1945  First fission bomb tested
1946  Big Bang cosmology (Gamow)
1946  Development of nuclear magnetic resonance (Bloch* and Purcell*)
1947  Development of radiocarbon dating (Libby*)
1947  First proton synchrocyclotron, 350 MeV (Berkeley)
1947  Discovery of $\pi$ meson (Powell*)
1948  First linear proton accelerator, 32 MeV (Alvarez*)
1949  Shell model of nuclear structure (Mayer*, Jensen*, Haxel, Suess)
1949  Development of scintillation counter (Kallmann, Coltman, Marshall)
1952  First proton synchrotron, 2.3 GeV (Brookhaven)
1952  First thermonuclear bomb tested
1953  Strangeness hypothesis (Gell-Mann*, Nishijima)
1953  Collective model of nuclear structure (A. Bohr*, Mottelson*, Rainwater*)
1953  First production of strange particles (Brookhaven)
1955  Discovery of antiproton (Chamberlain* and Segrè*)
1956  Experimental detection of neutrino (Reines and Cowan)
1956  Parity violation in weak interactions (Lee*, Yang*, Wu et al.)
1958  Recoilless emission of gamma rays (Mössbauer*)
1959  26-GeV proton synchrotron (CERN)
1964  Observation of CP violation in $K^0$ decay (Cronin* and Fitch*)
1964  Quark model of hadrons (Gell-Mann*, Zweig)
1967  Initial operation of SLAC accelerator for 20-GeV electrons (Stanford)
1967  Electroweak model proposed (Weinberg*, Salam*)
1970  Charm hypothesis (Glashow*)
1971  Proton-proton collider (CERN)
1972  500-GeV proton synchrotron (Fermilab)
1974  $J/\psi$ particle discovered and charmed quark confirmed (Richter*, Ting*)
1975  Discovery of $\tau$ lepton (Perl)
1977  $\Upsilon$ particle discovered and bottom quark inferred (Lederman)
1983  Operation of proton-antiproton collider at 300 GeV (CERN)
1983  Discovery of weak bosons $W^\pm$ and $Z^0$ (Rubbia*)

---

Names marked with an asterisk are Nobel laureates in physics or chemistry, although not necessarily for the work listed.

# INTRODUCTORY NUCLEAR PHYSICS

# INTRODUCTORY NUCLEAR PHYSICS

**Kenneth S. Krane**

Oregon State University

**JOHN WILEY & SONS**

New York · Chichester · Brisbane · Toronto · Singapore

*Library of Congress Cataloging in Publication Data:*

Krane, Kenneth S.
  Introductory nuclear physics.

  Rev. ed. of Introductory nuclear physics/David Halliday. 2nd. ed. 1955.
  1. Nuclear physics.   I. Halliday, David, 1916 –
Introductory nuclear physics.   II. Title.
QC777.K73   1987            539.7            87-10623
ISBN 0-471-80553-X

Printed in the United States of America

10 9

# PREFACE

This work began as a collaborative attempt with David Halliday to revise and update the second edition of his classic text *Introductory Nuclear Physics* (New York: Wiley, 1955). As the project evolved, it became clear that, owing to other commitments, Professor Halliday would be able to devote only limited time to the project and he therefore volunteered to remove himself from active participation, a proposal to which I reluctantly and regretfully agreed. He was kind enough to sign over to me the rights to use the material from the previous edition.

I first encountered Halliday's text as an undergraduate physics major, and it was perhaps my first real introduction to nuclear physics. I recall being impressed by its clarity and its readability, and in preparing this new version, I have tried to preserve these elements, which are among the strengths of the previous work.

**Audience**  This text is written primarily for an undergraduate audience, but could be used in introductory graduate surveys of nuclear physics as well. It can be used specifically for physics majors as part of a survey of modern physics, but could (with an appropriate selection of material) serve as an introductory course for other areas of nuclear science and technology, including nuclear chemistry, nuclear engineering, radiation biology, and nuclear medicine.

**Background**  It is expected that students have a previous background in quantum physics, either at the introductory level [such as the author's text *Modern Physics* (New York: Wiley, 1983)] or at a more advanced, but still undergraduate level. (A brief summary of the needed quantum background is given in Chapter 2.) The text is therefore designed in a "two-track" mode, so that the material that requires the advanced work in quantum mechanics, for instance, transition probabilities or matrix elements, can be separated from the rest of the text by skipping those sections that require such a background. This can be done without interrupting the logical flow of the discussion.

Mathematical background at the level of differential equations should be sufficient for most applications.

**Emphasis**  There are two features that distinguish the present book. The first is the emphasis on breadth. The presentation of a broad selection of material permits the instructor to tailor a curriculum to meet the needs of any particular student audience. The complete text is somewhat short for a full-year course, but too long for a course of quarter or semester length. The instructor is therefore able to select material that will provide students with the broadest possible introduction to the field of nuclear physics, consistent with the time available for the course.

The second feature is the unabashedly experimental and phenomenological emphasis and orientation of the presentation. The discussions of decay and reaction phenomena are accompanied with examples of experimental studies from the literature. These examples have been carefully selected following searches for papers that present data in the clearest possible manner and that relate most directly to the matter under discussion. These original experiments are discussed, often with accompanying diagrams of apparatus, and results *with uncertainties* are given, all in the attempt to convince students that progress in nuclear physics sprang not exclusively from the forehead of Fermi, but instead has been painstakingly won in the laboratory. At the same time, the rationale and motivation for the experiments are discussed, and their contributions to the theory are emphasized.

**Organization** The book is divided into four units: Basic Nuclear Structure, Nuclear Decay and Radioactivity, Nuclear Reactions, and Extensions and Applications. The first unit presents background material on nuclear sizes and shapes, discusses the two-nucleon problem, and presents an introduction to nuclear models. These latter two topics can be skipped without loss of continuity in an abbreviated course. The second unit on decay and radioactivity presents the traditional topics, with new material included to bring nuclear decay nearly into the current era (the recently discovered "heavy" decay modes, such as $^{14}$C, double $\beta$ decay, $\beta$-delayed nucleon emission, Mössbauer effect, and so on). The third unit surveys nuclear reactions, including fission and fusion and their applications. The final unit deals with topics that fall only loosely under the nuclear physics classification, including hyperfine interactions, particle physics, nuclear astrophysics, and general applications including nuclear medicine. The emphasis here is on the overlap with other physics and nonphysics specialties, including atomic physics, high-energy physics, cosmology, chemistry, and medicine. Much of this material, particularly in Chapters 18 and 19, represents accomplishments of the last couple of years and therefore, as in all such volatile areas, may be outdated before the book is published. Even if this should occur, however, the instructor is presented with a golden opportunity to make important points about progress in science. Chapter 20 features applications involving similarly recent developments, such as PET scans. The material in this last unit builds to a considerable degree on the previous material; it would be very unwise, for example, to attempt the material on meson physics or particle physics without a firm grounding in nuclear reactions.

**Sequence** Chapters or sections that can be omitted without loss of continuity in an abbreviated reading are indicated with asterisks (*) in the table of contents. An introductory short course in nuclear physics could be based on Chapters 1, 2, 3, 6, 8, 9, 10, and 11, which cover the fundamental aspects of nuclear decay and reactions, but little of nuclear structure. Fission and fusion can be added from

Chapters 13 and 14. Detectors and accelerators can be included with material selected from Chapters 7 and 15.

The last unit (Chapters 16 to 20) deals with applications and does not necessarily follow Chapter 15 in sequence. In fact, most of this material could be incorporated at any time after Chapter 11 (Nuclear Reactions). Chapter 16, covering spins and moments, could even be moved into the first unit after Chapter 3. Chapter 19 (Nuclear Astrophysics) requires background material on fission and fusion from Chapters 13 and 14.

Most of the text can be understood with only a minimal background in quantum mechanics. Chapters or sections that require a greater background (but still at the undergraduate level) are indicated in the table of contents with a dagger (†).

Many undergraduates, in my experience, struggle with even the most basic aspects of the quantum theory of angular momentum, and more abstract concepts, such as isospin, can present them with serious difficulties. For this reason, the introduction of isospin is delayed until it is absolutely necessary in Chapter 11 (Nuclear Reactions) where references to its application to beta and gamma decays are given to show its importance to those cases as well. No attempt is made to use isospin coupling theory to calculate amplitudes or cross sections. In an abbreviated coverage, it is therefore possible to omit completely any discussion of isospin, but it absolutely must be included before attempting Chapters 17 and 18 on meson and particle physics.

**Notation**   Standard notation has been adopted, which unfortunately overworks the symbol $T$ to represent kinetic energy, temperature, and isospin. The particle physicist's choice of $I$ for isospin and $J$ for nuclear spin leaves no obvious alternative for the total electronic angular momentum. Therefore, $I$ has been reserved for the total nuclear angular momentum, $J$ for the total electronic angular momentum, and $T$ for the isospin. To be consistent, the same scheme is extended into the particle physics regime in Chapters 17 and 18, even though it may be contrary to the generally accepted notation in particle physics. The lowercase $j$ refers to the total angular momentum of a single nucleon or atomic electron.

**References**   No attempt has been made to produce an historically accurate set of references to original work. This omission is done partly out of my insecurity about assuming the role of historian of science and partly out of the conviction that references tend to clutter, rather than illuminate, textbooks that are aimed largely at undergraduates. Historical discussions have been kept to a minimum, although major insights are identified with their sources. The history of nuclear physics, which so closely accompanies the revolutions wrought in twentieth-century physics by relativity and quantum theory, is a fascinating study in itself, and I encourage serious students to pursue it. In stark contrast to modern works, the classic papers are surprisingly readable. Many references to these early papers can be found in Halliday's book or in the collection by Robert T. Beyer, *Foundations of Nuclear Physics* (New York: Dover, 1949), which contains reprints of 13 pivotal papers and a classified bibliography of essentially every nuclear physics publication up to 1947.

Each chapter in this textbook is followed with a list of references for further reading, where more detailed or extensive treatments can be found. Included in the lists are review papers as well as popular-level books and articles.

Several of the end-of-chapter problems require use of systematic tabulations of nuclear properties, for which the student should have ready access to the current edition of the *Table of Isotopes* or to a complete collection of the *Nuclear Data Sheets*.

**Acknowledgments**   Many individuals read chapters or sections of the manuscript. I am grateful for the assistance of the following professional colleagues and friends: David Arnett, Carroll Bingham, Merle Bunker, H. K. Carter, Charles W. Drake, W. A. Fowler, Roger J. Hanson, Andrew Klein, Elliot J. Krane, Rubin H. Landau, Victor A. Madsen, Harvey Marshak, David K. McDaniels, Frank A. Rickey, Kandula S. R. Sastry, Larry Schecter, E. Brooks Shera, Richard R. Silbar, Paul Simms, Rolf M. Steffen, Gary Steigman, Morton M. Sternheim, Albert W. Stetz, and Ken Toth. They made many wise and valuable suggestions, and I thank them for their efforts. Many of the problems were checked by Milton Sagen and Paula Sonawala. Hundreds of original illustrations were requested of and generously supplied by nuclear scientists from throughout the world. Kathy Haag typed the original manuscript with truly astounding speed and accuracy and in the process helped to keep its preparation on schedule. The staff at John Wiley & Sons were exceedingly helpful and supportive, including physics editor Robert McConnin, copy editors Virginia Dunn and Deborah Herbert, and production supervisor Charlene Cassimire. Finally, without the kind support and encouragement of David Halliday, this work would not have been possible.

*Corvallis, Oregon*                                                      Kenneth S. Krane
*February 1987*

# CONTENTS

\*Denotes material that can be omitted without loss of continuity in an abbreviated reading.
† Denotes material that requires somewhat greater familiarity with quantum mechanics.

# INTRODUCTORY
# NUCLEAR
# PHYSICS

# UNIT I
## BASIC
## NUCLEAR
## STRUCTURE

# 1

# BASIC CONCEPTS

Whether we date the origin of nuclear physics from Becquerel's discovery of radioactivity in 1896 or Rutherford's hypothesis of the existence of the nucleus in 1911, it is clear that experimental and theoretical studies in nuclear physics have played a prominent role in the development of twentieth century physics. As a result of these studies, a chronology of which is given on the inside of the front cover of this book, we have today a reasonably good understanding of the properties of nuclei and of the structure that is responsible for those properties. Furthermore, techniques of nuclear physics have important applications in other areas, including atomic and solid-state physics. Laboratory experiments in nuclear physics have been applied to the understanding of an incredible variety of problems, from the interactions of quarks (the most fundamental particles of which matter is composed), to the processes that occurred during the early evolution of the universe just after the Big Bang. Today physicians use techniques learned from nuclear physics experiments to perform diagnosis and therapy in areas deep inside the body without recourse to surgery; but other techniques learned from nuclear physics experiments are used to build fearsome weapons of mass destruction, whose proliferation is a constant threat to our future. No other field of science comes readily to mind in which theory encompasses so broad a spectrum, from the most microscopic to the cosmic, nor is there another field in which direct applications of basic research contain the potential for the ultimate limits of good and evil.

Nuclear physics lacks a coherent theoretical formulation that would permit us to analyze and interpret all phenomena in a fundamental way; atomic physics has such a formulation in quantum electrodynamics, which permits calculations of some observable quantities to more than six significant figures. As a result, we must discuss nuclear physics in a phenomenological way, using a different formulation to describe each different type of phenomenon, such as $\alpha$ decay, $\beta$ decay, direct reactions, or fission. Within each type, our ability to interpret experimental results and predict new results is relatively complete, yet the methods and formulation that apply to one phenomenon often are not applicable to another. In place of a single unifying theory there are islands of coherent knowledge in a sea of seemingly uncorrelated observations. Some of the most fundamental problems of nuclear physics, such as the exact nature of the forces that hold the nucleus together, are yet unsolved. In recent years, much progress has been made toward understanding the basic force between the quarks that are the ultimate constituents of matter, and indeed attempts have been made at applying this knowledge to nuclei, but these efforts have thus far not contributed to the clarification of nuclear properties.

We therefore adopt in this text the phenomenological approach, discussing each type of measurement, the theoretical formulation used in its analysis, and the insight into nuclear structure gained from its interpretation. We begin with a summary of the basic aspects of nuclear theory, and then turn to the experiments that contribute to our knowledge of structure, first radioactive decay and then nuclear reactions. Finally, we discuss special topics that contribute to microscopic nuclear structure, the relationship of nuclear physics to other disciplines, and applications to other areas of research and technology.

## 1.1 HISTORY AND OVERVIEW

The search for the fundamental nature of matter had its beginnings in the speculations of the early Greek philosophers; in particular, Democritus in the fourth century B.C. believed that each kind of material could be subdivided into smaller and smaller bits until one reached the very limit beyond which no further division was possible. This *atom* of material, invisible to the naked eye, was to Democritus the basic constituent particle of matter. For the next 2400 years, this idea remained only a speculation, until investigators in the early nineteenth century applied the methods of *experimental science* to this problem and from their studies obtained the evidence needed to raise the idea of atomism to the level of a full-fledged scientific theory. Today, with our tendency toward the specialization and compartmentalization of science, we would probably classify these early scientists (Dalton, Avogadro, Faraday) as chemists. Once the chemists had elucidated the kinds of atoms, the rules governing their combinations in matter, and their systematic classification (Mendeleev's periodic table), it was only natural that the next step would be a study of the fundamental properties of individual atoms of the various elements, an activity that we would today classify as atomic physics. These studies led to the discovery in 1896 by Becquerel of the radioactivity of certain species of atoms and to the further identification of radioactive substances by the Curies in 1898. Rutherford next took up the study of these radiations and their properties; once he had achieved an understanding of the nature of the radiations, he turned them around and used them as probes of the atoms themselves. In the process he proposed in 1911 the existence of the atomic nucleus, the confirmation of which (through the painstaking experiments of Geiger and Marsden) provided a new branch of science, nuclear physics, dedicated to studying matter at its most fundamental level. Investigations into the properties of the nucleus have continued from Rutherford's time to the present. In the 1940s and 1950s, it was discovered that there was yet another level of structure even more elementary and fundamental than the nucleus. Studies of the particles that contribute to the structure at this level are today carried out in the realm of elementary particle (or high energy) physics.

Thus nuclear physics can be regarded as the descendent of chemistry and atomic physics and in turn the progenitor of particle physics. Although nuclear physics no longer occupies center stage in the search for the ultimate components of matter, experiments with nuclei continue to contribute to the understanding of basic interactions. Investigation of nuclear properties and the laws governing the structure of nuclei is an active and productive area of physical research in its own right, and practical applications, such as smoke detectors, cardiac pacemakers, and medical imaging devices, have become common. Thus nuclear physics has in reality three aspects: probing the fundamental particles and their interactions, classifying and interpreting the properties of nuclei, and providing technological advances that benefit society.

## 1.2  SOME INTRODUCTORY TERMINOLOGY

A nuclear species is characterized by the total amount of positive charge in the nucleus and by its total number of mass units. The net nuclear charge is equal to $+Ze$, where $Z$ is the *atomic number* and $e$ is the magnitude of the electronic charge. The fundamental positively charged particle in the nucleus is the *proton*, which is the nucleus of the simplest atom, hydrogen. A nucleus of atomic number $Z$ therefore contains $Z$ protons, and an electrically neutral atom therefore must contain $Z$ negatively charged electrons. Since the mass of the electrons is negligible compared with the proton mass ($m_p \simeq 2000 m_e$), the electron can often be ignored in discussions of the mass of an atom. The *mass number* of a nuclear species, indicated by the symbol $A$, is the integer nearest to the ratio between the nuclear mass and the fundamental mass unit, defined so that the proton has a mass of nearly one unit. (We will discuss mass units in more detail in Chapter 3.) For nearly all nuclei, $A$ is greater than $Z$, in most cases by a factor of two or more. Thus there must be other massive components in the nucleus. Before 1932, it was believed that the nucleus contained $A$ protons, in order to provide the proper mass, along with $A - Z$ *nuclear* electrons to give a net positive charge of $Ze$. However, the presence of electrons within the nucleus is unsatisfactory for several reasons:

1.  The nuclear electrons would need to be bound to the protons by a very strong force, stronger even than the Coulomb force. Yet no evidence for this strong force exists between protons and *atomic* electrons.

2.  If we were to confine electrons in a region of space as small as a nucleus ($\Delta x \sim 10^{-14}$ m), the uncertainty principle would require that these electrons have a momentum distribution with a range $\Delta p \sim \hbar/\Delta x = 20$ MeV/$c$. Electrons that are emitted from the nucleus in radioactive $\beta$ decay have energies generally less than 1 MeV; never do we see decay electrons with 20 MeV energies. Thus the existence of 20 MeV electrons in the nucleus is not confirmed by observation.

3.  The total intrinsic angular momentum (spin) of nuclei for which $A - Z$ is odd would disagree with observed values if $A$ protons and $A - Z$ electrons were present in the nucleus. Consider the nucleus of deuterium ($A = 2$, $Z = 1$), which according to the proton-electron hypothesis would contain 2 protons and 1 electron. The proton and electron each have intrinsic angular momentum (spin) of $\frac{1}{2}$, and the quantum mechanical rules for adding spins of particles would require that these three spins of $\frac{1}{2}$ combine to a total of either $\frac{3}{2}$ or $\frac{1}{2}$. Yet the observed spin of the deuterium nucleus is 1.

4.  Nuclei containing unpaired electrons would be expected to have magnetic dipole moments far greater than those observed. If a single electron were present in a deuterium nucleus, for example, we would expect the nucleus to have a magnetic dipole moment about the same size as that of an electron, but the observed magnetic moment of the deuterium nucleus is about $\frac{1}{2000}$ of the electron's magnetic moment.

Of course it is possible to invent all sorts of ad hoc reasons for the above arguments to be wrong, but the necessity for doing so was eliminated in 1932 when the *neutron* was discovered by Chadwick. The neutron is electrically neutral and has a mass about equal to the proton mass (actually about 0.1% larger). Thus a nucleus with $Z$ protons and $A - Z$ neutrons has the proper total mass *and* charge, without the need to introduce nuclear electrons. When we wish to indicate a specific nuclear species, or *nuclide*, we generally use the form $^{A}_{Z}X_{N}$, where X is the chemical symbol and $N$ is the *neutron number, $A - Z$*. The symbols for some nuclides are $^{1}_{1}H_{0}$, $^{238}_{92}U_{146}$, $^{56}_{26}Fe_{30}$. The chemical symbol and the atomic number $Z$ are redundant—every H nucleus has $Z = 1$, every U nucleus has $Z = 92$, and so on. It is therefore not necessary to write $Z$. It is also not necessary to write $N$, since we can always find it from $A - Z$. Thus $^{238}U$ is a perfectly valid way to indicate that particular nuclide; a glance at the periodic table tells us that U has $Z = 92$, and therefore $^{238}U$ has $238 - 92 = 146$ neutrons. You may find the symbols for nuclides written sometimes with $Z$ and $N$, and sometimes without them. When we are trying to balance $Z$ and $N$ in a decay or reaction process, it is convenient to have them written down; at other times it is cumbersome and unnecessary to write them.

Neutrons and protons are the two members of the family of *nucleons*. When we wish simply to discuss nuclear particles without reference to whether they are protons or neutrons, we use the term nucleons. Thus a nucleus of mass number $A$ contains $A$ nucleons.

When we analyze samples of many naturally occurring elements, we find that nuclides with a given atomic number can have several different mass numbers; that is, a nuclide with $Z$ protons can have a variety of different neutron numbers. Nuclides with the same proton number but different neutron numbers are called *isotopes*; for example, the element chlorine has two isotopes that are stable against radioactive decay, $^{35}Cl$ and $^{37}Cl$. It also has many other unstable isotopes that are artificially produced in nuclear reactions; these are the radioactive isotopes (or *radioisotopes*) of Cl.

It is often convenient to refer to a sequence of nuclides with the same $N$ but different $Z$; these are called *isotones*. The stable isotones with $N = 1$ are $^{2}H$ and $^{3}He$. Nuclides with the same mass number $A$ are known as *isobars*; thus stable $^{3}He$ and radioactive $^{3}H$ are isobars.

## 1.3  NUCLEAR PROPERTIES

Once we have identified a nuclide, we can then set about to measure its properties, among which (to be discussed later in this text) are mass, radius, relative abundance (for stable nuclides), decay modes and half-lives (for radioactive nuclides), reaction modes and cross sections, spin, magnetic dipole and electric quadrupole moments, and excited states. Thus far we have identified

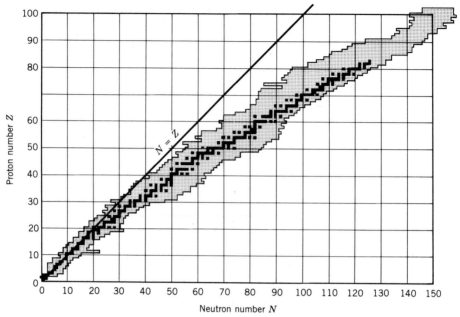

**Figure 1.1**  Stable nuclei are shown in dark shading and known radioactive nuclei are in light shading.

nuclides with 108 different atomic numbers (0 to 107); counting all the different isotopes, the total number of nuclides is well over 1000, and the number of carefully studied new nuclides is growing rapidly owing to new accelerators dedicated to studying the isotopes far from their stable isobars. Figure 1.1 shows a representation of the stable and known radioactive nuclides.

As one might expect, cataloging all of the measured properties of these many nuclides is a formidable task. An equally formidable task is the retrieval of that information: if we require the best current experimental value of the decay modes of an isotope or the spin and magnetic moment of another, where do we look?

Nuclear physicists generally publish the results of their investigations in journals that are read by other nuclear physicists; in this way, researchers from distant laboratories are aware of one another's activities and can exchange ideas. Some of the more common journals in which to find such communications are *Physical Review*, Section C (abbreviated *Phys. Rev. C*), *Physical Review Letters* (*Phys. Rev. Lett.*), *Physics Letters*, Section B (*Phys. Lett. B*), *Nuclear Physics*, Section A (*Nucl. Phys. A*), *Zeitschrift für Physik*, Section A (*Z. Phys. A*), and *Journal of Physics*, Section G (*J. Phys. G*). These journals are generally published monthly, and by reading them (or by scanning the table of contents), we can find out about the results of different researchers. Many college and university libraries subscribe to these journals, and the study of nuclear physics is often aided by browsing through a selection of current research papers.

Unfortunately, browsing through current journals usually does not help us to locate the specific nuclear physics information we are seeking, unless we happen to stumble across an article on that topic. For this reason, there are many sources of compiled nuclear physics information that summarize nuclear properties and give references to the literature where the original publication may be consulted. A one-volume summary of the properties of all known nuclides is the *Table of Isotopes*, edited by M. Lederer and V. Shirley (New York: Wiley, 1978). A copy of this indispensible work is owned by every nuclear physicist. A more current updating of nuclear data can be found in the *Nuclear Data Sheets*, which not only publish regular updated collections of information for each set of isobars, but also give an annual summary of all published papers in nuclear physics, classified by nuclide. This information is published in journal form and is also carried by many libraries. It is therefore a relatively easy process to check the recently published work concerning a certain nuclide.

Two other review works are the *Atomic Data and Nuclear Data Tables*, which regularly produces compilations of nuclear properties (for example, $\beta$ or $\gamma$ transition rates or fission energies), and the *Annual Review of Nuclear and Particle Science* (formerly called the *Annual Review of Nuclear Science*), which each year publishes a collection of review papers on current topics in nuclear and particle physics.

## 1.4 UNITS AND DIMENSIONS

In nuclear physics we encounter lengths of the order of $10^{-15}$ m, which is one femtometer (fm). This unit is colloquially known as one fermi, in honor of the pioneer Italian-American nuclear physicist, Enrico Fermi. Nuclear sizes range from about 1 fm for a single nucleon to about 7 fm for the heaviest nuclei.

The time scale of nuclear phenomena has an enormous range. Some nuclei, such as $^5$He or $^8$Be, break apart in times of the order of $10^{-20}$ s. Many nuclear reactions take place on this time scale, which is roughly the length of time that the reacting nuclei are within range of each other's nuclear force. Electromagnetic ($\gamma$) decays of nuclei occur generally within lifetimes of the order of $10^{-9}$ s (nanosecond, ns) to $10^{-12}$ s (picosecond, ps), but many decays occur with much shorter or longer lifetimes. $\alpha$ and $\beta$ decays occur with even longer lifetimes, often minutes or hours, but sometimes thousands or even millions of years.

Nuclear energies are conveniently measured in millions of electron-volts (MeV), where 1 eV = $1.602 \times 10^{-19}$ J is the energy gained by a single unit of electronic charge when accelerated through a potential difference of one volt. Typical $\beta$ and $\gamma$ decay energies are in the range of 1 MeV, and low-energy nuclear reactions take place with kinetic energies of order 10 MeV. Such energies are far smaller than the nuclear rest energies, and so we are justified in using nonrelativistic formulas for energy and momentum of the nucleons, but $\beta$-decay electrons must be treated relativistically.

Nuclear masses are measured in terms of the *unified atomic mass unit*, u, defined such that the mass of an *atom* of $^{12}$C is exactly 12 u. Thus the nucleons have masses of approximately 1 u. In analyzing nuclear decays and reactions, we generally work with mass energies rather than with the masses themselves. The conversion factor is 1 u = 931.502 MeV, so the nucleons have mass energies of approximately 1000 MeV. The conversion of mass to energy is of course done using the fundamental result from special relativity, $E = mc^2$; thus we are free to work either with masses or energies at our convenience, and in these units $c^2 = 931.502$ MeV/u.

## REFERENCES FOR ADDITIONAL READING

The following comprehensive nuclear physics texts provide explanations or formulations alternative to those of this book. Those at the introductory level are at about the same level as the present text; higher-level texts often form the basis for more advanced graduate courses in nuclear physics. No attempt has been made to produce a complete list of reference works; rather, these are the ones the author has found most useful in preparing this book.

These "classic" texts now mostly outdated but still containing much useful material are interesting for gaining historical perspective: R. D. Evans, *The Atomic Nucleus* (New York: McGraw-Hill, 1955) (For 20 years, since his graduate-student days, the most frequently used book on the author's shelves. Its binding has all but deteriorated, but its completeness and clarity remain.); David Halliday, *Introductory Nuclear Physics* (New York: Wiley, 1955); I. Kaplan, *Nuclear Physics* (Reading, MA: Addison-Wesley, 1955).

Introductory texts complementary to this text are: W. E. Burcham, *Nuclear Physics: An Introduction* (London: Longman, 1973); B. L. Cohen, *Concepts of Nuclear Physics* (New York: McGraw-Hill, 1971); Harald A. Enge, *Introduction to Nuclear Physics* (Reading, MA: Addison-Wesley, 1966); Robert A. Howard, *Nuclear Physics* (Belmont, CA: Wadsworth, 1963); Walter E. Meyerhof, *Elements of Nuclear Physics* (New York: McGraw-Hill, 1967); Haro Von Buttlar, *Nuclear Physics: An Introduction* (New York: Academic Press, 1968).

Intermediate texts, covering much the same material as the present one but distinguished primarily by a more rigorous use of quantum mechanics, are: M. G. Bowler, *Nuclear Physics* (Oxford: Pergamon, 1973); Emilio Segré, *Nuclei and Particles* (Reading, MA: W. A. Benjamin, 1977).

Advanced texts, primarily for graduate courses, but still containing much material of a more basic nature, are: Hans Frauenfelder and Ernest M. Henley, *Subatomic Physics* (Englewood Cliffs, NJ: Prentice-Hall, 1974); M. A. Preston, *Physics of the Nucleus* (Reading, MA: Addison-Wesley, 1962).

Advanced works, more monographs than texts in nature, are: John M. Blatt and Victor F. Weisskopf, *Theoretical Nuclear Physics* (New York: Wiley, 1952); A. Bohr and B. R. Mottelson, *Nuclear Structure* (New York: W. A. Benjamin, 1969); A. deShalit and H. Feshbach, *Theoretical Nuclear Physics* (New York: Wiley, 1974).

# 2

# ELEMENTS OF
# QUANTUM MECHANICS

Nucleons in a nucleus do not behave like classical particles, colliding like billiard balls. Instead, the *wave behavior* of the nucleons determines the properties of the nucleus, and to analyze this behavior requires that we use the mathematical techniques of quantum mechanics.

From a variety of scattering experiments, we know that the nucleons in a nucleus are in motion with kinetic energies of the order of 10 MeV. This energy is small compared with the nucleon rest energy (about 1000 MeV), and so we can with confidence use *nonrelativistic* quantum mechanics.

To give a complete introduction to quantum mechanics would require a text larger than the present one. In this chapter, we summarize some of the important concepts that we will need later in this book. We assume a previous introduction to the concepts of modern physics and a familiarity with some of the early experiments that could not be understood using classical physics; these experiments include thermal (blackbody) radiation, Compton scattering, and the photoelectric effect. At the end of this chapter is a list of several introductory modern physics texts for review. Included in the list are more advanced quantum physics texts, which contain more complete and rigorous discussions of the topics summarized in this chapter.

## 2.1 QUANTUM BEHAVIOR

Quantum mechanics is a mathematical formulation that enables us to calculate the wave behavior of material particles. It is not at all *a priori* evident that such behavior should occur, but the suggestion follows by analogy with the quantum behavior of light. Before 1900, light was generally believed to be a wave phenomenon, but the work of Planck in 1900 (analyzing blackbody radiation) and Einstein in 1905 (analyzing the photoelectric effect) showed that it was also necessary to consider light as if its energy were delivered not smoothly and continuously as a wave but instead in concentrated bundles or "quanta," in effect "particles of light."

The analogy between matter and light was made in 1924 by de Broglie, drawing on the previous work of Einstein and Compton. If light, which we generally regard as a wave phenomenon, also has particle aspects, then (so de Broglie argued) might not matter, which we generally regard as composed of particles, also have a wave aspect? Again proceeding by analogy with light, de Broglie postulated that associated with a "particle" moving with momentum $p$ is a "wave" of wavelength $\lambda = h/p$ where $h$ is Planck's constant. The wavelength defined in this way is generally called the de Broglie wavelength. Experimental confirmation of de Broglie's hypothesis soon followed in 1927 through the experiments of Thomson and of Davisson and Germer. They showed that electrons (particles) were diffracted like waves with the de Broglie wavelength.

The de Broglie theory was successful in these instances, but it is incomplete and unsatisfying for several reasons. For one, we seldom see particles with a unique momentum $p$; if the momentum of a particle changes, such as when it is acted upon by an external force, its wavelength must change, but the de Broglie relationship lacks the capability to enable computation of the dynamical behavior of the waves. For this we need a more complete mathematical theory, which was supplied by Schrödinger in 1925 and which we review in Section 2 of this chapter. A second objection to the de Broglie theory is its reliance on classical concepts and terminology. "Particle" and "wave" are mutually exclusive sorts of behaviors, but the de Broglie relationship involves classical particles with uniquely defined momenta and classical waves with uniquely defined wavelengths. A classical particle has a definite position in space. Now, according to de Broglie, that localized particle is to be represented by a pure wave that extends throughout all space and has no beginning, end, or easily identifiable "position."

The solution to this dilemma requires us to discard the classical idea of "particle" when we enter the domain of quantum physics. The size of a classical particle is the same in every experiment we may do; the "size" of a quantum particle varies with the experiment we perform. Quantum physics forces us to sacrifice the objective reality of a concept such as "size" and instead to substitute an operational definition that depends on the experiment that is being done. Thus an electron may have a certain size in one experiment and a very different size in another. Only through this coupling of the observing system and the observed object can we define observations in quantum physics. A particle, then, is *localized* within some region of space of dimension $\Delta x$. It is likely to be found in that region and unlikely to be found elsewhere. The dimension $\Delta x$ of an electron is determined by the kind of experiment we do—it may be the dimension of a block of material if we are studying electrical conduction in solids, or the dimension of a single atom if we are studying atomic physics, or of a nucleus if we are studying $\beta$ decay. The wave that characterizes the particle has large amplitude in the region $\Delta x$ and small amplitude elsewhere. The single de Broglie wave corresponding to the unique momentum component $p_x$ had a large amplitude everywhere; thus a definite momentum (wavelength) corresponds to a completely unlocalized particle. To localize the particle, we must add (superpose) other wavelengths corresponding to other momenta $p_x$, so that we make the resultant wave small outside the region $\Delta x$. *We improve our knowledge of $\Delta x$ at the expense of our knowledge of $p_x$. The very act of confining the particle to $\Delta x$ destroys the precision of our knowledge of $p_x$ and introduces a range of values $\Delta p_x$.* If we try to make a *simultaneous* determination of $x$ and $p_x$, our result will show that each is uncertain by the respective amounts $\Delta x$ and $\Delta p_x$, which are related by the Heisenberg uncertainty relationship

$$\Delta x \, \Delta p_x \geq \frac{\hbar}{2} \tag{2.1}$$

with similar expressions for the $y$ and $z$ components. (The symbol $\hbar$, read as "h-bar," is $h/2\pi$ where $h$ is Planck's constant.) The key word here is "simultaneous"—we can indeed measure $x$ with arbitrarily small uncertainty ($\Delta x = 0$) if we are willing to sacrifice all simultaneous knowledge of the momentum of the particle. Having made that determination, we could then make an arbitrarily precise measurement of the *new* momentum ($\Delta p_x = 0$), which would simultaneously destroy our previous precise knowledge of its position.

We describe the particle by a "wave packet," a collection of waves, representing a range of momenta $\Delta p_x$ around $p_x$, with an amplitude that is reasonably large only within the region $\Delta x$ about $x$. A particle is localized in a region of space defined by its wave packet; the wave packet contains all of the available information about the particle. Whenever we use the term "particle" we really mean "wave packet"; although we often speak of electrons or nucleons as if they had an independent existence, our knowledge of them is limited by the uncertainty relationship to the information contained in the wave packet that describes their particular situation.

These arguments about uncertainty hold for other kinds of measurements as well. The energy $E$ of a system is related to the frequency $\nu$ of its de Broglie wave according to $E = h\nu$. To determine $E$ precisely, we must observe for a sufficiently long time interval $\Delta t$ so that we can determine $\nu$ precisely. The uncertainty relationship in this case is

$$\Delta E \, \Delta t \geq \frac{\hbar}{2} \tag{2.2}$$

If a system lives for a time $\Delta t$, we cannot determine its energy except to within an uncertainty $\Delta E$. The energy of a system that is absolutely stable against decay can be measured with arbitrarily small uncertainty; for all decaying systems there is an uncertainty in energy, commonly called the energy "width."

A third uncertainty relationship involves the angular momentum. Classically, we can determine all three components $\ell_x, \ell_y, \ell_z$ of the angular momentum vector $\ell$. In quantum mechanics, when we try to improve our knowledge of one component, it is at the expense of our knowledge of the other two components. Let us choose to measure the $z$ component, and let the location of the projection of $\ell$ in the $xy$ plane be characterized by the azimuthal angle $\phi$. Then

$$\Delta \ell_z \, \Delta \phi \geq \frac{\hbar}{2} \tag{2.3}$$

If we know $\ell_z$ exactly, then we know nothing at all about $\phi$. We can think of $\ell$ as rotating or precessing about the $z$ axis, keeping $\ell_z$ fixed but allowing all possible $\ell_x$ and $\ell_y$, so that $\phi$ is completely uncertain.

## 2.2  PRINCIPLES OF QUANTUM MECHANICS

The mathematical aspects of nonrelativistic quantum mechanics are determined by solutions to the *Schrödinger equation*. In one dimension, the time-independent Schrödinger equation for a particle of mass $m$ with potential energy $V(x)$ is

$$-\frac{\hbar^2}{2m}\frac{d^2\psi}{dx^2} + V(x)\,\psi(x) = E\,\psi(x) \tag{2.4}$$

where $\psi(x)$ is the Schrödinger *wave function*. The wave function is the mathematical description of the wave packet. In general, this equation will have solutions only for certain values of the energy $E$; these values, which usually result from applying boundary conditions to $\psi(x)$, are known as the energy *eigenvalues*. The complete solution, including the time dependence, is

$$\Psi(x,t) = \psi(x)\,e^{-i\omega t} \tag{2.5}$$

where $\omega = E/\hbar$.

An important condition on the wave function is that $\psi$ and its first derivative $d\psi/dx$ must be continuous across any boundary; in fact, the same situation applies to classical waves. Whenever there is a boundary between two media, let us say at $x = a$, we must have

$$\lim_{\varepsilon \to 0} \left[ \psi(a+\varepsilon) - \psi(a-\varepsilon) \right] = 0 \tag{2.6a}$$

and

$$\lim_{\varepsilon \to 0} \left[ \left(\frac{d\psi}{dx}\right)_{x=a+\varepsilon} - \left(\frac{d\psi}{dx}\right)_{x=a-\varepsilon} \right] = 0 \tag{2.6b}$$

It is permitted to violate condition 2.6b if there is an *infinite* discontinuity in $V(x)$; however, condition 2.6a must always be true.

Another condition on $\psi$, which originates from the interpretation of probability density to be discussed below, is that $\psi$ must remain finite. Any solution for the Schrödinger equation that allows $\psi$ to become infinite must be discarded.

Knowledge of the wave function $\Psi(x, t)$ for a system enables us to calculate many properties of the system. For example, the probability to find the particle (the wave packet) between $x$ and $x + dx$ is

$$P(x)\,dx = \Psi^*(x,t)\,\Psi(x,t)\,dx \tag{2.7}$$

where $\Psi^*$ is the complex conjugate of $\Psi$. The quantity $\Psi^*\Psi$ is known as the *probability density*. The probability to find the particle between the limits $x_1$ and $x_2$ is the integral of all the infinitesimal probabilities:

$$P = \int_{x_1}^{x_2} \Psi^*\Psi\,dx \tag{2.8}$$

The total probability to find the particle must be 1:

$$\int_{-\infty}^{\infty} \Psi^*\Psi\,dx = 1 \tag{2.9}$$

This condition is known as the *normalization condition* and in effect it determines any multiplicative constants included in $\Psi$. All physically meaningful wave functions must be properly normalized.

Any function of $x$, $f(x)$, can be evaluated for this quantum mechanical system. The values that we measure for $f(x)$ are determined by the probability density, and the average value of $f(x)$ is determined by finding the contribution to the average for each value of $x$:

$$\langle f \rangle = \int \Psi^* f \, \Psi \, dx \qquad (2.10)$$

Average values computed in this way are called quantum mechanical *expectation values*.

We must be a bit careful how we interpret these expectation values. Quantum mechanics deals with statistical outcomes, and many of our calculations are really statistical averages. If we prepare a large number of identical systems and measure $f(x)$ for each of them, the average of these measurements will be $\langle f \rangle$. One of the unsatisfying aspects of quantum theory is its inability to predict with certainty the outcome of an experiment; all we can do is predict the statistical average of a large number of measurements.

Often we must compute the average values of quantities that are not simple functions of $x$. For example, how can we compute $\langle p_x \rangle$? Since $p_x$ is not a function of $x$, we cannot use Equation 2.10 for this calculation. The solution to this difficulty comes from the mathematics of quantum theory. Corresponding to each classical variable, there is a quantum mechanical *operator*. An operator is a symbol that directs us to perform a mathematical operation, such as exp or sin or $d/dx$. We adopt the convention that the operator acts only on the variable or function immediately to its right, unless we indicate otherwise by grouping functions in parentheses. This convention means that it is very important to remember the form of Equation 2.10; the operator is "sandwiched" between $\Psi^*$ and $\Psi$, and operates only on $\Psi$. Two of the most common operators encountered in quantum mechanics are the momentum operator, $p_x = -i\hbar\partial/\partial x$ and the energy, $E = i\hbar\partial/\partial t$. Notice that the first term on the left of the Schrödinger equation 2.4 is just $p_x^2/2m$, which we can regard as the kinetic energy operator. Notice also that the *operator* $E$ applied to $\Psi(x, t)$ in Equation 2.5 gives the *number* $E$ multiplying $\Psi(x, t)$.

We can now evaluate the expectation value of the $x$ component of the momentum:

$$\langle p_x \rangle = \int \Psi^* \left( -i\hbar \frac{\partial}{\partial x} \right) \Psi \, dx$$

$$= -i\hbar \int \Psi^* \frac{\partial \Psi}{\partial x} \, dx \qquad (2.11)$$

One very important feature emerges from these calculations: when we take the complex conjugate of $\Psi$ as given by Equation 2.5, the time-dependent factor becomes $e^{+i\omega t}$, and therefore the time dependence cancels from Equations 2.7–2.11. None of the observable properties of the system depend on the time. Such conditions are known for obvious reasons as *stationary states*; a system in a stationary state stays in that state for all times and all of the dynamical variables are constants of the motion. This is of course an idealization—no system lives forever, but many systems can be regarded as being in states that are approximately stationary. Thus an atom can make a transition from one "stationary" excited state to another "stationary" state.

Associated with the wave function $\Psi$ is the *particle current density j*:

$$j = \frac{\hbar}{2mi}\left(\Psi^* \frac{\partial \Psi}{\partial x} - \Psi \frac{\partial \Psi^*}{\partial x}\right) \tag{2.12}$$

This quantity is analogous to an electric current, in that it gives the number of particles per second passing any point $x$.

In three dimensions, the form of the Schrödinger equation depends on the coordinate system in which we choose to work. In Cartesian coordinates, the potential energy is a function of $(x, y, z)$ and the Schrödinger equation is

$$-\frac{\hbar^2}{2m}\left(\frac{\partial^2 \psi}{\partial x^2} + \frac{\partial^2 \psi}{\partial y^2} + \frac{\partial^2 \psi}{\partial z^2}\right) + V(x, y, z)\psi(x, y, z) = E\psi(x, y, z) \tag{2.13}$$

The complete time-dependent solution is again

$$\Psi(x, y, z, t) = \psi(x, y, z)\,e^{-i\omega t} \tag{2.14}$$

The probability density $\Psi^*\Psi$ now gives the probability per unit volume; the probability to find the particle in the volume element $dv = dx\,dy\,dz$ at $x, y, z$ is

$$P\,dv = \Psi^*\Psi\,dv \tag{2.15}$$

To find the total probability in some volume $V$, we must do a triple integral over $x$, $y$, and $z$. All of the other properties discussed above for the one-dimensional system can easily be extended to the three-dimensional system.

Since nuclei are approximately spherical, the Cartesian coordinate system is not the most appropriate one. Instead, we must work in spherical polar coordinates $(r, \theta, \phi)$, which are shown in Figure 2.1. In this case the Schrödinger equation is

$$-\frac{\hbar^2}{2m}\left[\frac{\partial^2 \psi}{\partial r^2} + \frac{2}{r}\frac{\partial \psi}{\partial r} + \frac{1}{r^2 \sin\theta}\frac{\partial}{\partial \theta}\left(\sin\theta \frac{\partial \psi}{\partial \theta}\right) + \frac{1}{r^2 \sin^2\theta}\frac{\partial^2 \psi}{\partial \phi^2}\right]$$
$$+ V(r, \theta, \phi)\psi(r, \theta, \phi) = E\psi(r, \theta, \phi) \tag{2.16}$$

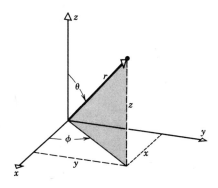

**Figure 2.1** Spherical polar coordinate system, showing the relationship to Cartesian coordinates.

All of the previous considerations apply in this case as well, with the volume element

$$dv = r^2 \sin\theta \, dr \, d\theta \, d\phi \tag{2.17}$$

The following two sections illustrate the application of these principles, first with the mathematically simpler one-dimensional problems and then with the more physical three-dimensional problems.

## 2.3 PROBLEMS IN ONE DIMENSION

### The Free Particle

For this case, no forces act and we take $V(x) = 0$ everywhere. We can then rewrite Equation 2.4 as

$$\frac{d^2\psi}{dx^2} = -\frac{2mE}{\hbar^2}\psi \tag{2.18}$$

The solution to this differential equation can be written

$$\psi(x) = A' \sin kx + B' \cos kx \tag{2.19}$$

or, equivalently

$$\psi(x) = A e^{ikx} + B e^{-ikx} \tag{2.20}$$

where $k^2 = 2mE/\hbar^2$ and where $A$ and $B$ (or $A'$ and $B'$) are constants.

The time-dependent wave function is

$$\Psi(x, t) = A e^{i(kx - \omega t)} + B e^{-i(kx + \omega t)} \tag{2.21}$$

The first term represents a wave traveling in the positive $x$ direction, while the second term represents a wave traveling in the negative $x$ direction. The intensities of these waves are given by the squares of the respective amplitudes, $|A|^2$ and $|B|^2$. Since there are no boundary conditions, there are no restrictions on the energy $E$; all values of $E$ give solutions to the equation. The normalization condition 2.9 cannot be applied in this case, because integrals of $\sin^2$ or $\cos^2$ do not converge in $x = -\infty$ to $+\infty$. Instead, we use a different normalization system for such constant potentials. Suppose we have a source such as an accelerator located at $x = -\infty$, emitting particles at a rate $I$ particles per second, with momentum $p = \hbar k$ in the positive $x$ direction. Since the particles are traveling in the positive $x$ direction, we can set $B$ to zero—the intensity of the wave representing particles traveling in the negative $x$ direction must vanish if there are no particles traveling in that direction. The particle current is, according to Equation 2.12,

$$j = \frac{\hbar k}{m}|A|^2 \tag{2.22}$$

which must be equal to the current of $I$ particles per second emitted by the source. Thus we can take $A = \sqrt{mI/\hbar k}$.

## Step Potential, $E > V_0$

The potential is

$$V(x) = 0 \qquad x < 0$$
$$= V_0 \qquad x > 0 \qquad (2.23)$$

where $V_0 > 0$. Let us call $x < 0$ region 1 and $x > 0$ region 2. Then in region 1, the Schrödinger equation is identical with Equation 2.18 and the solutions $\psi_1$ are given by Equation 2.20 with $k = k_1 = \sqrt{2mE/\hbar^2}$. In region 2, the Schrödinger equation is

$$\frac{d^2\psi_2}{dx^2} = -\frac{2m(E - V_0)}{\hbar^2}\psi_2 \qquad (2.24)$$

Since $E > V_0$, we can write the solution as

$$\psi_2 = C e^{ik_2 x} + D e^{-ik_2 x} \qquad (2.25)$$

where $k_2 = \sqrt{2m(E - V_0)/\hbar^2}$.

Applying the boundary conditions at $x = 0$ gives

$$A + B = C + D \qquad (2.26a)$$

from Equation 2.6a, and

$$k_1(A - B) = k_2(C - D) \qquad (2.26b)$$

from Equation 2.6b.

Let's assume that particles are incident on the step from a source at $x = -\infty$. Then the $A$ term in $\psi_1$ represents the *incident wave* (the wave in $x < 0$ traveling toward the step at $x = 0$), the $B$ term in $\psi_1$ represents the *reflected wave* (the wave in $x < 0$ traveling back toward $x = -\infty$), and the $C$ term in $\psi_2$ represents the *transmitted wave* (the wave in $x > 0$ traveling away from $x = 0$). The $D$ term cannot represent any part of this problem because there is no way for a wave to be moving toward the origin in region 2, and so we eliminate this term by setting $D$ to zero. Solving Equations 2.26a and 2.26b, we have

$$B = A\frac{1 - k_2/k_1}{1 + k_2/k_1} \qquad (2.27)$$

$$C = A\frac{2}{1 + k_2/k_1} \qquad (2.28)$$

The *reflection coefficient* $R$ is defined as the current in the reflected wave divided by the incident current:

$$R = \frac{j_{\text{reflected}}}{j_{\text{incident}}} \qquad (2.29)$$

and using Equation 2.22 we find

$$R = \frac{|B|^2}{|A|^2} = \left(\frac{1 - k_2/k_1}{1 + k_2/k_1}\right)^2 \qquad (2.30)$$

The *transmission coefficient* $T$ is similarly defined as the fraction of the incident

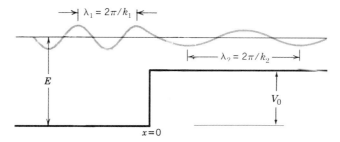

**Figure 2.2**  The wave function of a particle of energy $E$ encountering a step of height $V_0$ for the case $E > V_0$. The de Broglie wavelength changes from $\lambda_1$ to $\lambda_2$ when the particle crosses the step, but $\psi$ and $d\psi/dx$ are continuous at $x = 0$.

current that is transmitted past the boundary:

$$T = \frac{j_{\text{transmitted}}}{j_{\text{incident}}} \tag{2.31}$$

and thus

$$T = \frac{k_2}{k_1}\frac{|C|^2}{|A|^2} = \frac{4k_2/k_1}{\left(1 + k_2/k_1\right)^2} \tag{2.32}$$

Notice that $R + T = 1$, as expected. The resulting solution is illustrated in Figure 2.2.

This is a simple example of a *scattering* problem. In Chapter 4 we show how these concepts can be extended to three dimensions and applied to nucleon–nucleon scattering problems.

### Step Potential, $E < V_0$

In this case, the potential is still given by Equation 2.23, and the solution for region 1 ($x < 0$) is identical with the previous calculation. In region 2, the Schrödinger equation gives

$$\frac{d^2\psi_2}{dx^2} = \frac{2m}{\hbar^2}(V_0 - E)\psi_2 \tag{2.33}$$

which has the solution

$$\psi_2 = Ce^{k_2 x} + De^{-k_2 x} \tag{2.34}$$

where $k_2 = \sqrt{2m(V_0 - E)/\hbar^2}$. Note that for *constant* potentials, the solutions are either oscillatory like Equation 2.19 or 2.20 when $E > V_0$, or exponential like Equation 2.34 when $E < V_0$. Although the mathematical forms may be different for nonconstant potentials $V(x)$, the general behavior is preserved: oscillatory (though not necessarily sinusoidal) when $E > V(x)$ and exponential when $E < V(x)$.

This solution, Equation 2.34, must be valid for the entire range $x > 0$. Since the first term would become infinite for $x \to \infty$, we must set $C = 0$ to keep the wave function finite. The $D$ term in $\psi_2$ illustrates an important difference between classical and quantum physics, *the penetration of the wave function into*

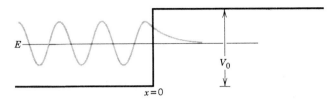

**Figure 2.3** The wave function of a particle of energy $E$ encountering a step of height $V_0$, for the case $E < V_0$. The wave function decreases exponentially in the classically forbidden region, where the classical kinetic energy would be negative. At $x = 0$, $\psi$ and $d\psi/dx$ are continuous.

*the classically forbidden region.* All (classical) *particles* are reflected at the boundary; the quantum mechanical *wave packet*, on the other hand, can penetrate a short distance into the forbidden region. The (classical) particle is never directly *observed* in that region; since $E < V_0$, the kinetic energy would be negative in region 2. The solution is illustrated in Figure 2.3

### Barrier Potential, $E > V_0$

The potential is

$$
\begin{aligned}
V(x) &= 0 & x < 0 \\
&= V_0 & 0 \le x \le a \\
&= 0 & x > a
\end{aligned}
\tag{2.35}
$$

In the three regions 1, 2, and 3, the solutions are

$$
\begin{aligned}
\psi_1 &= A\,e^{ik_1 x} + B\,e^{-ik_1 x} \\
\psi_2 &= C\,e^{ik_2 x} + D\,e^{-ik_2 x} \\
\psi_3 &= F\,e^{ik_3 x} + G\,e^{-ik_3 x}
\end{aligned}
\tag{2.36}
$$

where $k_1 = k_3 = \sqrt{2mE/\hbar^2}$ and $k_2 = \sqrt{2m(E - V_0)/\hbar^2}$.

Using the continuity conditions at $x = 0$ and at $x = a$, and assuming again that particles are incident from $x = -\infty$ (so that $G$ can be set to zero), after considerable algebraic manipulation we can find the transmission coefficient $T = |F|^2/|A|^2$:

$$
T = \frac{1}{1 + \dfrac{1}{4}\dfrac{V_0^2}{E(E - V_0)}\sin^2 k_2 a}
\tag{2.37}
$$

The solution is illustrated in Figure 2.4.

### Barrier Potential, $E < V_0$

For this case, the $\psi_1$ and $\psi_3$ solutions are as above, but $\psi_2$ becomes

$$
\psi_2 = C\,e^{k_2 x} + D\,e^{-k_2 x}
\tag{2.38}
$$

where now $k_2 = \sqrt{2m(V_0 - E)/\hbar^2}$. Because region 2 extends only from $x = 0$

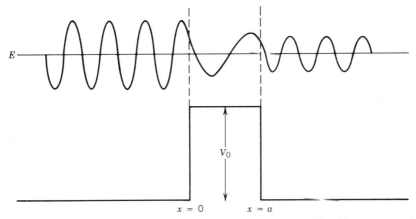

**Figure 2.4** The wave function of a particle of energy $E > V_0$ encountering a barrier potential. The particle is incident from the left. The wave undergoes reflections at both boundaries, and the transmitted wave emerges with smaller amplitude.

to $x = a$, the question of an exponential solution going to infinity does not arise, so we cannot set $C$ or $D$ to zero.

Again, applying the boundary conditions at $x = 0$ and $x = a$ permits the solution for the transmission coefficient:

$$T = \cfrac{1}{1 + \cfrac{1}{4} \cfrac{V_0^2}{E(V_0 - E)} \sinh^2 k_2 a} \tag{2.39}$$

Classically, we would expect $T = 0$—the particle is not permitted to enter the forbidden region where it would have negative kinetic energy. The quantum wave can penetrate the barrier and give a nonzero probability to find the particle beyond the barrier. The solution is illustrated in Figure 2.5.

This phenomenon of *barrier penetration* or quantum mechanical *tunneling* has important applications in nuclear physics, especially in the theory of $\alpha$ decay, which we discuss in Chapter 8.

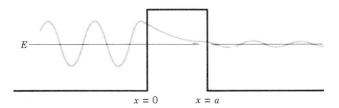

**Figure 2.5** The wave function of a particle of energy $E < V_0$ encountering a barrier potential (the particle would be incident from the left in the figure). The wavelength is the same on both sides of the barrier, but the amplitude beyond the barrier is much less than the original amplitude. The particle can never be observed, inside the barrier (where it would have negative kinetic energy) but it can be observed *beyond* the barrier.

**The Infinite Well**

The potential is (see Figure 2.6)

$$V(x) = \infty \qquad x < 0, \quad x > a$$
$$= 0 \qquad 0 \leq x \leq a \tag{2.40}$$

That is, the particle is trapped between $x = 0$ and $x = a$. The walls at $x = 0$ and $x = a$ are absolutely impenetrable; thus the particle is never outside the well and $\psi = 0$ for $x < 0$ and for $x > a$. Inside the well, the Schrödinger equation has the form of Equation 2.18, and we will choose a solution in the form of Equation 2.19:

$$\psi = A \sin kx + B \cos kx \tag{2.41}$$

The continuity condition on $\psi$ at $x = 0$ gives $\psi(0) = 0$, which is true only for $B = 0$. At $x = a$, the continuity condition on $\psi$ gives

$$A \sin ka = 0 \tag{2.42}$$

The solution $A = 0$ is not acceptable, for that would give $\psi = 0$ everywhere. Thus $\sin ka = 0$, or

$$ka = n\pi \qquad n = 1, 2, 3, \ldots \tag{2.43}$$

and

$$E_n = \frac{\hbar^2 k^2}{2m} = \frac{\hbar^2 \pi^2}{2ma^2} n^2 \tag{2.44}$$

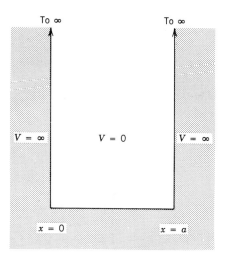

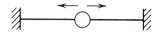

**Figure 2.6**  A particle moves freely in the one-dimensional region $0 \leq x \leq a$ but is excluded completely from $x < 0$ and $x > a$. A bead sliding without friction on a wire and bouncing elastically from the walls is a simple physical example.

Here the energy is *quantized*—only certain values of the energy are permitted. The energy spectrum is illustrated in Figure 2.7. These states are *bound states*, in which the potential confines the particle to a certain region of space.

The corresponding wave functions are

$$\psi_n = \sqrt{\frac{2}{a}} \, \sin \frac{n\pi x}{a} \qquad (2.45)$$

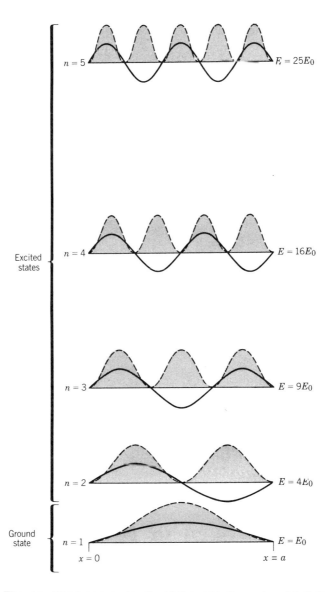

**Figure 2.7**   The permitted energy levels of the one-dimensional infinite square well. The wave function for each level is shown by the solid curve, and the shaded region gives the probability density for each level. The energy $E_0$ is $\hbar^2\pi^2/2ma^2$.

where the constant $A$ has been evaluated using Equation 2.9. The probability densities $|\psi|^2$ of some of the lower states are illustrated in Figure 2.7.

## The Finite Potential Well

For this case we assume the well has depth $V_0$ between $+a/2$ and $-a/2$:

$$V(x) = V_0 \qquad |x| > a/2$$
$$\phantom{V(x)} = 0 \qquad |x| < a/2 \qquad (2.46)$$

We look for *bound-state* solutions, with $E < V_0$. The solutions are

$$\psi_1 = A e^{k_1 x} + B e^{-k_1 x} \qquad x < -a/2$$
$$\psi_2 = C e^{ik_2 x} + D e^{-ik_2 x} \qquad -a/2 \le x \le a/2 \qquad (2.47)$$
$$\psi_3 = F e^{k_1 x} + G e^{-k_1 x} \qquad x > a/2$$

where $k_1 = \sqrt{2m(V_0 - E)/\hbar^2}$ and $k_2 = \sqrt{2mE/\hbar^2}$. To keep the wave function finite in region 1 when $x \rightarrow -\infty$, we must have $B = 0$, and to keep it finite in region 3 for $x \rightarrow +\infty$, we require $F = 0$.

Applying the continuity conditions at $x = -a/2$ and at $x = +a/2$, we find the following two relationships:

$$k_2 \tan \frac{k_2 a}{2} = k_1 \qquad (2.48a)$$

or

$$-k_2 \cot \frac{k_2 a}{2} = k_1 \qquad (2.48b)$$

These transcendental equations cannot be solved directly. They can be solved numerically on a computer, or graphically. The graphical solutions are easiest if we rewrite Equations 2.48 in the following form:

$$\alpha \tan \alpha = (P^2 - \alpha^2)^{1/2} \qquad (2.49a)$$

$$-\alpha \cot \alpha = (P^2 - \alpha^2)^{1/2} \qquad (2.49b)$$

where $\alpha = k_2 a/2$ and $P = (mV_0 a^2/2\hbar^2)^{1/2}$. The right side of these equations defines a circle of radius $P$, while the left side gives a tangentlike function with several discrete branches. The solutions are determined by the points where the circle intersects the tangent function, as shown in Figure 2.8. *Therefore, the number of solutions is determined by the radius $P$, and thus by the depth $V_0$ of the well.* (Contrast this with the infinite well, which had an infinite number of bound states.) For example, when $P < \pi/2$, there is only one bound state. For $\pi/2 < P < \pi$ there are two bound states. Conversely, if we studied a system of this sort and found only one bound state, we could deduce some limits on the depth of the well. As we will discuss in Chapter 4, a similar technique allows us to estimate the depth of the nuclear potential, because the deuteron, the simplest two-nucleon bound system, has only one bound state.

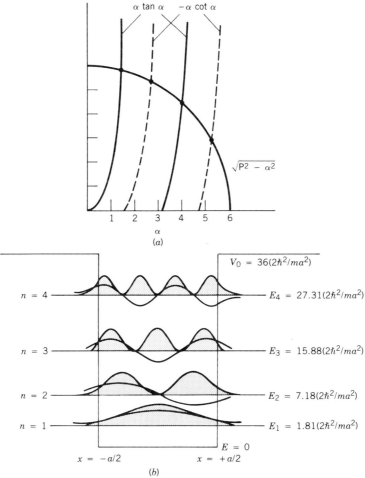

**Figure 2.8**   (*a*) The graphical solution of Equations 2.49*a* and 2.49*b*. For the case of $P = 6$ (chosen arbitrarily) there are four solutions at $\alpha = 1.345$, 2.679, 3.985, and 5.226. (*b*) The wave functions and probability densities (shaded) for the four states. (Compare with the infinite well shown in Figure 2.7.)

## The Simple Harmonic Oscillator

Any reasonably well-behaved potential can be expanded in a Taylor series about the point $x_0$:

$$V(x) = V(x_0) + \left(\frac{dV}{dx}\right)_{x=x_0}(x - x_0) + \frac{1}{2}\left(\frac{d^2V}{dx^2}\right)_{x=x_0}(x - x_0)^2 + \cdots$$

$$(2.50)$$

If $x_0$ is a potential minimum, the second term in the series vanishes, and since the first term contributes only a constant to the energy, the interesting term is the third term. Thus to a first approximation, near its minimum the system behaves

**Table 2.1**  Sample Wave Functions of the One-Dimensional Simple Harmonic Oscillator

| $n$ | $E_n$ | $\psi_n(x)$ |
|---|---|---|
| 0 | $\frac{1}{2}\hbar\omega_0$ | $\pi^{-1/4}e^{-\alpha^2x^2/2}$ |
| 1 | $\frac{3}{2}\hbar\omega_0$ | $2^{-1/2}\pi^{-1/4}(2\alpha x)\,e^{-\alpha^2x^2/2}$ |
| 2 | $\frac{5}{2}\hbar\omega_0$ | $2^{-3/2}\pi^{-1/4}(4\alpha^2x^2-2)\,e^{-\alpha^2x^2/2}$ |
| 3 | $\frac{7}{2}\hbar\omega_0$ | $(1/4\sqrt{3}\,\pi^{1/4})(8\alpha^3x^3-12\alpha x)\,e^{-\alpha^2x^2/2}$ |
| 4 | $\frac{9}{2}\hbar\omega_0$ | $(1/8\sqrt{6}\,\pi^{1/4})(16\alpha^4x^4-48\alpha^2x^2+12)\,e^{-\alpha^2x^2/2}$ |

$$E_n = \hbar\omega_0(n+\tfrac{1}{2})$$
$$\psi_n(x) = (2^n n!\sqrt{\pi}\,)^{-1/2}H_n(\alpha x)\,e^{-\alpha^2x^2/2}$$
where $H_n(\alpha x)$ is a Hermite polynomial

like a simple harmonic oscillator, which has the similar potential $\frac{1}{2}k(x-x_0)^2$. The study of the simple harmonic oscillator therefore is important for understanding a variety of systems.

For our system, we choose the potential energy

$$V(x) = \tfrac{1}{2}kx^2 \tag{2.51}$$

for all $x$. The Schrödinger equation for this case is solved through the substitution $\psi(x) = h(x)\,e^{-\alpha^2x^2/2}$, where $\alpha^2 = \sqrt{km}/\hbar$. The function $h(x)$ turns out to be a simple polynomial in $x$. The degree of the polynomial (the highest power of $x$ that appears) is determined by the quantum number $n$ that labels the energy states, which are also found from the solution to the Schrödinger equation:

$$E_n = \hbar\omega_0\left(n+\tfrac{1}{2}\right) \qquad n = 0,1,2,3,\dots \tag{2.52}$$

where $\omega_0 = \sqrt{k/m}$, the classical angular frequency of the oscillator. Some of the resulting wave functions are listed in Table 2.1, and the corresponding energy levels and probability densities are illustrated in Figure 2.9. Notice that the probabilities resemble those of Figure 2.8; where $E > V$, the solution oscillates somewhat sinusoidally, while for $E < V$ (beyond the classical turning points where the oscillator comes to rest and reverses its motion) the solution decays exponentially. This solution also shows penetration of the probability density into the classically forbidden region.

A noteworthy feature of this solution is that the energy levels are equally spaced. Also, because the potential is infinitely deep, there are infinitely many bound states.

## Summary

By studying these one-dimensional problems, we learn several important details about the wave properties of particles.

1. Quantum waves can undergo reflection and transmission when they encounter a potential barrier; this behavior is very similar to that of classical waves.

2. A wave packet can penetrate into the classically forbidden region and appear beyond a potential barrier that it does not have enough energy to overcome.

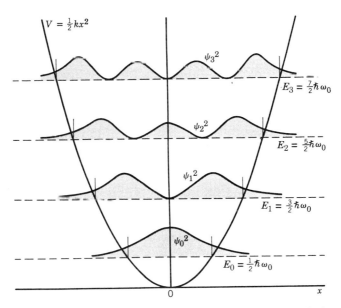

**Figure 2.9**  The lowest few energy levels and corresponding probability densities of the harmonic oscillator.

3.  Wave functions oscillate whenever $E > V(x)$ and decay exponentially whenever $E < V(x)$.

4.  When a potential confines a particle to a region of space, bound-state wave functions can result. The particle is permitted only a set of discrete energy values; the number of allowed energy values is determined by the depth of the potential well.

## 2.4  PROBLEMS IN THREE DIMENSIONS

### The Infinite Cartesian Well

We begin with a problem in Cartesian coordinates that illustrates an important feature present in three-dimensional problems but not in one-dimensional problems. The potential in this case is

$$V(x, y, z) = 0 \qquad 0 \le x \le a, \quad 0 \le y \le a, \quad 0 \le z \le a$$

$$\qquad\qquad = \infty \quad x < 0, \quad x > a, \quad y < 0, \quad y > a, \quad z < 0, \quad z > a$$

(2.53)

The particle is thus confined to a cubical box of dimension $a$. Beyond the impenetrable walls of the well, $\psi = 0$ as before. Inside the well, the Schrödinger equation is

$$-\frac{\hbar^2}{2m}\left(\frac{\partial^2 \psi}{\partial x^2} + \frac{\partial^2 \psi}{\partial y^2} + \frac{\partial^2 \psi}{\partial z^2}\right) = E\psi(x, y, z)$$

(2.54)

The usual procedure for solving partial differential equations is to try to find a

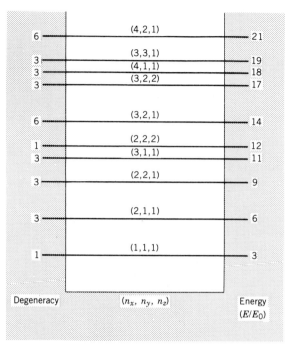

**Figure 2.10** Energy levels of a particle confined to a three-dimensional cubical box. The energy is given in units of $E_0 = \hbar^2\pi^2 / 2ma^2$.

separable solution, with $\psi(x, y, z) = X(x)\, Y(y)\, Z(z)$, where $X$, $Y$, and $Z$ are each functions of a single variable. We will skip the mathematical details and give only the result of the calculation:

$$\psi_{n_x n_y n_z}(x, y, z) = \sqrt{\left(\frac{2}{a}\right)^3}\, \sin\frac{n_x \pi x}{a}\, \sin\frac{n_y \pi y}{a}\, \sin\frac{n_z \pi z}{a} \qquad (2.55)$$

$$E_{n_x n_y n_z} = \frac{\hbar^2\pi^2}{2ma^2}\left(n_x^2 + n_y^2 + n_z^2\right) \qquad (2.56)$$

where $n_x$, $n_y$, and $n_z$ are independent integers greater than zero. The lowest state, the ground state, has quantum numbers $(n_x, n_y, n_z) = (1, 1, 1)$. Its probability distribution would show a maximum at the center of the box ($x = y = z = a/2$), falling gradually to zero at the walls like $\sin^2$.

The first excited state has three possible sets of quantum numbers: $(2, 1, 1)$, $(1, 2, 1)$, and $(1, 1, 2)$. Each of these distinct and independent states has a different wave function, and therefore a different probability density and different expectation values of the physical observables, *but they all have the same energy*. This situation is known as *degeneracy*; the first excited state is threefold degenerate. Degeneracy is extremely important for atomic structure since it tells us how many electrons can be put in each atomic subshell. We will soon discuss its similar role in the nuclear shell model.

Figure 2.10 shows the lower portion of the energy spectrum of excited states. Notice that the spacing and ordering do not have the regularity of the one-dimensional problem.

### The Infinite Spherical Well

If we work in spherical coordinates with a potential that depends only on $r$ (not on $\theta$ or $\phi$), another new feature arises that will be important in our subsequent investigations of nuclear structure. When we search for separable solutions, of the form $\psi(r, \theta, \phi) = R(r)\,\Theta(\theta)\,\Phi(\phi)$, the central potential $V(r)$ appears only in the radial part of the separation equation, and the angular parts can be solved directly. The differential equation for $\Phi(\phi)$ is

$$\frac{d^2\Phi}{d\phi^2} + m_\ell^2\, \Phi = 0 \tag{2.57}$$

where $m_\ell^2$ is the separation constant.

The solution is

$$\Phi_{m_\ell}(\phi) = \frac{1}{\sqrt{2\pi}}\, e^{im_\ell\phi} \tag{2.58}$$

where $m_\ell = 0, \pm 1, \pm 2, \ldots$ . The equation for $\Theta(\theta)$ is

$$\frac{1}{\sin\theta}\frac{d}{d\theta}\left(\sin\theta\,\frac{d\Theta}{d\theta}\right) + \left[\ell(\ell+1) - \frac{m_\ell^2}{\sin^2\theta}\right]\Theta = 0 \tag{2.59}$$

where $\ell = 0, 1, 2, 3, \ldots$ and $m_\ell = 0, \pm 1, \pm 2, \ldots, \pm\ell$. The solution $\Theta_{\ell m_\ell}(\theta)$ can be expressed as a polynomial of degree $\ell$ in $\sin\theta$ or $\cos\theta$. Together, and normalized, $\Phi_{m_\ell}(\phi)$ and $\Theta_{\ell m_\ell}(\theta)$ give the *spherical harmonics* $Y_{\ell m_\ell}(\theta, \phi)$, some examples of which are listed in Table 2.2. These functions give the angular part of the solution to the Schrödinger equation for *any central potential* $V(r)$. For example, it is these angular functions that give the spatial properties of atomic orbitals that are responsible for molecular bonds.

For each potential $V(r)$, all we need to do is to find a solution of the radial equation

$$-\frac{\hbar^2}{2m}\left(\frac{d^2R}{dr^2} + \frac{2}{r}\frac{dR}{dr}\right) + \left[V(r) + \frac{\ell(\ell+1)\hbar^2}{2mr^2}\right]R = ER \tag{2.60}$$

**Table 2.2**   Spherical Harmonics for Some Low $\ell$ Values

| $\ell$ | $m_\ell$ | $Y_{\ell m_\ell}(\theta, \phi) = \Theta_{\ell m_\ell}(\theta)\,\Phi_{m_\ell}(\phi)$ |
|---|---|---|
| 0 | 0 | $(1/4\pi)^{1/2}$ |
| 1 | 0 | $(3/4\pi)^{1/2}\cos\theta$ |
| 1 | $\pm 1$ | $\mp(3/8\pi)^{1/2}\sin\theta\, e^{\pm i\phi}$ |
| 2 | 0 | $(5/16\pi)^{1/2}(3\cos^2\theta - 1)$ |
| 2 | $\pm 1$ | $\mp(15/8\pi)^{1/2}\sin\theta\cos\theta\, e^{\pm i\phi}$ |
| 2 | $\pm 2$ | $(15/32\pi)^{1/2}\sin^2\theta\, e^{\pm 2i\phi}$ |

$$\Phi_{m_\ell}(\phi) = \frac{1}{\sqrt{2\pi}}\, e^{im_\ell\phi}$$

$$\Theta_{\ell m_\ell}(\theta) = \left[\frac{2\ell+1}{2}\frac{(\ell - m_\ell)!}{(\ell + m_\ell)!}\right]^{1/2} P_\ell^{m_\ell}(\theta)$$

where $P_\ell^{m_\ell}(\theta)$ is the associated Legendre polynomial

**Table 2.3** Spherical Bessel Functions —
Sample Expressions and Limits

$$j_0(kr) = \frac{\sin kr}{kr}$$

$$j_1(kr) = \frac{\sin kr}{(kr)^2} - \frac{\cos kr}{kr}$$

$$j_2(kr) = \frac{3 \sin kr}{(kr)^3} - \frac{3 \cos kr}{(kr)^2} - \frac{\sin kr}{kr}$$

$$j_\ell(kr) \cong \frac{(kr)^\ell}{1 \cdot 3 \cdot 5 \cdots (2\ell + 1)} \qquad kr \to 0$$

$$j_\ell(kr) \cong \frac{\sin(kr - \ell\pi/2)}{kr} \qquad kr \to \infty$$

$$j_\ell(kr) = \left(-\frac{r}{k}\right)^\ell \left(\frac{1}{r}\frac{d}{dr}\right)^\ell j_0(kr)$$

The $\ell(\ell + 1)$ term is generally written as an addition to the potential; it is called the "centrifugal potential" and it acts like a potential that keeps the particle away from the origin when $\ell > 0$.

As an example, we consider the case of the infinite spherical well,

$$V(r) = 0 \qquad r < a$$

$$= \infty \qquad r > a \qquad (2.61)$$

We require again that $R(r) = 0$ for $r > a$, since the walls of the infinite well are impenetrable. Inside the well, the solution to Equation 2.60 for $V = 0$ can be expressed in terms of the oscillatory functions known as *spherical Bessel functions* $j_\ell(kr)$, some of which are listed in Table 2.3. To find the energy eigenvalues, we proceed exactly as in the one-dimensional problem and apply the continuity condition on $\psi$ at $r = a$. This gives

$$j_\ell(ka) = 0 \qquad (2.62)$$

This is in effect a transcendental equation, which must be solved numerically. Tables of the spherical Bessel functions are available that can be consulted to find the zeros for any given value of $\ell$.* For example, we consider the case $\ell = 0$. From the tables we find $j_0(x) = 0$ at $x = 3.14, 6.28, 9.42, 12.57$, and so on. For $\ell = 1$, the first few zeros of $j_1(x)$ are at $x = 4.49, 7.73, 10.90, 14.07$. Since $E = \hbar^2 k^2 / 2m$, we can then solve for the allowed values of the energies. Repeating this process for $\ell = 2$, $\ell = 3$, and so on, we would be able to construct a spectrum of the energy states, as is shown in Figure 2.11. As in the case of the Cartesian well, the regularity of the one-dimensional problem is not present. Also, note that the levels are again degenerate—since the energy depends only on $\ell$, the wave functions with different $m_\ell$ values all have the same energy. Thus, in the case of the level with $\ell = 2$, the possible wave functions are $j_2(kr)Y_{22}(\theta, \phi)$,

*M. Abramowitz and I. A. Stegun, *Handbook of Mathematical Functions* (New York: Dover, 1965).

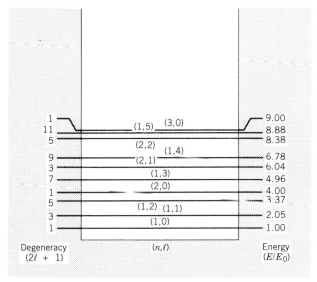

**Figure 2.11** Energy levels of a particle confined to a three-dimensional spherical container. The energy is given in units of $E_0 = \hbar^2\pi^2 / 2ma^2$. Compare with the spacings and degeneracies of Figure 2.10. The quantum number $n$ does not arise directly in the solution in this case; it serves to number the states of a given $\ell$.

$j_2(kr)Y_{21}(\theta, \phi)$, $j_2(kr)Y_{20}(\theta, \phi)$, $j_2(kr)Y_{2-1}(\theta, \phi)$, and $j_2(kr)Y_{2-2}(\theta, \phi)$, for a fivefold degeneracy. In fact, since $m_\ell$ is restricted to the values $0$, $\pm 1$, $\pm 2, \ldots, \pm \ell$, there are exactly $2\ell + 1$ possible $Y_{\ell m_\ell}$ for a given $\ell$, and thus each level has a degeneracy of $2\ell + 1$. (This situation is very similar to the case of electronic orbits in atoms, in which there is also a central potential. The capacity of each atomic subshell contains the factor of $2\ell + 1$, which likewise arises from the $m_\ell$ degeneracy.)

The probability to locate the particle in a volume $dv$ is given by $|\psi|^2\, dv$, where the volume element was given in Equation 2.17. Such three-dimensional distributions are difficult to represent graphically, so we often consider the radial and angular parts separately. To find the radial probability density, which gives the probability to find the particle between $r$ and $r + dr$ averaged over all angles, we integrate the probability density over $\theta$ and $\phi$:

$$P(r)\, dr = \int |\psi|^2\, dv$$

$$= r^2\, |R(r)|^2\, dr \int \sin\theta\, d\theta \int d\phi\, |Y_{\ell m_\ell}|^2 \qquad (2.63)$$

The spherical harmonics $Y_{\ell m_\ell}$ are themselves normalized, so that the integral gives 1, and thus

$$P(r) = r^2\, |R(r)|^2 \qquad (2.64)$$

Some sample radial probability distributions for the infinite well are shown in Figure 2.12.

The angular dependence of the probability density for any central potential is given by $|Y_{\ell m_\ell}(\theta, \phi)|^2$, some samples of which are illustrated in Figure 2.13.

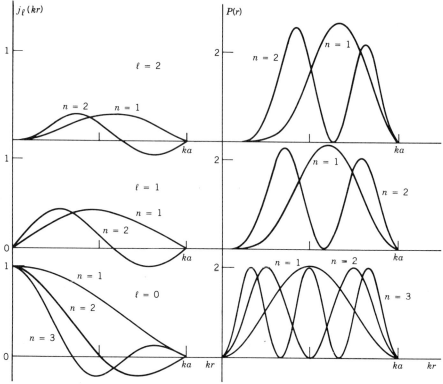

**Figure 2.12** The left side shows, for some of the lower energy levels, the unnormalized $j_\ell(kr)$, adjusted so that $j_\ell(ka) = 0$. The right side shows the corresponding normalized radial probability density, $r^2 R^2$. Note that all $j_\ell$ vanish at the origin except $j_0$, and that all probability densities vanish at $r = 0$. Also, note how the "centrifugal repulsion" pushes corresponding maxima in $P(r)$ away from the origin as $\ell$ increases.

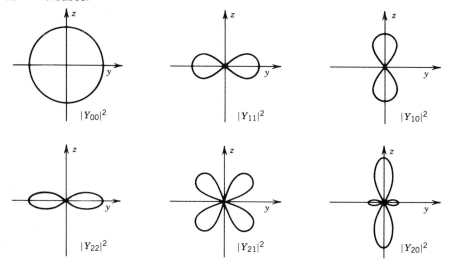

**Figure 2.13** Spatial probability distributions resulting from the $Y_{\ell m_\ell}$. The three-dimensional representations can be obtained by rotating each figure about the $z$ axis.

**Table 2.4**  Sample Radial Wave Functions for Three-Dimensional
Simple Harmonic Oscillator

| $n$ | $\ell$ | $E_n$ | $R(r)$ |
|---|---|---|---|
| 0 | 0 | $\frac{3}{2}\hbar\omega_0$ | $(2\alpha^{3/2}/\pi^{1/4})\,e^{-\alpha^2 r^2/2}$ |
| 1 | 1 | $\frac{5}{2}\hbar\omega_0$ | $(2\alpha^{3/2}\sqrt{2}/\sqrt{3}\,\pi^{1/4})(\alpha r)\,e^{-\alpha^2 r^2/2}$ |
| 2 | 0 | $\frac{7}{2}\hbar\omega_0$ | $(2\alpha^{3/2}\sqrt{2}/\sqrt{3}\,\pi^{1/4})(\frac{3}{2}-\alpha^2 r^2)\,e^{-\alpha^2 r^2/2}$ |
| 2 | 2 | $\frac{7}{2}\hbar\omega_0$ | $(4\alpha^{3/2}/\sqrt{15}\,\pi^{1/4})(\alpha^2 r^2)\,e^{-\alpha^2 r^2/2}$ |
| 3 | 1 | $\frac{9}{2}\hbar\omega_0$ | $(4\alpha^{3/2}/\sqrt{15}\,\pi^{1/4})(\frac{5}{2}\alpha r-\alpha^3 r^3)\,e^{-\alpha^2 r^2/2}$ |
| 3 | 3 | $\frac{9}{2}\hbar\omega_0$ | $(4\alpha^{3/2}\sqrt{2}/\sqrt{105}\,\pi^{1/4})(\alpha^3 r^3)\,e^{-\alpha^2 r^2/2}$ |
| 4 | 0 | $\frac{11}{2}\hbar\omega_0$ | $(4\alpha^{3/2}\sqrt{2}/\sqrt{15}\,\pi^{1/4})(\frac{15}{8}-\frac{5}{2}\alpha^2 r^2+\frac{1}{2}\alpha^4 r^4)\,e^{-\alpha^2 r^2/2}$ |
| 4 | 2 | $\frac{11}{2}\hbar\omega_0$ | $(4\alpha^{3/2}\sqrt{2}/\sqrt{105}\,\pi^{1/4})(\frac{7}{2}\alpha^2 r^2-\alpha^4 r^4)\,e^{-\alpha^2 r^2/2}$ |
| 4 | 4 | $\frac{11}{2}\hbar\omega_0$ | $(8\alpha^{3/2}/3\sqrt{105}\,\pi^{1/4})\,\alpha^4 r^4\,e^{-\alpha^2 r^2/2}$ |

Note the similarity in form (polynomial × exponential) between these solutions and those of the one-dimensional problem shown in Table 2.1. In this case the polynomials are called Laguerre polynomials. The solutions are discussed in J. L. Powell and B. Crasemann, *Quantum Mechanics* (Reading, MA: Addison-Wesley, 1961), Chapter 7.

## The Simple Harmonic Oscillator

We consider a central oscillator potential, $V(r) = \frac{1}{2}kr^2$. The angular part of the solution to the Schrödinger equation is $Y_{\ell m_\ell}(\theta,\phi)$ for all central potentials, so all we need to consider here is the solution to the radial equation. As in the one-dimensional case, the solution can be expressed as the product of an exponential and a finite polynomial. Some representative solutions are listed in Table 2.4, and the corresponding radial probability densities are illustrated in Figure 2.14. The general properties of the one-dimensional solutions are also present in this case: oscillation in the classically allowed region, and exponential decay in the classically forbidden region.

The energy levels are given by

$$E_n = \hbar\omega_0\left(n + \tfrac{3}{2}\right) \tag{2.65}$$

where $n = 0, 1, 2, 3, \ldots$ . The energy does not depend on $\ell$, but not all $\ell$ values are permitted. From the mathematical solution of the radial equation, the restrictions on $\ell$ are as follows: $\ell$ can be at most equal to $n$ and takes only even or only odd values as $n$ is even or odd. For $n = 5$, the permitted values of $\ell$ are 1, 3, and 5; for $n = 4$, the values of $\ell$ are 0, 2, and 4. Since the energies do not depend on $m_\ell$ either, there is an additional degeneracy of $2\ell + 1$ for each $\ell$ value. Thus the $n = 5$ level has a degeneracy of $[(2 \times 1 + 1) + (2 \times 3 + 1) + (2 \times 5 + 1)] = 21$, while the $n = 4$ level has a degeneracy of $[(2 \times 0 + 1) + (2 \times 2 + 1) + (2 \times 4 + 1)] = 15$. Figure 2.15 shows some of the energy levels and their degeneracies, which are equal to $\frac{1}{2}(n + 1)(n + 2)$.

## The Coulomb Potential

The attractive Coulomb potential energy also has a simple central form, $V(r) = -Ze^2/4\pi\epsilon_0 r$, for the interaction between electrical charges $+Ze$ and $-e$, such as in the case of a single electron in an atom of atomic number $Z$. The angular

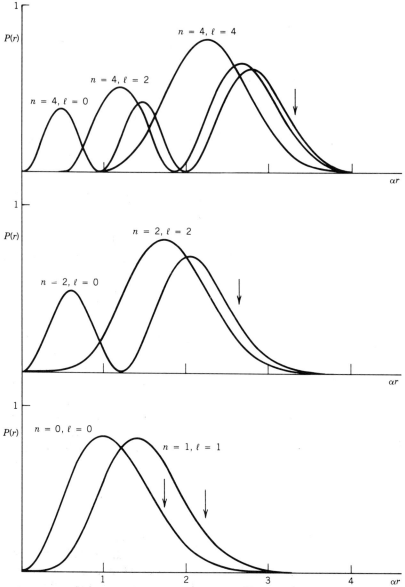

**Figure 2.14** Radial probabililty densities for some states of the three-dimensional harmonic oscillator. The vertical arrows mark the classical turning points. As in Figure 2.12, note that $P(r)$ vanishes for $r = 0$ (but note from Table 2.4 that $R(r)$ is nonvanishing at $r = 0$ only for $\ell = 0$). Also note the "centrifugal repulsion" for the larger $\ell$ values.

part of the wave function is again given by $Y_{\ell m_\ell}(\theta, \phi)$, and some of the radial wave functions $R(r)$ are listed in Table 2.5. The energy levels are $E_n = (-mZ^2e^4/32\pi^2\epsilon_0^2\hbar^2n^2)$ and are shown in Figure 2.16; the radial probability density is plotted for several states in Figure 2.17. The relationship between $n$ and $\ell$ is different from what it was for the oscillator potential: for each level $n$, the permitted values of $\ell$ are $0, 1, 2, \ldots, (n-1)$. The total degeneracy of each

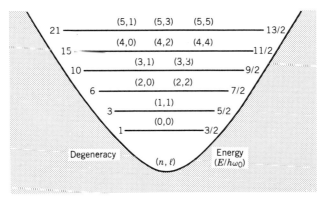

**Figure 2.15** Some of the lower energy levels of a particle in a central three-dimensional oscillator potential.

**Table 2.5**  Coulomb (Hydrogenic) Radial Wave Functions

| $n$ | $\ell$ | $R(r)$ |
|---|---|---|
| 1 | 0 | $2(Z/a_0)^{3/2}\, e^{-Zr/a_0}$ |
| 2 | 0 | $(Z/2a_0)^{3/2}(2 - Zr/a_0)\, e^{-Zr/2a_0}$ |
| 2 | 1 | $3^{-1/2}(Z/2a_0)^{3/2}(Zr/a_0)\, e^{-Zr/2a_0}$ |
| 3 | 0 | $\frac{2}{3}(Z/3a_0)^{3/2}[3 - 2Zr/a_0 + 2(Zr/3a_0)^2]\, e^{-Zr/3a_0}$ |
| 3 | 1 | $(4\sqrt{2}/9)(Z/3a_0)^{3/2}(Zr/a_0)(1 - Zr/6a_0)\, e^{-Zr/3a_0}$ |
| 3 | 2 | $(2\sqrt{2}/27\sqrt{5})(Z/3a_0)^{3/2}(Zr/a_0)^2\, e^{-Zr/3a_0}$ |

The radial wave functions have the mathematical form of associated Laguerre polynomials multiplied by exponentials. The Bohr radius $a_0$ is $4\pi\epsilon_0\hbar^2/me^2$. For a discussion of these solutions, see L. Pauling and E. B. Wilson, *Introduction to Quantum Mechanics* (New York: McGraw-Hill, 1935), Chapter 5.

energy level, including the various $\ell$ values and the $2\ell + 1$ degeneracy of each, is $n^2$.

## Summary

The mathematical techniques of finding and using solutions to the Schrödinger equation in three dimensions are similar to the techniques we illustrated previously for one-dimensional problems. There are two important new features in the three-dimensional calculations that do not arise in the one-dimensional calculations: (1) The energy levels are *degenerate*—several different wave functions can have the same energy. The degeneracies will have the same effect in the nuclear shell model that the $\ell$ and $m_\ell$ degeneracies of the energy levels of the Coulomb potential have in the atomic shell model—they tell us how many particles can occupy each energy level. (2) When the potential depends only on $r$ (not on $\theta$ or $\phi$), the wave functions can be assigned a definite angular momentum quantum number $\ell$. These new features will have important consequences when we discuss

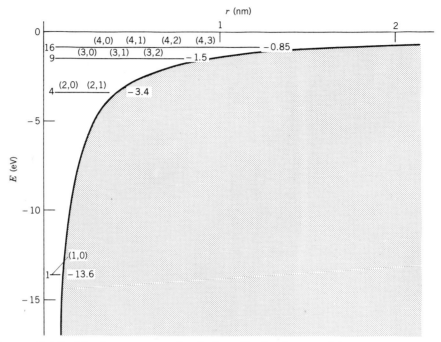

**Figure 2.16** The lower energy levels in a Coulomb potential, shown for $Z = 1$ (hydrogen atom). The states are labeled with $(n, \ell)$; the degeneracies are indicated on the left and the energy values on the right.

nuclear models in Chapter 5. The behavior of angular momentum in quantum theory is discussed in the next section.

## 2.5 QUANTUM THEORY OF ANGULAR MOMENTUM

In solutions of the Schrödinger equation for three-dimensional problems, the quantum number $\ell$ plays a prominent role. In atomic physics, for example, it serves to label different electron wave functions and to tell us something about the spatial behavior of the wave functions. This *angular momentum quantum number* has the same function in all three-dimensional problems involving central potentials, where $V = V(r)$.

In classical physics, the angular momentum $\ell$ of a particle moving with linear momentum $p$ at a location $r$ from a reference point is defined as

$$\ell = r \times p \tag{2.66}$$

In quantum mechanics, we can evaluate the expectation value of the angular momentum by analogy with Equation 2.10. We first consider the magnitude of the angular momentum, and for this purpose it is simplest to calculate $\ell^2$. We must first find a quantum mechanical *operator* for $\ell^2$, as we discussed in Section 2.2. This can be done simply by replacing the components of $p$ with their operator equivalents: $p_x = -i\hbar\, \partial/\partial x$, $p_y = -i\hbar\, \partial/\partial y$, $p_z = -i\hbar\, \partial/\partial z$. Evaluating the cross product then gives terms of the form $\ell_x = yp_z - zp_y$, and finally computing $\langle \ell^2 \rangle = \langle \ell_x^2 + \ell_y^2 + \ell_z^2 \rangle$ gives the remarkably simple result, which is

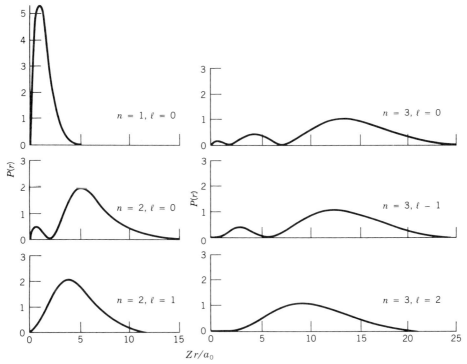

**Figure 2.17** Radial probability distributions for a particle in a Coulomb potential (hydrogenic atom). The probability vanishes at $r = 0$, but as before the $\ell = 0$ wave functions do not. This property becomes especially important for phenomena that depend on the overlap of atomic wave functions with the nucleus — only $\ell = 0$ states contribute substantially to such phenomena (electron capture, hyperfine structure, etc.). Why doesn't the "centrifugal repulsion" appear to occur in this case?

independent of the form of $R(r)$,

$$\langle \ell^2 \rangle = \hbar^2 \ell(\ell + 1) \tag{2.67}$$

That is, whenever we have a central potential, which gives a wave function $R(r)Y_{\ell m_\ell}(\theta, \phi)$, the magnitude of the angular momentum is fixed at the value given by Equation 2.67; *the angular momentum is a constant of the motion* (as it is in classical physics for central potentials). The atomic substates with a given $\ell$ value are labeled using *spectroscopic notation*; we use the same spectroscopic notation in nuclear physics: s for $\ell = 0$, p for $\ell = 1$, and so on. These are summarized in Table 2.6.

When we now try to find the direction of $\ell$, we run into a barrier imposed by the uncertainty principle: quantum mechanics permits us to know exactly only

**Table 2.6**  Spectroscopic Notation

| $\ell$ value | 0 | 1 | 2 | 3 | 4 | 5 | 6 |
|---|---|---|---|---|---|---|---|
| Symbol | s | p | d | f | g | h | i |

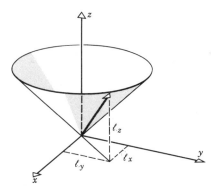

**Figure 2.18** The vector $\ell$ precesses rapidly about the $z$ axis, so that $\ell_z$ stays constant, but $\ell_x$ and $\ell_y$ are variable.

one component of $\ell$ at a time. Once we determine the value of one component, the other two components are completely indeterminate. (This is a fundamental limitation, and no amount of trickery can get us around it. It is the very act of measuring one component that *makes* the other two indeterminate. When we measure $\ell_x$, we force $\ell_y$ and $\ell_z$ into indeterminacy; when we then measure $\ell_y$ for the same system, our previous knowledge of $\ell_x$ is destroyed as $\ell_x$ is now forced into indeterminacy.) By convention, we usually choose the $z$ component of $\ell$ to be determined, and computing $\langle \ell_z \rangle$ as described above,

$$\langle \ell_z \rangle = \hbar m_\ell \qquad (2.68)$$

where $m_\ell = 0, \pm 1, \pm 2, \ldots, \pm \ell$. Notice that $|\langle \ell_z \rangle| < |\ell| = \hbar\sqrt{\ell(\ell+1)}$ —the $z$ component of the vector is always less than its length. If $|\langle \ell_z \rangle| = |\ell|$ were permitted, then we would have exact knowledge of all three components of $\ell$ ($\ell_x$ and $\ell_y$ would be zero if $\ell$ were permitted to align with the $z$ axis). The conventional vector representation of this indeterminacy is shown in Figure 2.18 —$\ell$ rotates or precesses about the $z$ axis keeping $\ell_z$ fixed but varying $\ell_x$ and $\ell_y$.

The complete description of an electronic state in an atom requires the introduction of a new quantum number, the *intrinsic angular momentum* or *spin*. For the electron, the spin quantum number is $s = \frac{1}{2}$. The spin can be treated as an angular momentum (although it cannot be represented in terms of classical variables, because it has no classical analog). Thus

$$\langle s^2 \rangle = \hbar^2 s(s+1) \qquad (2.69)$$

$$\langle s_z \rangle = \hbar m_s \qquad \left(m_s = \pm \tfrac{1}{2}\right) \qquad (2.70)$$

It is often useful to imagine the spin as a vector $s$ with possible $z$ components $\pm \frac{1}{2}\hbar$.

Nucleons, like electrons, have spin quantum numbers of $\frac{1}{2}$. A nucleon moving in a central potential with orbital angular momentum $\ell$ and spin $s$ has a *total angular momentum*

$$\boldsymbol{j} = \boldsymbol{\ell} + \boldsymbol{s} \qquad (2.71)$$

The total angular momentum $j$ behaves in a manner similar to $\ell$ and $s$:

$$\langle j^2 \rangle = \hbar^2 j(j+1) \qquad (2.72)$$

$$\langle j_z \rangle = \langle \ell_z + s_z \rangle = \hbar m_j \qquad (2.73)$$

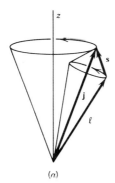
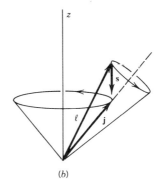

(a)                                          (b)

**Figure 2.19**  The coupling of orbital angular momentum $\ell$ to spin angular momentum $s$ giving total angular momentum $j$. (a) Coupling giving $j = \ell + \frac{1}{2}$. The vectors $\ell$ and $s$ have definite lengths, as does $j$. The combined $\ell$ and $s$ vectors rotate or precess about the direction of $j$; in this coupling the z components $\ell_z$ and $s_z$ thus do not have definite values. The vector $j$ precesses about the z direction so that $j_z$ has a definite value. (b) The similar case of $j = \ell - \frac{1}{2}$. In interpreting both figures, keep in mind that all such representations of vectors governed by the rules of quantum mechanics are at best symbolic and at worst misleading.

where $m_j = -j, -j + 1, \ldots, j - 1, j$ and where $j$ is the total angular momentum quantum number. From Equations 2.68, 2.70, and 2.73 it is apparent that

$$m_j = m_\ell + m_s = m_\ell \pm \tfrac{1}{2} \qquad (2.74)$$

Since $m_\ell$ is always an integer, $m_j$ must be half-integral ($\pm \frac{1}{2}, \pm \frac{3}{2}, \pm \frac{5}{2}, \ldots$) and thus $j$ must be half-integral. The vector coupling of Equation 2.71 suggests only two possible values for $j$: $\ell + \frac{1}{2}$ or $\ell - \frac{1}{2}$, which are illustrated in Figure 2.19.

Usually, we indicate the $j$ value as a subscript in spectroscopic notation. Thus, for $\ell = 1$ (p states), there are two possible $j$ values: $\ell + \frac{1}{2} = \frac{3}{2}$ and $\ell - \frac{1}{2} = \frac{1}{2}$. We would indicate these states as $p_{3/2}$ and $p_{1/2}$. When there is an additional quantum number, such as a principal quantum number $n$ (or perhaps just an index which counts the states in order of increasing energy), we indicate it as $2p_{3/2}$, $3p_{3/2}$, and so on.

In atoms, it is often useful for us to picture the electrons as moving in well defined orbits with definite $\ell$ and $j$. It is not at all obvious that a similar picture is useful for nucleons inside the nucleus, and thus it is not clear that $\ell$ and $j$ will be useful labels. We discuss this topic in detail when we consider the nuclear shell model in Chapter 5.

## 2.6  PARITY

The parity operation causes a reflection of all of the coordinates through the origin: $r \rightarrow -r$. In Cartesian coordinates, this means $x \rightarrow -x$, $y \rightarrow -y$, $z \rightarrow -z$; in spherical coordinates, $r \rightarrow r$, $\theta \rightarrow \pi - \theta$, $\phi \rightarrow \phi + \pi$. If a system is left unchanged by the parity operation, then we expect that none of the observable properties should change as a result of the reflection. Since the values we measure for the observable quantities depend on $|\psi|^2$, then we have the following reasonable assertion:

If $V(r) = V(-r)$,    then $|\psi(r)|^2 = |\psi(-r)|^2$.

This assertion, whose reverse is also true, has two important consequences for our work in nuclear physics:

1.  If $|\psi(r)|^2 = |\psi(-r)|^2$ then $\psi(-r) = \pm\psi(r)$. That is, the parity operation has either of two effects on a wave function. The case $\psi(-r) = +\psi(r)$ is known as *positive* or *even* parity, while the case $\psi(-r) = -\psi(r)$ is *negative* or *odd* parity. If the potential $V(r)$ is left unchanged by the parity operation, then the resulting stationary-state wave functions must be of either even or odd parity. Mixed-parity wave functions are not permitted. Recall our solutions for the one-dimensional harmonic oscillator. The potential $\frac{1}{2}kx^2$ is certainly invariant with respect to the parity operation $x \to -x$. The wave functions listed in Table 2.1 have either only odd powers of $x$, and therefore odd parity, or only even powers of $x$, and therefore even parity. Polynomials mixing odd and even powers do not occur. Also, review the solutions for the finite potential well. Since the well lies between $x = +a/2$ and $x = -a/2$, the potential is symmetric with respect to the parity operation: $V(x) = V(-x)$. Notice the solutions illustrated in Figure 2.8. For some of the solutions, $\psi(-x) = \psi(x)$ and their parity is even; the other solutions have $\psi(-x) = -\psi(x)$ and odd parity.

    In three dimensions, the parity operation applied to the $Y_{\ell m_\ell}$ gives a phase $(-1)^\ell$:

    $$Y_{\ell m_\ell}(\pi - \theta, \phi + \pi) = (-1)^\ell Y_{\ell m_\ell}(\theta, \phi) \qquad (2.75)$$

    Central potentials, which depend only on the magnitude of $r$, are thus invariant with respect to parity, and their wave functions have definite parity, odd if $\ell$ is odd and even if $\ell$ is even.

    The wave function for a system of many particles is formed from the product of the wave functions for the individual particles. The parity of the combined wave function will be even if the combined wave function represents any number of even-parity particles or an even number of odd-parity particles; it will be odd if there is an odd number of odd-parity particles. Thus nuclear states can be assigned a definite parity, odd or even. This is usually indicated along with the total angular momentum for that state, as for example, $\frac{5}{2}^+$ or $\frac{3}{2}^-$. In Chapter 10 we will discuss how the parity of a state can be determined experimentally.

2.  The second consequence of the parity rule is based on its converse. If we find a system for which $|\psi(r)|^2 \neq |\psi(-r)|^2$, then we must conclude that $V(r) \neq V(-r)$; that is, the system is *not* invariant with respect to parity. In 1957 it was discovered that certain nuclear processes ($\beta$ decays) gave observable quantities whose measured values did not respect the parity symmety. On the other hand, no evidence has yet been obtained that either the strong nuclear interaction or the electromagnetic interaction violate parity. The establishment of parity violation in $\beta$ decay was one of the most dramatic discoveries in nuclear physics and has had profound influences on the development of theories of fundamental interactions between particles. A description of these experiments is given in Section 9.9.

## 2.7   QUANTUM STATISTICS

When we group several particles together to make a larger quantum system (several nucleons in a nucleus, several electrons in an atom, several atoms in a molecule) a new quantum effect arises if the particles are indistinguishable from one another. Let us consider the case of two particles, for example, the two electrons in a helium atom. Suppose one electron is described by coordinates $r_1$ and is in the state $\psi_A$, while the other electron is described by coordinates $r_2$ and is in the state $\psi_B$. The combined wave function is the product of the two component wave functions; thus $\psi = \psi_A(r_1)\psi_B(r_2)$. Now suppose the two electrons are interchanged so that the new wave function is $\psi' = \psi_B(r_1)\psi_A(r_2)$. Is there any measurement we could do to detect whether this interchange had taken place?

If the electrons are truly *indistinguishable*, the answer to this question must be no. There is no *observational scheme* for distinguishing the "first electron" from the "second electron." Thus we have a result that is somewhat similar to our result for the parity operation: *Probability densities must be invariant with respect to exchange of identical particles.* That is, the exchanged wave function $\psi_{21}$ can at most differ only in sign from the original wave function $\psi_{12}$. We therefore have two cases. If the sign does not change upon exchange of the particles, we have a *symmetric* wave function; for symmetric wave functions, $\psi_{12} = \psi_{21}$. If the exchange changes the sign, we have an *antisymmetric* wave function, for which $\psi_{21} = -\psi_{12}$. *All combined wave functions representing identical particles must be either completely symmetric or completely antisymmetric.* No "mixed symmetry" wave functions are allowed.

When we turn to our laboratory experiments to verify these assertions, we find a further classification to which there are no known exceptions: all particles with integral spins $(0, 1, 2, \ldots)$ have symmetric combined wave functions, while all particles with half-integral spins $(\frac{1}{2}, \frac{3}{2}, \frac{5}{2}, \ldots)$ have antisymmetric combined wave functions.

The above two-particle functions $\psi$ and $\psi'$ will not do for combined wave functions because they are neither symmetric nor antisymmetric. That is, $\psi'$ does not at all look like either $\psi$ or $-\psi$. Instead, consider the following combined wave function:

$$\psi_{12} = \frac{1}{\sqrt{2}}\left[\psi_A(r_1)\psi_B(r_2) \pm \psi_B(r_1)\psi_A(r_2)\right] \tag{2.76}$$

If we choose the plus sign, then the combined wave function is symmetric with respect to interchange of the particles. If we choose the minus sign then the result is an antisymmetric wave function. The factor of $1/\sqrt{2}$ ensures that the resulting combination is normalized (assuming that each of the component wave functions is itself normalized).

A special case arises when we have identical quantum states $A$ and $B$. (We can regard $A$ and $B$ as representing a set of quantum numbers.) When $A$ is the same as $B$, the antisymmetric combination wave function vanishes identically, and so its probability density is always zero. *The probability to find two identical particles of half-integral spin in the same quantum state must always vanish.* This is of course

does not apply for integral spin

just the *Pauli exclusion principle*, which determines why atomic subshells fill in a certain way. This vanishing of the antisymmetric wave function is the mathematical basis of the Pauli principle. No such vanishing occurs for the symmetric combination, so there is nothing to prevent identical particles of integral spin from occupying the same quantum state.

Later in this text, we apply the Pauli principle to nucleons and show its importance in understanding the nuclear shell model. We also construct some simple antisymmetric wave functions for the quarks that make up nucleons and other similar particles.

## 2.8   TRANSITIONS BETWEEN STATES

A true stationary state lives forever. The expectation values of physical observables, computed from the wave function of a stationary state, do not change with time. In particular, the expectation value of the energy is constant in time. The energy of the state is precisely determined, and the uncertainty in the energy,

$$\Delta E = \sqrt{\langle E^2 \rangle - \langle E \rangle^2} \qquad (2.77)$$

vanishes, because $\langle E^2 \rangle = \langle E \rangle^2$ for this case. The Heisenberg relationship, $\Delta E \, \Delta t \geq \hbar/2$, then implies that $\Delta t = \infty$. Thus a state with an exact energy lives forever; its lifetime against decay (to lower excited states, for example) is infinite.

Now suppose our system is subject to a weak perturbing potential $V'$, in addition to the original potential $V$. In the absence of $V'$, we can solve the Schrödinger equation for the potential $V$ and find a set of eigenstates $\psi_n$ and corresponding eigenvalues $E_n$. If we now include the weak additional potential $V'$, we find that the states are approximately, but not exactly, the previous eigenstates $\psi_n$ of $V$. This weak additional potential permits the system to make transitions between the "approximate" eigenstates $\psi_n$. Thus, under the interaction with a weak electromagnetic field, a hydrogen atom can make transitions, such as 2p → 1s or 3d → 2p. We still describe the various levels as if they were eigenstates of the system.

Even though a system may make a transition from an initial energy state $E_i$ to a final state $E_f$, energy must be conserved. Thus the total decay energy must be constant. If the final state $E_f$ is of lower energy than $E_i$, the energy difference $E_i - E_f$ must appear as radiation emitted in the decay. In transitions between atomic or nuclear excited states, a photon is emitted to carry the energy $E_i - E_f$.

A nonstationary state has a nonzero energy uncertainty $\Delta E$. This quantity is often called the "width" of the state and is usually represented by $\Gamma$. The *lifetime* $\tau$ of this state (the mean or average time it lives before making a transition to a lower state) can be estimated from the uncertainty principle by associating $\tau$ with the time $\Delta t$ during which we are permitted to carry out a measurement of the energy of the state. Thus $\tau \simeq \hbar/\Gamma$. The *decay probability* or *transition probability* $\lambda$ (the number of decays per unit time) is inversely related to the mean lifetime $\tau$:

$$\lambda = \frac{1}{\tau} \qquad (2.78)$$

It would be very useful to have a way to calculate $\lambda$ or $\tau$ directly from the nuclear wave functions. We can do this if we have knowledge of (1) the initial and final wave functions $\psi_i$ and $\psi_f$, which we regard as approximate stationary states of the potential $V$; and (2) the interaction $V'$ that causes the transition between the states. The calculation of $\lambda$ is too detailed for this text, but can be found in any advanced text on quantum mechanics. We will merely state the result, which is known as *Fermi's Golden Rule*:

$$\lambda = \frac{2\pi}{\hbar} |V_{fi}'|^2 \rho(E_f) \qquad (2.79)$$

The quantity $V_{fi}'$ has the form of an expectation value:

$$V_{fi}' = \int \psi_f^* \, V' \psi_i \, dv \qquad (2.80)$$

Notice again the ordering of the states f and i in the integral. The integral $V_{fi}'$ is sometimes called the *matrix element* of the transition operator $V'$. This terminology comes from an alternative formulation of quantum mechanics based on matrices instead of differential equations. Be sure to take special notice that *the decay probability depends on the square of the transition matrix element*.

The quantity $\rho(E_f)$ is known as the *density of final states*. It is the number of states per unit energy interval at $E_f$, and it must be included for the following reason: if the final state $E_f$ is a single isolated state, then the decay probability will be much smaller than it would be in the case that there are many, many states in a narrow band near $E_f$. If there is a large density of states near $E_f$, there are more possible final states that can be reached by the transition and thus a larger transition probability. The density of final states must be computed based on the type of decay that occurs, and we shall consider examples when we discuss $\beta$ decay, $\gamma$ decay, and scattering cross sections.


## REFERENCES FOR ADDITIONAL READING

The following introductory (sophomore-junior) modern physics texts provide background material necessary for the study of quantum mechanics: A. Beiser, *Concepts of Modern Physics*, 3rd ed. (New York: McGraw-Hill, 1981); K. S. Krane, *Modern Physics* (New York: Wiley, 1983); P. A. Tipler, *Modern Physics* (New York: Worth, 1978); R. T. Weidner and R. L. Sells, *Elementary Modern Physics*, 3rd ed. (Boston: Allyn and Bacon, 1980).

Quantum mechanics references at about the same level as the present text are listed below: R. Eisberg and R. Resnick, *Quantum Physics of Atoms, Molecules, Solids, Nuclei, and Particles*, 2nd ed. (New York: Wiley, 1985); A. P. French and E. F. Taylor, *An Introduction to Quantum Physics* (New York: Norton, 1978); R. B. Leighton, *Principles of Modern Physics* (New York: McGraw-Hill, 1969); D. S. Saxon, *Elementary Quantum Mechanics* (San Francisco: Holden-Day, 1968).

Advanced quantum texts, which can be consulted to find more detailed discussions of topics discussed only briefly in this text, are the following: C. Cohen-Tannoudji, B. Diu, and F. Laloë, *Quantum Mechanics* (New York: Wiley-Interscience, 1977); D. Park, *Introduction to the Quantum Theory*, 2nd ed. (New York: McGraw-Hill, 1974); E. Merzbacher, *Quantum Mechanics*, 2nd ed. (New York: Wiley, 1970).

## PROBLEMS

1. Derive Equation 2.37 and plot the transmission coefficient as a function of the energy $E$ of the incident particle. Comment on the behavior of $T$.

2. Derive Equation 2.39 and plot the transmission coefficient as a function of $E$.

3. Solve the Schrödinger equation for the following potential:

$$V(x) = \infty \qquad\qquad x < 0$$
$$= -V_0 \qquad 0 < x < a$$
$$= 0 \qquad\qquad x > a$$

   Here $V_0$ is positive and solutions are needed for energies $E > 0$. Evaluate all undetermined coefficients in terms of a single common coefficient, but do not attempt to normalize the wave function. Assume particles are incident from $x = -\infty$.

4. Find the number of bound states and their energies in the finite one-dimensional square well when $P = 10$.

5. Find the solution to the "half" harmonic oscillator:

$$V(x) = \infty \qquad\quad x < 0$$
$$= \tfrac{1}{2}kx^2 \qquad x > 0$$

   Compare the energy values and wave functions with those of the full harmonic oscillator. Why are some of the full solutions present and some missing in the "half" problem?

6. For the ground state and first two excited states of the one-dimensional simple harmonic oscillator, find the probability for the particle to be beyond the classical turning points.

7. (a) For the ground state of the one-dimensional simple harmonic oscillator, evaluate $\langle x \rangle$ and $\langle x^2 \rangle$.
   (b) Find $\Delta x = [\langle x^2 \rangle - \langle x \rangle^2]^{1/2}$.
   (c) Without carrying out any additional calculations, evaluate $\langle p_x \rangle$ and $\langle p_x^2 \rangle$. (*Hint:* Find $\langle p_x^2/2m \rangle$.)
   (d) Evaluate $\Delta p_x$ and the product $\Delta x \cdot \Delta p_x$. A wave packet with this shape (called a Gaussian shape) is known as a "minimum-uncertainty" wave packet. Why?

8. (a) Find the wave functions and energy levels for a particle confined to a two-dimensional rectangular box, with

$$V(x, y) = 0 \qquad -a \le x \le +a, -b \le y \le +b$$
$$= \infty \qquad |x| > a, |y| > b$$

   (b) Make a diagram similar to Figure 2.10 showing the levels and degeneracies for $b = a$ and for $b = 2a$.

9. Continue Figure 2.10 upward to $50E_0$.

10. Continue Figure 2.11 upward to $20E_0$.

11. Carry out the separation procedure necessary to obtain the solution to Equation 2.54.

12. Show that the first four radial wave functions listed in Table 2.4 are solutions to the Schrödinger equation corresponding to the proper value of the energy, and also show that they are normalized.

13. Find the solutions to the one-dimensional infinite square well when the potential extends from $-a/2$ to $+a/2$ instead of from 0 to $+a$. Is the potential invariant with respect to parity? Are the wave functions? Discuss the assignment of odd and even parity to the solutions.

14. Find the angle between the angular momentum vector $\ell$ and the $z$ axis for all possible orientations when $\ell = 3$.

15. (a) What are the possible values of $j$ for f states?

    (b) What are the corresponding $m_j$?

    (c) How many total $m_j$ states are there?

    (d) How many states would there be if we instead used the labels $m_\ell$ and $m_s$?

16. A combined wave function representing the total spin $S$ of three electrons can be represented as $\psi = \psi_1(m_s)\psi_2(m_s)\psi_3(m_s)$ where $m_s = \pm \frac{1}{2}$ for a spin-$\frac{1}{2}$ electron. (a) List all such possible wave functions and their total projection $M_S$. (b) Identify the number of $M_S$ values with the number of different values of the total spin $S$. (Some values of $S$ may occur more than once.) (c) Draw simple vector diagrams showing how the different couplings of $s_1$, $s_2$, and $s_3$ can lead to the deduced number and values of $S$. (*Hint:* First couple two of the spins, and then couple the third to the possible resultants of the first two.) (d) In the coupling of four electrons, give the possible values of $S$ and their multiplicity, and show that the number of $M_S$ states agrees with what would be expected from tabulating the possible wave functions.

# 3

# NUCLEAR PROPERTIES

Like many systems governed by the laws of quantum mechanics, the nucleus is a somewhat enigmatic and mysterious object whose properties are much more difficult to characterize than are those of macroscopic objects. The list of instructions needed to build an exact replica of a French colonial house or a '57 Chevy is relatively short; the list of instructions necessary to characterize all of the mutual interactions of the 50 nucleons in a medium weight nucleus could contain as many as 50! or about $10^{64}$ terms! We must therefore select a different approach and try to specify the overall characteristics of the entire nucleus. Are there a few physical properties that can be listed to give an adequate description of any nucleus?

To a considerable extent, we can describe a nucleus by a relatively small number of parameters: electric charge, radius, mass, binding energy, angular momentum, parity, magnetic dipole and electric quadrupole moments, and energies of excited states. These are the *static properties* of nuclei that we consider in this chapter. In later chapters we discuss the *dynamic properties* of nuclei, including decay and reaction probabilities. To understand these static and dynamic properties in terms of the interaction between individual nucleons is the formidable challenge that confronts the nuclear physicist.

## 3.1 THE NUCLEAR RADIUS

Like the radius of an atom, the radius of a nucleus is not a precisely defined quantity; neither atoms nor nuclei are solid spheres with abrupt boundaries. Both the Coulomb potential that binds the atom and the resulting electronic charge distribution extend to infinity, although both become negligibly small at distances far beyond the atomic radius ($10^{-10}$ m). What is required is an "operational definition" of what we are to take as the value of the atomic radius. For instance, we might define the atomic radius to be the largest mean radius of the various electronic states populated in an atom. Such a property would be exceedingly difficult to measure, and so more practical definitions are used, such as the spacing between atoms in ionic compounds containing the atom of interest. This also leads to some difficulties since we obtain different radii for an atom when it is in different compounds or in different valence states.

For nuclei, the situation is better in some aspects and worse in others. As we will soon discuss, the density of nucleons and the nuclear potential have a similar spatial dependence—relatively constant over short distances beyond which they drop rapidly to zero. It is therefore relatively natural to characterize the nuclear shape with two parameters: the mean radius, where the density is half its central value, and the "skin thickness," over which the density drops from near its maximum to near its minimum. (In Section 5 of this chapter, we discuss a third parameter, which is necessary to characterize nuclei whose shape is not spherical.)

The problems we face result from the difficulty in determining just what it is that the distribution is describing; the radius that we measure depends on the kind of experiment we are doing to measure the nuclear shape. In some experiments, such as high-energy electron scattering, muonic X rays, optical and X-ray isotope shifts, and energy differences of mirror nuclei, we measure the Coulomb interaction of a charged particle with the nucleus. These experiments would then determine the *distribution of nuclear charge* (primarily the distribution of protons but also involving somewhat the distribution of neutrons, because of their internal structure). In other experiments, such as Rutherford scattering, $\alpha$ decay, and pionic X rays, we measure the strong nuclear interaction of nuclear particles, and we would determine the distribution of nucleons, called the *distribution of nuclear matter*.

## The Distribution of Nuclear Charge

Our usual means for determining the size and shape of an object is to examine the radiation scattered from it (which is, after all, what we do when we look at an object or take its photograph). To see the object and its details, the wavelength of the radiation must be smaller than the dimensions of the object; otherwise the effects of diffraction will partially or completely obscure the image. For nuclei, with a diameter of about 10 fm, we require $\lambda \leq 10$ fm, corresponding to $p \geq 100$ MeV/$c$. Beams of electrons with energies 100 MeV to 1 GeV can be produced with high-energy accelerators, such as the Stanford linear accelerator, and can be analyzed with a precise spectrometer to select only those electrons that are elastically scattered from the selected nuclear target. Figure 3.1 shows an example of the results of such an experiment. The first minimum in the diffractionlike pattern can clearly be seen; for diffraction by a circular disk of diameter $D$, the first minimum should appear at $\theta = \sin^{-1}(1.22\lambda/D)$, and the resulting estimates for the nuclear radii are 2.6 fm for $^{16}$O and 2.3 fm for $^{12}$C. These are, however, only rough estimates because potential scattering is a three-dimensional problem only approximately related to diffraction by a two-dimensional disk.

Figure 3.2 shows the results of elastic scattering from a heavy nucleus, $^{208}$Pb. Several minima in the diffractionlike pattern can be seen. These minima do not fall to zero like diffraction minima seen with light incident on an opaque disk, because the nucleus does not have a sharp boundary.

Let us try to make this problem more quantitative. The initial electron wave function is of the form $e^{i\mathbf{k}_i \cdot \mathbf{r}}$, appropriate for a free particle of momentum $\mathbf{p}_i = \hbar \mathbf{k}_i$. The scattered electron can also be regarded as a free particle of momentum $\mathbf{p}_f = \hbar \mathbf{k}_f$ and wave function $e^{i\mathbf{k}_f \cdot \mathbf{r}}$. The interaction $V(r)$ converts the initial wave into the scattered wave, and according to Equation 2.80 the probabil-

$p = \dfrac{h}{\lambda}$

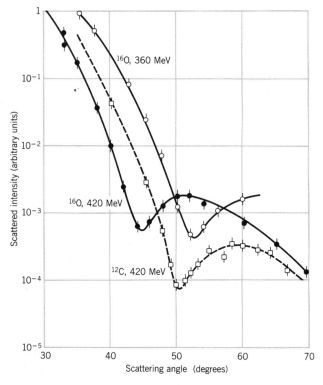

**Figure 3.1**   Electron scattering from $^{16}$O and $^{12}$C. The shape of the cross section is somewhat similar to that of diffraction patterns obtained with light waves. The data come from early experiments at the Stanford Linear Accelerator Center (H. F. Ehrenberg et al., *Phys. Rev.* **113**, 666 (1959)).

ity for the transition will be proportional to the square of the quantity

$$F(\boldsymbol{k}_i, \boldsymbol{k}_f) = \int \psi_f^* V(r) \psi_i \, dv \tag{3.1}$$

$$F(\boldsymbol{q}) = \int e^{i\boldsymbol{q}\cdot\boldsymbol{r}} V(r) \, dv \tag{3.2}$$

apart from a normalization constant, which is chosen so that $F(0) = 1$. Here $\boldsymbol{q} = \boldsymbol{k}_i - \boldsymbol{k}_f$, which is essentially the momentum change of the scattered electron. The interaction $V(r)$ depends on the nuclear charge density $Ze\rho_e(\boldsymbol{r}')$, where $\boldsymbol{r}'$ is a coordinate describing a point in the nuclear volume and $\rho_e$ gives the distribution of nuclear charge. That is, as indicated in Figure 3.3, an electron located at $\boldsymbol{r}$ feels a potential energy due to the element of charge $dQ$ located at $\boldsymbol{r}'$:

$$dV = -\frac{e \, dQ}{4\pi\epsilon_0 |\boldsymbol{r} - \boldsymbol{r}'|}$$

$$= -\frac{Ze^2 \rho_e(\boldsymbol{r}') \, dv'}{4\pi\epsilon_0 |\boldsymbol{r} - \boldsymbol{r}'|} \tag{3.3}$$

To find the complete interaction energy $V(r)$, we sum over all of the contribu-

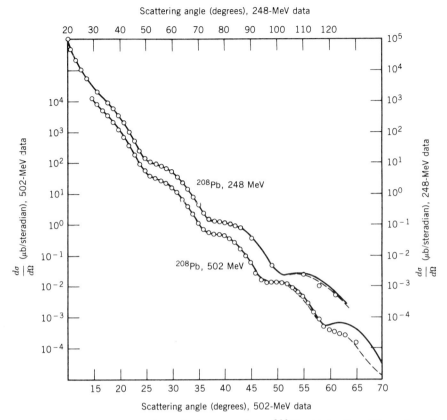

**Figure 3.2**  Elastic scattering of electrons from $^{208}$Pb. Note the different vertical and horizontal scales for the two energies. This also shows diffractionlike behavior, but lacks sharp minima. (J. Heisenberg et al., *Phys. Rev. Lett.* **23**, 1402 (1969).)

tions $dQ$ within the nucleus:

$$V(r) = -\frac{Ze^2}{4\pi\epsilon_0} \int \frac{\rho_e(r')\,dv'}{|r - r'|} \tag{3.4}$$

Writing $q \cdot r = qr \sin\theta$ in Equation 3.2 and integrating over $r$, the properly normalized result is

$$F(q) = \int e^{iq \cdot r'} \rho_e(r')\,dv' \tag{3.5}$$

and if $\rho_e(r')$ depends only on the magnitude $r'$ (not on $\theta'$ or $\phi'$) we obtain

$$F(q) = \frac{4\pi}{q} \int \sin qr' \,\rho_e(r')r'\,dr' \tag{3.6}$$

This quantity is a function of $q$, the magnitude of $q$. Since we originally assumed that the scattering was elastic, then $|p_i| = |p_f|$ and $q$ is merely a function of the scattering angle $\alpha$ between $p_i$ and $p_f$; a bit of vector manipulation shows $q = (2p/\hbar)\sin\alpha/2$ where $p$ is the momentum of the electron. Measuring the scattering probability as a function of the angle $\alpha$ then gives us the dependence

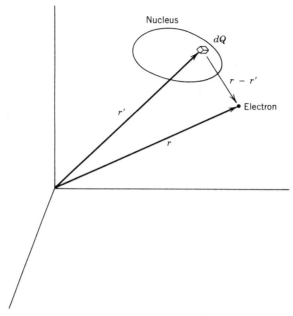

**Figure 3.3** The geometry of scattering experiments. The origin of coordinates is located arbitrarily. The vector **r'** locates an element of charge $dQ$ within the nucleus, and the vector **r** defines the position of the electron.

of Equation 3.6 on $q$. The quantity $F(q)$ is known as a form factor, and the numerical inversion of Equation 3.6, actually an inverse Fourier transformation, then gives us $\rho_e(r')$.

The results of this procedure for several different nuclei are shown in Figure 3.4. One remarkable conclusion is obvious—the central nuclear charge density is nearly the same for all nuclei. Nucleons do not seem to congregate near the center of the nucleus, but instead have a fairly constant distribution out to the surface. (The conclusion from measurements of the nuclear matter distribution is the same.) Thus *the number of nucleons per unit volume is roughly constant*:

$$\frac{A}{\frac{4}{3}\pi R^3} \sim \text{constant} \tag{3.7}$$

where $R$ is the mean nuclear radius. Thus $R \propto A^{1/3}$, and defining the proportionality constant $R_0$ gives

$$R = R_0 A^{1/3} \tag{3.8}$$

From electron scattering measurements, such as those in Figure 3.4, it is concluded that $R_0 \simeq 1.2$ fm. These measurements give the most detailed descriptions of the complete nuclear charge distribution.

Figure 3.4 also shows how diffuse the nuclear surface appears to be. The charge density is roughly constant out to a certain point and then drops relatively slowly to zero. The distance over which this drop occurs is nearly independent of the size of the nucleus, and is usually taken to be constant. We define the *skin*

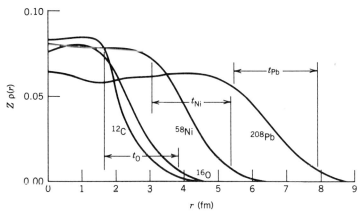

**Figure 3.4** The radial charge distribution of several nuclei determined from electron scattering. The skin thickness $t$ is shown for O, Ni, and Pb; its value is roughly constant at 2.3 fm. The central density changes very little from the lightest nuclei to the heaviest. These distributions were adapted from R. C. Barrett and D. F. Jackson, *Nuclear Sizes and Structure* (Oxford: Clarendon, 1977), which gives more detail on methods of determining $\rho(r)$.

*thickness* parameter $t$ as the distance over which the charge density falls from 90% of its central value to 10%. The value of $t$ is approximately 2.3 fm.

Figure 3.5 shows a more quantitative determination of the relationship between the nuclear radius and mass number, based on electron scattering results. The root mean square (rms) radius, $\langle r^2 \rangle^{1/2}$, is deduced directly from the distribution of scattered electrons; for a uniformly charged sphere $\langle r^2 \rangle = \frac{3}{5} R^2$, where $R$ is the radius of the sphere. Figure 3.5 shows that the dependence of $R$ on $A^{1/3}$ is approximately valid over the range from the lightest to the heaviest nuclei. From the slope of the line, we deduce $R_0 = 1.23$ fm.

The nuclear charge density can also be examined by a careful study of atomic transitions. In solving the Schrödinger equation for the case of the atom with a single electron, it is always assumed that the electron feels the Coulomb attraction of a *point* nucleus, $V(r) = -Ze^2/4\pi\epsilon_0 r$. Since real nuclei are not points, the electron wave function can penetrate to $r < R$, and thus the electron spends part of its time *inside* the nuclear charge distribution, where it feels a very different interaction. In particular, as $r \to 0$, $V(r)$ will not tend toward infinity for a nucleus with a nonzero radius. As a rough approximation, we can assume the nucleus to be a uniformly charged sphere of radius $R$, for which the potential energy of the electron for $r \leq R$ is

$$V'(r) = -\frac{Ze^2}{4\pi\epsilon_0 R}\left\{\frac{3}{2} - \frac{1}{2}\left(\frac{r}{R}\right)^2\right\} \tag{3.9}$$

while for $r \geq R$, the potential energy has the point–nucleus form.

The total energy $E$ of the electron in a state $\psi_n$ of a point nucleus depends in part on the expectation value of the potential energy

$$\langle V \rangle = \int \psi_n^* V \psi_n \, dv \tag{3.10}$$

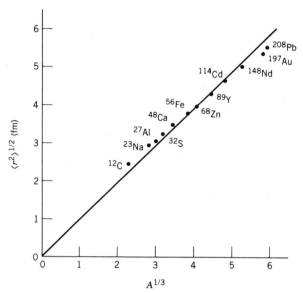

**Figure 3.5** The rms nuclear radius determined from electron scattering experiments. The slope of the straight line gives $R_0 = 1.23$ fm. (The line is not a true fit to the data points, but is forced to go through the origin to satisfy the equation $R = R_0 A^{1/3}$.) The error bars are typically smaller than the size of the points ($\pm 0.01$ fm). More complete listings of data and references can be found in the review of C. W. de Jager et al., *Atomic Data and Nuclear Data Tables* **14**, 479 (1974).

where $V$ is the point–nucleus Coulomb potential energy. If we assume (as a first approximation) that changing from a point nucleus to a uniformly charged spherical nucleus does not significantly change the electronic wave function $\psi_n$, then the energy $E'$ of the electron in a state of a uniformly charged spherical nucleus depends on the expectation value of the potential $V'$:

$$\langle V' \rangle = \int_{r < R} \psi_n^* V' \psi_n \, dv + \int_{r > R} \psi_n^* V \psi_n \, dv \qquad (3.11)$$

where the second integral involves only the $1/r$ potential energy. The effect of the spherical nucleus is thus to change the energy of the electronic states, relative to the point–nucleus value, by $\Delta E = E' - E = \langle V' \rangle - \langle V \rangle$; the latter step follows directly from our assumption that the wave functions are the same, in which case the kinetic energy terms in $E$ and $E'$ are identical. Using the 1s hydrogenic wave function from Table 2.5, we find

$$\Delta E = \frac{e^2}{4\pi\epsilon_0} \frac{4Z^4}{a_0^3} \int_0^R e^{-2Zr/a_0} \left\{ \frac{1}{r} - \frac{3}{2R} + \frac{1}{2} \frac{r^2}{R^3} \right\} r^2 \, dr \qquad (3.12)$$

The exponential factor in the integrand is nearly unity, because $R/a_0 \simeq 10^{-5}$, and evaluating the remaining terms gives

$$\Delta E = \frac{2}{5} \frac{Z^4 e^2}{4\pi\epsilon_0} \frac{R^2}{a_0^3} \qquad (3.13)$$

This $\Delta E$ is the difference between the energy of the 1s state in an atom with a "point" nucleus and the 1s energy in an atom with a uniformly charged nucleus of radius $R$. The latter is a fair approximation to real nuclei (as suggested by Figure 3.4); if we could only find a supply of atoms with "point" nuclei we could measure $\Delta E$ and deduce $R$! Since no such nuclei exist, the next best strategy would be to determine $E'$ from measurement (from K X rays, for example) and to calculate the point–nucleus value $E$ from the 1s atomic wave function. Unfortunately, the atomic wave functions are not known with sufficient precision to do this—$\Delta E$ is a very small difference, perhaps $10^{-4}$ of $E$, and the simple hydrogenlike 1s wave functions are not good enough to calculate $E$ to a precision of 1 part in $10^4$ (relativistic effects and the presence of other electrons are two factors that also influence the 1s energy). Thus a single measurement of the energy of a K X ray cannot be used to deduce the nuclear radius.

Let's instead measure and compare the K X-ray energies (resulting from $2p \rightarrow 1s$ electronic transitions) in two neighboring isotopes of mass numbers $A$ and $A'$. Letting $E_K(A)$ and $E_K(A')$ represent the observed K X-ray energies, we have

$$E_K(A) - E_K(A') = E_{2p}(A) - E_{1s}(A) - E_{2p}(A') + E_{1s}(A') \quad (3.14)$$

If we assume that the 2p energy difference is negligible (recall from Chapter 2 that p-electron wave functions vanish at $r = 0$), the remaining 1s energy difference reduces to the difference between the $\Delta E$ values of Equation 3.13, because $E_{1s} \equiv E' = E + \Delta E$ and the "point" nucleus values $E$ would be the same for the isotopes $A$ and $A'$. Thus

$$E_K(A) - E_K(A') = \Delta E(A') - \Delta E(A)$$

$$= -\frac{2}{5} \frac{Z^4 e^2}{4\pi\epsilon_0} \frac{1}{a_0^3} R_0^2 (A^{2/3} - A'^{2/3}) \quad (3.15)$$

The quantity $E_K(A) - E_K(A')$ is called the K X-ray *isotope shift*, and a plot against $A^{2/3}$ of a sequence of isotope shifts of different isotopes $A$ all compared with the same reference $A'$ should give a straight line from whose slope we could deduce $R_0$. Figure 3.6 is an example of such a plot for some isotopes of Hg. The agreement with the $A^{2/3}$ dependence is excellent. The slope, however, does not give a reasonable value of $R_0$, because the 1s wave function used in Equation 3.12 is not a very good representation of the true 1s wave function. The calculated K X-ray energies, for example, are about 10% lower than the observed values. Detailed calculations that treat the 1s electron relativistically and take into account the effect of other electrons give a more realistic relationship between the slope of Figure 3.6 and the value of $R_0$. The resulting values are in the range of 1.2 fm, in agreement with the results of electron scattering experiments.

It is also possible to measure isotope shifts for the *optical* radiations in atoms (those transitions among the outer electronic shells that produce visible light). Because these electronic orbits lie much further from the nucleus than the 1s orbit, their wave functions integrated over the nuclear volume, as in Equation 3.12, give much smaller shifts than we obtain for the inner 1s electrons. In Chapter 2 we showed that s states ($\ell = 0$ wave functions) give nonzero limits on

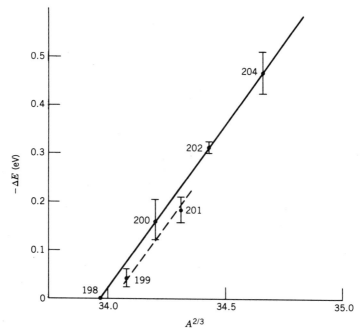

**Figure 3.6**   K X-ray isotope shifts in Hg. The energy of the K X ray in Hg is about 100 keV, so the relative isotope shift is of the order of $10^{-6}$. The data show the predicted dependence on $A^{2/3}$. There is an "odd-even" shift in radius of odd-mass nuclei relative to their even-$A$ neighbors, brought about by the orbit of the odd particle. For this reason, odd-$A$ isotopes must be plotted separately from even-$A$ isotopes. Both groups, however, show the $A^{2/3}$ dependence. The data are taken from P. L. Lee et al., *Phys. Rev. C* **17**, 1859 (1978).

$\psi$ at small $r$. If the optical transitions involve s states, their isotope shifts can be large enough to be measured precisely, especially using modern techniques of laser interferometry. Figure 3.7 shows an example of optical shifts in Hg isotopes; again the expected $A^{2/3}$ dependence is consistent with the data. Measurements across a large range of nuclei are consistent with $R_0 = 1.2$ fm.

These effects of the nuclear size on X-ray and optical transitions are very small, about $10^{-4}$ to $10^{-6}$ of the transition energy. The reason these effects are so small has to do with the difference in scale of $10^4$ between the Bohr radius $a_0$ and the nuclear radius $R$. For integrals of the form of Equation 3.12 to give large effects, the atomic wave function should be large at values of $r$ near $R$, but instead the atomic wave functions are large near $r = a_0/Z$, which is far greater than $R$. We can improve on this situation by using a *muonic atom*. The muon is a particle identical to the electron in all characteristics except its mass, which is 207 times the electronic mass. Since the Bohr radius depends inversely on the mass, the muonic orbits have $1/207$ the radius of the corresponding electronic orbits. In fact, in a heavy nucleus like Pb, the muonic 1s orbit has its mean radius *inside* the nuclear radius $R$; the effect of the nuclear size is a factor of 2 in the transition energies, a considerable improvement over the factor of $10^{-4}$ to $10^{-6}$ in electronic transitions.

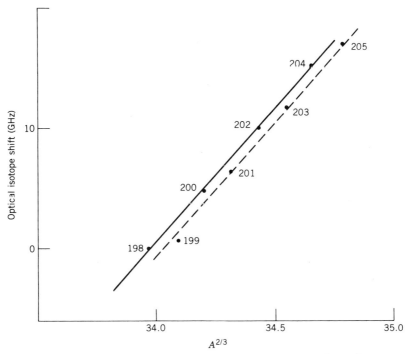

**Figure 3.7** Optical isotope shifts in Hg isotopes from 198 to 205, measured relative to 198. These data were obtained through laser spectroscopy; the experimental uncertainties are about $\pm 1\%$. The optical transition used for these measurements has a wavelength of 253.7 nm, and the isotope shift is therefore about one part in $10^7$. Compare these results with Figure 3.6. Data taken from J. Bonn et al., *Z. Phys. A* **276**, 203 (1976).

Muons are not present in ordinary matter, but must be made artificially using large accelerators that produce intense beams of $\pi$ mesons. The $\pi$ mesons then decay rapidly ($10^{-8}$ s) to muons. (The properties of muons and $\pi$ mesons are discussed in Chapters 17 and 18.) Beams of the resulting muons are then focused onto a suitably chosen target; atoms of the target capture the muons into orbits similar to electronic orbits. Initially the muon is in a state of very high principal quantum number $n$, and as the muon cascades down toward its 1s ground state, photons are emitted, in analogy with the photons emitted in electronic transitions between energy levels. The energy levels of atomic hydrogen depend directly on the electronic mass; we therefore expect the muonic energy levels and transition energies to be 207 times their electronic counterparts. Since ordinary K X rays are in the energy range of tens of keV, muonic K X rays will have energies of a few MeV. Figure 3.8 shows some typical muonic K X rays; the isotope shift is large compared with the isotope shift of electronic K X rays, which is typically $10^{-2}$ eV per unit change in $A$.

In contrast to the case with electronic K X rays, where uncertainties in atomic wave functions made it difficult to interpret the isotope shifts, we can use the observed muonic X-ray energies directly to compute the parameters of the nuclear charge distribution. Figure 3.9 shows the deduced rms radii, based once

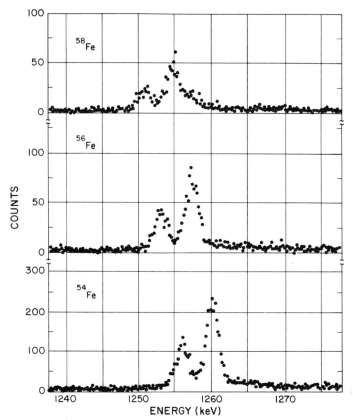

**Figure 3.8** The muonic K X rays in some Fe isotopes. The two peaks show the $2p_{3/2}$ to $1s_{1/2}$ and $2p_{1/2}$ to $1s_{1/2}$ transitions, which have relative intensities in the ratio $2:1$ determined by the statistical weight ($2j + 1$) of the initial state. The isotope shift can clearly be seen as the change in energy of the transitions. The effect is about 0.4%, which should be compared with the $10^{-6}$ effect obtained with electronic K X rays (Figure 3.6). From E. B. Shera et al., *Phys. Rev. C* **14**, 731 (1976).

again on the model of the uniformly charged sphere. The data are roughly consistent with $R_0 A^{1/3}$, with $R_0 = 1.25$ fm.

Yet another way to determine the nuclear charge radius is from direct measurement of the Coulomb energy differences of nuclei. Consider, for example, $_1^3H_2$ and $_2^3He_1$. To get from $^3He$ to $^3H$, we must change a proton into a neutron. As we discuss in Chapter 4, there is strong evidence which suggests that the *nuclear* force does not distinguish between protons and neutrons. Changing a proton into a neutron should therefore not affect the *nuclear* energy of the three-nucleon system; only the Coulomb energy should change, because the two protons in $^3He$ experience a repulsion that is not present in $^3H$. The energy difference between $^3He$ and $^3H$ is thus a measure of the Coulomb energy of the second proton, and the usual formula for the Coulomb repulsion energy can be used to calculate the distance between the protons and thus the size of the nucleus.

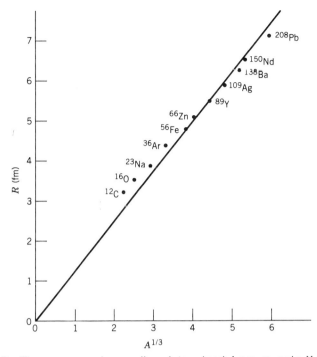

**Figure 3.9** The mean nuclear radius determined from muonic X-ray measurements. As in Figure 3.5, the data depend roughly linearly on $A^{1/3}$ (again forcing the line to go through the origin). The slope of the line gives $R_0 = 1.25$ fm. The data are taken from a review of muonic X-ray determinations of nuclear charge distributions by R. Engfer et al., *Atomic Data and Nuclear Data Tables* **14**, 509 (1974).

Consider now a more complex nucleus, such as $^{238}_{92}\text{U}_{146}$. If we try to change a proton to a neutron we now have a very different circumstance, because the 92nd proton would become the 147th neutron. Because neutrons and protons each must obey the Pauli principle, the orbital of the 92nd proton will differ from the orbital of the 147th neutron, and in general it is not possible to calculate this effect to sufficient accuracy to be able to extract the Coulomb energy. The situation is resolved if we choose a case (as with $^{3}\text{He}$–$^{3}\text{H}$) in which no change of orbital is involved, that is, in which the number of the last proton that makes the change is identical to the number of the last neutron after the change. The Z of the first nucleus must equal the N of the second (and thus the N of the first equals the Z of the second). Such pairs of nuclei are called *mirror nuclei* because one is changed into the other by "reflecting" in a mirror that exchanges protons and neutrons. Examples of such pairs of mirror nuclei are $^{13}_{7}\text{N}_6$ and $^{13}_{6}\text{C}_7$, or $^{39}_{20}\text{Ca}_{19}$ and $^{39}_{19}\text{K}_{20}$.

The Coulomb energy of a uniformly charged sphere of radius $R$ is

$$E_c = \frac{3}{5} \frac{1}{4\pi\epsilon_0} \frac{Q^2}{R} \qquad (3.16)$$

where $Q$ is the total charge of the sphere. The difference in Coulomb energy between the mirror pairs is thus

$$\Delta E_c = \frac{3}{5} \frac{e^2}{4\pi\epsilon_0 R} \left[ Z^2 - (Z-1)^2 \right]$$

$$= \frac{3}{5} \frac{e^2}{4\pi\epsilon_0 R} (2Z - 1) \qquad (3.17)$$

Since $Z$ represents the nucleus of higher atomic number, the $N$ of that nucleus must be $Z - 1$, and so $A = 2Z - 1$. With $R = R_0 A^{1/3}$,

$$\Delta E_c = \frac{3}{5} \frac{e^2}{4\pi\epsilon_0 R_0} A^{2/3} \qquad (3.18)$$

These Coulomb energy differences can be directly measured in two ways. One of the nuclei in the pair can decay to the other through *nuclear β decay*, in which a proton changes into a neutron with the emission of a positive electron (positron). The maximum energy of the positron is a measure of the energy difference between the nuclei. A second method of measuring the energy difference is through nuclear reactions; for example, when a nucleus such as $^{11}$B is bombarded with protons, occasionally a neutron will be emitted leaving the residual nucleus $^{11}$C. The minimum proton energy necessary to cause this reaction is a measure of the energy difference between $^{11}$B and $^{11}$C. (We discuss $\beta$ decay in Chapter 9 and reaction kinematics in Chapter 11.) The measured energy differences are plotted against $A^{2/3}$ in Figure 3.10. As expected from Equation 3.18, the dependence is very nearly linear. The slope of the line gives $R_0 = 1.22$ fm.

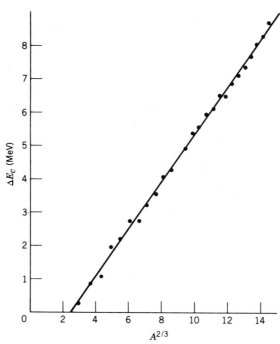

**Figure 3.10** Coulomb energy differences of mirror nuclei. The data show the expected $A^{2/3}$ dependence, and the slope of the line gives $R_0 = 1.22$ fm.

Even though these measurements of the nuclear charge radius use very different techniques, they all give essentially the same results: <u>the nuclear radius varies with mass number as $R_0 A^{1/3}$, with $R_0 = 1.2 - 1.25$ fm.</u>

## The Distribution of Nuclear Matter

An experiment that involves the nuclear force between two nuclei will often provide a measure of the nuclear radius. The determination of the spatial variation of the force between nuclei enables the calculation of the nuclear radii. In this case the radius is characteristic of the *nuclear*, rather than the *Coulomb*, force; these radii therefore reflect the distribution of all nucleons in a nucleus, not only the protons.

As an example of a measurement that determines the size of the nuclear matter distribution, we consider an experiment in which a $^4$He nucleus ($\alpha$ particle) is scattered from a much heavier target of $^{197}$Au. If the separation between the two nuclei is always greater than the sum of their radii, each is always beyond the range of the other's nuclear force, so only the Coulomb force acts. (This situation is known as Rutherford scattering and is discussed in Chapter 11.) The probability for scattering at a certain angle depends on the energy of the incident particle exactly as predicted by the Rutherford formula, when the energy of the incident particle is below a certain value. As the energy of the incident $\alpha$ particle is increased, the Coulomb repulsion of the nuclei is overcome and they may approach close enough to allow the nuclear force to act. In this case the Rutherford formula no longer holds. Figure 3.11 shows an example of this effect.

For another example, we consider the form of radioactive decay in which an $\alpha$ particle is emitted from the nucleus (see Chapter 8 for a complete discussion of $\alpha$ decay). The $\alpha$ particle must escape the nuclear potential and penetrate a Coulomb potential barrier, as indicated in Figure 3.12. The $\alpha$ decay probabilities can be calculated from a standard barrier-penetration approach using the Schrödinger equation. These calculated values depend on the nuclear matter radius $R$, and comparisons with measured decay probabilities permit values of $R$ to be deduced.

A third method for determining the nuclear matter radius is the measurement of the energy of $\pi$-mesic X rays. This method is very similar to the muonic X-ray technique discussed above for measuring the charge radius. The difference between the two techniques results from differences between muons and $\pi$ mesons: the muons interact with the nucleus through the Coulomb force, while the $\pi$ mesons interact with the nucleus through the nuclear and the Coulomb forces. Like the muons, the negatively charged $\pi$ mesons cascade through electronlike orbits and emit photons known as $\pi$-mesic X rays. When the $\pi$-meson wave functions begin to overlap with the nucleus, the energy levels are shifted somewhat from values calculated using only the Coulomb interaction. In addition, the $\pi$ mesons can be directly absorbed into the nucleus, especially from the inner orbits; thus there are fewer X-ray transitions among these inner levels. The "disappearance rate" of $\pi$ mesons gives another way to determine the nuclear radius.

All of these effects could in principle be used as a basis for deducing the nuclear radius. However, the calculations are very sensitive to the exact onset of

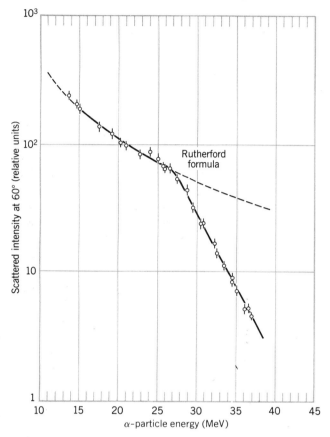

**Figure 3.11** The breakdown of the Rutherford scattering formula. When the incident α particle gets close enough to the target Pb nucleus so that they can interact through the nuclear force (in addition to the Coulomb force that acts when they are far apart) the Rutherford formula no longer holds. The point at which this breakdown occurs gives a measure of the size of the nucleus. Adapted from a review of α particle scattering by R. M. Eisberg and C. E. Porter, *Rev. Mod. Phys.* **33**, 190 (1961).

overlap between the probe particle and the nuclear matter distribution. For these calculations it is therefore very wrong to use the "uniform sphere" model of assuming a constant density out to $R$ and zero beyond $R$. We should instead use a distribution, such as those of Figure 3.4, with a proper tail beyond the mean radius.

We will not go into the details of the calculations, which are far more complicated than our previous calculations of the charge radius. We merely give the result, which may seem a bit surprising: *the charge and matter radii of nuclei are nearly equal, to within about 0.1 fm*. Both show the $A^{1/3}$ dependence with $R_0 \simeq 1.2$ fm. Because heavy nuclei have about 50% more neutrons than protons, we might have expected the neutron radius to be somewhat larger than the proton radius; however, the proton repulsion tends to push the protons outward and the neutron–proton force tends to pull the neutrons inward, until the

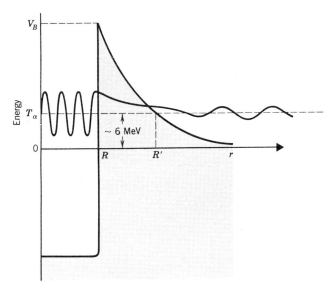

**Figure 3.12**   Barrier penetration in $\alpha$ decay. The half-life for $\alpha$ emission depends on the probability to penetrate the barrier, which in turn depends on its thickness. The measured half-lives can thus be used to determine the radius $R$ where the nuclear force ends and the Coulomb repulsion begins.

neutrons and protons are so completely intermixed that the charge and matter radii are nearly equal.

## 3.2   MASS AND ABUNDANCE OF NUCLIDES

In Appendix C is a table of measured values of the masses and abundances of neutral *atoms* of various stable and radioactive nuclei. Even though we must analyze the energy balance in nuclear reactions and decays using nuclear masses, it is conventional to tabulate the masses of neutral atoms. It may therefore be necessary to correct for the mass and binding energy of the electrons.

As we probe ever deeper into the constituents of matter, the binding energy becomes ever greater in comparison with the rest energy of the bound system. In a hydrogen atom, the binding energy of 13.6 eV constitutes only $1.4 \times 10^{-8}$ of the total rest energy of the atom. In a simple nucleus, such as deuterium, the binding energy of 2.2 MeV is $1.2 \times 10^{-3}$ of the total mass energy. The deuteron is relatively weakly bound and thus this number is rather low compared with typical nuclei, for which the fraction would be more like $8 \times 10^{-3}$. At a yet deeper level, three massive quarks make up a nucleon. The masses of the quarks are not known (no free quarks have yet been confirmed experimentally and quarks may not be permitted to exist in a free state), but it is possible that they may be greater than 100 GeV/$c^2$. If so, the binding energy of the quarks in a nucleon would be a fraction greater than 0.99 of the total mass of the quarks— 3 quarks of total rest energy of perhaps 300 GeV combine to produce a nucleon of rest energy 1 GeV!

It is therefore not possible to separate a discussion of nuclear mass from a discussion of nuclear binding energy; if it were, then nuclei would have masses given by $Zm_p + Nm_n$, and the subject would hardly be of interest. In this section, we confine our discussion to the experimental determination of nuclear masses, treating the nucleus as a simple object with no internal structure. In the next section we analyze the measured masses to determine the binding energy.

The measurement of nuclear masses occupies an extremely important place in the development of nuclear physics. Mass spectrometry was the first technique of high precision available to the experimenter, and since the mass of a nucleus increases in a regular way with the addition of one proton or neutron, measuring masses permitted the entire scheme of stable isotopes to be mapped. Not so with atomic physics—nineteenth-century measurements of average atomic weights led to discrepancies in the atomic periodic table, such as misordering of elements cobalt and nickel, cobalt being heavier but preceding nickel in the proper ordering based on atomic *number* not atomic *weight*. Equally as important, no understanding of nuclear structure can be successful unless we can explain the variation in nuclear properties from one isotope to another; before we can measure such properties, we must determine which isotopes are present and simultaneously attempt to separate them from one another for experimental investigations.

To determine the nuclear masses and relative abundances in a sample of ordinary matter, which even for a pure element may be a mixture of different isotopes, we must have a way to separate the isotopes from one another by their masses. The mere separation of isotopes does not require an instrument of great sensitivity—neighboring isotopes of medium-weight nuclei differ in mass by about 1%. To measure masses to precisions of order $10^{-6}$ requires instruments of much greater sophistication, known as *mass spectroscopes*. The separated masses may be focused to make an image on a photographic plate, in which case the instrument is called a *spectrograph*; or the masses may pass through a detecting slit and be recorded electronically (as a current, for instance), in which case we would have a *spectrometer*. A schematic diagram of a typical mass spectrograph is shown in Figure 3.13.

All mass spectroscopes begin with an *ion source*, which produces a beam of ionized atoms or molecules. Often a vapor of the material under study is bombarded with electrons to produce the ions; in other cases the ions can be formed as a result of a spark discharge between electrodes coated with the material. Ions emerging from the source have a broad range of velocities, as might be expected for a thermal distribution, and of course many different masses might be included.

The next element is a *velocity selector*, consisting of perpendicular electric and magnetic fields. The $E$ field would exert a force $qE$ that would tend to divert the ions upward in Figure 3.13; the $B$ field would exert a downward force $qvB$. Ions pass through undeflected if the forces cancel, for which

$$qE = qvB$$

$$v = \frac{E}{B} \tag{3.19}$$

The final element is a momentum selector, which is essentially a uniform magnetic field that bends the beam into a circular path with radius $r$ determined

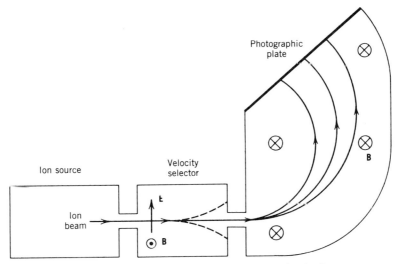

**Figure 3.13** Schematic diagram of mass spectrograph. An ion source produces a beam with a thermal distribution of velocities. A velocity selector passes only those ions with a particular velocity (others being deflected as shown), and momentum selection by a uniform magnetic field permits identification of individual masses.

by the momentum:

$$mv = qBr$$

$$r = \frac{mv}{qB} \tag{3.20}$$

Since $q$, $B$, and $v$ are uniquely determined, each different mass $m$ appears at a particular $r$. Often the magnetic fields of the velocity and momentum selectors are common, in which case

$$m = \frac{qrB^2}{E} \tag{3.21}$$

To determine masses to one part in $10^6$, we must know all quantities in Equation 3.21 to that precision, which hardly seems possible. In practice we could calibrate for one particular mass, and then determine all masses by relative measurements. The fixed point on the atomic mass scale is $^{12}C$, which is taken to be exactly 12.000000 u. To determine the mass of another atom, such as $^1H$, we would need to make considerable changes in $E$ and $B$, and it is perhaps questionable whether the calibration would be valid to one part in $10^6$ over such a range. It would be preferable to measure the smaller difference between two nearly equal masses. For example, let us set the apparatus for mass 128 and measure the difference between the molecular masses of $C_9H_{20}$ (nonane) and $C_{10}H_8$ (naphthalene). This difference is measured to be $\Delta = 0.09390032 \pm 0.00000012$ u. Neglecting corrections for the difference in the molecular binding energies of the two molecules (which is of the order of $10^{-9}$ u), we can write

$$\Delta = m(C_9H_{20}) - m(C_{10}H_8) = 12m(^1H) - m(^{12}C)$$

Thus

$$m(^1\mathrm{H}) = \tfrac{1}{12}\big[m(^{12}\mathrm{C}) + \Delta\big]$$

$$= 1.00000000 + \tfrac{1}{12}\Delta$$

$$= 1.00782503 \pm 0.00000001 \ \mathrm{u}$$

Given this accurate value we could then set the apparatus for mass 28 and determine the difference between $\mathrm{C_2H_4}$ and $\mathrm{N_2}$:

$$\Delta = m(\mathrm{C_2H_4}) - m(\mathrm{N_2}) = 2m(^{12}\mathrm{C}) + 4m(^1\mathrm{H}) - 2m(^{14}\mathrm{N})$$

$$= 0.025152196 \pm 0.000000030 \ \mathrm{u}$$

from which we find:

$$m(^{14}\mathrm{N}) = m(^{12}\mathrm{C}) + 2m(^1\mathrm{H}) - \tfrac{1}{2}\Delta = 14.00307396 \pm 0.00000002 \ \mathrm{u}$$

This system of measuring small differences between close-lying masses is known as the *mass doublet* method, and you can see how it gives extremely precise mass values. Notice in particular how the 1 part in $10^6$ uncertainties in the measured $\Delta$ values give uncertainties in the deduced atomic masses of the order of 1 part in $10^8$ or $10^9$.

It is also possible to determine mass differences by measuring the energies of particles in nuclear reactions. Consider the nuclear reaction $\mathrm{x} + \mathrm{X} \rightarrow \mathrm{y} + \mathrm{Y}$, in which a projectile x is incident on a stationary target X. By measuring the kinetic energies of the reacting particles, we can determine the difference in masses, which is known as the $Q$ value of the reaction:

$$Q = \big[m(\mathrm{x}) + m(\mathrm{X}) - m(\mathrm{y}) - m(\mathrm{Y})\big]c^2 \qquad (3.22)$$

(We consider reaction $Q$ values in detail in Section 11.2.) For example, consider the reaction $^1\mathrm{H} + {}^{14}\mathrm{N} \rightarrow {}^{12}\mathrm{N} + {}^3\mathrm{H}$. From mass doublet measurements we know that $m(^1\mathrm{H}) = 1.007825$ u, $m(^{14}\mathrm{N}) = 14.003074$ u, and $m(^3\mathrm{H}) = 3.016049$ u. The measured $Q$ value is $-22.1355 \pm 0.0010$ MeV. We thus deduce

$$m(^{12}\mathrm{N}) = m(^1\mathrm{H}) + m(^{14}\mathrm{N}) - m(^3\mathrm{H}) - Q/c^2$$

$$= 12.018613 \pm 0.000001 \ \mathrm{u}$$

The main contribution to the uncertainty of the deduced mass comes from the $Q$ value; the $^1\mathrm{H}$, $^3\mathrm{H}$, and $^{14}\mathrm{N}$ masses are known to much greater precision. The nuclide $^{12}\mathrm{N}$ is unstable and decays with a half-life of only 0.01 s, which is far too short to allow its mass to be measured with a mass spectrometer. The nuclear reaction method allows us to determine the masses of unstable nuclides whose masses cannot be measured directly.

**Nuclide Abundances**  The mass spectrometer also permits us to measure the relative abundances of the various isotopes of an element. Measuring the current passing through an exit slit (which replaces the photographic plate of Figure 3.13) as we scan the mass range by varying $E$ or $B$, we can produce results such as those shown in Figure 3.14. From the relative areas of the peaks, we can

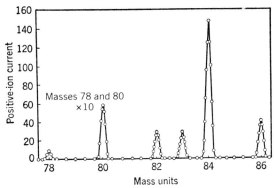

**Figure 3.14** A mass-spectrum analysis of krypton. The ordinates for the peaks at mass positions 78 and 80 should be divided by 10 to show these peaks in their true relation to the others.

determine the abundances of the stable isotopes of krypton:

| | | | |
|---|---|---|---|
| $^{78}$Kr | 0.356% | $^{83}$Kr | 11.5% |
| $^{80}$Kr | 2.27% | $^{84}$Kr | 57.0% |
| $^{82}$Kr | 11.6% | $^{86}$Kr | 17.3% |

The masses that do not appear in the scan ($^{79}$Kr, $^{81}$Kr, $^{85}$Kr, plus those below $^{78}$Kr and above $^{86}$Kr) are radioactive and are not present in natural krypton. A typical sample of natural krypton would consist of a mixture of the six stable isotopes with the above relative composition. If we add the measured masses of the six stable isotopes with the abundances as relative weighting factors, we can compute the "average" atomic mass of krypton

$$m = 0.00356m(^{78}\text{Kr}) + 0.0227m(^{80}\text{Kr}) + \cdots$$
$$= 83.8 \text{ u}$$

which is identical with the accepted atomic mass of Kr, such as is normally given in the periodic table of the elements.

**Separated Isotopes**   If we set the mass spectrometer on a single mass and collect for a very long time, we can accumulate a large quantity of a particular isotope, enough to use for laboratory experiments. Some mass spectrometers are designed to process large quantities of material (often at the expense of another characteristic of the equipment, such as its ability to resolve nearby masses as in Figure 3.14); the isotope separation facility at Oak Ridge National Laboratory is an example. Separated isotopes, which can be purchased from these facilities, are used for an enormous variety of experiments, not only in nuclear physics, where work with separated isotopes enables us to measure specific properties such as cross sections associated with a particular isotope, but also in other fields including chemistry or biology. For example, we can observe the ingestion of nutrients by plants using stable isotopes as an alternative to using radioactive tracers. Ordinary carbon is about 99% $^{12}$C and 1% $^{13}$C; nitrogen is 99.6% $^{14}$N and 0.4% $^{15}$N. If we surround a plant with an atmosphere of $CO_2$ made from

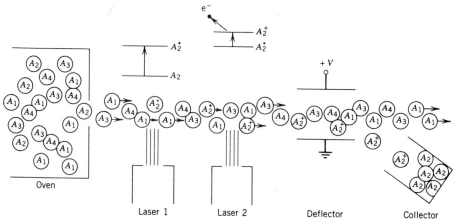

**Figure 3.15** Laser isotope separation. The beam of neutral atoms from the oven is a mixture of four isotopes $A_1$, $A_2$, $A_3$, and $A_4$. The first laser is tuned to the transition corresponding to the resonant excitation of isotope $A_2$ to a certain excited state; because of the sharpness of the laser energy and the isotope shift that gives that particular transition a different energy in the other isotopes, only $A_2$ is excited. The second laser has a broad energy profile, so that many free-electron states can be reached in the ionization of the atoms; but because only the $A_2$ isotopes are in the excited state, only the $A_2$ atoms are ionized. The $A_2$ ions are then deflected and collected.

$^{13}$C, and if we use fertilizers made with $^{15}$N instead of $^{14}$N, we can then study how these isotopes are incorporated into the plant. The longest-lived radioactive isotope of nitrogen has a half-life of 10 min; thus, long-term studies with radioactive tracers would not be possible, and in addition radioactive decays could adversely affect the plant and the personnel who must care for them.

**Laser Isotope Separation** A completely different technique for separating isotopes takes advantage of the extremely sharp (that is, monochromatic) beams available from lasers. As discussed in the last section, the optical radiations of different isotopes of the same element do not have exactly the same energy; the differences in nuclear size cause small variations in the transition energies, called the *isotope shift*. Laser beams are sufficiently sharp so that they can be tuned to excite electrons in one isotope of a mixture of isotopes but not in the others. A schematic representation of the process is shown in Figure 3.15. A beam of neutral atoms passes through a laser beam, which is tuned so that electrons in the desired isotope (but not in the others) will absorb the radiation and make a transition to a particular excited state. A second laser beam is set to a wavelength that corresponds to ionization of the excited atoms. The final energy states of the free electron are continuous rather than quantized, and hence the second laser should have a broad energy profile; this will not result in the ionization of the unwanted isotopes because only those that have been excited by the first laser have electrons in the excited state. After passing through the second laser, the beam consists of ionized atoms of one isotope and neutral atoms of all the others; the ionized atoms can be removed from the beam by an electric field and collected.

## 3.3 NUCLEAR BINDING ENERGY

The mass energy $m_N c^2$ of a certain nuclide is its atomic mass energy $m_A c^2$ less the total mass energy of $Z$ electrons and the total *electronic* binding energy:

$$m_N c^2 = m_A c^2 - Zm_e c^2 + \sum_{i=1}^{Z} B_i \qquad (3.23)$$

where $B_i$ is the binding energy of the $i$th electron. Electronic binding energies are of order 10–100 keV in heavy atoms, while atomic mass energies are of order $A \times 1000$ MeV; thus to a precision of about 1 part in $10^6$ we can neglect the last term of Equation 3.23. (Even this $10^{-6}$ precision does not affect measurements in nuclear physics because we usually work with *differences* in mass energies, such as in determining decay or reaction energies; the effects of electron binding energies tend to cancel in these differences.)

The *binding energy B* of a nucleus is the difference in mass energy between a nucleus $_Z^A X_N$ and its constituent $Z$ protons and $N$ neutrons:

$$B = \left\{ Zm_p + Nm_n - \left[ m(^A X) - Zm_e \right] \right\} c^2 \qquad (3.24)$$

where we have dropped the subscript from $m_A$—from now on, unless we indicate otherwise, we shall always be dealing with *atomic* masses.

Grouping the $Z$ proton and electron masses into $Z$ neutral hydrogen atoms, we can rewrite Equation 3.24 as

$$B = \left[ Zm(^1 H) + Nm_n - m(^A X) \right] c^2 \qquad (3.25)$$

With the masses generally given in atomic mass units, it is convenient to include the unit conversion factor in $c^2$, thus: $c^2 = 931.50$ MeV/u.

We occasionally find atomic mass tables in which, rather than $m(^A X)$, what is given is the *mass defect* $\Delta = (m - A)c^2$. Given the mass defect, it is possible to use Equation 3.25 to deduce the atomic mass.

Other useful and interesting properties that are often tabulated are the neutron and proton separation energies. The *neutron separation energy $S_n$* is the amount of energy that is needed to remove a neutron from a nucleus $_Z^A X_N$, equal to the difference in binding energies between $_Z^A X_N$ and $_Z^{A-1} X_{N-1}$:

$$S_n = B\left(_Z^A X_N\right) - B\left(_Z^{A-1} X_{N-1}\right)$$

$$= \left[ m\left(_Z^{A-1} X_{N-1}\right) - m\left(_Z^A X_N\right) + m_n \right] c^2 \qquad (3.26)$$

In a similar way we can define the *proton separation energy $S_p$* as the energy needed to remove a proton:

$$S_p = B\left(_Z^A X_N\right) - B\left(_{Z-1}^{A-1} X_N\right)$$

$$= \left[ m\left(_{Z-1}^{A-1} X_N\right) - m\left(_Z^A X_N\right) + m(^1 H) \right] c^2 \qquad (3.27)$$

The hydrogen mass appears in this equation instead of the proton mass, since we are always working with *atomic* masses; you can see immediately how the $Z$ electron masses cancel from Equations 3.26 and 3.27.

The neutron and proton separation energies are analogous to the ionization energies in atomic physics—they tell us about the binding of the outermost or

**Table 3.1**  Some Mass Defects and Separation Energies

| Nuclide | $\Delta$ (MeV) | $S_n$ (MeV) | $S_p$ (MeV) |
|---------|-------------|-------------|-------------|
| $^{16}$O | $-4.737$ | 15.66 | 12.13 |
| $^{17}$O | $-0.810$ | 4.14 | 13.78 |
| $^{17}$F | $+1.952$ | 16.81 | 0.60 |
| $^{40}$Ca | $-34.847$ | 15.64 | 8.33 |
| $^{41}$Ca | $-35.138$ | 8.36 | 8.89 |
| $^{41}$Sc | $-28.644$ | 16.19 | 1.09 |
| $^{208}$Pb | $-21.759$ | 7.37 | 8.01 |
| $^{209}$Pb | $-17.624$ | 3.94 | 8.15 |
| $^{209}$Bi | $-18.268$ | 7.46 | 3.80 |

valence nucleons. Just like the atomic ionization energies, the separation energies show evidence for nuclear shell structure that is similar to atomic shell structure. We therefore delay discussion of the systematics of separation energies until we discuss nuclear models in Chapter 5. Table 3.1 gives some representative values of mass defects and separation energies.

As with many other nuclear properties that we will discuss, we gain valuable clues to nuclear structure from a *systematic* study of nuclear binding energy. Since the binding energy increases more or less linearly with $A$, it is general practice to show the average binding energy per nucleon, $B/A$, as a function of $A$. Figure 3.16 shows the variation of $B/A$ with nucleon number. Several remarkable features are immediately apparent. First of all, the curve is relatively constant except for the very light nuclei. The average binding energy of most nuclei is, to within 10%, about 8 MeV per nucleon. Second, we note that the curve reaches a peak near $A = 60$, where the nuclei are most tightly bound. This suggests we can "gain" (that is, release) energy in two ways—below $A = 60$, by assembling lighter nuclei into heavier nuclei, or above $A = 60$, by breaking heavier nuclei into lighter nuclei. In either case we "climb the curve of binding energy" and liberate nuclear energy; the first method is known as *nuclear fusion* and the second as *nuclear fission*. These important subjects are discussed in Chapters 13 and 14.

Attempting to understand this curve of binding energy leads us to the *semiempirical mass formula*, in which we try to use a few general parameters to characterize the variation of $B$ with $A$.

The most obvious term to include in estimating $B/A$ is the constant term, since to lowest order $B \propto A$. The contribution to the binding energy from this "volume" term is thus $B = a_v A$ where $a_v$ is a constant to be determined, which should be of order 8 MeV. This linear dependence of $B$ on $A$ is in fact somewhat surprising, and gives us our first insight into the properties of the nuclear force. If every nucleon attracted all of the others, then the binding energy would be proportional to $A(A - 1)$, or roughly to $A^2$. Since $B$ varies linearly with $A$, this suggests that each nucleon attracts only its closest neighbors, and *not* all of the other nucleons. From electron scattering we learned that the nuclear density is roughly constant, and thus each nucleon has about the same number of neigh-

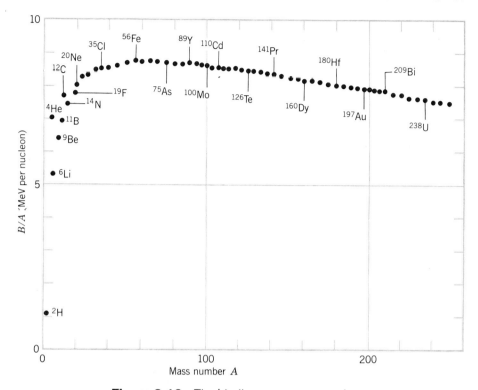

**Figure 3.16** The binding energy per nucleon.

bors; each nucleon thus contributes roughly the same amount to the binding energy.

An exception to the above argument is a nucleon on the nuclear surface, which is surrounded by fewer neighbors and thus less tightly bound than those in the central region. These nucleons do not contribute to $B$ quite as much as those in the center, and thus $B = a_v A$ overestimates $B$ by giving full weight to the surface nucleons. We must therefore subtract from $B$ a term proportional to the nuclear surface area. The surface area of the nucleus is proportional to $R^2$ or to $A^{2/3}$, since $R \propto A^{1/3}$. Thus the surface nucleons contribute to the binding energy a term of the form $-a_s A^{2/3}$.

Our binding energy formula must also include the Coulomb repulsion of the protons, which likewise tends to make the nucleus less tightly bound. Since each proton repels all of the others, this term is proportional to $Z(Z - 1)$, and we may do an exact calculation, assuming a uniformly charged sphere, to obtain $-\frac{3}{5}(e^2/4\pi\epsilon_0 R_0)Z(Z - 1)/A^{1/3}$ where the negative sign implies a reduction in binding energy. The constants evaluate to 0.72 MeV with $R_0 = 1.2$ fm; we can allow this constant to be adjustable by replacing it with a general Coulomb constant $a_c$.

We also note, from our study of the distribution of stable and radioactive isotopes (Figure 1.1), that stable nuclei have $Z \simeq A/2$. (The explanation for this effect will come from our discussion of the shell model in Chapter 5.) If our binding energy formula is to be realistic in describing the stable nuclei that are actually observed, it must take this effect into account. (Otherwise it would allow stable isotopes of hydrogen with hundreds of neutrons!) This term is very important for light nuclei, for which $Z \simeq A/2$ is more strictly observed. For heavy nuclei, this term becomes less important, because the rapid increase in the Coulomb repulsion term requires additional neutrons for nuclear stability. A possible form for this term, called the symmetry term because it tends to make the nucleus symmetric in protons and neutrons, is $-a_{sym}(A - 2Z)^2/A$ which has the correct form of favoring nuclei with $Z = A/2$ and reducing in importance for large $A$.

Finally, we must include another term that accounts for the tendency of like nucleons to couple pairwise to especially stable configurations. When we have an odd number of nucleons (odd $Z$ and even $N$, or even $Z$ and odd $N$), this term does not contribute. However, when both $Z$ and $N$ are odd, we gain binding energy by converting one of the odd protons into a neutron (or vice versa) so that it can now form a pair with its formerly odd partner. We find evidence for this *pairing force* simply by looking at the stable nuclei found in nature—there are only four nuclei with odd $N$ and $Z$ ($^2$H, $^6$Li, $^{10}$B, $^{14}$N), but 167 with even $N$ and $Z$. This pairing energy $\delta$ is usually expressed as $+a_p A^{-3/4}$ for $Z$ and $N$ even, $-a_p A^{-3/4}$ for $Z$ and $N$ odd, and zero for $A$ odd.

Combining these five terms we get the complete binding energy:

$$B = a_v A - a_s A^{2/3} - a_c Z(Z - 1)A^{-1/3}$$

$$-a_{sym}\frac{(A - 2Z)^2}{A} + \delta \qquad (3.28)$$

and using this expression for $B$ we have the *semiempirical mass formula*:

$$M(Z, A) = Zm(^1H) + Nm_n - B(Z, A)/c^2 \qquad (3.29)$$

The constants must be adjusted to give the best agreement with the experimental curve of Figure 3.16. A particular choice of $a_v = 15.5$ MeV, $a_s = 16.8$ MeV, $a_c = 0.72$ MeV, $a_{sym} = 23$ MeV, $a_p = 34$ MeV, gives the result shown in Figure 3.17, which reproduces the observed behavior of $B$ rather well.

The importance of the semiempirical mass formula is not that it allows us to predict any new or exotic phenomena of nuclear physics. Rather, it should be regarded as a first attempt to apply nuclear models to understand the systematic behavior of a nuclear property, in this case the binding energy. It includes several different varieties of nuclear models: the *liquid-drop model*, which treats some of the gross collective features of nuclei in a way similar to the calculation of the properties of a droplet of liquid (indeed, the first three terms of Equation 3.28 would also appear in a calculation of the energy of a charged liquid droplet), and the *shell model*, which deals more with individual nucleons and is responsible for the last two terms of Equation 3.28.

For constant $A$, Equation 3.29 represents a parabola of $M$ vs. $Z$. The parabola will be centered about the point where Equation 3.29 reaches a minimum. To compare this result with the behavior of actual nuclei, we must find the minimum, where $\partial M/\partial Z = 0$:

$$Z_{min} = \frac{\left[m_n - m(^1H)\right] + a_c A^{-1/3} + 4a_{sym}}{2a_c A^{-1/3} + 8a_{sym}A^{-1}} \qquad (3.30)$$

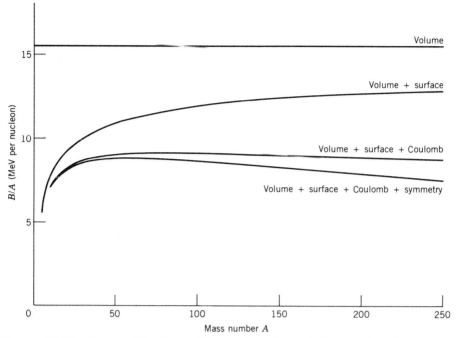

**Figure 3.17** The contributions of the various terms in the semiempirical mass formula to the binding energy per nucleon.

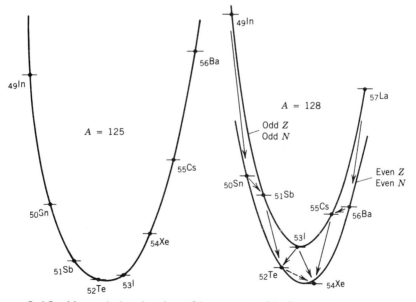

**Figure 3.18** Mass chains for $A = 125$ and $A = 128$. For $A = 125$, note how the energy differences between neighboring isotopes increase as we go further from the stable member at the energy minimum. For $A = 128$, note the effect of the pairing term; in particular, $^{128}$I can decay in either direction, and it is energetically possible for $^{128}$Te to decay directly to $^{128}$Xe by the process known as double $\beta$ decay.

With $a_c = 0.72$ MeV and $a_{sym} = 23$ MeV, it follows that the first two terms in the numerator are negligible, and so

$$Z_{min} \simeq \frac{A}{2} \frac{1}{1 + \frac{1}{4}A^{2/3}a_c/a_{sym}} \qquad (3.31)$$

For small $A$, $Z_{min} \simeq A/2$ as expected, but for large $A$, $Z_{min} < A/2$. For heavy nuclei, Equation 3.31 gives $Z/A \simeq 0.41$, consistent with observed values for heavy stable nuclei.

Figure 3.18 shows a typical odd-$A$ decay chain for $A = 125$, leading to the stable nucleus at $Z = 52$. The unstable nuclei approach stability by converting a neutron into a proton or a proton into a neutron by radioactive $\beta$ decay. Notice how the decay energy (that is, the mass difference between neighboring isobars) increases as we go further from stability. For even $A$, the pairing term gives two parabolas, displaced by $2\delta$. This permits two unusual effects, not seen in odd-$A$ decays: (1) some odd-$Z$, odd-$N$ nuclei can decay in either direction, converting a neutron to a proton or a proton to a neutron; (2) certain *double $\beta$ decays* can become energetically possible, in which the decay may change 2 protons to 2 neutrons. Both of these effects are discussed in Chapter 9.

## 3.4 NUCLEAR ANGULAR MOMENTUM AND PARITY

In Section 2.5 we discussed the coupling of orbital angular momentum $\ell$ and spin $s$ to give total angular momentum $j$. To the extent that the nuclear potential is central, $\ell$ and $s$ (and therefore $j$) will be constants of the motion. In the quantum mechanical sense, we can therefore label every nucleon with the corresponding quantum numbers $\ell$, $s$, and $j$. The total angular momentum of a nucleus containing $A$ nucleons would then be the vector sum of the angular momenta of all the nucleons. This total angular momentum is usually called the *nuclear spin* and is represented by the symbol $I$. The angular momentum $I$ has all of the usual properties of quantum mechanical angular momentum vectors: $I^2 = \hbar^2 I(I + 1)$ and $I_z = m\hbar$ ($m = -I, \ldots, +I$). For many applications involving angular momentum, the nucleus behaves as if it were a single entity with an intrinsic angular momentum of $I$. In ordinary magnetic fields, for example, we can observe the nuclear Zeeman effect, as the state $I$ splits up into its $2I + 1$ individual substates $m = -I, -I + 1, \ldots, I - 1, I$. These substates are equally spaced, as in the atomic normal Zeeman effect. If we could apply an incredibly strong magnetic field, so strong that the coupling between the nucleons were broken, we would see each individual $j$ splitting into its $2j + 1$ substates. Atomic physics also has an analogy here: when we apply large magnetic fields we can break the coupling between the electronic $\ell$ and $s$ and separate the $2\ell + 1$ components of $\ell$ and the $2s + 1$ components of $s$. No fields of sufficient strength to break the coupling of the nucleons can be produced. We therefore observe the behavior of $I$ as if the nucleus were only a single "spinning" particle. For this reason, the spin (total angular momentum) $I$ and the corresponding spin quantum number $I$ are used to describe nuclear states.

To avoid confusion, we will always use $I$ to denote the nuclear spin; we will use $j$ to represent the total angular momentum of a single nucleon. It will often be the case that a single valence particle determines all of the nuclear properties; in that case, $I = j$. In other cases, it may be necessary to consider two valence particles, in which case $I = j_1 + j_2$, and several different resultant values of $I$ may be possible. Sometimes the odd particle and the remaining core of nucleons each contribute to the angular momentum, with $I = j_{\text{particle}} + j_{\text{core}}$.

One important restriction on the allowed values of $I$ comes from considering the possible $z$ components of the total angular momentum of the individual nucleons. Each $j$ must be half-integral ($\frac{1}{2}, \frac{3}{2}, \frac{5}{2}, \ldots$) and thus its only possible $z$ components are likewise half-integral ($\pm\frac{1}{2}\hbar, \pm\frac{3}{2}\hbar, \pm\frac{5}{2}\hbar, \ldots$). If we have an even number of nucleons, there will be an even number of half-integral components, with the result that the $z$ component of the total $I$ can take only integral values. This requires that $I$ itself be an integer. If the number of nucleons is odd, the total $z$ component must be half-integral and so must the total $I$. We therefore require the following rules:

$$\left|\begin{array}{ll} \text{odd-}A \text{ nuclei:} & I = \text{half-integral} \\ \text{even-}A \text{ nuclei:} & I = \text{integral} \end{array}\right.$$

The measured values of the nuclear spin can tell us a great deal about the nuclear structure. For example, of the hundreds of known (stable and radioactive) even-$Z$, even-$N$ nuclei, all have spin-0 ground states. This is evidence for the nuclear pairing force we discussed in the previous section; the nucleons couple together in spin-0 pairs, giving a total $I$ of zero. As a corollary, the ground state spin of an odd-$A$ nucleus must be equal to the $j$ of the odd proton or neutron. We discuss this point further when we consider the nuclear shell model in Chapter 5.

Along with the nuclear spin, the *parity* is also used to label nuclear states. The parity can take either + (even) or − (odd) values. If we knew the wave function of every nucleon, we could determine the nuclear parity by multiplying together the parities of each of the $A$ nucleons, ending with a result $\pi$ either + or −: $\pi = \pi_1\pi_2 \cdots \pi_A$. However, in practice no such procedure is possible, for we generally cannot assign a definite wave function of known parity to every nucleon. Like the spin $I$, we regard the parity $\pi$ as an "overall" property of the whole nucleus. It can be directly measured using a variety of techniques of nuclear decays and reactions. The parity is denoted by a + or − superscript to the nuclear spin, as $I^\pi$. Examples are $0^+$, $2^-$, $\frac{1}{2}^-$, $\frac{5}{2}^+$. There is no direct theoretical relationship between $I$ and $\pi$; for any value of $I$, it is possible to have either $\pi = +$ or $\pi = -$.

## 3.5  NUCLEAR ELECTROMAGNETIC MOMENTS

Much of what we know about nuclear structure comes from studying not the strong nuclear interaction of nuclei with their surroundings, but instead the much weaker electromagnetic interaction. That is, the strong nuclear interaction establishes the distribution and motion of nucleons in the nucleus, and we probe that distribution with the electromagnetic interaction. In doing so, we can use electromagnetic fields that have less effect on the motion of nucleons than the strong force of the nuclear environment; thus our measurements do not seriously distort the object we are trying to measure.

Any distribution of electric charges and currents produces electric and magnetic fields that vary with distance in a characteristic fashion. It is customary to assign to the charge and current distribution an electromagnetic *multipole moment* associated with each characteristic spatial dependence—the $1/r^2$ electric field arises from the net charge, which we can assign as the zeroth or *monopole* moment; the $1/r^3$ electric field arises from the first or *dipole* moment; the $1/r^4$ electric field arises from the second or *quadrupole* moment, and so on. The magnetic multipole moments behave similarly, with the exception of the monopole moment; as far as we know, magnetic monopoles either do not exist or are exceedingly rare, and thus the magnetic monopole field ($\propto 1/r^2$) does not contribute. Electromagnetic theory gives us a recipe for calculating the various electric and magnetic multipole moments, and the same recipe can be carried over into the nuclear regime using quantum mechanics, by treating the multipole moments in operator form and calculating their expectation values for various nuclear states. These expectation values can then be directly compared with the experimental values we measure in the laboratory. Techniques for measuring the nuclear moments are discussed in Chapter 16.

The simplest distributions of charges and currents give only the lowest order multipole fields. A spherical charge distribution gives only a monopole (Coulomb) field; the higher order fields all vanish. A circular current loop gives only a magnetic dipole field. Nature has not been arbitrary in the construction of nuclei; if a simple, symmetric structure (consistent with the nuclear interaction) is possible, then nuclei tend to acquire that structure. It is therefore usually necessary to measure or calculate only the lowest order multipole moments to characterize the electromagnetic properties of the nucleus.

Another restriction on the multipole moments comes about from the symmetry of the nucleus, and is directly related to the parity of the nuclear states. Each electromagnetic multipole moment has a parity, determined by the behavior of the multipole operator when $r \rightarrow -r$. The parity of electric moments is $(-1)^L$, where $L$ is the order of the moment ($L = 0$ for monopole, $L = 1$ for dipole, $L = 2$ for quadrupole, etc.); for magnetic moments the parity is $(-1)^{L+1}$. When we compute the expectation value of a moment, we must evaluate an integral of the form $\int \psi^* \mathcal{O} \psi \, dv$, where $\mathcal{O}$ is the appropriate electromagnetic operator. The parity of $\psi$ itself is not important; because $\psi$ appears twice in the integral, whether $\psi \rightarrow +\psi$ or $\psi \rightarrow -\psi$ does not change the integrand. If, however, $\mathcal{O}$ has odd parity, then the integrand is an odd function of the coordinates and must vanish identically. *Thus all odd-parity static multipole moments must vanish*—electric dipole, magnetic quadrupole, electric octupole ($L = 3$), and so on.

The monopole electric moment is just the net nuclear charge $Ze$. The next nonvanishing moment is the *magnetic dipole moment* $\mu$. A circular loop carrying current $i$ and enclosing area $A$ has a magnetic moment of magnitude $|\mu| = iA$; if the current is caused by a charge $e$, moving with speed $v$ in a circle of radius $r$ (with period $2\pi r/v$), then

$$|\mu| = \frac{e}{(2\pi r/v)}\pi r^2 = \frac{evr}{2} = \frac{e}{2m}|\ell| \tag{3.32}$$

where $|\ell|$ is the classical angular momentum $mvr$. In quantum mechanics, we operationally *define* the observable magnetic moment to correspond to the direction of greatest component of $\ell$; thus we can take Equation 3.32 directly into the quantum regime by replacing $\ell$ with the expectation value relative to the axis where it has maximum projection, which is $m_\ell \hbar$ with $m_\ell = +\ell$. Thus

$$\mu = \frac{e\hbar}{2m}\ell \qquad \text{\scriptsize small } m, \text{ big } \mu \qquad (3.33)$$

where now $\ell$ is the angular momentum quantum number of the orbit.

The quantity $e\hbar/2m$ is called a *magneton*. For atomic motion we use the electron mass and obtain the *Bohr magneton* $\mu_B = 5.7884 \times 10^{-5}$ eV/T. Putting in the proton mass we have the *nuclear magneton* $\mu_N = 3.1525 \times 10^{-8}$ eV/T. Note that $\mu_N \ll \mu_B$ owing to the difference in the masses; thus under most circumstances atomic magnetism has much larger effects than nuclear magnetism. Ordinary magnetic interactions of matter (ferromagnetism, for instance) are determined by atomic magnetism; only in very special circumstances can we observe the effects of nuclear magnetism (see Chapter 16).

We can rewrite Equation 3.33 in a more useful form:

$$\mu = g_\ell \ell \mu_N \qquad (3.34)$$

where $g_\ell$ is the *g factor* associated with the orbital angular momentum $\ell$. For protons $g_\ell = 1$; because neutrons have no electric charge, we can use Equation 3.34 to describe the orbital motion of neutrons if we put $g_\ell = 0$.

We have thus far been considering only the orbital motion of nucleons. Protons and neutrons, like electrons, also have intrinsic or spin magnetic moments, which have no classical analog but which we write in the same form as Equation 3.34:

$$\mu = g_s s \mu_N \qquad (3.35)$$

where $s = \frac{1}{2}$ for protons, neutrons, and electrons. The quantity $g_s$ is known as the *spin g factor* and is calculated by solving a relativistic quantum mechanical equation. For a spin-$\frac{1}{2}$ point particle such as the electron, the Dirac equation gives $g_s = 2$, and measurement is quite consistent with that value for the electron: $g_s = 2.0023$. The difference between $g_s$ and 2 is quite small and can be very accurately computed using the higher order corrections of quantum electrodynamics. On the other hand, for free nucleons, the experimental values are far from the expected value for point particles:

$$\text{proton:} \quad g_s = \quad 5.5856912 \pm 0.0000022$$

$$\text{neutron:} \quad g_s = \ -3.8260837 \pm 0.0000018$$

(The measured magnetic moments, in nuclear magnetons, are just half the $g_s$ factors.) Not only is the proton value far from the expected value of 2 for a point particle, but the uncharged neutron has a nonzero magnetic moment! Here is perhaps our first evidence that the nucleons are not elementary point particles like the electron, but have an internal structure; the internal structure of the nucleons must be due to charged particles in motion, whose resulting currents give the observed spin magnetic moments. It is interesting to note that $g_s$ for the proton is greater than its expected value by about 3.6, while $g_s$ for the neutron is

**Table 3.2**  Sample Values of Nuclear Magnetic Dipole Moments

| Nuclide | $\mu(\mu_N)$ |
|---|---|
| n | $-1.9130418$ |
| p | $+2.7928456$ |
| $^2$H (D) | $+0.8574376$ |
| $^{17}$O | $-1.89379$ |
| $^{57}$Fe | $+0.09062293$ |
| $^{57}$Co | $+4.733$ |
| $^{93}$Nb | $+6.1705$ |

All values refer to the nuclear ground states; uncertainties are typically a few parts in the last digit. For a complete tabulation, see V. S. Shirley, in *Table of Isotopes* (Wiley: New York, 1978), Appendix VII.

less than its expected value (zero) by roughly the same amount. Formerly these differences between the expected and measured $g_s$ values were ascribed to the clouds of $\pi$ mesons that surround nucleons, with positive and neutral $\pi$ mesons in the proton's cloud, and negative and neutral $\pi$ mesons in the neutron's cloud. The equal and opposite contributions of the meson cloud are therefore not surprising. In present theories we consider the nucleons as composed of three quarks; adding the magnetic moments of the quarks gives the nucleon magnetic moments directly (see Chapter 18).

In nuclei, the pairing force favors the coupling of nucleons so that their orbital angular momentum and spin angular momentum each add to zero. Thus the paired nucleons do not contribute to the magnetic moment, and we need only consider a few valence nucleons. If this were not so, we might expect on statistical grounds alone to see a few heavy nuclei with very large magnetic moments, perhaps tens of nuclear magnetons. However, no nucleus has been observed with a magnetic dipole moment larger than about $6\mu_N$.

Table 3.2 gives some representative values of nuclear magnetic dipole moments. Because of the pairing force, we can analyze these magnetic moments to learn about the nuclear structure. In Chapter 4, we discuss the magnetic moment of the deuteron, and in Chapter 5 we consider how nuclear models predict the magnetic moments of heavier nuclei.

The next nonvanishing moment is the *electric quadrupole moment*. The quadrupole moment $eQ$ of a classical point charge $e$ is of the form $e(3z^2 - r^2)$. If the particle moves with spherical symmetry, then (on the average) $z^2 = x^2 = y^2 = r^2/3$ and the quadrupole moment vanishes. If the particle moves in a classical flat orbit, say in the $xy$ plane, then $z = 0$ and $Q = -r^2$. The quadrupole moment in quantum mechanics is

$$eQ = e \int \psi^*(3z^2 - r^2)\psi \, dv \qquad (3.36)$$

for a single proton; for an orbiting neutron, $Q = 0$. If $|\psi|^2$ is spherically symmetric, then $Q = 0$. If $|\psi|^2$ is concentrated in the $xy$ plane ($z \cong 0$), then

**Table 3.3**  Some Values of Nuclear Electric
Quadrupole Moments

| Nuclide | $Q$ (b) |
|---|---|
| $^2$H (D) | $+0.00288$ |
| $^{17}$O | $-0.02578$ |
| $^{59}$Co | $+0.40$ |
| $^{63}$Cu | $-0.209$ |
| $^{133}$Cs | $-0.003$ |
| $^{161}$Dy | $+2.4$ |
| $^{176}$Lu | $+8.0$ |
| $^{209}$Bi | $-0.37$ |

All values refer to nuclear ground states; uncertainties
are typically a few parts in the last digit. For a complete
tabulation, see V. S. Shirley, in *Table of Isotopes* (Wiley:
New York, 1978), Appendix VII.

$Q \sim -\langle r^2 \rangle$, while if $|\psi|^2$ is concentrated along the $z$ axis ($z \cong r$), we might
have $Q \sim +2\langle r^2 \rangle$. Here $\langle r^2 \rangle$ is the mean-square radius of the orbit. Once again
the pairing force is helpful, for if the paired nucleons move in spherically
symmetric orbits, they do not contribute to $Q$. We might therefore expect that for
many nuclei, the quadrupole moment can be estimated from the valence nucleon,
which we can assume to orbit near the surface, so $r = R_0 A^{1/3}$. We therefore
estimate $|eQ| \leq eR_0^2 A^{2/3}$, which ranges from about $6 \times 10^{-30}$ em$^2$ for light
nuclei to $50 \times 10^{-30}$ em$^2$ for heavy nuclei. The unit of $10^{-28}$ m$^2$ is used
frequently in nuclear reaction studies for cross sections, and is known as a *barn*
(b). This unit is also convenient for measuring quadrupole moments; thus the
expected maximum is from 0.06 to 0.5 eb. As you can see from Table 3.3, many
nuclei do fall within that range, but several, especially in the rare-earth
region, are far outside. Here the quadrupole moment is giving important information—the model of the single particle cannot explain the large observed quadrupole moments. Most or all of the protons must somehow collectively contribute
to have such a large $Q$. The assumption of a spherically symmetric core of paired
nucleons is not valid for these nuclei. The core in certain nuclei can take on a
static nonspherical shape that can give a large quadrupole moment. The properties of such strongly deformed nuclei are discussed in Chapter 5.

## 3.6  NUCLEAR EXCITED STATES

Just as we learn about atoms by studying their excited states, we study nuclear
structure in part through the properties of nuclear excited states. (And like
atomic excited states, the nuclear excited states are unstable and decay rapidly to
the ground state.) In atoms, we make excited states by moving individual
electrons to higher energy orbits, and we can do the same for individual
nucleons; thus the excited states can reveal something about the orbits of
individual nucleons. We have already several times in this chapter referred to the
complementary single-particle and collective structure of nuclei—we can also

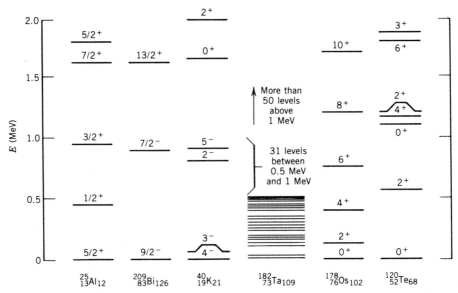

**Figure 3.19** Some sample level schemes showing the excited states below 2 MeV. Some nuclei, such as $^{209}$Bi, show great simplicity, while others, such as $^{182}$Ta, show great complexity. There is a regularity associated with the levels of $^{178}$Os that is duplicated in all even-$Z$, even-$N$ nuclei in the range $150 \le A \le 190$. Structures similar to $^{120}$Te are found in many nuclei in the range $50 \le A \le 150$.

produce excited states by adding energy to the core of paired nucleons. This energy can take the form of collective rotation or vibrations of the entire core, or it might even break one of the pairs, thereby adding two additional valence nucleons.

Part of the goal of nuclear spectroscopy is to observe the possible excited states and to measure their properties. The experimental techniques available include all manner of radioactive decay and nuclear reaction studies that we will consider in detail in subsequent sections. Among the properties we should like to measure for each excited state are: energy of excitation, lifetime and mode(s) of decay, spin and parity, magnetic dipole moment, and electric quadrupole moment. With more than 1000 individual nuclides, each of which may have hundreds of excited states, the tasks of measuring, tabulating, and interpreting these data are almost overwhelming.

Figure 3.19 shows some sample nuclear level schemes. A few of the excited states are identified as originating from excitations of the valence nucleons or the core; such identifications come about only after the properties listed above have been measured and have been compared with the predictions of calculations based on either single particle or collective core excitations, to see which agrees best with experiment. In subsequent chapters, we will explore the experimental techniques used to extract this information and nuclear models used to interpret it. Only through difficult and precise experiments, and through calculations involving the most powerful computers now available, can we obtain such detailed interpretations of nuclear structure.

## REFERENCES FOR ADDITIONAL READING

Two essential references on nuclear charge and mass distributions are Roger C. Barrett and Daphne F. Jackson, *Nuclear Sizes and Structure* (Oxford: Clarendon, 1977), and the collection of articles on nuclear charge and moment distributions in *Atomic Data and Nuclear Data Tables* **14**, 479–653 (1974).

A collection of reprints of articles relating to electron scattering is R. Hofstadter, *Nuclear and Nucleon Structure* (New York: Benjamin, 1963). The use of muonic atoms to determine nuclear charge distributions is reviewed by C. S. Wu and L. Wilets, *Ann. Rev. Nucl. Sci*, **19**, 527 (1969). General techniques of using lasers to do high-resolution optical spectroscopy to study nuclear properties are discussed in D. E. Murnick and M. S. Feld, *Ann. Rev. Nucl. Sci.* **29**, 411 (1979). For more detail on the semiempirical mass formula, see R. D. Evans, *The Atomic Nucleus* (New York: McGraw-Hill, 1955), Chapter 11.

## PROBLEMS

1. Show that the mean-square charge radius of a uniformly charged sphere is $\langle r^2 \rangle = 3R^2/5$.

2. (a) Derive Equation 3.9. (b) Fill in the missing steps in the derivation of Equation 3.13 beginning with Equation 3.9.

3. Compute the form factors $F(q)$ for the following charge distributions:
   (a) $\rho(r) = \rho_0, \quad r < R$     (b) $\rho(r) = \rho_0 e^{-(\ln 2)r^2/R^2}$
   $$= 0, \quad r > R$$

4. A nuclear charge distribution more realistic than the uniformly charged distribution is the Fermi distribution, $\rho(r) = \rho_0\{1 + \exp[(r - R)/a]\}^{-1}$. (a) Sketch this distribution and compare with Figure 3.4. (b) Find the value of $a$ if $t = 2.3$ fm. (c) What is the significance of the parameter $R$? (d) Evaluate $\langle r^2 \rangle$ according to this distribution.

5. Why is the electron screening correction, which is a great difficulty for analyzing electronic X rays, not a problem for muonic X rays?

6. (a) Using a one-electron model, evaluate the energies of the muonic K X rays in Fe assuming a point nucleus, and compare with the energies shown in Figure 3.8. (b) Evaluate the correction $\Delta E$ due to the finite nuclear size. Compare the corrected value with the measured energies.

7. (a) From the known masses of $^{15}$O and $^{15}$N, compute the difference in binding energy. (b) Assuming this difference to arise from the difference in Coulomb energy, compute the nuclear radius of $^{15}$O and $^{15}$N.

8. Given the following mass doublet values (in units of $10^{-6}$ u), compute the corresponding values for the atomic mass of $^{37}$Cl:

$$m(C_3H) - m(^{37}Cl) = 41922.2 \pm 0.3$$
$$m(C_2D_8) - m(^{37}ClH_3) = 123436.5 \pm 0.1$$
$$m(C_3H_6O_2) - m(^{37}Cl_2) = 104974.24 \pm 0.08$$

Here $D \equiv {}^2H$, $C \equiv {}^{12}C$, and $O \equiv {}^{16}O$. Include in your calculation the effect of uncertainties in the H, D, O, and C masses.

9. Compute the total binding energy and the binding energy per nucleon for (a) $^7$Li; (b) $^{20}$Ne; (c) $^{56}$Fe; (d) $^{235}$U.

10. For each of the following nuclei, use the semiempirical mass formula to compute the total binding energy and the Coulomb energy: (a) $^{21}$Ne; (b) $^{57}$Fe; (c) $^{209}$Bi; (d) $^{256}$Fm.

11. Compute the mass defects of (a) $^{32}$S; (b) $^{20}$F; (c) $^{238}$U.

12. Given the following mass defects, find the corresponding atomic mass: (a) $^{24}$Na: $-8.418$ MeV; (b) $^{144}$Sm: $-81.964$ MeV; (c) $^{240}$Pu: $+50.123$ MeV.

13. Evaluate (a) the neutron separation energies of $^7$Li, $^{91}$Zr, and $^{236}$U; (b) the proton separation energies of $^{20}$Ne, $^{55}$Mn, and $^{197}$Au.

14. Examine carefully the $S_n$ and $S_p$ values given in Table 3.1 and draw conclusions about the strength of the binding of the last proton or neutron in the mirror pairs ($^{17}$O, $^{17}$F) and ($^{41}$Ca, $^{41}$Sc). Try to account for general or systematic behavior. Compare the nucleon separation energies in nuclei with identical numbers of protons or neutrons (for example, $S_n$ in $^{16}$O and $^{17}$F or $S_p$ in $^{16}$O and $^{17}$O). Extend these systematics by evaluating and tabulating the $S_n$ and $S_p$ for $^4$He, $^5$He, $^5$Li and for $^{56}$Ni, $^{57}$Ni, and $^{57}$Cu. (*Note:* Nuclei with $Z$ or $N$ equal to 2, 8, 20, or 28 have unusual stability. We explore the reasons for this behavior in Chapter 5.)

15. Use the semiempirical mass formula to obtain an expression for the two-neutron separation energy when $A \gg 1$. (*Hint:* A differential method is far easier than an algebraic one for this problem.) Estimate the size of the various terms and discuss the $A$ dependence. Compare with the following data for Al and Te:

| $^{25}$Al | 31.82 MeV | $^{117}$Te | 18.89 MeV | $^{124}$Te | 16.36 MeV |
|---|---|---|---|---|---|
| $^{26}$Al | 28.30 MeV | $^{118}$Te | 18.45 MeV | $^{125}$Te | 16.00 MeV |
| $^{27}$Al | 24.42 MeV | $^{119}$Te | 18.17 MeV | $^{126}$Te | 15.69 MeV |
| $^{28}$Al | 20.78 MeV | $^{120}$Te | 17.88 MeV | $^{127}$Te | 15.41 MeV |
| $^{29}$Al | 17.16 MeV | $^{121}$Te | 17.46 MeV | $^{128}$Te | 15.07 MeV |
| $^{30}$Al | 15.19 MeV | $^{122}$Te | 17.04 MeV | $^{129}$Te | 14.86 MeV |
| $^{31}$Al | 13.03 MeV | $^{123}$Te | 16.80 MeV | $^{130}$Te | 14.50 MeV |

Why do we choose two-neutron, rather than one-neutron, separation energies for this comparison?

16. In analogy with the previous problem, use the semiempirical mass formula to find approximate expressions for the variation of $S_p$ with $A$ holding $Z$ constant. Obtain data for several sets of isotopes, plot the data, and compare with the predictions of the semiempirical mass formula.

17. The spin-parity of $^9$Be and $^9$B are both $\frac{3}{2}^-$. Assuming in both cases that the spin and parity are characteristic only of the odd nucleon, show how it is possible to obtain the observed spin-parity of $^{10}$B ($3^+$). What other spin-parity combinations could also appear? (These are observed as excited states of $^{10}$B.)

18. Let's suppose we can form $^3$He or $^3$H by adding a proton or a neutron to $^2$H, which has spin equal to 1 and even parity. Let $\ell$ be the orbital angular

momentum of the added nucleon relative to the $^2$H center of mass. What are the possible values of the total angular momentum of $^3$H or $^3$He? Given that the ground-state parity of $^3$H and $^3$He is even, which of these can be eliminated? What is the most likely value of the ground-state angular momentum of $^3$H or $^3$He? Can you make a similar argument based on removing a proton or a neutron from $^4$He? (What is the ground-state spin-parity of $^4$He?) How would you account for the spin-parity of $^5$Li and $^5$He $(\frac{3}{2}^-)$?

19. (a) Consider a neutron as consisting of a proton plus a negative $\pi$ meson in an $\ell = 1$ orbital state. What would be the orbital magnetic dipole moment of such a configuration? Express your result as a multiple of the proton's magnetic moment. (b) Is it possible to account for the observed neutron magnetic moment from such a model? Suppose the neutron wave function consisted of two pieces, one corresponding to a $g = 0$ "Dirac" neutron and the other to proton-plus-$\pi$ meson. What would be the relative sizes of the two pieces of the wave function? (Assume the proton also to behave like an ideal Dirac particle.) (c) Repeat the previous analysis for the proton magnetic moment; that is, consider the proton as part pure Dirac proton, plus part Dirac neutron with orbiting positive $\pi$ meson in $\ell = 1$ state.

20. Suppose the proton magnetic moment were to be interpreted as due to the rotational motion of a positive spherical uniform charge distribution of radius $R$ spinning about its axis with angular speed $\omega$. (a) Show that $\mu = e\omega R^2/5$ by integrating over the charge distribution. (b) Using the classical relationship between angular momentum and rotational speed, show that $\omega R^2 = s/0.4m$. (c) Finally, obtain $\mu = (e/2m)s$, which is analogous to Equation 3.32.

21. Calculate the electric quadrupole moment of a uniformly charged ellipsoid of revolution of semimajor axis $b$ and semiminor axis $a$.

# THE FORCE BETWEEN NUCLEONS

Even before describing any further experiments to study the force between two nucleons, we can already guess at a few of the properties of the nucleon–nucleon force:

1. At short distances it is stronger than the Coulomb force; the nuclear force can overcome the Coulomb repulsion of protons in the nucleus.
2. At long distances, of the order of atomic sizes, the nuclear force is negligibly feeble; the interactions among nuclei in a molecule can be understood based only on the Coulomb force.
3. Some particles are immune from the nuclear force; there is no evidence from atomic structure, for example, that electrons feel the nuclear force at all.

As we begin to do experiments specifically to explore the properties of the nuclear force, we find several other remarkable properties:

4. The nucleon–nucleon force seems to be nearly independent of whether the nucleons are neutrons or protons. This property is called *charge independence*.
5. The nucleon–nucleon force depends on whether the spins of the nucleons are parallel or antiparallel.
6. The nucleon–nucleon force includes a repulsive term, which keeps the nucleons at a certain average separation.
7. The nucleon–nucleon force has a noncentral or *tensor* component. This part of the force does not conserve orbital angular momentum, which is a constant of the motion under central forces.

In this chapter we explore these properties in detail, discuss how they are measured, and propose some possible forms for the basic nucleon–nucleon interaction.

## 4.1  THE DEUTERON

A *deuteron* ($^2$H nucleus) consists of a neutron and a proton. (A neutral atom of $^2$H is called *deuterium*.) It is the simplest bound state of nucleons and therefore gives us an ideal system for studying the nucleon–nucleon interaction. For nuclear physicists, the deuteron should be what the hydrogen atom is for atomic physicists. Just as the measured Balmer series of electromagnetic transitions between the excited states of hydrogen led to an understanding of the structure of hydrogen, so should the electromagnetic transitions between the excited states of the deuteron lead to an understanding of its structure. Unfortunately, there are *no excited states* of the deuteron—it is such a weakly bound system that the only "excited states" are unbound systems consisting of a free proton and neutron.

## Binding Energy

The binding energy of the deuteron is a very precisely measured quantity, which can be determined in three different ways. By spectroscopy, we can directly determine the mass of the deuteron, and we can use Equation 3.25 to find the binding energy. Using the mass doublet method described in Section 3.2, the following determinations have been made (we use the symbol D for $^2$H):

$$m(C_6H_{12}) - m(C_6D_6) = (9.289710 \pm 0.000024) \times 10^{-3} \text{ u}$$

and

$$m(C_5D_{12}) - m(C_6D_6) = (84.610626 \pm 0.000090) \times 10^{-3} \text{ u}.$$

From the first difference we find, using 1.007825037 u for the $^1$H mass,

$$m(^2H) = 2.014101789 \pm 0.000000021 \text{ u}$$

and from the second,

$$m(^2H) = 2.014101771 \pm 0.000000015 \text{ u}$$

These precise values are in very good agreement, and using the measured $^1$H and neutron masses we can find the binding energy

$$B = \left[ m(^1H) + m(n) - m(^2H) \right] c^2 = 2.22463 \pm 0.00004 \text{ MeV}$$

We can also determine this binding energy directly by bringing a proton and a neutron together to form $^2$H and measuring the energy of the $\gamma$-ray photon that is emitted:

$$^1H + n \rightarrow {}^2H + \gamma$$

The deduced binding energy, which is equal to the observed energy of the photon less a small recoil correction, is $2.224589 \pm 0.000002$ MeV, in excellent agreement with the mass spectroscopic value. A third method uses the reverse reaction, called *photodissociation*,

$$\gamma + {}^2H \rightarrow {}^1H + n$$

in which a $\gamma$-ray photon breaks apart a deuteron. The minimum $\gamma$-ray energy that accomplishes this process is equal to the binding energy (again, corrected for the recoil of the final products). The observed value is $2.224 \pm 0.002$ MeV, in good agreement with the mass spectroscopic value.

As we discussed in Section 3.3, the average binding energy per nucleon is about 8 MeV. The deuteron is therefore very weakly bound compared with typical nuclei. Let's see how we can analyze this result to study the properties of the deuteron.

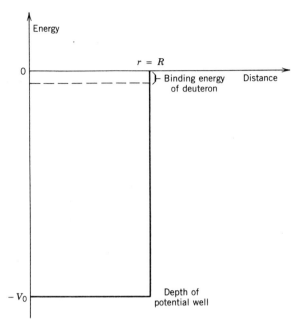

**Figure 4.1** The spherical square-well potential is an approximation to the nuclear potential. The depth is $-V_0$, where $V_0$ is deduced to be about 35 MeV. The bound state of the deuteron, at an energy of about $-2$ MeV, is very close to the top of the well.

To simplify the analysis of the deuteron, we will assume that we can represent the nucleon–nucleon potential as a three-dimensional square well, as shown in Figure 4.1:

$$V(r) = -V_0 \quad \text{for } r < R$$
$$= 0 \quad \text{for } r > R \tag{4.1}$$

This is of course an oversimplification, but it is sufficient for at least some qualitative conclusions. Here $r$ represents the separation between the proton and the neutron, so $R$ is in effect a measure of the *diameter* of the deuteron. Let's assume that the lowest energy state of the deuteron, just like the lowest energy state of the hydrogen atom, has $\ell = 0$. (We justify this assumption later in this section when we discuss the spin of the deuteron.) If we define the radial part of $\psi(\mathbf{r})$ as $u(r)/r$, then we can rewrite Equation 2.60 as

$$-\frac{\hbar^2}{2m}\frac{d^2 u}{dr^2} + V(r)u(r) = E u(r) \tag{4.2}$$

This expression looks exactly like the one-dimensional Equation 2.4, and the solutions can be written in analogy with Equations 2.47. For $r < R$,

$$u(r) = A \sin k_1 r + B \cos k_1 r \tag{4.3}$$

with $k_1 = \sqrt{2m(E + V_0)/\hbar^2}$, and for $r > R$,

$$u(r) = C e^{-k_2 r} + D e^{+k_2 r} \tag{4.4}$$

with $k_2 = \sqrt{-2mE/\hbar^2}$. (Remember, $E < 0$ for bound states.) To keep the wave

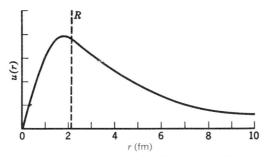

**Figure 4.2**   The deuteron wave function for $R$ = 2.1 fm. Note how the exponential joins smoothly to the sine at $r = R$, so that both $u(r)$ and $du/dr$ are continuous. If the wave function did not "turn over" inside $r = R$, it would not be possible to connect smoothly to a decaying exponential (negative slope) and there would be no bound state.

function finite for $r \to \infty$ we must have $D = 0$, and to keep it finite for $r \to 0$ we must have $B = 0$. ($\psi$ depends on $u(r)/r$; as $r \to 0$, $u(r)$ also must go to zero.) Applying the continuity conditions on $u$ and $du/dr$ at $r = R$, we obtain

$$k_1 \cot k_1 R = -k_2 \qquad (4.5)$$

This transcendental equation gives a relationship between $V_0$ and $R$. From electron scattering experiments, the rms charge radius of the deuteron is known to be about 2.1 fm, which provides a reasonable first estimate for $R$. Solving Equation 4.5 numerically (see Problem 6 at the end of this chapter) the result is $V_0 = 35$ MeV. This is actually quite a reasonable estimate of the strength of the nucleon–nucleon potential, even in more complex nuclei. (Note, however, that the proton and neutron are very likely to be found at separations greater than $R$; see Problem 4.)

We can see from Figure 4.1 how close the deuteron is to the top of the well. If the nucleon–nucleon force were just a bit weaker, the deuteron bound state would not exist (see Problem 3). We are fortunate that it does, however, because the formation of deuterium from hydrogen is the first step not only in the proton-proton cycle of fusion by which our sun makes its energy, but also in the formation of stable matter from the primordial hydrogen that filled the early universe. If no stable two-nucleon bound state existed, we would not be here to discuss it! (For more on the cosmological consequences of the formation of deuterium in the early universe, see Chapter 19.)

The deuteron wave function is shown in Figure 4.2. The weak binding means that $\psi(r)$ is just barely able to "turn over" in the well so as to connect at $r = R$ with the negative slope of the decaying exponential.

## Spin and Parity

The total angular momentum $I$ of the deuteron should have three components: the individual spins $s_n$ and $s_p$ of the neutron and proton (each equal to $\frac{1}{2}$), and the orbital angular momentum $\ell$ of the nucleons as they move about their common center of mass:

$$I = s_n + s_p + \ell \qquad (4.6)$$

When we solved the Schrödinger equation for the deuteron, we assumed $\ell = 0$ in analogy with the lowest bound state (the 1s state) in atomic hydrogen. The measured spin of the deuteron is $I = 1$ (how this is measured is discussed in Chapter 16). Since the neutron and proton spins can be either parallel (for a total of 1) or antiparallel (for a total of zero), there are four ways to couple $s_n$, $s_p$, and $\ell$ to get a total $I$ of 1:

(a)   $s_n$ and $s_p$ parallel with $\ell = 0$,
(b)   $s_n$ and $s_p$ antiparallel with $\ell = 1$,
(c)   $s_n$ and $s_p$ parallel with $\ell = 1$,
(d)   $s_n$ and $s_p$ parallel with $\ell = 2$.

Another property of the deuteron that we can determine is its *parity* (even or odd), the behavior of its wave function when $r \rightarrow -r$ (see Section 2.6). By studying the reactions involving deuterons and the property of the photon emitted during the formation of deuterons, we know that its parity is even. In Section 2.6 we discussed that the parity associated with orbital motion is $(-1)^\ell$, even parity for $\ell = 0$ (s states) and $\ell = 2$ (d states) and odd parity for $\ell = 1$ (p states). The observed even parity allows us to eliminate the combinations of spins that include $\ell = 1$, leaving $\ell = 0$ and $\ell = 2$ as possibilities. The spin and parity of the deuteron are therefore consistent with $\ell = 0$ as we assumed, but of course we cannot yet exclude the possibility of $\ell = 2$.

**Magnetic Dipole Moment**

In Section 3.5 we discussed the spin and orbital contributions to the magnetic dipole moment. If the $\ell = 0$ assumption is correct, there should be no orbital contribution to the magnetic moment, and we can assume the total magnetic moment to be simply the combination of the neutron and proton magnetic moments:

$$\mu = \mu_n + \mu_p$$

$$= \frac{g_{sn}\mu_N}{\hbar}s_n + \frac{g_{sp}\mu_N}{\hbar}s_p \tag{4.7}$$

where $g_{sn} = -3.826084$ and $g_{sp} = 5.585691$. As we did in Section 3.5, we take the observed magnetic moment to be the $z$ component of $\mu$ when the spins have their maximum value $(+\tfrac{1}{2}\hbar)$:

$$\mu = \tfrac{1}{2}\mu_N(g_{sn} + g_{sp}) \tag{4.8}$$

$$= 0.879804 \ \mu_N$$

The observed value is $0.8574376 \pm 0.0000004 \ \mu_N$, in good but not quite exact agreement with the calculated value. The small discrepancy can be ascribed to any of a number of factors, such as contributions from the mesons exchanged between the neutron and proton; in the context of the present discussion, we can assume the discrepancy to arise from a small mixture of d state ($\ell = 2$) in the deuteron wave function:

$$\psi = a_s \psi(\ell = 0) + a_d \psi(\ell = 2) \tag{4.9}$$

Calculating the magnetic moment from this wave function gives

$$\mu = a_s^2 \mu(\ell = 0) + a_d^2 \mu(\ell = 2) \qquad (4.10)$$

where $\mu(\ell = 0)$ is the value calculated in Equation 4.8 and $\mu(\ell = 2) = \frac{1}{4}(3 - g_{sp} - g_{sn})\mu_N$ is the value calculated for a d state. The observed value is consistent with $a_s^2 = 0.96$, $a_d^2 = 0.04$; that is, the deuteron is 96% $\ell = 0$ and only 4% $\ell = 2$. The assumption of the pure $\ell = 0$ state, which we made in calculating the well depth, is thus pretty good but not quite exact.

## Electric Quadrupole Moment

The bare neutron and proton have no electric quadrupole moment, and so any measured nonzero value for the quadrupole moment must be due to the orbital motion. Thus the pure $\ell = 0$ wave function would have a vanishing quadrupole moment. The observed quadrupole moment is

$$Q = 0.00288 \pm 0.00002 \text{ b}$$

which, while small by comparison with many other nuclei, is certainly not zero.

The mixed wave function of Equation 4.9, when used as in Equation 3.36 to evaluate $Q$, gives two contributions, one proportional to $a_d^2$ and another proportional to the cross-term $a_s a_d$. Performing the calculation we obtain

$$Q = \frac{\sqrt{2}}{10} a_s a_d \langle r^2 \rangle_{sd} - \frac{1}{20} a_d^2 \langle r^2 \rangle_{dd} \qquad (4.11)$$

where $\langle r^2 \rangle_{sd} = \int r^2 R_s(r) R_d(r) r^2 \, dr$ is the integral of $r^2$ over the radial wave functions; $\langle r^2 \rangle_{dd}$ is similarly defined. To calculate $Q$ we must know the deuteron d-state wave function, which is not directly measurable. Using the realistic phenomenological potentials discussed later in this chapter gives reasonable values for $Q$ with d-state admixtures of several percent, consistent with the value of 4% deduced from the magnetic moment.

This good agreement between the d-state admixtures deduced from $\mu$ and $Q$ should be regarded as a happy accident and not taken too seriously. In the case of the magnetic dipole moment, there is no reason to expect that it is correct to use the free-nucleon magnetic moments in nuclei. (In fact, in the next chapter we see that there is strong evidence to the contrary.) Unfortunately, a nucleon in a deuteron lies somewhere between a free nucleon and a strongly bound nucleon in a nucleus, and we have no firm clues about what values to take for the magnetic moments. Spin-orbit interactions, relativistic effects, and meson exchanges may have greater effects on $\mu$ than the d-state admixture (but may cancel one another's effects). For the quadrupole moment, the poor knowledge of the d-state wave function makes the deduced d-state admixture uncertain. (It would probably be more valid to regard the calculation of $Q$, using a known d-state mixture, as a test of the d-state wave function.) Other experiments, particularly scattering experiments using deuterons as targets, also give d-state admixtures in the range of 4%. Thus our conclusions from the magnetic dipole and electric quadrupole moments may be valid after all!

It is important that we have an accurate knowledge of the d-state wave function because the mixing of $\ell$ values in the deuteron is the best evidence we have for the noncentral (tensor) character of the nuclear force.

## 4.2 NUCLEON – NUCLEON SCATTERING

Although the study of the deuteron gives us a number of clues about the nucleon–nucleon interaction, the total amount of information available is limited. Because there are no excited states, we can only study the dynamics of the nucleon–nucleon interaction in the configuration of the deuteron: $\ell = 0$, parallel spins, 2-fm separation. (Excited states, if they were present, might have different $\ell$ values or spin orientations.) To study the nucleon–nucleon interaction in different configurations, we can perform nucleon–nucleon scattering experiments, in which an incident beam of nucleons is scattered from a target of nucleons. If the target is a nucleus with many nucleons, then there will be several target nucleons within the range of the nuclear potential of the incident nucleon; in this case the observed scattering of a single nucleon will include the complicated effects of multiple encounters, making it very difficult to extract the properties of the interaction between individual nucleons. We therefore select a target of hydrogen so that incident particles can scatter from the individual protons. (It is still possible to have multiple scattering, but in this case it must occur through scattering first from one proton, then from another that is quite far from the first on the scale of nuclear dimensions. If the probability for a single encounter is small, the probability for multiple encounters will be negligible. This is very different from the case of scattering from a heavier nucleus, in which each single encounter with a target nucleus consists of many nucleon–nucleon interactions.)

Before we discuss the nuclear scattering problem, let's look at an analogous problem in optics, the diffraction of waves at a small slit or obstacle, as shown in Figure 4.3. The diffraction pattern produced by an obstacle is very similar to that produced by a slit of the same size. Nuclear scattering more resembles diffraction by the obstacle, so we will concentrate our discussion on it. There are three features of the optical diffraction that are analogous to the scattering of nucleons:

1.  The incident wave is represented by a plane wave, while far from the obstacle the scattered wave fronts are spherical. The total energy content of any expanding spherical wave front cannot vary; thus its intensity (per unit area) must decrease like $r^{-2}$ and its amplitude must decrease like $r^{-1}$.
2.  Along the surface of any spherical scattered wave front, the diffraction is responsible for a variation in intensity of the radiation. The intensity thus depends on angular coordinates $\theta$ and $\phi$.
3.  A radiation detector placed at any point far from the obstacle would record both incident and scattered waves.

To solve the nucleon–nucleon scattering problem using quantum mechanics, we will again assume that we can represent the interaction by a square-well potential, as we did in the previous section for the deuteron. In fact, the only difference between this calculation and that of the deuteron is that we are concerned with free incident particles with $E > 0$. We will again simplify the Schrödinger equation by assuming $\ell = 0$. The justification for this assumption has nothing to do with that of the identical assumption made in the calculation for the deuteron. Consider an incident nucleon striking a target nucleon just at its surface; that is, the *impact parameter* (the perpendicular distance from the center of the target nucleus to the line of flight of the incident nucleon) is of the order of $R \simeq 1$ fm. If the incident particle has velocity $v$, its angular momentum

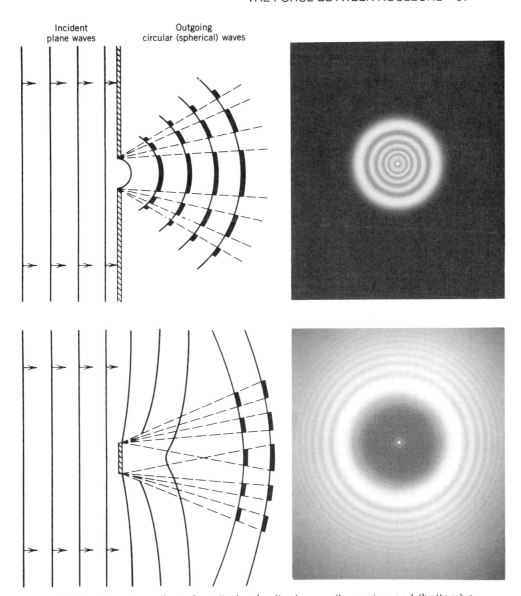

**Figure 4.3** Representation of scattering by (top) a small opening and (bottom) a small obstacle. The shading of the wavefronts shows regions of large and small intensity. On the right are shown photographs of diffraction by a circular opening and an opaque circular disk. Source of photographs: M. Cagnet, M. Francon, and J. C. Thrierr, *Atlas of Optical Phenomena* (Berlin: Springer-Verlag, 1962).

relative to the target is $mvR$. The relative angular momentum between the nucleons must be quantized in units of $\hbar$; that is, $mvR = \ell\hbar$ in semiclassical notation. If $mvR \ll \hbar$, then only $\ell = 0$ interactions are likely to occur. Thus $v \ll \hbar/mR$ and the corresponding kinetic energy is estimated as

$$T = \tfrac{1}{2}mv^2 \ll \frac{\hbar^2}{2mR^2} = \frac{\hbar^2 c^2}{2mc^2R^2} = \frac{(200 \text{ MeV} \cdot \text{fm})^2}{2(1000 \text{ MeV})(1 \text{ fm})^2} = 20 \text{ MeV}$$

If the incident energy is far below 20 MeV, the $\ell = 0$ assumption is justified. We will consider only low-energy scattering, for which the $\ell = 0$ assumption is valid.

The nucleon–nucleon scattering problem will be solved in the center-of-mass coordinate system (see Appendix B). The mass appearing in the Schrödinger equation is the reduced mass, which in this case is about half the nucleon mass.

The solution to the square-well problem for $r < R$ is given by Equation 4.3; as before, $B = 0$ in order that $u(r)/r$ remain finite for $r \to 0$. For $r > R$, the wave function is

$$u(r) = C' \sin k_2 r + D' \cos k_2 r \tag{4.12}$$

with $k_2 = \sqrt{2mE/\hbar^2}$. It is convenient to rewrite Equation 4.12 as

$$u(r) = C \sin(k_2 r + \delta) \tag{4.13}$$

where

$$C' = C \cos \delta \quad \text{and} \quad D' = C \sin \delta \tag{4.14}$$

The boundary conditions on $u$ and $du/dr$ at $r = R$ give

$$C \sin(k_2 R + \delta) = A \sin k_1 R \tag{4.15}$$

and

$$k_2 C \cos(k_2 R + \delta) = k_1 A \cos k_1 R \tag{4.16}$$

Dividing,

$$k_2 \cot(k_2 R + \delta) = k_1 \cot k_1 R \tag{4.17}$$

Again, we have a transcendental equation to solve; given $E$ (which we control through the energy of the incident particle), $V_0$, and $R$, we can in principle solve for $\delta$.

Before we discuss the methods for extracting the parameter $\delta$ from Equation 4.17, we examine how $\delta$ enters the solution to the Schrödinger equation. As $V_0 \to 0$ (in which case no scattering occurs), $k_1 \to k_2$ and $\delta \to 0$. This is just the free particle solution. The effect of $V_0$ on the wave function is indicated in Figure 4.4. The wave function at $r > R$ has the same form as the free particle, but it has experienced a *phase shift* $\delta$. The nodes (zeros) of the wave function are "pulled" toward the origin by the attractive potential. (A repulsive potential would "push" the nodes away from the origin and would give a negative phase shift.) We can analyze the incident waves into components according to their angular momentum relative to the target: $\ell = 0$ (which we have been considering so far), $\ell = 1$, and so on. Associated with each $\ell$ there will be a different solution to the Schrödinger equation and a different phase shift $\delta_\ell$.

Let us see how our square-well problem relates to more general scattering theory. The incident wave is (as in the optical analogy) a plane wave traveling in the $z$ direction:

$$\psi_{\text{incident}} = A e^{ikz} \tag{4.18}$$

Let the target be located at the origin. Multiplying by the time-dependent factor gives

$$\psi(z, t) = A e^{i(kz - \omega t)} \tag{4.19}$$

which always moves in the $+z$ direction (toward the target for $z < 0$ and away

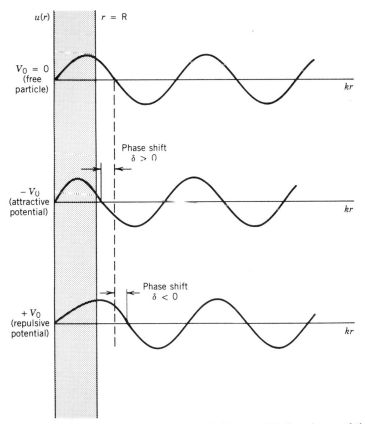

**Figure 4.4** The effect of a scattering potential is to shift the phase of the scattered wave at points beyond the scattering regions, where the wave function is that of a free particle.

from it for $z > 0$). It is mathematically easier to work with spherical waves $e^{ikr}/r$ and $e^{-ikr}/r$, and multiplying with $e^{-i\omega t}$ shows that $e^{ikr}$ gives an outgoing wave and $e^{-ikr}$ gives an incoming wave. (A more rigorous treatment of scattering theory, including terms with $\ell > 0$, is given in Chapter 11.) For $\ell = 0$ we can take

$$\psi_{\text{incident}} = \frac{A}{2ik}\left[\frac{e^{ikr}}{r} - \frac{e^{-ikr}}{r}\right] \qquad (4.20)$$

The minus sign between the two terms keeps $\psi$ finite for $r \to 0$, and using the coefficient $A$ for both terms sets the amplitudes of the incoming and outgoing waves to be equal. We assume that the scattering does not create or destroy particles, and thus the scattering cannot change the amplitudes of the $e^{ikr}$ or $e^{-ikr}$ terms (since the squared amplitudes give the probabilities to detect incoming or outgoing particles). *All that can result from the scattering is a change in phase of the outgoing wave*:

$$\psi(r) = \frac{A}{2ik}\left[\frac{e^{i(kr+\beta)}}{r} - \frac{e^{-ikr}}{r}\right] \qquad (4.21)$$

where $\beta$ is the change in phase.

Manipulation of Equation 4.13 gives the relationship between $\beta$ and $\delta_0$:

$$
\begin{aligned}
\psi(r) &= \frac{C}{r}\sin(kr + \delta_0) \\
&= \frac{C}{r}\frac{e^{i(kr+\delta_0)} - e^{-i(kr+\delta_0)}}{2i} \\
&= \frac{C}{2i}e^{-i\delta_0}\left[\frac{e^{i(kr+2\delta_0)}}{r} - \frac{e^{-ikr}}{r}\right]
\end{aligned}
\tag{4.22}
$$

Thus $\beta = 2\delta_0$ and $A = kCe^{-i\delta_0}$.

To evaluate the probability for scattering, we need the amplitude of the scattered wave. The wave function $\psi$ represents all waves in the region $r > R$, and to find the amplitude of only the scattered wave we must subtract away the incident amplitude:

$$
\begin{aligned}
\psi_{\text{scattered}} &= \psi - \psi_{\text{incident}} \\
&= \frac{A}{2ik}\left(e^{2i\delta_0} - 1\right)\frac{e^{ikr}}{r}
\end{aligned}
\tag{4.23}
$$

The current of scattered particles per unit area can be found using Equation 2.12 extended to three dimensions:

$$
\begin{aligned}
j_{\text{scattered}} &= \frac{\hbar}{2mi}\left(\psi^*\frac{\partial \psi}{\partial r} - \frac{\partial \psi^*}{\partial r}\psi\right) \tag{4.24} \\
&= \frac{\hbar|A|^2}{mkr^2}\sin^2\delta_0 \tag{4.25}
\end{aligned}
$$

and the incident current is, in analogy with Equation 2.22

$$
j_{\text{incident}} = \frac{\hbar k|A|^2}{m}
\tag{4.26}
$$

The scattered current is uniformly distributed over a sphere of radius $r$. An element of area $r^2 d\Omega$ on that sphere subtends a solid angle $d\Omega = \sin\theta\, d\theta\, d\phi$ at the scattering center; see Figure 4.5. The *differential cross section* $d\sigma/d\Omega$ is the probability per unit solid angle that an incident particle is scattered into the solid angle $d\Omega$; the probability $d\sigma$ that an incident particle is scattered into $d\Omega$ is the ratio of the scattered current through $d\Omega$ to the incident current:

$$
d\sigma = \frac{(j_{\text{scattered}})(r^2 d\Omega)}{j_{\text{incident}}}
\tag{4.27}
$$

Using Equations 4.25 and 4.26 for the scattered and incident currents, we obtain

$$
\frac{d\sigma}{d\Omega} = \frac{\sin^2\delta_0}{k^2}
\tag{4.28}
$$

The *total cross section* $\sigma$ is the total probability to be scattered in any direction:

$$
\sigma = \int \frac{d\sigma}{d\Omega}\, d\Omega
\tag{4.29}
$$

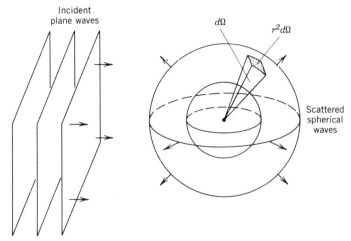

**Figure 4.5**   The basic geometry of scattering.

In general, $d\sigma/d\Omega$ varies with direction over the surface of the sphere; in the special case of $\ell = 0$ scattering, $d\sigma/d\Omega$ is constant and comes out of the integral:

$$\sigma = 4\pi \frac{d\sigma}{d\Omega}$$

$$= \frac{4\pi \sin^2 \delta_0}{k^2} \tag{4.30}$$

Thus the $\ell = 0$ phase shift is directly related to the probability for scattering to occur. That is, we can evaluate $\delta_0$ from our simple square-well model, Equation 4.17, find the total cross section from Equation 4.30, and compare with the experimental cross section.

We now return to the analysis of Equation 4.17. Let us assume the incident energy is small, say $E \leq 10$ keV. Then $k_1 = \sqrt{2m(V_0 + E)/\hbar^2} \simeq 0.92$ fm$^{-1}$, with $V_0 = 35$ MeV from our analysis of the deuteron bound state, and $k_2 = \sqrt{2mE/\hbar^2} \lesssim 0.016$ fm$^{-1}$. If we let the right side of Equation 4.17 equal $-\alpha$,

$$\alpha = -k_1 \cot k_1 R \tag{4.31}$$

then a bit of trigonometric manipulation gives

$$\sin^2 \delta_0 = \frac{\cos k_2 R + (\alpha/k_2)\sin k_2 R}{1 + \alpha^2/k_2^2} \tag{4.32}$$

and so

$$\sigma = \frac{4\pi}{k_2^2 + \alpha^2}\left(\cos k_2 R + \frac{\alpha}{k_2}\sin k_2 R\right) \tag{4.33}$$

Using $R \approx 2$ fm from the study of the $^2$H bound state gives $\alpha \approx 0.2$ fm$^{-1}$. Thus $k_2^2 \ll \alpha^2$ and $k_2 R \ll 1$, giving

$$\sigma \simeq \frac{4\pi}{\alpha^2}(1 + \alpha R) = 4.6 \text{ b} \tag{4.34}$$

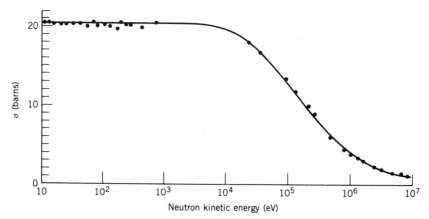

**Figure 4.6** The neutron–proton scattering cross section at low energy. Data taken from a review by R. K. Adair, *Rev. Mod. Phys.* **22**, 249 (1950), with additional recent results from T. L. Houk, *Phys. Rev. C* **3**, 1886 (1970).

where 1 barn (b) = $10^{-28}$ m². This result suggests that the cross section should be constant at low energy and should have a value close to 4–5 b.

Figure 4.6 shows the experimental cross sections for scattering of neutrons by protons. The cross section is indeed constant at low energy, and it decreases with $E$ at large energy as Equation 4.33 predicts, but the low-energy cross section, 20.4 b, is not in agreement with our calculated value of 4–5 b.

For the solution to this discrepancy, we must study the relative spins of the incident and scattered nucleons. The proton and neutron spins (each $\frac{1}{2}$) can combine to give a total spin $S = s_p + s_n$ that can have magnitude either 0 or 1. The $S = 1$ combination has three orientations (corresponding to $z$ components $+1, 0, -1$) and the $S = 0$ combination has only a single orientation. For that reason, the $S = 1$ combination is called a *triplet* state and the $S = 0$ combination is called a *singlet* state. Of the four possible relative spin orientations, three are associated with the triplet state and one with the singlet state. As the incident nucleon approaches the target, the probability of being in a triplet state is 3/4 and the probability of being in a singlet state is 1/4. If the scattering cross section is different for the singlet and triplet states, then

$$\sigma = \tfrac{3}{4}\sigma_t + \tfrac{1}{4}\sigma_s \qquad (4.35)$$

where $\sigma_t$ and $\sigma_s$ are the cross sections for scattering in the triplet and singlet states, respectively. In estimating the cross section in Equation 4.34, we used parameters obtained from the deuteron, which is in a $S = 1$ state. We therefore take $\sigma_t = 4.6$ b and using the measured value of $\sigma = 20.4$ b for the low-energy cross section, we deduce

$$\sigma_s = 67.8 \text{ b}$$

This calculation indicates that there is an enormous difference between the cross sections in the singlet and triplet states—that is, *the nuclear force must be spin dependent*.

Even from our investigation of the deuteron, we should have concluded that the force is spin dependent. If the neutron–proton force did *not* depend on the relative direction of the spins, then we would expect to find deuteron bound states with $S = 0$ and $S = 1$ at essentially the same energy. Because we find no $S = 0$ bound state, we conclude that the force must be spin dependent.

We can verify our conclusions about the singlet and triplet cross sections in a variety of ways. One method is to scatter very low energy neutrons from hydrogen *molecules*. Molecular hydrogen has two forms, known as orthohydrogen and parahydrogen. In orthohydrogen the two proton spins are parallel, while in parahydrogen they are antiparallel. The difference between the neutron scattering cross sections of ortho- and parahydrogen is evidence of the spin-dependent part of the nucleon–nucleon force.

Our discussion of the cross section for neutron–proton scattering is inadequate for analysis of scattering of neutrons from $H_2$ molecules. Very low energy neutrons ($E < 0.01$ eV) have a de Broglie wavelength larger than 0.05 nm, thus greater than the separation of the two protons in $H_2$. The uncertainty principle requires that the size of the wave packet that describes a particle be no smaller than its de Broglie wavelength. Thus the wave packet of the incident neutron overlaps simultaneously with both protons in $H_2$, even though the range of the nuclear force of the individual neutron–proton interactions remains of the order of 1 fm. The scattered neutron waves $\psi_1$ and $\psi_2$ from the two protons will therefore combine *coherently*; that is, they will interfere, and the cross section depends on $|\psi_1 + \psi_2|^2$, not $|\psi_1|^2 + |\psi_2|^2$. We cannot therefore simply add the cross sections from the two individual scatterings. (At higher energy, where the de Broglie wavelength would be small compared with the separation of the protons, the scattered waves would not interfere and we could indeed add the cross sections directly. The reason for choosing to work at very low energy is partly to observe the interference effect and partly to prevent the neutron from transferring enough energy to the $H_2$ molecule to start it rotating, which would complicate the analysis. The minimum rotational energy is about 0.015 eV, and so neutrons with energies in the range of 0.01 eV do not excite rotational states of the molecule.)

To analyze the interference effect in problems of this sort, we introduce the *scattering length a*, defined such that the low-energy cross section is equal to $4\pi a^2$:

$$\lim_{k \to 0} \sigma = 4\pi a^2 \tag{4.36}$$

Comparison with Equation 4.30 shows that

$$a = \pm \lim_{k \to 0} \frac{\sin \delta_0}{k} \tag{4.37}$$

The choice of sign is arbitrary, but it is conventional to choose the minus sign.

Even though the scattering length has the dimension of length, it is a parameter that represents the strength of the scattering, *not* its range. To see this, we note from Equation 4.37 that $\delta_0$ must approach 0 at low energy in order that $a$ remain finite. Equation 4.23 for the scattered wave function can be written for small $\delta_0$ as

$$\psi_{\text{scattered}} \simeq A \frac{\delta_0}{k} \frac{e^{ikr}}{r} = -Aa \frac{e^{ikr}}{r} \tag{4.38}$$

Thus $a$ gives in effect the amplitude of the scattered wave.

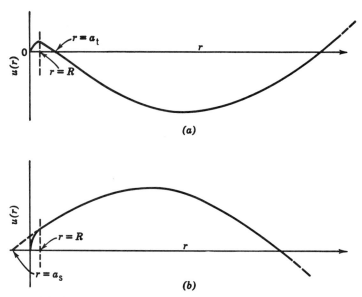

**Figure 4.7** (a) Wave function for triplet np scattering for a laboratory neutron energy of ~ 200 keV and a well radius of 2.1 fm. Note the positive scattering length. (b) Wave function exhibiting a negative scattering length. This happens to be the case for singlet np scattering.

The sign of the scattering length also carries physical information. Figure 4.7 shows representations of the triplet and singlet scattered wave functions $u(r)$. At low energy we can write $a \simeq -\delta_0/k$ and the scattered wave function, Equation 4.13, becomes

$$u(r) = C \sin k_2(r - a) \tag{4.39}$$

The value of $a$ is given by the point at which $u(r)$ passes through zero. The triplet wave function for $r < R$ looks just like the bound state wave function for the deuteron: $u(r)$ "turns over" for $r < R$ to form the bound state. The value of $a_t$ is therefore positive. Because there is no singlet bound state, $u(r)$ does not "turn over" for $r < R$, so it reaches the boundary at $r = R$ with positive slope. When we make the smooth connection at $r = R$ to the wave function beyond the potential and extrapolate to $u(r) = 0$, we find that $a_s$, the singlet scattering length, is negative.

Our estimate $\sigma_t = 4.6$ b from the properties of the deuteron leads to $a_t = +6.1$ fm, and the estimate of $\sigma_s = 67.8$ b needed to reproduce the observed total cross section gives $a_s = -23.2$ fm.

The theory of neutron scattering from ortho- and parahydrogen gives

$$\sigma_{\text{para}} = 5.7(3a_t + a_s)^2 \tag{4.40}$$

$$\sigma_{\text{ortho}} = \sigma_{\text{para}} + 12.9(a_t - a_s)^2 \tag{4.41}$$

where the numerical coefficients depend on the speed of the incident neutron. Equations 4.40 and 4.41 are written for neutrons of about 770 m/s, slower even than "thermal" neutrons (2200 m/s). The measured cross sections, corrected for absorption, for neutrons of this speed are $\sigma_{\text{para}} = 3.2 \pm 0.2$ b and $\sigma_{\text{ortho}} = 108 \pm 1$ b. If the nuclear force were independent of spin, we would have $\sigma_t = \sigma_s$ and $a_t = a_s$; thus $\sigma_{\text{para}}$ and $\sigma_{\text{ortho}}$ would be the same. The great difference between the measured values shows that $a_t \neq a_s$, and it also suggests that $a_t$ and $a_s$ must have different signs, so that $-3a_t \simeq a_s$ in order to make $\sigma_{\text{para}}$ small. Solving Equations 4.40 and 4.41 for $a_s$ and $a_t$ gives

$$a_s = -23.55 \pm 0.12 \text{ fm}$$

$$a_t = +5.35 \pm 0.06 \text{ fm}$$

consistent with the values deduced previously from $\sigma_t$ and $\sigma_s$. A description of these experiments can be found in G. L. Squires and A. T. Stewart, *Proc. Roy. Soc.* (*London*) **A230**, 19 (1955).

There are several other experiments that are sensitive to the singlet and triplet scattering lengths; these include neutron diffraction by crystals that contain hydrogen (such as hydrides) as well as the total reflection of neutron beams at small angles from hydrogen-rich materials (such as hydrocarbons). These techniques give results in good agreement with the above values for $a_s$ and $a_t$.

The theory we have outlined is valid only for $\ell = 0$ scattering of low-energy incident particles. The $\ell = 0$ restriction required particles of incident energies below 20 MeV, while our other low-energy approximations required eV or keV energies. As we increase the energy of the incident particle, we will violate Equation 4.36 long before we reach energies of 20 MeV. We therefore still have $\ell = 0$ scattering, but at these energies (of order 1 MeV) equations such as 4.38 are not valid. This case is generally treated in the *effective range approximation*, in which we take

$$k \cot \delta_0 = \frac{1}{a} + \frac{1}{2} r_0 k^2 + \cdots \qquad (4.42)$$

and where terms in higher powers of $k$ are neglected. The quantity $a$ is the zero-energy scattering length we already defined (and, in fact, this reduces to Equation 4.37 in the $k \to 0$ limit), and the quantity $r_0$ is a new parameter, the *effective range*. One of the advantages of this representation is that $a$ and $r_0$ characterize the nuclear potential independent of its shape; that is, we could repeat all of the calculations done in this section with a potential other than the square well, and we would deduce identical values of $a$ and $r_0$ from analyzing the experimental cross sections. Of course there is an accompanying disadvantage in that we can learn little about the shape of the nuclear potential from an analysis in which calculations with different potentials give identical results!

Like the scattering lengths, the effective range is different for singlet and triplet states. From a variety of scattering experiments we can deduce the best set of $\ell = 0$ parameters for the neutron–proton interaction:

$$a_s = -23.715 \pm 0.015 \text{ fm} \qquad a_t = 5.423 \pm 0.005 \text{ fm}$$

$$r_{0s} = \quad 2.73 \quad \pm 0.03 \text{ fm} \qquad r_{0t} = 1.748 \pm 0.006 \text{ fm}$$

As a final comment regarding the singlet and triplet neutron–proton interactions, we can try to estimate the energy of the singlet n-p state relative to the bound triplet state at $-2.22$ MeV. Using Equations 4.34, 4.31, and 4.5 we would deduce that the energy of the singlet state is about $+77$ keV. Thus the singlet state is only slightly unbound.

### 4.3 PROTON–PROTON AND NEUTRON–NEUTRON INTERACTIONS

There is one very important difference between the scattering of identical nucleons (proton–proton and neutron–neutron scattering) and the scattering of different nucleons (neutron–proton scattering). This difference comes about because the identical projectile and target nucleons must be described by a common wave function, as discussed in Section 2.7. Because nucleons have spin $\frac{1}{2}$, their wave functions must be antisymmetric with respect to interchange of the nucleons. If we again consider only low-energy scattering, so that $\ell = 0$, interchanging the spatial coordinates of the two particles gives no change in sign. (This situation is somewhat analogous to the parity operation described in Section 2.6.) Thus the wave function is symmetric with respect to interchange of spatial coordinates and must therefore be antisymmetric with respect to interchange of spin coordinates in order that the total (spatial times spin) wave function be antisymmetric. The antisymmetric spin wave function is of the form of Equation 2.76 and must correspond to a total combined spin of 0; that is, the spin orientations must be different. *Only singlet spin states can thus contribute to the scattering.* (At higher energies, the antisymmetric $\ell = 1$ spatial states can occur, accompanied by only the symmetric triplet spin states.)

The derivation of the differential cross section relies on another feature of quantum physics. Consider Figure 4.8, which represents the scattering of two identical particles in the center of mass reference frame. Since the particles are

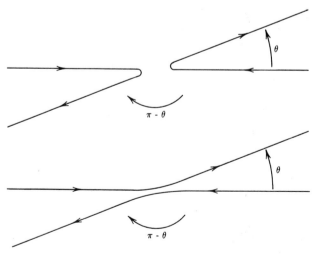

**Figure 4.8** Scattering of identical particles in the center-of-mass system. One particle emerges at the angle $\theta$ and the other at $\pi - \theta$; because the particles are identical, there is no way to tell which particle emerges at which angle, and therefore we cannot distinguish the two cases shown.

identical, there is no experimental way to distinguish the two situations in the figure. The scattered wave function must therefore include contributions for scattering at $\theta$ and at $\pi - \theta$. When we square the scattered wave function to calculate the cross section, there will be a term proportional to the interference between the parts of the wave function that give scattering at $\theta$ and at $\pi - \theta$. This interference is a purely quantum effect that has no classical analog.

Let's first consider scattering between two protons; the wave function must describe both Coulomb and nuclear scattering, and there will be an additional Coulomb-nuclear interference term in the cross section. (The scattered wave function must include one term resulting from Coulomb scattering and another resulting from nuclear scattering; the Coulomb term must vanish in the limit $e \rightarrow 0$, and the nuclear term must vanish as the nuclear potential vanishes, in which case $\delta_0 \rightarrow 0$. When we square the wave function to find the cross section, we get a term that includes both the Coulomb and nuclear scattering.) The derivation of the cross section is beyond the level of this text; for discussions of its derivation and of early work on proton–proton scattering, see J. D. Jackson and J. M. Blatt, *Rev. Mod. Phys.* **22**, 77 (1950). The differential cross section is

$$
\frac{d\sigma}{d\Omega} = \left( \frac{e^2}{4\pi\epsilon_0} \right)^2 \frac{1}{4T^2} \left\{ \frac{1}{\sin^4(\theta/2)} + \frac{1}{\cos^4(\theta/2)} - \frac{\cos\left[ \eta \ln \tan^2(\theta/2) \right]}{\sin^2(\theta/2)\cos^2(\theta/2)} \right.
$$

$$
- \frac{2}{\eta}(\sin \delta_0) \left( \frac{\cos\left[ \delta_0 + \eta \ln \sin^2(\theta/2) \right]}{\sin^2(\theta/2)} + \frac{\cos\left[ \delta_0 + \eta \ln \cos^2(\theta/2) \right]}{\cos^2(\theta/2)} \right)
$$

$$
\left. + \frac{4}{\eta^2} \sin^2 \delta_0 \right\} \tag{4.43}
$$

Here $T$ is the *laboratory* kinetic energy of the incident proton (assuming the target proton to be at rest), $\theta$ is the scattering angle in the center-of-mass system, $\delta_0$ the $\ell = 0$ phase shift for pure nuclear scattering, and $\eta = (e^2/4\pi\epsilon_0 \hbar c)\beta^{-1} = \alpha/\beta$, where $\alpha$ is the fine-structure constant (with a value of nearly $\frac{1}{137}$) and $\beta = v/c$ is the (dimensionless) relative velocity of the protons. The six terms in brackets in Equation 4.43 can be readily identified: (1) The $\sin^{-4}(\theta/2)$ is characteristic of Coulomb scattering, also known as Rutherford scattering. We discuss this further in Chapter 11. (2) Since the two protons are identical, we cannot tell the case in which the incident proton comes out at $\theta$ and the target proton at $\pi - \theta$ (in the center-of-mass system) from the case in which the incident proton comes out at $\pi - \theta$ and the target proton at $\theta$. Thus the scattering cross section must include a characteristic Coulomb (Rutherford) term $\sin^{-4}(\pi - \theta)/2 = \cos^{-4}(\theta/2)$. (3) This term describes the interference between Coulomb scattering at $\theta$ and at $\pi - \theta$. (4 and 5) These two terms result from the interference between Coulomb and nuclear scattering. (6) The last term is the pure nuclear scattering term. In the limit $e \rightarrow 0$ (pure nuclear scattering), only this term survives and Equation 4.43 reduces to Equation 4.28, as it should.

Although it may be complicated in practice, the procedure for studying the proton–proton interaction is simple in concept: since $\delta_0$ is the only unknown in Equation 4.43, we can measure the differential scattering cross section as a function of angle (for a specific incident kinetic energy) and extract $\delta_0$ from the

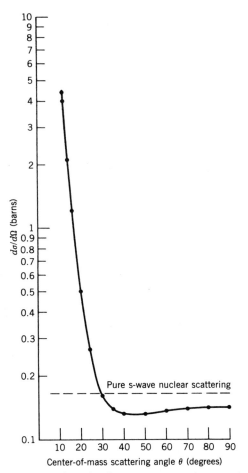

**Figure 4.9** The cross section for low-energy proton–proton scattering at an incident proton energy of 3.037 MeV. Fitting the data points to Equation 4.43 gives the s-wave phase shift $\delta_0 = 50.966°$. The cross section for pure nuclear scattering would be 0.165 b; the observation of values of the cross section *smaller* than the pure nuclear value is evidence of the interference between the Coulomb and nuclear parts of the wave function. Data from D. J. Knecht et al., *Phys. Rev.* **148**, 1031 (1966).

best fit of the results to Equation 4.43. Figure 4.9 shows an example of such data, from which it is deduced that $\delta_0 = 50.966°$ at $T = 3.037$ MeV. From many such experiments we can observe the dependence of $\delta_0$ on energy, as shown in Figure 4.10.

The next step in the interpretation of these data is to represent the scattering in terms of energy-independent quantities such as the scattering length and effective range, as we did in Equation 4.42. Unfortunately, this cannot easily be done because the Coulomb interaction has infinite range and even in the $k \to 0$ limit we cannot neglect the higher-order terms of Equation 4.42. With certain modifications, however, it is possible to obtain an expression incorporating the effects of Coulomb and nuclear scattering in a form similar to Equation 4.42 and thus to

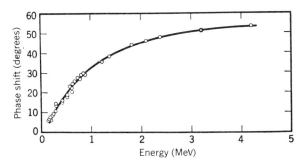

**Figure 4.10** The s-wave phase shift for pp scattering as deduced from the experimental results of several workers.

obtain values for the proton–proton scattering length and effective range:

$$a = -7.82 \pm 0.01 \text{ fm}$$
$$r_0 = \phantom{-}2.79 \pm 0.02 \text{ fm}$$

The effective range is entirely consistent with the *singlet* np values deduced in the previous section. The scattering length, which measures the strength of the interaction, includes Coulomb as well as nuclear effects and thus cannot be compared directly with the corresponding np value. (It is, however, important to note that $a$ is negative, suggesting that there is no pp bound state; that is, the nucleus $^2$He does not exist.) The comparison of the pp and np scattering lengths will be discussed further in the next section.

The study of neutron–neutron scattering should be free of the effects of the Coulomb interaction that made the analysis of proton–proton scattering so complicated. Here the difficulty is an experimental one—although beams of neutrons are readily available, targets of free neutrons are not. Measurement of neutron–neutron scattering parameters therefore requires that we use a nuclear reaction to create two neutrons in relative motion within the range of each other's nuclear force. As the two neutrons separate, we have in effect a scattering experiment. Unfortunately, such reactions must also create a third particle, which will have interactions with both of the neutrons (individually and collectively), but the necessary corrections can be calculated with sufficient precision to enable values to be extracted for the neutron–neutron scattering length and effective range. The experiments that have been reported include the breakup of a deuteron following capture of a negative $\pi$ meson ($\pi^- + {}^2\text{H} \rightarrow 2\text{n} + \gamma$) and following neutron scattering (n $+ {}^2\text{H} \rightarrow 2\text{n} + \text{p}$). It is also possible to deduce the nn parameters from comparison of mirror reactions such as $^3\text{He} + {}^2\text{H} \rightarrow {}^3\text{H} + 2\text{p}$ and $^3\text{H} + {}^2\text{H} \rightarrow {}^3\text{He} + 2\text{n}$, using known pp parameters as an aid in calculating the final-state effects of the three particles. The analysis of these (and other) experiments gives the neutron–neutron parameters

$$a = -16.6 \pm 0.5 \text{ fm}$$
$$r_0 = \phantom{-}2.66 \pm 0.15 \text{ fm}$$

As with the proton–proton interaction, the negative scattering length shows that the two neutrons do not form a stable bound state. (It is tempting, but incorrect, to explain the nonexistence of the di-proton as arising from Coulomb repulsion. No such temptation exists for the di-neutron, the nonexistence of which must arise from the spin dependence of the nuclear interaction. Reviewing the evidence, we first learned that the deuteron ground state is a spin triplet and that no bound spin singlet state exists. We then argued that, because identical fermions must have total antisymmetric wave functions and because the lowest state is expected to be a spatially symmetric $\ell = 0$ state, the di-proton and di-neutron systems must have antisymmetric, or singlet, spin states which are unbound.)

## 4.4 PROPERTIES OF THE NUCLEAR FORCE

Based on the low-energy properties described in the previous sections, we can learn many details about the nuclear force. When we include results from higher energy experiments, still more details emerge. In this section we summarize the main features of the internucleon force and in the next section we discuss a particular representation for the force that reproduces many of these details.

### The Interaction between Two Nucleons Consists to Lowest Order of an Attractive Central Potential

In this chapter we have used for this potential a square-well form, which simplifies the calculations and reproduces the observed data fairly well. Other more realistic forms could just as well have been chosen, but the essential conclusions would not change (in fact, the effective range approximation is virtually independent of the shape assumed for the potential). The common characteristic of these potentials is that they depend only on the internucleon distance $r$. We therefore represent this central term as $V_c(r)$. The experimental program to study $V_c(r)$ would be to measure the energy dependence of nucleon–nucleon parameters such as scattering phase shifts, and then to try to choose the form for $V_c(r)$ that best reproduces those parameters.

### The Nucleon – Nucleon Interaction is Strongly Spin Dependent

This observation follows from the failure to observe a singlet bound state of the deuteron and also from the measured differences between the singlet and triplet cross sections. What is the form of an additional term that must be added to the potential to account for this effect? Obviously the term must depend on the spins of the two nucleons, $s_1$ and $s_2$, but not all possible combinations of $s_1$ and $s_2$ are permitted. The nuclear force must satisfy certain symmetries, which restrict the possible forms that the potential could have. Examples of these symmetries are *parity* ($r \to -r$) and *time reversal* ($t \to -t$). Experiments indicate that, to a high degree of precision (one part in $10^7$ for parity and one part in $10^3$ for time reversal), the internucleon potential is invariant with respect to these operations. Under the parity operator, which involves spatial reflection, angular momentum vectors are unchanged. This statement may seem somewhat surprising, because upon inverting a coordinate system we would naturally expect all vectors defined in that coordinate system to invert. However, angular momentum is not a true or polar vector; it is a pseudo- or axial vector that does not invert when $r \to -r$. This follows directly from the definition $r \times p$ or can be inferred from a diagram of a spinning object. Under the time-reversal operation, all motions (including linear and angular momentum) are reversed. Thus terms such as $s_1$ or $s_2$ or a linear combination $As_1 + Bs_2$ in the potential would violate time-reversal invariance and cannot be part of the nuclear potential; terms such as $s_1^2$, $s_2^2$, or $s_1 \cdot s_2$ are invariant with respect to time reversal and are therefore allowed. (All of these terms are also invariant with respect to parity.) The simplest term involving both nucleon spins is $s_1 \cdot s_2$. Let's consider the value of $s_1 \cdot s_2$ for singlet and triplet states. To do this we evaluate the total spin $S = s_1 + s_2$

$$S^2 = S \cdot S = (s_1 + s_2) \cdot (s_1 + s_2)$$
$$= s_1^2 + s_2^2 + 2s_1 \cdot s_2$$

Thus

$$s_1 \cdot s_2 = \tfrac{1}{2}\left(S^2 - s_1^2 - s_2^2\right) \tag{4.44}$$

To evaluate this expression, we must remember that in quantum mechanics all squared angular momenta evaluate as $s^2 = \hbar^2 s(s+1)$; see Section 2.5 and Equation 2.69.

$$\langle s_1 \cdot s_2 \rangle = \tfrac{1}{2}\left[S(S+1) - s_1(s_1+1) - s_2(s_2+1)\right]\hbar^2 \tag{4.45}$$

With nucleon spins $s_1$ and $s_2$ of $\tfrac{1}{2}$, the value of $s_1 \cdot s_2$ is, for triplet ($S = 1$) states:

$$\langle s_1 \cdot s_2 \rangle = \tfrac{1}{2}\left[1(1+1) - \tfrac{1}{2}(\tfrac{1}{2}+1) - \tfrac{1}{2}(\tfrac{1}{2}+1)\right]\hbar^2 = \tfrac{1}{4}\hbar^2 \tag{4.46}$$

and for singlet ($S = 0$) states:

$$\langle s_1 \cdot s_2 \rangle = \tfrac{1}{2}\left[0(0+1) - \tfrac{1}{2}(\tfrac{1}{2}+1) - \tfrac{1}{2}(\tfrac{1}{2}+1)\right]\hbar^2 = -\tfrac{3}{4}\hbar^2 \tag{4.47}$$

Thus a spin-dependent expression of the form $s_1 \cdot s_2 V_s(r)$ can be included in the potential and will have the effect of giving different calculated cross sections for singlet and triplet states. The magnitude of $V_s$ can be adjusted to give the correct differences between the singlet and triplet cross sections and the radial dependence can be adjusted to give the proper dependence on energy.

We could also write the potential including $V_c$ and $V_s$ as

$$V(r) = -\left(\frac{s_1 \cdot s_2}{\hbar^2} - \tfrac{1}{4}\right)V_1(r) + \left(\frac{s_1 \cdot s_2}{\hbar^2} + \tfrac{3}{4}\right)V_3(r) \tag{4.48}$$

where $V_1(r)$ and $V_3(r)$ are potentials that separately give the proper singlet and triplet behaviors.

## The Internucleon Potential Includes a Noncentral Term, Known as a Tensor Potential

Evidence for the tensor force comes primarily from the observed quadrupole moment of the ground state of the deuteron. An s-state ($\ell = 0$) wave function is spherically symmetric; the electric quadrupole moment vanishes. Wave functions with mixed $\ell$ states must result from noncentral potentials. This tensor force must be of the form $V(\mathbf{r})$, instead of $V(r)$. For a single nucleon, the choice of a certain direction in space is obviously arbitrary; nucleons do not distinguish north from south or east from west. The only reference direction for a nucleon is its spin, and thus only terms of the form $\mathbf{s} \cdot \mathbf{r}$ or $\mathbf{s} \times \mathbf{r}$, which relate $\mathbf{r}$ to the direction of $\mathbf{s}$, can contribute. To satisfy the requirements of parity invariance, there must be an even number of factors of $\mathbf{r}$, and so for two nucleons the potential must depend on terms such as $(\mathbf{s}_1 \cdot \mathbf{r})(\mathbf{s}_2 \cdot \mathbf{r})$ or $(\mathbf{s}_1 \times \mathbf{r}) \cdot (\mathbf{s}_2 \times \mathbf{r})$. Using vector identities we can show that the second form can be written in terms of the first and the additional term $\mathbf{s}_1 \cdot \mathbf{s}_2$, which we already included in $V(r)$. Thus without loss of generality we can choose the tensor contribution to the internucleon potential to be of the form $V_{\mathrm{T}}(r) S_{12}$, where $V_{\mathrm{T}}(r)$ gives the force the proper radial dependence and magnitude, and

$$S_{12} = 3(\mathbf{s}_1 \cdot \mathbf{r})(\mathbf{s}_2 \cdot \mathbf{r})/r^2 - \mathbf{s}_1 \cdot \mathbf{s}_2 \qquad (4.49)$$

which gives the force its proper tensor character and also averages to zero over all angles.

### The Nucleon–Nucleon Force Is Charge Symmetric

This means that the proton–proton interaction is identical to the neutron–neutron interaction, after we correct for the Coulomb force in the proton–proton system. Here "charge" refers to the character of the nucleon (proton or neutron) and not to electric charge. Evidence in support of this assertion comes from the equality of the pp and nn scattering lengths and effective ranges. Of course, the pp parameters must first be corrected for the Coulomb interaction. When this is done, the resulting singlet pp parameters are

$$a = -17.1 \pm 0.2 \text{ fm}$$

$$r_0 = 2.84 \pm 0.03 \text{ fm}$$

These are in very good agreement with the measured nn parameters ($a = -16.6 \pm 0.5$ fm, $r_0 = 2.66 \pm 0.15$ fm), which strongly supports the notion of charge symmetry.

### The Nucleon–Nucleon Force Is Nearly Charge Independent

This means that (in analogous spin states) the three *nuclear* forces nn, pp, *and* pn are identical, again correcting for the pp Coulomb force. Charge independence is thus a stronger requirement than charge symmetry. Here the evidence is not so conclusive; in fact, the singlet np scattering length ($-23.7$ fm) seems to differ substantially from the pp and nn scattering lengths ($-17$ fm). However, we see from Figure 4.11 that large negative scattering lengths are extraordinarily sensitive to the nuclear wave function near $r = R$, and a very small change in $\psi$ can give a large change in the scattering length. Thus the large difference between the scattering lengths may correspond to a very small difference (of order 1%) between the potentials, which (as we see in the next section) is easily explained by the exchange force model.

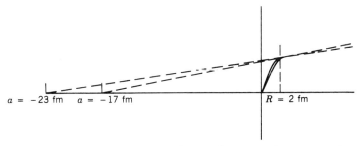

**Figure 4.11**  Very small changes in the nucleon–nucleon wave function near $r = R$ can lead to substantial differences in the scattering length when the extrapolation is made (compare Figure 4.7b).

### The Nucleon–Nucleon Interaction Becomes Repulsive at Short Distances

This conclusion follows from qualitative considerations of the nuclear density: as we add more nucleons, the nucleus grows in such a way that its central density remains roughly constant, and thus something is keeping the nucleons from crowding too closely together. More quantitatively, we can study nucleon–nucleon scattering at higher energies. Figure 4.12 shows the deduced singlet s-wave phase shifts for nucleon–nucleon scattering up to 500 MeV. (At these energies, phase shifts from higher partial waves, p and d for example, also contribute to the cross

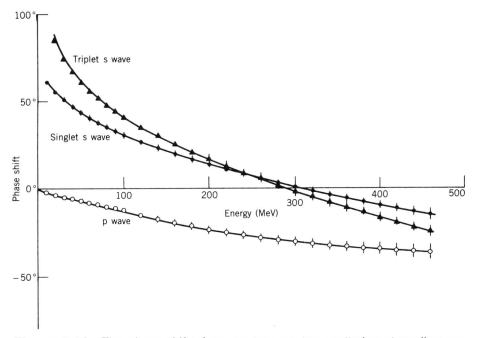

**Figure 4.12**  The phase shifts from neutron–proton scattering at medium energies. The change in the s-wave phase shift from positive to negative at about 300 MeV shows that at these energies the incident nucleon is probing a repulsive core in the nucleon–nucleon interaction. ▲, $^3S_1$; ●, $^1S_0$; O, $^1P_1$. Data from M. MacGregor et al., *Phys. Rev.* **182**, 1714 (1969).

sections. The s-wave phase shifts can be easily extracted from the differential scattering measurements of $d\sigma/d\Omega$ vs $\theta$ because they do not depend on $\theta$.) At about 300 MeV, the s-wave phase shift becomes *negative*, corresponding to a change from an attractive to a repulsive force. To account for the repulsive core, we must modify the potentials we use in our calculations. For example, again choosing a square-well form to simplify the calculation, we might try

$$V(r) = +\infty \qquad r < R_{\text{core}}$$
$$= -V_0 \qquad R_{\text{core}} \leq r \leq R$$
$$= 0 \qquad r > R \qquad (4.50)$$

and we can adjust $R_{\text{core}}$ until we get satisfactory agreement with the observed s-wave phase shifts. The value $R_{\text{core}} \simeq 0.5$ fm gives agreement with the observed phase shifts.

### The Nucleon – Nucleon Interaction May Also Depend on the Relative Velocity or Momentum of the Nucleons

Forces depending on velocity or momentum cannot be represented by a scalar potential, but we can include these forces in a reasonable manner by introducing terms linear in $p$, quadratic in $p$, and so on, with each term including a characteristic $V(r)$. Under the parity operation, $p \to -p$, and also under time reversal $p \to -p$. Thus any term simply linear in $p$ is unacceptable because it violates both parity and time-reversal invariance. Terms of the form $r \cdot p$ or $r \times p$ are invariant with respect to parity, but still violate time reversal. A possible structure for this term that is first order in $p$ and invariant with respect to both parity and time reversal is $V(r)(r \times p) \cdot S$, where $S = s_1 + s_2$ is the total spin of the two nucleons. The relative angular momentum of the nucleons is $\ell = r \times p$, and therefore this term, known as the *spin-orbit term* in analogy with atomic physics, is written $V_{\text{so}}(r)\ell \cdot S$. Although higher-order terms may be present, this is the only first-order term in $p$ that satisfies the symmetries of both parity and time reversal.

The experimental evidence in support of the spin-orbit interaction comes from the observation that scattered nucleons can have their spins aligned, or *polarized*, in certain directions. The polarization of the nucleons in a beam (or in a target) is defined as

$$P = \frac{N(\uparrow) - N(\downarrow)}{N(\uparrow) + N(\downarrow)} \qquad (4.51)$$

where $N(\uparrow)$ and $N(\downarrow)$ refer to the number of nucleons with their spins pointed up and down, respectively. Values of $P$ range from $+1$, for a 100% spin-up polarized beam, to $-1$, for a 100% spin-down polarized beam. An unpolarized beam, with $P = 0$, has equal numbers of nucleons with spins pointing up and down.

Consider the scattering experiment shown in Figure 4.13$a$, in which an unpolarized beam (shown as a mixture of spin-up and spin-down nucleons) is incident on a spin-up target nucleon. Let's suppose the nucleon–nucleon interaction causes the incident spin-up nucleons to be scattered to the left at angle $\theta$

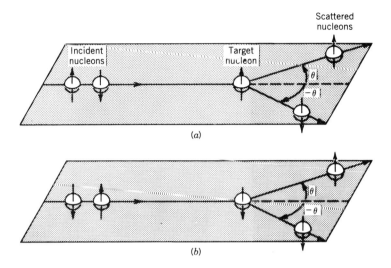

**Figure 4.13** An unpolarized beam (shown as a mixture of spin-up and spin-down nucleons) is scattered from a target that can have either spin up or spin down. In part *a*, the incident nucleons with spin up are scattered to the left at angle $\theta$, while those with spin down are scattered to the right at $-\theta$. Part *b* can be obtained from part *a* by viewing from below or by rotating 180° about the beam direction; it shows that the same conclusions follow in scattering from a spin-down polarized target.

and the incident spin-down nucleons to be scattered to the right at angle $-\theta$. Part *b* of the figure shows the *same experiment* viewed from below or else rotated 180° about the direction of the incident beam. We can also interpret Figure 4.13*b* as the scattering of an unpolarized beam from a spin-down target nucleon, and once again the spin-up incident nucleons scatter to the left and the spin-down nucleons scatter to the right. The results would be the same, even in an unpolarized target, which would contain a mixture of spin-up and spin-down nucleons: when an unpolarized beam is scattered from an unpolarized target, the spin-up scattered nucleons appear preferentially at $\theta$ and the spin-down scattered nucleons at $-\theta$.

Although this situation may appear superficially to violate reflection symmetry (parity), you can convince yourself that this is not so by sketching the experiment and its mirror image. Parity is conserved if at angle $\theta$ we observe a net polarization $P$, while at angle $-\theta$ we observe a net polarization of $-P$.

Let's now see how the spin-orbit interaction can give rise to this type of scattering with polarization. Figure 4.14 shows two nucleons with spin up incident on a spin up target, so that $S = 1$. (Scattering that includes only s waves must be spherically symmetric, and therefore there can be no polarizations. The p-wave ($\ell = 1$) scattering of identical nucleons has an antisymmetric spatial wave function and therefore a symmetric spin wave function.) Let's assume that $V_{so}(r)$ is negative. For incident nucleon 1, $\ell = r \times p$ is down (into the page), and therefore $\ell \cdot S$ is negative because $\ell$ and $S$ point in opposite directions. The combination $V_{so}(r)\ell \cdot S$ is positive and so there is a repulsive force between the

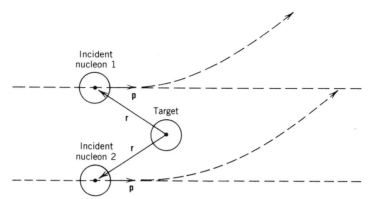

**Figure 4.14** Top view of nucleon–nucleon scattering experiment. All spins point up (out of the paper). Incident nucleon 1 has $r \times p$ into the paper, and thus $\ell \cdot S$ is negative, giving a repulsive force and scattering to the left. Incident nucleon 2 has $r \times p$ out of the paper, resulting in an attractive force and again scattering to the left.

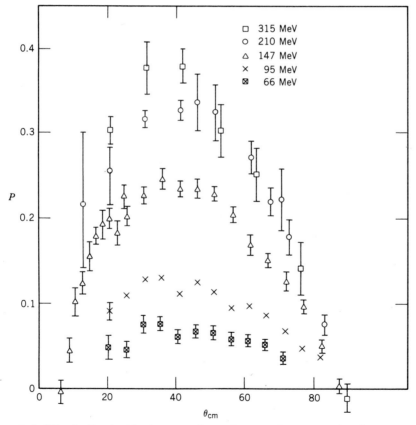

**Figure 4.15** As the incident energy in proton–proton scattering increases, the maximum polarization increases. From R. Wilson, *The Nucleon–Nucleon Interaction* (New York: Wiley-Interscience, 1963).

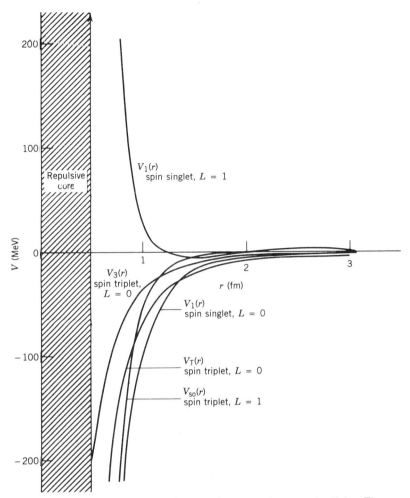

**Figure 4.16** Some representative nucleon–nucleon potentials. Those shown include the attractive singlet and triplet terms that contribute to s-wave scattering, the repulsive term that gives one type of p-wave ($L = 1$) scattering, and the attractive tensor and spin-orbit terms. All potentials have a repulsive core at $r = 0.49$ fm. These curves are based on an early set of functional forms proposed by T. Hamada and I. D. Johnston, *Nucl. Phys.* **34**, 382 (1962); other relatively similar forms are in current use.

target and incident nucleon 1, which is pushed to the left. For nucleon 2, $\ell$ points up, $\ell \cdot S$ is positive, and the interaction is attractive; incident nucleon 2 is pulled toward the target and also appears on the left side. Spin-up incident nucleons are therefore preferentially scattered to the left and (by a similar argument) spin-down nucleons to the right. Thus the spin-orbit force can produce polarized scattered beams when unpolarized particles are incident on a target.

At low energy, where s-wave scattering dominates, we expect no polarization. As the incident energy increases, the contribution of p-wave scattering increases and there should be a corresponding increase in the polarization. Figure 4.15

shows that this expectation is correct. From the variation of $P$ with $\theta$ and with energy, we can make deductions about the form of $V_{so}(r)$.

The general topic of polarization in nuclear reactions is far more complicated than we have indicated in this brief discussion. We should also consider the effect on the measured cross sections of using polarized beams and polarized targets, which we do in Chapter 11.

From this enormous set of experimental information (total and differential cross sections, spin dependence, polarizations), it is possible to propose a set of phenomenological potentials $V(r)$ that give reasonable agreement with the observed nucleon–nucleon data. These potentials can then be used in calculations for more complicated nuclei. As an example, Figure 4.16 illustrates one such set of potentials. As is usually the case, negative potentials give an attractive force and positive potentials give a repulsive force. Notice how the potentials incorporate such features as the range of the interactions, the repulsive core, the strong attractive s-wave phase shifts, the repulsive p-wave phase shifts, and charge independence (since no distinctions are made for the characters of the nucleons).

## 4.5 THE EXCHANGE FORCE MODEL

The phenomenological potentials discussed in the previous section have been fairly successful in accounting for a variety of measured properties of the nucleon–nucleon interaction. Of course, the ability of these potentials to give accurate predictions would be improved if we added more terms to the interaction. For example, we could have included a term that is second-order in the momentum dependence (proportional to $\ell^2$), we could write potentials that are different for each $\ell$ value, and so on. Each new term in the potential may improve the calculation, but it may be at the expense of simplicity. It also serves to make us lose sight of our main objective: to understand the nucleon–nucleon interaction. Simply because we have included enough potentials to do accurate calculations does not mean we have improved our understanding of the fundamental character of the nucleon–nucleon interaction. We therefore try to postulate a physical mechanism for the nucleon–nucleon force that will yield potentials similar to those that have already proven to be successful in calculations.

A successful mechanism is that of the *exchange force*. There are two principal arguments in support of the presence of exchange forces in nuclei. The first comes from the saturation of nuclear forces. The experimental support for saturation comes from the relatively constant nuclear density and binding energy per nucleon as we go to heavier nuclei. A given nucleon seems to attract only a small number of near neighbors, but it also repels at small distances to keep those neighbors from getting too close. (We explained this behavior in the previous section by choosing a central potential that was of finite range and had a repulsive core.) We encounter exactly the same sort of behavior in molecules. When we bring two atoms together to form a diatomic molecule, such as one with covalent bonding, electrons are shared or exchanged between the two atoms, and a stable molecule forms with the atoms in equilibrium separated by a certain

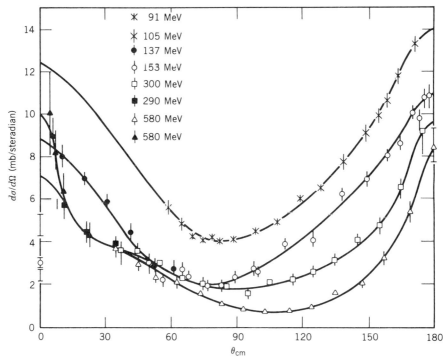

**Figure 4.17** The neutron–proton differential cross section at medium energies. The strong forward-scattering peak (near 0°) is expected; the equally strong backward peak (near 180°) is evidence for the exchange force. From R. Wilson, *The Nucleon–Nucleon Interaction* (New York: Wiley-Interscience, 1963).

distance. If we try to force the atoms closer together, the overlap of the filled electronic shells causes a strong repulsive force. Furthermore, approaching the molecule with a third atom may result only in very weak forces between the first two atoms and the third; if all of the valence electrons are occupied in the first set of bonds, none are available to form new bonds. Nuclear forces show a similar saturation character.

Another argument in favor of the exchange force model comes from the study of np scattering at high energies. Figure 4.17 shows the np differential cross section. There is a strong peak in the cross section at forward angles near 0°, corresponding to a small momentum transfer between the projectile and the target. We can estimate the extent of this forward peak by studying the maximum momentum transfer in the following way: For small deflection angles, $\sin \theta \simeq \theta = \Delta p/p$ where $p$ is the momentum of the incident particle and $\Delta p$ is the transverse momentum added during the collision. If $F$ is the average force that acts during the collision time $\Delta t$, then $\Delta p = F \Delta t$. The force $F$ is $-dV/dr$, and thus the average force should be of the order of $V_0/R$, where $V_0$ is the depth of the nucleon–nucleon square-well potential and $R$ is its range. (Even if the actual potential is not at all constant, such as the central term of Figure 4.16, the average value of $dV/dr$ should be of the order of $V_0/R$.) The collision time $\Delta t$

should be of the order of $R/v$, where $v$ is the projectile velocity. Thus

$$\theta \simeq \frac{\Delta p}{p} = \frac{F \Delta t}{p} = \frac{1}{p} \frac{V_0}{R} \frac{R}{v} = \frac{V_0}{pv} = \frac{V_0}{2T} \tag{4.52}$$

where $T$ is the projectile kinetic energy. For the energies shown in Figure 4.17, this gives values of $\theta$ in the range of $10°$ or smaller. We certainly do not expect to see a peak at $180°$! Although it is tempting to regard this "backward" peak in the center of mass frame as the result of a head-on collision in which the incident particle has its motion reversed, our estimate above indicates such an explanation is not likely to be correct.

A more successful explanation can be found in the exchange model if, during the collision, the neutron and proton exchange places. That is, the forward-moving neutron becomes a proton and the backward-moving (in the center-of-mass system) proton becomes a neutron. The incident nucleon then reappears in the laboratory as a forward-moving nucleon (now a proton), consistent with our estimate of the small deflection angle in nucleon–nucleon scattering.

In summary, both the saturation of nuclear forces and the strong backward peak in np scattering are explained by exchange forces. In the former case, "something" is exchanged between nucleons to produce a sort of saturated bond. In the second case, "something" is exchanged between nucleons and actually changes their character.

In the early development of classical physics, objects were said to interact by means of "action at a distance." Somehow one object mysteriously transmitted through space its force on the other object. The great development in nineteenth-century theoretical physics was the introduction of the concept of *fields*, according to which one object establishes throughout space a force field (electromagnetic and gravitational fields are examples) and the second object interacts only with the field, not directly with the first object. Maxwell showed in the case of electromagnetism how the fields were transmitted through space. The major development of twentieth-century physics is quantum mechanics, according to which all exchanges of energy must occur in bundles of a discrete size. The classical field is smooth and continuous, and to bring classical field theory into agreement with quantum theory, the field itself must be quantized. That is, according to quantum field theory, the first object does not set up a classical field throughout space but instead emits field quanta. The second object can then absorb those field quanta (and reemit them back to the first object). The two objects interact directly with the exchanged field quanta and therefore indirectly with each other.

In view of the preceding discussion, it is natural to associate the "something" that is exchanged in the nucleon–nucleon interaction with quanta of the nuclear field. For a spin-$\frac{1}{2}$ neutron to turn into a spin-$\frac{1}{2}$ proton, it is clear that the exchanged particle must have integral spin (0 or 1) and must carry electric charge. In addition, if we wish to apply the same exchange-force concepts to nn and pp interactions, there must also be an uncharged variety of exchanged particle. Based on the observed range of the nuclear force, we can estimate the mass of the exchanged particle. Let us assume that a nucleon (which we denote by N, to include both neutrons and protons) emits a particle x. A second nucleon absorbs the particle x:

$$N_1 \rightarrow N_1 + x$$
$$x + N_2 \rightarrow N_2$$

How is it possible for a nucleon to emit a particle of mass energy $m_xc^2$ and still remain a nucleon, without violating conservation of energy? It is not possible, unless the emission and reabsorption take place within a short enough time $\Delta t$ that we are unaware energy conservation has been violated. Since the limits of our ability to measure an energy (and therefore to determine whether energy is conserved) are restricted by the uncertainty principle, if $\Delta t < \hbar/(m_xc^2)$, we will be unaware that energy conservation has been violated by an amount $\Delta E = m_xc^2$. The maximum range of the force is determined by the maximum distance that the particle x can travel in the time $\Delta t$. If it moves at speeds of the order of $c$, then the range $R$ can be at most

$$R = c\,\Delta t = \frac{\hbar c}{m_xc^2} = \frac{200 \text{ MeV} \cdot \text{fm}}{m_xc^2} \tag{4.53}$$

where we have used a convenient approximation for $\hbar c$. Equation 4.53 gives a useful relationship between the mass energy of the exchanged particles and the range of the force. For nuclear forces with a range of about 1 fm, it is clear that we must have an exchanged particle with a mass energy of the order of 200 MeV.

Such particles that exist only for fleeting instants and allow us to violate conservation of energy (and momentum—the emitting and absorbing nucleons do not recoil) are known as *virtual* particles. We can observe the force that results from the exchange of virtual particles, but we cannot observe the particles themselves during the exchange. (Exchanged virtual particles can be identical with ordinary particles, however. According to field theory, the Coulomb interaction between electric charges can be regarded as the exchange of virtual photons, which have properties in common with ordinary real photons.)

The exchanged particles that carry the nuclear force are called *mesons* (from the Greek "meso" meaning middle, because the predicted mass was between the masses of the electron and the nucleon). The lightest of the mesons, the $\pi$-meson or simply *pion*, is responsible for the major portion of the longer range (1.0 to 1.5 fm) part of the nucleon–nucleon potential. To satisfy all the varieties of the exchanges needed in the two-nucleon system, there must be three pions, with electric charges of $+1$, 0, and $-1$. The pions have spin 0 and rest energies of 139.6 MeV (for $\pi^\pm$) and 135.0 MeV (for $\pi^0$). At shorter ranges (0.5–1.0 fm), two-pion exchange is probably responsible for the nuclear binding; at much shorter ranges (0.25 fm) the exchange of $\omega$ mesons ($mc^2 = 783$ MeV) may contribute to the repulsive core whereas the exchange of $\rho$ mesons ($mc^2 = 769$ MeV) may provide the spin-orbit part of the interaction. Further properties of these mesons are discussed in Chapter 17.

The differing masses for the charged and neutral pions may explain the possible small violation of charge independence we discussed previously. The single pion that is exchanged between two identical nucleons must be a $\pi^0$:

$$n_1 \rightarrow n_1 + \pi^0 \qquad \pi^0 + n_2 \rightarrow n_2$$

or

$$p_1 \rightarrow p_1 + \pi^0 \qquad \pi^0 + p_2 \rightarrow p_2$$

Charged pion exchange will not work:

$$n_1 \to p_1 + \pi^- \qquad \text{but } \pi^- + n_2 \to ?$$
$$p_1 \to n_1 + \pi^+ \qquad \text{but } \pi^+ + p_2 \to ?$$

because there are no nucleons with charges $-1$ or $+2$. (There are *excited states* of the nucleon with these charges, as we discuss in Chapters 17 and 18, but these high-energy states are unlikely to contribute substantially to the low-energy experiments we have discussed in this chapter.) However, the neutron–proton interaction can be carried by charged as well as neutral pions:

$$n_1 \to n_1 + \pi^0 \qquad \pi^0 + p_2 \to p_2$$
$$n_1 \to p_1 + \pi^- \qquad \pi^- + p_2 \to n_2$$

This additional term in the np interaction (and the difference in mass between the charged and neutral pions) may be responsible for the small difference in the potential that produces the observed difference in the scattering lengths.

The meson-exchange theory of nuclear forces was first worked out by Yukawa in 1935; some details of his work are summarized in Chapter 17. Meson exchange can be represented by a potential in the basic form of $r^{-1}e^{-r/R}$, where $R$ is the range of the force ($R = \hbar/m_\pi c = 1.5$ fm for pions). A more detailed form for the one-pion exchange potential (called OPEP in the literature) is

$$V(r) = \frac{g_\pi^2 (m_\pi c^2)^3}{3(Mc^2)^2 \hbar^2} \left[ s_1 \cdot s_2 + S_{12} \left( 1 + \frac{3R}{r} + \frac{3R^2}{r^2} \right) \right] \frac{e^{-r/R}}{r/R} \qquad (4.54)$$

Here $g_\pi^2$ is a dimensionless coupling constant that gives the strength of the interaction (just as $e^2$ gives the strength of the electromagnetic interaction) and $M$ is the nucleon mass. This particular potential describes only the long-range part of the nucleon–nucleon interaction; other aspects of the interaction are described by other potentials.

The exchange-force model enjoyed a remarkable success in accounting for the properties of the nucleon–nucleon system. The forces are based on the exchange of virtual mesons, all of which can be produced in the laboratory and studied directly. The pion is the lightest of the mesons and therefore has the longest range. Exploring the nucleus with higher energy probes (with shorter de Broglie wavelengths) allows us to study phenomena that are responsible for the finer details of the nuclear structure, such as those that occur only over very short distances. These phenomena are interpreted as arising from the exchange of heavier mesons. Studying the spatial and spin dependence of these detailed interactions allows us to deduce the properties of the hypothetical exchanged meson. On the other hand, particle physicists are able to observe a large variety of mesons from high-energy collisions done with large accelerators. Among the debris from those collisions they can observe many varieties of new particles and catalog their properties. Nuclear physicists are then able to choose from this list candidates for the mesons exchanged in various details of the nucleon–nucleon interaction. This slightly oversimplified view nevertheless emphasizes the close historical relationship between nuclear physics and elementary particle physics.

## REFERENCES FOR ADDITIONAL READING

For similar treatments of the nucleon–nucleon interaction, see H. Enge, *Introduction to Nuclear Physics* (Reading: Addison-Wesley, 1966), Chapters 2 and 3; R. D. Evans, *The Atomic Nucleus* (New York: McGraw-Hill, 1955), Chapter 10; E. Segrè, *Nuclei and Particles* (Reading, MA: Benjamin, 1977), Chapter 10.

Monographs devoted to the nucleon–nucleon interaction are H. A. Bethe and P. Morrison, *Elementary Nuclear Theory* (New York: Wiley, 1956); M. J. Moravcsik, *The Two Nucleon Interaction* (Oxford: Clarendon, 1963); R. Wilson, *The Nucleon Nucleon Interaction* (New York: Wiley, 1963).

A review of early work on nucleon–nucleon scattering can be found in R. K. Adair, *Rev. Mod. Phys.* **22**, 249 (1950). Analyses of scattering data to extract phase shifts and other parameters are reviewed by M. H. MacGregor, M. J. Moravcsik, and H. P. Stapp, *Ann. Rev. Nucl. Sci.* **10**, 291 (1960), and H P. Noyes, *Ann. Rev. Nucl. Sci.* **22**, 465 (1972). See also M. H. MacGregor, *Phys. Today* **22**, 21 (December 1969).

## PROBLEMS

1.  What is the minimum photon energy necessary to dissociate $^2$H? Take the binding energy to be 2.224589 MeV.

2.  (a) Use the continuity and normalization conditions to evaluate the coefficients $A$ and $C$ in the deuteron wave functions, Equations 4.3 and 4.4.

    (b) From the resulting wave function, evaluate the root-mean-square radius of the deuteron.

3.  The condition for the existence of a bound state in the square-well potential can be determined through the following steps:

    (a) Using the complete normalized wave function, Equations 4.3 and 4.4, show that the expectation value of the potential energy is

    $$\langle V \rangle = \int \psi^* V \psi \, dv = -V_0 A^2 \left[ \tfrac{1}{2} R - \frac{1}{4k_1} \sin 2k_1 R \right]$$

    (b) Show that the expectation value of the kinetic energy is

    $$\langle T \rangle = \frac{\hbar^2}{2m} \int_0^\infty \left| \frac{\partial \psi}{\partial r} \right|^2 dv$$

    $$= \frac{\hbar^2}{2m} A^2 \left[ \tfrac{1}{2} k_1^2 R + \tfrac{1}{4} k_1 \sin 2k_1 R + \frac{k_2}{2} \sin^2 k_1 R \right]$$

    (c) Show that, for a bound state to exist, it must be true that $\langle T \rangle < -\langle V \rangle$.

    (d) Finally, show that a bound state will exist only for $V_0 \geq \pi^2 \hbar^2 / 8mR^2$ and evaluate the minimum depth of the potential that gives a bound state of the deuteron.

    (*Note:* This calculation is valid only in three-dimensional problems. In the one-dimensional square well (indeed, in all reasonably well-behaved attractive one-dimensional potentials) there is always at least one bound state. Only in three-dimensional problems is there a critical depth for the existence of a bound state. See C. A. Kocher, *Am. J. Phys.* **45**, 71 (1977).)

4.   What fraction of the time do the neutron and proton in the deuteron spend beyond the range of their nuclear force?

5.   From Equation 4.5, plot $V_0$ against $R$ for $R$ in the range of 1.0 to 3.0 fm. Discuss the sensitivity of $V_0$ to $R$.

6.   (a) Show that Equation 4.5 can be written in the transcendental form $x = -\tan bx$, where $x = \sqrt{-(V_0 + E)/E}$. Evaluate the parameter $b$ for $R = 2$ fm. Note that in Equation 4.2, $m$ is the reduced mass $m_p m_n/(m_p + m_n)$, which is approximately $m_p/2$. (b) Solve the transcendental equation in two ways: graphically and iteratively using a programmable calculator or computer. For a review of using iterative techniques on similar equations, see K. S. Krane, *Am. J. Phys.* **50**, 521 (1982).

7.   Assuming a deuteron wave function of the form of Equation 4.9, deduce why there is a cross-term (that is, a term proportional to $a_s a_d$) in the expression for the electric quadrupole moment, Equation 4.11, but not in the expression for the magnetic dipole moment, Equation 4.10.

8.   Evaluate the energy of the magnetic dipole–dipole interaction in the deuteron, and compare with the nuclear binding energy. Consider separately the cases in which the nucleon spins are perpendicular to and parallel to the line joining the nucleons.

9.   Find the scattering cross section of the "hard sphere":

$$V(r) = \infty \qquad \text{for } r < R$$
$$= 0 \qquad \text{for } r > R$$

10.   Suppose the binding energy of the deuteron were much weaker, say 10 keV. Evaluate the resulting neutron–proton s-wave cross section.

11.   Show that the singlet neutron–proton state is unbound and evaluate its energy.

12.   Antiprotons ($\bar{p}$) and antineutrons ($\bar{n}$) can be produced at several accelerator facilities throughout the world. Discuss the properties of the following systems in comparison with those discussed in this chapter: (a) $\bar{n}p$ bound state; (b) $n\bar{p}$ bound state; (c) $\bar{n}p$ s-wave scattering; (d) $\bar{p}p$ scattering.

13.   Solve the Schrödinger equation for the potential given in Equation 4.50 for s-wave neutron–proton scattering. Find an expression that relates the s-wave phase shift to the core radius, and find the value of $R_{core}$ that causes the phase shift to go negative, as in Figure 4.12.

14.   In a measurement of the pp differential cross section, the result $d\sigma/d\Omega = 0.111$ b/steradian was obtained at a laboratory energy of 4.2 MeV and a laboratory scattering angle of 30°. What is the corresponding s-wave phase shift?

15.   Suppose the nucleon–nucleon force were stronger, so that the deuteron had the following bound states:

State A is the "well-known" ground state with the properties discussed in this chapter. State B is very close to state A. At a large energy gap $\Delta E$

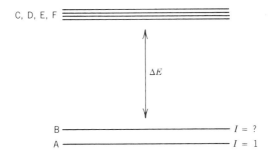

above states A and B are four states, C, D, E, and F. There are no other states near C, D, E, and F.

(a) What is the most likely value of the relative orbital angular momentum of the nucleons in state B? What is the relative orientation of the nucleon intrinsic spins in state B, and what is the resulting value of the total angular momentum $I$ in state B?

(b) In states C, D, E, and F the nucleons have the same relative orbital angular momentum (but different from state A). Make a reasonable guess at the value of the orbital angular momentum and justify your choice.

(c) By considering the possible couplings of the orbital angular momentum and the intrinsic spins of the two nucleons, show that there must be only four states in the excited multiplet, and give the four possible values of the total angular momentum $I$ and the parity.

(d) Assuming the energy gap $\Delta E$ to result primarily from the "centripetal" contribution to the potential, estimate $\Delta E$. Be sure to justify any choice of parameters you use in your estimate.

16. Low-energy (s-wave) neutrons are scattered from protons, and the distribution of "recoil" protons is to be observed and analyzed. Let the neutron scattering angle be $\theta$ in the laboratory coordinate system, and let the incident neutron kinetic energy be $T_n$. (a) Show that the protons emerge at an angle $\phi = 90° - \theta$ with respect to the direction of the incident neutrons. (b) Show that $T'_n = T_n \cos^2 \theta$ and $T'_p = T_n \sin^2 \theta$, where $T'$ signifies the energy after the scattering. (c) Show that the laboratory and center-of-mass cross sections are related by $(d\sigma/d\Omega)_{lab} = (4 \cos \theta)(d\sigma/d\Omega)_{cm}$. (d) Given that the scattering is independent of direction in the center-of-mass system, show that (in the laboratory system) $d\sigma/dT'_p = \sigma/T_n$, where $\sigma$ is the total cross section. This latter result shows that the number of recoil protons observed at any particular energy $T'_p$ ($0 \le T'_p \le T_n$) is independent of $T'_p$. (e) What is the angular distribution of the recoil protons in the laboratory?

# 5

# NUCLEAR MODELS

At this point it is tempting to try to extend the ideas of the previous chapter to heavier nuclei. Unfortunately, we run into several fundamental difficulties when we do. One difficulty arises from the mathematics of solving the many-body problem. If we again assume an oversimplified form for the nuclear potential, such as a square well or an harmonic oscillator, we could in principle write down a set of coupled equations describing the mutual interactions of the $A$ nucleons. These equations cannot be solved analytically, but instead must be attacked using numerical methods. A second difficulty has to do with the nature of the nuclear force itself. There is evidence to suggest that the nucleons interact not only through mutual two-body forces, but through three-body forces as well. That is, the force on nucleon 1 not only depends on the individual positions of nucleons 2 and 3, it contains an *additional* contribution that arises from the correlation of the positions of nucleons 2 and 3. Such forces have no classical analog.

In principle it is possible to do additional scattering experiments in the three-body system to try (in analogy with the two-body studies described in Chapter 4) to extract some parameters that describe the three-body forces. However, we quickly reach a point at which such a microscopic approach obscures, rather than illuminates, the essential physics of the nucleus. It is somewhat like trying to obtain a microscopic description of the properties of a gas by studying the interactions of its atoms and then trying to solve the dynamical equations that describe the interatomic forces. Most of the physical insight into the properties of a gas comes from a few general parameters such as pressure and temperature, rather than from a detailed microscopic theory.

We therefore adopt the following approach for nuclei. We choose a deliberately oversimplified theory, but one that is mathematically tractable and rich in physical insight. If that theory is fairly successful in accounting for at least a few nuclear properties, we can then improve it by adding additional terms. Through such operations we construct a *nuclear model*, a simplified view of nuclear structure that still contains the essentials of nuclear physics. A successful model must satisfy two criteria: (1) it must reasonably well account for previously measured nuclear properties, and (2) it must predict additional properties that can be measured in new experiments. This system of modeling complex processes is a common one in many areas of science; biochemists model the complex processes such as occur in the replication of genes, and atmospheric scientists model the complex dynamics of air and water currents that affect climate.

## 5.1 THE SHELL MODEL

Atomic theory based on the shell model has provided remarkable clarification of the complicated details of atomic structure. Nuclear physicists therefore attempted to use a similar theory to attack the problem of nuclear structure, in the hope of similar success in clarifying the properties of nuclei. In the atomic shell model, we fill the shells with electrons in order of increasing energy, consistent with the requirement of the Pauli principle. When we do so, we obtain an inert core of filled shells and some number of valence electrons; the model then assumes that atomic properties are determined primarily by the valence electrons. When we compare some measured properties of atomic systems with the predictions of the model, we find remarkable agreement. In particular, we see regular and smooth variations of atomic properties *within* a subshell, but rather sudden and dramatic changes in the properties when we fill one subshell and enter the next. Figure 5.1 shows the effects of a change in subshell on the ionic radius and ionization energy of the elements.

When we try to carry this model over to the nuclear realm, we immediately encounter several objections. In the atomic case, the potential is supplied by the Coulomb field of the nucleus; the subshells ("orbits") are established by an external agent. We can solve the Schrödinger equation for this potential and calculate the energies of the subshells into which electrons can then be placed. In the nucleus, there is no such external agent; the nucleons move in a potential that they themselves create.

Another appealing aspect of atomic shell theory is the existence of spatial orbits. It is often very useful to describe atomic properties in terms of spatial orbits of the electrons. The electrons can move in those orbits relatively free of collisions with other electrons. Nucleons have a relatively large diameter compared with the size of the nucleus. How can we regard the nucleons as moving in well defined orbits when a single nucleon can make many collisions during each orbit?

First let's examine the experimental evidence that supports the existence of nuclear shells. Figure 5.2 shows measured proton and neutron separation energies, plotted as deviations from the predictions of the semiempirical mass formula, Equation 3.28. (The gross changes in nuclear binding are removed by plotting the data in this form, allowing the shell effects to become more apparent.) The similarity with Figure 5.1 is striking—the separation energy, like the atomic ionization energy, increases gradually with $N$ or $Z$ except for a few sharp drops that occur at the same neutron and proton numbers. We are led to guess that the sharp discontinuities in the separation energy correspond (as in the atomic case) to the filling of major shells. Figure 5.3 shows some additional evidence from a variety of experiments; the sudden and discontinuous behavior occurs at the same proton or neutron numbers as in the case of the separation energies. These so-called "magic numbers" ($Z$ or $N = 2, 8, 20, 28, 50, 82,$ and $126$) represent the effects of filled major shells, and any successful theory must be able to account for the existence of shell closures at those occupation numbers.

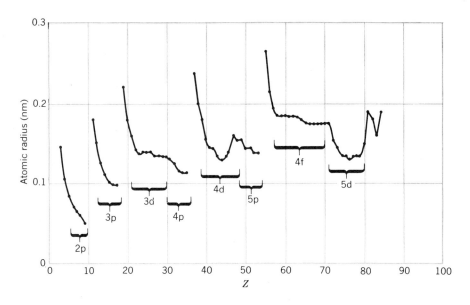

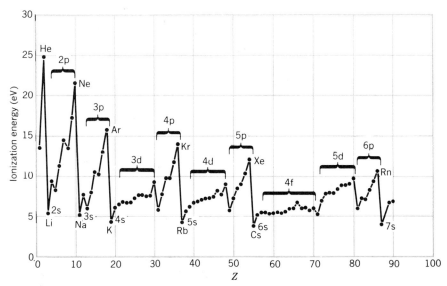

**Figure 5.1** Atomic radius (top) and ionization energy (bottom) of the elements. The smooth variations in these properties correspond to the gradual filling of an atomic shell, and the sudden jumps show transitions to the next shell.

The question of the existence of a nuclear potential is dealt with by the fundamental assumption of the shell model: the motion of a single nucleon is governed by a potential caused by all of the other nucleons. If we treat each individual nucleon in this way, then we can allow the nucleons in turn to occupy the energy levels of a series of subshells.

The existence of definite spatial orbits depends on the Pauli principle. Consider in a heavy nucleus a collision between two nucleons in a state near the very

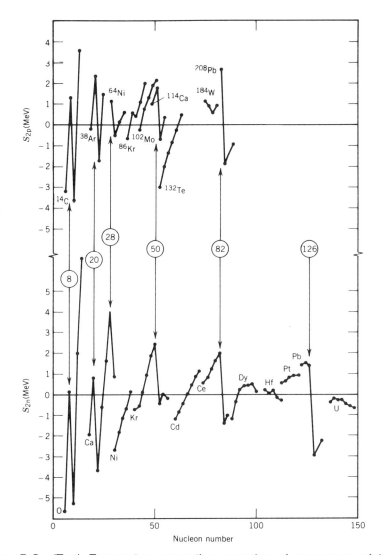

**Figure 5.2** (Top) Two-proton separation energies of sequences of isotones (constant $N$). The lowest $Z$ member of each sequence is noted. (Bottom) Two-neutron separation energies of sequences of isotopes. The sudden changes at the indicated "magic numbers" are apparent. The data plotted are differences between the measured values and the predictions of the semiempirical mass formula. Measured values are from the 1977 atomic mass tables (A. H. Wapstra and K. Bos, *Atomic Data and Nuclear Data Tables* **19**, 215 (1977)).

bottom of the potential well. When the nucleons collide they will transfer energy to one another, but if all of the energy levels are filled up to the level of the valence nucleons, there is no way for one of the nucleons to gain energy except to move up to the valence level. The other levels near the original level are filled and cannot accept an additional nucleon. Such a transfer, from a low-lying level to the valence band, requires more energy than the nucleons are likely to transfer in

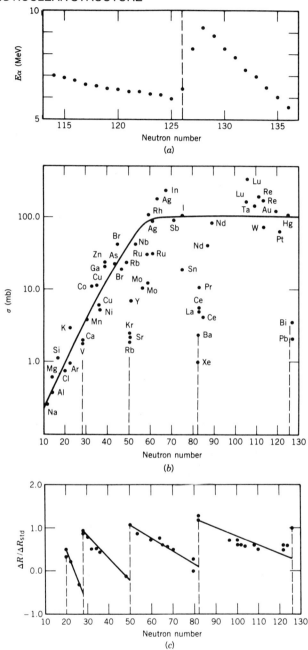

**Figure 5.3** Additional evidence for nuclear shell structure. (a) Energies of α particles emitted by isotopes of Rn. Note the sudden increase when the *daughter* has N = 126 (i.e., when the parent has N = 128). If the daughter nucleus is more tightly bound, the α decay is able to carry away more energy. (b) Neutron-capture cross sections of various nuclei. Note the decreases by roughly two orders of magnitude near N = 50, 82, and 126. (c) Change in the nuclear charge radius when ΔN = 2. Note the sudden jumps at 20, 28, 50, 82, and 126 and compare with Figure 5.1. To emphasize the shell effects, the radius difference ΔR has been divided by the standard ΔR expected from the $A^{1/3}$ dependence. From E. B. Shera et al., *Phys. Rev. C* **14**, 731 (1976).

a collision. Thus the collisions cannot occur, and the nucleons can indeed orbit as if they were transparent to one another!

### Shell Model Potential

The first step in developing the shell model is the choice of the potential, and we begin by considering two potentials for which we solved the three-dimensional Schrödinger equation in Chapter 2: the infinite well and the harmonic oscillator. The energy levels we obtained are shown in Figure 5.4. As in the case of atomic physics, the degeneracy of each level is the the number of nucleons that can be put in each level, namely $2(2\ell + 1)$. The factor of $(2\ell + 1)$ arises from the $m_\ell$ degeneracy, and the additional factor of 2 comes from the $m_s$ degeneracy. As in atomic physics, we use spectroscopic notation to label the levels, with one important exception: the index $n$ is *not* the principal quantum number, but simply counts the number of levels with that $\ell$ value. Thus 1d means the first (lowest) d state, 2d means the second, and so on. (In atomic spectroscopic notation, there are no 1d or 2d states.) Figure 5.4 also shows the occupation

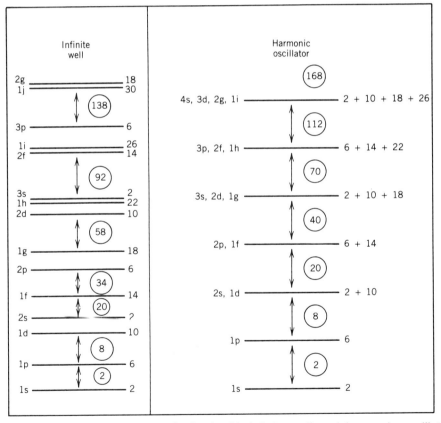

**Figure 5.4** Shell structure obtained with infinite well and harmonic oscillator potentials. The capacity of each level is indicated to its right. Large gaps occur between the levels, which we associate with closed shells. The circled numbers indicate the total number of nucleons at each shell closure.

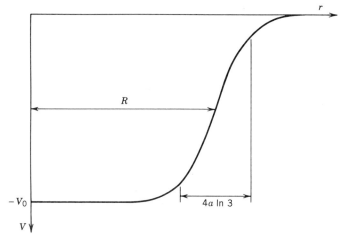

**Figure 5.5** A realistic form for the shell-model potential. The "skin thickness" $4a \ln 3$ is the distance over which the potential changes from $0.9V_0$ to $0.1V_0$.

number of each level and the cumulative number of nucleons that would correspond to the filling of major shells. (Neutrons and protons, being nonidentical particles, are counted separately. Thus the 1s level can hold 2 protons as well as 2 neutrons.) It is encouraging to see the magic numbers of 2, 8, and 20 emerging in both of these schemes, but the higher levels do not correspond at all to the observed magic numbers.

As a first step in improving the model, we try to choose a more realistic potential. The infinite well is not a good approximation to the nuclear potential for several reasons: To separate a neutron or proton, we must supply enough energy to take it out of the well—an infinite amount! In addition, the nuclear potential does not have a sharp edge, but rather closely approximates the nuclear charge and matter distribution, falling smoothly to zero beyond the mean radius $R$. The harmonic oscillator, on the other hand, does not have a sharp enough edge, and it also requires infinite separation energies. Instead, we choose an intermediate form:

$$V(r) = \frac{-V_0}{1 + \exp\left[(r - R)/a\right]} \qquad (5.1)$$

which is sketched in Figure 5.5. The parameters $R$ and $a$ give, respectively, the mean radius and skin thickness, and their values are chosen in accordance with the measurements discussed in Chapter 3: $R = 1.25A^{1/3}$ fm and $a = 0.524$ fm. The well depth $V_0$ is adjusted to give the proper separation energies and is of order 50 MeV. The resulting energy levels are shown in Figure 5.6; the effect of the potential, as compared with the harmonic oscillator (Figure 5.4) is to remove the $\ell$ degeneracies of the major shells. As we go higher in energy, the splitting becomes more and more severe, eventually becoming as large as the spacing between the oscillator levels themselves. Filling the shells in order with $2(2\ell + 1)$ nucleons, we again get the magic numbers 2, 8, and 20, but the higher magic numbers do not emerge from the calculations.

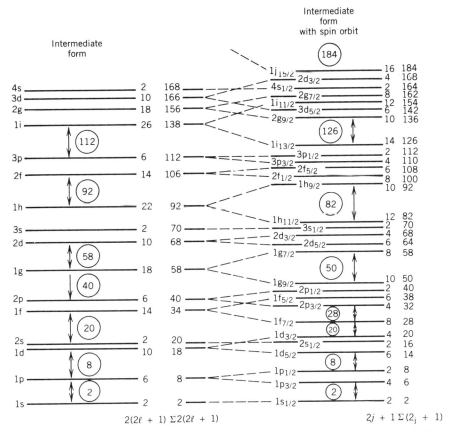

**Figure 5.6** At the left are the energy levels calculated with the potential of Figure 5.5. To the right of each level are shown its capacity and the cumulative number of nucleons up to that level. The right side of the figure shows the effect of the spin-orbit interaction, which splits the levels with $\ell > 0$ into two new levels. The shell effect is quite apparent, and the magic numbers are exactly reproduced.

## Spin-Orbit Potential

How can we modify the potential to give the proper magic numbers? We certainly cannot make a radical change in the potential, because we do not want to destroy the physical content of the model—Equation 5.1 is already a very good guess at how the nuclear potential *should* look. It is therefore necessary to add various terms to Equation 5.1 to try to improve the situation. In the 1940s, many unsuccessful attempts were made at finding the needed correction; success was finally achieved by Mayer, Haxel, Suess, and Jensen who showed in 1949 that the inclusion of a *spin-orbit* potential could give the proper separation of the subshells.

Once again, we are borrowing an idea from our colleagues, the atomic physicists. In atomic physics the spin-orbit interaction, which causes the observed fine structure of spectral lines, comes about because of the electromagnetic interaction of the electron's magnetic moment with the magnetic field generated by its motion about the nucleus. The effects are typically very small, perhaps one part in $10^5$ in the spacing of atomic levels. No such electromagnetic interaction would be strong enough to give the substantial changes in the *nuclear* level spacing needed to generate the observed magic numbers. Nevertheless we adopt the concept of a *nuclear* spin-orbit force of the same form as the atomic spin-orbit force but certainly *not* electromagnetic in origin. In fact, we know from the scattering experiments discussed in Chapter 4 that there is strong evidence for a nucleon–nucleon spin-orbit force.

The spin-orbit interaction is written as $V_{so}(r)\boldsymbol{\ell} \cdot \boldsymbol{s}$, but the form of $V_{so}(r)$ is not particularly important. It is the $\boldsymbol{\ell} \cdot \boldsymbol{s}$ factor that causes the reordering of the levels. As in atomic physics, in the presence of a spin-orbit interaction it is appropriate to label the states with the *total angular momentum* $\boldsymbol{j} = \boldsymbol{\ell} + \boldsymbol{s}$. A single nucleon has $s = \frac{1}{2}$, so the possible values of the total angular momentum quantum number are $j = \ell + \frac{1}{2}$ or $j = \ell - \frac{1}{2}$ (except for $\ell = 0$, in which case only $j = \frac{1}{2}$ is allowed). The expectation value of $\boldsymbol{\ell} \cdot \boldsymbol{s}$ can be calculated using a common trick. We first evaluate $\boldsymbol{j}^2 = (\boldsymbol{\ell} + \boldsymbol{s})^2$:

$$\boldsymbol{j}^2 = \boldsymbol{\ell}^2 + 2\boldsymbol{\ell} \cdot \boldsymbol{s} + \boldsymbol{s}^2$$
$$\boldsymbol{\ell} \cdot \boldsymbol{s} = \tfrac{1}{2}(\boldsymbol{j}^2 - \boldsymbol{\ell}^2 - \boldsymbol{s}^2) \tag{5.2}$$

Putting in the expectation values gives

$$\langle \boldsymbol{\ell} \cdot \boldsymbol{s} \rangle = \tfrac{1}{2}[j(j+1) - \ell(\ell+1) - s(s+1)]\hbar^2 \tag{5.3}$$

Consider a level such as the 1f level ($\ell = 3$), which has a degeneracy of $2(2\ell + 1) = 14$. The possible $j$ values are $\ell \pm \frac{1}{2} = \frac{5}{2}$ or $\frac{7}{2}$. Thus we have the levels $1f_{5/2}$ and $1f_{7/2}$. The degeneracy of each level is $(2j + 1)$, which comes from the $m_j$ values. (With spin-orbit interactions, $m_s$ and $m_\ell$ are no longer "good" quantum numbers and can no longer be used to label states or to count degeneracies.) The capacity of the $1f_{5/2}$ level is therefore 6 and that of $1f_{7/2}$ is 8, giving again 14 states (the number of possible states must be preserved; we are only grouping them differently). For the $1f_{5/2}$ and $1f_{7/2}$ states, which are known as a spin-orbit pair or doublet, there is an energy separation that is proportional to the value of $\langle \boldsymbol{\ell} \cdot \boldsymbol{s} \rangle$ for each state. Indeed, for any pair of states with $\ell > 0$, we can compute the energy difference using Equation 5.3:

$$\langle \boldsymbol{\ell} \cdot \boldsymbol{s} \rangle_{j=\ell+1/2} - \langle \boldsymbol{\ell} \cdot \boldsymbol{s} \rangle_{j=\ell-1/2} = \tfrac{1}{2}(2\ell + 1)\hbar^2 \tag{5.4}$$

The energy splitting increases with increasing $\ell$. Consider the effect of choosing $V_{so}(r)$ to be negative, so that the member of the pair with the larger $j$ is pushed downward. Figure 5.6 shows the effect of this splitting. The $1f_{7/2}$ level now appears in the gap between the second and third shells; its capacity of 8 nucleons gives the magic number 28. (The p and d splittings do not result in any major regrouping of the levels.) The next major effect of the spin-orbit term is on the 1g level. The $1g_{9/2}$ state is pushed down all the way to the next lower major shell; its capacity of 10 nucleons adds to the previous total of 40 for that shell to give the magic number of 50. A similar effect occurs at the top of each major shell. In each case the lower energy member of the spin-orbit pair from the next shell is pushed down into the lower shell, and the remaining magic numbers follow exactly as expected. (We even predict a new one, at 184, which has not yet been seen.)

As an example of the application of the shell model, consider the filling of levels needed to produce $^{15}_{8}$O and $^{17}_{8}$O. The 8 protons fill a major shell and do not

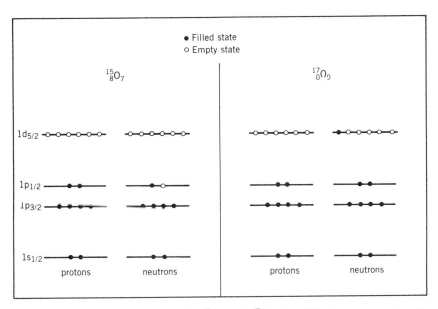

**Figure 5.7** The filling of shells in $^{15}$O and $^{17}$O. The filled proton shells do not contribute to the structure; the properties of the ground state are determined primarily by the odd neutron.

contribute to the structure. Figure 5.7 shows the filling of levels. The extreme limit of the shell model asserts that only the single unpaired nucleon determines the properties of the nucleus. In the case of $^{15}$O, the unpaired neutron is in the $p_{1/2}$ shell; we would therefore predict that the ground state of $^{15}$O has spin $\frac{1}{2}$ and odd parity, since the parity is determined by $(-1)^{\ell}$. The ground state of $^{17}$O should be characteristic of a d$_{5/2}$ neutron with spin $\frac{5}{2}$ and even parity. These two predictions are in exact agreement with the observed spin-parity assignments, and in fact similar agreements are found throughout the range of odd-$A$ nuclei where the shell model is valid (generally $A < 150$ and $190 < A < 220$, for reasons to be discussed later in this chapter). This success in accounting for the observed ground-state spin-parity assignments was a great triumph for the shell model.

**Magnetic Dipole Moments**

Another case in which the shell model gives a reasonable (but not so exact) agreement with observed nuclear properties is in the case of magnetic dipole moments. You will recall from Chapter 3 that the magnetic moment is computed from the expectation value of the magnetic moment operator in the state with maximum $z$ projection of angular momentum. Thus, including both $\ell$ and $s$ terms, we must evaluate

$$\mu = \mu_N (g_\ell \ell_z + g_s s_z)/\hbar \tag{5.5}$$

when $j_z = j\hbar$. This cannot be evaluated directly, since $\ell_z$ and $s_z$ do not have precisely defined values when we work in a system in which $j$ is precisely defined. We can rewrite this expression, using $j = \ell + s$, as

$$\mu = \left[ g_\ell j_z + (g_s - g_\ell)s_z \right] \mu_N/\hbar \tag{5.6}$$

and, taking the expectation value when $j_z = j\hbar$, the result is

$$\langle \mu \rangle = \left[ g_\ell j + (g_s - g_\ell)\langle s_z \rangle/\hbar \right] \mu_N \tag{5.7}$$

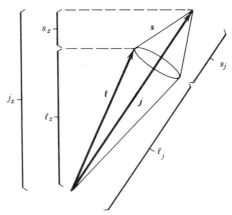

**Figure 5.8** As the total angular momentum $j$ precesses about the $z$ axis keeping $j_z$ constant, the vectors $\ell$ and $s$ precess about $j$. The components of $\ell$ and $s$ along $j$ remain constant, but $\ell_z$ and $s_z$ vary.

The expection value of $\langle s_z \rangle$ can be quickly computed by recalling that $j$ is the only vector of interest in this problem—the $\ell$ and $s$ vectors are meaningful only in their relationship to $j$. Specifically, when we compute $\langle s_z \rangle$ the only surviving part will be from the component of $s$ along $j$, as suggested by the vector diagram of Figure 5.8. The instantaneous value of $s_z$ varies, but its component along $j$ remains constant. We therefore need an expression for the vector $s_j$, the component of $s$ along $j$. The unit vector along $j$ is $j/|j|$, and the component of $s$ along $j$ is $|s \cdot j|/|j|$. The vector $s_j$ is therefore $j|s \cdot j|/|j|^2$, and replacing all quantities by their expectation values gives

$$\langle s_z \rangle = \frac{j}{2j(j+1)} \left[ j(j+1) - \ell(\ell+1) + s(s+1) \right] \hbar \qquad (5.8)$$

where $s \cdot j = s \cdot (\ell + s)$ is computed using Equation 5.3. Thus for $j = \ell + \frac{1}{2}$, $\langle s_z \rangle = \hbar/2$, while for $j = \ell - \frac{1}{2}$ we have $\langle s_z \rangle = -\hbar j/2(j+1)$. The corresponding magnetic moments are

$$j = \ell + \tfrac{1}{2} \qquad \langle \mu \rangle = \left[ g_\ell \left( j - \tfrac{1}{2} \right) + \tfrac{1}{2} g_s \right] \mu_N$$

$$j = \ell - \tfrac{1}{2} \qquad \langle \mu \rangle = \left[ g_\ell \frac{j \left( j + \tfrac{3}{2} \right)}{(j+1)} - \frac{1}{2} \frac{1}{j+1} g_s \right] \mu_N \qquad (5.9)$$

Figure 5.9 shows a comparison of these calculated values with measured values for shell-model odd-$A$ nuclei. The computed values are shown as solid lines and are known as the Schmidt lines; this calculation was first done by Schmidt in 1937. The experimental values fall within the limits of the Schmidt lines, but are generally smaller in magnitude and have considerable scatter. One defect of this theory is the assumption that $g_s$ for a nucleon in a nucleus is the same as $g_s$ for a free nucleon. We discussed in Chapter 3 how the spin $g$ factors of nucleons differ considerably from the value of 2 expected for "elementary" spin-$\frac{1}{2}$ particles. If we regard the substantial differences as arising from the "meson cloud" that surrounds the nucleon, then it is not at all surprising that the meson cloud in nuclei, where there are other surrounding nucleons and mesons, differs from what it is for free nucleons. It is customary to account for this effect by (somewhat

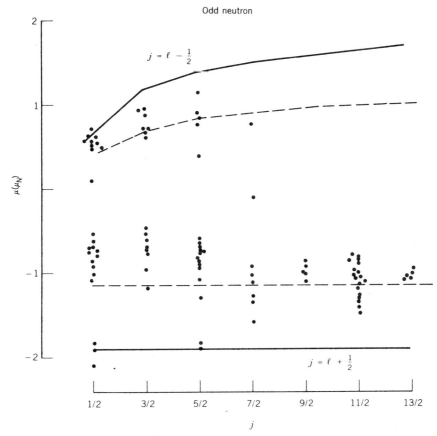

**Figure 5.9** Experimental values for the magnetic moments of odd-neutron and odd-proton shell-model nuclei. The Schmidt lines are shown as solid for $g_s = g_s$(free) and dashed for $g_s = 0.6g_s$(free).

arbitrarily) reducing the $g_s$ factor; for example, the lines for $g_s = 0.6g_s$(free) are shown in Figure 5.9. The overall agreement with experiment is better, but the scatter of the points suggests that the model is oversimplifying the calculation of magnetic moments. Nevertheless, the success in indicating the general trend of the observed magnetic moments suggests that the shell model gives us at least an approximate understanding of the structure of these nuclei.

## Electric Quadrupole Moments

The calculation of electric quadrupole moments in the shell model is done by evaluating the electric quadrupole operator, $3z^2 - r^2$, in a state in which the total angular momentum of the odd particle has its maximum projection along the $z$ axis (that is, $m_j = +j$). Let's assume for now that the odd particle is a proton. If its angular momentum is aligned (as closely as quantum mechanics allows) with the $z$ axis, then it must be orbiting mostly in the $xy$ plane. As we indicated in the discussion following Equation 3.36, this would give a negative quadrupole

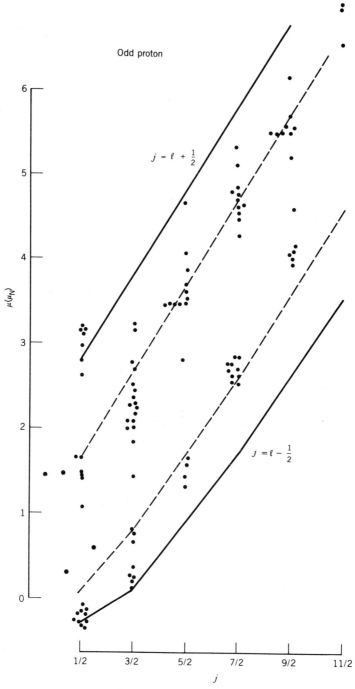

**Figure 5.9** Continued.

**Table 5.1**    Shell-Model Quadrupole Moments

| Shell-Model State | Calculated $Q$ (single proton) | Measured $Q$ | | | |
|---|---|---|---|---|---|
| | | Single Particle | | Single Hole | |
| | | p | n | p | n |
| $1p_{3/2}$ | $-0.013$ | $-0.0366(^{7}\text{Li})$ | | $+0.0407(^{11}\text{B})$ | $+0.053(^{9}\text{Be})$ |
| $1d_{5/2}$ | $-0.036$ | $-0.12(^{19}\text{F})$ | $-0.026(^{17}\text{O})$ | $+0.140(^{27}\text{Al})$ | $+0.201(^{25}\text{Mg})$ |
| $1d_{3/2}$ | $-0.037$ | $-0.08249(^{35}\text{Cl})$ | $-0.064(^{33}\text{S})$ | $+0.056(^{39}\text{K})$ | $+0.45(^{35}\text{S})$ |
| $1f_{7/2}$ | $-0.071$ | $-0.26(^{43}\text{Sc})$ | $-0.080(^{41}\text{Ca})$ | $+0.40(^{59}\text{Co})$ | $+0.24(^{49}\text{Ti})$ |
| $2p_{3/2}$ | $-0.055$ | $-0.209(^{63}\text{Cu})$ | $-0.0285(^{53}\text{Cr})$ | $+0.195(^{67}\text{Ga})$ | $+0.20(^{57}\text{Fe})$ |
| $1f_{5/2}$ | $-0.086$ | | $-0.20(^{61}\text{Ni})$ | $+0.274(^{85}\text{Rb})$ | $+0.15(^{67}\text{Zn})$ |
| $1g_{9/2}$ | $-0.13$ | $-0.32(^{93}\text{Nb})$ | $-0.17(^{73}\text{Ge})$ | $+0.86(^{115}\text{In})$ | $+0.45(^{85}\text{Kr})$ |
| $1g_{7/2}$ | $-0.14$ | $-0.49(^{123}\text{Sb})$ | | $+0.20(^{139}\text{La})$ | |
| $2d_{5/2}$ | $-0.12$ | $-0.36(^{121}\text{Sb})$ | $-0.236(^{91}\text{Zr})$ | | $+0.44(^{111}\text{Cd})$ |

Data for this table are derived primarily from the compilation of V. S. Shirley in the *Table of Isotopes*, 7th ed. (New York: Wiley, 1978). The uncertainties in the values are typically a few parts in the last quoted significant digit.

moment of the order of $Q \approx -\langle r^2 \rangle$. Some experimental values of quadrupole moments of nuclei that have one proton beyond a filled subshell are listed in Table 5.1. Values of $\langle r^2 \rangle$ range from 0.03 b for $A = 7$ to 0.3 b for $A = 209$, and thus the measured values are in good agreement with our expectations.

A more refined quantum mechanical calculation gives the single-particle quadrupole moment of an odd proton in a shell-model state $j$:

$$\langle Q_{sp} \rangle = -\frac{2j-1}{2(j+1)} \langle r^2 \rangle \tag{5.10}$$

For a uniformly charged sphere, $\langle r^2 \rangle = \frac{3}{5}R^2 = \frac{3}{5}R_0^2 A^{2/3}$. Using these results, we can compute the quadrupole moments for the nuclei shown in Table 5.1. The calculated values have the correct sign but are about a factor of 2–3 too small.

A more disturbing difficulty concerns nuclei with an odd neutron. An uncharged neutron outside a filled subshell should have no quadrupole moment at all. From Table 5.1 we see that the odd-neutron values are generally smaller than the odd-proton values, but they are most definitely not zero.

When a subshell contains more than a single particle, all of the particles in the subshell can contribute to the quadrupole moment. Since the capacity of any subshell is $2j + 1$, the number of nucleons in an unfilled subshell will range from 1 to $2j$. The corresponding quadrupole moment is

$$\langle Q \rangle = \langle Q_{sp} \rangle \left[ 1 - 2\frac{n-1}{2j-1} \right] \tag{5.11}$$

where $n$ is the number of nucleons in the subshell ($1 \leq n \leq 2j$) and $Q_{sp}$ is the

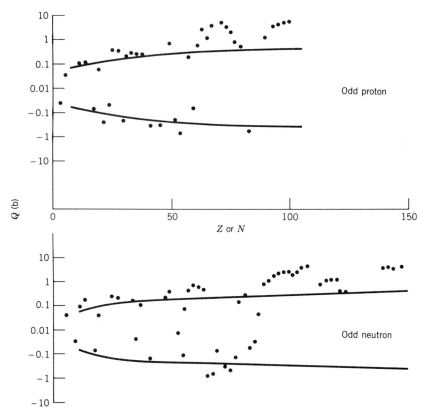

**Figure 5.10** Experimental values of electric quadrupole moments of odd-neutron and odd-proton nuclei. The solid lines show the limits $Q \sim \langle r^2 \rangle$ expected for shell-model nuclei. The data are within the limits, except for the regions $60 < Z < 80$, $Z > 90$, $90 < N < 120$, and $N > 140$, where the experimental values are more than an order of magnitude larger than predicted by the shell model.

single-particle quadrupole moment given in Equation 5.10. When $n = 1$, $Q = Q_{sp}$, but when $n = 2j$ (corresponding to a subshell that lacks only one nucleon from being filled), $Q = -Q_{sp}$. Table 5.1 shows the quadrupole moments of these so-called "hole" states, and you can see that to a very good approximation, $Q(\text{particle}) = -Q(\text{hole})$. In particular, the quadrupole moments of the hole states are positive and opposite in sign to the quadrupole moments of the particle states.

Before we are overcome with enthusiasm with the success of this simple model, let us look at the entire systematic behavior of the quadrupole moments. Figure 5.10 summarizes the measured quadrupole moments of the ground states of odd-mass nuclei. There is some evidence for the change in sign of $Q$ predicted by Equation 5.11, but the situation is not entirely symmetric—there are far more positive than negative quadrupole moments. Even worse, the model fails to predict the extremely large quadrupole moments of several barns observed for certain heavy nuclei. The explanations for these failures give us insight into other aspects of nuclear structure that cannot be explained within the shell model. We discuss these new features in the last two sections of this chapter.

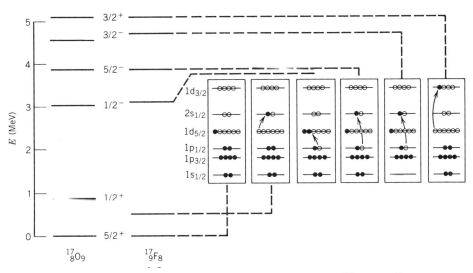

**Figure 5.11** Shell-model interpretation of the levels of $^{17}$O and $^{17}$F. All levels below about 5 MeV are shown, and the similarity between the levels of the two nuclei suggests they have common structures, determined by the valence nucleons. The even-parity states are easily explained by the excitation of the single odd nucleon from the $d_{5/2}$ ground state to $2s_{1/2}$ or $1d_{3/2}$. The odd-parity states have more complicated structures; one possible configuration is shown, but others are also important.

## Valence Nucleons

The shell model, despite its simplicity, is successful in accounting for the spins and parities of nearly all odd-$A$ ground states, and is somewhat less successful (but still satisfactory) in accounting for magnetic dipole and electric quadrupole moments. The particular application of the shell model that we have considered is known as the *extreme independent particle model*. The basic assumption of the extreme independent particle model is that all nucleons but one are paired, and the nuclear properties arise from the motion of the single unpaired nucleon. This is obviously an oversimplification, and as a next better approximation we can treat all of the particles in the unfilled subshell. Thus in a nucleus such as $^{43}_{20}$Ca$_{23}$, with three neutrons beyond the closed shell at $N = 20$, the extreme version of the shell model considers only the 23rd neutron, but a more complete shell model calculation should consider all three valence neutrons. For $^{45}_{22}$Ti$_{23}$, we should take into account all five particles (2 protons, 3 neutrons) beyond the closed shells at $Z = 20$ and $N = 20$.

If the extreme independent particle model were valid, we should be able to reproduce diagrams like Figure 5.6 by studying the excited states of nuclei. Let's examine some examples of this procedure. Figure 5.11 shows some of the excited states of $^{17}_{8}$O$_{9}$ and $^{17}_{9}$F$_{8}$, each of which has only one nucleon beyond a doubly magic ($Z = 8$, $N = 8$) core. The ground state is $\frac{5}{2}^{+}$, as expected for the $d_{5/2}$ shell-model state of the 9th nucleon. From Figure 5.6 we would expect to find excited states with spin-parity assignments of $\frac{1}{2}^{+}$ and $\frac{3}{2}^{+}$, corresponding to the $1s_{1/2}$ and $1d_{3/2}$ shell-model levels. According to this assumption, when we add

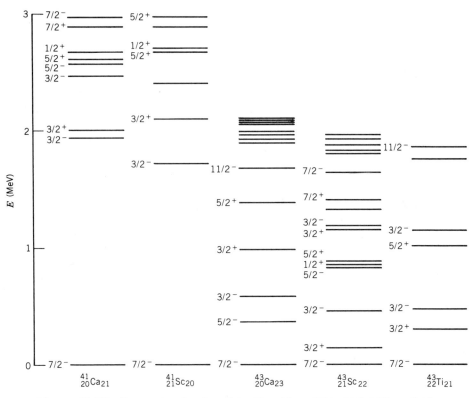

**Figure 5.12** Energy levels of nuclei with odd particles in the $1f_{7/2}$ shell.

energy to the nucleus, the core remains inert and the odd particle absorbs the energy and moves to higher shell-model levels. The expected $\frac{1}{2}^{+}$ shell-model state appears as the first excited state, and the $\frac{3}{2}^{+}$ state is much higher, but how can we account for $\frac{1}{2}^{-}$, $\frac{3}{2}^{-}$, and $\frac{5}{2}^{-}$? (The negative parity $2p_{1/2}$, $2p_{3/2}$, and $1f_{5/2}$ shell-model states are well above the $1d_{3/2}$ state, which should therefore appear lower.) Figure 5.11 shows one possible explanation for the $\frac{1}{2}^{-}$ state: instead of exciting the odd nucleon to a higher state, we break the pair in the $1p_{1/2}$ level and excite one of the nucleons to pair with the nucleon in the $d_{5/2}$ level. The odd nucleon is now in the $1p_{1/2}$ state, giving us a $\frac{1}{2}^{-}$ excited state. (Because the pairing energy increases with $\ell$, it is actually energetically favorable to break an $\ell = 1$ pair and form an $\ell = 2$ pair.) Verification of this hypothesis requires that we determine by experiment whether the properties of the $\frac{1}{2}^{-}$ state agree with those expected for a $p_{1/2}$ shell-model state. A similar assumption might do as well for the $\frac{3}{2}^{-}$ state (breaking a $p_{3/2}$ pair), but that still does not explain the $\frac{5}{2}^{-}$ state or the many other excited states.

In Figure 5.12, we show a similar situation for nuclei in the $1f_{7/2}$ shell. The $\frac{7}{2}^{-}$ ground state ($1f_{7/2}$) and the $\frac{3}{2}^{-}$ excited state ($2p_{3/2}$) appear as expected in the nuclei $^{41}$Ca and $^{41}$Sc, each of which has only a single nucleon beyond a doubly magic ($Z = 20$, $N = 20$) core. In $^{43}$Ca, the structure is clearly quite different from that of $^{41}$Ca. Many more low-lying states are present. These states come from the coupling of three particles in the $1f_{7/2}$ shell and illustrate the difference between the complete shell model and its extreme independent particle limit. If only the odd particle were important, $^{43}$Ca should be similar to $^{41}$Ca. In $^{43}$Sc, you can see how the 21st and 22nd neutrons, which would be ignored in the extreme independent particle limit, have a great effect on the structure. Similarly, the level scheme of $^{43}$Ti shows that the 21st and 22nd protons have a great effect on the shell-model levels of the 21st neutron.

In addition to spin-parity assignments, magnetic dipole moments, electric quadrupole moments, and excited states, the shell model can also be used to calculate the probability of making a transition from one state to another as a result of a radioactive decay or a nuclear reaction. We examine the shell model predictions for these processes in later chapters.

Let's conclude this discussion of the shell model with a brief discussion of the question we raised at the beginning—how can we be sure that the very concept of a nucleon with definite orbital properties remains valid deep in the nuclear interior? After all, many of the tests of the shell model involve such nuclear properties as the spin and electromagnetic moments of the valence particles, all of which are concentrated near the nuclear surface. Likewise, many experimental probes of the nucleus, including other particles that feel the nuclear force, tell us mostly about the surface properties. To answer the question we have proposed,

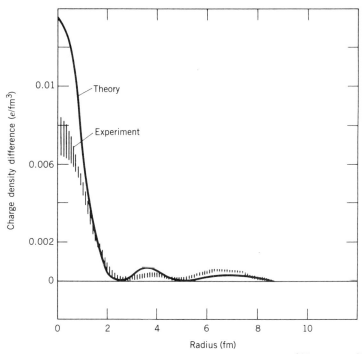

**Figure 5.13**  The difference in charge density between $^{205}$Tl and $^{206}$Pb, as determined by electron scattering. The curve marked "theory" is just the square of a harmonic oscillator 3s wave function. The theory reproduces the variations in the charge density extremely well. Experimental data are from J. M. Cavedon et al., *Phys. Rev. Lett.* **49**, 978 (1982).

what is needed is a probe that reaches deep into the nucleus, and we must use that probe to measure a nuclear property that characterizes the interior of the nucleus and not its surface. For a probe we choose high-energy electrons, as we did in studying the nuclear charge distribution in Chapter 3. The property that is to be measured is the charge density of a single nucleon in its orbit, which is equivalent to the square of its wave function, $|\psi|^2$. Reviewing Figure 2.12 we recall that only s-state wave functions penetrate deep into the nuclear interior; for other states $\psi \to 0$ as $r \to 0$. For our experiment we therefore choose a nucleus such as $^{205}_{81}\text{Tl}_{124}$, which lacks a single proton in the $3s_{1/2}$ orbit from filling all subshells below the $Z = 82$ gap. How can we measure the contribution of just the $3s_{1/2}$ proton to the charge distribution and ignore the other protons? We can do so by measuring the *difference* in charge distribution between $^{205}\text{Tl}$ and $^{206}_{82}\text{Pb}_{124}$, which has the filled proton shell. Any difference between the charge distributions of these two nuclei must be due to the extra $3s_{1/2}$ proton in $^{206}\text{Pb}$. Figure 5.13 shows the experimentally observed difference in the charge distributions as measured in a recent experiment. The comparison with $|\psi|^2$ for a 3s wave function is very successful (using the same harmonic oscillator wave function plotted in Figure 2.12, except that here we plot $|\psi|^2$, *not $r^2 R^2$*), thus confirming the validity of the assumption about nucleon orbits retaining their character deep in the nuclear interior. From such experiments we gain confidence that the independent-particle description, so vital to the shell model, is not just a convenience for analyzing measurements near the nuclear surface, but instead is a valid representation of the behavior of nucleons throughout the nucleus.

## 5.2 EVEN-*Z*, EVEN-*N* NUCLEI AND COLLECTIVE STRUCTURE

Now let's try to understand the structure of nuclei with even numbers of protons and neutrons (known as *even-even* nuclei). As an example, consider the case of $^{130}\text{Sn}$, shown in Figure 5.14. The shell model predicts that all even-even nuclei will have $0^+$ (spin 0, even parity) ground states, because all of the nucleons are paired. According to the shell model, the 50 protons of $^{130}\text{Sn}$ fill the $g_{9/2}$ shell and the 80 neutrons lack 2 from filling the $h_{11/2}$ shell to complete the magic number of $N = 82$. To form an excited state, we can break one of the pairs and excite a nucleon to a higher level; the coupling between the two odd nucleons then determines the spin and parity of the levels. Promoting one of the $g_{9/2}$ protons or $h_{11/2}$ neutrons to a higher level requires a great deal of energy, because the gap between the major shells must be crossed (see Figure 5.6). We therefore expect that the major components of the wave functions of the lower excited states will consist of neutron excitation within the last occupied major shell. For example, if we assume that the ground-state configuration of $^{130}\text{Sn}$ consists of filled $s_{1/2}$ and $d_{3/2}$ subshells and 10 neutrons (out of a possible 12) occupying the $h_{11/2}$ subshell, then we could form an excited state by breaking the $s_{1/2}$ pair and promoting one of the $s_{1/2}$ neutrons to the $h_{11/2}$ subshell. Thus we would have one neutron in the $s_{1/2}$ subshell and 11 neutrons in the $h_{11/2}$ subshell. The properties of such a system would be determined mainly by the coupling of the $s_{1/2}$ neutron with the unpaired $h_{11/2}$ neutron. Coupling angular momenta $j_1$ and $j_2$ in quantum mechanics gives values from the sum $j_1 + j_2$ to

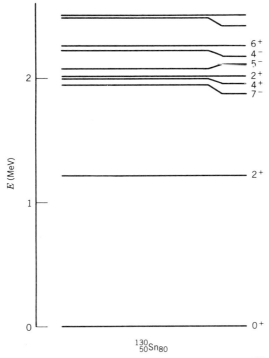

**Figure 5.14** The low-lying energy levels of $^{130}$Sn.

the difference $|j_1 - j_2|$ in integer steps. In this case the possible resultants are $\frac{11}{2} + \frac{1}{2} = 6$ and $\frac{11}{2} - \frac{1}{2} = 5$. Another possibility would be to break one of the $d_{3/2}$ pairs and again place an odd neutron in the $h_{11/2}$ subshell. This would give resulting angular momenta ranging from $\frac{11}{2} + \frac{3}{2} = 7$ to $\frac{11}{2} - \frac{3}{2} = 4$. Because the $s_{1/2}$ and $d_{3/2}$ neutrons have even parity and the $h_{11/2}$ neutron has odd parity, all of these couplings will give states with odd parity. If we examine the $^{130}$Sn level scheme, we do indeed see several odd parity states with spins in the range of 4–7 with energies about 2 MeV. This energy is characteristic of what is needed to break a pair and excite a particle within a shell, and so we have a strong indication that we understand those states. Another possibility to form excited states would be to break one of the $h_{11/2}$ pairs and, keeping both members of the pair in the $h_{11/2}$ subshell, merely recouple them to a spin different from 0; according to the angular momentum coupling rules, the possibilities would be anything from $\frac{11}{2} + \frac{11}{2} = 11$ to $\frac{11}{2} - \frac{11}{2} = 0$. The two $h_{11/2}$ neutrons must be treated as identical particles and must therefore be described by a properly symmetrized wave function. This requirement restricts the resultant coupled spin to even values, and thus the possibilities are $0^+, 2^+, 4^+, 6^+, 8^+, 10^+$. There are several candidates for these states in the 2-MeV region, and here again the shell model seems to give us a reasonable description of the level structure.

A major exception to this successful interpretation is the $2^+$ state at about 1.2 MeV. Restricting our discussion to the neutron states, what are the possible ways to couple two neutrons to get $2^+$? As discussed above, the two $h_{11/2}$ neutrons can couple to $2^+$. We can also excite a pair of $d_{3/2}$ neutrons to the $h_{11/2}$ subshell (thus filling that shell and making an especially stable configuration), then break the coupling of the two remaining $d_{3/2}$ neutrons and recouple them to $2^+$. Yet another possibility would be to place the pair of $s_{1/2}$ neutrons into the $h_{11/2}$ subshell, and excite one of the $d_{3/2}$ neutrons to the $s_{1/2}$ subshell. We would then have an odd neutron in each of the $d_{3/2}$ and $s_{1/2}$ subshells, which could couple to $2^+$. However, in all these cases we must first break a pair, and thus the resulting states would be expected at about 2 MeV.

Of course, the shell-model description is only an approximation, and it is unlikely that "pure" shell-model states will appear in a complex level scheme. A better approach is to recognize that if we wish to use the shell model as a means to interpret the structure, then the physical states must be described as combinations of shell-model states, thus:

$$\psi(2^+) = a\psi\left(\nu h_{11/2} \oplus \nu h_{11/2}\right) + b\psi\left(\nu d_{3/2} \oplus \nu d_{3/2}\right)$$

$$+ c\psi\left(\nu d_{3/2} \oplus \nu s_{1/2}\right) + \cdots \qquad (5.12)$$

where $\nu$ stands for neutron and the $\oplus$ indicates that we are doing the proper angular momentum coupling to get the $2^+$ resultant. The puzzle of the low-lying $2^+$ state can now be rephrased as follows: Each of the constituent states has an energy of about 2 MeV. What is it about the nuclear interaction that gives the right mixture of expansion coefficients $a, b, c, \ldots$ to force the state down to an energy of 1.2 MeV?

Our first thought is that this structure may be a result of the particular shell-model levels occupied by the valence particles of $^{130}$Sn. We therefore examine other even-even nuclei and find this remarkable fact: of the hundreds of known even-even nuclei in the shell-model region, *each one* has an "anomalous" $2^+$ state at an energy at or below one-half of the energy needed to break a pair. In all but a very few cases, this $2^+$ state is the lowest excited state. The occurrence of this state is thus not an accident resulting from the shell-model structure of $^{130}$Sn. Instead it is a *general property* of even-$Z$, even-$N$ nuclei, valid throughout the entire mass range, independent of which particular shell-model states happen to be occupied. We will see that there are other general properties that are common to all nuclei, and it is reasonable to identify those properties not with the motion of a few valence nucleons, but instead with the entire nucleus. Such properties are known as *collective properties* and their origin lies in the nuclear collective motion, in which many nucleons contribute cooperatively to the nuclear properties. The collective properties vary smoothly and gradually with mass number and are mostly independent of the number and kind of valence nucleons outside of filled subshells (although the valence nucleons may contribute shell structure that couples with the collective structure).

In Figures 5.15 and 5.16 are shown four different properties of even-even nuclei that reveal collective behavior. The energy of the first $2^+$ excited state (Figure 5.15a) seems to decrease rather smoothly as a function of $A$ (excepting the regions near closed shells). The region from about $A = 150$ to $A = 190$ shows values of $E(2^+)$ that are both exceptionally small and remarkably constant. Again excepting nuclei near closed shells, the ratio $E(4^+)/E(2^+)$ (Figure 5.15b)

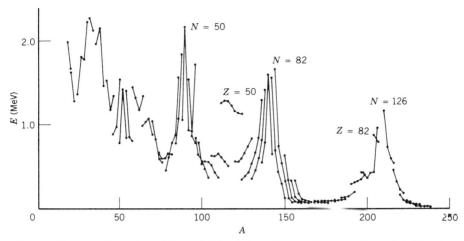

**Figure 5.15a** Energies of lowest $2^+$ states of even-$Z$, even-$N$ nuclei. The lines connect sequences of isotopes.

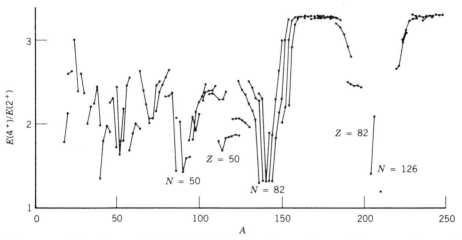

**Figure 5.15b** The ratio $E(4^+)/E(2^+)$ for the lowest $2^+$ and $4^+$ states of even-$Z$, even-$N$ nuclei. The lines connect sequences of isotopes.

is roughly 2.0 for nuclei below $A = 150$ and very constant at 3.3 for $150 < A < 190$ and $A > 230$. The magnetic moments of the $2^+$ states (Figure 5.16a) are fairly constant in the range 0.7–1.0, and the electric quadrupole moments (Figure 5.16b) are small for $A < 150$ and much larger for $A > 150$. These illustrations suggest that we must consider two types of collective structure, for the nuclei with $A < 150$ seem to show one set of properties and the nuclei with $150 < A < 190$ show quite a different set.

The nuclei with $A < 150$ are generally treated in terms of a model based on vibrations about a spherical equilibrium shape, while nuclei with $A$ between 150 and 190 show structures most characteristic of rotations of a nonspherical system. Vibrations and rotations are the two major types of collective nuclear

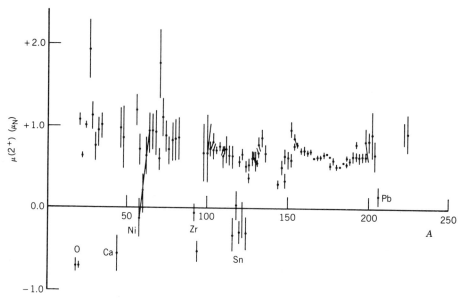

**Figure 5.16a**  Magnetic moments of lowest $2^+$ states of even-$Z$, even-$N$ nuclei. Shell-model nuclei showing noncollective behavior are indicated.

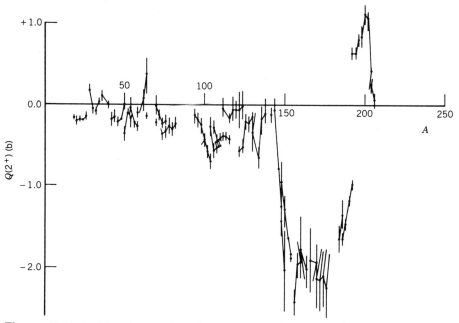

**Figure 5.16b**  Electric quadrupole moments of lowest $2^+$ states of even-$Z$, even-$N$ nuclei. The lines connect sequences of isotopes.

motion, and we will consider each in turn. The collective nuclear model is often called the "liquid drop" model, for the vibrations and rotations of a nucleus resemble those of a suspended drop of liquid and can be treated with a similar mathematical analysis.

## Nuclear Vibrations

Imagining a liquid drop vibrating at high frequency, we can get a good idea of the physics of nuclear vibrations. Although the average shape is spherical, the instantaneous shape is not. It is convenient to give the instantaneous coordinate $R(t)$ of a point on the nuclear surface at $(\theta, \phi)$, as shown in Figure 5.17, in terms of the spherical harmonics $Y_{\lambda\mu}(\theta, \phi)$. Each spherical harmonic component will have an amplitude $\alpha_{\lambda\mu}(t)$:

$$R(t) = R_{av} + \sum_{\lambda \geq 1} \sum_{\mu=-\lambda}^{+\lambda} \alpha_{\lambda\mu}(t) Y_{\lambda\mu}(\theta, \phi) \tag{5.13}$$

The $\alpha_{\lambda\mu}$ are not completely arbitrary; reflection symmetry requires that $\alpha_{\lambda\mu} = \alpha_{\lambda-\mu}$, and if we assume the nuclear fluid to be incompressible, further restrictions apply. The constant ($\lambda = 0$) term is incorporated into the average radius $R_{av}$, which is just $R_0 A^{1/3}$. A typical $\lambda = 1$ vibration, known as a *dipole* vibration, is shown in Figure 5.18. Notice that this gives a net displacement of the center of mass and therefore cannot result from the action of internal nuclear forces. We therefore consider the next lowest mode, the $\lambda = 2$ (quadrupole) vibration. In analogy with the quantum theory of electromagnetism, in which a unit of electromagnetic energy is called a photon, a quantum of vibrational energy is

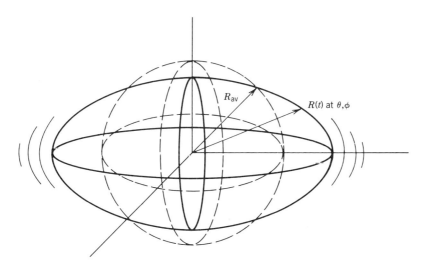

**Figure 5.17** A vibrating nucleus with a spherical equilibrium shape. The time-dependent coordinate $R(t)$ locates a point on the surface in the direction $\theta, \phi$.

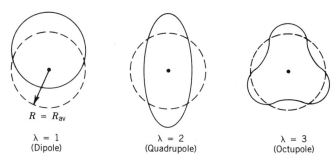

$R = R_{av}$

| $\lambda = 1$ | $\lambda = 2$ | $\lambda = 3$ |
| (Dipole) | (Quadrupole) | (Octupole) |

**Figure 5.18** The lowest three vibrational modes of a nucleus. The drawings represent a slice through the midplane. The dashed lines show the spherical equilibrium shape and the solid lines show an instantaneous view of the vibrating surface.

called a *phonon*. Whenever we produce mechanical vibrations, we can equivalently say that we are producing vibrational phonons. A single unit of $\lambda = 2$ nuclear vibration is thus a *quadrupole phonon*.

Let's consider the effect of adding one unit of vibrational energy (a quadrupole phonon) to the $0^{+}$ ground state of an even-even nucleus. The $\lambda = 2$ phonon carries 2 units of angular momentum (it adds a $Y_{2\mu}$ dependence to the nuclear wave function, just like a $Y_{\ell m}$ with $\ell = 2$) and even parity, since the parity of a $Y_{\ell m}$ is $(-1)^{\ell}$. Adding two units of angular momentum to a $0^{+}$ state gives only a $2^{+}$ state, in exact agreement with the observed spin-parity of first excited states of spherical even-$Z$, even-$N$ nuclei. (The energy of the quadrupole phonon is not predicted by this theory and must be regarded as an adjustable parameter.) Suppose now we add a second quadrupole phonon. There are 5 possible components $\mu$ for each phonon and therefore 25 possible combinations of the $\lambda\mu$ for the two phonons, as enumerated in Table 5.2. Let's try to examine the resulting sums. There is one possible combination with total $\mu = +4$. It is natural to associate this with a transfer of 4 units of angular momentum (a $Y_{\ell m}$ with $m = +4$ and therefore $\ell = 4$). There are two combinations with total $\mu = +3$: ($\mu_{1} = +1$, $\mu_{2} = +2$) and ($\mu_{1} = +2$, $\mu_{2} = +1$). However, when we make the proper symmetric combination of the phonon wave functions (phonons, with integer spins, must have symmetric total wave functions; see Section 2.7), only

**Table 5.2** Combinations of z Projections of Two Quadrupole Phonons into a Resultant Total z Component[a]

| | $\mu_1$ | | | | |
|---|---|---|---|---|---|
| $\mu_2$ | $-2$ | $-1$ | $0$ | $+1$ | $+2$ |
| $-2$ | $-4$ | $-3$ | $-2$ | $-1$ | $0$ |
| $-1$ | $-3$ | $-2$ | $-1$ | $0$ | $+1$ |
| $0$ | $-2$ | $-1$ | $0$ | $+1$ | $+2$ |
| $+1$ | $-1$ | $0$ | $+1$ | $+2$ | $+3$ |
| $+2$ | $0$ | $+1$ | $+2$ | $+3$ | $+4$ |

[a] The entries show $\mu = \mu_1 + \mu_2$.

one combination appears. There are three combinations that give $\mu = +2$: $(\mu_1, \mu_2) = (+2, 0), (+1, +1)$, and $(0, +2)$. The first and third must be combined into a symmetric wave function; the $(+1, +1)$ combination is already symmetric. Continuing in this way, we would find not 25 but 15 possible allowed combinations: one with $\mu = +4$, one with $\mu = +3$, two with $\mu = +2$, two with $\mu = +1$, three with $\mu = 0$, two with $\mu = -1$, two with $\mu = -2$, one with $\mu = -3$ and one with $\mu = -4$. We can group these in the following way:

$$\ell = 4 \qquad \mu = +4, +3, +2, +1, 0, -1, -2, -3, -4$$
$$\ell = 2 \qquad \mu = +2, +1, 0, -1, -2$$
$$\ell = 0 \qquad \mu = 0$$

Thus we expect a triplet of states with spins $0^+, 2^+, 4^+$ at twice the energy of the first $2^+$ state (since two identical phonons carry twice as much energy as one). This $0^+, 2^+, 4^+$ triplet is a common feature of vibrational nuclei and gives strong support to this model. The three states are never exactly at the same energy, owing to additional effects not considered in this simple model. A similar calculation for three quadrupole phonons gives states $0^+, 2^+, 3^+, 4^+, 6^+$ (see Problem 10).

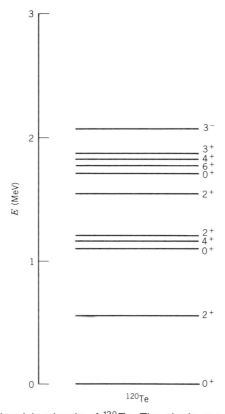

**Figure 5.19** The low-lying levels of $^{120}$Te. The single quadrupole phonon state (first $2^+$), the two-phonon triplet, and the three-phonon quintuplet are obviously seen. The $3^-$ state presumably is due to the octupole vibration. Above 2 MeV the structure becomes quite complicated, and no vibrational patterns can be seen.

The next highest mode of vibration is the $\lambda = 3$ *octupole* mode, which carries three units of angular momentum and negative parity. Adding a single octupole phonon to the $0^+$ ground state gives a $3^-$ state. Such states are also commonly found in vibrational nuclei, usually at energies somewhat above the two-phonon triplet. As we go higher in energy, the vibrational structure begins to give way to particle excitation corresponding to the breaking of a pair in the ground state. These excitations are very complicated to handle and are not a part of the collective structure of nuclei.

The vibrational model makes several predictions that can be tested in the laboratory. If the equilibrium shape is spherical, the quadrupole moments of the first $2^+$ state should vanish; Figure 5.16b showed they are small and often vanishing in the region $A < 150$. The magnetic moments of the first $2^+$ states are predicted to be $2(Z/A)$, which is in the range 0.8–1.0 for the nuclei considered; this is also in reasonable agreement with experiment. The predicted ratio $E(4^+)/E(2^+)$ is 2.0, if the $4^+$ state is a member of the two-phonon triplet and the $2^+$ state is the first excited state; Figure 5.15b shows reasonable agreement with this prediction in the range $A < 150$. In Chapter 10 we show the good agreement with $\gamma$-ray transition probabilities as well. Figure 5.19 shows an example of the low-lying level structure of a typical "vibrational" nucleus, and many of the predicted features are readily apparent. Thus the spherical vibrational model gives us quite an accurate picture of the structure of these nuclei.

## Nuclear Rotations

Rotational motion can be observed only in nuclei with nonspherical equilibrium shapes. These nuclei can have substantial distortions from spherical shape and are often called *deformed nuclei*. They are found in the mass ranges $150 < A < 190$ and $A > 220$ (rare earths and actinides). Figure 5.10 showed that the odd-mass nuclei in these regions also have quadrupole moments that are unexpectedly large. A common representation of the shape of these nuclei is that of an ellipsoid of revolution (Figure 5.20), the surface of which is described by

$$R(\theta, \phi) = R_{av}\left[1 + \beta Y_{20}(\theta, \phi)\right] \qquad (5.14)$$

which is independent of $\phi$ and therefore gives the nucleus cylindrical symmetry. The *deformation parameter* $\beta$ is related to the eccentricity of the ellipse as

$$\beta = \frac{4}{3}\sqrt{\frac{\pi}{5}}\frac{\Delta R}{R_{av}} \qquad (5.15)$$

where $\Delta R$ is the difference between the semimajor and semiminor axes of the ellipse. It is customary (although not quite exact) to take $R_{av} = R_0 A^{1/3}$. The approximation is not exact because the volume of the nucleus described by Equation 5.14 is not quite $\frac{4}{3}\pi R_{av}^3$; see Problem 11. The axis of symmetry of Equation 5.14 is the reference axis relative to which $\theta$ is defined. When $\beta > 0$, the nucleus has the elongated form of a *prolate* ellipsoid; when $\beta < 0$, the nucleus has the flattened form of an *oblate* ellipsoid.

One indicator of the stable deformation of a nucleus is a large electric quadrupole moment, such as those shown in Figure 5.10. The relationship

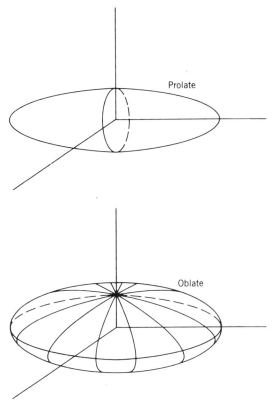

**Figure 5.20** Equilibrium shapes of nuclei with permanent deformations. These sketches differ from Figures 5.17 and 5.18 in that these do not represent snapshots of a moving surface at a particular instant of time, but instead show the static shape of the nucleus.

between the deformation parameter and the quadrupole moment is

$$Q_0 = \frac{3}{\sqrt{5\pi}} R_{\text{av}}^2 Z\beta(1 + 0.16\beta) \qquad (5.16)$$

The quadrupole moment $Q_0$ is known as the *intrinsic* quadrupole moment and would only be observed in a frame of reference in which the nucleus were at rest. In the laboratory frame of reference, the nucleus is rotating and quite a different quadrupole moment $Q$ is measured. In fact, as indicated in Figure 5.21, rotating a prolate intrinsic distribution about an axis perpendicular to the symmetry axis (no rotations can be observed parallel to the symmetry axis) gives a time-averaged oblate distribution. Thus for $Q_0 > 0$, we would observe $Q < 0$. The relationship between $Q$ and $Q_0$ depends on the nuclear angular momentum; for $2^+$ states, $Q = -\frac{2}{7}Q_0$. Figure 5.16$b$ shows $Q \simeq -2$ b for nuclei in the region of stable permanent deformations ($150 \leq A \leq 190$), and so $Q_0 \simeq +7$ b. From Equation 5.16, we would deduce $\beta \simeq 0.29$. This corresponds to a substantial deviation from a spherical nucleus; the difference in the lengths of the semimajor and semiminor axes is, according to Equation 5.15, about 0.3 of the nuclear radius.

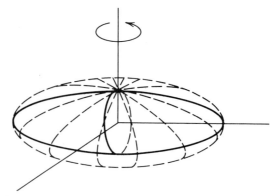

**Figure 5.21** Rotating a static prolate distribution about an axis perpendicular to its symmetry axis gives in effect a smeared-out oblate (flattened) distribution.

The kinetic energy of a rotating object is $\frac{1}{2}\mathscr{I}\omega^2$, where $\mathscr{I}$ is the moment of inertia. In terms of the angular momentum $\ell = \mathscr{I}\omega$, the energy is $\ell^2/2\mathscr{I}$. Taking the quantum mechanical value of $\ell^2$, and letting $I$ represent the angular momentum quantum number, gives

$$E = \frac{\hbar^2}{2\mathscr{I}}I(I+1) \tag{5.17}$$

for the energies of a rotating object in quantum mechanics. Increasing the quantum number $I$ corresponds to adding rotational energy to the nucleus, and the nuclear excited states form a sequence known as a *rotational band*. (Excited states in molecules also form rotational bands, corresponding to rotations of the molecule about its center of mass.) The ground state of an even-$Z$, even-$N$ nucleus is always a $0^+$ state, and the mirror symmetry of the nucleus restricts the sequence of rotational states in this special case to *even* values of $I$. We therefore expect to see the following sequence of states:

$$E(0^+) = 0$$
$$E(2^+) = 6(\hbar^2/2\mathscr{I})$$
$$E(4^+) = 20(\hbar^2/2\mathscr{I})$$
$$E(6^+) = 42(\hbar^2/2\mathscr{I})$$
$$E(8^+) = 72(\hbar^2/2\mathscr{I})$$

and so on.

Figure 5.22 shows the excited states of a typical rotational nucleus. The first excited state is at $E(2^+) = 91.4$ keV, and thus we have $\hbar^2/2\mathscr{I} = 15.2$ keV. The energies of the next few states in the ground-state rotational band are computed to be

$$E(4^+) = 20(\hbar^2/2\mathscr{I}) = 305 \text{ keV} \qquad \text{(measured 300 keV)}$$
$$E(6^+) = 42(\hbar^2/2\mathscr{I}) = 640 \text{ keV} \qquad \text{(measured 614 keV)}$$
$$E(8^+) = 72(\hbar^2/2\mathscr{I}) = 1097 \text{ keV} \qquad \text{(measured 1025 keV)}$$

The calculated energy levels are not quite exact (perhaps because the nucleus behaves somewhat like a fluid of nucleons and not quite like a rigid object with a fixed moment of inertia), but are good enough to give us confidence that we have at least a rough idea of the origin of the excited levels. In particular, the predicted

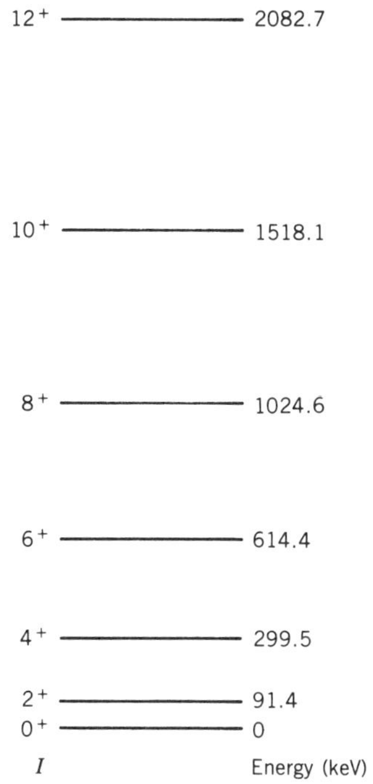

$12^+$ ———————— 2082.7

$10^+$ ———————— 1518.1

$8^+$ ———————— 1024.6

$6^+$ ———————— 614.4

$4^+$ ———————— 299.5

$2^+$ ———————— 91.4

$0^+$ ———————— 0

$I$                 Energy (keV)

**Figure 5.22** The excited states resulting from rotation of the ground state in $^{164}$Er.

ratio $E(4^+)/E(2^+)$ is 3.33, in remarkable agreement with the systematics of nuclear levels for $150 < A < 190$ and $A > 230$ shown in Figure 5.15*b*.

We can gain some insight into the structure of deformed nuclei by considering the moment of inertia in two extreme cases. A solid ellipsoid of revolution of mass $M$ whose surface is described by Equation 5.14 has a rigid moment of inertia

$$\mathscr{I}_{\text{rigid}} = \tfrac{2}{5}MR_{\text{av}}^2(1 + 0.31\beta) \qquad (5.18)$$

which of course reduces to the familiar value for a sphere when $\beta = 0$. For a typical nucleus in the deformed region ($A \approx 170$), this gives a rotational energy constant

$$\frac{\hbar^2}{2\mathscr{I}_{\text{rigid}}} \cong 6 \text{ keV}$$

which is of the right order of magnitude but is too small compared with the observed values (about 15 keV for $E(2^+) = 90$ keV). That is, the rigid moment of inertia is too large by about a factor of 2–3. We can take the other extreme and regard the nucleus as a fluid inside a rotating ellipsoidal vessel, which would give a moment of inertia

$$\mathscr{I}_{\text{fluid}} = \frac{9}{8\pi}MR_{\text{av}}^2\beta \qquad (5.19)$$

from which we would estimate

$$\frac{\hbar^2}{2\mathscr{I}_{\text{fluid}}} \cong 90 \text{ keV}$$

The fluid moment inertia is thus too small, and we conclude $\mathscr{I}_{\text{rigid}} > \mathscr{I} > \mathscr{I}_{\text{fluid}}$. The rotational behavior is thus intermediate between a rigid object, in which the particles are tightly bonded together, and a fluid, in which the particles are only weakly bonded. (We probably should have guessed this result, based on our studies of the nuclear force. Strong forces exist between a nucleon and its immediate neighbors only, and thus a nucleus does not show the long-range structure that would characterize a rigid solid.)

Another indication of the lack of rigidity of the nucleus is the increase in the moment of inertia that occurs at high angular momentum or rotational frequency. This effect, called "centrifugal stretching," is seen most often in heavy-ion reactions, to be discussed in Section 11.13.

Of course, the nucleus has no "vessel" to define the shape of the rotating fluid; it is the potential supplied by the nucleons themselves which gives the nucleus its shape. The next issue to be faced is whether the concept of a shape has any meaning for a rotating nucleus. If the rotation is very fast compared with the speed of nucleons in their "orbits" defined by the nuclear potential (as seen in a frame of reference in which the nucleus is at rest), then the concept of a static nuclear shape is not very meaningful because the motion of the nucleons will be dominated by the rotation. The average kinetic energy of a nucleon in a nucleus is of the order of 20 MeV, corresponding to a speed of approximately $0.2c$. This is a reasonable estimate for the speed of internal motion of the nucleons. The angular velocity of a rotating state is $\omega = \sqrt{2E/\mathscr{I}}$, where $E$ is the energy of the state. For the first rotational state, $\omega \simeq 1.1 \times 10^{20}$ rad/s and a nucleon near the surface would rotate with a tangential speed of $v \simeq 0.002c$. The rotational motion is therefore far slower than the internal motion. The correct picture of a rotating deformed nucleus is therefore a stable equilibrium shape determined by nucleons in rapid internal motion in the nuclear potential, with the entire resulting distribution rotating sufficiently slowly that the rotation has little effect on the nuclear structure or on the nucleon orbits. (The rotational model is sometimes described as "adiabatic" for this reason.)

It is also possible to form other kinds of excited states upon which new rotational bands can be built. Examples of such states, known as *intrinsic states* because they change the intrinsic structure of the nucleus, are vibrational states (in which the nucleus vibrates about a deformed equilibrium shape) and pair-breaking particle excitations. If the intrinsic state has spin different from zero, the rotational band built on that state will have the sequence of spins $I, I + 1, I + 2, \ldots$. The vibrational states in deformed nuclei are of two types: $\beta$ vibrations, in which the deformation parameter $\beta$ oscillates and the nucleus preserves its cylindrical symmetry, and $\gamma$ vibrations, in which the cylindrical symmetry is violated. (Picture a nucleus shaped like a football. $\beta$ vibrations correspond to pushing and pulling on the ends of the football, while $\gamma$ vibrations correspond to pushing and pulling on its sides.) Both the vibrational states and the particle excitations occur at energies of about 1 MeV, while the rotational spacing is much smaller (typically $\hbar^2/2\mathscr{I} \simeq 10$–$20$ keV).

Figure 5.23 shows the complete low-energy structure of $^{164}$Er. Although the entire set of excited states shows no obvious patterns, knowing the spin-parity assignments helps us to group the states into rotational bands, which show the characteristic $I(I + 1)$ spacing. Other properties of the excited states (for example, $\gamma$-ray emission probabilities) also help us to identify the structure.

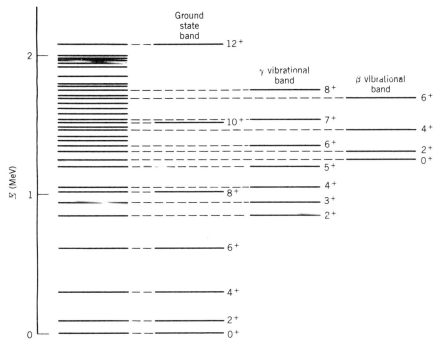

**Figure 5.23**  The states of $^{164}$Er below 2 MeV. Most of the states are identified with three rotational bands: one built on the deformed ground state, a second built on a γ-type vibration (in which the surface vibrates transverse to the symmetry axis), and a third built on a β-type vibration (in which the surface vibrates along the symmetry axis). Many of the other excited states originate from pair-breaking particle excitations and their associated rotational bands.

Both the vibrational and rotational collective motions give the nucleus a magnetic moment. We can regard the movement of the protons as an electric current, and a single proton moving with angular momentum quantum number $\ell$ would give a magnetic moment $\mu = \ell\mu_N$. However, the entire angular momentum of a nuclear state does not arise from the protons alone; the neutrons also contribute, and if we assume that the protons and neutrons move with identical collective motions (a reasonable but not quite exact assumption), we would predict that the protons contribute to the total nuclear angular momentum a fraction of nearly $Z/A$. (We assume that the collective motion of the neutrons does not contribute to the magnetic moment, and we also assume that the protons and neutrons are all coupled pairwise so that the spin magnetic moments do not contribute.) The collective model thus predicts for the magnetic moment of a vibrational or rotational state of angular momentum $I$

$$\mu(I) = I\frac{Z}{A}\mu_N \tag{5.20}$$

For light nuclei, $Z/A \simeq 0.5$ and $\mu(2) \simeq +1\mu_N$, while for heavier nuclei, $Z/A \simeq 0.4$ and $\mu(2) \simeq +0.8\mu_N$. Figure 5.16a shows that, with the exception of closed-shell nuclei (for which the collective model is not valid), the magnetic moments of the $2^+$ states are in very good agreement with this prediction.

As a final point in this brief introduction to nuclear collective motion, we must try to justify the origin of collective behavior based on a more microscopic approach to nuclear structure. This is especially true for rotational nuclei with permanent deformations. We have already seen how well the shell model with a spherically symmetric potential works for many nuclei. We can easily allow the shell-model potential to vibrate about equilibrium when energy is added to the nucleus, and so the vibrational motion can be handled in a natural way in the shell model. As we learned in our discussion of the $^{130}$Sn structure at the beginning of this section, we can analyze the collective vibrational structure from an even more microscopic approach; for example, we consider all valence nucleons (those outside closed shells), find all possible couplings (including those that break pairs) giving $2^+$ resultant spins, and try to find the correct mixture of wave functions that gives the observed $2^+$ first excited state. If there are many possible couplings, this procedure may turn out to be mathematically complex, but the essentials of the shell model on which it is based are not significantly different from the extreme independent particle model we considered in the previous section. This approach works for spherical nuclei, but it does not lead naturally to a rotational nucleus with a permanent deformation.

Here is the critical question: How do shell-model orbits, calculated using a spherical potential, result in a nonspherical nucleus? We get a clue to the answer to this question by superimposing a diagram showing the "magic numbers" on a

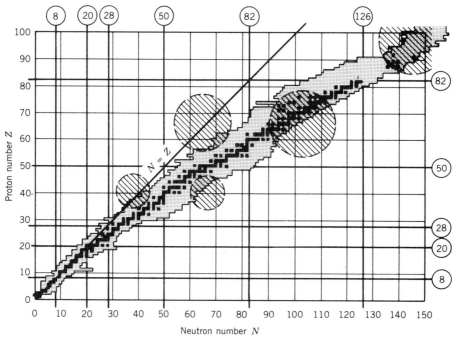

**Figure 5.24** The crosshatched areas show the regions far from closed shells where we expect that the cooperative effects of many single-particle shell-model states may combine to produce a permanent nuclear deformation. Such deformed nuclei have been identified in all of the regions where the crosshatched areas overlap the known nuclei.

chart of the known nuclear species, as shown in Figure 5.24. The deformed nuclei exist only in regions far from filled neutron *and* proton shells. Just as the cooperative effect of a few nucleon pairs outside of a filled shell was responsible for the microscopic structure of the vibrations of spherical nuclei, the cooperative effect of many valence nucleon pairs can distort the "core" of nucleons until the equilibrium shape becomes strongly deformed.

## 5.3  MORE REALISTIC NUCLEAR MODELS

Both the shell model for odd-*A* nuclei and the collective model for even-even nuclei are idealizations that are only approximately valid for real nuclei, which are far more complex in their structure than our simple models suggest. Moreover, in real nuclei we cannot "turn off" one type of structure and consider only the other. Thus even very collective nuclei show single-particle effects, while the core of nucleons in shell-model nuclei may contribute collective effects that we have ignored up to this point. The structure of most nuclei cannot be quite so neatly divided between single-particle and collective motion, and usually we must consider a combination of both. Such a *unified nuclear model* is mathematically too complicated to be discussed here, and hence we will merely illustrate a few of the resulting properties of nuclei and try to relate them to the more elementary aspects of the shell and collective models.

### Many-Particle Shell Model

In our study of the shell model, we considered only the effects due to the last unpaired single particle. A more realistic approach for odd-*A* nuclei would be to include all particles outside of closed shells. Let us consider for example the nuclei with odd $Z$ or $N$ between 20 and 28, so that the odd nucleons are in the $f_{7/2}$ shell. For simplicity, we shall confine our discussion to one kind of nucleon only, and thus we require not only that there be an even number of the other kind of nucleon, but also that it be a magic number. Figure 5.25 shows the lower excited states of several such nuclei. The nuclei whose structure is determined by a *single* particle ($^{41}$Ca and $^{55}$Co) show the expected levels—a $\frac{7}{2}^-$ ground state, corresponding to the single odd $f_{7/2}$ particle (or vacancy, in the case of $^{55}$Co, since a single vacancy or hole in a shell behaves like a single particle in the shell), and a $\frac{3}{2}^-$ excited state at about 2 MeV, corresponding to exciting the single odd particle to the $p_{3/2}$ state. The nuclei with 3 or 5 particles in the $f_{7/2}$ level show a much richer spectrum of states, and in particular the very low negative-parity states cannot be explained by the extreme single-particle shell model. If the $\frac{5}{2}^-$ state, for instance, originated from the excitation of a single particle to the $f_{5/2}$ shell, we would expect it to appear above 2 MeV because the $f_{5/2}$ level occurs above the $p_{3/2}$ level (see Figure 5.6); the lowest $\frac{5}{2}^-$ level in the single-particle nuclei occurs at 2.6 MeV (in $^{41}$Ca) and 3.3 MeV (in $^{55}$Co).

We use the shorthand notation $(f_{7/2})^n$ to indicate the configuration with $n$ particles in the $f_{7/2}$ shell, and we consider the possible resultant values of $I$ for the configuration $(f_{7/2})^3$. (From the symmetry between particles and holes, the levels of three holes, or five particles, in the $f_{7/2}$ shell will be the same.) Because

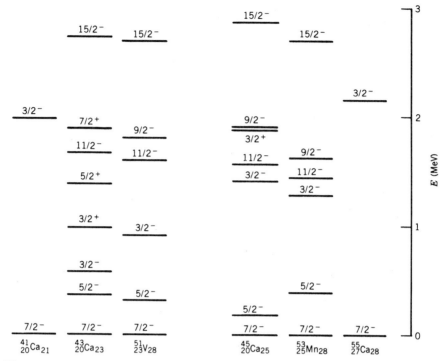

**Figure 5.25**  Excited states of some nuclei with valence particles in the $f_{7/2}$ shell. All known levels below about 2 MeV are shown, and in addition the $\frac{15}{2}^-$ state is included.

the nucleons have half-integral spins, they must obey the Pauli principle, and thus no two particles can have the same set of quantum numbers. Each particle in the shell model is described by the angular momentum $j = \frac{7}{2}$, which can have the projections or $z$ components corresponding to $m = \pm\frac{1}{2}, \pm\frac{3}{2}, \pm\frac{5}{2}, \pm\frac{7}{2}$. The Pauli principle requires that each of the 3 particles have a different value of $m$. Immediately we conclude that the maximum value of the total projection, $M = m_1 + m_2 + m_3$, for the three particles is $+\frac{7}{2} + \frac{5}{2} + \frac{3}{2} = +\frac{15}{2}$. (Without the Pauli principle, the maximum would be $\frac{21}{2}$.) We therefore expect to find no state in the configuration $(f_{7/2})^3$ with $I$ greater than $\frac{15}{2}$; the maximum resultant angular momentum is $I = \frac{15}{2}$, which can have all possible $M$ from $+\frac{15}{2}$ to $-\frac{15}{2}$. The next highest possible $M$ is $\frac{13}{2}$, which can only be obtained from $+\frac{7}{2} + \frac{5}{2} + \frac{1}{2}$ ($+\frac{7}{2} + \frac{3}{2} + \frac{3}{2}$ is not permitted, nor is $+\frac{7}{2} + \frac{7}{2} - \frac{1}{2}$). This single $M = \frac{13}{2}$ state must belong to the $M$ states we have already assigned to the $I = \frac{15}{2}$ configuration; thus we have no possibility to have a $I = \frac{13}{2}$ resultant. Continuing in this way, we find two possibilities to obtain $M = +\frac{11}{2}$ ($+\frac{7}{2} + \frac{3}{2} + \frac{1}{2}$ and $+\frac{7}{2} + \frac{5}{2} - \frac{1}{2}$); there are thus two possible $M = +\frac{11}{2}$ states, one for the $I = \frac{15}{2}$ configuration and another that we can assign to $I = \frac{11}{2}$. Extending this reasoning, we expect to find the following states for $(f_{7/2})^3$ or $(f_{7/2})^5$: $I = \frac{15}{2}, \frac{11}{2}, \frac{9}{2}, \frac{7}{2}, \frac{5}{2}$, and $\frac{3}{2}$. Because each of the three or five particles has negative parity, the resultant parity is $(-1)^3$. The nuclei shown in Figure 5.25 show low-lying negative-parity states with the expected spins (and also with the expected absences—no low-lying $\frac{1}{2}^-$ or $\frac{13}{2}^-$ states appear).

Although this analysis is reasonably successful, it is incomplete—if we do indeed treat all valence particles as independent and equivalent, then the energy of a level should be independent of the orientation of the different $m$'s—that is, all of the resultant $I$'s should have the same energy. This is obviously not even approximately true; in the case of the $(f_{7/2})^3$ multiplet, the energy splitting between the highest and lowest energy levels is 2.7 MeV, about the same energy as pair-breaking and particle-exciting interactions. We can analyze these energy splittings in terms of a *residual interaction* between the valence particles, and thus the level structure of these nuclei gives us in effect a way to probe the nucleon–nucleon interaction in an environment different from the free-nucleon studies we discussed in Chapter 4.

As a final comment, we state without proof that the configurations with $n$ particles in the same shell have another common feature that lends itself to experimental test—their magnetic moments should all be proportional to $I$. That is, given two different states 1 and 2 belonging to the same configuration, we expect

$$\frac{\mu_1}{\mu_2} = \frac{I_1}{I_2} \tag{5.21}$$

Unfortunately, few of the excited-state magnetic moments are well enough known to test this prediction. In the case of $^{51}$V, the ground-state moment is $\mu = +5.1514 \pm 0.0001 \ \mu_N$ and the moment of the first excited state is $\mu = +3.86 \pm 0.33 \ \mu_N$. The ratio of the moments is thus $1.33 \pm 0.11$, in agreement with the expected ratio $\frac{7}{2}/\frac{5}{2} = 1.4$. In the case of $^{53}$Mn, the ratio of moments of the same states is $5.024 \pm 0.007 \ \mu_N/3.25 \pm 0.30 \ \mu_N = 1.55 \pm 0.14$. The evidence from the magnetic moments thus supports our assumption about the nature of these states.

### Single-Particle States in Deformed Nuclei

The calculated levels of the nuclear shell model are based on the assumption that the nuclear potential is spherical. We know, however, that this is not true for nuclei in the range $150 \leq A \leq 190$ and $A > 230$. For these nuclei we should use a shell-model potential that approximates the actual nuclear shape, specifically a rotational ellipsoid. In calculations using the Schrödinger equation with a non-spherical potential, the angular momentum $\ell$ is no longer a "good" quantum number; that is, we cannot identify states by their spectroscopic notation (s, p, d, f, etc.) as we did for the spherical shell model. To put it another way, the states that result from the calculation have mixtures of different $\ell$ values (but based on consideration of parity, we expect mixtures of only even or only odd $\ell$ values).

In the spherical case, the energy levels of each single particle state have a degeneracy of $(2j + 1)$. (That is, relative to any arbitrary axis of our choice, all $2j + 1$ possible orientations of $j$ are equivalent.) If the potential has a deformed shape, this will no longer be true—*the energy levels in the deformed potential depend on the spatial orientation of the orbit*. More precisely, the energy depends on the component of $j$ along the symmetry axis of the core. For example, an $f_{7/2}$

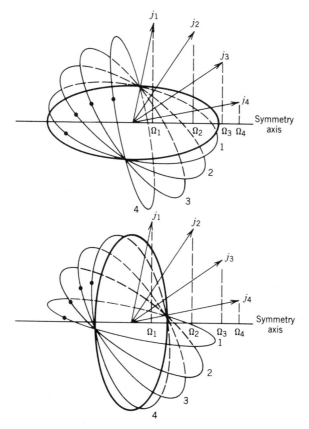

**Figure 5.26** Single-particle orbits with $j = \frac{7}{2}$ and their possible projections along the symmetry axis, for prolate (top) and oblate (bottom) deformations. The possible projections are $\Omega_1 = \frac{1}{2}$, $\Omega_2 = \frac{3}{2}$, $\Omega_3 = \frac{5}{2}$, and $\Omega_4 = \frac{7}{2}$. (For clarity, only the positive projections are shown.) Note that in the prolate case, orbit 1 lies closest (on the average) to the core and will interact most strongly with the core; in the oblate case, it is orbit 4 that has the strongest interaction with the core.

nucleon can have eight possible components of $j$, ranging from $-\frac{7}{2}$ to $+\frac{7}{2}$. This component of $j$ along the symmetry axis is generally denoted by $\Omega$. Because the nuclei have reflection symmetry for either of the two possible directions of the symmetry axis, the components $+\Omega$ and $-\Omega$ will have the same energy, giving the levels a degeneracy of 2. That is, what we previously called the $f_{7/2}$ state splits up into four states if we deform the central potential; these states are labeled $\Omega = \frac{1}{2}, \frac{3}{2}, \frac{5}{2}, \frac{7}{2}$ and all have negative parity. Figure 5.26 indicates the different possible "orbits" of the odd particle for prolate and oblate deformations. For prolate deformations, the orbit with the smallest possible $\Omega$ (equal to $\frac{1}{2}$) interacts most strongly with the core and is thus more tightly bound and lowest in energy. The situation is different for oblate deformations, in which the orbit with maximum $\Omega$ (equal to $j$) has the strongest interaction with the core and the lowest energy. Figure 5.27 shows how the $f_{7/2}$ states would split as the deformation is increased.

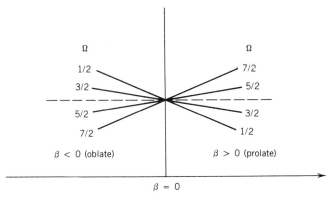

**Figure 5.27**   This shows directly the effect of the various orientations of the $f_{7/2}$ orbit. As shown in Figure 5.26, the orbit with component $\Omega = \frac{1}{2}$ along the symmetry axis has the strongest interaction with the *prolate* core and therefore lies lowest in energy. For an *oblate* core, it is the $\Omega = \frac{7}{2}$ component that lies lowest.

Of course, we must keep in mind that Figures 5.26 and 5.27 are not strictly correct because the spherical single-particle quantum numbers $\ell$ and $j$ are not valid when the potential is deformed. The negative parity state with $\Omega = \frac{5}{2}$, for example, cannot be identified with the $f_{7/2}$ state, even though it approaches that state as $\beta \to 0$. The wave function of the $\Omega = \frac{5}{2}$ state can be expressed as a mixture (or linear combination) of many different $\ell$ and $j$ (but only with $j \geq \frac{5}{2}$, in order to get a component of $\frac{5}{2}$). It is customary to make the approximation that states from different major oscillator shells (see Figures 5.4 and 5.6) do not mix. Thus, for example, the $\Omega = \frac{5}{2}$ state that approaches the $2f_{7/2}$ level as $\beta \to 0$ will include contributions from only those states of the 5th oscillator shell ($2f_{5/2}$, $2f_{7/2}$, $1h_{9/2}$, $1h_{11/2}$). The 4th and 6th oscillator shells have the opposite parity and so will not mix, and the next odd-parity shells are far away and do not mix strongly. Writing the spherical wave functions as $\psi_{N\ell j}$, we must have

$$\psi'(\Omega) = \sum_{\ell j} a(N\ell j)\psi_{N\ell j} \tag{5.22}$$

where $\psi'(\Omega)$ represents the wave function of the deformed state $\Omega$ and where $a(N\ell j)$ are the expansion coefficients. For the $\Omega = \frac{5}{2}$ state

$$\psi'(\Omega) = a\left(53\tfrac{5}{2}\right)\psi_{53\frac{5}{2}} + a\left(53\tfrac{7}{2}\right)\psi_{53\frac{7}{2}} + a\left(55\tfrac{9}{2}\right)\psi_{55\frac{9}{2}} + a\left(55\tfrac{11}{2}\right)\psi_{55\frac{11}{2}} \tag{5.23}$$

The coefficients $a(N\ell j)$ can be obtained by solving the Schrödinger equation for the deformed potential, which was first done by S. G. Nilsson in 1955. The coefficients will vary with $\beta$, and of course for $\beta \to 0$ we expect $a(53\frac{7}{2})$ to approach 1 while the others all approach 0. For $\beta = 0.3$ (a typical prolate deformation), Nilsson calculated the values

$$a\left(53\tfrac{5}{2}\right) = 0.267 \qquad a\left(53\tfrac{7}{2}\right) = 0.832$$
$$a\left(55\tfrac{9}{2}\right) = 0.415 \qquad a\left(55\tfrac{11}{2}\right) = -0.255$$

for the $\Omega = \frac{5}{2}$ level we have been considering.

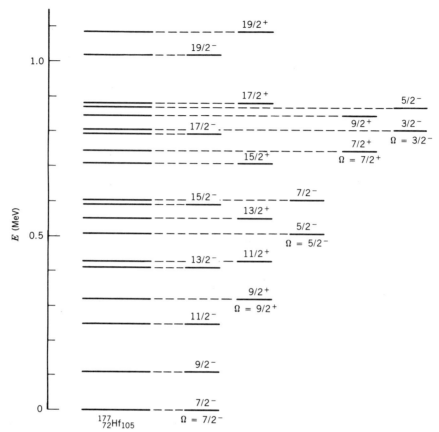

**Figure 5.28** The energy levels of $^{177}$Hf. As in the case of $^{164}$Er (Figure 5.23), knowledge of the spin-parity assignments helps us to group the states into rotational bands. The lowest state in each band has $I = \Omega$, and the higher states follow the $I(I + 1)$ energy spacing.

Given such wave functions for single-particle states in deformed nuclei, we can then allow the nuclei to rotate, and we expect to find a sequence of rotational states, following the $I(I + 1)$ energy spacing, built on each single-particle state. The lowest state of the rotational band has $I = \Omega$, and as rotational energy is added the angular momentum increases in the sequence $I = \Omega, \Omega + 1, \Omega + 2, \ldots$. Figure 5.28 shows the energy levels of the nucleus $^{177}$Hf, in which two well-developed rotational bands have been found and several other single-particle states have been identified.

To interpret the observed single-particle levels, we require a diagram similar to Figure 5.27 but which shows all possible single-particle states and how their energies vary with deformation. Such a diagram is shown in Figure 5.29 for the neutron states that are appropriate to the $150 \leq A \leq 190$ region. Recalling that the degeneracy of each deformed single-particle level is 2, we proceed exactly as we did in the spherical shell model, placing two neutrons in each state up to $N = 105$ and two protons in each state up to $Z = 72$. We can invoke the pairing argument to neglect the single-particle states of the protons and examine the possible levels of the 105th neutron for the typical deformation of $\beta \approx 0.3$. You

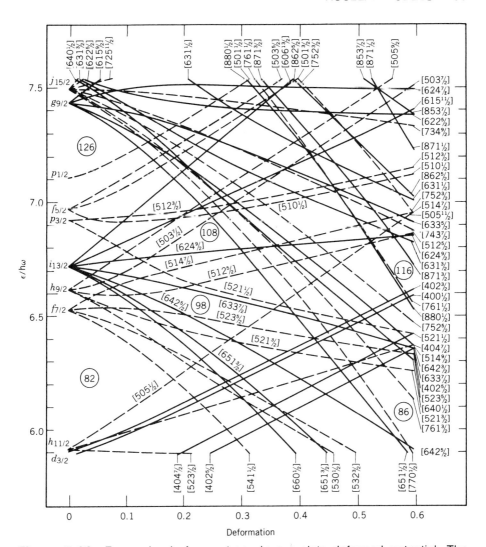

**Figure 5.29** Energy levels for neutrons in a prolate deformed potential. The deformation is measured essentially by the parameter $\beta$. The numbers in the brackets label the states; for our purposes, we are interested only in the first number, which is the principal quantum number $N$ of the oscillator shell and therefore tells us the parity of the state $(-1)^N$, and the last number, which is the component $\Omega$. Solid lines show states with even parity, and dashed lines show odd parity. For a deformation between 0.2 and 0.3 (typical for nuclei in this region) the 105th neutron of $^{177}$Hf would go into the state $[514\frac{7}{2}]$, that is, an odd-parity state with $\Omega = \frac{7}{2}$. A small excitation takes it into the state $[624\frac{9}{2}]$, an even parity state with $\Omega = \frac{9}{2}$. Both intrinsic states (and their associated rotational bands) can be seen in Figure 5.28. Other observed states in $^{177}$Hf result from breaking a pair of neutrons in a lower state and exciting one to pair with the $\frac{7}{2}^-$ neutron. In this way, for instance, we could produce a single neutron in the state $[512\frac{5}{2}]$, which gives the odd-parity $\Omega = \frac{5}{2}$ state in $^{177}$Hf. From C. Gustafson et al., *Ark. Fys.* **36**, 613 (1967).

can see from the diagram that the expected single-particle levels correspond exactly with the observed levels of $^{177}$Hf.

The general structure of the odd-$A$ deformed nuclei is thus characterized by rotational bands built on single-particle states calculated from the deformed shell-model potential. The proton and neutron states are filled (two nucleons per state), and the nuclear properties are determined in the extreme single-particle limit by the properties of the odd particle. This model, with the wave functions calculated by Nilsson, has had extraordinary success in accounting for the nuclear properties in this region. In general, the calculations based on the properties of the odd particle have been far more successful in the deformed region than have the analogous calculations in the spherical region.

In this chapter we have discussed evidence for types of nuclear structure based on the *static* properties of nuclei—energy levels, spin-parity assignments, magnetic dipole and electric quadrupole moments. The wave functions that result from solving the Schrödinger equation for these various models permit many other features of nuclear structure to be calculated, particularly the transitions between different nuclear states. Often the evidence for collective structure, for instance, may be inconclusive based on the energy levels alone, while the transition probabilities between the excited states may give definitive proof of collective effects. It may also be the case that a specific excited state may have alternative interpretations—for example, a vibrational state or a 2-particle coupling. Studying the transition probabilities will usually help us to discriminate between these competing interpretations. The complete study of nuclear structure therefore requires that we study *radioactive decays*, which give spontaneous transitions between states, and *nuclear reactions*, in which the experimenter can select the initial and final states. In both cases, we can compare calculated decay and reaction probabilities with their experimental values in order to draw conclusions about the structure of nuclear states. The methods of each of these areas of *nuclear spectroscopy* will occupy us for most of the remainder of this text.

## REFERENCES FOR ADDITIONAL READING

Nuclear physics texts providing more detailed and advanced information on nuclear models are the following: B. L. Cohen, *Concepts of Nuclear Physics* (New York: McGraw-Hill, 1971), Chapters 4–6; H. Frauenfelder and E. M. Henley, *Subatomic Physics* (Englewood Cliffs, NJ: Prentice-Hall, 1974), Chapters 14–16; M. A. Preston, *Physics of the Nucleus* (Reading, MA: Addison-Wesley, 1962), Chapters 7–10.

Two essential monographs, covering respectively the shell model and the collective model, have been written by recipients of the Nobel Prize for their work on nuclear models: M. G. Mayer and J. H. D. Jensen, *Elementary Theory of Nuclear Shell Structure* (New York: Wiley, 1955) and A. Bohr and B. R. Mottelson, *Nuclear Structure* (Reading, MA: Benjamin, 1975), Vol. 2 (Vol. 1 treats the shell model). Another comprehensive but advanced work is J. M. Eisenberg and W. Greiner, *Nuclear Models* (Amsterdam: North-Holland, 1970).

## PROBLEMS

1. Give the expected shell-model spin and parity assignments for the ground states of (a) $^7$Li; (b) $^{11}$B; (c) $^{15}$C; (d) $^{17}$F; (e) $^{31}$P; (f) $^{141}$Pr.

2. The low-lying levels of $^{13}$C are ground state, $\frac{1}{2}^-$; 3.09 MeV, $\frac{1}{2}^+$; 3.68 MeV, $\frac{3}{2}^-$; 3.85 MeV, $\frac{5}{2}^+$. The next states are about 7 MeV and above. Interpret these four states according to the shell model.

3. The level scheme of Figure 5.6 would lead us to expect $I^\pi = \frac{11}{2}^-$ for the ground state of $^{203}$Tl ($Z = 81$), while the observed value is $\frac{1}{2}^+$. A similar case occurs in $^{207}$Pb ($N = 125$) and $^{199}$Hg ($N = 119$), where $\frac{13}{2}^+$ is expected but $\frac{1}{2}^-$ is observed. Given that the pairing force increases strongly with $\ell$, give the shell-model configurations for these nuclei that are consistent with the observed spin-parity assignments.

4. Figure 5.6 is only a schematic, average representation of the shell-model single-particle states. The energies of the states will vary with the proton number and neutron number. To illustrate this effect, consider the available states of the 51st proton in Sb isotopes. Make a chart in the style of Figure 5.25 showing the $\frac{5}{2}^+$ and $\frac{7}{2}^+$ states in $^{113}$Sb to $^{133}$Sb. (Consult the *Table of Isotopes* and the Nuclear Data Sheets for information on the energy levels.) Discuss the relative positions of the $g_{7/2}$ and $d_{5/2}$ *proton* states as a function of *neutron* number.

5. In the single-particle shell model, the ground state of a nucleus with an odd proton and an odd neutron is determined from the coupling of the proton and neutron shell-model states: $I = j_p + j_n$. Consider the following nuclei: $^{16}$N $- 2^-$; $^{12}$B $- 1^+$; $^{34}$P $- 1^+$; $^{28}$Al $- 3^+$. Draw simple vector diagrams illustrating these couplings, then replace $j_p$ and $j_n$, respectively, by $\ell_p + s_p$ and $\ell_n + s_n$. Examine your four diagrams and deduce an empirical rule for the relative orientation of $s_p$ and $s_n$ in the ground state. Finally, use your empirical rule to predict the $I^\pi$ assignments of $^{26}$Na and $^{28}$Na.

6. (a) If the energy of a single-particle state in the absence of spin-orbit splitting is $E_0$, find the energies of the two members of the spin-orbit doublet whose difference is given by Equation 5.4. (b) Show that the "center of gravity" of the doublet is $E_0$.

7. Compute the expected shell-model quadrupole moment of $^{209}$Bi ($\frac{9}{2}^-$) and compare with the experimental value, $-0.37$ b.

8. Compute the values of the magnetic dipole moments expected from the shell model, and compare with the experimental values:

| Nuclide | $I^\pi$ | $\mu$(exp) ($\mu_N$) |
|---------|---------|----------------------|
| $^{75}$Ge | $\frac{1}{2}^-$ | $+0.510$ |
| $^{87}$Sr | $\frac{9}{2}^+$ | $-1.093$ |
| $^{91}$Zr | $\frac{5}{2}^+$ | $-1.304$ |
| $^{47}$Sc | $\frac{7}{2}^-$ | $+5.34$ |
| $^{147}$Eu | $\frac{11}{2}^-$ | $+6.06$ |

9. Compute the vibrational frequency associated with typical quadrupole vibrations. Taking typical values for decay lifetimes of $2^+$ states in vibrational nuclei (you can find these in the *Table of Isotopes*, for example), comment on whether the decays generally occur on a shorter or longer time scale than the nuclear vibrations. If $\alpha$ represents the vibrational amplitude, as in Equation 5.13, could we observe quantities dependent on $\langle \alpha \rangle$? On $\langle \alpha^2 \rangle$?

10. By tabulating the possible $m$ states of three quadrupole ($\ell = 2$) phonons, and their symmetrized combinations, show that the permitted resultant states are $0^+$, $2^+$, $3^+$, $4^+$, and $6^+$.

11. Find the volume of the nucleus whose surface is described by Equation 5.14.

12. Consider a uniformly charged ellipsoidal nucleus whose surface is described by Equation 5.14. Show that the electric quadrupole moment, defined by Equation 3.36, is given by Equation 5.16.

13. The levels of $^{174}$Hf show two similar rotational bands, with energies given as follows (in MeV):

| | $E(0^+)$ | $E(2^+)$ | $E(4^+)$ | $E(6^+)$ | $E(8^+)$ | $E(10^+)$ | $E(12^+)$ |
|---|---|---|---|---|---|---|---|
| Band 1 | 0 | 0.091 | 0.297 | 0.608 | 1.010 | 1.486 | 2.021 |
| Band 2 | 0.827 | 0.900 | 1.063 | 1.307 | 1.630 | 2.026 | 2.489 |

Compare the moments of inertia of these two bands, and comment on any difference.

14. The low-lying levels of $^{17}$O and $^{19}$O differ primarily in the presence of states of $I^\pi = \frac{3}{2}^+$ and $\frac{9}{2}^+$ in $^{19}$O; these two states have no counterparts in $^{17}$O. Show that these two states could result from the configuration $(d_{5/2})^3$ and thus are not expected in $^{17}$O.

15. The nucleus $^{24}$Mg has a $2^+$ first excited state at 1.369 MeV and a $4^+$ second excited state at 4.123 MeV. The $2^+$ state has a magnetic dipole moment of 1.02 $\mu_N$ and an electric quadrupole moment of $-0.27$ b. Which model would be most likely to provide an accurate description of these states? Justify your choice by calculating parameters appropriate to your choice of model.

# UNIT II
## NUCLEAR
## DECAY
## AND
## RADIOACTIVITY

# 6

# RADIOACTIVE DECAY

The radioactive decays of naturally occurring minerals containing uranium and thorium are in large part responsible for the birth of the study of nuclear physics. These decays have half-lives that are of the order of the age of the Earth, suggesting that the materials are survivors of an early period in the creation of matter by aggregation of nucleons; the shorter-lived nuclei have long since decayed away, and we observe today the remaining long-lived decays. Were it not for the extremely long half-lives of $^{235}$U and $^{238}$U, we would today find no uranium in nature and would probably have no nuclear reactors or nuclear weapons.

In addition to this naturally occurring radioactivity, we also have the capability to produce radioactive nuclei in the laboratory through nuclear reactions. This was first done in 1934 by Irene Curie and Pierre Joliot, who used $\alpha$ particles from the natural radioactive decay of polonium to bombard aluminum, thereby producing the isotope $^{30}$P, which they observed to decay through positron emission with a half-life of 2.5 min. In their words:

Our latest experiments have shown a very striking fact: when an aluminum foil is irradiated on a polonium preparation, the emission of positrons does not cease immediately when the active preparation is removed. The foil remains radioactive and the emission of radiation decays exponentially as for an ordinary radioelement.

For this work on artificially produced radioactivity the Joliot-Curie team was awarded the 1935 Nobel Prize in Chemistry (following a family tradition—Irene's parents, Pierre and Marie Curie, shared with Becquerel the 1903 Nobel Prize in Physics for their work on the natural radioactivity of the element radium, and Marie Curie became the first person twice honored, when she was awarded the 1911 Nobel Prize in Chemistry).

In this chapter we explore the physical laws governing the production and decay of radioactive materials, which we take to mean those substances whose nuclei spontaneously emit radiations and thereby change the state of the nucleus.

## 6.1 THE RADIOACTIVE DECAY LAW

Three years following the 1896 discovery of radioactivity it was noted that the decay rate of a pure radioactive substance decreases with time according to an exponential law. It took several more years to realize that radioactivity represents changes in the individual atoms and not a change in the sample as a whole. It took another two years to realize that the decay is statistical in nature, that it is impossible to predict when any given atom will disintegrate, and that this hypothesis leads directly to the exponential law. This lack of predictability of the behavior of single particles does not bother most scientists today, but this early instance of it, before the development of quantum theory, was apparently difficult to accept. Much labor was required of these dedicated investigators to establish what now may seem like evident facts.

If $N$ radioactive nuclei are present at time $t$ and if no new nuclei are introduced into the sample, then the number $dN$ decaying in a time $dt$ is proportional to $N$, and so

$$\lambda = -\frac{(dN/dt)}{N} \tag{6.1}$$

in which $\lambda$ is a constant called the *disintegration or decay constant*. The right side of Equation 6.1 is the probability per unit time for the decay of an atom. *That this probability is constant, regardless of the age of the atoms, is the basic assumption of the statistical theory of radioactive decay*. (Human lifetimes do not follow this law!)

Integrating Equation 6.1 leads to the *exponential law of radioactive decay*

$$N(t) = N_0 e^{-\lambda t} \tag{6.2}$$

where $N_0$, the constant of integration, gives the original number of nuclei present at $t = 0$. The *half-life* $t_{1/2}$ gives the time necessary for half of the nuclei to decay. Putting $N = N_0/2$ in Equation 6.2 gives

$$t_{1/2} = \frac{0.693}{\lambda} \tag{6.3}$$

It is also useful to consider the *mean lifetime* (sometimes called just the lifetime) $\tau$, which is defined as the average time that a nucleus is likely to survive before it decays. The number that survive to time $t$ is just $N(t)$, and the number that decay between $t$ and $t + dt$ is $|dN/dt|\, dt$. The mean lifetime is then

$$\tau = \frac{\int_0^\infty t\, |dN/dt|\, dt}{\int_0^\infty |dN/dt|\, dt} \tag{6.4}$$

where the denominator gives the total number of decays. Evaluating the integrals gives

$$\tau = \frac{1}{\lambda} \tag{6.5}$$

Thus the mean lifetime is simply the inverse of the decay constant.

Equation 6.2 allows us to predict the number of undecayed nuclei of a given type remaining after a time $t$. Unfortunately, the law in that form is of limited usefulness because $N$ is a very difficult quantity to measure. Instead of counting the number of undecayed nuclei in a sample, it is easier to count the number of decays (by observing the emitted radiations) that occur between the times $t_1$ and $t_2$. If we deduce a change $\Delta N$ in the number of nuclei between $t$ and $t + \Delta t$, then

$$|\Delta N| = N(t) - N(t + \Delta t) = N_0 e^{-\lambda t}(1 - e^{-\lambda \Delta t}) \tag{6.6}$$

If the interval $\Delta t$ during which we count is much smaller than $\lambda^{-1}$ (and thus, in effect, $\Delta t \ll t_{1/2}$), we can ignore higher order terms in the expansion of the second exponential, and

$$|\Delta N| = \lambda N_0 e^{-\lambda t} \Delta t \tag{6.7}$$

Going over to the differential limit gives

$$\left| \frac{dN}{dt} \right| = \lambda N_0 e^{-\lambda t} \tag{6.8}$$

Defining the *activity* $\mathscr{A}$ to be the rate at which decays occur in the sample,

$$\mathscr{A}(t) \equiv \lambda N(t) = \mathscr{A}_0 e^{-\lambda t} \tag{6.9}$$

The initial activity at $t = 0$ is $\mathscr{A}_0 = \lambda N_0$.

Actually, we could have obtained Equation 6.8 by differentiating Equation 6.2 directly, but we choose this more circuitous path to emphasize an important but often overlooked point: *Measuring the number of counts $\Delta N$ in a time interval $\Delta t$ gives the activity of the sample only if $\Delta t \ll t_{1/2}$.* The number of decays in the interval from $t_1$ to $t_2$ is

$$\Delta N = \int_{t_1}^{t_2 = t_1 + \Delta t} \mathscr{A}\, dt \tag{6.10}$$

which equals $\mathscr{A}\Delta t$ only if $\Delta t \ll t_{1/2}$. (Consider an extreme case—if $t_{1/2} = 1$ s, we observe the same number of counts in 1 min as we do in 1 h.) See Problem 1 at the end of this chapter for more on the relation between $\mathscr{A}$ and $\Delta N$.

The activity of a radioactive sample is exactly the number of decays of the sample per unit time, and decays/s is a convenient unit of measure. Another unit for measuring activity is the *curie* (Ci), which originally indicated the activity of one gram of radium but is now defined simply as

$$1 \text{ Ci} = 3.7 \times 10^{10} \text{ decays/s}$$

Most common radioactive sources of strengths typically used in laboratories have activities in the range of microcuries to millicuries. The SI unit for activity is the becquerel (Bq), equal to one decay per second; however, the curie is so firmly in place as a unit of activity that the becquerel has not yet become the commonly used unit.

Note that the activity tells us only the number of disintegrations per second; it says nothing about the *kind* of radiations emitted or their *energies*. If we want to know about the effects of radiation on a biological system, the activity is not a useful quantity since different radiations may give different effects. In Section 6.8 we discuss some alternative units for measuring radiation that take into account their relative biological effects.

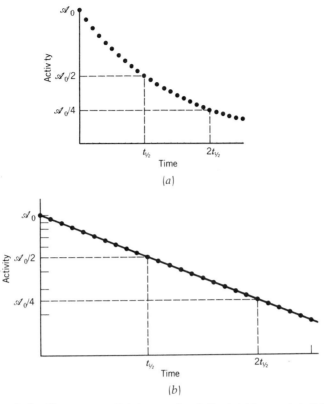

**Figure 6.1**   The exponential decay of activity. (*a*) Linear plot. (*b*) Semilog plot.

$$A = A_0 e^{-\lambda t}$$
$$\ln A = \ln A_0 - \lambda t$$

Equation 6.9 shows that the activity decays exponentially with time. We can thus measure the activity as a function of time by counting the number of decays in a sequence of short time intervals $\Delta t$. Plotting these data on a semilog graph (that is, $\ln \mathscr{A}$ vs $t$) should give a straight line of slope $-\lambda$. Figure 6.1 is an example of this kind of experiment, from which one can determine the half-life of a radioactive decay.

This method of measurement is useful only for half-lives that are neither too short nor too long. The half-life must be short enough that we can see the sample decaying—for half-lives far greater than a human lifetime, we would not be able to observe any substantial reduction in activity. For such cases, we can use Equation 6.1 directly, by measuring $dN/dt$ (which is just the activity in this simple decay process) and by determining the number of atoms (such as by weighing a sample whose chemical composition is accurately known).

For half-lives that are very short (say, small compared with 1 s), observing the successive disintegration rates is also not useful, for the activity decays to nothing in the time that it would take to switch the counting apparatus on and off. For these cases we use a more precise technique, described in Chapter 7, that permits the routine measurement of half-lives down to nanoseconds ($10^{-9}$ s) and even picoseconds ($10^{-12}$ s).

It is important to keep in mind that the simple exponential law of radioactive decay applies only in a limited set of circumstances—a given initial quantity of a substance decays (by emitting radiation) to a stable end product. Under these circumstances, when radioactive nucleus 1 decays with decay constant $\lambda_1$ to stable nucleus 2, the number of nuclei present is

$$N_1 = N_0 e^{-\lambda_1 t} \qquad (6.11a)$$

$$N_2 = N_0(1 - e^{-\lambda_1 t}) \qquad (6.11b)$$

Note that the number of nuclei of type 2 starts out at 0 and approaches $N_0$ as $t \to \infty$ (all of type 1 eventually end as type 2) and also note that $N_1 + N_2 = N_0$ (the total number of nuclei is constant). If nuclei of type 2 are themselves radioactive, or if nuclei of type 1 are being produced (as a result of a nuclear reaction, for instance) then Equations 6.11 do not apply. We consider these cases in Sections 6.3 and 6.4.

Often it will happen that a given initial nucleus can decay in two or more different ways, ending with two different final nuclei. Let's call these two decay modes a and b. The rate of decay into mode a, $(dN/dt)_a$, is determined by the *partial decay constant* $\lambda_a$, and the rate of decay into mode b, $(dN/dt)_b$, by $\lambda_b$:

$$\lambda_a = \frac{-(dN/dt)_a}{N}$$
$$\qquad (6.12)$$
$$\lambda_b = \frac{-(dN/dt)_b}{N}$$

The *total decay rate* $(dN/dt)_t$ is

$$-\left(\frac{dN}{dt}\right)_t = -\left(\frac{dN}{dt}\right)_a - \left(\frac{dN}{dt}\right)_b = N(\lambda_a + \lambda_b) = N\lambda_t \qquad (6.13)$$

where $\lambda_t = \lambda_a + \lambda_b$ is the *total decay constant*. The nuclei therefore decay according to $N = N_0 e^{-\lambda_t t}$, and the activity $|dN/dt|$ decays with decay constant $\lambda_t$. *Whether we count the radiation leading to final states a or b, we observe only the total decay constant $\lambda_t$*; we never observe an exponential decay of the activity with constants $\lambda_a$ or $\lambda_b$. The relative decay constants $\lambda_a$ and $\lambda_b$ determine the probability for the decay to proceed by mode a or b. Thus a fraction $\lambda_a/\lambda_t$ of the nuclei decay by mode a and a fraction $\lambda_b/\lambda_t$ decay by mode b, so that

$$N_1 = N_0 e^{-\lambda_{1,t} t}$$
$$N_{2,a} = (\lambda_a/\lambda_t) N_0(1 - e^{-\lambda_{1,t} t}) \qquad (6.14)$$
$$N_{2,b} = (\lambda_b/\lambda_t) N_0(1 - e^{-\lambda_{1,t} t})$$

The separate factors $\lambda_a$ or $\lambda_b$ never appear in any exponential term; we cannot "turn off" one decay mode to observe the exponential decay of the other.

Another special case is that of a sample with two or more radionuclei with genetically unrelated decay schemes. Consider a mixture of $^{64}$Cu (12.7 h) and $^{61}$Cu (3.4 h); such mixtures cannot be chemically separated of course. The activity of a particular mixture is plotted against time on semilog paper in Figure 6.2. At the right end of the curve we assume (because the curve is linear) that only one activity is present; the limiting slope shows a 12.7-h half-life. By (1)

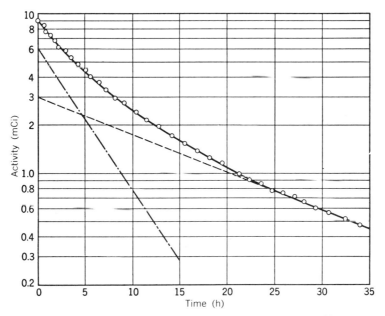

**Figure 6.2**   Decay curve for a sample containing a mixture of $^{64}$Cu (12.7 h) and $^{61}$Cu (3.4 h).

extending this limiting slope backward, (2) taking differences between the curve and this straight line at various abscissas, and (3) plotting these differences on the same scale, we get the dot-dashed straight line that represents the 3.4-h half-life. The intercepts of both straight lines on the vertical axis give the initial counting rates for each component. This method can be extended to mixtures with more than two components, if the half-lives are sufficiently different from one another.

## 6.2   QUANTUM THEORY OF RADIATIVE DECAYS

The energy levels we obtain by solving the Schrödinger equation for various time-independent potentials share one property—they are *stationary states*. A quantum system that is originally in a particular stationary state will remain in that state for all times and will *not* make transitions to (i.e., decay to) other states. We can allow a quantum system to be found sometimes in one state and sometimes in another by making a mixture of two or more states, such as $\psi = c_1\psi_1 + c_2\psi_2$ which has the probability $|c_1|^2$ to be found in state 1 and $|c_2|^2$ to be found in state 2. For time-independent potentials, $c_1$ and $c_2$ are independent of time, which does not correspond with observations for decaying states, in which the probability to find one state decays with time. Moreover, on a philosophical level, we should be forced to abandon the notion of pure states with well-defined wave functions, making the interpretation of nuclear structure very difficult indeed.

We therefore adopt the following approach: The potential is assumed to be of the form $V + V'$, where $V$ is the nuclear potential that gives the stationary states and $V'$ is a very weak additional potential that can cause transitions between the states. For the moment neglecting $V'$, we solve the Schrödinger equation for the potential $V$ and obtain the static nuclear wave functions. We then use those wave functions to calculate the transition probability between the "stationary states" under the influence of $V'$. This transition probability is just the decay constant $\lambda$, which is given by Fermi's Golden Rule as discussed in Section 2.8:

$$\lambda = \frac{2\pi}{\hbar} |V'_{fi}|^2 \, \rho(E_f) \tag{6.15}$$

where

$$V'_{fi} = \int \psi_f^* V' \psi_i \, dv \tag{6.16}$$

Given the initial and final wave functions $\psi_i$ and $\psi_f$, we can evaluate the "matrix element" of $V'$ and thus calculate the transition probability (which can then be compared with its experimental value).

The transition probability is also influenced by the *density of final states* $\rho(E_f)$ —within an energy interval $dE_f$, the number of final states accessible to the system is $dn_f = \rho(E_f) \, dE_f$. The transition probability will be large if there is a large number of final states accessible for the decay. There are two contributions to the density of final states because the final state after the decay includes two components—the final nuclear state and the emitted radiation. Let's consider in turn each of these two components, beginning with the nuclear state.

Solving the Schrödinger equation for the time-independent potential $V$ gives us the stationary states of the nucleus, $\psi_a(r)$. The time-dependent wave function $\Psi_a(r, t)$ for the state a is

$$\Psi_a(r, t) = \psi_a(r) \, e^{-iE_a t/\hbar} \tag{6.17}$$

where $E_a$ is the energy of the state. The probability of finding the system in the state a is $|\Psi_a(r, t)|^2$, which is independent of time for a stationary state. To be consistent with the radioactive decay law, we would like the probability of finding our decaying system in the state a to decrease with time like $e^{-t/\tau_a}$:

$$|\Psi_a(t)|^2 = |\Psi_a(t = 0)|^2 \, e^{-t/\tau_a} \tag{6.18}$$

where $\tau_a = 1/\lambda_a$ is the mean lifetime of the state whose decay constant is $\lambda_a$. We should therefore have written Equation 6.17 as

$$\Psi_a(r, t) = \psi_a(r) \, e^{-iE_a t/\hbar} \, e^{-t/2\tau_a} \tag{6.19}$$

The price we pay for including the *real* exponential term in $\Psi_a$ is the loss of the ability to determine exactly the energy of the state—we no longer have a stationary state. (Recall the energy-time uncertainty relationship, Equation 2.2. If a state lives forever, $\Delta t \to \infty$ and we can determine its energy exactly, since $\Delta E = 0$. If a state lives on the average for a time $\tau$, we cannot determine its energy except to within an uncertainty of $\Delta E \sim \hbar/\tau$.) We can make this discussion more rigorous by calculating the distribution of energy states (actually the Fourier transform of $e^{-t/2\tau_a}$). The probability to observe the system in the energy interval between $E$ and $E + dE$ in the vicinity of $E_a$ is given by the

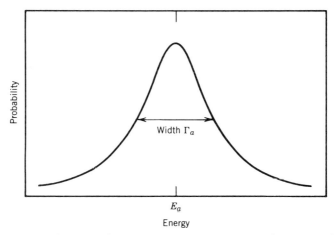

**Figure 6.3**  Probability to observe the energy of an unstable state of width $\Gamma_a$.

square of this distribution:

$$P(E)\,dE = \frac{dE}{(E - E_a)^2 + \Gamma_a^2/4} \tag{6.20}$$

where $\Gamma_a = \hbar/\tau_a$ is the *width* of the state a. Figure 6.3 shows the function $P(E)$. If we measure the energy of this system, we may no longer find the value $E_a$ (although the average of many measurements gives $E_a$). The width $\Gamma_a$ is a measure of our inability to determine precisely the energy of the state (through no fault of our own—nature imposes the limit of uncertainty, not our measuring instruments; as indicated by Figure 6.3, a state with the "exact" energy $E_a$ cannot be observed).

If nuclear states do not have exact energies, can we speak of transitions between distinct levels? We can, because *the widths of the low-lying nuclear levels are small compared with their energy spacing*. Nuclear states typically have lifetimes greater than $10^{-12}$ s, corresponding to $\Gamma < 10^{-10}$ MeV. The discrete low-lying nuclear states that are populated in ordinary decays (and many nuclear reactions, as well) have typical separations of the order of $10^{-3}$ MeV and larger. Thus if we were to measure the energy of a final nuclear state after a decay process (by measuring the energy of the emitted radiation, for example), it is very unlikely that the overlap of the energy distributions of two different final states a and b could cause confusion as to the final "stationary" state resulting from the decay (see Figure 6.4).

We therefore conclude that it *is* reasonable to speak of discrete pseudo-stationary states because their separation is far greater than their width, and we also conclude that such nuclear states do not contribute to the density of final states because there is only one nuclear state that can be reached in a given decay process.

It is thus only the radiation field that contributes to the density of states, and we must consider the properties of the emitted radiations in calculating $\rho(E_f)$. For the present, we will only make some general comments regarding $\rho(E_f)$. If

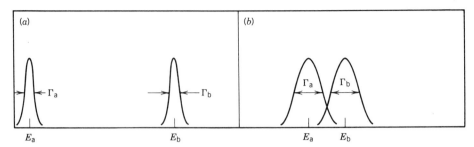

**Figure 6.4** When the widths of unstable states are small compared with their separation, as in (a), the states are distinct and observable. In (b), the states a and b overlap and are strongly mixed; these states do not have distinctly observable wave functions.

we observe only the probability to form the nuclear state $E_f$, then we must consider all possible radiations of energy $E_i - E_f$. Specifically, the radiation can be emitted in any direction and in any state of polarization (if the radiation consists of a particle with spin, the spin may have any possible orientation), assuming of course that we do not observe the direction of the radiation or its polarization. It is this process of counting the number of accessible final states that gives the density of states, which we consider further when we discuss specific radiation types in Chapters 8–10.

In solving the differential equation (6.1) to obtain the radioactive decay law, we assumed the decay probability $\lambda$ to be (1) small and (2) constant in time, which happen to be the same assumptions made in deriving Fermi's Golden Rule. If $V'$ is independent of time, then $\lambda$ calculated according to Equation 6.15 will also be independent of time. Under such a condition, the effect of $V'$ on the stationary states a and b of $V$ is

$$\psi_a \rightarrow \psi_a + \frac{V'_{ba}}{E_b - E_a}\psi_b$$

and the system formerly in the state a has a probability proportional to $|V'_{ba}|^2$ to be found in the state b. We observe this as a "decay" from a to b.

To apply Fermi's Golden Rule, the probability for decay must also be small, so that the amplitude of $\psi_b$ in the above expression is small. It is this requirement that gives us a decay process. If the decay probability were large, then there would be enough radiation present to induce the reverse transition b → a through the process of resonant absorption. The system would then oscillate between the states a and b, in analogy with a classical system of two coupled oscillators.

The final connection between the effective decay probability for an ensemble of a large number of nuclei and the microscopic decay probability computed from the quantum mechanics of a single nucleus requires the assumption that each nucleus of the ensemble emits its radiation independently of all the others. We assume that the decay of a given nucleus is independent of the decay of its neighbors. This assumption then permits us to have confidence that the decay constant we measure in the laboratory can be compared with the result of our quantum mechanical calculation.

## 6.3  PRODUCTION AND DECAY OF RADIOACTIVITY

It quite frequently happens that a basic condition imposed in deriving the exponential law, that no new nuclei are introduced into the sample, is not valid. In solving Equation 6.1 we obtained a fixed number $N_0$ of nuclei present at $t = 0$. In many applications, however, we produce activity continuously, such as by a nuclear reaction. In this case, Equation 6.2 is no longer valid and we must consider in more detail the processes that occur in the production and decay of the activity.

Let's assume that we place a target of stable nuclei into a reactor or an accelerator such as a cyclotron. The nuclei of the target will capture a neutron or a charged particle, possibly leading to the production of a radioactive species. The rate $R$ at which this occurs will depend on the number $N_0$ of target atoms present, on the flux or current $I$ of incident particles, and on the reaction cross section $\sigma$ (which measures the probability for one incident particle to react with one target nucleus). A typical flux of particles in a reactor or cyclotron might be of the order of $10^{14}/s \cdot cm^2$, and typical cross sections are at most of the order of barns ($10^{-24}\ cm^2$). Thus the probability to convert a target particle from stable to radioactive is about $10^{-10}/s$. Even if the reaction is allowed to continue for hours, the absolute number of converted target particles is small (say, less than $10^{-6}$ of the original number). *We can therefore, to a very good approximation, regard the number of target nuclei as constant*, and under this approximation the rate $R$ is constant. (As we "burn up" target nuclei, $N_0$ will decrease by a small amount and the rate may therefore similarly decrease with time. Obviously $N_0$ must go to zero as $t \rightarrow \infty$, but for ordinary reaction times and typical cross sections we ignore this very small effect.) Thus

$$R = N_0 \sigma I \qquad (6.21)$$

is taken to be a constant giving the rate at which the radioactive product nuclei are formed.

Let's denote by $N_1$ the number of radioactive nuclei that are formed as a result of the reaction. These nuclei decay with decay constant $\lambda_1$ to the stable nuclei denoted by $N_2$. Thus the number of nuclei $N_1$ present *increases* owing to the production at the rate $R$ and *decreases* owing to the radioactive decay:

$$dN_1 = R\,dt - \lambda_1 N_1\,dt \qquad (6.22)$$

and the solution to this equation is easily obtained

$$N_1(t) = \frac{R}{\lambda_1}(1 - e^{-\lambda_1 t}) \qquad (6.23)$$

and

$$\mathscr{A}_1(t) = \lambda_1 N_1(t) = R(1 - e^{-\lambda_1 t}) \qquad (6.24)$$

If the irradiation time is short compared with one half-life, then we can expand the exponential and keep only the term linear in $t$:

$$\mathscr{A}_1(t) \cong R\lambda_1 t \qquad t \ll t_{1/2} \qquad (6.25)$$

For small times, the activity thus increases at a constant rate. This corresponds to the linear (in time) accumulation of product nuclei, whose number is not yet seriously depleted by radioactive decays.

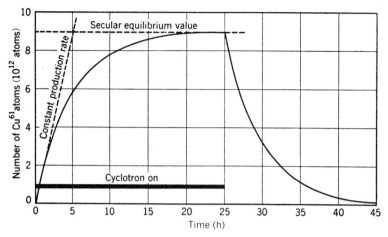

**Figure 6.5** A plot of the number of radioactive $^{61}$Cu atoms present in a Ni target at various times during and after bombardment with deuterons in a cyclotron.

For times long compared with the half-life the exponential approaches zero and the activity is approximately constant:

$$\mathscr{A}_1(t) \cong R \qquad t \gg t_{1/2} \qquad (6.26)$$

In this case new activity is being formed at the same rate at which the older activity decays. This is an example of *secular equilibrium* which we discuss in more detail in the next section.

If we irradiate the sample for a time $t_1$ and then remove it from the accelerator or reactor, it will decay according to the simple exponential law, since no new activity is being formed. Figure 6.5 shows the activity resulting from the deuteron bombardment of $^{61}$Ni to form $^{61}$Cu ($t_{1/2} = 3.4$ h).

From Equation 6.24 we see that we produce 75% of the maximum possible activity by irradiating for two half-lives and 87.5% by irradiating for three half-lives. Further irradiation increases the activity by a steadily diminishing amount, so that we gain relatively little additional activity by irradiating for more than 2–3 half-lives. In fact, since the cost of using a reactor or accelerator is usually in direct proportion to the irradiation time, the best value (maximum activity per unit cost) is obtained by remaining close to the linear regime ($t \ll t_{1/2}$).

## 6.4 GROWTH OF DAUGHTER ACTIVITIES

Another common situation occurs when a radioactive decay results in a product nucleus that is also radioactive. It is thus possible to have series or chains of radioactive decays $1 \rightarrow 2 \rightarrow 3 \rightarrow 4 \ldots$, and it has become common to refer to the original nucleus (type 1) as the parent and the succeeding "generations" as daughter (type 2), granddaughter (type 3), and so on.

We assume that we begin with $N_0$ atoms of the parent at $t = 0$ and that no atoms of the decay products are originally present:

$$N_1(t = 0) = N_0$$
$$N_2(t = 0) = N_3(t - 0) = \cdots = 0 \qquad (6.27)$$

The various decay constants are represented by $\lambda_1, \lambda_2, \lambda_3, \ldots$ . For the present calculation, we will assume that the granddaughter is the stable end-product of the decay. The number of parent nuclei decreases with time according to the usual form

$$dN_1 = -\lambda_1 N_1 \, dt \qquad (6.28)$$

The number of daughter nuclei increases as a result of decays of the parent and decreases as a result of its own decay:

$$dN_2 = \lambda_1 N_1 \, dt - \lambda_2 N_2 \, dt \qquad (6.29)$$

The number of parent nuclei can be found directly from integrating Equation 6.28:

$$N_1(t) = N_0 e^{-\lambda_1 t} \qquad (6.30)$$

To solve Equation 6.29, we try a solution of the form $N_2(t) = A e^{-\lambda_1 t} + B e^{-\lambda_2 t}$ and by substituting into Equation 6.29 and using the initial condition $N_2(0) = 0$ we find

$$N_2(t) = N_0 \frac{\lambda_1}{\lambda_2 - \lambda_1} (e^{-\lambda_1 t} - e^{-\lambda_2 t}) \qquad (6.31)$$

$$\mathscr{A}_2(t) \equiv \lambda_2 N_2(t) = N_0 \frac{\lambda_2 \lambda_1}{\lambda_2 - \lambda_1} (e^{-\lambda_1 t} - e^{-\lambda_2 t}) \qquad (6.32)$$

Note that Equation 6.31 reduces to Equation 6.11b if nuclei of type 2 are stable ($\lambda_2 \rightarrow 0$). We can also include the results of the previous section as a special case of Equation 6.31. Let's suppose that $\lambda_1$ is very small (but not quite zero), so that $N_1 \simeq N_0 - N_0\lambda_1 t$. In a nuclear reaction, the number of target nuclei decreases at the rate $R$ according to $N_0 - Rt$, and thus identifying $N_0\lambda_1$ with $R$ and neglecting $\lambda_1$ in comparison with $\lambda_2$, Equation 6.31 reduces to Equation 6.24 for the activity of type 2.

## $\lambda_1 \ll \lambda_2$

In this case the parent is so long-lived that it decays at an essentially constant rate; for all practical times $e^{-\lambda_1 t} \simeq 1$ and

$$N_2(t) \cong N_0 \frac{\lambda_1}{\lambda_2} (1 - e^{-\lambda_2 t}) \qquad (6.33)$$

which is of the same form as Equation 6.24. Thus the activity $\mathscr{A}_2$ approaches the limiting value $N_0\lambda_1$ as was shown in Figure 6.5.

This is another example of *secular equilibrium*, where as $t$ becomes large nuclei of type 2 are decaying at the same rate at which they are formed: $\lambda_2 N_2 = \lambda_1 N_1$. (Note that Equation 6.29 shows immediately that $dN_2/dt = 0$ in this case.) Figure 6.6 shows an example of approximate secular equilibrium.

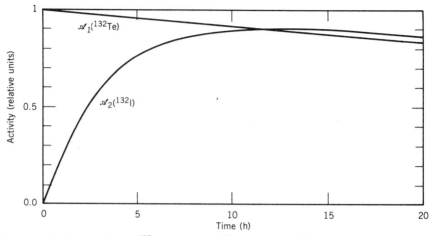

**Figure 6.6**   In the decay $^{132}$Te (78 h) $\rightarrow$ $^{132}$I (2.28 h) $\rightarrow$ $^{132}$Xe, approximate secular equilibrium is reached at about 12 h.

### $\lambda_1 < \lambda_2$

From Equations 6.30 and 6.31 we can calculate the ratio of the two activities:

$$\frac{\lambda_2 N_2}{\lambda_1 N_1} = \frac{\lambda_2}{\lambda_2 - \lambda_1}\left(1 - e^{-(\lambda_2 - \lambda_1)t}\right) \qquad (6.34)$$

As $t$ increases, the exponential term becomes smaller and the ratio $\mathscr{A}_2/\mathscr{A}_1$ approaches the limiting constant value $\lambda_2/(\lambda_2 - \lambda_1)$. The activities themselves are not constant, but the nuclei of type 2 decay (in effect) with the decay constant

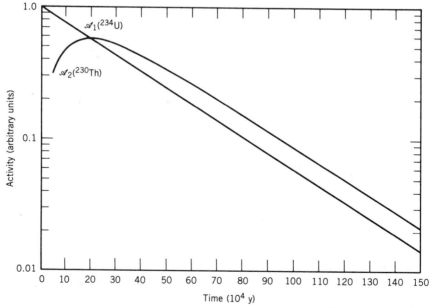

**Figure 6.7**   An example of equilibrium in the decays of $^{234}$U (2.45 $\times$ 10$^5$ y) to $^{230}$Th (8.0 $\times$ 10$^4$ y). The ratio $\mathscr{A}_2/\mathscr{A}_1$ approaches the constant value 1.48.

of type 1. This situation is known as *transient equilibrium* and is illustrated in Figure 6.7.

## $\lambda_1 > \lambda_2$

In this case the parent decays quickly, and the daughter activity rises to a maximum and then decays with its characteristic decay constant. When this occurs the number of nuclei of type 1 is small and nearly insignificant. If $t$ is so long that $e^{-\lambda_1 t}$ effectively vanishes, then Equation 6.31 becomes

$$N_2(t) \cong N_0 \frac{\lambda_1}{\lambda_1 - \lambda_2} e^{-\lambda_2 t} \tag{6.35}$$

which reveals that the type 2 nuclei decay approximately according to the exponential law.

### Series of Decays

If we now assume that there are several succeeding generations of radioactive nuclei (that is, the granddaughter nuclei type 3 are themselves radioactive, as are types $4, 5, 6, \ldots$), we can then easily generalize Equation 6.29 since each species is populated by the preceding one:

$$dN_i = \lambda_{i-1} N_{i-1}\, dt - \lambda_i N_i\, dt \tag{6.36}$$

A general solution, for the case of $N_0$ nuclei of type 1 and none of the other types initially present, is given by the *Bateman equations*, in which the activity of the $n$th member of the chain is given in terms of the decay constants of all preceding members:

$$
\begin{aligned}
\mathcal{A}_n &= N_0 \sum_{i=1}^{n} c_i e^{-\lambda_i t} \\
&= N_0 \left( c_1 e^{-\lambda_1 t} + c_2 e^{-\lambda_2 t} + \cdots + c_n e^{-\lambda_n t} \right)
\end{aligned} \tag{6.37}
$$

where

$$
\begin{aligned}
c_m &= \frac{\displaystyle\prod_{i=1}^{n} \lambda_i}{\displaystyle\prod_{i=1}^{n}{}' (\lambda_i - \lambda_m)} \\
&= \frac{\lambda_1 \lambda_2 \lambda_3 \cdots \lambda_n}{(\lambda_1 - \lambda_m)(\lambda_2 - \lambda_m) \cdots (\lambda_n - \lambda_m)}
\end{aligned} \tag{6.38}
$$

where the prime on the lower product indicates we omit the term with $i = m$.

It is also possible to have secular equilibrium in this case, with $\lambda_1 N_1 = \lambda_2 N_2 = \cdots = \lambda_n N_n$.

## 6.5  TYPES OF DECAYS

The three primary decay types, to be discussed in greater detail in Chapters 8, 9, and 10, are $\alpha$, $\beta$, and $\gamma$ decays. In $\alpha$- and $\beta$-decay processes, an unstable nucleus emits an $\alpha$ or a $\beta$ particle as it tries to become a more stable nucleus (that is, to approach the most stable isobar for the resulting mass number). In $\gamma$-decay processes, an excited state decays toward the ground state without changing the nuclear species.

## $\alpha$ Decay

In this process, a nucleus emits an $\alpha$ particle (which Rutherford and his co-workers showed to be a nucleus of helium, $^4_2\text{He}_2$). The $^4\text{He}$ nucleus is chosen as the agent for this process because it is such a tightly bound system, and thus the kinetic energy released in the decay is maximized. Such decays are favored, as we shall discuss in Chapter 8. The decay process is

$$^A_Z\text{X}_N \rightarrow ^{A-4}_{Z-2}\text{X}'_{N-2} + ^4_2\text{He}_2$$

where X and X' represent the chemical symbols of the initial and final nuclei. Notice that the number of protons and the number of neutrons must separately be conserved in the decay process. An example of an $\alpha$-decay process is

$$^{226}_{88}\text{Ra}_{138} \rightarrow ^{222}_{86}\text{Rn}_{136} + \alpha$$

in which the half-life is 1600 years and the $\alpha$ particle appears with a kinetic energy of about 4.8 MeV.

## $\beta$ Decay

Here the nucleus can correct a proton or a neutron excess by directly converting a proton into a neutron or a neutron into a proton. This process can occur in three possible ways, each of which must involve another charged particle to conserve electric charge (the charged particle, originally called a $\beta$ particle, was later shown to be identical with ordinary electrons).

$$\text{n} \rightarrow \text{p} + \text{e}^- \qquad \beta^- \text{ decay}$$

$$\text{p} \rightarrow \text{n} + \text{e}^+ \qquad \beta^+ \text{ decay}$$

$$\text{p} + \text{e}^- \rightarrow \text{n} \qquad \text{electron capture } (\varepsilon)$$

The first process is known as negative $\beta$ decay or negatron decay and involves the creation and emission of an ordinary electron. The second process is positive $\beta$ decay or positron decay, in which a positively charged electron is emitted. In the third process, an atomic electron that strays too close to the nucleus is swallowed, allowing the conversion of a proton to a neutron.

In all three processes, yet another particle called a *neutrino* is also emitted, but since the neutrino has no electric charge, its inclusion in the decay process does not affect the identity of the other final particles.

Note that in positive and negative $\beta$ decay, a particle is created (out of the decay energy, according to $m = E/c^2$). The electron or positron did not exist inside the nucleus before the decay. (Contrast the case of $\alpha$ decay, in which the emitted nucleons were inside the nucleus before the decay.)

Some representative $\beta$-decay processes are

$$\underset{53}{^{131}}\text{I}_{78} \underset{\beta^-}{\rightarrow} \underset{54}{^{131}}\text{Xe}_{77} \qquad t_{1/2} = 8.0 \text{ d}$$

$$\underset{13}{^{25}}\text{Al}_{12} \underset{\beta^+}{\rightarrow} \underset{12}{^{25}}\text{Mg}_{13} \qquad t_{1/2} = 7.2 \text{ s}$$

$$\underset{25}{^{54}}\text{Mn}_{29} \underset{\varepsilon}{\rightarrow} \underset{24}{^{54}}\text{Cr}_{30} \qquad t_{1/2} = 312 \text{ d}$$

In these processes, $Z$ and $N$ each change by one unit, but the total mass number $Z + N$ remains constant.

### $\gamma$ Decay

Radioactive $\gamma$ emission is analogous to the emission of atomic radiations such as optical or X-ray transitions. An excited state decays to a lower excited state or possibly the ground state by the emission of a photon of $\gamma$ radiation of energy equal to the difference in energy between the nuclear states (less a usually negligible correction for the "recoil" energy of the emitting nucleus). Gamma emission is observed in all nuclei that have excited bound states ($A > 5$), and usually follows $\alpha$ and $\beta$ decays since those decays will often lead to excited states in the daughter nucleus.

The half-lives for $\gamma$ emission are usually quite short, generally less than $10^{-9}$ s, but occasionally we find half-lives for $\gamma$ emission that are significantly longer, even hours or days. These transitions are known as *isomeric transitions* and the long-lived excited states are called *isomeric states* or *isomers* (or sometimes *metastable* states). There is no clear criterion for classifying a state as isomeric or not; the distinction was originally taken to be whether or not the half-life was directly measurable, but today we can measure half-lives well below $10^{-9}$ s. Clearly a state with $t_{1/2} = 10^{-6}$ s is an isomer and one with $t_{1/2} = 10^{-12}$ s is not, but in between the boundary is rather fuzzy. We usually indicate metastable states with a superscript m, thus: $^{110}\text{Ag}^m$ or $^{110m}\text{Ag}$.

A process that often competes with $\gamma$ emission is *internal conversion*, in which the nucleus de-excites by transferring its energy directly to an *atomic* electron, which then appears in the laboratory as a free electron. (This is very different from $\beta$ decay in that no change of $Z$ or $N$ occurs, although the atom becomes ionized in the process.)

### Spontaneous Fission

We usually think of fission as occurring under very unnatural and artificial conditions, such as in a nuclear reactor. There are, however, some nuclei that fission spontaneously, as a form of radioactive decay. This process is similar to the neutron-induced fission that occurs in reactors, with the exception that no previous neutron capture is needed to initiate the fission. In the process, a heavy nucleus with an excess of neutrons splits roughly in half into two lighter nuclei; the final nuclei are not rigidly determined, as they are in $\alpha$ or $\beta$ decay, but are statistically distributed over the entire range of medium-weight nuclei. Examples of spontaneously fissioning nuclei are $^{256}$Fm ($t_{1/2} = 2.6$ h) and $^{254}$Cf ($t_{1/2} = 60.5$ days).

## Nucleon Emission

As we move further and further from the "valley" of stable nuclei, the energy differences between neighboring isobars increases (recall the mass parabolas of constant $A$ of Figure 3.18). Eventually the difference exceeds the nucleon binding energy (about 8 MeV, on the average) and it becomes possible to have radioactive decay by nucleon emission. This type of decay occurs most frequently in fission products, which have a very large neutron excess, and it is responsible for the "delayed" neutrons (that is, delayed by the half-life of the decay) that are used to control nuclear reactors. For example, $^{138}$I $\beta$ decays with a half-life of 6.5 s to $^{138}$Xe. Most of the $\beta$ decays populate low excited states in $^{138}$Xe, but about 5% of the $^{138}$I decays populate states in $^{138}$Xe at about 6.5 MeV; these states decay by direct neutron emission to $^{137}$Xe. Similarly, 0.7% of the $^{73}$Kr $\beta^+$ decays ($t_{1/2} = 27$ s) go to states in $^{73}$Br at about 5 MeV; these states decay by proton emission to states in $^{72}$Se.

## Branching Ratios and Partial Half-lives

Figure 6.8 summarizes a variety of different decay processes, and Figure 6.9 shows a small section of the chart of stable and radioactive nuclei (Figure 1.1) with several decay processes indicated. Some nuclei may decay only through a single process, but more often decay schemes are very complicated, involving the emission of $\alpha$'s, $\beta$'s, and $\gamma$'s in competing modes. We specify the relative intensities of the competing modes by their *branching ratios*. Thus $^{226}$Ra $\alpha$ decays to the ground state of $^{222}$Rn with a branching ratio of 94% and to the first excited state with a branching ratio of 6%. Often different decay modes can compete: $^{226}$Ac decays by $\alpha$ emission (0.006%), $\beta^-$ emission (83%), and $\varepsilon$ (17%); $^{132}$Cs decays by $\beta^-$ emission (2%) and by $\beta^+$ and $\varepsilon$ (98%); the metastable state $^{95m}$Nb decays by $\beta^-$ emission (2.5%) or by an isomeric transition (97.5%). The isomeric transition itself includes a 27% branch by $\gamma$ emission and a 73% branch by internal conversion.

Frequently, we specify the branching ratio by giving the partial decay constant or partial half-life. For example, we consider the decay of $^{226}$Ac ($t_{1/2} = 29$ h). The total decay constant is

$$\lambda_t = \frac{0.693}{t_{1/2}} = 0.024 \text{ h}^{-1} = 6.6 \times 10^{-6} \text{ s}^{-1}$$

The partial decay constants are

$$\lambda_\beta = 0.83\lambda_t = 5.5 \times 10^{-6} \text{ s}^{-1}$$

$$\lambda_\varepsilon = 0.17\lambda_t = 1.1 \times 10^{-6} \text{ s}^{-1}$$

$$\lambda_\alpha = 6 \times 10^{-5}\lambda_t = 4 \times 10^{-10} \text{ s}^{-1}$$

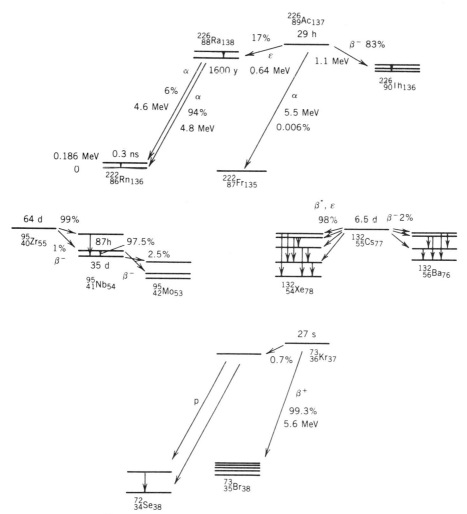

**Figure 6.8**   A variety of different decay processes.

and the partial half-lives are

$$t_{1/2,\beta} = \frac{0.693}{\lambda_\beta} = 1.3 \times 10^5 \, \text{s} = 35 \, \text{h}$$

$$t_{1/2,\varepsilon} = \frac{0.693}{\lambda_\varepsilon} = 6.1 \times 10^5 \, \text{s} = 170 \, \text{h}$$

$$t_{1/2,\alpha} = \frac{0.693}{\lambda_\alpha} = 1.7 \times 10^9 \, \text{s} = 55 \, \text{y}$$

The partial half-life is merely a convenient way to represent branching ratios; a glance at the above figures shows that $\alpha$ emission is far less probable than $\beta$ emission for $^{226}$Ac. *However, the activity would be observed to decay only with the total half-life.* Even if we were to observe the decay of $^{226}$Ac by its $\alpha$ emission, the

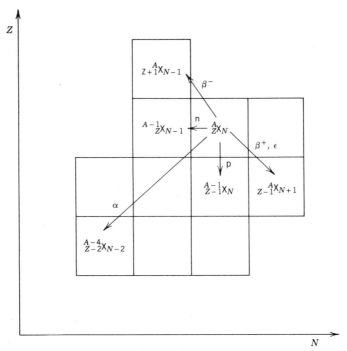

**Figure 6.9** The initial nucleus ${}^A_Z X_N$ can reach different final nuclei through a variety of possible decay processes.

activity would decay with time according to a half-life of 29 h. (Imagine if this were not so, and two observers were studying the decay of ${}^{226}$Ac, one by observing the $\beta$'s and the other by observing the $\alpha$'s. Since the radioactive decay law gives the number of undecayed nuclei, the $\beta$ observer would conclude that half of the original ${}^{226}$Ac nuclei remained after 35 h, while the $\alpha$ observer would have to wait 55 years similarly to observe half of the nuclei undecayed! In reality, half of the nuclei decay every 29 h, no matter what method we use to observe those decays.)

## 6.6 NATURAL RADIOACTIVITY

The Earth and the other planets of our solar system formed about $4.5 \times 10^9$ y ago out of material rich in iron, carbon, oxygen, silicon, and other medium and heavy elements. These elements in turn were created from the hydrogen and helium that resulted from the Big Bang some $15 \times 10^9$ y ago. During the $10 \times 10^9$ y from the Big Bang until the condensation of the solar system, the hydrogen and helium were "cooked" into heavier elements in stellar interiors, novas, and supernovas; we are made of the recycled debris of these long dead stars. Most of the elements thus formed were radioactive, but have since decayed to stable nuclei. A few of the radioactive elements have half-lives that are long compared with the age of the Earth, and so we can still observe their radioactivity. This radioactivity forms the major portion of our natural radioactive environment, and is also probably responsible for the inner heating of the terrestrial planets.

**Table 6.1** Some Characteristics of the Disintegration Series
of the Heavy Elements

| Name of Series | Type[a] | Final Nucleus (Stable) | Longest-Lived Member | |
|---|---|---|---|---|
| | | | Nucleus | Half-Life, (y) |
| Thorium | $4n$ | $^{208}$Pb | $^{232}$Th | $1.41 \times 10^{10}$ |
| Neptunium | $4n + 1$ | $^{209}$Bi | $^{237}$Np | $2.14 \times 10^{6}$ |
| Uranium | $4n + 2$ | $^{206}$Pb | $^{238}$U | $4.47 \times 10^{9}$ |
| Actinium | $4n + 3$ | $^{207}$Pb | $^{235}$U | $7.04 \times 10^{8}$ |

[a]$n$ is an integer.

Although there are long-lived natural radioactive elements of other varieties, most of those observed today originate with the very heavy elements, which have no stable isotopes at all. These nuclides decay by $\alpha$ and $\beta$ emission, decreasing $Z$ and $A$ until a lighter, stable nucleus is finally reached. Alpha decay changes $A$ by four units and $\beta$ decay does not change $A$ at all, and so therefore we have four independent decay chains with mass numbers $4n$, $4n + 1$, $4n + 2$, and $4n + 3$, where $n$ is an integer. The decay processes will tend to concentrate the nuclei in the longest-lived member of the chain, and if the lifetime of that nuclide is at least of the order of the age of the Earth, we will observe that activity today. The four series are listed in Table 6.1. Notice that the longest-lived member of the neptunium series has far too short a half-life to have survived since the formation of the Earth; this series is not observed in natural material.

Consider, for example, the thorium series illustrated in Figure 6.10. Let us assume that we had created, in a short period of time, a variety of plutonium (Pu) isotopes. The isotopes $^{232}$Pu and $^{236}$Pu decay rapidly to 72-y $^{232}$U and other species of much shorter half-lives. Thus in a time long compared with 72 y (say, $10^3$ y), all traces of these isotopes have vanished, leaving only the stable end product $^{208}$Pb. The isotopes $^{240}$Pu and $^{244}$Pu decay much more slowly, the former comparatively quickly and the latter very slowly to $^{236}$U, which in turn decays to the longest-lived member of the series, $^{232}$Th. In a time greater than $81 \times 10^6$ y but less than $14 \times 10^9$ y, the original $^{240}$Pu and $^{244}$Pu (and the intermediate $^{236}$U) will all have decayed to $^{232}$Th, the decay of which we still observe today.

These radioactive isotopes are present in material all around us, especially in rocks and minerals that condensed with the Earth $4.5 \times 10^9$ y ago. (In fact, their decays provide a reliable technique for determining the time since the condensation of the rocks and thus the age of the Earth; see Section 6.7 and Chapter 19 for discussions of these techniques.) In general the radioactive elements are tightly bound to the minerals and are not hazardous to our health, but all of the natural radioactive series involve the emission of a gaseous radioactive element, radon. This element, if formed deep within rocks, normally has little chance to migrate to the surface and into the air before it decays. However, when rocks are fractured, the radon gas can escape (in fact the presence of radon gas has in recent years been observed as a precursor of earthquakes). There is also the possibility of escape of radon from the surface of minerals, and particularly those that are used in the construction of buildings. Inhalation of this radioactive gas

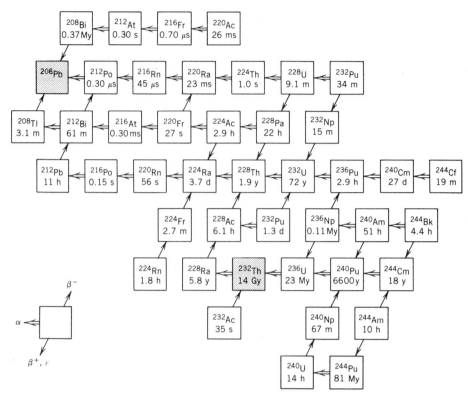

**Figure 6.10** The thorium series of naturally occurring radioactive decays. Some half-lives are indicated in My ($10^6$ y) and Gy ($10^9$ y). The shaded members are the longest-lived radioactive nuclide in the series (Th, after which the series is named) and the stable end product.

**Table 6.2** Some Natural Radioactive Isotopes

| Isotope | $t_{1/2}$ (y) |
|---------|---------------|
| $^{40}$K | $1.28 \times 10^9$ |
| $^{87}$Rb | $4.8 \times 10^{10}$ |
| $^{113}$Cd | $9 \times 10^{15}$ |
| $^{115}$In | $4.4 \times 10^{14}$ |
| $^{138}$La | $1.3 \times 10^{11}$ |
| $^{176}$Lu | $3.6 \times 10^{10}$ |
| $^{187}$Re | $5 \times 10^{10}$ |

could possibly be responsible for many lung cancers, and there is a current suspicion that smoking may accelerate this process by causing the accumulation of these radioactive products in the lungs. It is perhaps ironic that the recent trends toward well insulated and tightly sealed buildings to conserve energy may be responsible for an increased concentration of radon gas, and as of this writing there is active research on the problem, including measurement of radon gas accumulation in buildings.

The heavy element series are not the only sources of naturally occurring radioactive isotopes of half-lives long enough to be present in terrestrial matter. Table 6.2 gives a partial list of others, some of which can also be used for radioactive dating.

There are also other natural sources of radioactivity of relatively short half-lives, which are not remnants of the production of elements before the Earth formed, but instead are being formed continuously today. These elements include $^3$II and $^{14}$C, which are formed in the upper atmosphere when cosmic rays (high-energy protons) strike atoms of the atmosphere and cause nuclear reactions. The isotope $^{14}$C has had important applications in radioactive dating.

## 6.7  RADIOACTIVE DATING

Although we cannot predict with certainty when an individual nucleus will decay, we can be very certain how long it will take for half of a large number of nuclei to decay. These two statements may seem inconsistent; their connection has to do with the statistical inferences that we can make by studying random processes. If we have a room containing a single gas molecule, we cannot predict with certainty whether it will be found in the left half of the room or the right half. If however we have a room containing a large number $N$ of molecules ($N \sim 10^{24}$), then we expect to find on the average $N/2$ molecules in each half. Furthermore, the fluctuations of the number in each half about the value $N/2$ are of the order of $\sqrt{N}$; thus the deviation of the fraction in each half from the value 0.5 is about $\sqrt{N}/N \simeq 10^{-12}$. The fraction in each half is thus $0.500000000000 \pm 0.0000000000001$. This extreme (and unreasonable) precision comes about because $N$ is large and thus the fractional error $N^{-1/2}$ is small.

A similar situation occurs for radioactive decay. (The laws of counting statistics are discussed in detail in Chapter 7.) If we had at $t = 0$ a collection of a large number $N_0$ of radioactive nuclei, then after a time equal to one half-life, we should find that the remaining fraction is $\frac{1}{2} \pm N_0^{-1/2}$. Thus despite the apparently random nature of the decay process, the decay of radioactive nuclei gives us a very accurate and entirely reliable clock for recording the passage of time. That is, if we know the decay constant $\lambda$, the exponential decrease in activity of a sample can be used to measure time.

The difficulty in using this process occurs when we try to apply it to decays that occur over geological times ($\sim 10^9$ y) because in this case we do not measure the activity as a function of time. Instead, we use the relative number of parent and daughter nuclei observed at time $t_1$ (now) compared with the relative number at time $t_0$ (when the "clock" started ticking, usually when the material such as a rock or mineral condensed, trapping the parent nuclei in their present sites). In principle this process is rather simple. Given the decay of parent isotope P to daughter isotope D, we merely count (by chemical means, for instance) the present numbers of P and D atoms, $N_P(t_1)$ and $N_D(t_1)$:

$$N_D(t_1) + N_P(t_1) = N_P(t_0) \tag{6.39}$$

$$N_P(t_1) = N_P(t_0) \, e^{-\lambda(t_1 - t_0)} \tag{6.40}$$

$$\Delta t \equiv t_1 - t_0 = \frac{1}{\lambda} \ln \frac{N_P(t_0)}{N_P(t_1)}$$

$$\Delta t = \frac{1}{\lambda} \ln \left( 1 + \frac{N_D(t_1)}{N_P(t_1)} \right) \tag{6.41}$$

Given the decay constant (which we can measure in the laboratory) and the present ratio of daughter to parent nuclei, the age of the sample is directly found, with a precision determined by our knowledge of $\lambda$ and by the counting statistics for $N_P$ and $N_D$.

Equations 6.39 and 6.40 contain assumptions that must be carefully tested before we can apply Equation 6.41 to determine the age of a sample. Equation 6.39 assumes that $N_D(t_0) = 0$—no daughter atoms are present at $t_0$—and also that the total number of atoms remains constant—no parent or daughter atoms escape from the mineral or solid in which they were originally contained. As we discuss below, we can modify the derivation of $\Delta t$ to account for the daughter atoms present at $t_0$ (even though when we analyze the sample today at time $t_1$, we cannot tell which daughter atoms were originally present and which resulted from decays during $\Delta t$). Equation 6.40 assumes that the variation in $N_P$ comes only from the decay—no new parent atoms are introduced (by a previous decay or by nuclear reactions induced by cosmic rays, for example).

Let's relax the assumption of Equation 6.39 and permit daughter nuclei to be present at $t = t_0$. These daughter nuclei can be formed from the decay of parent nuclei at times before $t_0$ or from the process that formed the original parent nuclei (a supernova explosion, for example); the means of formation of these original daughter nuclei is of no importance for our calculation. We therefore take

$$N_D(t_1) + N_P(t_1) = N_D(t_0) + N_P(t_0) \tag{6.42}$$

Because we have introduced another unknown, $N_D(t_0)$, we can no longer solve directly for the age $\Delta t$. If, however, there is also present in the sample a different isotope of the daughter, D′, which is neither radioactive nor formed from the decay of a long-lived parent, we can again find the age of the sample. The population of this stable isotope is represented by $N_{D'}$, and if D′ is stable then $N_{D'}(t_1) = N_{D'}(t_0)$, in which case

$$\frac{N_D(t_1) + N_P(t_1)}{N_{D'}(t_1)} = \frac{N_D(t_0) + N_P(t_0)}{N_{D'}(t_0)} \tag{6.43}$$

which can be written as

$$\frac{N_D(t_1)}{N_{D'}(t_1)} = \frac{N_P(t_1)}{N_{D'}(t_1)} [e^{\lambda(t_1 - t_0)} - 1] + \frac{N_D(t_0)}{N_{D'}(t_0)} \tag{6.44}$$

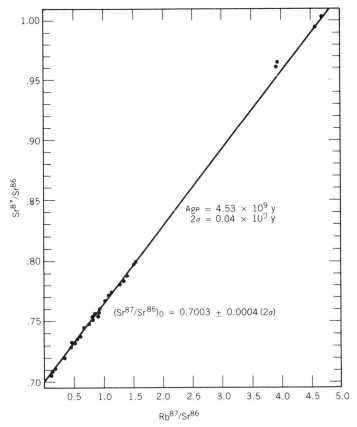

**Figure 6.11** The Rb-Sr dating method, allowing for the presence of some initial $^{87}$Sr. The linear behavior is consistent with Equation 6.44. From G. W. Wetherill, *Ann. Rev. Nucl. Sci.* **25**, 283 (1975).

The ratios $N_D(t_1)/N_{D'}(t_1)$ and $N_P(t_1)/N_{D'}(t_1)$ can be measured in the laboratory, but that still leaves two unknowns in Equation 6.44: the age $\Delta t$ and the initial isotopic ratio $N_D(t_0)/N_{D'}(t_0)$. Minerals that crystallize from a common origin should show identical ages and identical isotopic ratios $N_D(t_0)/N_{D'}(t_0)$, even though the original $N_P(t_0)$ may be very different (from differing chemical compositions, for example). If these hypotheses are correct, we expect to observe today minerals with various ratios $N_D(t_1)/N_{D'}(t_1)$ and $N_P(t_1)/N_{D'}(t_1)$ corresponding to common values of $\Delta t$ and $N_D(t_0)/N_{D'}(t_0)$. We can test these assumptions by plotting $y = N_D(t_1)/N_{D'}(t_1)$ against $x = N_P(t_1)/N_{D'}(t_1)$ for a variety of minerals. Equation 6.44 is of the form $y = mx + b$, a straight line with slope $m = e^{\lambda(t_1 - t_0)} - 1$ and intercept $b = N_D(t_0)/N_{D'}(t_0)$. Figure 6.11 is an example of such a procedure for the decay $^{87}\text{Rb} \rightarrow {}^{87}\text{Sr}$ ($t_{1/2} = 4.8 \times 10^{10}$ y), in which the comparison is done with stable $^{86}$Sr. Even though the present ratio of $^{87}$Rb to $^{86}$Sr varies by more than an order of magnitude, the data indicate a common age of the Earth, $\Delta t = 4.5 \times 10^9$ y. The good linear fit is especially important, for it justifies our assumptions of no loss of parent or daughter nuclei.

Other similar methods for dating minerals from the Earth, Moon, and meteorites give a common age of $4.5 \times 10^9$ y. These methods include the decay of $^{40}$K to $^{40}$Ar, the decay of $^{235}$U and $^{238}$U to $^{207}$Pb and $^{206}$Pb, and the spontaneous fission of $^{238}$U and $^{244}$Pu, which are analyzed either by chemical separation of the fission products or by microscopic observation of the tracks left in the minerals by the fission fragments.

For dating more recent samples of organic matter, the $^{14}$C dating method is used. The $CO_2$ that is absorbed by organic matter consists almost entirely of stable $^{12}$C (98.89%), with a small mixture of stable $^{13}$C (1.11%). Radioactive $^{14}$C is being continuously formed in the upper atmosphere as a result of cosmic-ray bombardment of atmospheric nitrogen, and thus all living matter is slightly radioactive owing to its $^{14}$C content. Because the production rate of $^{14}$C by cosmic rays has been relatively constant for thousands of years, living organic material reaches equilibrium of its carbon with atmospheric carbon, with about 1 atom of $^{14}$C for every $10^{12}$ atoms of $^{12}$C. The half-life of $^{14}$C is 5730 y, and thus each gram of carbon shows an activity of about 15 decays per minute. When an organism dies, it goes out of equilibrium with atmospheric carbon; it stops acquiring new $^{14}$C and its previous content of $^{14}$C decreases according to the radioactive decay law. We can therefore determine the age of samples by measuring the *specific activity* (activity per gram) of their carbon content. This method applies as long as we have enough $^{14}$C intensity to determine the activity; from matter that has decayed for 10 or more half-lives, the decay is so weak that the $^{14}$C method cannot be used. Recent techniques using accelerators as mass spectrometers have the potential to exceed this limit by counting $^{14}$C atoms directly; these techniques are discussed in Chapter 20.

The major assumption of this method is the relatively constant production of $^{14}$C by cosmic rays over the last 50,000 y or so. We can test this assumption by comparing the ages determined by radiocarbon dating with ages known or determined by independent means (historical records or tree-ring counting, for example). These comparisons show very good agreement and support the assumption of a relatively uniform flux of cosmic rays.

In later millennia, the radiocarbon method may no longer be applicable. During the last 100 years, the burning of fossil fuels has upset the atmospheric balance by diluting the atmosphere with stable carbon (the hydrocarbons of fossil fuels are old enough for all of their $^{14}$C to have decayed away). During the 1950s and 1960s, atmospheric testing of nuclear weapons has placed additional $^{14}$C in the atmosphere, perhaps doubling the concentration over the equilibrium value from cosmic-ray production alone.

## 6.8   UNITS FOR MEASURING RADIATION

The activity of a radioactive sample (in curies or in decays per second) does not depend on the type of radiation or on its energy. Thus the activity may be a useful means to compare two different sources of the same decaying isotope (10 mCi of $^{60}$Co is stronger than 1 mCi of $^{60}$Co), but how can we compare different decays? For instance, how does a 10-mCi source of $^{60}$Co compare in strength with 10 mCi of $^{14}$C, or how does a 10-$\mu$Ci $\gamma$ emitter compare in strength with a 10-mCi $\alpha$ emitter? And just what exactly do we mean by the "strength" of a source of radiation?

One common property of nuclear radiations is their ability to *ionize* (knock electrons from) atoms with which they interact. (For this reason, nuclear radiation is often called *ionizing radiation*.) We begin by considering the passage of X-ray and γ-ray photons through air. The photons interact many times with atoms in the air through a variety of processes (Compton scattering, photoelectric effect, electron-positron pair production), each of which creates a free electron, often of reasonably high energy. These secondary electrons can themselves produce ionization (and additional electrons). The total electric charge $Q$ on the ions produced in a given mass $m$ of air is called the *exposure X*, and we may take γ-ray sources as being of the same strength if they result in the same exposure, even though the energies of the γ rays and the activities of the sources may be quite different. Specifically, the exposure is

$$X = \frac{Q}{m} \tag{6.45}$$

and is measured in the SI units of coulomb per kilogram. More frequently we encounter the *roentgen* unit (R), which is defined as the exposure resulting in an ionization charge of 1 electrostatic unit (the cgs unit of electric charge, in terms of which the electronic charge $e$ is $4.80 \times 10^{-10}$ electrostatic unit) in 1 cm³ of air at 0°C and 760 mm pressure (corresponding to a mass of 0.001293 g). Thus

$$1 \text{ R} = \frac{1 \text{ esu}}{0.001293 \text{ g}} = 2.58 \times 10^{-4} \text{ C/kg}$$

Assigning one unit of electric charge to each ion, an exposure of 1 R means that $(2.58 \times 10^{-4} \text{ C/kg})/1.60 \times 10^{-19} \text{ C} = 1.61 \times 10^{15}$ ions are formed per kg of air, or $2.08 \times 10^{9}$ ions per cm³. It takes on the average about 34 eV to form an ion in air, and thus an exposure of 1 R results in an energy absorption by the air of $7.08 \times 10^{10}$ eV/cm³ or 0.113 erg/cm³, or 88 erg/g.

The ionization produced by a γ ray depends on its energy. With about 34 eV needed to produce each ion in air, a 1-MeV γ ray can be expected to produce, on the average, about 30,000 ions. A radioactive source of a given activity will generally produce many different γ rays with different intensities and energies. The exposure resulting from this source will depend on the number of decays and also on the intensities and energies of each of the γ rays, and the exposure rate (exposure per unit time) will depend on the activity of the source. It will also depend on how far we are from the source; if we imagine that we are to measure the ionization produced in 1 cm³ of air, that ionization will obviously depend on whether we hold that volume of air very close to the radioactive source or very far away. We can therefore write

$$\frac{\Delta X}{\Delta t} = \Gamma \frac{\mathscr{A}}{d^2} \tag{6.46}$$

where $\Delta X/\Delta t$ is the exposure rate, $\mathscr{A}$ is the activity, $d$ is the distance from the source, and $\Gamma$ is a constant, the *specific γ-ray constant*, which depends on the details of γ-ray emission of each radionuclide (the fraction of γ rays with each particular energy and the ionizing ability of photons of that particular energy). It is customary to take $d = 1$ m as a standard distance for measuring the relation-

**Table 6.3**  Specific γ-Ray Constants for Various Radioisotopes[a]

| Nuclide | $t_{1/2}$ | γ-Ray Energy (MeV) and Abundance (%) | Γ |
|---|---|---|---|
| $^{22}$Na | 2.6 y | 0.511 (181), 1.275 (100) | 1.20 |
| $^{24}$Na | 15.02 h | 1.369 (100), 2.754 (100) | 1.84 |
| $^{59}$Fe | 44.6 d | 0.143 (1), 0.192 (3), 1.099 (56), 1.292 (44) | 0.60 |
| $^{57}$Co | 270 d | 0.014 (9), 0.122 (85), 0.136 (11) | 0.059 |
| $^{60}$Co | 5.27 y | 1.173 (100), 1.333 (100) | 1.28 |
| $^{131}$I | 8.06 d | 0.08 (2), 0.284 (6), 0.364 (82), 0.637 (7), 0.723 (2) | 0.22 |
| $^{137}$Cs | 30.1 y | 0.032 (8), 0.662 (85) | 0.32 |
| $^{198}$Au | 2.7 d | 0.412 (95), 0.676 (1) | 0.23 |
| $^{226}$Ra and daughters | | | 0.84 |

[a]Units for Γ are R · m$^2$/h · Ci. Note the relationship between Γ and the energy and intensity of the γ rays.

ship between exposure rate and activity, and thus Γ has units of $(R/h)/(Ci/m^2)$. Some representative values of Γ are given in Table 6.3.

Materials other than air exposed to ionizing radiation will differ in their rate of energy absorption. It is therefore necessary to have a standard for defining the energy absorption by ionization in different materials. This quantity is called the *absorbed dose D* of the material and measures the energy deposited by ionizing radiation per unit mass of material. The commonly used unit of absorbed dose is the *rad* (*r*adiation *a*bsorbed *d*ose) equal to an energy absorption of 100 ergs per gram of material. (Thus 1 R = 0.88 rad in air.) The SI unit for absorbed dose is the *gray* (Gy), equal to the absorption of 1 joule per kilogram of material, and so 1 Gy = 100 rad.

To define standards for radiation protection of human beings, it is necessary to have some measure of the biological effects of different kinds of radiations. That is, some radiations may deposit their energy over a very long path, so that relatively little energy is deposited over any small interval (say, of the size of a typical human cell); β and γ rays are examples of such radiations. Other types of radiations, α particles for instance, lose energy more rapidly and deposit essentially all of their energy over a very short path length. The probability of cell damage from 1 rad of α radiation is thus far greater than that from 1 rad of γ radiation. To quantify these differences, we define the *relative biological effectiveness* (RBE), as the ratio of the dose of a certain radiation to the dose of X rays *that produces the same biological effect*. Values of the RBE range from 1 to about 20 for α radiation. Since the RBE is a relatively difficult quantity to measure, it is customary to work instead with the *quality factor* (QF), which is calculated for a given type (and energy) of radiation according to the energy deposited per unit path length. Radiations that deposit relatively little energy per unit length (β's and γ's) have QF near 1, while radiations that deposit more energy per unit length (α's) have QF ranging up to about 20. Table 6.4 shows some representative values of QF.

**Table 6.4**  Quality Factors for Absorbed Radiation

| Radiation | QF |
|---|---|
| X rays, $\beta$, $\gamma$ | 1 |
| Low-energy p, n ($\sim$ keV) | 2–5 |
| Energetic p, n ($\sim$ MeV) | 5–10 |
| $\alpha$ | 20 |

**Table 6.5**  Quantities and Units for Measuring Radiation

| Quantity | Measure of | Traditional Unit | SI Unit |
|---|---|---|---|
| Activity ($\mathscr{A}$) | Decay rate | curie (Ci) | becquerel (Bq) |
| Exposure ($X$) | Ionization in air | roentgen (R) | coulomb per kilogram (C/kg) |
| Absorbed dose ($D$) | Energy absorption | rad | gray (Gy) |
| Dose equivalent (DE) | Biological effectiveness | rem | sievert (Sv) |

The effect of a certain radiation on a biological system then depends on the absorbed dose $D$ and on the quality factor QF of the radiation. The *dose equivalent* DE is obtained by multiplying these quantities together:

$$DE = D \cdot QF \tag{6.47}$$

The dose equivalent is measured in units of *rem* (*r*oentgen *e*quivalent *m*an) when the dose $D$ is in rads. When the SI unit of gray is used for $D$, then the dose equivalent is in *sievert* (Sv). Previously we noted that 1 Gy = 100 rad, and so it follows that 1 Sv = 100 rem.

We therefore see that "strength" of radiation has many different ways of being defined, depending on whether we wish to merely count the rate at which the decays occur (activity) or to measure the effect on living systems (dose equivalent). Table 6.5 summarizes these various measures and the traditional and SI units in which they are expressed.

Standards for radiation exposure of the general public and of radiation workers are specified in rems over a certain period of time (usually per calendar quarter and per year). From natural background sources (cosmic rays and naturally occurring radioactive isotopes, such as the uranium and thorium series and $^{40}$K) we receive about 0.1–0.2 rem per year. The International Commission on Radiation Protection (ICRP) has recommended limiting annual whole-body absorbed dose to 0.5 rem per year for the general public and 5 rem per year for those who work with radiation. By way of contrast, the dose absorbed by a particularly sensitive area of the body, the bone marrow, is about 0.05 rem for a typical chest X ray and 0.002 rem for dental X rays. Unfortunately, the physiological effects of radiation exposure are difficult to calculate and to measure, and so the guideline must be to keep the exposure as low as possible. (For this reason, many physicians no longer recommend chest X rays as a part of the regular annual physical examination, and dentists often place a lead apron over the sensitive areas of a patient's body while taking X-ray pictures of the mouth.) Although the evidence is not conclusive, there is reason to believe that the risk of radiation-induced cancers and genetic damage remains at even very low doses while other effects, such as cataracts and loss of fertility, may show a definite threshold of exposure below which there is no risk at all. Much of our knowledge in this area comes from studies of the survivors of the nuclear weapons exploded over Hiroshima and Nagasaki in World War II, from which we know that there is virtual certainty of death following a short-term dose of 100 rem, but the evidence regarding the linear relationship between dose and risk is less clear. The effects of long-term, low-level doses are still under active debate, with serious consequences for standards of radiation protection and for the health of the general public.

## REFERENCES FOR ADDITIONAL READING

The quantum mechanics of decay processes is treated in more detail in M. G. Bowler, *Nuclear Physics* (Oxford: Pergamon, 1973); see especially Sections 3.1 and 3.2.

A more complete treatment of radioactive decay series can be found in R. D. Evans, *The Atomic Nucleus* (New York: McGraw-Hill, 1955), Chapter 15.

Radioactive dating of the solar system has been reviewed by L. T. Aldrich and G. W. Wetherill, *Ann. Rev. Nucl. Sci.* **8**, 257 (1958), and more recently by G. W. Wetherill, *Ann. Rev. Nucl. Sci.* **25**, 283 (1975).

For more information on radioactivity in the atmosphere and in the oceans, see D. Lal and H. E. Suess, *Ann. Rev. Nucl. Sci.* **18**, 407 (1968).

Additional information on radiation exposure can be found in many references on health physics. See, for example, E. Pochin, *Nuclear Radiation: Risks and Benefits* (Oxford: Clarendon, 1983).

## PROBLEMS

1. Three radioactive sources each have activities of 1.0 $\mu$Ci at $t = 0$. Their half-lives are, respectively, 1.0 s, 1.0 h, and 1.0 d. (a) How many radioactive nuclei are present at $t = 0$ in each source? (b) How many nuclei of each source decay between $t = 0$ and $t = 1$ s? (c) How many decay between $t = 0$ and $t = 1$ h?

2. Naturally occurring samarium includes 15.1% of the radioactive isotope $^{147}$Sm, which decays by $\alpha$ emission. One gram of natural Sm gives $89 \pm 5$ $\alpha$ decays per second. From these data calculate the half-life of the isotope $^{147}$Sm and give its uncertainty.

3. Among the radioactive products emitted in the 1986 Chernobyl reactor accident were $^{131}$I ($t_{1/2} = 8.0$ d) and $^{137}$Cs ($t_{1/2} = 30$ y). There are about five times as many $^{137}$Cs atoms as $^{131}$I atoms produced in fission. (a) Which

isotope contributes the greater activity to the radiation cloud? Assume the reactor had been operating continuously for several days before the radiation was released. (b) How long after the original incident does it take for the two activities to become equal? (c) About 1% of fission events produce $^{131}I$, and each fission event releases an energy of about 200 MeV Given a reactor of the Chernobyl size (1000 MW), calculate the activity in curies of $^{131}I$ after 24 h of operation.

4. Consider a chain of radioactive decays $1 \rightarrow 2 \rightarrow 3$, where nuclei of type 3 are stable. (a) Show that Equation 6.31 is the solution to Equation 6.29. (b) Write a *differential equation* for the number of nuclei of type 3 and solve the differential equation for $N_3(t)$. (c) Evaluate $N_1(t) + N_2(t) + N_3(t)$ and interpret. (d) Examine $N_1$, $N_2$, and $N_3$ at small $t$, keeping only linear terms. Interpret the results. (e) Find the limits of $N_1$, $N_2$, and $N_3$ as $t \rightarrow \infty$ and interpret.

5. The human body contains on the average about 18% carbon and 0.2% potassium. Compute the intrinsic activity of the average person from $^{14}C$ and $^{40}K$.

6. A radioactive isotope is prepared by a nuclear reaction in a cyclotron. At the conclusion of the irradiation, which lasts a very short time in comparison with the decay half-life, a chemical procedure is used to extract the radioactive isotope. The chemical procedure takes 1 h to perform and is 100% efficient in recovering the activity. After the chemical separation, the sample is counted for a series of 1-min intervals, with the following results ($t = 0$ is taken to be the conclusion of the irradiation):

| $t$ (min) | Decays / min | $t$ (min) | Decays / min | $t$ (min) | Decays / min |
|---|---|---|---|---|---|
| 62.0 | 592 | 112.0 | 290 | 163.0 | 125 |
| 68.0 | 527 | 120.0 | 242 | 170.0 | 110 |
| 73.0 | 510 | 125.0 | 215 | 175.0 | 109 |
| 85.0 | 431 | 130.0 | 208 | 180.0 | 100 |
| 90.0 | 380 | 138.0 | 187 | | |
| 97.0 | 353 | 144.0 | 177 | | |
| 101.0 | 318 | 149.0 | 158 | | |
| 105.0 | 310 | 156.0 | 142 | | |

(a) Plot these data on semilog paper and determine the half-life and initial ($t = 0$) activity from your graph. Show the range of uncertainty of each data point, and try to estimate the resulting uncertainty in the half-life. (b) Use an analytic procedure to do a linear least-squares fit of the data (in the form of log $N$ vs $t$) and determine the half-life and its uncertainty. Formulas for linear least-squares fits can be found in K. S. Krane and L. Schecter, *Am. J. Phys.* **50**, 82 (1982).

7. A sample of a certain element with two naturally occurring isotopes becomes activated by neutron capture. After 1 h in the reactor, it is placed

in a counting room, in which the total number of decays in 1 h is recorded at daily intervals. Here is a summary of the recorded data:

| Time (d) | No. Decays | Time (d) | No. Decays |
|---|---|---|---|
| 0 | 102,515 | 20 | 2372 |
| 1 | 79,205 | 40 | 1421 |
| 2 | 61,903 | 60 | 1135 |
| 3 | 48,213 | 80 | 862 |
| 4 | 37,431 | 100 | 725 |
| 5 | 29,367 | 120 | 551 |
| 6 | 23,511 | 140 | 462 |
| 7 | 18,495 | 160 | 359 |
| 8 | 14,829 | 180 | 265 |
| 9 | 11,853 | 200 | 225 |
| 10 | 9,595 | | |

From these data, determine the half-lives and initial activities of the two components. What is the element?

8.  Consider a simple decay process in which an initial number $N_0$ of radioactive nuclei of type A decay to stable nuclei of type B. In a time interval from $t_1$ to $t_1 + \Delta t$, how many decays will occur? Solve this problem in two ways: (1) use Equation 6.10, and (2) use the difference between $N(t_1)$ and $N(t_1 + \Delta t)$. *Note:* Only the first of these methods is correct in general; see the next problem.

9.  Consider a decay process A → B → C, in which $N_A(t = 0) = N_0$, and $N_B(t = 0) = N_C(t = 0) = 0$. How many decays of type B nuclei will be observed between $t_1$ and $t_1 + \Delta t$? (*Hint:* See the previous problem, and explain why method (2) will not work in this case. Figures 6.6 and 6.7 may also provide convincing evidence.)

10. Nuclei of type A, produced at a constant rate R in a nuclear reaction, decay to type B which in turn decay to stable nuclei C. (a) Set up and solve the differential equations for $N_A$, $N_B$, and $N_C$ as functions of the time during which the reaction occurs. (b) Evaluate the sum $N_A + N_B + N_C$ and interpret.

11. The radioactive isotope $^{233}$Pa ($t_{1/2} = 27.0$ d) can be produced following neutron capture by $^{232}$Th. The resulting $^{233}$Th decays to $^{233}$Pa with a half-life of 22.3 min. Neutron capture in 1 g of $^{232}$Th, in the neutron flux from a typical reactor, produces $^{233}$Th at a rate of $2.0 \times 10^{11}$ s$^{-1}$. (a) At the end of 1 h of irradiation, what are the resulting activities of $^{233}$Th and $^{233}$Pa? (b) After 1 h of irradiation, the sample is placed in storage so that the $^{233}$Th activity can decay away. What are the $^{233}$Th and $^{233}$Pa activities after 24 h and after 48 h of storage? (c) The $^{233}$Pa decay results in $^{233}$U, which is itself radioactive ($t_{1/2} = 1.6 \times 10^5$ y). After the above sample has been stored for 1 year, what is the $^{233}$U activity? (*Hint:* It should not be necessary to set up an additional differential equation to find the $^{233}$U activity.)

**12.** An initial activity of nucleus A decays to B, which in turn decays to stable nucleus C. (a) Discuss qualitatively why it must be true at short times that $\mathscr{A}_A > \mathscr{A}_B$, while at long times $\mathscr{A}_B > \mathscr{A}_A$. (b) There must therefore be a time $T$ at which $\mathscr{A}_A = \mathscr{A}_B$. Find $T$ in terms of the decay constants $\lambda_A$ and $\lambda_B$.

**13.** The decay chain $^{139}Cs \rightarrow {}^{139}Ba \rightarrow {}^{139}La$ is observed from an initially pure sample of 1 mCi of $^{139}Cs$. The half-lives are $^{139}Cs$, 9.5 min; $^{139}Ba$, 82.9 min; $^{139}La$, stable. What is the maximum $^{139}Ba$ activity and when does it occur?

**14.** In the decay process $^{235}U \rightarrow {}^{231}Th \rightarrow {}^{231}Pa$ ($t_{1/2} = 7.04 \times 10^8$ y for $^{235}U$; $t_{1/2} = 25.5$ h for $^{231}Th$) plot the $^{235}U$ and $^{231}Th$ activities as a function of time from $t = 0$ to $t = 100$ h. Assume the sample initially consists of 1.0 mCi of $^{235}U$. Discuss the condition of secular equilibrium in this decay process.

**15.** The $\alpha$ decay of $^{238}U$ ($t_{1/2} = 4.47 \times 10^9$ y) leads to 24.1-d $^{234}Th$. A sample of uranium ore should reveal $^{234}Th$ activity in secular equilibrium with the parent. What would be the $^{234}Th$ activity in each gram of uranium?

**16.** Prepare a diagram similar to Figure 6.10 showing the $4n + 2$ natural radioactive series.

**17.** The radioactive decay of $^{232}Th$ leads eventually to stable $^{208}Pb$. A rock is determined to contain 3.65 g of $^{232}Th$ and 0.75 g of $^{208}Pb$. (a) What is the age of the rock, as deduced from the Th/Pb ratio? (b) If the rock is large, the $\alpha$ particles emitted in the decay processes remain trapped. If the rock were pulverized, the $\alpha$'s could be collected as helium gas. At 760 mm and 0°C, what volume of helium gas could be collected from this rock?

**18.** It is desired to determine the age of a wood timber used to construct an ancient shelter. A sample of the wood is analyzed for its $^{14}C$ content and gives 2.1 decays per minute. Another sample of the same size from a recently cut tree of the same type gives 5.3 decays per minute. What is the age of the sample?

**19.** Show that the present $^{14}C$ content of organic material gives it an activity of about 15 decays per minute per gram of carbon.

**20.** What is the probability of a $^{14}C$ decay taking place in the lungs during a single breath? The atmosphere is about 0.03% $CO_2$, and in an average breath we inhale about 0.5 liter of air and exhale it about 3.5 s later.

**21.** (a) What is the $\gamma$-ray flux ($\gamma$'s per unit area) a distance of 1.0 m from a 7.5-mCi source of $^{60}Co$?

(b) How many ions per minute are produced in a cubic centimeter of air at that distance?

# 7

# DETECTING NUCLEAR RADIATIONS

In their basic principles of operation, most detectors of nuclear radiations follow similar characteristics: the radiation enters the detector, interacts with the atoms of the detector material (losing part or all of its energy), and releases a large number of relatively low-energy electrons from their atomic orbits. These electrons are then collected and formed into a voltage or current pulse for analysis by electronic circuitry. The choice of material to use for radiation detectors depends on the type of radiation we are trying to detect and on the information about that radiation we are trying to gather. For $\alpha$ particles from radioactive decays or charged particles from nuclear reactions at low (MeV) energies, very thin detectors are sufficient, as the maximum range of these particles in most solids is typically less than 100 $\mu$m. For electrons, such as those emitted in $\beta$ decay, a detector of thickness 0.1 to 1 mm is required, while for $\gamma$ rays the range is large and even detectors of 5-cm thickness may not be sufficient to convert energetic photons (MeV or above) into an electronic pulse. Merely to show the presence of radiation, the familiar click of the Geiger counter may be sufficient; all incident radiations give the same output. To measure the energy of the radiation, we should select a detector in which the output pulse amplitude is proportional to the energy of the radiation; here we must choose a material in which the number of released electrons is large, so that if we experience statistical fluctuations or fail to count a few, it does not substantially affect our ability to determine the energy. To determine the time at which the radiation was emitted, we must choose a material in which the electrons can be gathered quickly into the pulse; the number of electrons gathered is of considerably less importance. To determine the type of particle (such as in a nuclear reaction, in which many varieties of particles may be produced), we must choose a material in which the mass or charge of the particle gives a distinctive signature. To measure the spin or polarization of the radiation, we must choose a detector that can resolve or separate different spin or polarization states. If we expect unusually high counting rates, we must choose a detector that can recover quickly from one radiation before counting the next; for very low counting rates, we must be concerned about detecting every event and about reducing the influence of background radiations. Finally, if we are interested in reconstructing the trajectory of the detected radiations, we must have a detector that is sensitive to the location at which the radiation enters the detector.

In this chapter we discuss various types of detectors that satisfy one or another of these requirements; no single detector can satisfy all of them. We limit our discussion to radiations that are likely to be encountered in most nuclear decay and reaction studies: heavy charged particles (protons, $\alpha$'s) of nonrelativistic energies, relativistic electrons (of typically MeV energies), and photons in the X-ray and $\gamma$-ray regions. Neutron detectors are considered separately in Chapter 12.

## 7.1 INTERACTIONS OF RADIATION WITH MATTER

### Heavy Charged Particles

Although Coulomb scattering of charged particles by nuclei (called Rutherford scattering) is an important process in nuclear physics, it has very little influence on the loss in energy of a charged particle as it travels through the detector material. Because the nuclei of the detector material occupy only about $10^{-15}$ of the volume of their atoms, it is (crudely) $10^{15}$ times more probable for the particle to collide with an electron than with a nucleus. The dominant mechanism for energy loss by charged particles is therefore Coulomb scattering by the atomic electrons of the detector.

Conservation of energy and momentum in a head-on elastic collision between a heavy particle of mass $M$ and an electron of mass $m$ (which we assume to be at rest for the sake of this simplified discussion) gives for the loss in kinetic energy of the particle

$$\Delta T = T\left(\frac{4m}{M}\right) \tag{7.1}$$

For a 5-MeV $\alpha$ particle (typical of those emitted in radioactive decay), this amounts to 2.7 keV. Four conclusions follow immediately:

1.  It takes many thousands of such events before the particle loses all its energy. (A head-on collision gives the *maximum* energy transfer to the electron; in most collisions, the energy loss of the particle will be much smaller.)
2.  In a glancing collision between an electron and a heavy particle, the heavy particle is deflected by a negligible angle, and so the particle follows very nearly a straight-line path.
3.  Because the Coulomb force has infinite range, the particle interacts simultaneously with many electrons and thus loses energy gradually but continuously along its path. After traveling a certain distance, it has lost all of its energy; this distance is called the *range* of the particle. The range is determined by the type of particle, type of material, and energy of the particle. Figure 7.1 shows cloud-chamber tracks of $\alpha$ particles; there is a rather well-defined distance beyond which there are no particles. Usually we work with the mean range, defined so that one-half the particles have longer ranges and one-half shorter; the variation about the mean is very small, at most a few percent, so the mean range is a useful and precisely defined quantity.

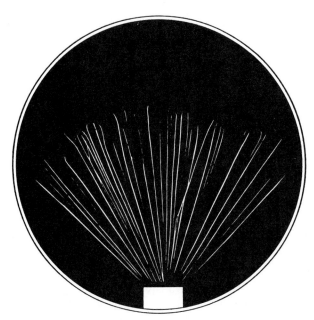

**Figure 7.1**  Cloud chamber tracks of $\alpha$ particles from the decay of $^{210}$Po.

4.  The energy needed to ionize an atom (i.e., to remove an electron) is of the order of 10 eV; thus many collisions will transfer enough energy to an electron to ionize the atom. (If the electron is not given enough energy to produce an ion, the atom is placed into an excited state, which quickly de-excites back to the ground state.) Furthermore, electrons given energies in the keV region (which are known as *delta* rays) can themselves produce ions by collisions, resulting in even more *secondary* electrons. To determine the energy lost by the particle, we must include the primary and secondary electrons as well as the atomic excitations.

Figure 7.2 shows the relationship between range and energy for air and for some other commonly encountered materials. For materials that are not shown, an estimate of the range can be made using a semiempirical relationship known as the Bragg-Kleeman rule:

$$\frac{R_1}{R_0} \cong \frac{\rho_0\sqrt{A_1}}{\rho_1\sqrt{A_0}} \tag{7.2}$$

where $R$ is the range, $\rho$ the density, and $A$ the atomic weight. The subscripts 0 and 1 refer, for instance, to the known and unknown ranges and materials, respectively.

The theoretical relationship between range and energy can be obtained from a quantum mechanical calculation of the collision process, which was first done in 1930 by Hans Bethe. The calculation gives the magnitude of the energy loss per unit length (sometimes called the *stopping power*):

$$\frac{dE}{dx} = \left(\frac{e^2}{4\pi\epsilon_0}\right)^2 \frac{4\pi z^2 N_0 Z \rho}{mc^2\beta^2 A}\left[\ln\left(\frac{2mc^2\beta^2}{I}\right) - \ln(1 - \beta^2) - \beta^2\right] \tag{7.3}$$

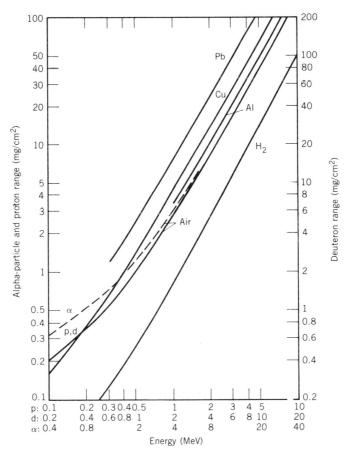

**Figure 7.2** The range-energy relationship in various materials. Because the particles lose energy through scattering by atomic electrons, the range depends inversely on the density. It is therefore convenient to plot the product range × density, in units of mg/cm². Unfortunately, this product is also called "range" in the literature. *From* A. H. Wapstra et al., *Nuclear Spectroscopy Tables* (Amsterdam: North-Holland, 1959).

where $v = \beta c$ is the velocity of the particle, $ze$ is its electric charge, $Z$, $A$, and $\rho$ are the atomic number, atomic weight, and density of the stopping material, $N_0$ is Avogadro's number, and $m$ is the electron mass. The parameter $I$ represents the mean excitation energy of the atomic electrons, which could in principle be computed by averaging over all atomic ionization and excitation processes. In practice, $I$ is regarded as an empirical constant, with a value in eV of the order of $10Z$. In air, for instance, $I = 86$ eV, while for Al, $I = 163$ eV.

The range can be calculated by integrating Equation 7.3 over the energies of the particle

$$R = \int_T^0 \left( -\frac{dE}{dx} \right)^{-1} dE \qquad (7.4)$$

However, Equation 7.3 fails at low energy near the end of the range, primarily because it does not take into account the capture of electrons by the now slow-moving particle. It is possible to write Equation 7.4 in the following form:

$$R = Mz^{-2} \int f(v) \, dv \tag{7.5}$$

where $f(v)$ is a function of the velocity of the particle that is independent of its mass and charge. We can therefore compare ranges in the same material for different particles of the same initial velocity:

$$\frac{R_1}{R_0} = \frac{M_1}{M_0} \frac{z_0^2}{z_1^2} \tag{7.6}$$

## Electrons

Electrons (positive and negative) interact through Coulomb scattering from atomic electrons, just like heavy charged particles. There are, however, a number of important differences: (1) Electrons, particularly those emitted in $\beta$ decay, travel at relativistic speeds. (2) Electrons will suffer large deflections in collisions with other electrons, and therefore will follow erratic paths. The range (defined as the linear distance of penetration into the material) will therefore be very different from the length of the path that the electron follows. (3) In head-on collisions of one electron with another, a large fraction of the initial energy may be transferred to the struck electron. (In fact, in electron–electron collisions we must take into account the identity of the two particles; after the collision, we cannot tell which electron was incident and which was struck.) (4) Because the electron may suffer rapid changes in the direction and the magnitude of its velocity, it is subject to large accelerations, and accelerated charged particles must radiate electromagnetic energy. Such radiation is called *bremsstrahlung* (German for "braking radiation").

The expressions for the energy loss per unit path length for electrons were also derived by Bethe, and can be written in a form similar to Equation 7.3:

$$\left(\frac{dE}{dx}\right)_c = \left(\frac{e^2}{4\pi\epsilon_0}\right)^2 \frac{2\pi N_0 Z \rho}{mc^2 \beta^2 A} \left[ \ln \frac{T(T + mc^2)^2 \beta^2}{2I^2 mc^2} + (1 - \beta^2) \right.$$

$$\left. - \left(2\sqrt{1 - \beta^2} - 1 + \beta^2\right) \ln 2 + \tfrac{1}{8}\left(1 - \sqrt{1 - \beta^2}\right)^2 \right] \tag{7.7}$$

$$\left(\frac{dE}{dx}\right)_r = \left(\frac{e^2}{4\pi\epsilon_0}\right)^2 \frac{Z^2 N_0 (T + mc^2)\rho}{137 m^2 c^4 A} \left[ 4 \ln \frac{2(T + mc^2)}{mc^2} - \frac{4}{3} \right] \tag{7.8}$$

where $T$ is the kinetic energy of the electron. The subscripts c and r stand for the energy losses due to collisions and radiation, respectively. The expression for the radiative loss is valid only for relativistic energies; below 1 MeV, the radiation losses are negligible.

The total energy loss is just the sum of these two contributions:

$$\frac{dE}{dx} = \left(\frac{dE}{dx}\right)_c + \left(\frac{dE}{dx}\right)_r \tag{7.9}$$

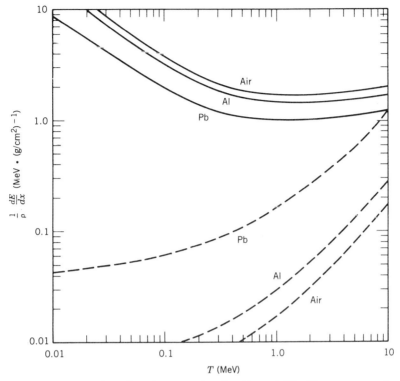

**Figure 7.3** Energy loss by electrons in air, Al, and Pb. To suppress the large variation in $dE/dx$ arising from the number of electrons of the material, the quantity $\rho^{-1}(dE/dx)$ is plotted. Solid lines are for collisions; dashed lines are for radiation. For additional tabulated data on energy losses, see L. Pages et al., *Atomic Data* **4**, 1 (1972).

To estimate the relative contributions of the two terms we can form their ratio, which in the relativistic region is approximately

$$\frac{(dE/dx)_r}{(dE/dx)_c} \approx \frac{T + mc^2}{mc^2}\frac{Z}{1600} \tag{7.10}$$

The radiative term is thus significant only at high energy and in heavy materials. Figure 7.3 shows the relative contributions for air, aluminum, and lead; for most materials used as electron detectors, the radiative contribution is small. Moreover, there is very little variation of the collisional losses with electron energy.

Calculation of the range of electrons could in principle be done by integrating Equations 7.7 and 7.8 over the path of the electrons; however, because of the random nature of the path, this is a difficult process. Instead, we use empirical data on absorption of beams of monoenergetic electrons to generate the range-energy relationship for electrons; Figure 7.4 is an example of this relationship. Based on the comparison of $\rho^{-1}(dE/dx)$ values in Figure 7.3, we conclude that the variation with type of absorber is small, and we can thus use Figure 7.4 to estimate ranges (in mg/cm$^2$, which is in reality range times density) in other materials as well.

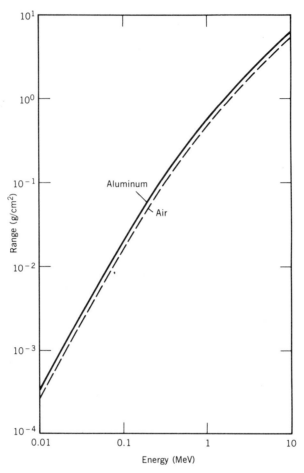

**Figure 7.4** Range-energy relationship for electrons in air and in aluminum.

## Electromagnetic Radiation

Gamma rays and X rays interact with matter primarily through three processes: photoelectric absorption, Compton scattering, and pair production. In the photoelectric effect, a photon is absorbed by an atom and one of the atomic electrons, known in this case as a photoelectron, is released. (Free electrons cannot absorb a photon and recoil. Energy and momentum cannot both be conserved in such a process; a heavy atom is necessary to absorb the momentum at little cost in energy.) The kinetic energy of the electron is equal to the photon energy less the binding energy of the electron:

$$T_e = E_\gamma - B_e \tag{7.11}$$

The probability for photoelectric absorption is difficult to calculate, but from experimental studies we know several features: it is most significant for low-energy photons ($\sim 100$ keV), it increases rapidly with the atomic number $Z$ of the absorber atoms (roughly as $Z^4$), and it decreases rapidly with increasing

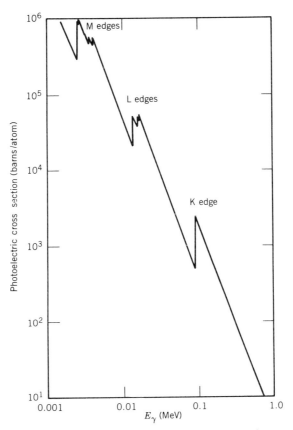

**Figure 7.5** Photoelectric cross section in Pb. The discrete jumps correspond to the binding energies of various electron shells; the K-electron binding energy, for example, is 88 keV. To convert the cross section to the linear absorption coefficient $\tau$ in $cm^{-1}$, multiply by 0.033.

photon energy (roughly as $E_\gamma^{-3}$). Furthermore, there are in the probability for photoelectric absorption discontinuous jumps at energies corresponding to the binding energies of particular electronic shells. That is, the binding energy of a K-shell electron in Pb is 88 keV. Incident photons of energy less than 88 keV cannot release K-shell photoelectrons (although they can release less tightly bound electrons from higher shells). When the photon energy is increased above 88 keV, the availability of the K-shell electrons to participate in the photoelectric absorption process causes a sudden increase in the absorption probability, known as the K-absorption edge or simply K edge. Figure 7.5 shows a sample of the photoelectric absorption cross section.

Compton scattering is the process by which a photon scatters from a nearly free atomic electron, resulting in a less energetic photon and a scattered electron carrying the energy lost by the photon. Figure 7.6 shows a schematic view of the process. If we regard the struck electron as free and at rest (a good approximation, since the photon energy is usually large compared with the orbital energies of the loosely bound outer atomic electrons), then conservation of linear

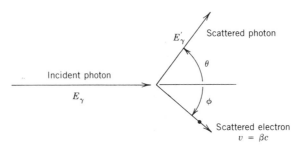

**Figure 7.6**   The geometry of Compton scattering.

momentum and total energy (using relativistic dynamics) gives

$$\frac{E_\gamma}{c} = \frac{E'_\gamma}{c} \cos\theta + \frac{mc\beta\cos\phi}{\sqrt{1-\beta^2}} \tag{7.12}$$

$$0 = \frac{E'_\gamma}{c}\sin\theta - \frac{mc\beta\sin\phi}{\sqrt{1-\beta^2}} \tag{7.13}$$

$$E_\gamma + mc^2 = E'_\gamma + \frac{mc^2}{\sqrt{1-\beta^2}} \tag{7.14}$$

If we observe the scattered photon, then we may eliminate the unobserved variables $\beta$ and $\phi$, giving the Compton-scattering formula

$$E'_\gamma = \frac{E_\gamma}{1 + (E_\gamma/mc^2)(1 - \cos\theta)} \tag{7.15}$$

The scattered photons range in energy from $E_\gamma$ for $\theta = 0°$ (forward scattering, corresponding to no interaction) to a minimum of roughly $mc^2/2 \approx 0.25$ MeV for $\theta = 180°$ when the photon energy is large.

The probability for Compton scattering at an angle $\theta$ can be determined through a quantum mechanical calculation of the process. The result is the *Klein-Nishina formula* for the differential cross section per electron:

$$\frac{d\sigma_c}{d\Omega} = r_0^2 \left[ \frac{1}{1 + \alpha(1 - \cos\theta)} \right]^3 \left[ \frac{1 + \cos\theta}{2} \right]$$

$$\times \left[ 1 + \frac{\alpha^2(1 - \cos\theta)^2}{(1 + \cos^2\theta)[1 + \alpha(1 - \cos\theta)]} \right] \tag{7.16}$$

Here $\alpha$ is the photon energy in units of the electron rest energy ($\alpha = E_\gamma/mc^2$) and $r_0$ is a parameter called the *classical electron radius*, $r_0 = e^2/4\pi\epsilon_0 mc^2 = 2.818$ fm. (This is merely a convenient parameter and has nothing whatever to do with the "size" of the electron.) Figure 7.7 shows a polar plot of the cross section.

If we are interested in the absorption of photons (that is, their removal from the incident beam of photons), we must integrate Equation 7.16 over all angles since we do not observe the scattered photon. The result is

$$\sigma_c = \frac{\pi r_0^2}{\alpha} \left\{ \left[ 1 - \frac{2(\alpha + 1)}{\alpha^2} \right] \ln(2\alpha + 1) + \frac{1}{2} + \frac{4}{\alpha} - \frac{1}{2(2\alpha + 1)^2} \right\} \tag{7.17}$$

for each electron in the scatterer.

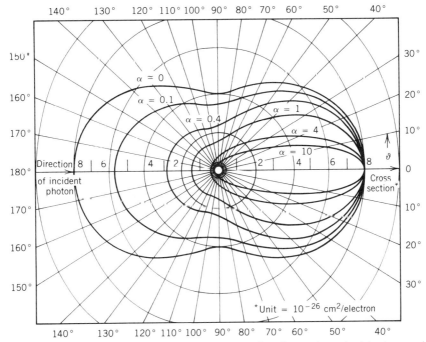

140° 130° 120° 110° 100° 90° 80° 70° 60° 50° 40°

150°                                         30°

$\alpha = 0$

160°                 $\alpha = 0.1$                  20°

$\alpha = 0.4$      $\alpha = 1$

170°                      $\alpha = 4$            10°

$\alpha = 10$            $\vartheta$

180° Direction 8 | 6 | | 4 | 2 | | 2 | 4 | 6 | 8     0
of incident                             Cross
photon                              section*

170°                                       10°

160°                                       20°

150°                                       30°

*Unit $= 10^{-26}$ cm$^2$/electron

140° 130° 120° 110° 100° 90° 80° 70° 60° 50° 40°

**Figure 7.7** The Compton-scattering cross section for various incident energies. The polar plot shows the intensity of the scattered radiation as a function of the scattering angle $\theta$. From R. D. Evans, *The Atomic Nucleus* (New York: McGraw-Hill, 1955).

The third interaction process is pair production, in which a photon creates an electron-positron pair; the photon disappears in this process. The energy balance is

$$E_\gamma = T_+ + mc^2 + T_- + mc^2 \qquad (7.18)$$

where $T_+$ and $T_-$ are the energies of the positron and electron. Like photoelectric absorption, this process requires the nearby presence of a massive atom for momentum conservation, but the recoil energy given to the atom is negligible compared with the other terms of Equation 7.18.

There is obviously a threshold of $2mc^2$, or 1.022 MeV, for this process, and in general pair production is important only for photons of high energy. Figure 7.8 shows the importance of pair production relative to the other two processes; pair production becomes dominant only for energies above 5 MeV.

Let us consider a highly collimated beam of monoenergetic photons incident on a slab of material of thickness $t$ (Figure 7.9). The photon may undergo photoelectric absorption or pair production and thus disappear, or Compton scattering and be deflected from reaching the detector. Those photons that reach the detector are those that have no interactions at all; there are simply fewer of them than there were in the incident beam. (Contrast this situation with the case of heavy charged particles, where if $t$ is less than the range, the number is unchanged but the energy is decreased.) The total probability *per unit length* $\mu$ for removal of a photon is called the *total linear attenuation coefficient*; it is simply the sum of the respective probabilities for photoelectric absorption ($\tau$),

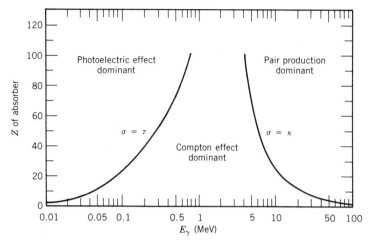

**Figure 7.8**   The three γ-ray interaction processes and their regions of dominance.

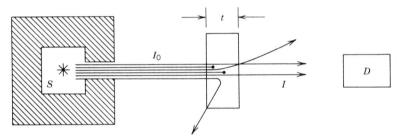

**Figure 7.9**   An experiment to measure absorption of radiation in a slab of material of thickness $t$. A beam of radiation from the source S is collimated and then is scattered or absorbed by the material. The remaining intensity $I$ reaches the detector D.

Compton scattering ($\sigma$), and pair production ($\kappa$):

$$\mu = \tau + \sigma + \kappa \tag{7.19}$$

The linear Compton absorption coefficient $\sigma$ is related to the calculated cross section per electron, $\sigma_c$ of Equation 7.17, according to

$$\sigma = \sigma_c NZ \tag{7.20}$$

where $Z$ and $N$ again represent the atomic number of the scattering material and the number of atoms per unit volume. All quantities in Equation 7.19 have dimensions of $(\text{length})^{-1}$.

The fractional loss in intensity in crossing any thickness $dx$ of material is

$$\frac{dI}{I} = -\mu \, dx \tag{7.21}$$

and thus

$$I = I_0 e^{-\mu t} \tag{7.22}$$

in passing through a thickness $t$.

Figure 7.10 shows some representative values for the energy dependence of the attenuation coefficients.

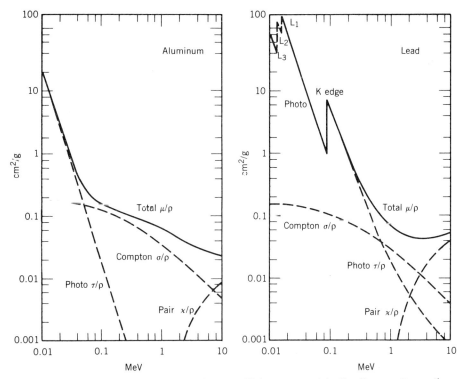

**Figure 7.10** Photon mass attenuation coefficients, equal to the linear attenuation coefficients divided by the density (to suppress effects due simply to the number of electrons in the material) for the three processes in Al and Pb.

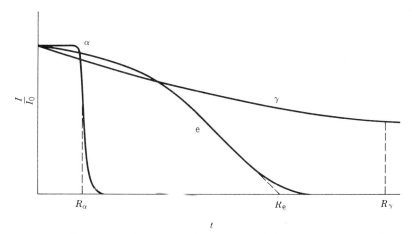

**Figure 7.11** The transmitted intensity measured in a geometry such as that shown in Figure 7.9. For $\alpha$'s, the value of $t$ such that $I/I_0 = 0.5$ is the mean range; for photons, with their simple exponential dependence, we can define the mean range similarly. For electrons, it is customary to define the *extrapolated* range by extending the linear portion of the absorption curve as shown. The horizontal scale is not at all linear; the range for $\gamma$'s may be $10^4$ that for $\alpha$'s.

If we were to study the loss in intensity of monoenergetic beams of 1-MeV $\alpha$'s, electrons, and $\gamma$ radiations in a geometry such as that shown in Figure 7.9, the results might look somewhat like Figure 7.11. The $\alpha$ intensity is undiminished until the thickness is very close to the mean range, and then drops very quickly to zero; the range of 1-MeV $\alpha$'s in aluminum is about 0.0003 cm. The electron intensity begins to decrease slowly even for thicknesses much less than the range, owing to electrons that are scattered out of the beam. The extrapolated range for electrons is about 0.18 cm. The $\gamma$ intensity decreases exponentially; the mean range (defined as the thickness such that $I = 0.5I_0$) is about 4.3 cm for a 1-MeV $\gamma$ in aluminum.

Note the somewhat different uses of the concept of range in these cases, and be sure to note that for $\alpha$'s and $\beta$'s but not for $\gamma$'s the energy per particle observed in the geometry of Figure 7.9 would decrease as the beam traveled through the material.

## 7.2 GAS-FILLED COUNTERS

The function of many detectors of nuclear radiation is to use an electric field to separate and count the ions (or electrons) formed as a result of the passage of radiation through the detector. The simplest type of detector that accomplishes this is the *ionization chamber*, which we can regard as a parallel-plate capacitor in which the region between the plates is filled with a gas, often air. The electric field in this region keeps the ions from recombining with the electrons, and it is useful to picture a cloud of electrons drifting toward the plate held at positive potential, while the positively charged ions drift toward the other plate. In air, the average energy needed to produce an ion is about 34 eV; thus a 1-MeV radiation produces a maximum of about $3 \times 10^4$ ions and electrons. For a medium-sized chamber, say $10 \times 10$ cm with a plate separation of 1 cm, the capacitance is $8.9 \times 10^{-12}$ F and the resulting voltage pulse is about

$$C \approx \frac{Q}{V} \qquad V \approx \frac{Q}{C} \qquad \frac{(3 \times 10^4 \text{ ions})(1.6 \times 10^{-19} \text{ C/ion})}{8.9 \times 10^{-12} \text{ F}} \cong 0.5 \text{ mV}$$

This is a rather small signal, which must be considerably amplified (by a factor of roughly $10^4$) before we can analyze it using standard electronics.

The amplitude of the signal is proportional to the number of ions formed (and thus to the energy deposited by the radiation), and is independent of the voltage between the plates. The applied voltage does determine the speed at which the electron and ion clouds drift to their respective electrodes. For a typical voltage of roughly 100 V, the ions move at speeds of about 1 m/s and will take roughly 0.01 s to travel across the 1-cm chamber. (Electrons are more mobile and travel about 1000 times faster.) This is an exceedingly long time by standards of nuclear counting (a weak radioactive source of 1 $\mu$Ci activity gives on the average one decay every 30 $\mu$s), and thus the ion chamber is of no use in counting individual pulses. It does find wide use as a radiation monitor, and many commercial radiation monitors are in fact ion chambers. The radiation intensity is recorded as a *current* representing the interaction of many radiations during the response time of the chamber. The current output is proportional both to the activity of

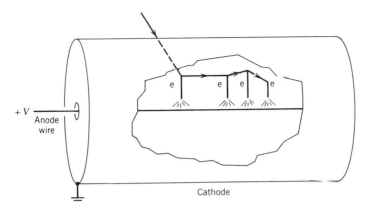

**Figure 7.12** The geometry of a cylindrical proportional counter. The incoming radiation creates many electron-ion pairs. The electrons drift relatively slowly until they reach the neighborhood of the anode wire, where they are accelerated rapidly and create many secondary ionizations.

the source and to the energy of the radiations—higher energy radiations give more ionization and thus a greater response.

To use a gas-filled detector to observe individual pulses, we must provide considerable amplification. One way of accomplishing this is to increase the voltage to larger values, usually in excess of 1000 V. The larger electric field is able to accelerate the electrons that result from the ionization process; rather than drifting slowly toward the anode, making occasional elastic collisions with gas atoms, the accelerated electrons can acquire enough energy to make inelastic collisions and even to create new ionized atoms (and more electrons to be accelerated in turn). This rapid amplification through production of *secondary* ionizations is called a *Townsend avalanche*. Even though there is a large number (perhaps $10^3$–$10^5$) of secondary events for each original ion, the chamber is always operated such that the number of secondary events is proportional to the number of primary events, and the device is therefore known as a *proportional counter*.

The geometry of the proportional counter is usually cylindrical, as shown in Figure 7.12. The electric field in this geometry at a radius $r$ is

$$E(r) = \frac{V}{r \ln(b/a)} \tag{7.23}$$

where $b$ is the inner radius of the cathode and $a$ is the outer radius of the anode wire. The avalanche will obviously occur in the high-field region near the anode wire. This region, however, represents only a very small fraction of the volume of the chamber. The vast majority of the original ions are formed far from this central region, and the electrons drift slowly until they begin the avalanche process. (A primary event that occurs within the high-field region would experience a somewhat smaller amplification factor because it would not have the opportunity to make as many collisions.)

Because the output signal of a proportional counter comes mainly from the avalanche process, which occurs very rapidly, the timing is determined by the *drift time* of the primary electrons from the point of the original ion formation to the vicinity of the anode wire where the avalanche occurs. This time is of the order of microseconds, and thus the counter can be operated in a pulse mode at counting rates of the order of $10^6/s$.

If the electric field is increased to even larger values, *secondary avalanches* can occur. These can be triggered by photons emitted by atoms excited in the original (or in a subsequent) avalanche. These photons can travel relatively far from the region of the original avalanche, and soon the entire tube is participating in the process. The amplification factor is perhaps as large as $10^{10}$. Because the entire tube participates for every incident event, there is no information at all on the energy of the original radiation—all incident radiations produce identical output pulses. This region of operation is called the *Geiger-Müller region* and counters based on this principle are usually known as *Geiger counters.* Geiger counters also are popular as portable radiation monitoring instruments.

The output signal from a Geiger counter consists of the collected electrons from the many avalanche processes; the signal is of the order of 1 V, so no further amplification is usually required. The collection time is of the order of $10^{-6}$ s, during which time the positive ions do not move far from the avalanche region. There is thus surrounding the anode wire a positively charged ion cloud that reduces the electric field intensity and eventually terminates the avalanche process.

The cycle would then be completed after the positive ions have drifted to the cathode and become neutralized (which takes $10^{-4}$–$10^{-3}$ s), but during their travel they can be accelerated and strike the cathode with enough energy to release electrons (from the cathode) and to begin the process again (and because of the nature of the multiple avalanche process in the Geiger tube, all it takes is *one electron* to create an output pulse). To prevent this from occurring, a second type of gas, called the *quenching* gas, is added to the tube. The quenching gas is usually one with complex organic molecules such as ethanol; the primary gas is generally one with simple molecules, such as argon, and a typical mixture might be 90% argon and 10% ethanol. As the space charge, consisting mostly of argon ions, begins to drift toward the cathode, collisions will occur with the quenching gas in which there is a high probability of the transfer of an electron, so that the argon is neutralized and the ionized ethanol begins to drift toward the cathode. When it arrives there and is neutralized, the energy that formerly went into releasing a free electron can now be absorbed in the dissociation of the molecule (a process not possible for a simple argon atom). The quenching gas is thus gradually used up, and the Geiger tube must periodically be replaced. Other Geiger tube designs use halogens as the quenching gas; the subsequent recombination of the dissociated molecule eliminates the need to replace the tube.

The various regions of operation of gas-filled counters are summarized in Figure 7.13. For low applied voltages, recombination of the primary electrons and ions occurs. As $V$ is increased, we reach the ion chamber region, where the pulse output is proportional to the primary ionization produced by the radiation and thus to its energy, but is independent of $V$. In the proportional region, the pulse amplitude rises with $V$ to make analysis easier, but the output pulse is still proportional to the energy of the radiation through the ionization produced. Finally, the Geiger plateau is reached, where all radiations give the same output

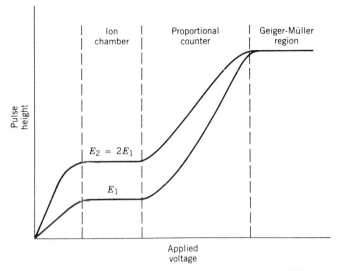

**Figure 7.13**  The pulse heights produced by different gas-filled counters as a function of the applied voltage, for two radiations differing in energy by a factor of 2. In the Geiger-Müller region, all radiations give the same output pulse height; in the other regions, the output pulse height is proportional to the energy deposited by the radiation through primary ionization.

pulse, irrespective of the amount of initial ionization or the energy of the radiation.

## 7.3  SCINTILLATION DETECTORS

A disadvantage of gas-filled counters is their low efficiency for many radiations of interest in nuclear physics—the range in air of a 1-MeV $\gamma$ ray is of the order of 100 m. Solid detectors have the higher densities that give reasonable absorption probabilities for detectors of reasonable size. To make a workable solid detector, however, we must satisfy two contradictory criteria: (1) The material must be able to support a large electric field, so that the electrons and ions can be collected and formed into an electronic pulse, and little or no current must flow in the absence of radiation, so that the background noise is low. (2) Electrons must be easily removed from atoms in large numbers by the radiation, and the electrons and ionized atoms must be able to travel easily through the material. (Actually, the ions themselves do not move in a solid; instead, the electronic vacancy or "hole" is filled by successive electron transfers from one atom to the next, so that the "hole" appears to travel.) The first condition supports the choice of an insulating material, while the second suggests using a conductor. The obvious compromise is a semiconductor, and we consider such devices in the next section. Bulk semiconducting materials in sizes large enough to make useful radiation detectors (tens of $cm^3$) did not become available until the late 1960s, and to fill the need for nuclear spectroscopic devices of high efficiency and reasonable resolution, *scintillation counters* were developed during the 1950s.

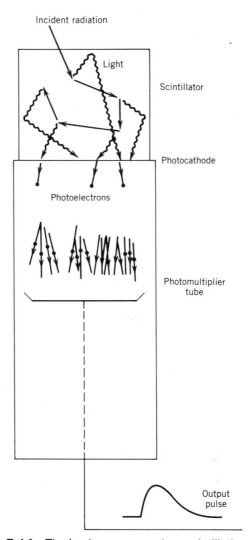

**Figure 7.14** The basic processes in a scintillation detector.

Scintillation counters solve our dilemma over the choice of materials in a clever way: The electrons that are formed in the ionization process are not the same electrons that form the electronic pulse. The intermediary between the ionization electrons and the pulse electrons is ordinary light. The complete process can be understood as follows, with reference to Figure 7.14: (1) The incident radiation enters the detector and suffers a large number of interactions, which result in the raising of the atoms to excited states. (2) The excited states rapidly emit visible (or near-visible) light; the material is said to *fluoresce*. (3) The light strikes a photosensitive surface, releasing at most one photoelectron per photon. (4) These secondary electrons are then multiplied, accelerated, and formed into the output pulse in the *photomultiplier* (PM) tube.

Many different varieties of scintillators and PM tubes are available, depending on the application in which they will be used. Properties that are usually

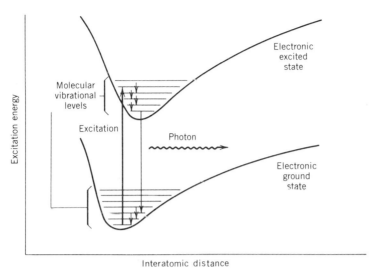

**Figure 7.15** Electronic structure in an organic scintillator. The electronic states are represented as a potential minimum, resulting from the combined effects of the molecular attraction that keeps us from separating the atoms to greater distances and the repulsion that keeps us from forcing the atoms closer together (because the Pauli principle does not let the atomic wave functions overlap). Inside the electronic potential minimum is a sequence of levels that result from the atoms of the molecule vibrating against one another.

considered in making the choice of a material include light output (the fraction of the incident energy that appears as light), efficiency (the probability for the radiation to be absorbed), timing, and energy resolution. Other criteria may have to do with ease of working with the material: one common scintillator, crystalline NaI, is hygroscopic; exposure to water vapor causes a transparent crystal to become an opaque powder, and NaI must be kept sealed. On the other hand, many plastic scintillators can be cut with an ordinary saw and formed into any desired size and shape.

To understand the operation of a scintillator, we must consider the mechanisms by which energy can be absorbed in raising electrons to excited states. There are two basic types of detectors, those composed of organic materials and those composed of inorganic materials.

In organic scintillators (which can be liquid or solid), the interactions between the molecules are relatively weak, and we can discuss their properties in terms of the discrete excited states of the molecules. There are two ways in which a molecule can absorb energy: the electrons can be excited to higher excited states, and the atoms in the molecule can vibrate against one another. Typical spacing of vibrational energies is about 0.1 eV, while the electronic excitation energies are of the order of a few eV. The resulting structure may look something like that of Figure 7.15. The excited electrons are generally those not strongly involved in the bonding of the material. In aromatic hydrocarbons, such as those typified by the ring structure of benzene, three of the four valence electrons of carbon are in the hybridized orbitals called $\sigma$ orbitals; these are strongly localized between each carbon, its two carbon neighbors, and a single hydrogen. The fourth electron, which is in the so-called $\pi$ orbital, is not as well localized and does not participate in the bonding process as strongly as the $\sigma$ electrons. It is this $\pi$ electron that is most responsible for the scintillation process.

The incoming radiation interacts with many molecules, losing a few eV at each interaction as it excites the molecule. Many possible vibrational states can be excited (and also many possible electronic excited states; for simplicity only the lowest electronic excited state is shown). These decay quickly ($\sim$ 1 ps) to the lowest vibrational state of the electronic excited state, which then decays (in a time of order 10 ns) to one of the vibrational states of the electronic ground state. These in turn decay quickly to the vibrational ground state.

Under normal circumstances, at room temperature all of the molecules of the scintillator are in the lowest vibrational state of the electronic ground state. The thermal energy $kT$ at room temperature is 0.025 eV, and thus according to the Boltzmann population distribution $e^{-E/kT}$, it is unlikely to find any population of the vibrational states above the electronic ground state. Thus only one of the many emitted photon transitions has any probability to be absorbed. This represents an important property of a scintillator: *it must be transparent to its own radiation*.

Of the inorganic scintillators, the most common variety is the single crystal of an alkali halide; NaI is the most frequently used. A single crystal is needed to obtain transparency; reflections and absorption at the crystal faces would make a polycrystalline scintillator useless. The cooperative interactions of the atoms in a crystal cause the discrete energy levels to "smear out" into a series of energy bands. The two highest bands are the *valence band* and the *conduction band* (Figure 7.16). In an insulating material such as NaI, the valence-band states are generally full and the conduction-band states are empty. An incoming radiation can excite an electron across the *energy gap* (about 4 eV) and into the conduction band; eventually, it loses energy by emission of a photon and drops back into the valence band.

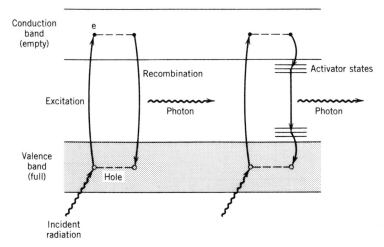

**Figure 7.16** Energy bands in a crystal. At the left are shown processes characteristic of a pure crystal such as NaI. At the right are shown the processes in the presence of an activator, such as Tl in NaI(Tl).

**Table 7.1** Properties of Some Common Scintillators

| Name | Type | Density (g/cm³) | Index of Refraction | Wavelength of Maximum Emission (nm) | Relative Output[a] | Time Constant (ns) |
|---|---|---|---|---|---|---|
| Anthracene | Organic solid | 1.25 | 1.62 | 447 | 0.43 | 30 |
| Pilot B | Plastic (organic solid) | 1.03 | 1.58 | 408 | 0.30 | 1.8 |
| NE 213 | Organic liquid | 0.87 | 1.508 | 425 | 0.34 | 3.7 |
| NaI(Tl) | Inorganic crystal | 3.67 | 1.85 | 410 | 1.00 | 230 |
| CsF | Inorganic crystal | 4.11 | 1.48 | 390 | 0.05 | 5 |

[a] The relative output includes the typical efficiency for photon absorption and the resulting light output.

To increase the probability for photon emission and to reduce self-absorption of the light, small amounts of impurities called *activators* are added to the crystal. A commonly used activator is thallium, and so these detectors are indicated as, for instance, NaI(Tl). The activator provides states in the energy gap and the light emission takes place between the activator states. In the case of NaI, the wavelength of maximum emission is shifted from 303 nm in pure NaI to 410 nm in NaI(Tl). Absorption at this energy cannot occur in NaI(Tl), because the activator ground states are not populated, and the change in wavelength from the ultraviolet to the visible gives a better overlap with the maximum sensitivity of most photomultiplier tubes.

Table 7.1 shows some properties of some commonly used scintillation detectors. The actual choice of scintillator will depend on the type of experiment that is being done. For example, where high efficiency for γ rays is concerned,

**Figure 7.17** (Left) A selection of NaI scintillators, some with photomultiplier tubes attached. (Right) A large NaI detector viewed by seven photomultipliers. Reproduced with permission of Harshaw / Filtrol Partnership.

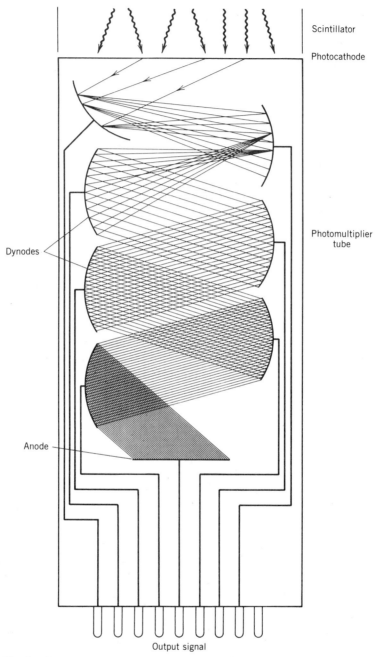

Scintillator

Photocathode

Photomultiplier
tube

Dynodes

Anode

Output signal

**Figure 7.18** Schematic of photomultiplier operation. Electrons released from the cathode are attracted to the first dynode and multiplied. Each successive dynode is at a higher potential that the previous one; a typical tube might have 10 or 14 dynodes. At each stage, the number of electrons increases by a factor of the order of 5.

NaI(Tl) is usually the choice, the large $Z$ (53) of I giving a high probability for photon absorption. However, for precise timing, NaI(Tl) is not very good and the relatively less efficient plastic scintillator may be a better choice.

The coupling of a scintillator to a photomultiplier tube can be done in a variety of ways. Some detector-tube combinations are purchased as a sealed unit. NaI(Tl) detectors can be placed into direct contact with the glass of the PM tube using transparent "optical grease" to provide a relatively uniform change in index of refraction and minimize internal reflection. Sometimes the photomultiplier geometry is very different from the scintillator geometry or it must be located far away from the scintillator (to eliminate the effects of magnetic fields, for instance). In this case a "light pipe" is used; light pipes can be cut to any size or shape out of any ordinary transparent material such as Lucite. Both the scintillator and the light pipe must be wrapped with reflective material to improve the efficiency of light collection. Figure 7.17 shows a selection of scintillation detectors, light pipes, and PM tubes.

A schematic diagram of a PM tube is shown in Figure 7.18. A small number of electrons (smaller than the number of incident photons) is released at the photocathode, then multiplied and focused by a series of electrodes called *dynodes*. The dynodes are connected to a voltage chain produced by a high-voltage supply and a series of voltage dividers. The typical potential difference between adjacent dynodes is about 100 V, and thus electrons strike the dynodes with about 100 eV of energy. The dynodes are constructed of materials with a high probability of secondary electron emission; it may take 2–3 eV to release an electron and thus a gain in the number of electrons of factors of 30–50 is possible. However, because the electrons are released in random directions in the material, relatively few will actually be released at the surface, and a gain of 5 at each dynode is more typical. Even so, with a 10-dynode tube, the overall gain would be $5^{10}$ (about $10^7$). For energy spectrometers, two important characteristics are *linearity* and *stability*. Linearity means that the amplitude of the eventual output pulse must be directly proportional to the original number of scintillation events and thus in turn to the energy deposited in the detector by the radiation. Because the gain of each dynode stage depends on the voltage difference, any change in the high voltage will cause a variation in the output pulse; thus it is often necessary to stabilize the high-voltage supply.

A wide variety of photomultiplier tubes is available; choices may be determined by such parameters as physical size, response of photocathode to different incident wavelengths, sensitivity of photocathode, gain, noise level, and timing characteristics.

## 7.4 SEMICONDUCTOR DETECTORS

As we discussed in the previous section, solid semiconducting materials (germanium and silicon) are alternatives to scintillators for radiation detectors. Both Ge and Si form solid crystals in which the valence-4 atoms form four covalent bonds with neighboring atoms. All valence electrons thus participate in covalent bonds, and the band structure shows a filled valence band and an empty conduction band. The difference between an insulator and a semiconductor is in the size of the energy gap, which is perhaps 5 eV in an insulator and 1 eV in a semiconductor. At room temperature a small number of electrons (perhaps 1 in $10^9$) is thermally excited across the gap and into the conduction band, leaving a valence-band vacancy known as a "hole." As an electron from a neighboring atom fills the hole (creating in the process a new hole) the hole appears to migrate through the crystal (but of course the positively charged atoms do not move).

To control the electrical conduction in semiconductors, small amounts of materials called _dopants_ are added. In the process of doping, atoms with valence 3 or 5 are introduced into the lattice. In the case of valence-5 atoms (P, As, Sb),

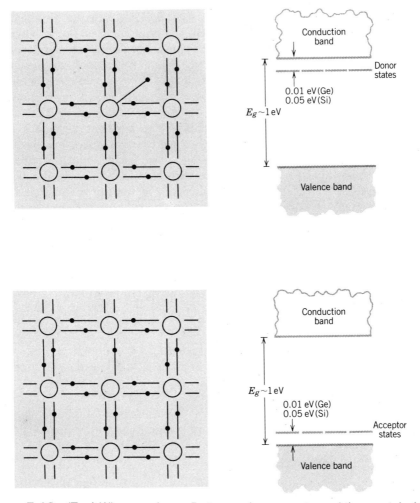

**Figure 7.19**   (Top) When a valence-5 atom replaces an atom of the crystal of Ge or Si, there is an extra electron that does not share in a covalent bond and that is easily excited into the conduction band. These atoms form donor states just below the conduction band. The material doped with the valence-5 impurity becomes an _n_-type semiconductor. (Bottom) When a valence-3 atom is used as the dopant, a vacancy or hole is formed which can easily accept an electron from the valence band. The hole can migrate through the material, which becomes _p_-type semiconductor.

four of the electrons form covalent bonds with neighboring Si or Ge; the fifth electron moves easily through the lattice and forms a set of discrete *donor states* just below the conduction band. Because there is an excess of negative charge carriers (electrons) this material is called *n-type semiconductor*. Alternatively, we could use valence-3 atoms, in which attempts to form covalent bonds with four neighboring atoms in the crystal would produce an excess of holes. These form *acceptor states* just above the valence band, and the material is called *p-type semiconductor* because the primary charge carriers are the positively charged holes. Figure 7.19 shows the band structure of the types of semiconductors.

It is useful to remember that *n*-type and *p*-type refer to the sign of the charge of the primary carriers of electric current. The materials themselves are electrically neutral.

When *p*-type and *n*-type materials are brought into contact, the electrons from the *n*-type material can diffuse across the junction into the *p*-type material and

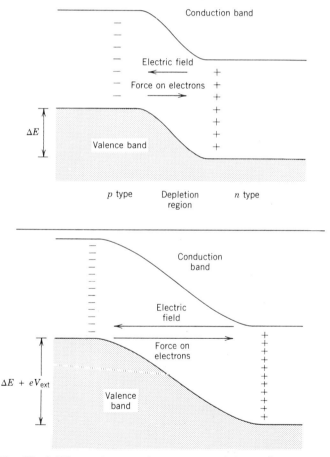

**Figure 7.20**  (Top) When *n*-type and *p*-type materials are brought into contact, electrons and holes near the junction can recombine to create a region that is depleted of charge carriers. (Bottom) Under reverse bias (when the − terminal of an external battery is connected to the *p*-type side), the depleted region becomes larger and the magnitude of the electric field increases.

combine with the holes; in the vicinity of the junction, the charge carriers are neutralized, creating a region called the *depletion region*. The diffusion of electrons from the *n*-type region leaves behind ionized fixed donor sites, while the similar diffusion of holes from the *p*-type region leaves behind negatively charged fixed acceptor sites. The space charge from the fixed sites creates an electric field which eventually halts further migration. The result, shown in Figure 7.20, is a *junction diode*.

If radiation enters the depletion region and creates electron-hole pairs, the result is very similar to that of an ionization chamber (in fact, the depletion region looks very much like a parallel-plate capacitor). Electrons flow in one direction, holes in the other, and the total number of electrons collected can form an electronic pulse whose amplitude is proportional to the energy of the radiation.

In practice, these detectors are operated with large reverse bias voltages (1000–3000 V), which has two effects: it increases the magnitude of the electric field in the depletion region, making charge collection more efficient, and it increases the dimensions of the depletion region (thereby increasing the sensitive volume of the detector) by forcing more carriers to drift across from one type of material to the other.

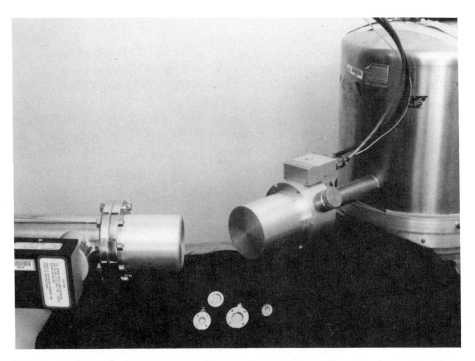

**Figure 7.21** In the foreground are shown four different sizes of silicon surface-barrier detectors. Two large Ge γ-ray detectors are housed in the evacuated aluminum cryostats of diameter about 8 cm; the detectors are cylinders about 4 cm long and 4 cm in diameter. Each detector is mounted on a copper rod that is in thermal contact with a dewar of liquid nitrogen, one of which is visible at right. The small boxes mounted along the cryostats are preamplifiers.

One common fabrication procedure for Ge and Si detectors is to begin with a sample of *p*-type material and diffuse into its surface a concentration of Li atoms, which tend to form donor states and thereby create a thin *n*-type region. Under reverse bias, and at a slightly elevated temperature, the Li drifts into the *p*-type region, making quite a large depletion region. Such detectors are known as *lithium-drifted* Ge and Si detectors, or Ge(Li) and Si(Li) detectors (pronounced "jelly" and "silly"). Following the Li drift, the Ge(Li) detector must be kept cold (usually at liquid nitrogen temperature of 77 K); otherwise the Li will migrate out of its lattice sites in the depletion region and destroy the effectiveness of the detector. Keeping the detector cold also reduces the thermal excitation of electrons across the energy gap, thereby reducing the background electrical noise produced by the detector. Recently, large-volume, high-purity Ge detectors have become available, owing to advances in the techniques of refining Ge crystals. These detectors do not need to be kept at 77 K, but are operated at that temperature to keep the noise level low. Figure 7.21 shows a selection of solid-state detectors.

The *n*-type layer created in producing Ge(Li) or Si(Li) detectors is of order 1 mm thick, which is easily penetrable by medium-energy $\gamma$ rays (the range of a 100-keV photon in Ge is about 4 mm and in Si about 2 cm). However, for charged particles the range is much smaller (for a 1-MeV electron, the range is about 1 mm in Si and Ge; for a 5-MeV $\alpha$ particle, the range is only 0.02 mm in Si and Ge), and such a layer would prevent the particles from reaching the depletion region. For charged-particle work, the preferred choice is the *surface-barrier* detector, in which an extremely thin *p*-type layer is etched or deposited on the surface of *n*-type Si. A thin layer of gold is then evaporated onto the front surface to serve as one electrical contact. The total thickness that the particles must penetrate to reach the depletion region is thus made to be about 0.1 $\mu$m.

The time necessary to collect the charge from a large-volume detector may be in the range of 10–100 ns; variations occur depending on the geometry of the detector (planar or coaxial) and on the point of entry of the radiation relative to the electrodes.

## 7.5  COUNTING STATISTICS

All laboratory measurements contain sources of uncertainty or error. Some originate with the properties of the measuring instrument (such as trying to estimate fractions of a millimeter on a meter stick graduated in millimeters). Others, of which radioactive decay is one, originate with the inherent statistical variations of a process whose occurrence is essentially random. If we make a single measurement of a phenomenon governed by a random, statistical process, then the outcome of the measurement is useful to us only if we can provide the answers to two questions: (1) How well does the measurement predict the outcome of future measurements? (2) How close to the "true" value is the outcome of a single measurement likely to be? To answer these questions, we must know how the various possible outcomes are distributed statistically.

Let us suppose we have a sample of *N* nuclei, and we wish to compute the probability $P(n)$ that *n* of them decay in a certain time interval. (For the moment we assume that we can measure time with arbitrarily small uncertainty.)

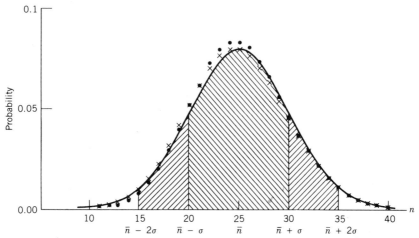

**Figure 7.22** Comparison of binomial (●), Poisson (×), and Gaussian (—) distributions for $\bar{n} = 25$. The approximation of using the Gaussian distribution improves as $\bar{n}$ increases. The area under the Gaussian distribution between $\bar{n} - \sigma$ and $\bar{n} + \sigma$ is 68% of the total area under the curve; thus any single measurement has a 68% probability to be within $\pm\sigma$ of the true mean $\bar{n}$. The area between the limits $\bar{n} \pm 2\sigma$ is 95% of the total area.

The probability for decay of a single nucleus is $p$. We assume that $p$ is constant —each nucleus decays independently of the state of the other nuclei. The desired probability can be found from the *binomial distribution*

$$P(n) = \frac{N!}{n!(N-n)!} p^n (1-p)^{N-n} \tag{7.24}$$

The binomial distribution is most often encountered in simple random experiments, such as tossing a coin or rolling dice. The distribution is shown in Figure 7.22 and is characterized by its mean $\bar{n}$, which is equal to $pN$ (as one would guess —the decay probability per nucleus times the total number of nuclei gives the number of decays) and also by the *variance* $\sigma^2$:

$$\bar{n} \equiv \sum_{n=0}^{N} nP(n) = pN \tag{7.25}$$

$$\sigma^2 \equiv \sum_{n=0}^{N} (n - \bar{n})^2 P(n) = \bar{n}(1-p) \tag{7.26}$$

The *standard deviation* $\sigma$ is a rough measure of the "width" of the statistical distribution.

When $n$ and $N$ are small, the binomial distribution is quite trivial to use, but when $n$ and $N$ are very large, as might be typical in decay processes, the distribution becomes less useful as the computations become difficult. We can obtain a less cumbersome approximation in the case when $p \ll 1$ (which will usually be true for radioactive decays):

$$P(n) = \frac{\bar{n}^n e^{-\bar{n}}}{n!} \tag{7.27}$$

where $\bar{n} = pN$ as before. This is known as the *Poisson distribution*. Note that the probability to observe $n$ decays depends only on the mean value $\bar{n}$. For this distribution, $\sigma = \sqrt{\bar{n}}$.

Another useful approximation occurs in the case of small $p$ and large $\bar{n}$ ($\gg 1$), in which case the *normal or Gaussian distribution* can be used:

$$P(n) = \frac{1}{\sqrt{2\pi\bar{n}}} e^{-(n-\bar{n})^2/2\bar{n}} \qquad (7.28)$$

and once again, $\sigma = \sqrt{\bar{n}}$.

For most practical purposes, we can use the Gaussian distribution for our statistical analysis. As shown in Figure 7.22, the Gaussian distribution has the property that 68% of the probability lies within $\pm\sigma$ of the mean value $\bar{n}$. Unfortunately the mean value $\bar{n}$ is not available for measurement; it results only from an *infinite* number of trials. Clearly the "true" value we are seeking is represented by $\bar{n}$, and our single measurement $n$ has a 68% chance of falling within $\pm\sigma$ of $\bar{n}$. We therefore take $n$ as the best guess for $\bar{n}$ and we quote the error limit of $n$ as $\pm\sigma$, or $\pm\sqrt{n}$.

If we were to repeat the measurement a large number of times, we could trace out a histogram that should come fairly close to the Gaussian distribution, and the fractional error would be reduced each time. Suppose we made $m$ independent measurements and recorded a total of $M$ counts ($M = \sum_{i=1}^{m} n_i$). Then

$$\sigma_M = \sqrt{\sum_{i=1}^{m} \sigma_i^2}$$

$$= \sqrt{\sum n_i}$$

$$= \sqrt{M} \qquad (7.29)$$

$$\bar{n} = \frac{M}{m}$$

$$\sigma_{\bar{n}} = \frac{\sigma_M}{m} = \frac{\sqrt{M}}{m} = \sqrt{\frac{\bar{n}}{m}} \qquad (7.30)$$

The standard deviation (error) of our best estimate for the true value $\bar{n}$ has been reduced by the factor $1/\sqrt{m}$. Each successive independent measurement reduces the uncertainty of the mean.

Any quantity that has a random statistical distribution of this type (the probability for any one event is small and independent of other events; the number of observed events is large) has the fundamental property that the expected uncertainty in the number of observations can be estimated as the square root of the number of observations. This applies not only to counts observed in a detector of radioactive decays, but also to the production of ionizing events in the detector itself. As we see in the next section, the energy resolution of a detector is determined by the square root of the number of ionizing events.

Finally, a word of caution—the square-root relationship applies only to raw numbers of events, not to calculated quantities such as rates or differences between numbers of counts. That is, if a detector records $N$ counts in time $t$, the counting rate is $N/t$ with an uncertainty of $N^{1/2}/t$. If we wish to subtract $B$ background counts, the uncertainty in the difference $N - B$ is $(N + B)^{1/2}$, following the usual rules for propagation of error.

## 7.6 ENERGY MEASUREMENTS

A schematic diagram of the equipment that might be used to measure the energy of nuclear radiations is shown in Figure 7.23. The electronic signal from the detector usually goes directly to a *preamplifier* (preamp), which converts the charge pulse from the detector to a voltage pulse (for example, by charging a capacitor) and then drives the pulse to the next element in the circuit. The *amplifier* provides the voltage gain to bring the millivolt preamp pulse to the range of a few volts where it can conveniently be processed. The amplifier must be *linear*, so that the proportionality of radiation energy and pulse height can be preserved. The many pulse heights that may be produced by a complex decay process can conveniently be displayed on a *multichannel analyzer* (MCA) in histogram style, that is, with pulse height on the horizontal scale and number of pulses on the vertical. The input pulses are digitized, and the digital pulse height is stored in a memory location referred to as a *channel*; hence, the horizontal axis is often labeled as *channel number*. The resulting pulse-height *spectrum* can then be used to determine the energies of the radiations emitted by the source (from their locations on the horizontal scale) and their relative intensities (from the areas of the various peaks in the spectrum). This is most frequently done for γ radiation, which is now discussed in more detail.

Figure 7.24 shows some of the processes that can occur when a γ-ray photon enters a solid detector. The photon can Compton scatter several times; after each scattering, the photon loses some energy and a free electron is produced. Gradually the photon suffers either of two fates: it continues the repeated Compton scattering, eventually becoming so low in energy that photoelectric absorption occurs and the photon vanishes, or else it wanders too close to the edge of the crystal and scatters out of the crystal. The energy of the photon is converted into electrons (photoelectrons or Compton-scattered electrons), which have a very short range in the crystal, and which therefore lose energy very rapidly, by creating light photons in a scintillator or electron-hole pairs in a

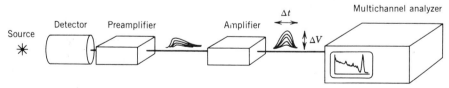

**Figure 7.23** Schematic diagram of electronic equipment that might be used in a measurement of the energies of radiations emitted by a source. The pulses between the preamplifier and amplifier generally have a short (ns) rise time and a long (ms) decay time, with an amplitude of millivolts. The amplifier pulses are more symmetric, with a width $\Delta t$ of μs and a pulse height $\Delta V$ of a few volts. The multichannel analyzer display shows $\Delta V$ on the horizontal axis.

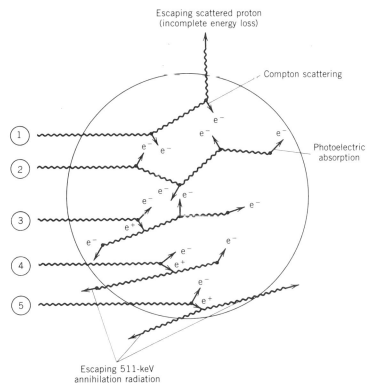

**Figure 7.24** Processes occurring in γ-ray detection. (1) The photon Compton scatters a few times and eventually leaves the detector before depositing all its energy. (2) Multiple Compton scattering is followed by photoelectric absorption, and complete energy deposition occurs. (3) Pair production followed by positron annihilation, Compton scattering, and photoelectric absorption; again, complete energy loss occurs. (4) One of the annihilation photons leaves the detector, and the γ ray deposits its full energy less 511 keV. (5) Both annihilation photons leave the detector, resulting in energy deposition of the full energy less 1022 keV. Processes (4) and (5) occur only if the γ-ray energy exceeds 1022 keV.

semiconductor detector. We can assume that all of this energy is absorbed, and we will refer to this quantity as the energy deposited in the detector by the original photon. If the original photon eventually suffers photoelectric absorption, the energy deposited is equal to the original γ-ray energy. If it scatters out of the crystal, the energy deposited is less than the original photon energy.

Let's consider how much energy is given to the scattered electron in a single Compton event. From Equation 7.15, we can find the electron kinetic energy:

$$T_e = E_\gamma - E'_\gamma = \frac{E_\gamma^2(1 - \cos\theta)}{mc^2 + E_\gamma(1 - \cos\theta)} \tag{7.31}$$

Since all scattering angles can occur in the detector, the scattered electron ranges in energy from 0 for $\theta = 0°$ to $2E_\gamma^2/(mc^2 + 2E_\gamma)$ for $\theta = 180°$. These electrons will normally be totally absorbed in the detector, and (if the scattered photons escape) they contribute to the energy response of the detector a continuum called

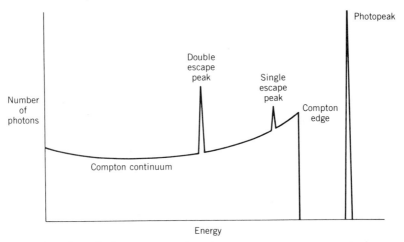

**Figure 7.25** A typical response of a detector to monoenergetic γ rays. The photopeak results from the γ ray losing all its energy in the detector, as in events 2 and 3 in Figure 7.24. The Compton continuum consists of many events of type 1, while the single- and double-escape peaks result from processes 4 and 5. The detector energy resolution might tend to broaden all peaks more than they are shown here, and multiple Compton scattering will fill in the gap between the Compton edge and the photopeak. The escape peaks appear only if the γ-ray energy is above 1.022 MeV.

the *Compton continuum*, ranging from zero up to a maximum known as the *Compton edge*. (This continuum is not flat; the Klein-Nishina formula, Equation 7.16, shows how the Compton scattering probability varies with angle.) The peak at $E = E_\gamma$ corresponding to complete photoelectric absorption (called the *full-energy peak* or *photopeak*) and the Compton continuum are shown in Figure 7.25.

We have so far neglected the third process of γ-ray interactions in the detector, that of pair production. The positron and electron are created with a total kinetic energy of $E_\gamma - 2mc^2$, as in Equation 7.18, and loss of that energy in the detector would result in a peak at the full energy. However, once the positron slows down to an energy near that of the atomic electron, *annihilation* takes place, in which the positron and an atomic electron disappear and are replaced by two photons of energy $mc^2$ or 511 keV. These two photons can travel out of the detector with no interactions, or can be totally or partially absorbed, through Compton scattering processes. We therefore expect to see peaks at $E_\gamma - 2mc^2$ (when both photons escape), $E_\gamma - mc^2$ (when one escapes and the other is totally absorbed), and $E_\gamma$ (when both are totally absorbed). These single- and double-escape peaks are shown in Figure 7.25.

The relative amplitudes of the photopeak, Compton continuum, and escape peaks depend on the size and shape of the detector. In general, the larger the detector, the smaller is the Compton continuum relative to the photopeak, for there is a smaller chance of a Compton-scattered photon surviving from the center to the surface without interacting again. Similarly, in a large detector, there is a greater chance of capturing one or both of the 511-keV annihilation photons.

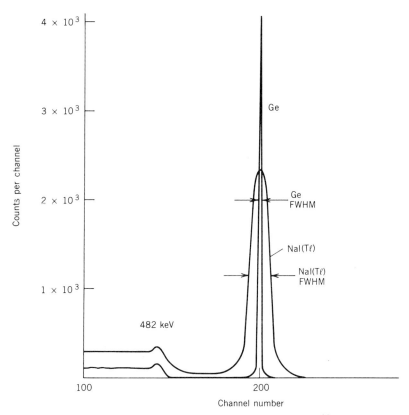

**Figure 7.26** Comparison of NaI(Tl) and Ge(Li) spectra of $^{137}$Cs. The energy of the photopeak is 662 keV. The resolution (FWHM) of the NaI(Tl) is about 40 keV, while that of the Ge is about 1 keV. The intensity (peak area) of the Ge is about 11% that of NaI(Tl).

Figure 7.26 shows MCA spectra of the decay of $^{137}$Cs, such as might be obtained with Ge and NaI(Tl) detectors. Only a single $\gamma$ ray, of energy 662 keV, is emitted in this decay. The Compton continuum is easily seen, as is the Compton edge. The "valley" between the Compton edge and the photopeak does not go quite to zero and the Compton edge itself is not sharp; in our previous discussion we assumed the Compton continuum to originate from a single-scattering event, and multiple-scattering events will distort the simple picture of Figure 7.25. The Compton edge is expected at $E = 478$ keV, in agreement with the observed spectrum.

What is striking about the two spectra shown are the differences in efficiency (essentially, the area of photopeak) and in resolution (the width of the photopeak) between Ge and NaI(Tl). NaI detectors have higher efficiencies than Ge detectors, and in addition have the advantages of lower cost (1/10 or less of the Ge cost) and simpler operating conditions (no cooling is required for NaI). Because the present demands of nuclear spectroscopy require the study of ever more complex decays, resolution has become of critical importance; to study these decays carefully, one must be able to determine all $\gamma$-ray energies and intensities, which would not be possible if all peaks had widths characteristic of NaI(Tl) detectors. Early work with Ge detectors suffered a considerable loss in efficiency to obtain the good resolution; these detectors had efficiencies only a few percent of NaI(Tl) detectors. Improvements in refining techniques of Ge have enabled the production of large-volume Ge detectors that are now only a factor of 2–3 below NaI(Tl) in efficiency. Because Ge has a lower atomic number than NaI, it will always have smaller interaction probabilities for photons and therefore smaller relative efficiency, the photoelectric absorption coefficient varying roughly as $Z^4$.

Let's now try to understand the reasons for the observed resolutions based on the statistics of the detection process.

The full-energy peaks of both NaI(Tl) and Ge detectors can be approximated as Gaussian shapes, and the width is characterized by the parameter $\sigma$ in the general form of the Gaussian distribution

$$f(E) = A\,e^{-(E-\bar{E})^2/2\sigma^2} \tag{7.32}$$

where $A$ is a normalization constant. The relationship between the mean $\bar{E}$ and $\sigma$ cannot be used because we do not know the number of events $\bar{n}$ that are represented in the mean. (Recall that $\sigma = \sqrt{\bar{n}}$ holds only for counting events.)

Generally the width is specified in terms of the full width at half maximum (FWHM), that is, the distance $\Delta E$ between the two points $E_1$ and $E_2$ where $f(E_1) = f(E_2) = A/2$. A bit of manipulation gives

$$\Delta E = 2\sigma\sqrt{2\ln 2} \cong 2.35\sigma \tag{7.33}$$

Often the FWHM is expressed as a ratio $\Delta E/\bar{E}$.

To estimate $\sigma$, we must estimate $\bar{n}$, the number of statistical events associated with the production of the detector signal. The incident $\gamma$-ray energy is 662 keV. In NaI(Tl) the scintillation efficiency (the fraction of the incident radiation converted into light) is about 13%; thus 86 keV of photons appear. The energy per photon in NaI is about 4 eV, and the number of light photons is thus (on the average) about 21,000. The contribution to the resolution from this is about $2.35\sqrt{21,000}/21,000$ or about 1.6%. At the photocathode, the number is reduced even further. The transmission of the light through the glass end of the photomultiplier tube is typically about 85%, and the *quantum efficiency* of a typical photocathode (that is, the number of photoelectrons emitted per incident photon) is typically 23%. Thus the number of photoelectrons is only about 20% of the number of incident photons, or about 4200, and the contribution to the resolution is about $2.35\sqrt{4200}/4200$, or 3.6%. The electron multiplication process in the photomultiplier will increase the number of events and give a smaller contribution to the FWHM. *In the entire NaI(Tl) detection process, the smallest number of events corresponds to the production of electrons at the photocathode, and this is therefore the most substantial contributor to the energy resolution.*

This simple calculation has ignored a number of effects in the crystal, photomultiplier, and amplifier, all of which can make nonstatistical contributions to the resolution. A figure of 6% (40 keV) might be more typical at 662 keV. The absolute FWHM (the width of the energy peak) increases with energy, roughly as $E^{1/2}$, but the ratio $\Delta E/E$ decreases like $E^{-1/2}$; thus at 1 MeV, the resolution should be about 5% (50 keV).

In a Ge detector, there is only a single event contributing to the statistics—the creation by the photon of electron-hole pairs. In Ge, it takes on the average about 3 eV to create an electron-hole pair, and thus the mean number of statistical events when a 662-keV photon is fully absorbed would be about 220,000. The contribution to the resolution is then about 3.3 keV (by convention, NaI resolutions are usually expressed as percentages, and Ge resolutions in keV). This gives a resolution more than an order of magnitude better than NaI(Tl), and this difference can be understood simply on the basis of the properties of the absorption of radiation in the detectors.

We have failed to include a number of factors in our estimate for the Ge resolution. The absorption is in fact not well described by Poisson statistics; proper consideration of the statistical nature of the process (done empirically) reduces the calculated value to about 1.0 keV. Nonstatistical processes (collection of charges by the electric field, electronic noise in the preamp and amplifier) will tend to increase the value somewhat. A typical value for a good detector today is 1.7 keV at 1332 keV (the energy of a $^{60}$Co $\gamma$-ray, taken as the standard for resolution measurement), which would correspond to about 1.2 keV at 662 keV, if the $E^{1/2}$ dependence is valid.

In $\gamma$-ray (or other radiation) spectroscopy measurements, the goal is usually to determine the energy and the intensity of the radiation. To find the energy, the centroid of a peak must be determined. For isolated, well-resolved peaks, the centroid can be determined by a simple numerical procedure. First it is necessary to subtract the background. (The peak may be on the Compton continuum of other peaks higher in energy.) This is usually done by drawing a straight line between groups of background channels above and below the peak (Figure 7.27). In this case a linear background is assumed, and the background counts can be

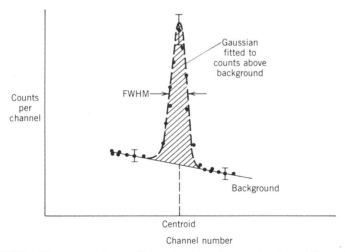

**Figure 7.27** The area of a well-resolved peak can be found by subtracting a linear background and then either adding the counts above background or fitting a Gaussian function to the counts above background. Because the number of counts in each channel results from a simple counting process, its uncertainty is just the square root of the number; this applies only to the total number in each channel and *not* to the number above background.

subtracted directly. The centroid and area can then be determined by

$$\text{area} = \sum y_i \tag{7.34}$$

$$\text{centroid} = \frac{\sum x_i y_i}{\sum y_i} \tag{7.35}$$

where $y_i$ represents the net number of counts above background in channel $i$. A slightly more sophisticated method is to fit a Gaussian function to the peak. This can be done most easily by assuming the functional dependence of the form of Equation 7.32 and taking the logarithm:

$$\ln y_i = \ln A - \frac{(x_i - \bar{x})^2}{2\sigma^2} \tag{7.36}$$

A least-squares fit to this form gives the parameters $\bar{x}$, $\sigma$, and $A$. Integrating the Gaussian form gives

$$\text{area} = \sigma A \sqrt{2\pi} \tag{7.37}$$

In the case of complicated spectra, this simple procedure will not work. Often backgrounds are not at all well represented by the linear approximation. Close-lying peaks that may overlap cannot be fitted in this way. Even the assumption of a Gaussian shape is not always valid, as there may be exponential "tails" on the high-energy or low-energy sides of the Gaussian. In this case, there are sophisticated fitting programs that can handle many parameters, including centroids and areas of many peaks and backgrounds of nonlinear shape.

Once we have values for the centroid and area of a peak, we would like to use those values to obtain the energy and counting rate for the radiation. To find the energy, it is necessary to calibrate the MCA so that channel number can be converted to energy. This is usually done with two or more radiations of known energy $E_1$ and $E_2$, which would be found to have centroids $\bar{x}_1$ and $\bar{x}_2$. It is then

**Table 7.2** Commonly Used Energy Calibration Standards

| Nuclide | $t_{1/2}$ | Radiation | Energy (keV) |
|---------|-----------|-----------|--------------|
| $^{109}$Cd | 453 d | $\gamma$ | $88.037 \pm 0.005$ |
| $^{57}$Co | 271 d | $\gamma$ | $122.06135 \pm 0.00013$ |
| | | | $136.47434 \pm 0.00030$ |
| $^{198}$Au | 2.696 d | $\gamma$ | $411.80441 \pm 0.00015$ |
| $^{137}$Cs | 30.17 y | $\gamma$ | $661.661 \pm 0.003$ |
| $^{60}$Co | 5.271 y | $\gamma$ | $1173.238 \pm 0.015$ |
| | | | $1332.513 \pm 0.018$ |
| $^{207}$Bi | 38 y | $e^-$ | $481.65 \pm 0.01$ |
| | | | $975.63 \pm 0.01$ |
| $^{241}$Am | 433 y | $\alpha$ | $5485.74 \pm 0.12$ |
| $^{226}$Ra | 1600 y | $\alpha$ | $4784.50 \pm 0.25$ |
| | | | $5489.66 \pm 0.30$ |
| | | | $6002.55 \pm 0.09$ |
| | | | $7687.09 \pm 0.06$ |

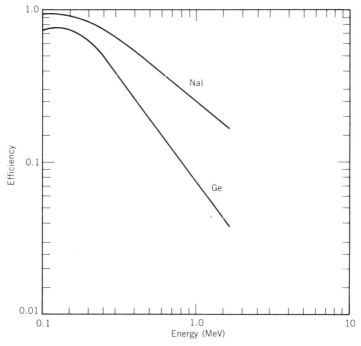

**Figure 7.28** Relative efficiencies of NaI and Ge detectors. Here "efficiency" means the probability that a photon *striking the detector* will appear in the photopeak; it does not take into account the differing sizes of the detectors. The curves are drawn for a 3-in.- diameter by 3-in.-long NaI(Tl) and a 4.2-cm-diameter by 4.2-cm long Ge, with the source at 10 cm from the detector. In the terms that are usually used to specify relative efficiencies, the relative probability for a 1.332-MeV photon (emitted by [60]Co) to appear in the Ge photopeak is 8% that of the NaI(Tl) photopeak, including the solid angle factor.

easy to find a linear relationship between $\bar{x}$ and $E$. Because MCAs (and other parts of the system, including the detector and amplifier) may be slightly nonlinear, it is advisable to choose calibration energies $E_1$ and $E_2$ as close as possible to the unknown energy $E$. It is necessary to use two radiations for calibration because channel zero of the MCA will not necessarily correspond to a pulse height of zero. Table 7.2 shows some commonly used calibration standards.

To convert the peak area to an absolute counting rate, we must know something about the efficiency of the detector—how large a solid angle does it subtend at the source of the radiation and what is the probability that an incident radiation will be absorbed into the photopeak? The detector efficiency depends on the energy rather dramatically for γ rays; Figure 7.28 shows the absolute efficiency of NaI(Tl) and Ge detectors as a function of the γ-ray energy.

## 7.7 COINCIDENCE MEASUREMENTS AND TIME RESOLUTION

When we wish to study radiations that follow one another in cascade—for example, γ rays that follow a particular β decay—equipment of the type shown in Figure 7.29 is often used. One primary object of this equipment is to determine

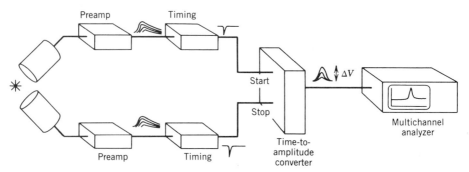

**Figure 7.29** Schematic diagram of equipment to determine whether two radiations from the source are in time coincidence (that is, whether they come close enough together in time to originate from the same nucleus in a sequential or cascade emission). The short rise time of each preamp signal triggers a timing circuit; the fast timing signals start and stop a time-to-amplitude converter (TAC), the output of which has a pulse height $\Delta V$ that is proportional to the time difference between the start and stop pulses. The spectrum of pulse heights (and therefore of times) can be displayed on a MCA.

whether the second radiation is in true time coincidence with the first. That is, do the two radiations come close enough together in time to originate from the same nucleus in a sequential or cascade emission? The time-to-amplitude converter (TAC) produces a time spectrum of output pulses whose pulse height is proportional to the difference in arrival times of the two input pulses; this in turn, after correcting for known delays in the intervening equipment and signal cables, must be related to the difference in emission times of the two radiations from the source.

A time coincidence of two radiations from the same nucleus is called a *true* or *real* coincidence. It is also possible for radiations from different nuclei to trigger the timing circuits; this produces an unwanted *chance* or *accidental* coincidence. In principle, discriminating between true and chance coincidences is relatively easy. After accepting the first radiation in the start channel of the TAC, we wait only a very short time (perhaps nanoseconds) for the second radiation to arrive and trigger the stop channel. The longer we must wait, the greater is the possibility for a radiation from a different nucleus to produce a chance coincidence. (We are assuming in this discussion that the time interval between the emission of the two radiations in the nucleus is negligibly short. The case in which this is not true is treated in the next section.) The pulse height spectrum provides a relatively trivial means to distinguish between true and chance coincidences. The pulse heights, and thus the time differences, corresponding to start and stop signals from true coincidences have a definite fixed time relationship. Chance coincidences, representing emissions from different nuclei, have no definite time relationship; if the decay rate from the source does not change, it will be just as probable to find the second radiation emitted (from a different nucleus) at any time following the first radiation. Chance coincidences produce a uniform range of pulse heights whereas true coincidences produce a single unique pulse height. Figure 7.30 illustrates the MCA display of the TAC spectrum that might result.

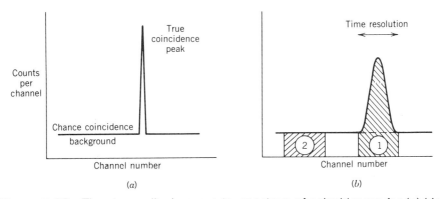

**Figure 7.30** Time-to-amplitude converter spectrum of coincidences for (a) ideal detectors and (b) real detectors. The pulse height (running along the horizontal axis, as in all MCA spectra) gives the time difference between the two pulses. The time jitter in real detectors broadens the true-coincidence peak. The area in pulse-height region 1 gives true + chance coincidences; if the background is flat, region 2 gives the chance coincidences, and the difference gives the true coincidences only.

Figure 7.30a illustrates the situation for ideal detectors and electronics, in which the timing signal is derived from the radiation entering the detector with no ambiguities or uncertainties. Actual detectors and electronics can introduce distortions (from electronic noise, for example, or from difficulties in triggering the timing circuit from the preamp pulse), and the resulting time spectrum is shown in Figure 7.30b. The sharp peak of true coincidences is broadened; it has a width that is characteristic of the *time resolution* $\tau$ of the detector and electronics system. To determine the rate of true coincidence emissions, we could electronically select regions 1 and 2 in Figure 7.30b, and the difference in their areas (if we are confident that the chance coincidence background is flat) gives the true coincidence peak.

By using this procedure, we can always correct for the chance coincidences. But if the chance rate is too high, the statistical uncertainties introduced by subtracting two large and nearly equal quantities can result in a deduced true coincidence rate with a relatively large uncertainty. It is therefore usually advisable to try to reduce the chance coincidence rate to the smallest possible level. This can be done in two ways. (1) Because the chance coincidence rate depends on the random overlap in time of two pulses, it increases as the square of the activity of the source. Doubling the activity, for example, increases the true coincidence rate by a factor of two but the chance rate increases by a factor of four. Reducing the activity therefore reduces the rate of chance coincidences. Reducing it too much, on the other hand, can reduce the rate of true coincidences so much that the statistical precision is degraded; therefore, compromises are necessary in adjusting the activity of the source. (2) Optimizing the detector and electronics can reduce the time resolution of the true coincidence peak in Figure 7.30b, which means that the corresponding chance coincidence background that must be counted is similarly reduced. The time resolution ultimately depends on the influence of random noise in the detector and electronics or in the variations in the time necessary to collect the electronic charges in the detector that form the preamp pulse. With Ge detectors, especially the large-volume crystals that are presently used for $\gamma$ ray detection, the charge collection time over the volume of the detector can limit the resolution to about 10 ns. For small NaI detectors with fast photomultipliers, values in the range 1 ns are possible, and plastic scintillators permit even smaller values, down to perhaps 100 ps.

## 7.8 MEASUREMENT OF NUCLEAR LIFETIMES

Techniques for measuring nuclear lifetimes are as varied as the lifetimes themselves—from more than $10^{15}$ y for some naturally occurring radioisotopes to below $10^{-15}$ s for short-lived excited states. Searches for double-$\beta$ decay (see Chapter 9) are conducted in the range of lifetimes of about $10^{20}$ y, and the breakup of highly unstable nuclei such as $^8$Be and $^5$He occurs with lifetimes in the range of $10^{-16}$ to $10^{-20}$ s. We will not discuss here the unusual techniques needed for these special cases, but instead we shall give examples of a few techniques for measuring lifetimes that are more commonly encountered in studying nuclear decays.

The most straightforward technique is of course to observe the exponential decay of activity as a function of time, such as was shown in Figure 6.1. Plotting $\mathscr{A}$ against $t$ on a semilog scale gives the decay constant $\lambda$ directly, and thus the half-life. For half-lives in the range of minutes to hours, this is a particularly easy laboratory exercise. We can count a certain type of radiation and display the accumulated counts on a scaler. This same technique can be used for half-lives in the range of years, although it is less convenient to follow the decay for several half-lives. As we get to significantly longer half-lives, this method becomes impractical. Figure 7.31 shows an example of this direct technique.

For very long half-lives, it is better to use the method of *specific activity*. (Specific activity indicates the quantity of activity per unit of substance, such as Ci per gram.) Here we can use the expression $\mathscr{A} = \lambda N$ directly. We can determine $\mathscr{A}$ by counting the number of decays emitted in a certain period of time, and we can determine $N$ (the number of radioactive nuclei) by such techniques as chemical analysis or mass spectrometry.

When we go to half-lives shorter than minutes, we must deal with our inability as experimenters to perform the measurement of $\mathscr{A}$ vs $t$. That is, suppose we had a decay with a 10-s half-life. We would hardly have time to take one data point and clear the scaler for the next before the sample would have decayed away. For this range, we can take advantage of a capability built into most multichannel analyzers, called *multiscaling*. In this mode of operation, the MCA accepts a logic pulse rather than a linear pulse. A variable *dwell time* can be set by the experimenter. The MCA starts in channel 1 and counts the number of logic pulses that arrive during the dwell time. It stores that number in channel 1, then moves to channel 2 and repeats the cycle. The MCA thus does all the reading and recording of the scalers for us, and we should be able simply to watch the display screen as decay curves similar to Figure 6.1 appear.

In principle we could extend this technique to shorter and shorter half-lives, but we face a limitation at about $10^{-3}$ s. Most radiation detectors cannot

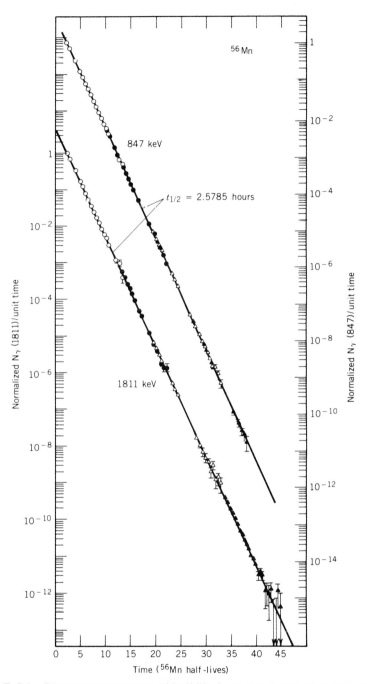

**Figure 7.31** Direct measurement of half-life from the decay of activity with time. Data from two different $\gamma$ rays in the decay of $^{56}$Mn are plotted against time on a semi-log scale over 45 half-lives. During this period the count rate changes by more than 13 orders of magnitude. From E. B. Norman, S. B. Gazes, and S. G. Crane, *Bull. Am. Phys. Soc.* **30**, 1273 (1985); courtesy E. B. Norman.

accommodate counting rates in excess of $10^5/s$. If we set the dwell time at $10^{-3}$ s, then we would accumulate a maximum of only about 100 counts per channel, and each channel would have an uncertainty of the order of $1/\sqrt{100}$, or 10%, resulting in a large uncertainty of the half-life.

We can measure half-lives shorter than $10^{-3}$ s if we have a precise way of determining the time interval between the formation of a nuclear state and its decay. The formation of a state is signaled by the observation of a radiation that leads to that state. We thus can do a *coincidence* experiment between a radiation that populates a state and a radiation from its decay. The easiest way to accomplish this is with the time-to-amplitude converter (TAC) as was shown in Figure 7.30 for cases in which the nuclear state has a half-life that is very short compared with the time resolution. If the half-life of the state is comparable to or greater than the time resolution, the TAC spectrum will show evidence of the usual sort of exponential decay that is shown by all radioactive systems. That is, the probability for the state to survive for a time $t$ after its formation decreases exponentially as the time $t$ increases. It is thus most probable that we will observe the second radiation within a short time after the formation of the state; the longer we wait, the less probable it is for the state to have survived and the

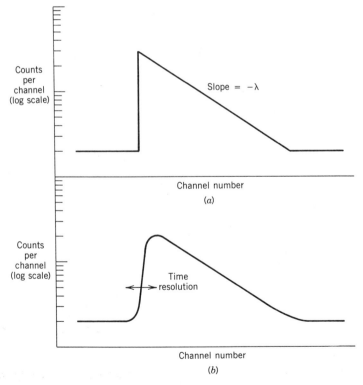

**Figure 7.32** If the nuclear state between the first and second radiations has a half-life that is *not* negligibly short compared with the time resolution, we can observe its exponential decay. These TAC spectra should be compared with those of Figure 7.30 to see the effect of the decay of the state. (*a*) Ideal detectors, (*b*) real detectors.

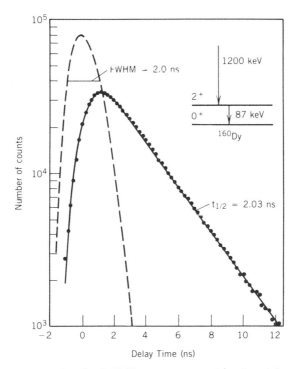

**Figure 7.33**  An example of a half-life measurement by the delayed coincidence technique. The dashed line shows the response expected for two radiations emitted essentially instantaneously, that is, within a time that is small compared with the time resolution of the system (2 ns); this is the so-called "prompt" curve. The delayed curve shows the rounding, as in Figure 7.32, owing to the finite time resolution. From the linear portion of the semilog plot, a half-life of $2.03 \pm 0.02$ ns is deduced for the 87-keV first excited state in $^{160}$Dy. Data from P. C. Lopiparo, R. L. Rasera, and M. E. Caspari, *Nucl. Phys.* **A178**, 577 (1972).

fewer of the second radiations we count. The resulting TAC spectrum is shown in Figure 7.32.

This type of experiment is an example of the *delayed coincidence* technique. An example of experimental results using this technique is shown in Figure 7.33.

The limitation on the use of the delayed coincidence technique is the ability to distinguish the prompt from the delayed curves as in Figure 7.33. That is, the time resolution must be smaller than the half-life. Typical time resolutions are of order 10 ns with Ge, 1 ns with NaI(Tl), and 0.1 ns with plastic scintillators. With careful measurement technique, the delayed coincidence method can be extended to half-lives below 10 ps, but its primary range of applicability is from $10^{-3}$ to $10^{-11}$ s.

For shorter half-lives, coincidence techniques are not applicable, and we must use a variety of other approaches. Some of these involve measuring the probability to excite a nuclear state by *absorbing* electromagnetic radiation from the ground state. Examples of such experiments will be considered in Sections 10.9 and 11.6. Other techniques are more applicable to nuclei produced in nuclear reactions. The product nuclei are allowed to recoil out of the reaction target; if

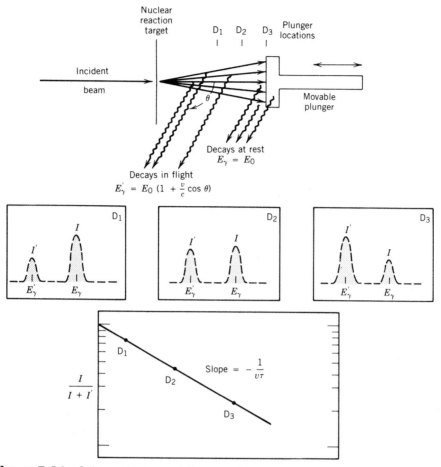

**Figure 7.34** Schematic views of the Doppler-recoil method of lifetime measurement. Decaying nuclei are observed by a detector whose axis makes an angle $\theta$ with the recoil direction. Moving the plunger changes the relative numbers of decays in flight and at rest, which can be used to determine the mean life of the decay.

they are formed in an excited state, then the $\gamma$ rays can be emitted while the nucleus is in flight. Such $\gamma$ rays will be Doppler shifted. If the beam of recoiling nuclei is allowed then to be stopped (by striking a solid target), any remaining nuclei in the excited state will decay at rest and their energies will not be Doppler shifted. We therefore see two $\gamma$-ray peaks, one at the shifted energy and one at the unshifted energy. The relative intensities of the two peaks will depend on the distance through which the recoiling nuclei travel before stopping. (If the distance is short, fewer nuclei will decay in flight.) Figure 7.34 shows a schematic view of the experiment and its outcome. The fraction of $\gamma$ rays in the unshifted peak shows an exponential dependence on the recoil distance, which in turn gives the half-life of the level. This method is useful in the range of $10^{-10}$ to $10^{-12}$ s. Below $10^{-12}$ s, the recoil distance will be so short that the technique cannot be easily used. (For a typical nonrelativistic nucleon, with $v = 0.1c$, the distance traveled in $10^{-12}$ s is 0.03 mm.) Instead, the recoiling nucleus is allowed to penetrate a solid backing directly after the reaction. The nucleus immediately begins slowing down and eventually comes to rest; the velocity decreases continuously, and thus the $\gamma$ emission varies continuously in energy from the shifted to the unshifted value. The profile of this energy distribution can be used to deduce the lifetime, once we understand the mechanism for energy loss by collisions

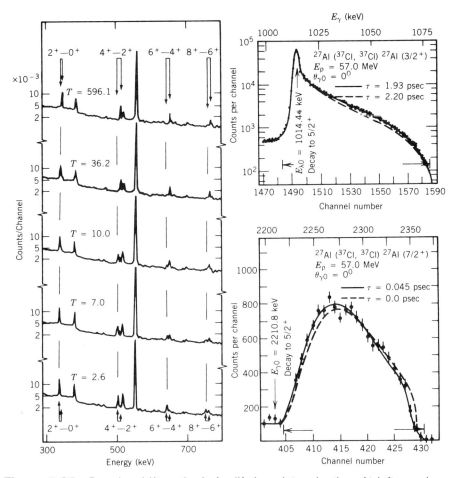

**Figure 7.35** Doppler-shift methods for lifetime determination. At left are shown raw data for transitions in the ground-state rotational band ($2^+$, $4^+$, $6^+$, $8^+$) of $^{122}$Xe. The locations of the shifted and unshifted transitions are marked at the top and bottom. These data were taken with the detector at $0°$ in the geometry of Figure 7.34; thus the shifted line is at higher, rather than lower, energy. The various spectra were taken at plunger distances corresponding to flight times of 2.6 to 596.1 ps. The deduced half-lives are: $2^+ \rightarrow 0^+$, $61.9 \pm 5.6$ ps; $4^+ \rightarrow 2^+$, $5.7 \pm 0.8$ ps; $6^+ \rightarrow 4^+$, $2.7 \pm 0.5$ ps; $8^+ \rightarrow 6^+$, $< 2.4$ ps. At the right are shown two cases in which the lifetimes are so short that the recoil technique cannot be used. The line is broadened over the entire region from the shifted to the unshifted energy. The upper case yields a half-life of $1.33 \pm 0.07$ ps and the lower case gives $0.025 \pm 0.013$ ps. From W. Kutschera et al., *Phys. Rev. C* **5**, 1658 (1972); D. Schwalm et al., *Nucl. Phys. A* **293**, 425 (1977).

in the backing material. This technique is applicable down to about $10^{-15}$ s. Figure 7.35 shows applications of both techniques.

## 7.9 OTHER DETECTOR TYPES

### Magnetic Spectrometers

In optical measurements, we use diffraction gratings to disperse light into its constituent wavelengths. We thus achieve a *spatial* separation of wavelength— different wavelengths appear at different locations, and can be recorded on a photographic film (as in a spectrograph) or recorded in intensity with a slit and a photoelectric device (as in a spectrometer). The goal in the design of an instrument to perform these measurements, as discussed in many introductory texts, is to achieve a high *resolving power* (defined as $\lambda/\Delta\lambda$, the ability to separate two nearby wavelengths at $\lambda$ differing by $\Delta\lambda$) and a large *dispersion* (defined as $\Delta\theta/\Delta\lambda$, the angular separation $\Delta\theta$ per unit wavelength interval $\Delta\lambda$).

The goals in the design of magnetic spectrometers for charged particles are similar. We would like to have a device that has a large dispersion and a small resolution. (The definition of resolution is one of many areas of disagreement in terminology between atomic and nuclear physicists. The resolving power of a grating, $\lambda/\Delta\lambda$, should be a large number; the resolution of a nuclear radiation detector, $\Delta E/E$, should be as small as possible. These two statements are equivalent, but are stated in different terms. Both state that monochromatic radiation incident on the device should emerge with the smallest possible spread in energy or wavelength.) We would also like to have the highest possible efficiency; in the case of charged particles, we would like to be able to focus monoenergetic particles traveling in different directions to a common location on the output device.

The design of magnetic spectrometers shares many features with the design of mass spectrometers, which were discussed in Section 3.2 and illustrated in Figure 3.13.

The basic operation of all magnetic spectrometers is sufficiently similar that we will provide only the broad general details of operation and leave the discussion of design and construction to more detailed works. Figure 7.36 illustrates the basic principles. Let's assume the radioactive source emits two distinct radiations of energies $E_1$ and $E_2$ ($\alpha$ particles, for instance). These are of course emitted in many different directions. In the uniform magnetic field they follow circular trajectories in which the product $Br$ determines the momentum of the particle (see Equation 3.20). Striking a recording device such as a photographic film, two distinct images are produced. All remaining details concern the design of the magnetic field to maximize the focusing effect and to improve the resolution.

A typical spectrometer for electrons is shown in Figure 7.37. The magnetic field is produced by a set of coils. For a particular value of the current in the coils (and thus of the field), electrons of one energy enter the output slit while others do not. A detector registers the electron intensity for different output currents. The resolution $\Delta E/E$ that can be obtained from such a device is typically below 0.1%, whereas the best resolution obtainable for electrons in a Si(Li) detector

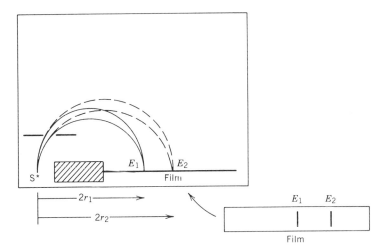

**Figure 7.36** A simple magnetic spectrometer. There is a uniform magnetic field *B* perpendicular to the plane of the paper. The momentum of the particle determines the radius of curvature *r* of its path. There is also a focusing effect, as particles emitted in a narrow range of angles are focused to a common point on the film.

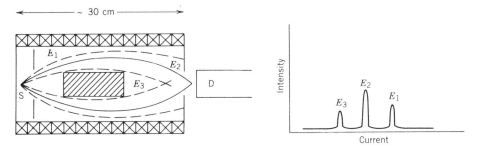

**Figure 7.37** A magnetic "lens" spectrometer designed for electrons. The operation is very similar to that of an optical lens. The coils produce a magnetic field along the axis of the system. Particles of a unique energy $E_2$ are focussed on the exit slit and reach the detector; particles of different energies are not recorded. Changing the current in the coils allows different energy groups to be brought into focus and observed by the detector.

might be 0.5%. This improvement in resolution is often critical in studies of the intensities of internal conversion electrons from different subshells (see Section 10.6).

For heavy particles, such as alphas or protons, the design principles are similar but the construction must take into account the larger mass of the particles. Larger fields are required to bend the paths of these heavy particles, and generally this calls for designs using magnetic iron rather than current-carrying coils. The resulting radii of curvature are still large, and so the physical size of these devices is much larger than that of electron spectrometers. A typical size might be several meters, and the total weight of so much iron may be 100 metric tons! Figure 7.38 shows an example of a magnetic spectrometer designed for

**Figure 7.38** High resolution proton spectrometer at the Los Alamos Meson Physics Facility. The incident proton beam enters through the pipe near the bottom left and scatters in the target chamber at the center. The scattered protons are deflected twice by the 75° vertical bending magnets and are detected at the top of the structure. The energy resolution is about 30 keV for 800-MeV protons. Photo courtesy Los Alamos National Laboratory.

heavy charged particles. The resolution is again about 0.1% or better, which is an improvement of a factor of 3–5 over Si(Li) or surface-barrier detectors.

## Counter Telescopes

A counter telescope consists of two or more counters, in which the radiation to be observed passes in sequence through the counters and is usually totally absorbed in the last counter. Generally the last counter has the largest volume, in order to achieve complete absorption of the energy of the particle; the remaining counters in the telescope are very thin, so that the particle loses only a small amount of energy $\Delta E$ in each. The $\Delta E$ counters are usually chosen for their timing properties; plastic scintillators are the most popular choice, for they have

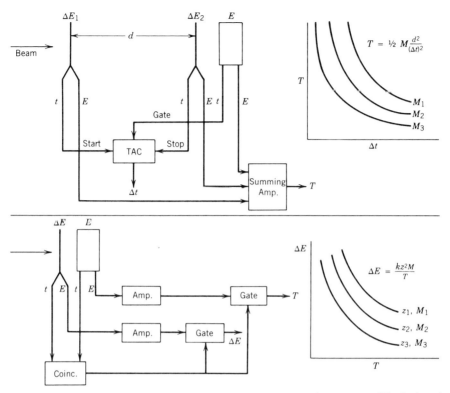

**Figure 7.39** Two different examples of counter telescopes. (Top) In the time-of-flight technique, a TAC measures the time that it takes the particle to travel the distance $d$ between the two $\Delta E$ detectors; a summing amplifier adds together the three energy losses to determine the energy of the particle. Plotting energy against $\Delta t$ gives a family of hyperbolas that determine the mass of the particle. (Bottom) The $\Delta E \cdot T$ technique also gives a family of hyperbolas, which determine both $z$ and $M$.

excellent timing and can be easily constructed in the variety of sizes and shapes needed for experiments. Other systems may use proportional counters, which can record the trajectory of the particle.

Counter telescopes have many different uses and can be constructed of a variety of different detectors. We consider only one application—their use as *particle identifiers*. (We discuss position-sensitive proportional counters in the next subsection.)

Figure 7.39 shows two simple telescopes that can be used for identifying particles. In the first method, we use two thin counters to extract timing signals; these timing signals can be used with a TAC to deduce the time that it took the particle to travel the distance between the two counters. We thus determine its velocity, and since its kinetic energy is determined from the $E$ counter, we can deduce its mass. This technique is called the *time-of-flight* method and has applications other than particle identification; it is often used in measuring energies of neutrons, for which $E$ counters do not give a photopeak and therefore give a relatively poor measure of the energy (see Chapter 12). Figure 7.40 shows a sample of the use of this technique to identify reaction products.

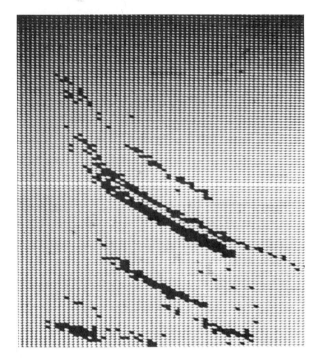

**Figure 7.40** Time-of-flight particle identification. The data appear plotted as $T$ against $\Delta t$, exactly as in Figure 7.39. From the top, the hyperbolas show $^{16}$O, $^{13}$C, $^{12}$C, $^{7}$Li, $^{6}$Li, and $^{4}$He. From W. F. W. Schneider et al., *Nucl. Instrum. Methods* **87**, 253 (1970).

The second technique involves the measurement of the energy loss in the thin counter. From the Bethe equation for stopping power, Equation 7.3, we see that $\Delta E \propto v^{-2}$ to a good approximation since the factors in the bracket will be small for nonrelativistic particles. Thus the product $\Delta E \cdot T$ is equal to $kz^2M$, where $ze$ is the charge of the particle, $M$ is its mass, and $k$ is a constant (depending on the absorbing material). Graphing $\Delta E$ against $T$ should yield a family of hyperbolas corresponding to the different values of $z^2M$. (Equation 7.7 for electrons can be written in a similar form for light relativistic particles.) Figure 7.41 shows the result of performing such a $\Delta E \cdot T$ analysis for a beam of particles.

**Multiwire Proportional Counters**

We include the multiwire proportional counter (MWPC) as an example of a detector that is sensitive to the position at which a particle interacts. We can use these detectors in a telescope arrangement to map the trajectory of a particle that results from a nuclear reaction. The basic MWPC (Figure 7.42) usually consists of two planes of individual anode and cathode wires, spaced perhaps 2–3 mm apart. The counter itself may be as large as 1 m². A charged particle passing through the chamber creates ionizing events that result in the avalanche occurring primarily in the vicinity of one of the wires. The wires are observed individually, and the output signals enable us to deduce the position of the particle within an uncertainty of the wire spacing (2–3 mm).

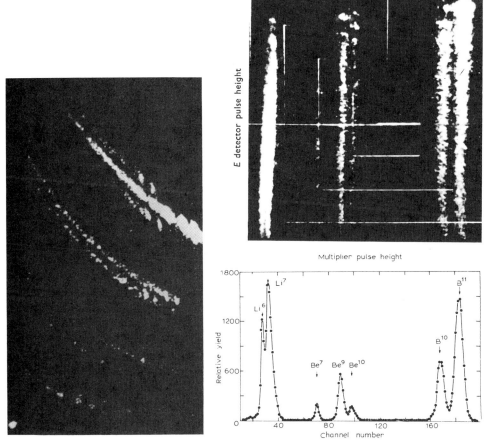

**Figure 7.41** $\Delta E \cdot T$ method of particle identification. To the left is shown the hyperbola plots in the same form as in Figure 7.39. At the top right is shown data plotted in a slightly different format: $T$ on the vertical scale and $\Delta E \cdot T$ on the horizontal scale. This way of plotting the data shows that $\Delta E \cdot T$ is indeed constant. At bottom right is shown the spectrum corresponding to the photograph; channel number gives the pulse height of the product $\Delta E \cdot T$. Notice that this method separates $^7$Li from $^7$Be and $^{10}$Be from $^{10}$B; these would not be separated in the time-of-flight technique, which is sensitive only to mass and not to charge. From M. W. Sachs et al., *Nucl. Instrum. Methods* **41**, 213 (1966).

## Polarimeters

Often we wish to measure the polarization of the observed radiation. For spin-$\frac{1}{2}$ particles, such as electrons or nucleons, the polarization means the component of the spin (up or down) relative to a particular axis. For photons, we are usually interested in the classical linear polarization as determined by the **E** vector of the electromagnetic radiation field.

The Compton scattering of photons is dependent on polarization, as shown in Figure 7.43. If we observe the intensity of the scattered radiation at different positions about the scatterer, we can deduce the degree of linear polarization of the incident radiation. A second method, used most often for measurement of

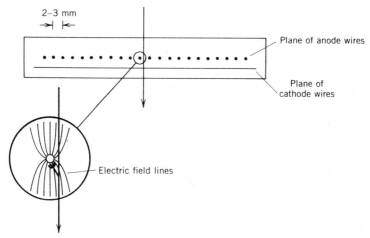

**Figure 7.42** Schematic view of multiwire proportional counter. The passage of a charged particle near one of the anode wires causes an avalanche that is read out as a signal on that wire only. The particle can thus be located to within an uncertainty of the wire spacing. The wires and the entrance and exit windows are extremely thin, so that the particle loses very little energy. These detectors are usually operated in a telescope with a thick $E$ counter at the end to determine the energy of the particle.

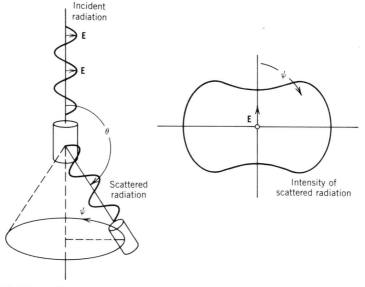

**Figure 7.43** A Compton polarimeter. As the detector travels in the cylindrical geometry over various values of $\psi$, the scattered intensity varies because of the polarization dependence of the Compton-scattering process. Measuring the intensity of the scattered radiation at two or more angles $\psi$ will enable the plane of polarization (direction of **E**) to be deduced.

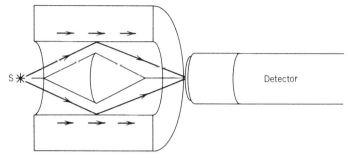

**Figure 7.44**  The Compton-scattering process also depends on the polarization of the electrons of the scattering material. Here the scattering material is magnetized iron. Reversing the direction of magnetization (by reversing the current in an electromagnet) will cause a change in the intensity of the radiation observed by the detector, from which the circular polarization can be deduced.

circular polarization, is based on the probability of scattering of photons from polarized electrons such as in a magnetic material. From the difference in intensity of the scattered radiation when the field is reversed, we can deduce the polarization (Figure 7.44).

Measuring the polarization of spin- $\frac{1}{2}$ particles also involves doing a scattering experiment. It is well known that the scattering of polarized electrons from polarized electrons (again, as in magnetic material) depends strongly on whether the spins of the electrons are parallel or antiparallel. The Pauli principle inhibits scattering in the parallel configuration, and thus scattering in the antiparallel arrangement is strongly preferred. (The ratio of cross sections for parallel and antiparallel scattering is in the range 0 to 0.1 at low energies.) For positron-electron scattering, the Pauli principle does not apply and the cross sections become equal at low energy.

Polarimeters for nucleons depend on the spin dependence of nuclear scattering, which was discussed in Chapter 4. Here we do not need to scatter from spins that are themselves polarized (as we did with electrons, for instance). The nuclear spin-orbit interaction gives the scattering cross section its spin dependence, and it is perfectly acceptable to determine polarizations by scattering of neutrons or protons by a spinless target such as $^4$He. Generally we use this technique to analyze the polarization of the products of nuclear reactions, which we discuss in Chapter 11.

## REFERENCES FOR ADDITIONAL READING

For comprehensive and up-to-date references on nuclear radiation detectors, see Glenn F. Knoll, *Radiation Detection and Measurement* (New York: Wiley, 1979), and J. B. A. England, *Techniques in Nuclear Structure Physics* (New York: Wiley, 1974). These two works contain extensive references to the original literature, and the latter work has a particularly good review of magnetic spectrometers.

Other general texts and references on this topic include G. G. Eichholz and J. S. Poston, *Principles of Nuclear Radiation Detection* (London: Butterworths, 1980); R. E. Lapp and H. L. Andrews, *Nuclear Radiation Physics*, 4th ed.

(Englewood Cliffs, NJ: Prentice-Hall, 1972); P. W. Nicholson, *Nuclear Electronics* (London: Wiley, 1974); R. D. Evans, *The Atomic Nucleus* (New York: McGraw-Hill, 1955), Chapters 18–25; G. Dearnaley and D. C. Northrop, *Semiconductor Counters for Nuclear Radiations* (New York: Barnes and Noble, 1966); J. R. Birks, *The Theory and Practice of Scintillation Counting* (New York: MacMillan, 1964).

Many topics concerning detectors, electronics, and techniques are covered in *Alpha-, Beta-, and Gamma-Ray Spectroscopy*, edited by K. Siegbahn (Amsterdam: North-Holland, 1965). See especially Chapters 1 and 2 on interactions of radiation with matter, Chapter 3 on spectrometers, Chapter 5 on scintillators, Chapter 6 on particle detectors, Chapters 7 and 8 on general experimental techniques, and Chapter 17 on measuring short lifetimes.

Photon cross sections are tabulated in E. Storm and H. I. Israel, *Nuclear Data Tables A* **7**, 565 (1970). Electron ranges in matter can be found in L. Pages, E. Bertel, H. Joffre, and L. Sklavenitis, *Atomic Data* **4**, 1 (1972).

For reviews of techniques for measuring nuclear lifetimes, see A. Z. Schwarzschild and E. K. Warburton, *Ann. Rev. Nucl. Sci.* **18**, 265 (1968). Techniques of particle identification are reviewed by F. S Goulding and B. G Harvey, *Ann. Rev. Nucl. Sci.* **25**, 167 (1975).

## PROBLEMS

1. Use the ranges given in Figure 7.2 to compute the range of (a) a 10-MeV $\alpha$ particle in gold; (b) a 5-MeV proton in beryllium; (c) a 1-MeV proton in water. Express the range in centimeters.

2. Find the range (in centimeters) in aluminum of 4.0-MeV $^3$He and $^3$H.

3. Calculate the energy loss $\Delta T$ of protons, deuterons, and $\alpha$'s between 10 and 200 MeV in passing through a 2-mm thickness of solid plastic scintillator. Plot $\Delta T$ against $T$.

4. Calculate and sketch the range of protons in NaI for energies between 1 and 100 MeV. (Extrapolate as necessary from Figure 7.2.)

5. The immediate environment of an accelerator or reactor contains large fluxes of $\gamma$ rays of energies in the vicinity of 5–10 MeV. What thickness of lead is required to reduce the photon intensity by a factor of $10^{12}$?

6. An $\alpha$-particle point source of 25 $\mu$Ci is placed in contact with one face of a large ionization chamber. The source emits a single $\alpha$ particle of energy 6.20 MeV. If the $\alpha$ particles that enter the chamber lose all their energy in the chamber, what is the current produced at the output of the chamber? (Assume 100% efficiency for collecting the charges in the chamber.)

7. A Geiger counter used in several applications over the course of a typical day produces on the average 100 counts per second. The tube is in the form of a cylinder 2 cm in diameter by 10 cm long and is filled with a mixture of 90% argon and 10% ethanol to a pressure of 0.1 atmosphere. In the Geiger-Müller region, each output count results from the formation of about $10^{10}$ ion-electron pairs. How long will it take for one-third of the quenching gas to be used up, thus necessitating replacement of the tube?

8.  The pulse-height spectrum of a radioactive source, known to emit only monoenergetic photons of fairly high energy, shows three prominent peaks, at pulse heights of 7.38, 6.49, and 5.60 V. What is the $\gamma$-ray energy?

9.  The 662-keV photon in the decay of $^{137}$Cs is observed by a NaI detector with an energy resolution (FWHM) of 53 keV. What will be the resolution for a measurement of the 1.836-MeV photon in the decay of $^{88}$Y?

10. Find the energy of the Compton edges in a $\gamma$-ray spectrum of the decay of $^{60}$Co (see Table 7.2).

11. In the decay of $^{88}$Y, two photons are emitted with energies of 0.898 MeV (92% of decays) and 1.836 MeV (100% of decays). Sketch the expected $\gamma$-ray spectrum, as in Figure 7.26, when a source of $^{88}$Y is placed in front of a NaI detector and a Ge detector.

12. For the binomial, Poisson, and Gaussian distributions, derive the expressions for the variance $\sigma^2$.

13. A certain radioactive source gives 3861 counts in a 10-min counting period. When the source is removed, the background alone gives 2648 counts in 30 min. Determine the net source counting rate (counts per second) and its uncertainty.

14. A technician is given the job of determining the strength of a radioactive source to the greatest possible precision. The counting equipment is in great demand, and only 1 h of total measuring time is available. How should the available time be scheduled if: (a) the net source counting rate is about 5 times the background rate; (b) the net source counting rate is about equal to the background rate; (c) the net source counting rate is about one-fifth of the background rate?

15. A coincidence counting experiment consists of two $\gamma$-ray detectors; one is fixed at a position defined as $\theta = 0°$ and the other can move in a horizontal plane to various values of $\theta$. The source, located on the axis of rotation of the second detector and equidistant from the two detectors, emits a $\gamma$ ray of energy 750 keV. It is possible for the photon to enter one detector and produce a single Compton-scattering event, following which the scattered photon can then travel to the other detector to be absorbed. Plot the energy of the radiation observed in the moveable detector against its position when this scattering occurs. (*Note*: In this geometry, $\theta$ is *not* the Compton-scattering angle.)

16. The thickness of a metallic foil is regulated during the manufacture by observing the attenuation of a beam of photons passing through the foil. A source of photons is placed above the foil as it emerges from the rollers, and a detector is placed below the foil. The photon energy is chosen so that the attenuation is exactly 50% at the desired thickness of 0.10 mm. The detection efficiency with no foil in place is 1%. To regulate the rollers, it is necessary to make a determination of the thickness in at most 1 s, and the thickness must be regulated to $\pm 5\%$. Calculate the required source strength.

17. What would be the appearance of a time-to-amplitude spectrum (such as Figure 7.32) if the two radiations were so close in energy that the detectors were unable to distinguish between them?

# 8

# ALPHA DECAY

Alpha particles were first identified as the least penetrating of the radiations emitted by naturally occurring materials. In 1903, Rutherford measured their charge-to-mass ratio by deflecting $\alpha$ particles from the decay of radium in electric and magnetic fields. Despite the difficulty of these early experiments, Rutherford's result was only about 25% higher than the presently accepted value. In 1909 Rutherford showed that, as suspected, the $\alpha$ particles were in fact helium nuclei; in his experiments the particles entered an evacuated thin-walled chamber by penetrating its walls, and after several days of collecting, atomic spectroscopy revealed the presence of helium gas inside the chamber.

Many heavy nuclei, especially those of the naturally occurring radioactive series, decay though $\alpha$ emission. Only exceedingly rarely does any other spontaneous radioactive process result in the emission of nucleons; we do not, for example, observe deuteron emission as a natural decay process. There must therefore be a special reason that nuclei choose $\alpha$ emission over other possible decay modes. In this chapter we examine this question and study the $\alpha$ decay process in detail. We also show how $\alpha$ spectroscopy can help us to understand nuclear structure.

## 8.1 WHY $\alpha$ DECAY OCCURS

Alpha emission is a Coulomb repulsion effect. It becomes increasingly important for heavy nuclei because the disruptive Coulomb force increases with size at a faster rate (namely, as $Z^2$) than does the specific nuclear binding force, which increases approximately as $A$.

Why is the $\alpha$ particle chosen as the agent for the spontaneous carrying away of positive charge? When we call a process *spontaneous* we mean that some kinetic energy has suddenly appeared in the system for no apparent cause; this energy must come from a decrease in the mass of the system. The $\alpha$ particle, because it is a very stable and tightly bound structure, has a relatively small mass compared with the mass of its separate constituents. It is particularly favored as an emitted particle if we hope to have the disintegration products as light as possible and thus get the largest possible release of kinetic energy.

**Table 8.1**   Energy Release ($Q$ value) for Various Modes of Decay of $^{232}$U[a]

| Emitted Particle | Energy Release (MeV) | Emitted Particle | Energy Release (MeV) |
|---|---|---|---|
| n | −7.26 | $^4$He | +5.41 |
| $^1$H | −6.12 | $^5$He | −2.59 |
| $^2$H | −10.70 | $^6$He | −6.19 |
| $^3$H | −10.24 | $^6$Li | −3.79 |
| $^3$He | −9.92 | $^7$Li | −1.94 |

[a] Computed from known masses.

For a typical $\alpha$ emitter $^{232}$U (72 y) we can compute, from the known masses, the energy release for various emitted particles. Table 8.1 summarizes the results. Of the particles considered, spontaneous decay is energetically possible *only* for the $\alpha$ particle. A positive disintegration energy results for some slightly heavier particles than those listed, $^8$Be or $^{12}$C, for example. We will show, however (Section 8.4), that the partial disintegration constant for emission of such heavy particles is normally vanishingly small compared with that for $\alpha$ emission. Such decays would be so rare that in practice they would almost never be noticed. This suggests that if a nucleus is to be recognized as an alpha emitter it is not enough for $\alpha$ decay to be energetically possible. The disintegration constant must also not be too small or else $\alpha$ emission will occur so rarely that it may not be detected. With present techniques this means that the half-life must be less than about $10^{16}$ y. Also, $\beta$ decay, if it has a much higher partial disintegration constant, can mask the $\alpha$ decay. Most nuclei with $A > 190$ (and many with $150 < A < 190$) are energetically unstable against $\alpha$ emission but only about one-half of them can meet these other requirements.

## 8.2   BASIC $\alpha$ DECAY PROCESSES

The spontaneous emission of an $\alpha$ particle can be represented by the following process:

$$^A_Z X_N \rightarrow ^{A-4}_{Z-2} X'_{N-2} + \alpha$$

The $\alpha$ particle, as was shown by Rutherford, is a nucleus of $^4$He, consisting of two neutrons and two protons. To understand the decay process, we must study the conservation of energy, linear momentum, and angular momentum.

Let's first consider the conservation of energy in the $\alpha$ decay process. We assume the initial decaying nucleus X to be at rest. Then the energy of the initial system is just the rest energy of X, $m_X c^2$. The final state consists of X' and $\alpha$, each of which will be in motion (to conserve linear momentum). Thus the final total energy is $m_{X'} c^2 + T_{X'} + m_\alpha c^2 + T_\alpha$, where $T$ represents the kinetic energy of the final particles. Thus conservation of energy gives

$$m_X c^2 = m_{X'} c^2 + T_{X'} + m_\alpha c^2 + T_\alpha \tag{8.1}$$

or

$$(m_X - m_{X'} - m_\alpha)c^2 = T_{X'} + T_\alpha \tag{8.2}$$

The quantity on the left side of Equation 8.2 is the net energy released in the decay, called the $Q$ value:

$$Q = (m_X - m_{X'} - m_\alpha)c^2 \qquad (8.3)$$

and the decay will occur spontaneously only if $Q > 0$. (The decay $Q$ values for $^{232}$U were listed in Table 8.1.) $Q$ values can be calculated from _atomic mass_ tables because even though Equation 8.3 represents a _nuclear_ process, the electron masses will cancel in the subtraction. When the masses are in atomic mass units (u), expressing $c^2$ as 931.502 MeV/u gives $Q$ values directly in MeV.

The $Q$ value is also equal to the total kinetic energy given to the decay fragments:

$$Q = T_{X'} + T_\alpha \qquad (8.4)$$

If the original nucleus X is at rest, then its linear momentum is zero, and conservation of linear momentum then requires that X' and $\alpha$ move with equal and opposite momenta in order that the final momentum also be zero:

$$p_\alpha = p_{X'} \qquad (8.5)$$

$\alpha$ decays typically release about 5 MeV of energy. Thus for both X' and $\alpha$, $T \ll mc^2$ and we may safely use nonrelativistic kinematics. Writing $T = p^2/2m$ and using Equations 8.4 and 8.5 gives the kinetic energy of the $\alpha$ particle in terms of the $Q$ value:

$$T_\alpha = \frac{Q}{(1 + m_\alpha/m_{X'})} \qquad (8.6)$$

Because the mass ratio is small compared with 1 (recall that X' represents a heavy nucleus), it is usually sufficiently accurate to express this ratio simply as $4/(A - 4)$, which gives, with $A \gg 4$,

$$T_\alpha = Q(1 - 4/A) \qquad (8.7)$$

Typically, the $\alpha$ particle carries about 98% of the $Q$ value, with the much heavier nuclear fragment X' carrying only about 2%. (This _recoil_ energy of the heavy fragment is not entirely negligible. For a typical $Q$ value of 5 MeV, the recoiling nucleus has an energy of the order of 100 keV. This energy is far in excess of that which binds atoms in solids, and thus the recoiling nucleus, if it is near the surface of the radioactive source, escapes from the source and can spread to the surroundings. If the $\alpha$ decay is part of a decay chain, then the recoiling daughter nucleus may itself be radioactive, and these recoils can result in the spread of radioactive material. Fortunately, the heavy recoil nuclei have an extremely short range in matter and their spread can be prevented by a thin coating, such as Mylar or lacquer, placed over the radioactive sample.)

The kinetic energy of an $\alpha$ particle can be measured directly with a magnetic spectrometer, and so the $Q$ value of a decay can be determined. This gives us a way to determine atomic masses, such as in a case in which we might know the mass of long-lived X as a result of direct measurement but X' is so short-lived that its mass cannot be determined by direct measurement.

## 8.3  α DECAY SYSTEMATICS

One feature of $\alpha$ decay is so striking that it was noticed as long ago as 1911, the year that Rutherford "discovered" the nucleus. Geiger and Nuttall noticed that $\alpha$ emitters with large disintegration energies had short half-lives and conversely. The variation is astonishingly rapid as we may see from the limiting cases of $^{232}$Th ($1.4 \times 10^{10}$ y; $Q = 4.08$ MeV) and $^{218}$Th ($1.0 \times 10^{-7}$ s; $Q = 9.85$ MeV). A factor of 2 in energy means a factor of $10^{24}$ in half-life! The theoretical explanation of this Geiger-Nuttall rule in 1928 was one of the first triumphs of quantum mechanics.

A plot of log $t_{1/2}$ against $Q$ in which all $\alpha$ emitters are included shows a considerable scatter about the general Geiger-Nuttall trend. Very smooth curves result, however, if we plot only $\alpha$ emitters with the same $Z$ and if further we select from this group only those with $Z$ and $N$ both even (Figure 8.1). Even-odd, odd-even, and odd-odd nuclei obey the general trend but do not plot into quite such smooth curves; their periods are 2–1000 times longer than those for even-even types with the same $Z$ and $Q$.

It is interesting that $^{235}$U (even $Z$, odd $N$) is one of these "extra-long-life" types. If its half-life were 1000 times shorter, this important nucleus would not occur in nature, and we probably would not have nuclear reactors today! We see in Chapter 13 that the same feature that apparently accounts for the long life against $\alpha$ decay, namely the odd neutron, also makes $^{235}$U very susceptible to fission by thermal neutrons.

Figure 8.2 shows another important systematic relationship for $\alpha$ emitters. Looking for the moment only at the data for $A > 212$, we see that adding neutrons to a nucleus reduces the disintegration energy, which, because of the

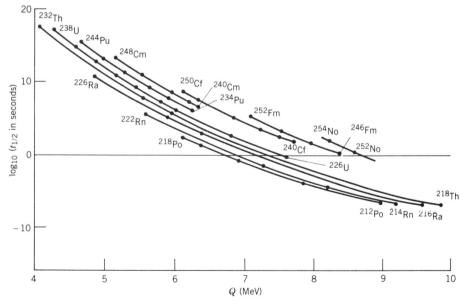

**Figure 8.1** The inverse relationship between $\alpha$-decay half-life and decay energy, called the Geiger-Nuttall rule. Only even-$Z$, even-$N$ nuclei are shown. The solid lines connect the data points.

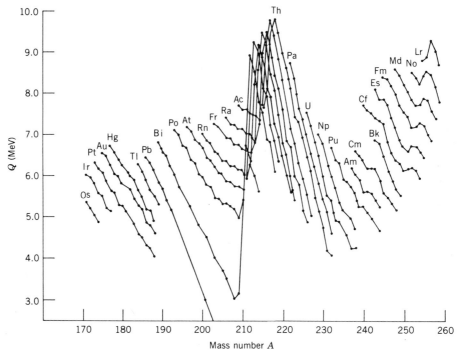

**Figure 8.2** Energy released in α decay for various isotopic sequences of heavy nuclei. In contrast to Figure 8.1, both odd-A and even-A isotopes are shown, and a small amount of odd-even staggering can be seen. The effects of the shell closures at $N = 126$ (large dip in data) and $Z = 82$ (larger than average spacing between Po, Bi, and Pb sequences) are apparent.

Geiger-Nuttall rule, increases the half-life. The nucleus becomes more stable. A striking discontinuity near $A = 212$ occurs where $N = 126$ and is another example of nuclear shell structure.

We can compare the systematic dependence of $Q$ on $A$ with the prediction of the semiempirical mass formula, Equation 3.28.

$$Q = B(^4\text{He}) + B(Z - 2, A - 4) - B(Z, A) \tag{8.8}$$

$$\cong 28.3 - 4a_v + \tfrac{8}{3}a_s A^{-1/3} + 4a_c Z A^{-1/3}(1 - Z/3A)$$

$$-4a_{\text{sym}}(1 - 2Z/A)^2 + 3a_p A^{-7/4} \tag{8.9}$$

where the approximation in Equation 8.9 is $Z, A \gg 1$. For $^{226}\text{Th}$, this formula gives $Q = 6.75$ MeV, not too far from the measured value of 6.45 MeV. What is perhaps more significant is that the general trend of Figure 8.2 is reproduced: for $^{232}\text{Th}$, Equation 8.9 gives $Q = 5.71$ MeV (to be compared with $Q = 4.08$ MeV), while for $^{220}\text{Th}$ the formula gives $Q = 7.77$ MeV (compared with $Q = 8.95$ MeV). Keep in mind that the parameters of the semiempirical mass formula are chosen to give rough agreement with observed binding energies across the entire range of nuclei. It is important that the formula gives us rough agreement with the decay $Q$ values and that it correctly gives $Q > 0$ for the heavy nuclei. It also correctly predicts the decrease of $Q$ with increasing $A$ for a sequence of isotopes such as those of thorium, although it gives too small a change of $Q$ with $A$ (the formula gives $\Delta Q = -0.17$ MeV per unit change in $A$, while for Th the observed average change is $\Delta Q = -0.40$ MeV per unit change in $A$).

## 8.4 THEORY OF α EMISSION

The general features of Figure 8.1 can be accounted for by a quantum mechanical theory developed in 1928 almost simultaneously by Gamow and by Gurney and Condon. In this theory an α particle is assumed to move in a spherical region determined by the *daughter* nucleus. The central feature of this *one-body model* is that the α particle is preformed inside the parent nucleus. Actually there is not much reason to believe that α particles do exist separately within heavy nuclei; nevertheless, the theory works quite well, especially for even-even nuclei. This success of the theory does not prove that α particles are preformed, but merely that they behave as if they were.

Figure 8.3 shows a plot, suitable for purposes of the theory, of the potential energy between the α particle and the residual nucleus for various distances between their centers. The horizontal line $Q$ is the disintegration energy. Note that the Coulomb potential is extended inward to a radius $a$ and then arbitrarily cut off. The radius $a$ can be taken as the sum of the radius of the residual nucleus and of the α particle. There are three regions of interest. In the spherical region $r < a$ we are inside the nucleus and speak of a potential well of depth $-V_0$, where $V_0$ is taken as a positive number. Classically the α particle can move in this region, with a kinetic energy $Q + V_0$ but it cannot escape from it. The annular-shell region $a < r < b$ forms a potential barrier because here the potential energy

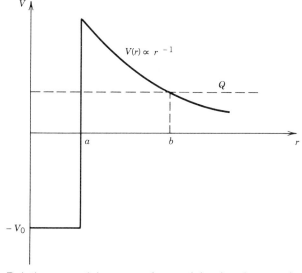

**Figure 8.3** Relative potential energy of α-particle, daughter-nucleus system as a function of their separation. Inside the nuclear surface at $r = a$, the potential is represented as a square well; beyond the surface, only the Coulomb repulsion operates. The α particle tunnels through the Coulomb barrier from $a$ to $b$.

is more than the total available energy $Q$. Classically the $\alpha$ particle cannot enter this region from either direction, just as a tennis ball dropped from a certain height cannot rebound higher; in each case the kinetic energy would have to be negative. The region $r > b$ is a classically permitted region outside the barrier.

From the classical point of view, an $\alpha$ particle in the spherical potential well would sharply reverse its motion every time it tried to pass beyond $r = a$. Quantum mechanically, however, there is a chance of "leakage" or "tunnelling" through such a barrier. This barrier accounts for the fact that $\alpha$-unstable nuclei do not decay immediately. The $\alpha$ particle within the nucleus must present itself again and again at the barrier surface until it finally penetrates. In $^{238}$U, for example, the leakage probability is so small that the $\alpha$ particle, on the average, must make $\sim 10^{38}$ tries before it escapes ($\sim 10^{21}$ per second for $\sim 10^{9}$ years)!

The barrier also operates in reverse, in the case of $\alpha$-particle scattering by nuclei (see Sections 3.1 and 11.6). Alpha particles incident on the barrier from outside the nucleus usually scatter in the Coulomb field if the incident energy is well below the barrier height. Tunnelling through the barrier, so that the nuclear force between the particle and target can cause nuclear reactions, is a relatively improbable process at low energy. The theoretical analysis of nuclear reactions induced by charged particles uses a formalism similar to that of $\alpha$ decay to calculate the barrier penetration probability. Fusion reactions, such as those responsible for the energy released in stars, also are analyzed using the barrier penetration approach (see Section 14.2).

The disintegration constant of an $\alpha$ emitter is given in the one-body theory by

$$\lambda = fP \tag{8.10}$$

where $f$ is the frequency with which the $\alpha$ particle presents itself at the barrier and $P$ is the probability of transmission through the barrier.

Equation 8.10 suggests that our treatment is going to be semiclassical in that our discussion of the situation for $r < a$ is very "billiard-ballish." A rigorous wave-mechanical treatment, however, gives about the same results for this problem. The quantity $f$ is roughly of the order of $v/a$ where $v$ is the relative velocity of the $\alpha$ particle as it rattles about inside the nucleus. We can find $v$ from the kinetic energy of the $\alpha$ particle for $r < a$. Estimating $V_0 \approx 35$ MeV for a typical well depth gives $f \approx 6 \times 10^{21}$/s for $Q \approx 5$ MeV. We will see later that we do not need to know $f$ very precisely to check the theory.

The barrier penetration probability $P$ must be obtained from a quantum mechanical calculation similar to the one-dimensional problem discussed in Section 2.3. Let's first use the result of that calculation, Equation 2.39, to estimate the probability $P$. Of course, the calculation that led to Equation 2.39 was based on a one-dimensional rectangular barrier, which is not directly applicable to the $1/r$ Coulomb potential, but we can at least find a rough order-of-magnitude estimate. The result, Equation 2.39, depends on the width of the barrier and on its height (called $V_0$ for the rectangular barrier) above the energy $E$ of the particle. The Coulomb barrier of Figure 8.3 has height $B$ at $r = a$, where

$$B = \frac{1}{4\pi\epsilon_0} \frac{zZ'e^2}{a} \tag{8.11}$$

In this expression the $\alpha$ particle has charge $ze$ and the daughter nucleus, which provides the Coulomb repulsion, has charge $Z'e = (Z - z)e$. The height of the barrier thus varies from $(B - Q)$ above the particle's energy at $r = a$ to zero at $r = b$, and we can take a representative average height to be $\frac{1}{2}(B - Q)$. We can similarly choose a representative average width to be $\frac{1}{2}(b - a)$. The factor $k_2$ of Equation 2.39 then becomes $\sqrt{(2m/\hbar^2)} \cdot \frac{1}{2}(B - Q)$. For a typical heavy nucleus ($Z = 90$, $a = 7.5$ fm), the barrier height $B$ is about 34 MeV, so the factor $k_2$ is about 1.6 fm$^{-1}$. The radius $b$ at which the $\alpha$ particle "leaves" the barrier is found from the equality of the particle's energy and the potential energy:

$$b = \frac{1}{4\pi\epsilon_0} \frac{zZ'e^2}{Q} \tag{8.12}$$

and for a typical case of a heavy nucleus with $Q \approx 6$ MeV, $b \approx 42$ fm. Thus $k_2 \cdot \frac{1}{2}(b - a) \gg 1$ and we can approximate Equation 2.39 as

$$P \cong e^{-2k_2 \cdot (1/2)(b-a)} \tag{8.13}$$

since the factors in front of the exponential are of unit order of magnitude. For the case we are estimating here, $P \sim 2 \times 10^{-25}$ and thus $\lambda \sim 10^{-3}$/s and $t_{1/2} \sim 700$ s. A slight change of $Q$ to 5 MeV changes $P$ to $1 \times 10^{-30}$ and $t_{1/2} \sim 10^8$ s. Even this very crude calculation is able to explain the many orders of magnitude change of $t_{1/2}$ between $Q = 5$ MeV and $Q = 6$ MeV, as illustrated in Figure 8.1.

The exact quantum mechanical calculation is very similar in spirit to the crude estimate above. We can think of the Coulomb barrier as made up of a sequence of infinitesimal rectangular barriers of height $V(r) = zZ'e^2/4\pi\epsilon_0 r$ and width $dr$. The probability to penetrate each infinitesimal barrier, which extends from $r$ to $r + dr$, is

$$dP = \exp\left\{-2\,dr\sqrt{(2m/\hbar^2)[V(r) - Q]}\right\} \tag{8.14}$$

The probability to penetrate the complete barrier is

$$P = e^{-2G} \tag{8.15}$$

where the *Gamow factor G* is

$$G = \sqrt{\frac{2m}{\hbar^2}} \int_a^b [V(r) - Q]^{1/2}\,dr \tag{8.16}$$

which can be evaluated as

$$G = \sqrt{\frac{2m}{\hbar^2 Q}} \frac{zZ'e^2}{4\pi\epsilon_0}\left[\arccos\sqrt{x} - \sqrt{x(1 - x)}\right] \tag{8.17}$$

where $x = a/b = Q/B$. The quantity in brackets in Equation 8.17 is approximately $\pi/2 - 2x^{1/2}$ when $x \ll 1$, as is the case for most decays of interest. Thus the result of the quantum mechanical calculation for the half-life of $\alpha$ decay is

$$t_{1/2} = 0.693\frac{a}{c}\sqrt{\frac{mc^2}{2(V_0 + Q)}}\exp\left\{2\sqrt{\frac{2mc^2}{(\hbar c)^2 Q}}\frac{zZ'e^2}{4\pi\epsilon_0}\left(\frac{\pi}{2} - 2\sqrt{\frac{Q}{B}}\right)\right\} \tag{8.18}$$

**Table 8.2** Calculated $\alpha$-Decay Half-lives for Th Isotopes

| | | $t_{1/2}$ (s) | |
|---|---|---|---|
| $A$ | $Q$ (MeV) | Measured | Calculated |
| 220 | 8.95 | $10^{-5}$ | $3.3 \times 10^{-7}$ |
| 222 | 8.13 | $2.8 \times 10^{-3}$ | $6.3 \times 10^{-5}$ |
| 224 | 7.31 | 1.04 | $3.3 \times 10^{-2}$ |
| 226 | 6.45 | 1854 | $6.0 \times 10^{1}$ |
| 228 | 5.52 | $6.0 \times 10^{7}$ | $2.4 \times 10^{6}$ |
| 230 | 4.77 | $2.5 \times 10^{12}$ | $1.0 \times 10^{11}$ |
| 232 | 4.08 | $4.4 \times 10^{17}$ | $2.6 \times 10^{16}$ |

The results of this calculation for the even isotopes of Th are shown in Table 8.2. The agreement is not exact, but the calculation is able to reproduce the trend of the half-lives within 1–2 orders of magnitude over a range of more than 20 orders of magnitude. We have neglected several important details in the calculation: we did not consider the initial and final nuclear wave functions (Fermi's Golden Rule, Equation 2.79, must be used to evaluate the decay probability), we did not consider the angular momentum carried by the $\alpha$ particle, and we assumed the nucleus to be spherical with a mean radius of $1.25A^{1/3}$ fm. The latter approximation has a very substantial influence on the calculated half-lives. The nuclei with $A \gtrsim 230$ have strongly deformed shapes, and the calculated half-lives are extremely sensitive to small changes in the assumed mean radius. For instance, changing the mean radius to $1.20A^{1/3}$ (a 4% change in $a$) changes the half-lives by a factor of 5! In fact, because of this extreme sensitivity, the procedure is often reversed—the measured half-lives are used to deduce the nuclear radius; what actually comes out of the calculation is more like the sum of the radii of the nucleus $X'$ and the $\alpha$ particle, if we assume their charge distributions to have a sharp edge. This result can then be used to obtain an estimate of the nuclear radius; see, for example, L. Marquez, *J. Phys. Lett.* **42**, 181 (1981).

Even though this oversimplified theory is not strictly correct, it gives us a good estimate of the decay half-lives. It also enables us to understand why other decays into light particles are not commonly seen, even though they may be allowed by the $Q$ value. For example, the decay $^{220}$Th $\rightarrow$ $^{12}$C + $^{208}$Po would have a $Q$ value of 32.1 MeV, and carrying through the calculation using Equation 8.18 gives $t_{1/2} = 2.3 \times 10^{6}$ s for the $^{220}$Th decay into $^{12}$C. This is a factor of $10^{13}$ longer than the $\alpha$-decay half-life and thus the decay will not easily be observable.

Recently, just such a decay mode has in fact been observed, the first example of a spontaneous decay process involving emission of a particle heavier than an $\alpha$. The decay of $^{223}$Ra normally proceeds by $\alpha$ emission with a half-life of 11.2 d, but there has now been discovered the decay process $^{223}$Ra $\rightarrow$ $^{14}$C + $^{209}$Pb. The probability for this process is very small, about $10^{-9}$ relative to the $\alpha$ decay. Figure 8.4 indicates the heroic efforts that are necessary to observe the process. To confirm that the emitted particle is $^{14}$C requires the $\Delta E \cdot T$ technique discussed in Chapter 7. Figure 8.4 shows a portion of the high-energy end of the tail of the hyperbola expected for observation of carbon. From the mass tables,

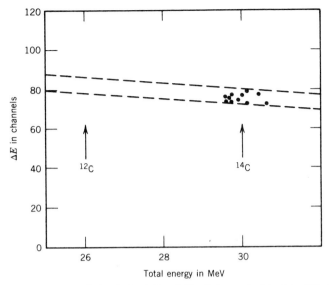

**Figure 8.4** A portion of the tail of the $\Delta E \cdot T$ hyperbola showing the observed $^{14}$C events from the decay of $^{223}$Ra. The dashed lines show the limits expected for carbon. The 11 $^{14}$C events result from 6 months of counting. From H. J. Rose and G. A. Jones, *Nature* **307**, 245 (1984). Reprinted by permission, copyright © Macmillan Journals Limited.

the decay $Q$ value is calculated to be 31.8 MeV, which (when corrected for the recoil) gives a $^{14}$C kinetic energy of 29.8 MeV. By contrast, the calculated energy for $^{12}$C emission would be about 26 MeV. The total of 11 events observed represents about *six months* of counting with a source of 3.3 $\mu$Ci of $^{223}$Ra in secular equilibrium with 21-y $^{227}$Ac, a member of the naturally occurring actinium series beginning with $^{235}$U.

Calculating the Gamow factor for $^{14}$C emission gives a decay probability of about $10^{-3}$ relative to $\alpha$ emission; the discrepancy between the calculated and observed ($10^{-9}$) values results from the assumptions about the preformation of the particle inside the nucleus. You will recall that our theory of $\alpha$ decay is based on the assumption that the $\alpha$ is preformed inside the nucleus. What the experiment tells us is that the probability for forming $^{14}$C clusters inside the nucleus is about $10^{-6}$ relative to the probability to preform $\alpha$'s.

For a description of the experiment, see H. J. Rose and G. A. Jones, *Nature* **307**, 245 (1984). Emission of $^{14}$C from several other nuclei in this region has also been observed, and emission of heavier decay fragments, including $^{24}$Ne, has been reported.

Going in the opposite direction, we can use Equation 8.18 with $z = 1$ to evaluate the half-life for proton decay—that is, the spontaneous emission of a proton by an unstable nucleus. In this case the Coulomb barrier will be only half as high as it is for $\alpha$ decay, but these decays are inhibited for a stronger reason: the $Q$ values for proton decay are generally negative and so the decays are absolutely forbidden by energy conservation. Such decays have recently been observed for a few proton-rich unstable nuclei, which are formed in nuclear reactions by bombarding a target with $N \approx Z$ using a projectile having $N \approx Z$.

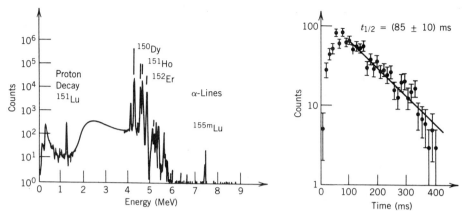

**Figure 8.5** (Left) Charged-particle spectrum emitted in the radioactive decays of products of the reaction $^{96}$Ru + $^{58}$Ni. The peaks above 4 MeV represent $\alpha$ decays; the 1.2-MeV peak is from proton emission. (Right) The decay with time of the proton peak gives a half-life of 85 ms. From S. Hofmann et al., *Z. Phys. A* **305**, 111 (1982).

This creates a heavy nucleus with $N \approx Z$, a very unstable configuration, and proton emission may be energetically possible, as the nucleus tries to relieve itself of its proton excess. The $Q$ value for proton decay can be found by a slight modification of Equation 8.3, which gives exactly the negative of the proton separation energy, Equation 3.27. Proton decay will be energetically possible when the $Q$ value is positive and therefore when the separation energy is negative. A glance at the mass tabulations (see A. H. Wapstra and G. Audi, *Nucl. Phys. A* **432**, 1 (1985)) shows only a few very rare cases in which the proton separation energy is negative, and even these are not directly measured but instead obtained by extrapolations from more stable nuclei.

In an experiment reported by Hofmann *et al.*, *Z. Phys. A* **305**, 111 (1982), a target of $^{96}$Ru was bombarded with $^{58}$Ni projectiles. Figure 8.5 shows the spectrum of light particles emitted following the reaction. The more energetic peaks are identified as $\alpha$ decays from unstable nuclei in the neighborhood of $A = 150$ produced in the reaction. The peak at 1.239 MeV was identified as a proton using $\Delta E \cdot T$ techniques as described in Chapter 7. Its half-life was measured as 85 ms, as shown in Figure 8.5. The decay was assigned to the isotope $^{151}$Lu based on a series of indirect arguments; unfortunately, reactions such as this produce many different products, and it is often a difficult task to identify the source of the observed radiations. This experiment thus provides evidence for the decay $^{151}$Lu $\rightarrow$ $^{150}$Yb + p.

Study of decays such as this enables us to extend our knowledge of nuclear mass systematics far beyond the previous limits; for instance, at the time of this work $^{151}$Lu was three protons further from stability than the previous last known isobar ($^{151}$Er). Figure 8.6 shows the $Q_p$ values deduced from known masses and from extrapolations based on systematics. The value for $^{151}$Lu lies right on the theoretical calculation, giving confidence to both the identification of the isotope and to the theoretical calculation.

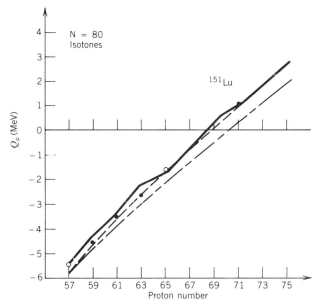

**Figure 8.6**   Proton-decay energies of $N = 80$ isotones. The solid lines are theoretical calculations based on nuclear mass formulas (somewhat like the semiempirical mass formula). Only for $^{151}$Lu is the decay energy positive. From S. Hofmann et al., *Z. Phys. A* **305**, 111 (1982).

Using Equation 8.18 for the half-life gives a value of about 1.7 $\mu$s, too small by nearly 5 orders of magnitude. For this reason, it has been proposed that the decay is inhibited by differences in the nuclear structure of the initial and final states (or possibly by a large angular momentum change in the decay, examples of which are discussed in the next section).

## 8.5   ANGULAR MOMENTUM AND PARITY IN $\alpha$ DECAY

We have up to this point neglected to discuss the angular momentum carried by the $\alpha$ particle. In a transition from an initial nuclear state of angular momentum $I_i$ to a final state $I_f$, the angular momentum of the $\alpha$ particle can range between $I_i + I_f$ and $|I_i - I_f|$. The nucleus $^4$He consists of two protons and two neutrons, all in 1s states and all with their spins coupled pairwise to 0. The spin of the $\alpha$ particle is therefore zero, and the total angular momentum carried by an $\alpha$ particle in a decay process is purely orbital in character. We will designate this by $\ell_\alpha$. The $\alpha$ particle wave function is then represented by a $Y_{\ell m}$ with $\ell = \ell_\alpha$; thus the parity change associated with $\alpha$ emission is $(-1)^{\ell_\alpha}$, and we have a parity *selection rule*, indicating which transitions are permitted and which are absolutely forbidden by conservation of parity: if the initial and final parities are the same, then $\ell_\alpha$ must be even; if the parities are different, then $\ell_\alpha$ must be odd.

To study the applications of these rules, we must recognize that we have also neglected one very significant feature of $\alpha$ decay—a given initial state can populate many different final states in the daughter nucleus. This property is

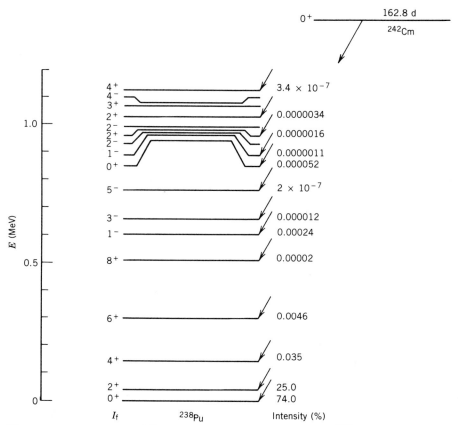

**Figure 8.7** $\alpha$ decay of $^{242}$Cm to different excited states of $^{238}$Pu. The intensity of each $\alpha$-decay branch is given to the right of the level.

sometimes known as the "fine structure" of $\alpha$ decay, but of course has nothing whatever to do with atomic fine structure. Figure 8.7 shows the $\alpha$ decay of $^{242}$Cm. The initial state is spin zero, and thus the angular momentum of the $\alpha$ particle $\ell_\alpha$ is equal to the angular momentum of the final nuclear state $I_f$. You can see that many different states of $^{238}$Pu are populated. The $\alpha$ decays have different $Q$ values (given by the $Q$ value for decay to the ground state, 6.216 MeV, less the excitation energy of the excited state) and different intensities. The intensity depends on the wave functions of the initial and final states, and also depends on the angular momentum $\ell_\alpha$. In Equation 2.60, it was shown how the "centrifugal potential" $\ell(\ell+1)\hbar^2/2mr^2$ must be included in spherical coordinates. This term, which is always positive, has the effect of raising the potential energy for $a < r < b$ and thus increasing the thickness of the barrier which must be penetrated. Consider for example the $0^+$, $2^+$, $4^+$, $6^+$, and $8^+$ states of the ground-state rotational band. The decay to the $2^+$ state has less intensity than the decay to the ground state for two reasons—the "centrifugal potential" raises the barrier by about 0.5 MeV, and the excitation energy lowers $Q$ by 0.044 MeV. The decay intensity continues to decrease for these same reasons as we go up the band to the $8^+$ state. If we use our previous theory for the decay rates, taking into account the increasing effective $B$ and decreasing $Q$, we obtain the following estimates for the relative decay branches: $0^+$, 76%; $2^+$, 23%; $4^+$, 1.5%; $6^+$, 0.077%; $8^+$, $8.4 \times 10^{-5}$%. These results are not in exact agreement with the observed decay intensities, but they do give us a rough idea of the origin of the decrease in intensity.

Once we go above the ground-state band, the $\alpha$ decay intensities become very small, of the order of $10^{-6}$% of the total decay intensity. This situation results from the poor match of initial and final wave functions—many of these excited states originate with vibrations or pair-breaking particle excitations, which are not at all similar to the paired, vibrationless $0^+$ ground state of $^{242}$Cm. You should note that there are some states for which there is no observed decay intensity at all. These include the $2^-$ states at 0.968 and 0.986 MeV, the $3^+$ state at 1.070 MeV, and the $4^-$ state at 1.083 MeV. Alpha decay to these states is absolutely forbidden by the parity selection rule. For example, a $0 \to 3$ decay must have $\ell_\alpha = 3$, which must give a change in parity between initial and final states. Thus $0^+ \to 3^-$ is possible, but not $0^+ \to 3^+$. Similarly, $0 \to 2$ and $0 \to 4$

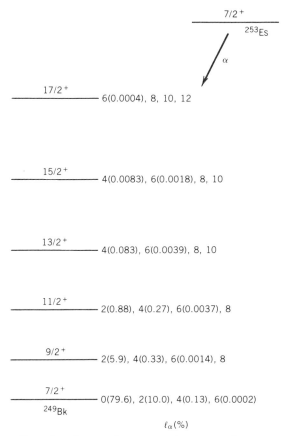

**Figure 8.8** Intensities of various $\alpha$-decay angular momentum components in the decay of $^{253}$Es. For $\ell_\alpha = 8$ and higher, the intensities are not known but are presumably negligibly small. From the results of a study of spin-aligned $\alpha$ decays by A. J. Soinski et al., *Phys. Rev. C* **2**, 2379 (1970).

decays cannot change the parity, and so $0^+ \rightarrow 2^-$ and $0^+ \rightarrow 4^-$ are not permitted.

When neither the initial nor the final states have spin 0, the situation is not so simple and there are no absolutely forbidden decays. For example, the decay $2^- \rightarrow 2^+$ must have odd $\ell_\alpha$ (because of the change in parity), and the angular momentum coupling rules require $0 \le \ell_\alpha \le 4$. Thus it is possible to have this decay with $\ell_\alpha = 1$ or 3. The next question that occurs is whether $\ell_\alpha = 1$ or $\ell_\alpha = 3$ is favored and by how much. Our previous discussion would lead us to guess that the $\ell_\alpha = 1$ intensity is roughly an order of magnitude greater than the $\ell_\alpha = 3$ intensity. However, measuring only the energy or the intensity of the decay gives us no information about how the total decay intensity is divided among the possible values of $\ell_\alpha$. To make the determination of the relative contributions of the different $\ell$ values, it is necessary to measure the angular distribution of the $\alpha$ particles. The emission of an $\ell = 1$ $\alpha$ particle is governed by a $Y_1(\theta, \phi)$, while an $\ell = 3$ $\alpha$ decay is emitted with a distribution according to $Y_3(\theta, \phi)$. If we determine the spatial distribution of these decays, we could in principle determine the relative amounts of the different $\ell$ values.

To do this experiment we must first align the spins of our $\alpha$-radioactive nuclei, such as by aligning their magnetic dipole or electric quadrupole moments in a magnetic field or in a crystalline electric field gradient. Keeping the spins aligned requires that the nuclei must be cooled to a temperature at which the thermal motion is not sufficient to destroy the alignment; generally temperatures below 0.01 K are required (that is, less than 0.01 degree above the absolute zero of temperature!).

As an example of such an experiment, we consider the decay of $^{253}$Es to states of the ground-state rotational band of $^{249}$Bk. The possible $\ell$ values are indicated in Figure 8.8, and the results of measuring the $\alpha$-particle angular distributions help us to determine the relative contribution of the different values of $\ell_\alpha$.

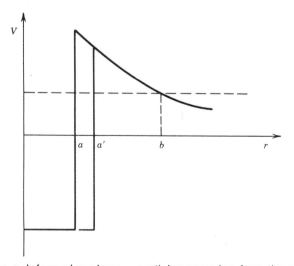

**Figure 8.9** In a deformed nucleus, $\alpha$ particles escaping from the poles enter the Coulomb barrier at the larger separation $a'$, and must therefore penetrate a lower, thinner barrier. It is therefore more probable to observe emission from the poles than from the equator.

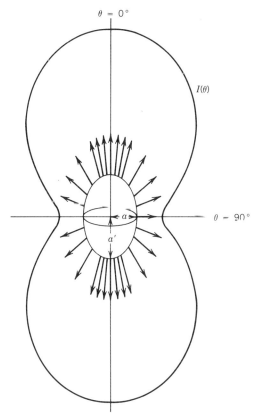

**Figure 8.10** Intensity distribution of $\alpha$ particles emitted from the deformed nucleus at the center of the figure. The polar plot of intensity shows a pronounced angular distribution effect.

Since many $\alpha$-emitting nuclei are deformed, these angular distribution measurements can also help us to answer another question: if we assume a stable prolate (elongated) nucleus, will more $\alpha$'s be emitted from the poles or from the equator? Figure 8.9 suggests a possible answer to this question: at the larger radius of the poles, the $\alpha$ particle feels a weaker Coulomb potential and must therefore penetrate a thinner and lower barrier. We therefore expect that polar emission ought to be more likely than equatorial emission. Figure 8.10 shows the angular distribution of $\alpha$ emission relative to the symmetry axis. You can see that emission from the poles is 3–4 times more probable than emission from the equator, exactly as we expect on the basis of the potential.

### 8.6 $\alpha$ DECAY SPECTROSCOPY

The final topic in our discussion of $\alpha$ decay is this: What can we learn about the energy levels of nuclei by studying $\alpha$ decay?

Let's consider, for example, the 5.3-h decay of $^{251}$Fm to levels of $^{247}$Cf. (The levels of $^{247}$Cf are also populated in the beta decay of $^{247}$Es, but the half-life of that decay is so short, 4.7 min, that it is more difficult to use as a detailed probe of the level structure of $^{247}$Cf.)

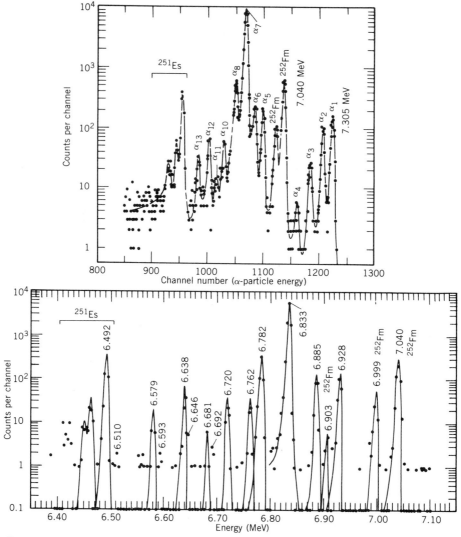

**Figure 8.11** $\alpha$ spectrum from the decay of $^{251}$Fm. The top portion shows the spectrum as observed with a Si detector. The bottom shows a portion of the same spectrum observed with a magnetic spectrometer, whose superior energy resolution enables observation of the 6.762-MeV decay, which would be missed in the upper spectrum. From Ahmad et al., *Phys. Rev. C* **8**, 737 (1973).

Figure 8.11 shows the energy spectrum of $\alpha$ decays from the decay of $^{251}$Fm. As you can see, there are 13 distinct groups of $\alpha$ particles; each group presumably represents the decay to a different excited state of $^{247}$Cf. How can we use this information to construct the level scheme of $^{247}$Cf? Based on the $\alpha$ spectrum, we first must find the energy and intensity of each $\alpha$ group. The energy is found by comparing with decays of known energy (the impurity decays from the $^{252}$Fm contaminant are helpful for this) and the intensity is found from the area of each peak. The result of this analysis is shown in Table 8.3, along with the uncertainties that come mostly from the counting statistics for each peak. (Notice that the

**Table 8.3** α Decays from $^{251}$Fm

| α Group | α Energy (keV) | Decay Energy (keV) | Excited-State Energy (keV) | α Intensity (%) |
|---------|----------------|--------------------|-----------------------------|------------------|
| $\alpha_1$ | 7305 ± 3 | 7423 | 0 | 1.5 ± 0.1 |
| $\alpha_2$ | 7251 ± 3 | 7368 | 55 | 0.93 ± 0.08 |
| $\alpha_3$ | 7184 ± 3 | 7300 | 123 | 0.29 ± 0.03 |
| $\alpha_4$ | 7106 ± 5 | 7221 | 202 | ~ 0.05 |
| $\alpha_5$ | 6928 ± 2 | 7040 | 383 | 1.8 ± 0.1 |
| $\alpha_6$ | 6885 ± 2 | 6996 | 427 | 1.7 ± 0.1 |
| $\alpha_7$ | 6833 ± 2 | 6944 | 479 | 87.0 ± 0.9 |
| $\alpha_8$ | 6782 ± 2 | 6892 | 531 | 4.8 ± 0.2 |
| $\alpha_9$ | 6762 ± 3 | 6872 | 552 | 0.38 ± 0.06 |
| $\alpha_{10}$ | 6720 ± 3 | 6829 | 594 | 0.44 ± 0.04 |
| $\alpha_{11}$ | 6681 ± 4 | 6789 | 634 | 0.07 ± 0.03 |
| $\alpha_{12}$ | 6638 ± 3 | 6745 | 678 | 0.56 ± 0.06 |
| $\alpha_{13}$ | 6579 ± 3 | 6686 | 738 | 0.26 ± 0.04 |

strongest peaks have the smallest *relative* uncertainties.) To find the decay energies (that is, the relative energies of the nuclear states), we must use Equation 8.7, since the measured α energies are only the kinetic energies. These are also shown in Table 8.3.

The different $^{247}$Cf excited states will quickly decay to the ground state by emitting γ-ray photons, so in constructing the decay scheme it is helpful to have the energies and intensities of the γ rays as well. Figure 8.12 shows the observed γ rays and Table 8.4 shows the deduced energies and intensities.

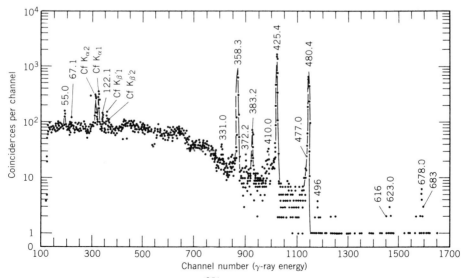

**Figure 8.12** γ-ray spectrum of $^{251}$Fm in coincidence with all α decays in the range 6.0 to 7.7 MeV. The spectrum was obtained with a Ge(Li) detector.

**Table 8.4** γ Rays in $^{247}$Cf following α Decay of $^{251}$Fm

| Energy (keV) | Intensity (% of decays) | Energy (keV) | Intensity (% of decays) |
|---|---|---|---|
| 55.0 ± 0.2 | 0.58 ± 0.08 | 425.4 ± 0.1 | 51 ± 4 |
| 67.1 ± 0.2 | 0.28 ± 0.05 | 477.0 ± 0.3 | 0.54 ± 0.08 |
| 122.1 ± 0.2 | 0.28 ± 0.05 | 480.4 ± 0.1 | 21 ± 2 |
| 331.0 ± 0.3 | 0.35 ± 0.07 | 496 ± 1 | ~ 0.08 |
| 358.3 ± 0.1 | 17 ± 1.5 | 616 ± 1 | ~ 0.05 |
| 372.2 ± 0.4 | 0.25 ± 0.05 | 623.0 ± 0.8 | 0.07 ± 0.02 |
| 382.2 ± 0.3 | 1.2 ± 0.13 | 678.0 ± 0.8 | 0.26 ± 0.06 |
| 410.0 ± 0.3 | 0.50 ± 0.07 | 683 ± 1 | ~ 0.04 |

Now the detective work comes. Let's assume (and here we must be very careful, as we see in the next example) that the highest energy α decay populates the ground state of $^{247}$Cf. (In an even-even nucleus, this would be a very good assumption, because $0^+ \rightarrow 0^+$ α decays are very strong and not inhibited by any differences between the wave functions of the initial and final nuclear states. In an odd-$A$ nucleus, the initial and final ground states may have very different characters so that the decay to the ground state may be very weak or even vanishing.) The decay just lower in energy differs from the ground-state decay by about 55 keV. Assuming this to populate the first excited state, we are pleased to find among the γ transitions one of energy 55 keV, which presumably represents the transition between the first excited state and the ground state. The next α decay populates a state at $123 \pm 3$ keV above the ground state, and we find among the γ rays one of energy 122.1 keV, which corresponds to a transition from the second excited state to the ground state. We also find a transition of energy 67.1 $(= 122.1 - 55.0)$ keV, which results from transitions between the second and first excited states.

Let's guess that these three states (with assumed energies 0, 55.0 keV, 122.1 keV) form a rotational band whose states, we recall from the discussion of odd-$A$ deformed nuclei in Section 5.3, have angular momenta $I = \Omega, \Omega + 1, \Omega + 2, \ldots,$ where $\Omega$ is the component of the angular momentum of the odd particle along the symmetry axis. The energy difference between the first excited state and ground state should then be

$$\Delta E_{21} \equiv E_2 - E_1 = \frac{\hbar^2}{2\mathscr{I}}[(\Omega + 1)(\Omega + 2) - \Omega(\Omega + 1)]$$

$$= \frac{\hbar^2}{2\mathscr{I}}2(\Omega + 1) \tag{8.19}$$

where we have used $E = (\hbar^2/2\mathscr{I})I(I + 1)$ for the energy of rotational states. Similarly, the difference between the ground state and second excited state is

$$\Delta E_{31} \equiv E_3 - E_1 = \frac{\hbar^2}{2\mathscr{I}}[(\Omega + 2)(\Omega + 3) - \Omega(\Omega + 1)]$$

$$= \frac{\hbar^2}{2\mathscr{I}}2(2\Omega + 3) \tag{8.20}$$

Combining these results with the experimental values, $\Delta E_{21} = 55.0$ keV and $\Delta E_{31} = 122.1$ keV, we conclude $\Omega = 3.5 \pm 0.2$ (that is, $\Omega = \frac{7}{2}$) and $\hbar^2/2\mathscr{I} = 6.11 \pm 0.02$ keV. These three states thus seem to form a rotational band with $I = \frac{7}{2}, \frac{9}{2}, \frac{11}{2}$. With our deduced values we can predict the energy of the $\frac{13}{2}$ state:

$$\Delta E_{41} = \frac{\hbar^2}{2\mathscr{I}}\left[\frac{13}{2}\cdot\frac{15}{2} - \frac{7}{2}\cdot\frac{9}{2}\right] = 201.6 \text{ keV}$$

and the $\frac{15}{2}$ state

$$\Delta E_{51} = \frac{\hbar^2}{2\mathscr{I}}\left[\frac{15}{2}\cdot\frac{17}{2} - \frac{7}{2}\cdot\frac{9}{2}\right] = 293.3 \text{ keV}$$

Apparently, the $\frac{13}{2}$ state is populated by the very weak $\alpha_4$ decay, but its $\gamma$ decays may be too weak to be seen in the spectrum of Figure 8.12. The decay to the $\frac{15}{2}$ state is not observed.

The interpretation of the remaining states is aided by $\alpha$-$\gamma$ coincidence studies, in which we electronically select only those $\gamma$ transitions that follow a given $\alpha$ decay within a certain short time interval (in this case 110 ns). Since this time is long compared with typical lifetimes of nuclear states, all $\gamma$ rays that follow the $\alpha$ decay will be recorded, even those that follow indirectly (such as the case in which two $\gamma$'s are emitted in cascade, one following the other). The following coincidence relationships were observed:

| Coincidence Gate | $\gamma$ Rays (keV) |
|---|---|
| $\alpha_5$ | 383.2 |
| $\alpha_6$ | 372.2, 383.2 |
| $\alpha_7$ | 55.0, 67.1, 122.1, 358.3, 425.4, 480.4 |
| $\alpha_8$ | 331.0, 358.3, 410.0, 425.4, 477.0, 480.4 |
| $\alpha_{12}$ | 623.0, 678.0 |

The decay $\alpha_5$ goes to a state at 383.2 keV, which then goes directly to the ground state by emitting a single $\gamma$ ray. The decay $\alpha_6$ populates a state at about 427 keV. There is no coincident $\gamma$ ray of that energy, which indicates no direct transition to the ground state, but there is a transition of energy 372.2 keV which, when added to 55.0 keV, gives 427.2 keV, very close to the energy of the state. We therefore conclude that this state, at 427.2 keV, decays to the first excited state at 55.0 keV. There is also a coincident transition at 383.2 keV, and thus this state at 427.2 keV must decay to the previously established state at 383.2 keV, by emitting a $\gamma$ ray of energy 427.2 − 383.2 = 44.0 keV; this $\gamma$ ray is not observed. The decay $\alpha_7$ to the state at 480.4 keV shows decays to the ground state and to the 55.0 and 122.1 keV states (425.4 + 55.0 = 480.4; 358.3 + 122.1 = 480.4). Similarly, the decay $\alpha_8$ to a state of 532.0 keV shows direct transitions to the lower states (331.0 + 201.0 = 532.0; 410.0 + 122.1 = 532.1; 477.0 + 55.0 = 532.0) but not directly to the ground state. It also shows coincident transitions that originate from the 480.4-keV level, so there must be a transition of energy 51.6 keV (= 532.0 − 480.4). In a similar fashion we analyze the other $\alpha$ and $\gamma$ decays, and Figure 8.13 shows the resulting decay scheme.

For the states above the ground-state band, the assignment of spins and intrinsic angular momentum $\Omega$ is not as easy as it was for the states of the

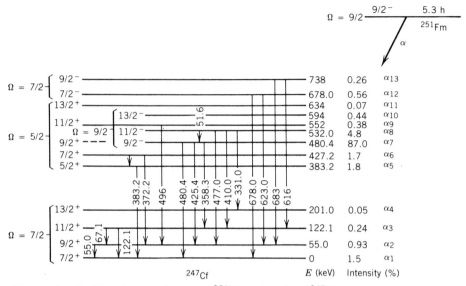

**Figure 8.13** The decay scheme of $^{251}$Fm to levels of $^{247}$Cf deduced from $\alpha$ and $\gamma$ spectroscopy. The spin assignments for the higher levels are deduced using $\gamma$-ray and internal conversion techniques described in Chapter 10.

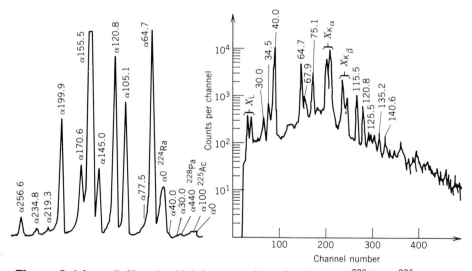

**Figure 8.14** $\alpha$ (left) and $\gamma$ (right) spectra from the decay of $^{229}$Pa to $^{225}$Ac. The $\alpha$ peaks are labeled according to the excited state populated in $^{225}$Ac; thus $\alpha$105.1 indicates the decay leading to the excited state at 105.1 keV. Prominent peaks from impurities are also indicated. The $\gamma$ spectrum is taken in coincidence with all $\alpha$'s. From P. Aguer et al., *Nucl. Phys. A* **202**, 37 (1973).

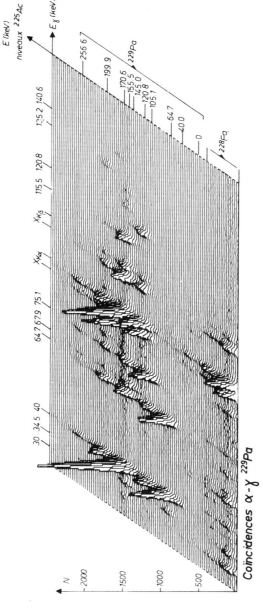

**Figure 8.15** Three-dimensional (sometimes called two-parameter) representation of α-γ coincidences in the decay of $^{229}$Pa. The horizontal axis shows γ-ray energies, labeled along the top. The oblique axis gives α-decay energies, labeled to indicate the $^{225}$Ac state populated in the decay. The vertical axis gives the intensity of the coincidence relationship.

ground-state rotational band. To make these assignments, we need additional information from the γ decays; these measurements are discussed in Chapter 10.

Notice the strong α branch to the state at 480.4 keV. This occurs because the wave functions of the initial and final states are identical—both come from the same $\Omega = \frac{9}{2}$ deformed single-particle state—and the result is that more than 93% of the decay intensity goes to states of that so-called "favored" band. The observed decay rates can be compared with values calculated for various deformed single-particle states using the Nilsson wave functions, and in general there is good agreement between the measured and calculated results, both for the favored and unfavored decays. It is such comparisons between theory and experiment that allow us to assign the single-particle states because the intrinsic $\Omega$ and Nilsson assignments are not directly measurable.

The data for this discussion were taken from I. Ahmad, J. Milsted, R. K. Sjoblom, J. Lerner, and P. R. Fields, *Phys. Rev. C* **8**, 737 (1973). Theoretical calculations of α transition amplitudes for states in even-*A* and odd-*A* deformed nuclei of the actinide region can be found in J. K. Poggenburg, H. J. Mang, and J. O. Rasmussen, *Phys. Rev.* **181**, 1697 (1969).

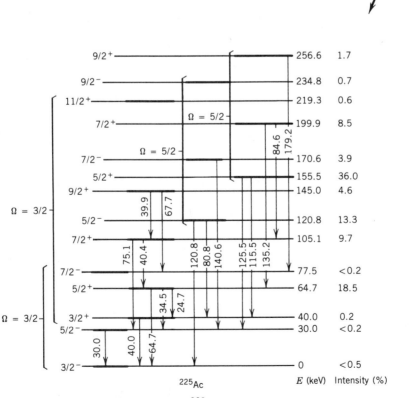

**Figure 8.16** Decay scheme of $^{229}$Pa deduced from α and γ spectroscopy.

Another example of the study of nuclear spectroscopy through $\alpha$ decay is illustrated in Figures 8.14–8.16. Figure 8.14 shows the $\alpha$ and $\gamma$ spectra from the decay $^{229}$Pa $\rightarrow$ $^{225}$Ac, and you can see that the decay to the ground state (labeled $\alpha_0$) cannot be verified. Again, the $\alpha$-$\gamma$ coincidences help to elucidate the decay scheme, and a particularly instructive way to illustrate the coincidences is shown in Figure 8.15. Each peak in this three-dimensional spectrum represents a definite coincidence relationship between the $\alpha$ and the $\gamma$ that label the axes. The information derived from the coincidence studies is used to make the decay scheme shown in Figure 8.16. Four rotational bands are identified in $^{225}$Ac, positive and negative parity bands with $\Omega = \frac{3}{2}$ and $\frac{5}{2}$. The decaying $^{229}$Pa is assigned $\frac{5}{2}^{+}$, so in this case the favored decay to the $\frac{5}{2}^{+}$ band in the daughter has about 46% of the decay intensity. The decay to the $\frac{3}{2}^{-}$ ground-state rotational band is strongly inhibited by the nuclear wave functions, resulting in the very weak (and possibly nonexistent) decay to the ground state. In this case it would lead to errors if we had assumed that the highest energy observed $\alpha$ group ($\alpha_{64.7}$, or $\alpha_{40.0}$ if we looked carefully) corresponded to transitions to the ground state.

The data for the $^{229}$Pa decay come from P. Aguer, A. Peghaire, and C. F. Liang, *Nucl. Phys. A* **202**, 37 (1973).

## REFERENCES FOR ADDITIONAL READING

Somewhat more extensive discussions of $\alpha$ decay can be found in Chapter 16 of R. D. Evans, *The Atomic Nucleus* (New York: McGraw-Hill, 1955), and in Chapter 13 of I. Kaplan, *Nuclear Physics* (Reading, MA: Addison-Wesley, 1955). For surveys of $\alpha$-decay theory, see H. J. Mang, *Ann. Rev. Nucl. Sci.* **14**, 1 (1964), and J. O. Rasmussen, "Alpha Decay," in *Alpha-, Beta- and Gamma-Ray Spectroscopy*, edited by K. Siegbahn (Amsterdam: North-Holland, 1965), Chapter XI. A discussion of the use of $\alpha$ decay for nuclear spectroscopy is that of F. S. Stephens, in *Nuclear Spectroscopy*, part A, edited by F. Ajzenberg-Selove (New York: Academic, 1959), Section I.E.2.

## PROBLEMS

1. Find the $Q$ values of the following decays:
   (a) $^{247}$Bk $\rightarrow$ $^{243}$Am $+ \alpha$; (b) $^{251}$Cf $\rightarrow$ $^{247}$Cm $+ \alpha$; (c) $^{230}$Th $\rightarrow$ $^{226}$Ra $+ \alpha$.

2. For each decay given in Problem 1, calculate the kinetic energy and velocity of the daughter nucleus after the decay.

3. From the known atomic masses, compute the $Q$ values of the decays:
   (a) $^{242}$Pu $\rightarrow$ $^{238}$U $+ \alpha$
   (b) $^{208}$Po $\rightarrow$ $^{204}$Pb $+ \alpha$
   (c) $^{208}$Po $\rightarrow$ $^{196}$Pt $+ ^{12}$C
   (d) $^{210}$Bi $\rightarrow$ $^{208}$Pb $+ ^{2}$H

4. In the decay of $^{242}$Cm to $^{238}$Pu, the maximum $\alpha$ energy is 6112.9 $\pm$ 0.1 keV. Given the mass of $^{238}$Pu, find the mass of $^{242}$Cm.

5. The highest energy $\alpha$ particle emitted in the decay of $^{238}$U to $^{234}$Th is 4196 $\pm$ 4 keV. From this information and the known mass of $^{238}$U, compute the mass of $^{234}$Th.

6. Use the uncertainty principle to estimate the minimum speed and kinetic energy of an $\alpha$ particle confined to the interior of a heavy nucleus.

7. (a) Compute the $Q$ values for the decays $^{224}$Ra $\rightarrow$ $^{212}$Pb $+$ $^{12}$C and $^{224}$Ra $\rightarrow$ $^{210}$Pb $+$ $^{14}$C. (b) Estimate the half-lives for these two possible decay processes. $^{224}$Ra is an $\alpha$ emitter with a half-life of 3.66 days.

8. The $Q$ value for the $\alpha$ decay of $^{203}$Tl is calculated to be 0.91 MeV from the masses of the initial and final nuclei. Estimate the half-life of $^{203}$Tl and explain why we call $^{203}$Tl a "stable" nucleus.

9. Use the semiempirical mass formula to estimate the $\alpha$-decay energy of $^{242}$Cf and compare with the measured value (see Figure 8.1).

10. For the $\alpha$ decay of $^{226}$Ra to $^{222}$Rn ($Q = 4.869$ MeV), compute the expected half-lives for values of the $^{222}$Rn radius of 7.0, 8.0, 9.0, and 10.0 fm. Estimate the nuclear radius required to give the measured half-life (1602 years); deduce the corresponding value of $R_0$ and interpret.

11. Using a scale similar to that of Figure 8.2, plot Equation 8.9 and show that it reproduces the general trend of the $Q$ values. Choose appropriate values of $Z$ corresponding to each $A$.

12. Make a drawing to scale of the Coulomb potential barrier encountered in the $\alpha$ decay of $^{242}$Cm ($Z = 96$) to $^{238}$Pu ($Z = 94$), for which $Q = 6.217$ MeV. Assume $R_0 = 1.5$ fm, to account for the diffuseness of the nuclear surface. Show also the Coulomb-plus-centrifugal barrier for the $\ell = 2$ decay to the first excited state (44 keV). Now use a method analogous to that of Equation 8.13 to estimate the reduction in the decay probability caused by the additional barrier, and correspondingly estimate the ratio of the $\alpha$ branching intensities to the ground and first excited states of $^{238}$Pu. (Don't forget to reduce the $Q$ value for decays to the excited state.) Compare your estimate with the actual intensities given in Figure 8.7.

13. Equations 8.1 to 8.5 are strictly correct even using relativistic mechanics; however, Equations 8.6 and 8.7 were obtained by assuming a nonrelativistic form for the kinetic energy. Using relativistic expressions for $p$ and $T$, derive relativistic versions of Equations 8.6 and 8.7 and calculate the error made by neglecting these corrections for a 6-MeV $\alpha$ particle.

14. Consider the strongly distorted nucleus $^{252}$Fm with $\varepsilon = 0.3$. That is, the nucleus is shaped like an ellipsoid of revolution with semimajor axis $a' = R(1 + \varepsilon)$ and semiminor axis $a = R(1 + \varepsilon)^{-1/2}$, where $R$ is the mean radius. Using a potential of the form of Figure 8.9, estimate the relative probabilities of polar and equatorial emission of $\alpha$ particles.

15. In a semiclassical picture, an $\ell = 0$ $\alpha$ particle is emitted along a line that passes through the center of the nucleus. (a) How far from the center of the nucleus must $\ell = 1$ and $\ell = 2$ $\alpha$ particles be emitted? Assume $Q = 6$ MeV for a heavy nucleus ($A = 230$). (b) What would be the recoil rotational kinetic energy if all of the recoil went into rotational motion of the daughter nucleus?

16. Use data from available reference material (*Table of Isotopes, Atomic Mass Tables*) to plot a series of curves showing $\alpha$-decay $Q$ values against $Z$ for constant $A$. Use Equation 8.9 to analyze the graph.

**17.** In the decay $^{228}\text{Th} \rightarrow {}^{224}\text{Ra} + \alpha$, the highest energy $\alpha$ particle has an energy 5.423 MeV and the next highest energy is 5.341 MeV. (a) The highest energy decay populates the $^{224}\text{Ra}$ ground state. Why is it natural to expect this to be so? (b) Compute the $Q$ value for the decay from the measured $\alpha$ energy. (c) Compute the energy of the first excited state of $^{224}\text{Ra}$

**18.** The $Q$ value of the decay $^{233}\text{U} \rightarrow {}^{229}\text{Th} + \alpha$ is 4.909 MeV. Excited states of $^{229}\text{Th}$ at 29, 42, 72, and 97 keV are populated in the decay. Compute the energies of the five most energetic $\alpha$ groups emitted in the $^{233}\text{U}$ decay.

**19.** The five highest energy $\alpha$'s emitted by $^{242}\text{Cm}$ (Figure 8.7) have energies (in MeV) of 6.113, 6.070, 5.972, 5.817, 5.609. Each state is connected with the state directly below it by a $\gamma$ transition. Calculate the energies of the $\gamma$ rays.

**20.** The $\alpha$ decay of a nucleus near mass 200 has two components of energies 4.687 and 4.650 MeV. Neither populates the ground state of the daughter, but each is followed by a $\gamma$ ray, of respective energy 266 and 305 keV. No other $\gamma$ rays are seen. (a) From this information, construct a decay scheme. (b) The decaying parent state has spin 1 and negative parity, and the ground state of the daughter has spin zero and also negative parity. Explain why there is no direct $\alpha$ decay to the ground state.

**21.** The $\alpha$ decay of $^{244}\text{Cm}$ populates a $0^+$ excited state in $^{240}\text{Pu}$ at 0.861 MeV with an intensity of $1.6 \times 10^{-4}\%$, while the $0^+$ ground state is populated with an intensity of 76.7%. Estimate the ratio between these decay intensities from the theory of $\alpha$ decay and compare with the experimental value.

**22.** In a certain decay process, a nucleus in the vicinity of mass 240 emits $\alpha$ particles with the following energies (in MeV): 5.545 ($\alpha_0$), 5.513 ($\alpha_1$), 5.486 ($\alpha_2$), 5.469 ($\alpha_3$), 5.443 ($\alpha_4$), 5.417 ($\alpha_5$), and 5.389 ($\alpha_6$). The following $\gamma$ rays in the daughter nucleus are seen (energies in keV): 26 ($\gamma_1$), 33 ($\gamma_2$), 43 ($\gamma_3$), 56 ($\gamma_4$), 60 ($\gamma_5$), 99 ($\gamma_6$), 103 ($\gamma_7$), and 125 ($\gamma_8$). Construct a decay scheme from this information, assuming $\alpha_0$ populates the ground state of the daughter.

**23.** For the decay of $^{253}\text{Es}$ to $^{249}\text{Bk}$ illustrated in Figure 8.8: (a) Estimate the intensities of the $\ell_\alpha = 0, 2, 4,$ and 6 contributions to the decay to the ground state and compare with the measured values. (b) Assuming the $\ell_\alpha = 2$ component to be dominant, estimate the relative intensities of the decays to the $\frac{9}{2}^+$ and $\frac{11}{2}^+$ states. The ground-state $Q$ value is 6.747 MeV, and the excited states are at 42 keV ($\frac{9}{2}^+$) and 94 keV ($\frac{11}{2}^+$).

**24.** The decay of $^{253}\text{Es}$ ($I = \frac{7}{2}$, $\pi = +$) leads to a sequence of negative-parity states in $^{249}\text{Bk}$ with $I = \frac{3}{2}, \frac{5}{2}, \frac{7}{2}, \frac{9}{2}, \frac{11}{2}, \frac{13}{2}$. For each state, find the permitted values of $\ell_\alpha$.

# 9

# BETA DECAY

The emission of ordinary negative electrons from the nucleus was among the earliest observed radioactive decay phenomena. The inverse process, capture by a nucleus of an electron from its atomic orbital, was not observed until 1938 when Alvarez detected the characteristic X rays emitted in the filling of the vacancy left by the captured electron. The Joliot-Curies in 1934 first observed the related process of positive electron (positron) emission in radioactive decay, only two years after the positron had been discovered in cosmic rays. These three nuclear processes are closely related and are grouped under the common name *beta* ($\beta$) *decay*.

The most basic $\beta$ decay process is the conversion of a proton to a neutron or of a neutron into a proton. In a nucleus, $\beta$ decay changes both $Z$ and $N$ by one unit: $Z \to Z \pm 1$, $N \to N \mp 1$ so that $A = Z + N$ remains constant. Thus $\beta$ decay provides a convenient way for an unstable nucleus to "slide down" the mass parabola (Figure 3.18, for example) of constant $A$ and to approach the stable isobar.

In contrast with $\alpha$ decay, progress in understanding $\beta$ decay has been achieved at an extremely slow pace, and often the experimental results have created new puzzles that challenged existing theories. Just as Rutherford's early experiments showed $\alpha$ particles to be identical with $^4$He nuclei, other early experiments showed the negative $\beta$ particles to have the same electric charge and charge-to-mass ratio as ordinary electrons. In Section 1.2, we discussed the evidence against the presence of electrons as nuclear constituents, and so we must regard the $\beta$ decay process as "creating" an electron from the available decay energy at the instant of decay; this electron is then immediately ejected from the nucleus. This situation contrasts with $\alpha$ decay, in which the $\alpha$ particle may be regarded as having a previous existence in the nucleus.

The basic decay processes are thus:

$$n \to p + e^- \quad \text{negative beta decay } (\beta^-)$$

$$p \to n + e^+ \quad \text{positive beta decay } (\beta^+)$$

$$p + e^- \to n \quad \text{orbital electron capture } (\varepsilon)$$

These processes are not complete, for there is yet another particle (a neutrino or antineutrino) involved in each. The latter two processes occur only for protons

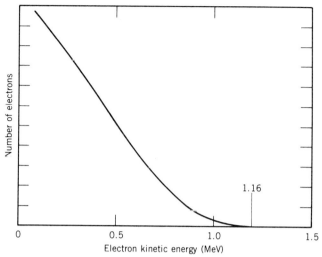

**Figure 9.1** The continuous electron distribution from the $\beta$ decay of $^{210}$Bi (also called RaE in the literature).

bound in nuclei; they are energetically forbidden for free protons or for protons in hydrogen atoms.

## 9.1 ENERGY RELEASE IN β DECAY

The continuous energy distribution of the $\beta$-decay electrons was a confusing experimental result in the 1920s. Alpha particles are emitted with sharp, well-defined energies, equal to the difference in mass energy between the initial and final states (less the small recoil corrections); all $\alpha$ decays connecting the same initial and final states have exactly the same kinetic energies. Beta particles have a continuous distribution of energies, from zero up to an upper limit (the endpoint energy) which is equal to the energy difference between the initial and final states. If $\beta$ decay were, like $\alpha$ decay, a two-body process, we would expect all of the $\beta$ particles to have a unique energy, but virtually all of the emitted particles have a smaller energy. For instance, we might expect on the basis of nuclear mass differences that the $\beta$ particles from $^{210}$Bi would be emitted with a kinetic energy of 1.16 MeV, yet we find a continuous distribution from 0 up to 1.16 MeV (Figure 9.1).

An early attempt to account for this "missing" energy hypothesized that the $\beta$'s are actually emitted with 1.16 MeV of kinetic energy, but lose energy, such as by collisions with atomic electrons, before they reach the detection system. Such a possibility was eliminated by very precise calorimetric experiments that confined a $\beta$ source and measured its decay energy by the heating effect. If a portion of the energy were transferred to the atomic electrons, a corresponding rise in temperature should be observed. These experiments showed that the shape of the spectrum shown in Figure 9.1 is a characteristic of the decay electrons themselves and not a result of any subsequent interactions.

To account for this energy release, Pauli in 1931 proposed that there was emitted in the decay process a second particle, later named by Fermi the *neutrino*. The neutrino carries the "missing" energy and, because it is highly penetrating radiation, it is not stopped within the calorimeter, thus accounting for the failure of those experiments to record its energy. Conservation of electric charge requires the neutrino to be electrically neutral, and angular momentum conservation and statistical considerations in the decay process require the neutrino to have (like the electron) a spin of $\frac{1}{2}$. Experiment shows that there are in fact two different kinds of neutrinos emitted in $\beta$ decay (and yet other varieties emitted in other decay processes; see Chapter 18). These are called the *neutrino* and the *antineutrino* and indicated by $\nu$ and $\bar{\nu}$. It is the antineutrino which is emitted in $\beta^-$ decay and the neutrino which is emitted in $\beta^+$ decay and electron capture. In discussing $\beta$ decay, the term "neutrino" is often used to refer to both neutrinos and antineutrinos, although it is of course necessary to distinguish between them in writing decay processes; the same is true for "electron."

To demonstrate $\beta$-decay energetics we first consider the decay of the free neutron (which occurs with a half-life of about 10 min),

$$\text{n} \rightarrow \text{p} + \text{e}^- + \bar{\nu}.$$

As we did in the case of $\alpha$ decay, we define the $Q$ value to be the difference between the initial and final *nuclear* mass energies.

$$Q = (m_\text{n} - m_\text{p} - m_\text{e} - m_{\bar{\nu}})c^2 \qquad (9.1)$$

and for decays of neutrons at rest,

$$Q = T_\text{p} + T_\text{e} + T_{\bar{\nu}} \qquad (9.2)$$

For the moment we will ignore the proton recoil kinetic energy $T_\text{p}$, which amounts to only 0.3 keV. The antineutrino and electron will then share the decay energy, which accounts for the continuous electron spectrum. The maximum-energy electrons correspond to minimum-energy antineutrinos, and when the antineutrinos have vanishingly small energies, $Q \simeq (T_\text{e})_{\max}$. The measured maximum energy of the electrons is $0.782 \pm 0.013$ MeV. Using the measured neutron, electron, and proton masses, we can compute the $Q$ value:

$$Q = m_\text{n}c^2 - m_\text{p}c^2 - m_\text{e}c^2 - m_{\bar{\nu}}c^2$$

$$= 939.573 \text{ MeV} - 938.280 \text{ MeV} - 0.511 \text{ MeV} - m_{\bar{\nu}}c^2$$

$$= 0.782 \text{ MeV} - m_{\bar{\nu}}c^2$$

Thus to within the precision of the measured maximum energy (about 13 keV) we may regard the antineutrino as massless. Other experiments provide more stringent upper limits, as we discuss in Section 9.6, and for the present discussion we take the masses of the neutrino and antineutrino to be identically zero.

Conservation of linear momentum can be used to identify $\beta$ decay as a three-body process, but this requires measuring the momentum of the recoiling nucleus in coincidence with the momentum of the electron. These experiments are difficult, for the low-energy nucleus ($T \lesssim$ keV) is easily scattered, but they have been done in a few cases, from which it can be deduced that the vector sum of the linear momenta of the electron and the recoiling nucleus is consistent with an unobserved third particle carrying the "missing" energy and having a rest mass of zero or nearly zero. Whatever its mass might be, the existence of the additional particle is absolutely required by these experiments, for the momenta of the electron and nucleus certainly do not sum to zero, as they would in a two-body decay.

Because the neutrino is massless, it moves with the speed of light and its total relativistic energy $E_\nu$ is the same as its kinetic energy; we will use $E_\nu$ to represent neutrino energies. (A review of the concepts and formulas of relativistic kinematics may be found in Appendix A.) For the electron, we will use both its kinetic energy $T_e$ and its total relativistic energy $E_e$, which are of course related by $E_e = T_e + m_e c^2$. (Decay energies are typically of order MeV; thus the nonrelativistic approximation $T \ll mc^2$ is certainly not valid for the decay electrons, and we *must* use relativistic kinematics.) The nuclear recoil is of very low energy and can be treated nonrelativistically.

Let's consider a typical negative $\beta$-decay process in a nucleus:

$$_Z^A X_N \rightarrow {}_{Z+1}^A X'_{N-1} + e^- + \bar{\nu}$$

$$Q_{\beta^-} = \left[ m_N(_Z^A X) - m_N(_{Z+1}^A X') - m_e \right] c^2 \tag{9.3}$$

where $m_N$ indicates *nuclear* masses. To convert nuclear masses into the tabulated neutral atomic masses, which we denote as $m(^A X)$, we use

$$m(^A X) c^2 = m_N(^A X) c^2 + Z m_e c^2 - \sum_{i=1}^{Z} B_i \tag{9.4}$$

where $B_i$ represents the binding energy of the $i$th electron. In terms of atomic masses,

$$Q_{\beta^-} = \left\{ \left[ m(^A X) - Z m_e \right] - \left[ m(^A X') - (Z+1) m_e \right] - m_e \right\} c^2$$

$$+ \left\{ \sum_{i=1}^{Z} B_i - \sum_{i=1}^{Z+1} B_i \right\} \tag{9.5}$$

Notice that the electron masses cancel in this case. Neglecting the differences in electron binding energy, we therefore find

$$Q_{\beta^-} = \left[ m(^A X) - m(^A X') \right] c^2 \tag{9.6}$$

where the masses are neutral atomic masses. The $Q$ value represents the energy shared by the electron and neutrino:

$$Q_{\beta^-} = T_e + E_{\bar{\nu}} \tag{9.7}$$

and it follows that each has its maximum when the other approaches zero:

$$(T_e)_{\max} = (E_{\bar{\nu}})_{\max} - Q_{\beta^-} \tag{9.8}$$

In the case of the $^{210}\text{Bi} \rightarrow {}^{210}\text{Po}$ decay, the mass tables give

$$Q_{\beta^-} = \left[ m(^{210}\text{Bi}) - m(^{210}\text{Po}) \right] c^2$$

$$= (209.984095 \text{ u} - 209.982848 \text{ u})(931.502 \text{ MeV/u})$$

$$= 1.161 \text{ MeV}$$

Figure 9.1 showed $(T_e)_{\max} = 1.16$ MeV, in agreement with the value expected from $Q_{\beta^-}$. Actually, this is really not an agreement between two independent values. The value of $Q_{\beta^-}$ is used in this case to *determine* the mass of $^{210}\text{Po}$, with the mass of $^{210}$Bi determined from that of $^{209}$Bi using neutron capture. Equation 9.6 is used with the measured $Q_{\beta^-}$ to obtain $m(^AX')$.

In the case of positron decay, a typical decay process is

$$^A_Z X_N \rightarrow \, _{Z-1}^{A}X'_{N+1} + e^+ + \nu$$

and a calculation similar to the previous one shows

$$Q_{\beta^+} = \left[ m(^AX) - m(^AX') - 2m_e \right] c^2 \tag{9.9}$$

again using *atomic* masses. Notice that the electron masses *do not* cancel in this case.

For electron-capture processes, such as

$$^A_Z X_N + e^- \rightarrow \, _{Z-1}^{A}X'_{N+1} + \nu$$

the calculation of the $Q$ value must take into account that the atom X' is in an *atomic* excited state immediately after the capture. That is, if the capture takes place from an inner shell, the K shell for instance, an electronic vacancy in that shell results. The vacancy is quickly filled as electrons from higher shells make downward transitions and emit characteristic X rays. Whether one X ray is emitted or several, the total X-ray energy is equal to the binding energy of the captured electron. Thus the atomic mass of X' immediately after the decay is greater than the mass of X' in its atomic ground state by $B_n$, the binding energy of the captured $n$-shell electron ($n = K, L, \ldots$). The $Q$ value is then

$$Q_\varepsilon = \left[ m(^AX) - m(^AX') \right] c^2 - B_n \tag{9.10}$$

Positive beta decay and electron capture both lead from the initial nucleus $^A_Z X_N$ to the final nucleus $_{Z-1}^{A}X'_{N+1}$, but note that both may not always be energetically possible ($Q$ must be positive for any decay process). Nuclei for which $\beta^+$ decay is energetically possible may also undergo electron capture, but the reverse is not true—it is possible to have $Q > 0$ for electron capture while $Q < 0$ for $\beta^+$ decay. The atomic mass energy difference must be at least $2m_e c^2 = 1.022$ MeV to permit $\beta^+$ decay.

In positron decay, expressions of the form of Equations 9.7 and 9.8 show that there is a continuous distribution of neutrino energies up to $Q_{\beta^+}$ (less the usually negligible nuclear recoil). In electron capture, however, the two-body final state results in unique values for the recoil energy and $E_\nu$. Neglecting the recoil, a monoenergetic neutrino with energy $Q_\varepsilon$ is emitted.

All of the above expressions refer to decays between nuclear ground states. If the final nuclear state X' is an excited state, the $Q$ value must be accordingly

**Table 9.1** Typical $\beta$-Decay Processes

| Decay | Type | Q (MeV) | $t_{1/2}$ |
|---|---|---|---|
| $^{23}$Ne $\rightarrow$ $^{23}$Na + e$^-$ + $\bar{\nu}$ | $\beta^-$ | 4.38 | 38 s |
| $^{99}$Tc $\rightarrow$ $^{99}$Ru + e$^-$ + $\bar{\nu}$ | $\beta^-$ | 0.29 | $2.1 \times 10^5$ y |
| $^{25}$Al $\rightarrow$ $^{25}$Mg + e$^+$ + $\nu$ | $\beta^+$ | 3.26 | 7.2 s |
| $^{124}$I $\rightarrow$ $^{124}$Te + e$^+$ + $\nu$ | $\beta^+$ | 2.14 | 4.2 d |
| $^{15}$O + e$^-$ $\rightarrow$ $^{15}$N + $\nu$ | $\varepsilon$ | 2.75 | 1.22 s |
| $^{41}$Ca + e$^-$ $\rightarrow$ $^{41}$K + $\nu$ | $\varepsilon$ | 0.43 | $1.0 \times 10^5$ y |

decreased by the excitation energy of the state:

$$Q_{ex} = Q_{ground} - E_{ex} \qquad (9.11)$$

Table 9.1 shows some typical $\beta$ decay processes, their energy releases, and their half-lives.

## 9.2 FERMI THEORY OF $\beta$ DECAY

In our calculation of $\alpha$-decay half-lives in Chapter 8, we found that the barrier penetration probability was the critical factor in determining the half-life. In negative $\beta$ decay there is no such barrier to penetrate and even in $\beta^+$ decay, it is possible to show from even a rough calculation that the exponential factor in the barrier penetration probability is of order unity. There are other important differences between $\alpha$ and $\beta$ decay which suggest to us that we must use a completely different approach for the calculation of transition probabilities in $\beta$ decay: (1) The electron and neutrino do not exist before the decay process, and therefore we must account for the formation of those particles. (2) The electron and neutrino must be treated relativistically. (3) The continuous distribution of electron energies must result from the calculation.

In 1934, Fermi developed a successful theory of $\beta$ decay based on Pauli's neutrino hypothesis. The essential features of the decay can be derived from the basic expression for the transition probability caused by an interaction that is weak compared with the interaction that forms the quasi-stationary states. This is certainly true for $\beta$ decay, in which the characteristic times (the half-lives, typically of order seconds or longer) are far longer than the characteristic nuclear time ($10^{-20}$ s). The result of this calculation, treating the decay-causing interaction as a weak perturbation, is Fermi's Golden Rule, a general result for any transition rate previously given in Equation 2.79:

$$\lambda = \frac{2\pi}{\hbar} |V_{fi}|^2 \rho(E_f) \qquad (9.12)$$

The matrix element $V_{fi}$ is the integral of the interaction $V$ between the initial and final quasi-stationary states of the system:

$$V_{fi} = \int \psi_f^* V \psi_i \, dv \qquad (9.13)$$

The factor $\rho(E_f)$ is the density of final states, which can also be written as $dn/dE_f$, the number $dn$ of final states in the energy interval $dE_f$. A given transition is more likely to occur if there is a large number of accessible final states.

Fermi did not know the mathematical form of $V$ for $\beta$ decay that would have permitted calculations using Equations 9.12 and 9.13. Instead, he considered all possible forms consistent with special relativity, and he showed that $V$ could be replaced with one of five mathematical operators $O_X$, where the subscript $X$ gives the form of the operator $O$ (that is, its transformation properties): $X = V$ (vector), A (axial vector), S (scalar), P (pseudoscalar), or T (tensor). Which of these is correct for $\beta$ decay can be revealed only through experiments that study the symmetries and the spatial properties of the decay products, and it took 20 years (and several mistaken conclusions) for the correct V–A form to be deduced.

The final state wave function must include not only the nucleus but also the electron and neutrino. For electron capture or neutrino capture, the forms would be similar, but the appropriate wave function would appear in the initial state. For $\beta$ decay, the interaction matrix element then has the form

$$V_{\text{fi}} = g \int [\psi_f^* \varphi_e^* \varphi_\nu^*] O_X \psi_i \, dv \tag{9.14}$$

where now $\psi_f$ refers only to the final nuclear wave function and $\varphi_e$ and $\varphi_\nu$ give the wave functions of the electron and neutrino. The quantity in square brackets represents the entire final system after the decay. The value of the constant $g$ determines the strength of the interaction; the electronic charge $e$ plays a similar role in the interaction between an atom and the electromagnetic field.

The density of states factor determines (to lowest order) the shape of the beta energy spectrum. To find the density of states, we need to know the number of final states accessible to the decay products. Let us suppose in the decay that we have an electron (or positron) emitted with momentum $p$ and a neutrino (or antineutrino) with momentum $q$. We are interested at this point only in the *shape* of the energy spectrum, and thus the directions of $p$ and $q$ are of no interest. If we imagine a coordinate system whose axes are labeled $p_x$, $p_y$, $p_z$, then the locus of the points representing a specific value of $|p| = (p_x^2 + p_y^2 + p_z^2)^{1/2}$ is a sphere of radius $p = |p|$. More specifically, the locus of points representing momenta in the range $dp$ at $p$ is a spherical shell of radius $p$ and thickness $dp$, thus having volume $4\pi p^2 \, dp$. If the electron is confined to a box of volume $V$ (this step is taken only for completeness and to permit the wave function to be normalized; the actual volume will cancel from the final result), then the number of final electron states $dn_e$, corresponding to momenta in the range $p$ to $p + dp$, is

$$dn_e = \frac{4\pi p^2 \, dp \, V}{h^3} \tag{9.15}$$

where the factor $h^3$ is included to make the result a dimensionless pure number.* Similarly, the number of neutrino states is

$$dn_\nu = \frac{4\pi q^2 \, dq \, V}{h^3} \tag{9.16}$$

and the number of final states which have simultaneously an electron and a neutrino with the proper momenta is

$$d^2n = dn_e \, dn_\nu = \frac{(4\pi)^2 V^2 p^2 \, dp \, q^2 \, dq}{h^6} \tag{9.17}$$

---

*The available spatial and momentum states are counted in six-dimensional $(x, y, z, p_x, p_y, p_z)$ *phase space*; the unit volume in phase space is $h^3$.

The electron and neutrino wave functions have the usual free-particle form, normalized within the volume $V$:

$$\varphi_e(r) = \frac{1}{\sqrt{V}} e^{\,i\mathbf{p}\cdot\mathbf{r}/\hbar}$$

$$\varphi_\nu(r) = \frac{1}{\sqrt{V}} e^{\,i\mathbf{q}\cdot\mathbf{r}/\hbar} \tag{9.18}$$

For an electron with 1 MeV kinetic energy, $p = 1.4$ MeV/$c$ and $p/\hbar = 0.007$ fm$^{-1}$. Thus over the nuclear volume, $pr \ll 1$ and we can expand the exponentials, keeping only the first term:

$$e^{\,i\mathbf{p}\cdot\mathbf{r}/\hbar} = 1 + \frac{i\mathbf{p}\cdot\mathbf{r}}{\hbar} + \cdots \cong 1$$

$$e^{\,i\mathbf{q}\cdot\mathbf{r}/\hbar} = 1 + \frac{i\mathbf{q}\cdot\mathbf{r}}{\hbar} + \cdots \cong 1 \tag{9.19}$$

This approximation is known as the *allowed* approximation.

In this approximation, the only factors that depend on the electron or neutrino energy come from the density of states. Let's assume we are trying to calculate the momentum and energy distributions of the emitted electrons. The partial decay rate for electrons and neutrinos with the proper momenta is

$$d\lambda = \frac{2\pi}{\hbar} g^2 |M_{\mathrm{fi}}|^2 (4\pi)^2 \frac{p^2\, dp\, q^2}{h^6} \frac{dq}{dE_{\mathrm{f}}} \tag{9.20}$$

where $M_{\mathrm{fi}} = \int \psi_{\mathrm{f}}^* O_X \psi_{\mathrm{i}}\, dv$ is the *nuclear matrix element*. The final energy $E_{\mathrm{f}}$ is just $E_e + E_\nu = E_e + qc$, and so $dq/dE_{\mathrm{f}} = 1/c$ at fixed $E_e$. As far as the shape of the electron spectrum is concerned, all of the factors in Equation 9.20 that do not involve the momentum (including $M_{\mathrm{fi}}$, which for the present we assume to be independent of $p$) can be combined into a constant $C$, and the resulting distribution gives the number of electrons with momentum between $p$ and $p + dp$:

$$N(p)\, dp = Cp^2 q^2\, dp \tag{9.21}$$

If $Q$ is the decay energy, then ignoring the negligible nuclear recoil energy,

$$q = \frac{Q - T_e}{c} = \frac{Q - \sqrt{p^2c^2 + m_e^2 c^4} + m_e c^2}{c} \tag{9.22}$$

and the spectrum shape is given by

$$N(p) = \frac{C}{c^2} p^2 (Q - T_e)^2 \tag{9.23}$$

$$= \frac{C}{c^2} p^2 \left[ Q - \sqrt{p^2 c^2 + m_e^2 c^4} + m_e c^2 \right]^2 \tag{9.24}$$

This function vanishes at $p = 0$ and also at the endpoint where $T_e = Q$; its shape is shown in Figure 9.2.

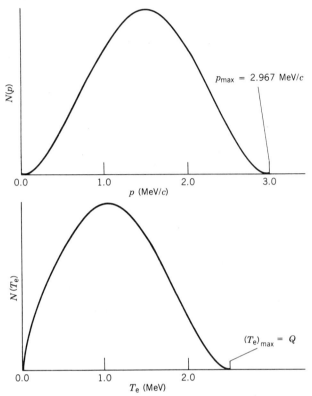

**Figure 9.2** Expected electron energy and momentum distributions, from Equations 9.24 and 9.25. These distributions are drawn for $Q = 2.5$ MeV.

More frequently we are interested in the energy spectrum, for electrons with kinetic energy between $T_e$ and $T_e + dT_e$. With $c^2 p \, dp = (T_e + m_e c^2) \, dT_e$, we have

$$N(T_e) = \frac{C}{c^5}\left(T_e^2 + 2T_e m_e c^2\right)^{1/2}(Q - T_e)^2(T_e + m_e c^2) \qquad (9.25)$$

This distribution, which also vanishes at $T_e = 0$ and at $T_e = Q$, is shown in Figure 9.2.

In Figure 9.3, the $\beta^+$ and $\beta^-$ decays of $^{64}$Cu are compared with the predictions of the theory. As you can see, the general shape of Figure 9.2 is evident, but there are systematic differences between theory and experiment. These differences originate with the Coulomb interaction between the $\beta$ particle and the daughter nucleus. Semiclassically, we can interpret the shapes of the momentum distributions of Figure 9.3 as a Coulomb repulsion of $\beta^+$ by the nucleus, giving fewer low-energy positrons, and a Coulomb attraction of $\beta^-$, giving more low-energy electrons. From the more correct standpoint of quantum mechanics, we should instead refer to the change in the electron plane wave, Equation 9.19, brought about by the Coulomb potential inside the nucleus. The quantum mechanical calculation of the effect of the nuclear Coulomb field on the electron wave function is beyond the level of this text. It modifies the spectrum by introducing an additional factor, the *Fermi function* $F(Z', p)$ or $F(Z', T_e)$, where $Z'$ is the atomic number of the daughter nucleus. Finally, we must

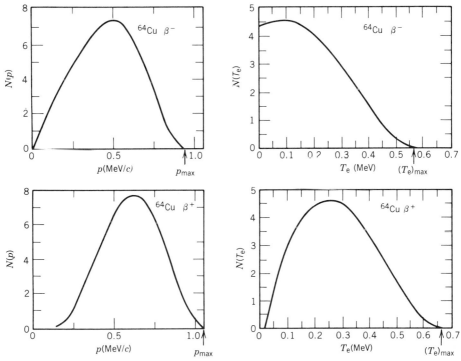

**Figure 9.3** Momentum and kinetic energy spectra of electrons and positrons emitted in the decay of $^{64}$Cu. Compare with Figure 9.2; the differences arise from the Coulomb interactions with the daughter nucleus. From R. D. Evans, *The Atomic Nucleus* (New York: McGraw-Hill, 1955).

consider the effect of the nuclear matrix element, $M_{fi}$, which we have up to now assumed not to influence the shape of the spectrum. This approximation (also called the allowed approximation) is often found to be a very good one, but there are some cases in which it is very bad—in fact, there are cases in which $M_{fi}$ vanishes in the allowed approximation, giving no spectrum at all! In such cases, we must take the next terms of the plane wave expansion, Equations 9.19, which introduce yet another momentum dependence. Such cases are called, somewhat incorrectly, *forbidden* decays; these decays are not absolutely forbidden, but as we will learn subsequently, they are less likely to occur than allowed decays and therefore tend to have longer half-lives. The degree to which a transition is forbidden depends on how far we must take the expansion of the plane wave to find a nonvanishing nuclear matrix element. Thus the first term beyond the 1 gives first-forbidden decays, the next term gives second-forbidden, and so on. We will see in Section 9.4 how the angular momentum and parity selection rules restrict the kinds of decay that can occur.

The complete $\beta$ spectrum then includes three factors:

1. The *statistical factor* $p^2(Q - T_e)^2$, derived from the number of final states accessible to the emitted particles.

2. The Fermi function $F(Z', p)$, which accounts for the influence of the nuclear Coulomb field.

3.  The nuclear matrix element $|M_{fi}|^2$, which accounts for the effects of particular initial and final nuclear states and which may include an additional electron and neutrino momentum dependence $S(p, q)$ from forbidden terms:

$$N(p) \propto p^2 (Q - T_e)^2 F(Z', p)|M_{fi}|^2 S(p, q) \qquad (9.26)$$

## 9.3 THE "CLASSICAL" EXPERIMENTAL TESTS OF THE FERMI THEORY

### The Shape of the β Spectrum

In the allowed approximation, we can rewrite Equation 9.26 as

$$(Q - T_e) \propto \sqrt{\frac{N(p)}{p^2 F(Z', p)}} \qquad (9.27)$$

and plotting $\sqrt{N(p)/p^2 F(Z', p)}$ against $T_e$ should give a straight line which intercepts the $x$ axis at the decay energy $Q$. Such a plot is called a *Kurie plot* (sometimes a Fermi plot or a Fermi-Kurie plot). An example of a Kurie plot is shown in Figure 9.4. The linear nature of this plot gives us confidence in the theory as it has been developed, and also gives us a convenient way to determine the decay endpoint energy (and therefore the $Q$ value).

In the case of forbidden decays, the standard Kurie plot does not give a straight line, but we can restore the linearity of the plot if we instead graph $\sqrt{N(p)/p^2 F(Z', p)S(p, q)}$ against $T_e$, where $S$ is the momentum dependence that results from the higher-order term in the expansion of the plane wave. The function $S$ is known as the *shape factor*; for certain first-forbidden decays, for example, it is simply $p^2 + q^2$.

Including the shape factor gives a linear plot, as Figure 9.5 shows.

### The Total Decay Rate

To find the total decay rate, we must integrate Equation 9.20 over all values of the electron momentum $p$, keeping the neutrino momentum at the value determined by Equation 9.22, which of course also depends on $p$. Thus, for allowed decays,

$$\lambda = \frac{g^2 |M_{fi}|^2}{2\pi^3 \hbar^7 c^3} \int_0^{p_{max}} F(Z', p) p^2 (Q - T_e)^2 \, dp \qquad (9.28)$$

The integral will ultimately depend only on $Z'$ and on the maximum electron total energy $E_0$ (since $cp_{max} = \sqrt{E_0^2 - m_e^2 c^4}$), and we therefore represent it as

$$f(Z', E_0) = \frac{1}{(m_e c)^3 (m_e c^2)^2} \int_0^{p_{max}} F(Z', p) p^2 (E_0 - E_e)^2 \, dp \qquad (9.29)$$

where the constants have been included to make $f$ dimensionless. The function $f(Z', E_0)$ is known as the *Fermi integral* and has been tabulated for values of $Z'$ and $E_0$.

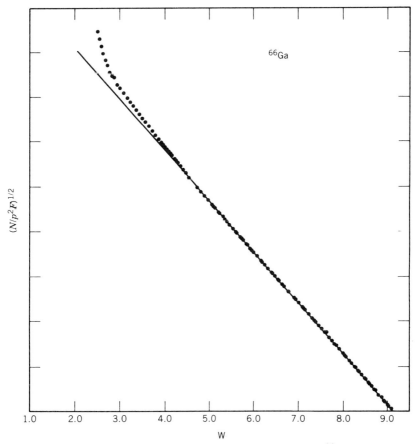

**Figure 9.4** Fermi – Kurie plot of allowed $0^+ \rightarrow 0^+$ decay of $^{66}$Ga. The horizontal scale is the relativistic *total* energy ($T_e + m_e c^2$) in units of $m_e c^2$. The deviation from the straight line at low energy arises from the scattering of low-energy electrons within the radioactive source. From D. C. Camp and L. M. Langer, *Phys. Rev.* **129**, 1782 (1963).

With $\lambda = 0.693/t_{1/2}$, we have

$$ft_{1/2} = 0.693 \frac{2\pi^3 \hbar^7}{g^2 m_e^5 c^4 |M_{fi}|^2} \tag{9.30}$$

The quantity on the left side of Equation 9.30 is called the *comparative half-life* or *ft value*. It gives us a way to compare the $\beta$ decay probabilities in different nuclei—Equation 9.28 shows that the decay rate depends on $Z'$ and on $E_0$, and this dependence is incorporated into $f$, so that *differences in ft values must be due to differences in the nuclear matrix element* and thus to differences in the nuclear wave function.

As in the case of $\alpha$ decay, there is an enormous range of half-lives in $\beta$ decay —$ft$ values range from about $10^3$ to $10^{20}$ s. For this reason, what is often quoted is the value of $\log_{10} ft$ (with $t$ given in seconds). The decays with the shortest comparative half-lives ($\log ft \simeq 3$–4) are known as *superallowed* decays. Some of

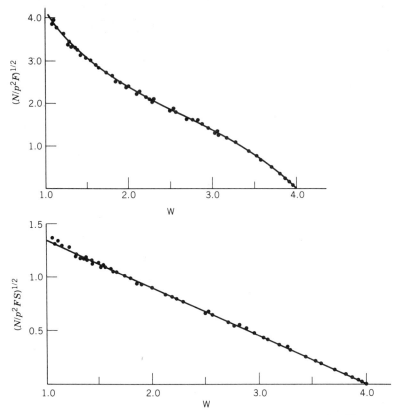

**Figure 9.5**   Uncorrected Fermi–Kurie plot in the $\beta$ decay of $^{91}$Y (top). The linearity is restored if the shape factor $S(p, q)$ is included; for this type of first-forbidden decay, the shape factor $p^2 + q^2$ gives a linear plot (bottom). Data from L. M. Langer and H. C. Price, *Phys. Rev.* **75**, 1109 (1949).

the superallowed decays have $0^+$ initial and final states, in which case the nuclear matrix element can be calculated quite easily: $M_{\mathrm{fi}} = \sqrt{2}$. The log $ft$ values for $0^+ \to 0^+$ decays should all be identical. Table 9.2 shows the log $ft$ values of all known $0^+ \to 0^+$ superallowed transitions, and within experimental error the values appear to be quite constant. Moreover, with $M_{\mathrm{fi}} = \sqrt{2}$, we can use Equation 9.30 to find a value of the $\beta$-decay strength constant

$$g = 0.88 \times 10^{-4} \text{ MeV} \cdot \text{fm}^3$$

To make this constant more comparable to other fundamental constants, we should express it in a dimensionless form. We can then compare it with dimensionless constants of other interactions (the fine structure constant which characterizes the electromagnetic interaction, for instance). Letting $M$, $L$, and $T$ represent, respectively, the dimensions of mass, length, and time, the dimensions of $g$ are $M^1L^5T^{-2}$, and no combinations of the fundamental constants $\hbar$ (dimension $M^1L^2T^{-1}$) and $c$ (dimension $L^1T^{-1}$) can be used to convert $g$ into a dimensionless constant. (For instance, $\hbar c^3$ has dimension $M^1L^5T^{-5}$, and so $g/\hbar c^3$ has dimension $T^3$.) Let us therefore introduce an arbitrary mass $m$ and

**Table 9.2**  *ft* Values for $0^+ \rightarrow 0^+$ Superallowed Decays

| Decay | *ft* (s) |
|---|---|
| $^{10}\text{C} \rightarrow {}^{10}\text{B}$ | $3100 \pm 31$ |
| $^{14}\text{O} \rightarrow {}^{14}\text{N}$ | $3092 \pm 4$ |
| $^{18}\text{Ne} \rightarrow {}^{18}\text{F}$ | $3084 \pm 76$ |
| $^{22}\text{Mg} \rightarrow {}^{22}\text{Na}$ | $3014 \pm 78$ |
| $^{26}\text{Al} \rightarrow {}^{26}\text{Mg}$ | $3081 \pm 4$ |
| $^{26}\text{Si} \rightarrow {}^{26}\text{Al}$ | $3052 \pm 51$ |
| $^{30}\text{S} \rightarrow {}^{30}\text{P}$ | $3120 \pm 82$ |
| $^{34}\text{Cl} \rightarrow {}^{34}\text{S}$ | $3087 \pm 9$ |
| $^{34}\text{Ar} \rightarrow {}^{34}\text{Cl}$ | $3101 \pm 20$ |
| $^{38}\text{K} \rightarrow {}^{38}\text{Ar}$ | $3102 \pm 8$ |
| $^{38}\text{Ca} \rightarrow {}^{38}\text{K}$ | $3145 \pm 138$ |
| $^{42}\text{Sc} \rightarrow {}^{42}\text{Ca}$ | $3091 \pm 7$ |
| $^{42}\text{Ti} \rightarrow {}^{42}\text{Sc}$ | $3275 \pm 1039$ |
| $^{46}\text{V} \rightarrow {}^{46}\text{Ti}$ | $3082 \pm 13$ |
| $^{46}\text{Cr} \rightarrow {}^{46}\text{V}$ | $2834 \pm 657$ |
| $^{50}\text{Mn} \rightarrow {}^{50}\text{Cr}$ | $3086 \pm 8$ |
| $^{54}\text{Co} \rightarrow {}^{54}\text{Fe}$ | $3091 \pm 5$ |
| $^{62}\text{Ga} \rightarrow {}^{62}\text{Zn}$ | $2549 \pm 1280$ |

try to choose the exponents $i$, $j$, and $k$ so that $g/m^i \hbar^j c^k$ is dimensionless. A solution immediately follows with $i = -2$, $j = 3$, $k = -1$. Thus the desired ratio, indicated by $G$, is

$$G = \frac{g}{m^{-2}\hbar^3 c^{-1}} = g\frac{m^2 c}{\hbar^3} \qquad (9.31)$$

There is no clear indication of what value to use for the mass in Equation 9.31. If we are concerned with the nucleon–nucleon interaction, it is appropriate to use the nucleon mass, in which case the resulting dimensionless strength constant is $G = 1.0 \times 10^{-5}$. The comparable constant describing the pion—nucleon interaction, denoted by $g_\pi^2$ in Chapter 4, is of order unity. We can therefore rank the four basic nucleon–nucleon interactions in order of strength:

| | |
|---|---|
| pion-nucleon ("strong") | 1 |
| electromagnetic | $10^{-2}$ |
| $\beta$ decay ("weak") | $10^{-5}$ |
| gravitational | $10^{-39}$ |

(The last entry follows from a similar conversion of the universal gravitational constant into dimensionless form also using the nucleon mass.) The $\beta$-decay interaction is one of a general class of phenomena known collectively as *weak interactions*, all of which are characterized by the strength parameter $g$. The Fermi theory is remarkably successful in describing these phenomena, to the extent that they are frequently discussed as examples of the *universal Fermi interaction*. Nevertheless, the Fermi theory fails in several respects to account for some details of the weak interaction (details which are unimportant for the present discussion of $\beta$ decay). A theory that describes the weak interaction in terms of exchanged particles (just as the strong nuclear force was described in Chapter 4) is more successful in explaining these properties. The recently discovered exchanged particles (with the unfortunate name *intermediate vector bosons*) are discussed in more detail in Chapter 18.

### The Mass of the Neutrino

The Fermi theory is based on the *assumption* that the rest mass of the neutrino is zero. Superficially, it might seem that the neutrino rest mass would be a reasonably easy quantity to measure in order to verify this assumption. Looking back at Equations 9.1 and 9.2, or their equivalents for nuclei with $A > 1$, we immediately see a method to test the assumption. We can calculate the decay $Q$ value (including a possible nonzero value of the neutrino mass) from Equation 9.6 or 9.9, and we can measure the $Q$ value, as in Equation 9.8, from the maximum energy of the $\beta$ particles. Comparison of these two values then permits a value for the neutrino mass to be deduced.

From this procedure we can conclude that the neutrino rest mass is smaller than about 1 keV/$c^2$, but we cannot extend far below that limit because the measured atomic masses used to compute $Q$ have precisions of the order of keV, and the deduced endpoint energies also have experimental uncertainties of the order of keV. A superior method uses the shape of the $\beta$ spectrum near the upper limit. If $m_\nu \neq 0$ then Equation 9.22 is no longer strictly valid. However, if $m_\nu c^2 \ll Q$, then over most of the observed $\beta$ spectrum $E_\nu \gg m_\nu c^2$ and the neutrino can be treated in the extreme relativistic approximation $E_\nu \approx qc$. In this case, Equation 9.22 will be a very good approximation and the neutrino mass will have a negligible effect. Near the endpoint of the $\beta$ spectrum, however, the neutrino energy approaches zero and at some point we would expect $E_\nu \sim m_\nu c^2$, in which case our previous calculation of the statistical factor for the spectrum shape is incorrect. Still closer to the endpoint, the neutrino kinetic energy becomes still smaller and we may begin to treat it nonrelativistically, so that $q^2 = 2m_\nu T_\nu$ and

$$N(p) \propto p^2 \left[ Q - \sqrt{p^2 c^2 + m_e^2 c^4} + m_e c^2 \right]^{1/2} \tag{9.32}$$

which follows from a procedure similar to that used to obtain Equation 9.24, except that for $m_\nu > 0$ we must use $dq/dE_f = m_\nu/q$ in the nonrelativistic limit. Also,

$$N(T_e) \propto \left( T_e^2 + 2T_e m_e c^2 \right)^{1/2} (Q - T_e)^{1/2} \left( T_e + m_e c^2 \right) \tag{9.33}$$

The quantity in square brackets in Equations 9.32 and 9.24, which is just $(Q - T_e)$, vanishes at the endpoint. Thus at the endpoint $dN/dp \rightarrow 0$ if $m_\nu = 0$, while $dN/dp \rightarrow \infty$ if $m_\nu > 0$. That is, the momentum spectrum approaches the

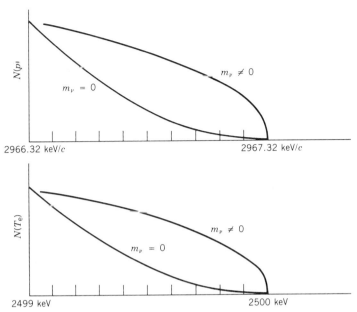

**Figure 9.6** Expanded view of the upper 1-keV region of the momentum and energy spectra of Figure 9.2. The normalizations are arbitrary; what is significant is the difference in the shape of the spectra for $m_\nu = 0$ and $m_\nu \neq 0$. For $m_\nu = 0$, the slope goes to zero at the endpoint; for $m_\nu \neq 0$, the slope at the endpoint is infinite.

endpoint with zero slope for $m_\nu = 0$ and with infinite slope for $m_\nu > 0$. The slope of the energy spectrum, $dN/dT_e$, behaves identically. We can therefore study the limit on the neutrino mass by looking at the slope at the endpoint of the spectrum, as suggested by Figure 9.6. Unfortunately $N(p)$ and $N(T_e)$ also approach zero here, and we must study the slope of a continuously diminishing (and therefore statistically worsening) quantity of data.

The most attractive choice for an experimental measurement of this sort would be a decay with a small $Q$ (so that the relative magnitude of the effect is larger) and one in which the atomic states before and after the decay are well understood, so that the important corrections for the influence of different atomic states can be calculated. (The effects of the atomic states are negligible in most $\beta$-decay experiments, but in this case in which we are searching for a very small effect, they become important.) The decay of $^3$H (tritium) is an appropriate candidate under both criteria. Its $Q$ value is relatively small (18.6 keV), and the one-electron atomic wave functions are well known. (In fact, the calculation of the state of the resulting $^3$He ion is a standard problem in first-year quantum mechanics.) Figure 9.7 illustrates some of the more precise experimental results. Langer and Moffat originally reported an upper limit of $m_\nu c^2 < 200$ eV, while two decades later, Bergkvist reduced the limit to 60 eV. One recent result may indicate a nonzero mass with a probable value between 14 and 46 eV, while others suggest an upper limit of about 20 eV. Several experiments are currently being performed to resolve this question and possibly to reduce the upper limit.

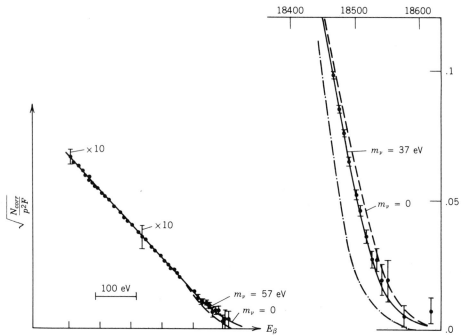

**Figure 9.7** Experimental determination of the neutrino mass from the $\beta$ decay of tritium ($^3$H). The data at left, from K.-E. Bergkvist, *Nucl. Phys. B* **39**, 317 (1972), are consistent with a mass of zero and indicate an upper limit of around 60 eV. The more recent data of V. A. Lubimov et al., *Phys. Lett. B* **94**, 266 (1980), seem to indicate a nonzero value of about 30 eV; however, these data are subject to corrections for instrumental resolution and atomic-state effects and may be consistent with a vanishing mass.

Why is so much effort expended to pursue these measurements? The neutrino mass has very important implications for two areas of physics that on the surface may seem to be unrelated. If the neutrinos have mass, then the "electroweak" theoretical formulism, which treats the weak and electromagnetic interactions as different aspects of the same basic force, permits electron-type neutrinos, those emitted in $\beta$ decay, to convert into other types of neutrinos, called muon and $\tau$ neutrinos (see Chapter 18). This conversion may perhaps explain why the number of neutrinos we observe coming from the sun is only about one-third of what it is expected to be, based on current theories of solar fusion. At the other end of the scale, there seems to be more matter holding the universe together than we can observe with even the most powerful telescopes. This matter is nonluminous, meaning it is not observed to emit any sort of radiation. The Big Bang cosmology, which seems to explain nearly all of the observed astronomical phenomena, predicts that the present universe should be full of neutrinos from the early universe, with a present concentration of the order of $10^8/m^3$. If these neutrinos were massless, they could not supply the necessary gravitational attraction to "close" the universe (that is, to halt and reverse the expansion), but with rest masses as low as 5 eV, they would provide sufficient mass-energy density. The study of the neutrino mass thus has direct and immediate bearing not only on nuclear and particle physics, but on solar physics and cosmology as well.

## 9.4 ANGULAR MOMENTUM AND PARITY SELECTION RULES

### Allowed Decays

In the allowed approximation, we replaced the electron and neutrino wave functions with their values at the origin; that is, we regard the electron and neutrino to have been created at $r = 0$. In this case they cannot carry any *orbital angular momentum*, and the only change in the angular momentum of the nucleus must result from the spins of the electron and neutrino, each of which has the value $s = \frac{1}{2}$. These two spins can be parallel (total $S = 1$) or antiparallel (total $S = 0$). If the spins are antiparallel (which is known as a *Fermi* decay) then in the allowed approximation ($\ell = 0$) there can be no change in the nuclear spin: $\Delta I = |I_i - I_f| = 0$. If the electron and neutrino spins are parallel (which is called a *Gamow-Teller* decay) in the allowed approximation, they carry a total angular momentum of 1 unit and thus $I_i$ and $I_f$ must be coupled through a vector of length 1: $\mathbf{I}_i = \mathbf{I}_f + \mathbf{1}$. This is possible only if $\Delta I = 0$ or 1 (except for $I_i = 0$ and $I_f = 0$, in which case only the Fermi transition can contribute).

   If the electron and neutrino carry no orbital angular momentum, then the parities of the initial and final states must be identical since the parity associated with orbital angular momentum $\ell$ is $(-1)^\ell$.

   We therefore have the following *selection rules* for *allowed $\beta$ decay*:

$$\Delta I = 0, 1 \qquad \Delta\pi \text{ (parity change)} = \text{no}$$

Some examples of allowed $\beta$ decay are

$^{14}O \rightarrow {}^{14}N^*$   This $0^+ \rightarrow 0^+$ decay to an excited state of $^{14}N$ must be pure Fermi type (because $0^+ \rightarrow 0^+$ decays cannot be accomplished through a Gamow-Teller decay, which must carry one unit of angular momentum). Other examples include $^{34}Cl \rightarrow {}^{34}S$ and $^{10}C \rightarrow {}^{10}B^*$, both of which are $0^+ \rightarrow 0^+$.

$^6He \rightarrow {}^6Li$   This decay is $0^+ \rightarrow 1^+$, which must be a pure Gamow-Teller transition. Other allowed pure Gamow-Teller decays include $^{13}B \rightarrow {}^{13}C$ ($\frac{3}{2}^- \rightarrow \frac{1}{2}^-$), $^{230}Pa \rightarrow {}^{230}Th^*$ ($2^- \rightarrow 3^-$), and $^{111}Sn \rightarrow {}^{111}In$ ($\frac{7}{2}^+ \rightarrow \frac{9}{2}^+$).

$n \rightarrow p$   In this case $\Delta I = 0$ ($\frac{1}{2}^+ \rightarrow \frac{1}{2}^+$), and so both the Fermi (F) and Gamow-Teller (GT) selection rules are satisfied. This is an example of a "mixed" F + GT transition, in which the exact proportions of F and GT are determined by the initial and final nuclear wave functions. It is convenient to define the ratio $y$ of the Fermi and Gamow-Teller *amplitudes* (that is, matrix elements):

$$y = \frac{g_F M_F}{g_{GT} M_{GT}} \tag{9.34}$$

where $M_F$ and $M_{GT}$ are the actual Fermi and Gamow-Teller nuclear matrix elements. We allow for the possibility that the Fermi and Gamow-Teller strength constants may differ by defining $g_F$ and $g_{GT}$ as the constants analogous to the single constant $g$ that appears in Equation 9.28. (In the decay rate, we should replace $g^2|M_{fi}|^2$ with $g_F^2|M_F|^2 + g_{GT}^2|M_{GT}|^2$.) We assume $g_F$ to be identical to the value $g$ deduced from the superallowed ($0^+ \rightarrow 0^+$) Fermi decays. For neutron decay, the Fermi matrix element can be simply calculated: $|M_F| = 1$. Since the decay rate is proportional to $g_F^2 M_F^2 (1 + y^{-2})$, the neutron decay rate permits a calculation of the ratio $y$, which yields the value $0.467 \pm 0.003$. That is, the decay is 82% Gamow-Teller and 18% Fermi.

In general, the initial and final nuclear wave functions make calculating $M_F$ and $M_{GT}$ a complicated and difficult process, but in one special group of decays the calculation is simplified. That group is the *mirror decays*, which we previously considered in Section 3.1. In decays such as $^{41}_{21}\text{Sc}_{20} \rightarrow ^{41}_{20}\text{Ca}_{21}$, where the 21st proton becomes the 21st neutron, no change of wave function is involved. Except for minor differences due to the Coulomb interaction, the initial and final wave functions are identical, and the calculation of $M_F$ and $M_{GT}$ is easily done. For these nuclei, $g_F$ and $M_F$ have the same values as they do for the decay of the free neutron.

This result may seem somewhat surprising because in a nucleus, a nucleon does not behave at all like a free nucleon, primarily because of the cloud of mesons that surrounds a nucleon as it participates in exchange interactions with its neighbors. The hypothesis that Fermi interactions of nucleons in nuclei are unchanged by the surrounding mesons is called the *conserved vector current* (CVC) hypothesis. (The term "vector" refers to the transformation properties of the operator that causes the Fermi part of the decay; the Gamow-Teller part arises from an "axial vector" type of interaction.) The CVC hypothesis can be understood by analogy with the electromagnetic interaction. The electric charge is not changed by the transformation $\text{p} \leftrightarrow \text{n} + \pi^+$ which is part of the exchange interaction in which a proton may participate. Electric charge is conserved in this process and the Coulomb interaction is unchanged. (The electrons bound to the nucleus by Coulomb forces are unaware of the transformation.) On the other hand, magnetic interactions are substantially changed by $\text{p} \leftrightarrow \text{n} + \pi^+$, as we discussed when we considered shell-model magnetic moments in Section 5.1. In $\beta$ decay, $g_F$ (like electric charge) is unaffected by the surrounding mesons, while $g_{GT}$ (like magnetic moments) may be affected by the meson cloud. In some nuclei, the change amounts to 20–30%. The matrix element $M_{GT}$ also varies with the particular shell model state of the nucleon that makes the transition.

**Table 9.3** Ratio of Fermi to Gamow-Teller Matrix Elements

|  | Decay | $y = g_F M_F / g_{GT} M_{GT}$ | %F | %GT |
|---|---|---|---|---|
| Mirror | $\text{n} \rightarrow \text{p}$ | $0.467 \pm 0.003$ | 18 | 82 |
| decays | $^3\text{H} \rightarrow ^3\text{He}$ | $0.479 \pm 0.001$ | 19 | 81 |
|  | $^{13}\text{N} \rightarrow ^{13}\text{C}$ | $1.779 \pm 0.006$ | 76 | 24 |
|  | $^{21}\text{Na} \rightarrow ^{21}\text{Ne}$ | $1.416 \pm 0.012$ | 67 | 33 |
|  | $^{41}\text{Sc} \rightarrow ^{41}\text{Ca}$ | $0.949 \pm 0.003$ | 47 | 53 |
| Nonmirror | $^{24}\text{Na} \rightarrow ^{24}\text{Mg}$ | $-0.021 \pm 0.007$ | 0.044 | 99.956 |
| decays | $^{41}\text{Ar} \rightarrow ^{41}\text{K}$ | $+0.027 \pm 0.011$ | 0.073 | 99.927 |
|  | $^{46}\text{Sc} \rightarrow ^{46}\text{Ti}$ | $-0.023 \pm 0.005$ | 0.053 | 99.947 |
|  | $^{52}\text{Mn} \rightarrow ^{52}\text{Cr}$ | $-0.144 \pm 0.006$ | 2 | 98 |
|  | $^{65}\text{Ni} \rightarrow ^{65}\text{Cu}$ | $-0.002 \pm 0.019$ | $< 0.04$ | $> 99.96$ |

Table 9.3 shows a summary of values of the ratio $y$ of the Fermi and Gamow-Teller amplitudes for some mirror nuclei, assuming the CVC hypothesis ($g_F$ is unchanged from its value for neutron decay) and taking $|M_F| = 1$. These values are derived from decay rates.

For decays in which the initial and final wave functions are very different, the Fermi matrix element vanishes, and so measuring the ratio $y$ for these decays is a way to test the purity of the wave functions. Table 9.3 includes some representative values of $y$ for transitions in other than mirror nuclei. These values come from measuring the angular distribution of the $\beta$ particles relative to a particular direction (similar to studies with $\alpha$ decays discussed in Chapter 8). You can see that the values are in general quite small, showing that the Fermi transitions are inhibited and thus that the wave functions are relatively pure.

## Forbidden Decays

The designation of decays as "forbidden" is really somewhat of a misnomer. These decays are usually less probable than allowed decays (and have generally longer half-lives, as we discuss in the next section), but if the allowed matrix elements happen to vanish, then the forbidden decays are the only ones that can occur.

The most frequent occurrence of a forbidden decay is when the initial and final states have opposite parities, and thus the selection rule for allowed decay is violated. To accomplish the change in parity, the electron and neutrino must be emitted with an odd value of the orbital angular momentum relative to the nucleus. Let us consider, for example, a 1-MeV decay process. If the electron is given all the decay energy, its momentum is 1.4 MeV/$c$ and the maximum angular momentum it can carry relative to the nucleus is $pR = 8.4$ MeV·fm/$c$ taking $R = 6$ fm as a typical nuclear radius. In units of $\hbar$, this is equivalent to $pR/\hbar = 0.04$. Thus, while it is less likely to have $\ell = 1$ decays relative to $\ell = 0$, decays with $\ell = 3, 5, 7, \ldots$ are extremely unlikely, and we can for the moment consider only those forbidden decays with $\ell = 1$. These are called *first-forbidden* decays, and like the allowed decays they have Fermi types, with the electron and neutrino spins opposite ($S = 0$), and Gamow-Teller types, with the spins parallel ($S = 1$). The coupling of $S = 0$ with $\ell = 1$ for the Fermi decays gives total angular momentum of one unit carried by the beta decay, so that $\Delta I = 0$ or 1 (but not $0 \rightarrow 0$). Coupling $S = 1$ with $\ell = 1$ for the Gamow-Teller decays gives 0, 1, or 2 units of total angular momentum, so that $\Delta I = 0$, 1, or 2. Thus the *selection rules* for first-forbidden decays are

$$\Delta I = 0, 1, 2 \qquad \Delta \pi = \text{yes}$$

In contrast to the relative simplicity of allowed decays, there are six different matrix elements for first-forbidden decays, and the analysis of decay rates or angular distributions becomes very complicated. We will merely cite some of the many examples of first-forbidden decays:

$$^{17}\text{N} \rightarrow {}^{17}\text{O} \qquad (\tfrac{1}{2}^- \rightarrow \tfrac{5}{2}^+)$$

$$^{76}\text{Br} \rightarrow {}^{76}\text{Se} \qquad (1^- \rightarrow 0^+)$$

$$^{122}\text{Sb} \rightarrow {}^{122}\text{Sn}^* \qquad (2^- \rightarrow 2^+)$$

Transitions with $\Delta I \geq 2$, but with no change in parity, are permitted by neither the allowed nor the first-forbidden selection rules. For these transitions we must look to the $\ell = 2$ $\beta$ emission, and consequently these are known as *second-forbidden* decays. When we couple $S = 0$ or 1 to $\ell = 2$, we can in principle change the nuclear spin by any amount from $\Delta I = 0$ to $\Delta I = 3$ (with certain exceptions, such as $0 \to 0$ and $\frac{1}{2} \to \frac{1}{2}$). The $\Delta I = 0$ and 1 cases fall within the selection rules for allowed decays, and we expect that the contribution of the second-forbidden terms to those decays will be negligible (perhaps $10^{-3}$ to $10^{-4}$ in angular distributions, and $10^{-6}$ to $10^{-8}$ in the spectrum shape). Excepting these cases, the selection rules for the second-forbidden decays are

$$\Delta I = 2, 3 \qquad \Delta\pi = \text{no}$$

Examples of second-forbidden decays are

$$^{22}\text{Na} \to {}^{22}\text{Ne} \quad (3^+ \to 0^+)$$

$$^{137}\text{Cs} \to {}^{137}\text{Ba} \quad \left(\tfrac{7}{2}^+ \to \tfrac{3}{2}^+\right)$$

Continuing this process, we would find third-forbidden decays ($\ell = 3$), in which the selection rules not also satisfied by first-forbidden processes are $\Delta I = 3$ or 4 and $\Delta\pi = \text{yes}$:

$$^{87}\text{Rb} \to {}^{87}\text{Sr} \quad \left(\tfrac{3}{2}^- \to \tfrac{9}{2}^+\right)$$

$$^{40}\text{K} \to {}^{40}\text{Ca} \quad (4^- \to 0^+)$$

In very unusual circumstances, even fourth-forbidden decays ($\ell = 4$) may occur, with $\Delta I = 4$ or 5 and $\Delta\pi = \text{no}$:

$$^{115}\text{In} \to {}^{115}\text{Sn} \left(\tfrac{9}{2}^+ \to \tfrac{1}{2}^+\right)$$

We will learn in the next section that the higher the order of forbiddenness, the more unlikely is the decay. Given the chance, a nucleus prefers to decay by allowed or first-forbidden decays, and higher orders are generally too weak to observe. Only when no other decay mode is possible can we observe these extremely rare third- and fourth-forbidden decays.

## 9.5 COMPARATIVE HALF-LIVES AND FORBIDDEN DECAYS

Beta-decay half-lives encompass an enormous range, from the order of milliseconds to about $10^{16}$ years. Part of this variation must be due to the poor match-up of the initial and final nuclear wave functions, but it is hard to imagine that nuclear wave functions are so purely one configuration or another that this effect can account for any but a relatively small part of this variation over 26 orders of magnitude.

The true source of the variation in half-lives is the relative difficulty of creating a $\beta$ particle and a neutrino in an angular momentum state with $\ell > 0$. As we found in the previous section, a typical (classical) angular momentum for a 1-MeV $\beta$ particle has a maximum value of the order of $\ell \sim 0.04\hbar$. That is, the probability is very small for the electron and neutrino to be emitted in a state with quantum number $\ell > 0$.

We can make this qualitative estimate more quantitative by considering the wave functions of the electron and neutrino, which are taken to be of the form of plane waves, $e^{i\mathbf{p}\cdot\mathbf{r}/\hbar}$. Expanding the exponential gives $1 + (i\mathbf{p}\cdot\mathbf{r})/\hbar + \frac{1}{2}[(i\mathbf{p}\cdot\mathbf{r})/\hbar]^2 + \cdots$. The first term (after sandwiching between the initial and final nuclear wave functions and including the appropriate spin terms) is responsible for *allowed* decays. In the event that the nuclear wave functions cause this term to vanish (they may be of opposite parity, for instance) then we must go to the next term, in which the nuclear part (excepting the spin) is $\int \psi_f^* \mathbf{r} \psi_i \, dv$. Such terms are responsible for the *first-forbidden* decays. The average value of $\mathbf{p}\cdot\mathbf{r}/\hbar$, integrated over the nuclear volume, is of order 0.01, as we found above. The transition probability is proportional to the square of the integral, and so the probability for first-forbidden decays is only about $10^{-4}$ that for allowed decays.

The integral also vanishes unless the initial and final states have opposite parities, which can be shown, for example, by writing $\mathbf{r}$ in terms of $Y_1(\theta, \phi)$. This gives again the selection rule $\Delta\pi = $ yes for first-forbidden decays, as discussed in the previous section.

Each succeeding term in the expansion of the exponential form of the plane wave gives a higher order of forbiddenness, and each gives a transition probability smaller than that of the previous term by a factor of the order of $(\mathbf{p}\cdot\mathbf{r}/\hbar)^2$, or about $10^{-4}$.

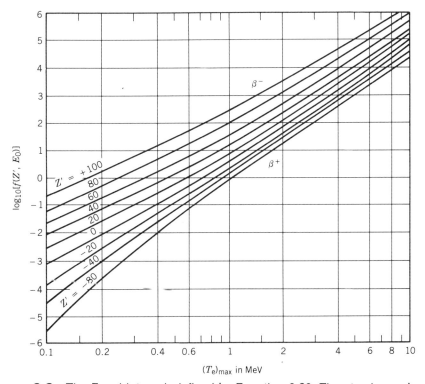

**Figure 9.8** The Fermi integral, defined by Equation 9.29. The atomic number $Z'$ refers to the daughter nucleus; the curves for positive $Z'$ are for $\beta^-$ decay, while negative $Z'$ is for $\beta^+$ decay. From R. D. Evans, *The Atomic Nucleus* (New York: McGraw-Hill, 1955).

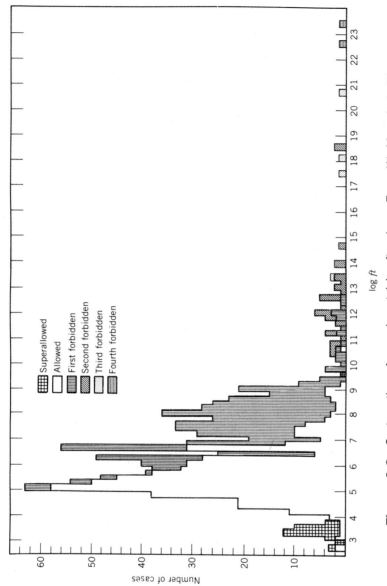

**Figure 9.9** Systematics of experimental log *ft* values. From W. Meyerhof, *Elements of Nuclear Physics* (New York: McGraw-Hill, 1967).

To compare the half-lives of different $\beta$ transitions, we must first correct for the variation in the $\beta$ decay probability that results from differences in the daughter atomic number $Z'$ or in the endpoint energy $E_0$. This is done through the *Fermi integral function* $f(Z', E_0)$, which was defined in Section 9.3. If we know the *partial* half-life for a certain decay process, we can find $f(Z', E_0)$ from curves such as those of Figure 9.8. The product $ft_{1/2}$ is the *comparative half-life* or *ft* value, which is usually given as $\log_{10} ft$, where $t_{1/2}$ is always in seconds.

As an example, we consider the $\beta^-$ decay of $^{203}$Hg. The half-life is 46.8 days, so $\log_{10} t_{1/2} = 6.6$. The $Q$ value for the decay to $^{203}$Tl is 0.491 MeV. However, essentially 100% of the decay goes to the 279-keV excited state of $^{203}$Tl, and so the $\beta$ endpoint energy will be $0.491 - 0.279 = 0.212$ MeV. From Figure 9.8 we estimate $\log_{10} f = -0.1$, and thus

$$\log_{10} ft = \log_{10} f + \log_{10} t_{1/2} = -0.1 + 6.6 = 6.5$$

For a second example, we take the $\beta^+$ decay of $^{22}$Na to the ground state of $^{22}$Ne ($Z' = 10$). The half-life is 2.60 years but the branching ratio to the ground state is only 0.06%. Thus the partial half-life is 2.60 years/$6 \times 10^{-4}$, so that $\log_{10} t_{1/2} = 11.1$. The $Q$ value for $\beta^+$ decay is 1.8 MeV, so from Figure 9.8 we estimate $\log_{10} f = 1.6$, and $\log_{10} ft = 11.1 + 1.6 = 12.7$.

In compilations of nuclear decay information, the log $ft$ values are given directly. We can determine the type of decay (allowed, $n$th-forbidden) based on the angular momentum and parity selection rules, and we can then try to relate the experimental log $ft$ values with the order of forbiddenness. Figure 9.9 summarizes the experimental values of log $ft$ for different types of decays, and you can see that there is indeed an effect of the order we estimated—each additional degree of forbiddenness increases the log $ft$ value by about 3.5, representing a reduction in the transition probability by $3 \times 10^{-4}$. (There is also a great deal of scatter within each type of decay, a large part of which is probably due to the effects of the particular initial and final nuclear wave functions.)

Most allowed decays have log $ft$ values in the range 3.5 to 7.5, and first-forbidden decays generally fall in the range 6.0 to 9.0. There are relatively fewer known second-forbidden decays, which have log $ft$ values from about 10 to 13, and the third-forbidden decays (four cases) range from about 14 to 20. There are two known fourth-forbidden decays, with log $ft$ about 23.

The value of summaries of this kind of information is in their *predictive ability*; for example, if we are studying a previously unknown decay scheme for which we measure log $ft = 5.0$, the decay is most probably of the allowed type, which permits us to assign the initial and final states the same parity and to conclude that their spins differ by at most one unit. We shall see the value of such deductions when we discuss $\beta$ spectroscopy in Section 9.10.

## 9.6 NEUTRINO PHYSICS

A process closely related to $\beta$ decay is capture of a neutrino (or an antineutrino) by a nucleon:

$$\bar{\nu} + p \rightarrow n + e^+$$
$$\nu + n \rightarrow p + e^-$$

which sometimes is called *inverse $\beta$ decay*.

Let's first discuss why only these processes occur and not others such as capture of a neutrino by a proton or of an antineutrino by a neutron. Electrons and neutrinos belong to a class of particles called *leptons*; the antiparticles $e^+$ and $\bar{\nu}$ are *antileptons*. Based on observations of many processes and failure to observe certain others, the *law of lepton conservation* is deduced: the total number of leptons minus antileptons on each side of a decay or reaction process must be the same. Many sensitive searches have been made to find violations of this law, but none have yet been found. The reaction $\nu + p \rightarrow n + e^+$, which conserves electric charge and nucleon number, does not conserve lepton number and is therefore, according to our present understanding of fundamental processes, absolutely forbidden.

The failure to observe such reactions is in fact one of our best indicators that $\nu$ and $\bar{\nu}$ are really different particles. The electron and positron differ in the sign of their electric charge (and in properties that depend on electric charge, such as magnetic moment). But $\nu$ and $\bar{\nu}$ are uncharged (and as uncharged point particles, have vanishing magnetic moments). They are thus immune from the electromagnetic interaction, which is often used to distinguish particles from antiparticles.

As we discussed in Section 9.1, the existence of the neutrino was inferred from the failure of $\beta$ decay to conform to the well-established conservation laws of energy and momentum conservation. Direct observation of the neutrinos did not occur until 25 years following Pauli's original proposal. To understand the difficulty of catching the elusive neutrino, we can try to estimate the probability for the basic neutrino capture reaction. Let us, in analogy with Equation 4.27, define the cross section for the reaction $\bar{\nu} + p \rightarrow n + e^+$ as

$$\sigma = \frac{\text{probability per target atom for the reaction to occur}}{\text{incident flux of } \bar{\nu}} \quad (9.35)$$

The reaction probability can be calculated using Fermi's Golden Rule, as in Equation 9.12. For the matrix element $V_{fi}$ we can take $(1/V)gM_{fi}$, as we did in the calculation based on the allowed approximation leading to Equation 9.20. Neglecting the recoil of the neutron, the density of final states comes only from the electron and is given by Equation 9.15. Finally, we can adapt the form of Equation 4.26 for the incident flux of $\bar{\nu}$, using the plane wave form of Equation 9.18 and recalling that the quantity $\hbar k/m$ came from the velocity of the incident particle, which is $c$ for neutrinos. The resulting cross section is thus

$$\sigma = \frac{\dfrac{2\pi}{\hbar} \dfrac{g^2}{V^2}|M_{fi}|^2 \dfrac{4\pi p^2\, dp}{h^3\, dE} V}{c/V} \quad (9.36)$$

Using $dp/dE = E/c^2p$ gives

$$\sigma = \frac{2\pi}{\hbar c} g^2 |M_{fi}|^2 \frac{4\pi pE}{c^2 h^3} \quad (9.37)$$

To make a numerical estimate, let us use the nuclear matrix element we found for the case of neutron $\beta$ decay in Section 9.4, $g^2|M_{fi}|^2 = g_F^2(1 + y^{-2}) \simeq 5.6g_F^2$; for $g_F$ we take the value deduced from the superallowed $\beta$ decays. We choose an incident antineutrino energy of 2.5 MeV, somewhat above the minimum energy of 1.8 MeV needed to initiate the reaction (because $m_pc^2 < m_nc^2$, we must supply the additional needed mass energy through the incident antineutrino), and thus the electron energy is 1.21 MeV. Putting in all of the numerical factors, the resulting cross section is $1.2 \times 10^{-19}$ b = $1.2 \times 10^{-43}$ cm$^2$. We can appreciate this incredibly small cross section (compare with the low-energy nucleon–nucleon scattering cross section of 20 b!) by evaluating the probability for a neutrino to be captured in passing through a typical solid, which contains of the order of $10^{24}$ protons per cm$^3$. The neutrino has a reaction cross section of about $10^{-43}$ cm$^2$ for each proton it encounters, and in passing through 1 cm$^3$ of material a neutrino will encounter about $10^{24}$ protons. The net reaction probability is $(10^{-43}$ cm$^2)(10^{24}$ cm$^{-3}) = 10^{-19}$ cm$^{-1}$; that is, the reaction probability is about $10^{-19}$ for each cm of material through which the neutrino passes. To have a reasonable probability to be captured, the neutrino must pass through about $10^{19}$ cm of material, or about 10 light-years. No wonder it took 25 years to find one!

The actual experimental detection was done through an ingenious and painstaking series of experiments carried out in the 1950s by Reines and Cowan. As a source of $\bar{\nu}$ they used a nuclear reactor, since the neutron-rich fission products undergo negative $\beta$ decay and consequently emit $\bar{\nu}$. The average emission rate is about 6 $\bar{\nu}$ per fission, and the net flux of $\bar{\nu}$ was about $10^{13}$ per cm$^2$ per second. For their neutrino detector, Reines and Cowan used a liquid scintillator (rich in free protons) into which a Cd compound had been introduced. The capture of $\bar{\nu}$ by a proton gives a neutron and a positron; the positron quickly annihilates $(e^- + e^+ \rightarrow 2\gamma)$ in the scintillator and gives a flash of light. The neutron travels through the solution and is gradually slowed, until finally it is captured by a Cd nucleus, which has a large neutron-capture cross section. Following the neutron capture, $^{114}$Cd is left in a highly excited state, which quickly emits a 9.1-MeV $\gamma$ ray. The characteristic signal of a $\bar{\nu}$ is thus a light signal from the positron annihilation radiation (0.511-MeV photons) followed about 10 $\mu$s later (the time necessary for the neutron to be slowed and captured) by the 9.1-MeV neutron capture $\gamma$ ray. Using a tank containing of the order of $10^6$ cm$^3$ of scintillator, Reines and Cowan observed a few events per hour that were candidates for $\bar{\nu}$ captures. To determine conclusively that these were indeed $\bar{\nu}$ captures, many additional experiments were necessary, following which the conclusion was inescapable—the $\bar{\nu}$ is a real particle, and not just a figment of Pauli's and Fermi's fertile imaginations.

To demonstrate that $\bar{\nu}$ capture by *neutrons* is not possible, a related experiment was done by Davis and co-workers. They used a large tank of CCl$_4$ in an attempt to observe $\bar{\nu} + {}^{37}\text{Cl} \rightarrow e^- + {}^{37}\text{Ar}$, again with reactor antineutrons. By purging the tank periodically and searching for the presence of radioactive $^{37}$Ar in the removed gas, Davis was able to conclude that the reaction was not observed, indicating that $\nu$ and $\bar{\nu}$ are in fact different particles.

We are thus resolved to the fact that $\nu$ and $\bar{\nu}$ are different particles, but we have not yet specified just what is the fundamental property that distinguishes $\nu$ from $\bar{\nu}$. Experimentally, there is one property: all $\bar{\nu}$ have their spin vectors parallel to their momentum vectors, while all $\nu$ have spin opposite to momentum. This property is called the *helicity* and is defined to be

$$ h = \frac{\mathbf{s} \cdot \mathbf{p}}{|\mathbf{s} \cdot \mathbf{p}|} \tag{9.38} $$

which has the value of $+1$ for $\bar{\nu}$ and $-1$ for $\nu$. (It is often said that $\bar{\nu}$ is "right-handed" and $\nu$ is "left-handed" because the precession of $s$ about $p$ traces out a pattern analogous to the threads of a right-handed screw for $\bar{\nu}$ and of a left-handed screw for $\nu$.) Electrons from $\beta$ decay have a similar property, with $h = -v/c$ for $e^-$ and $h = +v/c$ for $e^+$, but this is not an intrinsic property of all $e^+$ and $e^-$, only those emitted in $\beta$ decay. Electrons in atoms have no definite helicities, nor do positrons that originate from pair production ($\gamma \rightarrow e^+ + e^-$). *All $\nu$ and $\bar{\nu}$, however, have definite helicities, right-handed for $\bar{\nu}$ and left-handed for $\nu$.*

Davis has used a similar technique to observe $\nu$ emitted by the sun as a result of fusion processes. (Fusion of light nuclei tends to produce neutron-deficient products, which undergo $\beta^+$ decay and thus emit $\nu$ rather than $\bar{\nu}$.) To shield against events produced by cosmic rays (a problem in his earlier experiments), Davis has placed his $CCl_4$ tank at the bottom of a 1500-m deep mine, and has spent more than 10 years counting these solar neutrinos. These are especially important because they come to us directly from the core of the sun, where the nuclear reactions occur; the light we see, on the other hand, comes from the sun's surface and contains relatively little direct information about processes that are now going on in the core.) The expected rate of conversion of $^{37}$Cl to $^{37}$Ar by solar neutrinos in Davis' tank is about *one atom per day*; yet despite years of heroic efforts, the observed rate is only about one-third of the expected value, which represents either an error in the assumptions made regarding the rate of neutrino emission by the sun (and thus a shortcoming of our present theory of solar processes) or an error in our present theories of properties of the neutrino.

## 9.7 DOUBLE-$\beta$ DECAY

Consider the decay of $^{48}$Ca (Figure 9.10). The $Q$ value for $\beta^-$ decay to $^{48}$Sc is 0.281 MeV, but the only $^{48}$Sc states accessible to the decay would be the $4^+$, $5^+$, and $6^+$ states, which would require either fourth- or sixth-forbidden decays. If we take our previous empirical estimate of $\log ft \sim 23$ for fourth-forbidden decays, then (with $\log f \simeq -2$ from Figure 9.8) we estimate $\log t \sim 25$ or $t_{1/2} \sim 10^{25}$ s ($10^{18}$ y). It is thus not surprising that we should regard $^{48}$Ca as a "stable" nucleus.

An alternative possible decay is the *double-$\beta$ ($\beta\beta$) decay* $^{48}$Ca $\rightarrow$ $^{48}$Ti $+ 2e^- + 2\bar{\nu}$. This is a direct process, which does not require the $^{48}$Sc intermediate state. (In fact, as we shall discuss, in most of the possible $\beta\beta$ decays, the intermediate state is of greater energy then the initial state and is energetically impossible to reach.) The advantage of this process over the single $\beta$ decay (in this case) is the $0^+ \rightarrow 0^+$ nature of the transition, which would place it in the superallowed, rather than the fourth-forbidden, category.

We can make a rough estimate of the probability for such decay by rewriting Equation 9.30 for single-$\beta$ decay as

$$\lambda_\beta = \left( \frac{m_e c^2}{\hbar} \right) \left\{ fg^2 \frac{m_e^4 c^2 |M_{fi}|^2}{2\pi^3 \hbar^6} \right\} \tag{9.39}$$

The first term has a value of approximately $0.8 \times 10^{21}$ s$^{-1}$ and can be considered

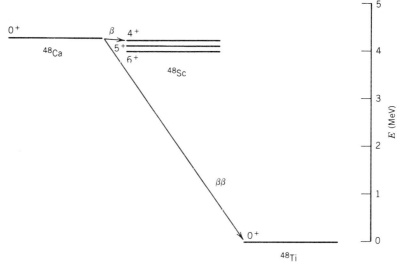

**Figure 9.10** The decay of $^{48}$Ca. The superallowed $\beta\beta$ decay to $^{48}$Ti is an alternative to the fourth-forbidden single-$\beta$ decay to $^{48}$Sc.

the dimensional scaling factor. The remaining term is dimensionless and contains all of the information on the $\beta$ decay and nuclear transition probabilities. It has a value of $1.5 \times 10^{-25} f$ (using $|M_{fi}| = \sqrt{2}$).

The decay rate for $\beta\beta$ decay then ought to be approximately given by

$$\lambda_{\beta\beta} = \left(\frac{m_e c^2}{\hbar}\right)\left\{ fg^2 \frac{m_e^4 c^2 |M_{fi}|^2}{2\pi^3 \hbar^6}\right\}^2 \tag{9.40}$$

which gives a half-life of the order of $10^{17}$ years, comparable with the value for single-$\beta$ decay (although this simple calculation should not be taken too seriously).

Double-beta decay can also occur in cases in which the intermediate state cannot be reached by the single decay mode. Consider the case of $^{128}$Te, shown in Figure 9.11. The decay $^{128}$Te $\rightarrow$ $^{128}$I has a *negative* $Q$ value of $-1.26$ MeV, and is therefore not possible. Yet the $\beta\beta$ decay $^{128}$Te $\rightarrow$ $^{128}$Xe is energetically possible, with $Q = 0.87$ MeV. In fact, such situations provide the most likely candidates

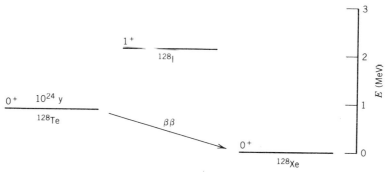

**Figure 9.11** Single-$\beta$ decay of $^{128}$Te is energetically forbidden, but $\beta\beta$ decay to $^{128}$Xe is possible. See Figure 3.18 to understand the relative masses of these nuclei.

for observing $\beta\beta$ decays because we do *not* want to study the case of two successive decays through an energetically accessible intermediate state.

There are two basic approaches to the observation of $\beta\beta$ decay. The first is the mass spectroscopic method, in which we search for the stable daughter nuclei in minerals of known geological age. If, for example, we were to find an excess abundance of $^{128}$Xe (relative to its abundance in atmospheric Xe, for example) in a tellurium-bearing rock, we could deduce an estimate for the $\beta\beta$-decay half-life of $^{128}$Te to $^{128}$Xe. Making the reasonable assumption that the $\beta\beta$-decay half-life is long compared with the age $T$ of the rock, the number of Xe resulting from the $\beta\beta$ decay is

$$N_{Xe} = N_{Te}(1 - e^{-\lambda T}) \cong N_{Te}\frac{0.693T}{t_{1/2}} \tag{9.41}$$

and so

$$t_{1/2} \cong 0.693T\frac{N_{Te}}{N_{Xe}} \tag{9.42}$$

The number of Te and Xe nuclei can be determined using mass spectroscopic techniques, and thus the $\beta\beta$-decay half-life may be found. Some typical values obtained using this method are

$$^{128}\text{Te} \rightarrow {}^{128}\text{Xe} \qquad (3.5 \pm 1.0) \times 10^{24} \text{ y}$$

$$^{130}\text{Te} \rightarrow {}^{130}\text{Xe} \qquad (2.2 \pm 0.6) \times 10^{21} \text{ y}$$

$$^{82}\text{Se} \rightarrow {}^{82}\text{Kr} \qquad (1.7 \pm 0.3) \times 10^{20} \text{ y}$$

The direct detection of $\beta\beta$ decay is obviously frustrated by the long half-lives —from one mole of sample, we would expect of the order of one decay per year in the worst case above and one per day in the best case. Experiments with such low count rates always suffer from spurious background counts, such as those from natural radioactivity or cosmic rays, and shielding against these unwanted counts severely taxes the skill of the experimenter. For example, one experiment was done under 4000 m of rock in a tunnel under Mont Blanc on the border between France and Italy!

A recent experiment reported by Moe and Lowenthal used strips of $^{82}$Se in a cloud chamber to search for evidence of $\beta\beta$ decays. Figure 9.12 shows examples of a typical event in which two electrons were emitted. Also shown for comparison is another event in which an $\alpha$ particle track originates from the same location as the two electrons; this event probably results from the natural radioactive background, most likely from the decay of $^{214}$Bi in the uranium series. A more sensitive search for $\beta\beta$ events by Elliott, Hahn, and Moe, reported in *Phys. Rev. Lett.* **56**, 2582 (1986), showed approximately 30 events possibly associated with $\beta\beta$ decays in more than 3000 h of measuring time. The deduced lower limit on the $\beta\beta$ half-life is $1.0 \times 10^{20}$ y, in agreement with the geochemical result listed above.

Although the direct method is exceedingly difficult and subject to many possible systematic uncertainties, it is extremely important to pursue these studies because they are sensitive to the critical question of *lepton conservation* (which we

 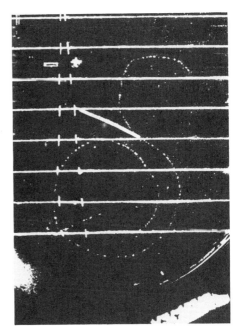

**Figure 9.12** Cloud chamber photograph of a suspected $\beta\beta$-decay event from $^{82}$Se. The horizontal lines are strips of $^{82}$Se source material. The $\beta\beta$-decay event is the pair of curved tracks originating from one of the strips in the exact center of the photograph at left. There are also background events due to natural radioactivity; these produce two $\beta$-decay electrons in succession (as in the natural radioactive chain of decays, Figure 6.10) and an $\alpha$ particle. Note the two electron tracks and the heavy $\alpha$ track originating from a common point near the center of the photograph on the right. A magnetic field perpendicular to the plane of the photos curves the tracks, so that the electron momentum can be deduced. From M. K. Moe and D. D. Lowenthal, *Phys. Rev. C* **22**, 2186 (1980).

discuss in greater detail in Chapter 18). If $\nu$ and $\bar{\nu}$ are not really distinct particles (that is, if they are coupled together or are linear combinations of yet other fundamental particles), then "neutrinoless" $\beta\beta$ decay would be possible:

$$^{A}_{Z}X_N \rightarrow ^{A}_{Z+2}X''_{N-2} + 2e^-$$

(In essence, we can think of this process as follows: the first $\beta$ decay proceeds through the virtual and energetically inaccessible intermediate state $^{A}_{Z+1}X'_{N-1}$. The emitted $\bar{\nu}$ turns into a $\nu$ and is reabsorbed by the virtual intermediate state giving $\nu + ^{A}_{Z+1}X'_{N-1} \rightarrow e^- + ^{A}_{Z+2}X''_{N-2}$. The net process therefore results in the emission of two $\beta$'s and no $\nu$'s.)

An experiment designed to search for neutrinoless $\beta\beta$ decay has been done in the case of $^{76}$Ge. Here a Ge detector is used both as the source of the decaying nuclei and as the detector of the decays. The total available decay energy is 2.04 MeV, and if the two electrons stop within the detector, it should record a single event with an energy of 2.04 MeV. The difficulty here is to reduce the background (from natural radioactivity, man-made radioactive contaminants, and cosmic rays) to a low enough level so that the 2.04-MeV region can be searched for a peak. The Mont Blanc experiment mentioned above was of this type and obtained a lower limit on the half-life of $5 \times 10^{21}$ y. In another underground experiment, reported recently by Avignone et al., *Phys. Rev. C* **34**, 666 (1986), extraordinary measures were taken to surround the detector only with material that would not contribute substantially to the background (stainless steel screws, which showed contamination from $^{60}$Co, were replaced with brass, and rubber O-rings were replaced with indium). After 9 months of counting, there was no visible peak at 2.04 MeV, and the half-life was deduced to be greater than $10^{23}$ y. These experiments are continuing, in the hope that continued improvements in sensitivity will enable both the two-neutrino and the neutrinoless $\beta\beta$ decays to be observed directly.

Although the theoretical interpretations are difficult, it may be that the search for evidence of neutrinoless $\beta\beta$ decay will be an important source of information on the fundamental character of the neutrinos. The emission-reabsorption process described above, for instance, is impossible for massless neutrinos with definite helicities ($\pm 1$), and so the observation of the neutrinoless $\beta\beta$ decay would immediately suggest that the "classical" properties of the neutrino are not correct.

## 9.8 BETA-DELAYED NUCLEON EMISSION

Gamma rays are not the only form of radiation that can be emitted from nuclear *excited* states that are populated following $\beta$ decay. Occasionally the states are unstable against the emission of one or more nucleons. The nucleon emission itself occurs rapidly (so that it competes with $\gamma$ emission), and thus overall the nucleon emission occurs with a half-life characteristic of the $\beta$ decay.

For decays of nuclei only one or two places from the most stable isobar of each mass number $A$, the decay energies are small (1–2 MeV), and nucleon emission is forbidden by energetics. Far from the stable nuclei, the decay energies may become large enough to populate highly excited states, which may then decay through nucleon emission. A schematic diagram of this process for proton emission is shown in Figure 9.13. The original $\beta$-decaying parent is called the *precursor*; the nucleons themselves come from the *emitter* and eventually lead to states in the *daughter*.

Interest in delayed nucleon emission has increased in recent years in concert with experimental studies of nuclei far from stability. Additional interest comes from the importance of delayed neutrons in the control of nuclear reactors (see Chapter 13). However, the discovery of the phenomenon dates from the early history of nuclear physics—Rutherford in 1916 reported "long-range alpha particles" following the $\beta$ decay of $^{212}$Bi. The main branch in this $\beta$ decay goes to the ground state of $^{212}$Po, which in turn emits $\alpha$ particles with an energy of 8.784 MeV. (Since the $\alpha$-decaying state is a $0^+$ ground state of an even-even nucleus, the decay proceeds virtually 100% to the ground state of $^{208}$Pb.) A small number of $\alpha$'s, however, were observed with *higher* energies (9.495 MeV, 0.0035%; 10.422 MeV, 0.0020%; 10.543 MeV, 0.017%). Lower energies would have indicated decays *to* excited states of $^{208}$Pb, but higher energies must indicate decays *from* excited states of $^{212}$Po. Similar behavior was observed in the decay of $^{214}$Bi.

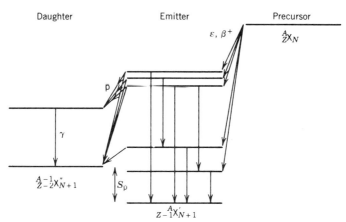

**Figure 9.13** Schematic of β-delayed nucleon emission The β decay of the precursor populates highly excited states in the emitter that are unstable with respect to nucleon emission. Note that the energy of the excited state in the emitter equals the sum of the energy of the emitted nucleon plus the nucleon separation energy between X' and X" (plus the small correction for the recoil of the emitting nucleus).

The calculation of the energy spectra of the emitted nucleons is a complicated process, requiring knowledge of the spectrum of excited states in the emitter, the probabilities for β decay from the precursor to each state of the emitter, and the probabilities of nucleon decay for each state of the emitter to the accessible states of the daughter. The difficulty is compounded in heavy nuclei by the large density of excited states—the average spacing between excited states at high energy may be of order eV, far smaller than our ability to resolve individual proton or neutron groups; thus all we observe in such cases is a broad distribution, similar in structure to the continuous distribution in β decay but originating from a very different effect. Because of these difficulties, we shall not discuss the theory of delayed nucleon emission; rather, we shall give some examples of experimental studies and their significance.

The energetics of β-delayed nucleon emission are relatively simple. Reference to Figure 9.13 shows immediately that the process can occur as long as the β-decay energy is greater than the nucleon separation energy: $Q_\beta > S_N$ (where N = n or p). Whenever this process is energetically permitted, there will always be competing processes; for example, γ decay of the emitting state or β decays to lower levels in the emitter that cannot decay by particle emission.

The information that we derive from β-delayed nucleon emission is mainly of two types: (1) Since the decay is a two-body process (emitted nucleon plus daughter nucleus), the nucleons emerge with a distinct energy, which gives directly the energy difference between the initial and final states. The energy levels in the daughter are usually well known, and so the energy of the emitted nucleon is in effect a measure of the energy of the excited state of the emitter. (2) From the relative probability of nucleon emission from different states in the emitter, we can deduce the relative population of these states in the β decay of the precursor. This provides information on the β-decay matrix elements. Because the highly excited states in the emitter are so close together, they nearly form a continuum, and it is more appropriate to consider a $\beta$-decay strength function $S_\beta(E_x)$, which gives the average $\beta$-decay intensity leading to excited states in the vicinity of excitation energy $E_x$. Usually there are few selection rules inhibiting $\beta$ decay to states at this high excitation, and so the $\beta$-decay strength function is rather featureless and is roughly proportional to the density of states $\rho(E_x)$. However, there is always one particular state that is so similar in character to the precursor that the majority of the $\beta$ decays populate that state (it has a particularly large Fermi-type matrix element). The state is known as the *isobaric analog state* (or simply, analog state) because its structure is analogous to the original decaying state in the neighboring isobar. The $\beta$-decay strength leading to the analog state (and its energy) can often be determined only through the technique of $\beta$-delayed nucleon emission.

As an example of a typical experiment, we consider the $\beta$-delayed neutron emission from $^{17}$N, which decays by negative $\beta$ emission to $^{17}$O. Figure 9.14 shows three readily identifiable neutron groups, with energies 383, 1171, and 1700 keV; we assume that three excited states of $^{17}$O are populated in the $\beta$ decay and that each emits a neutron to form $^{16}$O. Let us assume that these decays go directly to the ground state of $^{16}$O. (This is certainly not going to be true in general, but $^{16}$O has its first excited state at more than 6 MeV; we will see that it is not possible that the $^{17}$N $\beta$ decay could have enough energy to populate such a highly excited state.)

To analyze the energy transfer in the decay, we first need the neutron separation energy of $^{17}$O; using Equation 3.26:

$$S_n = \left[ m(^{16}O) - m(^{17}O) + m_n \right] c^2$$
$$- (15.99491464\,u - 16.9991306\,u + 1.008664967\,u)931.502\,\text{MeV}/u$$
$$= 4.144\,\text{MeV}$$

This is the energy that must be supplied to remove a neutron from $^{17}$O. Let's regard the initial state of the system as $^{17}$O in an excited state with energy $E_x$. The initial energy is therefore $m(^{17}O)c^2 + E_x$. The final energy is $m(^{16}O)c^2 + E_x'$ $+ m_n c^2 + T_n + T_R$, where $T_n$ is the neutron kinetic energy and $T_R$ is the energy of the $^{16}$O recoil, which must occur to conserve momentum. We have included a possible excitation energy of $^{16}$O in the term $E_x'$; later we will show it must be zero in this case. Energy conservation gives

$$m(^{17}O)c^2 + E_x = m(^{16}O)c^2 + E_x' + m_n c^2 + T_n + T_R$$

or

$$E_x = E_x' + T_n + T_R + S_n \qquad (9.43)$$

which is a general result. The recoil correction is obtained by application of conservation of momentum, yielding

$$T_R = T_n \left( \frac{m_n}{m_R} \right) \approx T_n \frac{1}{A-1} \qquad (9.44)$$

where $m_R$ is the mass of the recoiling nucleus. Since this is a small correction, we can approximate $m_n/m_R$ by $1/(A-1)$. The final result is

$$E_x = E_x' + \frac{A}{A-1} T_n + S_n \qquad (9.45)$$

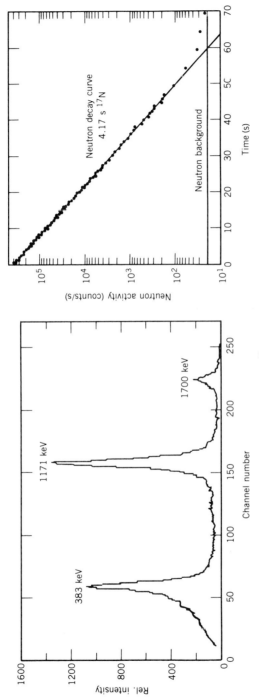

**Figure 9.14** Beta-delayed neutrons following the decay of $^{17}$N. The neutron energy spectrum is shown at the left; the decay of the neutron activity with time is at the right. From H. Ohm et al., *Nucl. Phys. A* **274**, 45 (1976).

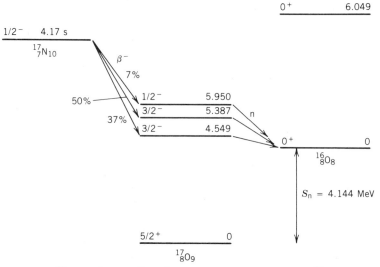

**Figure 9.15** The $\beta$-delayed neutron decay of $^{17}$N.

Assuming $E'_x = 0$ for $^{16}$O, the three measured $^{17}$N $\beta$-delayed neutron energies give excitation energies of 4.551, 5.388, and 5.950 MeV. Nuclear reactions can also be used to measure the energies of the $^{17}$O excited states, and three states are found in reaction studies with the energies we have just calculated. If we were to consider the possibility to reach excited states in $^{16}$O (that is, $E'_x \geq 6.049$ MeV, the first excited state in $^{16}$O), then the lowest possible excitation in $^{17}$O would be 10.6 MeV, which is greater than the $Q$ value of the $^{17}$N $\beta$ decay (8.68 MeV). Excited states in $^{16}$O are therefore not populated in this decay.

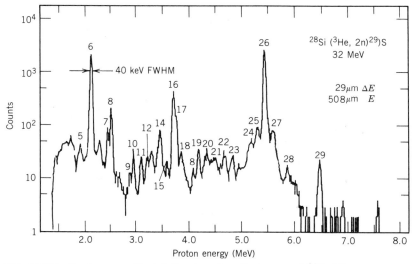

**Figure 9.16** Protons emitted following the $\beta$ decay of $^{29}$S. The protons were observed using a $\Delta E \cdot E$ telescope of Si detectors. The numbers refer to specific proton decays of excited state of $^{29}$P. Data from D. J. Vieira et al., *Phys. Rev. C* **19**, 177 (1979).

Figure 9.14 also shows the rate of neutron emission as a function of time, which gives the half-life of $^{17}$N to be 4.17 s. This half-life is far too long for the decay to be a direct neutron emission process and it must therefore be a $\beta$-delayed emission process. The resulting decay is shown in Figure 9.15.

Proton emission will occur most easily from nuclei with an excess of protons, which is certainly the case for $^{29}$S ($Z = 16$, $N = 13$). The activity is formed through the reaction $^{28}$Si $+ \,^3$He $\rightarrow \,^{29}$S $+ 2$n, which essentially adds two protons and removes a neutron from the stable initial nucleus ($Z = 14$, $N = 14$). The

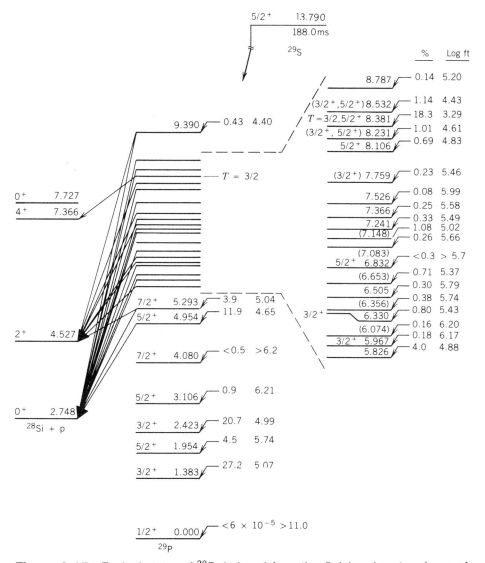

**Figure 9.17** Excited states of $^{29}$P deduced from the $\beta$-delayed proton decay of $^{29}$S. The ft values are deduced from the intensity of the observed protons. Note the strong decay branch (small ft value) in the decay to the state at 8.381 MeV, which is the analog state of the $^{29}$S ground state.

precursor $^{29}$S decays by $\beta^+$ emission to states in the emitter $^{29}$P, which then emits protons leading to final states in $^{28}$Si. Figure 9.16 shows the observed proton spectrum, and Figure 9.17 illustrates the assignment of these proton groups to known initial and final states in $^{29}$P and $^{28}$Si. Many of the arguments for placing these decays proceed indirectly; for example, the energy difference between the $0^+$ ground state and $2^+$ first excited state in $^{28}$Si is known to be 1.778 MeV, and thus two proton groups differing in energy by 1.778 MeV can be assumed to lead from the same state in the emitter to these two different final states in the daughter (groups 16 and 26, 18 and 27, 22 and 29). The analog state is associated with the strong groups 16 and 26; its $\log ft$ value of 3.29 is characteristic of superallowed decays, as expected for this strongly favored transition.

As we go to heavier nuclei, the density of excited states in the emitter becomes so large that the spacing between levels is smaller than the energy resolution of the detector. When this occurs, it is no longer possible to make the above identification of decays from specific states in the emitter, and only broad, average features of the decay can be discussed (Figure 9.18).

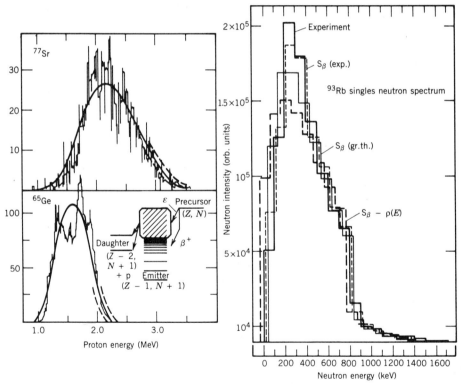

**Figure 9.18** Proton (left) and neutron (right) emission following $\beta$ decay in heavy nuclei. The spacing between excited states in the emitter is so small that we observe only a broad distribution, rather than the individual peaks of Figures 9.14 and 9.16. Attempts to fit the experimental data are based on statistical models, rather than on detailed calculations of individual nuclear states. Proton data from J. C. Hardy et al., *Phys. Lett. B* **63**, 27 (1976); neutron data from K.-L. Kratz et al., *Z. Phys. A* **306**, 239 (1982).

## 9.9 NONCONSERVATION OF PARITY

The parity operation (as distinguished from the parity quantum number) consists of reflecting all of the coordinates of a system: $r \rightarrow -r$. If the parity operation gives us a physical system or set of equations that obeys the same laws as the original system, we conclude that the system is invariant with respect to parity. The original and reflected systems would both represent possible states of nature, and in fact we could not distinguish in any fundamental way the original system from its reflection.

Of course, the macroscopic world *does* show a definite preference for one direction over another—for example, we humans tend to have our hearts on the left side of our bodies. There is no law of nature that demands that this be so, and we could construct a perfectly acceptable human with the heart on the right side. It is the reflection symmetry of the laws of nature themselves with which we are concerned, not the accidental arrangement of the objects governed by the laws.

In fact, there are three different "reflections" with which we frequently work. The first is the spatial reflection $r \rightarrow -r$, which is the parity (P) operation. The second "reflection" consists of replacing all particles with their antiparticles; this operation is called *charge conjugation* (C), although there are properties in addition to electric charge that are reversed in this operation. The third operation is *time reversal* (T), in which we replace $t$ by $-t$ and in effect reverse the direction in time of all of the processes in the system. Figure 9.19 shows how three basic processes would appear under the P, C, and T operations. Notice especially that there are some vectors that change sign under P (coordinates, velocity, force, electric field) and some that do not (angular momentum, magnetic field, torque). The former are called true or polar vectors and the latter are pseudo- or axial vectors. Figure 9.20 shows a complete view of a rotating object reflected through the origin. You can see quite clearly that the angular momentum vector does not change direction upon reflection.

In each case shown in Figure 9.19, the reflected image represents a real physical situation that we could achieve in the laboratory, and we believe that gravity and electromagnetism are invariant with respect to P, C, and T.

One way of testing the invariance of the nuclear interaction to P, C, and T would be to perform the series of experiments described in Figure 9.21. In the original experiment, a reaction between particles A and B produces C and D. We could test P by interchanging the particles (for example, have projectile B incident on target A, instead of projectile A incident on target B). We could test C by doing the reaction with antiparticles and T by reacting particles C and D to produce A and B. In each case we could compare the probability of the reversed reaction with that of the original, and if the probabilities proved to be identical, we could conclude that P, C, and T were invariant operations for the nuclear interaction.

In the case of decays A → B + C, we could perform the same type of tests, as shown in the figure, and could again study the invariance of P, C, and T in decay processes.

We must take care how we test the P operation because, as shown in the figure, the reflected experiment is identical to what we would observe if we turned the page around or stood on our heads to observe the decay or reaction process.

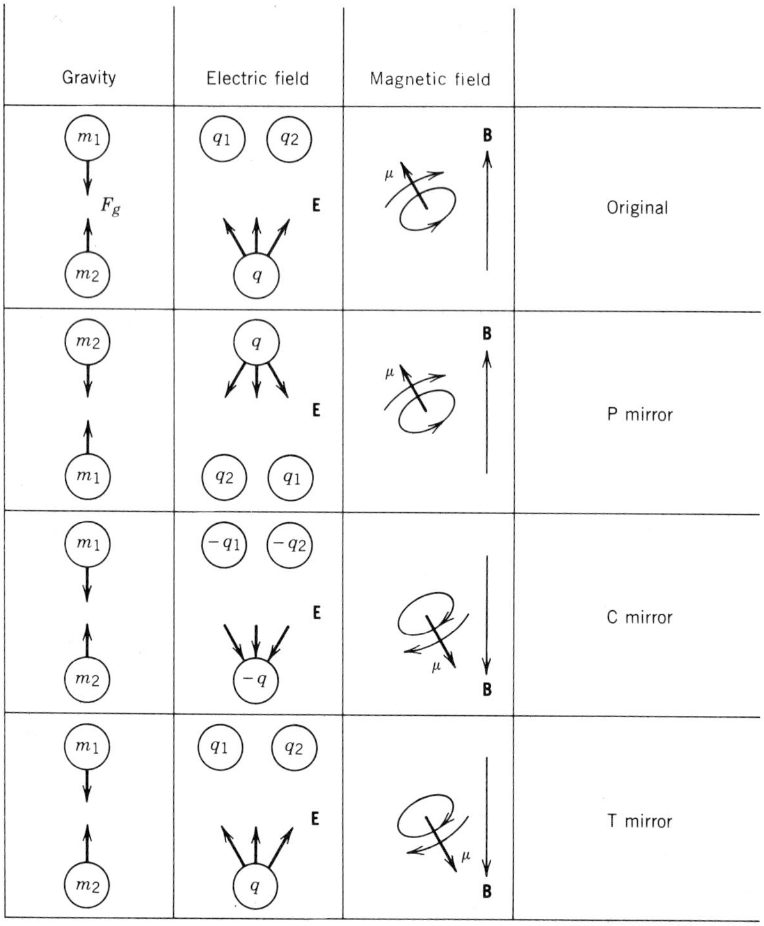

**Figure 9.19** The effect of P, C, and T reversal on gravitational and electromagnetic interactions. In all cases the reversed diagrams represent possible physical situations, and thus these interactions are invariant under P, C, and T.

Since our goal is not to test the invariance of the laws of nature to physicists standing on their heads, we must have some way more clearly to identify the reflected process.

One way is to assume the decaying particle A to have a spin vector that is pointing in a specific direction. The spin does not change direction under P, but it certainly does if we view it upside down. Thus the original experiment shows particle B emitted in the same direction as the spin of A, while the reflected experiment shows B emitted opposite to the spin of A. Quite clearly the experiment differs from its reflection. If, however, we have a large number of A nuclei, all with spins aligned in the same direction, and if they tend to emit B's in equal numbers along the spin and opposite to the spin, then once again the experiment looks like its image. Here then is a way to test P directly—we simply

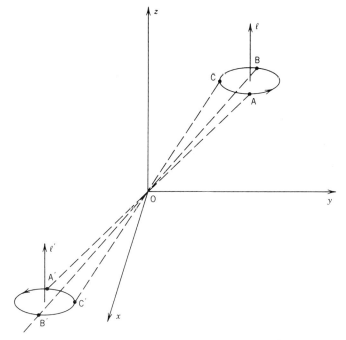

**Figure 9.20**  The effect of the P operation on a rotating object. If coordinates A, B, and C in the orbit are reflected through the origin ($r \rightarrow -r$), coordinates A', B', and C' result. As the original particle travels from A to B to C, the reflected particle travels from A' to B' to C', and using the right-hand rule to define the direction of the angular momentum indicates that both $\ell$ and $\ell'$ point upward. Thus $\ell$ is a vector that does not change sign under P; such vectors are called *axial vectors*. (Vectors such $r$ that do change sign under P are called *polar vectors*.)

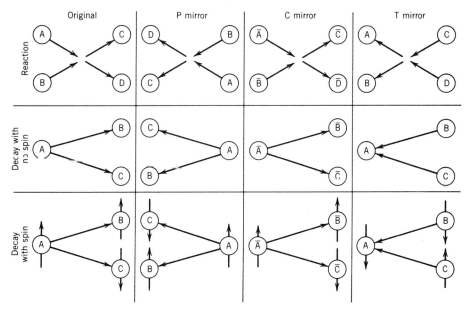

**Figure 9.21**  Nuclear physics tests of P, C, and T.

align the spins of some decaying nuclei and look to see if the decay products are emitted equally in both directions or preferentially in one direction.

In 1956, T. D. Lee and C. N. Yang pointed out that P had not yet been tested in $\beta$ decay, even though it had been well tested in other nuclear decay and reaction processes. They were led to this assertion by an unusual situation called the $\theta$-$\tau$ puzzle. At that time there were two particles, called $\theta$ and $\tau$, which appeared to have identical spins, masses, and lifetimes; this suggested that $\theta$ and $\tau$ were in fact the same particle. Yet the decays of these particles lead to final states of different parities. Since the decays were governed by a process similar to nuclear $\beta$ decay, Lee and Yang suggested that $\theta$ and $\tau$ were the same particle (today called a K meson) which could decay into final states of differing parities *if the P operation were not an invariant process for $\beta$ decay.*

Several experimental groups set out to test the suggestion of Lee and Yang, and a successful experiment was soon done by C. S. Wu and her co-workers using the $\beta$ decay of $^{60}$Co. They aligned the $^{60}$Co spins by aligning their magnetic dipole moments in a magnetic field at very low temperature ($T \sim 0.01$ K, low enough so that thermal motion would not destroy the alignment). Reversing the magnetic field direction reversed the spins and in effect accomplished the reflection. If $\beta$ particles would have been observed in equal numbers along and opposite to the magnetic field, then $\beta$ decay would have been invariant with respect to the P operation. What was observed in fact was that at least 70% of the $\beta$ particles were emitted opposite to the nuclear spin. Figure 9.22 shows the original data of Wu and colleagues, and you can see quite clearly that the $\beta$ counting rate reverses as the magnetic field direction is reversed.

Twenty-five years after the original experiment, Wu's research group repeated the $^{60}$Co experiment with new apparatus that represented considerably advanced technology for cooling the nuclei, polarizing their spins, and detecting the $\beta$ particles. Figure 9.23 shows the result of this new experiment, which demonstrates quite clearly the parity-violating effect.

Figure 9.24 shows schematically the $^{60}$Co experiment and its reflection in the P mirror. In the P-reflected experiment, the electrons are emitted preferentially along, rather than opposite to, the direction of the magnetic field. Since this represents a state of affairs that is not observed in nature, it must be concluded that, at least as far as $\beta$ decay is concerned, the P operation is not a valid symmetry. There is yet another surprising result that follows from this experiment. Consider the reflection of the original experiment in the C mirror, also shown in Figure 9.24. The electrons flowing in the wires that produce the magnetic field become positrons, so that the magnetic field reverses. In the C-reflected experiment, the $\beta$ particles are now emitted preferentially *along* the magnetic field. Thus matter and antimatter behave differently in beta decay, which is a violation of the C symmetry. (In his book *The Ambidextrous Universe*, Martin Gardner discusses how this experiment can be used to try to decide whether an extraterrestrial civilization, with whom we may someday be in communication, is composed of matter or antimatter.)

If, however, we reflect the experiment in a mirror that simultaneously performs both the P and C operations, as shown in Figure 9.24, the original experiment is restored. Even though the separate C and P operations are not valid symmetries, the CP combination is. (We discuss in Chapter 17 that certain decays of the K

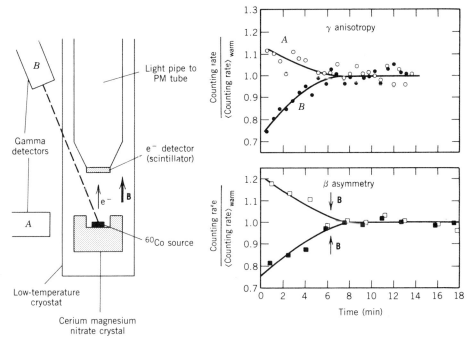

**Figure 9.22** Schematic arrangement of experimental test of parity violation in the decay of $^{60}$Co. At left is shown the apparatus; the cerium magnesium nitrate crystal is used to cool the radioactive source to about 0.01 K. At bottom right is shown the observed $\beta$ counting rates; reversing the magnetic field direction is equivalent to subjecting the nuclei to the P operation (see Figures 9.21 and 9.24). If P were *not* violated, there would be no asymmetry and the field-up and field-down curves would coincide. The vanishing of the asymmetry at about 8 min is due to the gradual warming of the source and the corresponding loss in polarization of the $^{60}$Co nuclei, as is demonstrated by the observed $\gamma$-ray counting rates. Data from C. S. Wu et al., *Phys. Rev.* **105**, 1413 (1957).

meson, which are analogous to $\beta$ decays, even violate to a small extent the CP invariance. There is as yet no evidence that the CP symmetry is violated in ordinary nuclear $\beta$ decay.)

Before we leave this topic, we should discuss the effect of the P nonconservation on nuclear spectroscopy. The interaction between nucleons in a nucleus consists of two parts: the "strong" part, which arises primarily from $\pi$ meson exchange and which respects the P symmetry, and the "weak" part, which comes from the same interaction responsible for $\beta$ decay:

$$V_{\text{nuclear}} = V_{\text{strong}} + V_{\text{weak}} \qquad (9.46)$$

Typically, the effects of $V_{\text{weak}}$ on nuclear spectroscopy are very small compared with those of $V_{\text{strong}}$, but $V_{\text{weak}}$ has a property that $V_{\text{strong}}$ lacks—it violates the P symmetry. As far as nuclear states are concerned, the effect of $V_{\text{weak}}$ is to add to the nuclear wave function a small contribution of the "wrong" parity:

$$\psi = \psi^{(\pi)} + F\psi^{(-\pi)} \qquad (9.47)$$

where $F$ is of order $10^{-7}$. Under most circumstances, this small addition to the

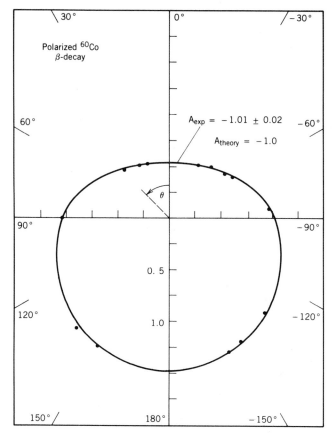

**Figure 9.23** Improved results of C. S. Wu and co-workers on the parity violation in the $^{60}$Co $\beta$ decay. The data points (plotted on a polar diagram) give the observed $\beta$ intensity at an angle $\theta$ with respect to the direction of polarization (spin direction) of the decaying $^{60}$Co nuclei. The solid curve represents the prediction of the Fermi theory, according to which the intensity should vary as $1 + AP\cos\theta$, where $P$ is a parameter that depends primarily on the nuclear polarization. If parity were *not* violated in $\beta$ decay, the intensities at 0 and 180° would be equal. From L. M. Chirovsky et al., *Phys. Lett. B* **94**, 127 (1980).

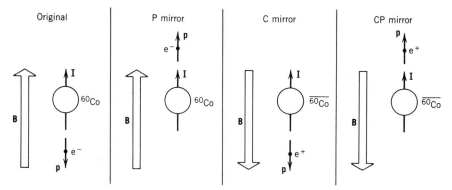

**Figure 9.24** The polarized $^{60}$Co experiment in the P, C, and combined CP mirrors.

wave function has no observable consequences for nuclear spectroscopy, but there are two cases in particular in which the effects can be observable. In the first, nuclear $\gamma$ radiation emitted by a polarized nucleus acquires a small difference in intensity between the directions along and opposite to the polarizing magnetic field. This is exactly analogous to the $^{60}$Co experiment, but is generally a very small effect (of the order of one part in $10^7$) because it arises only from the small part of the wave function and the regular part $\psi^{(\pi)}$ gives no difference in the two intensities. In one very favorable case in the decay of the $^{180}$Hf isomeric state, described in K. S. Krane et al., *Phys. Rev. C* **4**, 1906 (1971), the difference is about 2%, but in general it is much smaller and probably beyond our ability to measure. A second type of observation involves the search for a process that would ordinarily be absolutely forbidden if $F$ were zero. For example, consider the $\alpha$ decay of the $2^-$ level of $^{16}$O to the $0^+$ ground state of $^{12}$C. The selection rules for $\alpha$ decay absolutely forbid $2^- \to 0^+$ decays (see Section 8.5), but if the $2^-$ state includes a small piece of $2^+$ state, the decay is permitted to occur with a very small intensity proportional to $F^2$. Based on a careful study of the $\alpha$ decay of the excited states of $^{16}$O, Neubeck et al. discovered a weak branch which they assigned to the parity-violating $2^- \to 0^+$ transition. The partial half-life for this transition was deduced to be $7 \times 10^{-7}$ s. By way of comparison, Equation 8.18 gives for the half-life of an ordinary $\alpha$ transition (with $Q = 1.7$ MeV, $B = 3.8$ MeV) the value $2 \times 10^{-21}$ s. The $\alpha$ decay intensity is thus indeed of order $F^2$ ($10^{-14}$), as expected for this P-violating process. A description of this difficult experiment can be found in *Phys. Rev. C* **10**, 320 (1974).

## 9.10  BETA SPECTROSCOPY

In this section we explore some techniques for deducing the properties of nuclear states (especially excitation energies and spin-parity assignments) through measurements of $\beta$ decays. This process is complicated by two features of the $\beta$-decay process (as compared with $\alpha$ decay, for instance): (1) The $\beta$ spectrum is continuous. The study of decay processes such as those discussed in Section 8.6 is

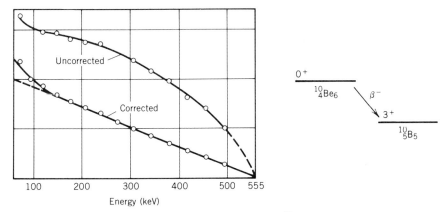

**Figure 9.25**  Uncorrected Fermi–Kurie plot for $^{10}$Be decay and correction for shape factor for second-forbidden transition. Data from L. Feldman and C. S. Wu, *Phys. Rev.* **87**, 1091 (1952).

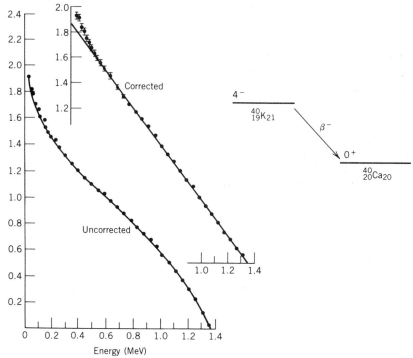

**Figure 9.26** Uncorrected Fermi–Kurie plot for the decay of $^{40}$K and correction for shape factor of third-forbidden transition. Data from W. H. Kelly et al., *Nucl. Phys.* **11**, 492 (1959).

not possible for $\beta$ decays with many branches, for we cannot reliably "unfold" the various components. (2) The $\beta$ selection rules are not absolute—the ranges of $ft$ values often overlap and cannot be used to make absolute deductions of decay types, and the measurement of the angular momentum carried by the $\beta$ particle is not sufficient to fix the relative parities of the initial and final states. There are, however, many cases in which it is possible to derive spectroscopic information from the decays.

Although the shape of the $\beta$ spectrum and the half-life (actually, the $ft$ value) of the decay are not absolute indicators of the decay type, they do give us strong clues about the type of decay (and therefore the relative spin-parity assignments of the initial and final nuclear states). As discussed in Section 9.3, a linear Fermi–Kurie plot with no shape factor strongly suggests a decay of the allowed type, and we would therefore expect $\Delta I = 0$ or $1$ and $\Delta\pi = $ no. A nonlinear Fermi–Kurie plot that is linearized by the shape factor $S = p^2 + q^2$ is, as shown in Figure 9.5, most likely of a first-forbidden type.

Figures 9.25 and 9.26 show additional examples of the use of the shape of the decay to deduce the properties of the initial state. The Fermi–Kurie plot for the decay of $^{10}$Be is linearized by a shape factor characteristic of $\Delta I = 3$, $\Delta\pi = $ no second-forbidden decays. Like all even-$Z$, even-$N$ nuclei, $^{10}$Be has a $0^+$ ground state, and so we immediately deduce the assignment of $3^+$ for the $^{10}$B final state.

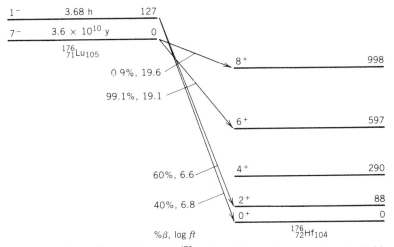

**Figure 9.27**  The $\beta$ decay of $^{176}$Lu. Level energies are given in keV.

Moreover, the log $ft$ value of 13.4 is consistent with that expected for second-forbidden decays. The $^{40}$K decay proceeds primarily by $\beta^-$ decay to $^{40}$Ca. The Fermi–Kurie plot is linearized by a shape factor characteristic of a $\Delta I = 4$, $\Delta\pi = $ yes third-forbidden decay. The final state is the $^{40}$Ca ground state, which is certainly $0^+$. The initial state is therefore $4^-$. The log $ft$ of 18.1 also suggests a third-forbidden process.

On the other hand, we must be careful not to rely too much on the empirical rules for log $ft$ values, which are merely based on systematics and not on any theory. In the decay of $^{176}$Lu (Figure 9.27) the log $ft$ is 19.1, while from the known spin-parity assignments we expect a first-forbidden decay (the log $ft$ values for which usually fall in the range 8–12). It is useful to remind ourselves that a log $ft$ of 19 means that the $\beta$ decay is slowed or inhibited by a factor of $10^7$ relative to a decay with log $ft$ of 12. The extreme effect in the $^{176}$Lu decay comes about from the unusually poor match of the initial and final nuclear wave functions.

A case in which two different $\beta$ groups contribute to the decay is illustrated by the decay of $^{72}$Zn (Figure 9.28). The weaker group can be reliably seen only through $\beta$-$\gamma$ coincidence measurements. The linear Fermi–Kurie plots and the small log $ft$ values suggest allowed decays, consistent with the $1^+$ assignments to both final states.

A more extreme example comes from the decay of $^{177}$Lu (Figure 9.29) in which through careful measurement it is possible to deduce four separate groups. The unfolding procedure begins with the highest group, which is assumed to have a nearly linear Fermi–Kurie plot. Extrapolating the linear high-energy portion backward and subtracting, the remaining spectrum shows an endpoint of 385 keV, and repeating the process reveals two additional components.

We cannot tell directly whether the highest-energy component represents a decay to the ground state of $^{177}$Hf, but we can show that it does by computing the $Q$ value for the decay to the ground state. Because $^{177}$Lu is radioactive, its mass cannot be measured directly, but we can deduce it through measuring the

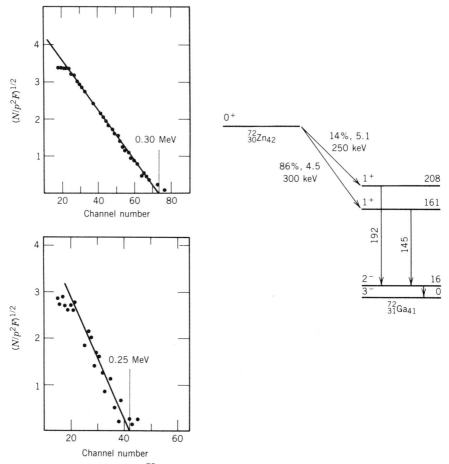

**Figure 9.28** $\beta$ decay of $^{72}$Zn. In this case, there are two decays, but the stripping method cannot reliably be used to find the much weaker, lower energy component. At left are shown Fermi–Kurie plots of $\beta$'s in coincidence with the 192-keV (bottom) and 145-keV (top) $\gamma$ rays. The coincidences enable the unambiguous separation of the two decays, the determination of their endpoints, and the demonstration of the linearity of their Fermi–Kurie plots. Data from M. Ishihara, *J. Phys. Soc. Jpn.* **18**, 1111 (1963).

energy released when a low-energy neutron is added to stable $^{176}$Lu to form $^{177}$Lu. This energy is determined to be $7.0726 \pm 0.0006$ MeV, and thus

$$m(^{177}\text{Lu}) = m(^{176}\text{Lu}) + m(\text{n}) - 7.0726 \text{ MeV}/c^2$$
$$= 176.943766 \pm 0.000006 \text{ u}$$

where the latter step is made using the known $^{176}$Lu mass. We can now find

$$Q_\beta = m(^{177}\text{Lu}) - m(^{177}\text{Hf})$$
$$= 496 \pm 8 \text{ keV}$$

in excellent agreement with the energy of the highest $\beta$ group. We therefore conclude that this group populates the $^{177}$Hf ground state, with the lower groups

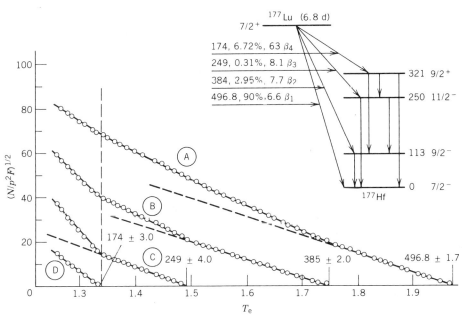

**Figure 9.29** Fermi–Kurie plot of the β decay of ¹⁷⁷Lu. Curve A represents the complete β spectrum. Extrapolating the high-energy portion (which presumably includes only a single component) gives the dashed line, and the difference between the extrapolated line and curve A gives curve B. The linear portion of curve B gives the endpoint of the next component, and repeating the procedure gives curves C and D. The resulting decay scheme is shown in the inset. Data from M. S. El-Nesr and E. Bashandy, *Nucl. Phys.* **31**, 128 (1962).

populating successively higher states at energies of 112 keV (= 497 − 385), 248 keV (= 497 − 249), and 323 keV (= 497 − 174). The γ spectrum shows results consistent with these deductions, as we discuss in Section 10.8. The 113-keV γ ray, for example, represents the transition from the first excited state to the ground state, and the β spectrum in coincidence with the 113-keV γ ray shows only the 385-keV component.

As a final example of a spectroscopic study, we consider the decay of ¹²⁶I, which can occur either through negative or positive β emission. The Fermi–Kurie plot (Figure 9.30) is definitely nonlinear at the high end, but when the upper end is corrected by the shape factor for a $\Delta I = 2$, $\Delta \pi$ = yes first-forbidden decay, it becomes linear and the stripping reveals three groups. Only the two lower groups are in coincidence with γ radiation, suggesting that the highest group populates the ¹²⁶Xe ground state ($0^+$) and thus that the decaying state must be $2^-$ (because the highest group is $\Delta I = 2$, $\Delta \pi$ = yes). The other groups must populate excited states at 385 keV (= 1250 − 865) and 865 keV (= 1250 − 385). (It is coincidental that the numbers happen to be interchangeable.) The positron spectrum (Figure 9.31) similarly shows two groups which by the same argument populate the ground and first excited (670-keV) state of ¹²⁶Te. The γ spectrum shows strong transitions of energies 389, 492, 666, 754, 880, and 1420 keV, which can be placed as shown in Figure 9.32, based on the observed β endpoints. The spins of the first excited states are $2^+$, and the second excited states must be $2^+$ as well,

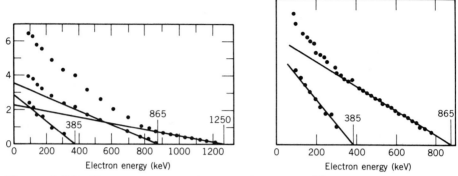

**Figure 9.30** Electron spectrum from the $\beta$ decay of $^{126}$I. The singles spectrum (left) shows three components, using a stripping procedure as in Figure 9.29. The coincidence spectrum with gamma radiation (right) does not show the highest energy component; if it is coincident with no $\gamma$ radiations, it must be a transition to the ground state. Data from L. Koerts et al., *Phys. Rev.* **98**, 1230 (1955).

based on the similarity of the log *ft* values and the observed "crossover" transition to the ground state.

The calculation of $\beta$ matrix elements from nuclear wave functions is a difficult process, and so we usually are content to compare experimental results from different but similar decays. For example, consider the mirror decays of $^{12}$B and $^{12}$N to $^{12}$C (Figure 9.33). The *ft* values to the different excited states are virtually identical for the $\beta^-$ and $\beta^+$ decays. The transition of the 7th proton into the 6th

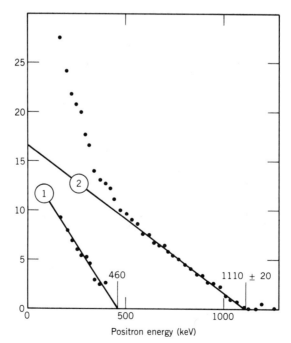

**Figure 9.31** Positron spectrum from the $\beta$ decay of $^{126}$I. Stripping reveals two components. Data from L. Koerts et al., *Phys. Rev.* **98**, 1230 (1955).

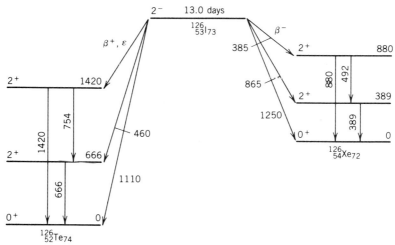

**Figure 9.32** Decay scheme of $^{126}$I, showing $\beta$ and $\gamma$ information. Energies of $\beta$'s, $\gamma$'s, and levels are given in keV.

neutron would be expected to involve initial and final nuclear wave functions (based on the shell model) identical with those in the transition of the 7th neutron into the 6th proton. The $ft$ values are consistent with this expectation. The transition of the proton into a neutron that leaves the neutron in the same shell-model state as the initial proton results in the population of the 15.11-MeV excited state of $^{12}$C. This state therefore has the same nuclear wave functions as the $^{12}$B and $^{12}$N ground states (except for the difference between protons and neutrons) and is the *analog state* of $^{12}$B and $^{12}$N. The particularly small $ft$ value in the decay of $^{12}$N to this state emphasizes its interpretation as the analog state.

Finally, let's look at the information on nuclear wave functions that can be obtained from $\beta$ decay. In particular, we examine the transitions between odd neutrons and protons within the $f_{7/2}$ shell. (That is, one $f_{7/2}$ nucleon is transformed into another.) Let us look specifically at cases in odd-$A$ nuclei involving $\Delta I = 0$, allowed decays between states of spin-parity $\frac{7}{2}^{-}$. The simplest example is

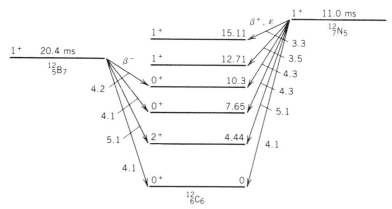

**Figure 9.33** Beta decays of $^{12}$B and $^{12}$N to $^{12}$C. Note the similarities in the log $ft$ values for the $\beta^{+}$ and $\beta^{-}$ decays leading to the same final state in $^{12}$C.

**Table 9.4**  $\beta$ Decays in the $f_{7/2}$ Shell ($\frac{7}{2}^- \to \frac{7}{2}^-$)

| $_{Z}^{A}X_N \to _{Z'}^{A}X'_{N'}$ | $\|Z - N'\| = \|N - Z'\|$ | $\log ft$ |
|---|---|---|
| $_{21}^{41}Sc_{20} \to _{20}^{41}Ca_{21}$ | 0 | 3.5 |
| $_{22}^{43}Ti_{21} \to _{21}^{43}Sc_{22}$ | 0 | 3.5 |
| $_{23}^{45}V_{22} \to _{22}^{45}Ti_{23}$ | 0 | 3.6 |
| $_{27}^{53}Co_{26} \to _{26}^{53}Fe_{27}$ | 0 | 3.6 |
| $_{21}^{43}Sc_{22} \to _{20}^{43}Ca_{23}$ | 2 | 5.0 |
| $_{22}^{45}Ti_{23} \to _{21}^{45}Sc_{24}$ | 2 | 4.6 |
| $_{26}^{53}Fe_{27} \to _{25}^{53}Mn_{28}$ | 2 | 5.2 |
| $_{20}^{45}Ca_{25} \to _{21}^{45}Sc_{24}$ | 4 | 6.0 |
| $_{21}^{47}Sc_{26} \to _{22}^{47}Ti_{25}$ | 4 | 5.3 |
| $_{23}^{49}V_{26} \to _{22}^{49}Ti_{27}$ | 4 | 6.2 |
| $_{24}^{51}Cr_{27} \to _{23}^{51}V_{28}$ | 4 | 5.4 |
| $_{20}^{47}Ca_{27} \to _{21}^{47}Sc_{26}$ | 6 | 8.5 |
| $_{21}^{49}Sc_{28} \to _{22}^{49}Ti_{27}$ | 6 | 5.7 |

the decay of $^{41}$Sc to $^{41}$Ca, in which a single proton outside the doubly magic $^{40}$Ca core changes into a single neutron. No change of nuclear wave function is involved, and the observed $\log ft$ for this decay is 3.5, placing it in the superallowed category. (This is an example of a *mirror* decay.) In the extreme independent particle shell model, all odd particles are treated equivalently, and we might therefore expect the decay $^{47}$Ca to $^{47}$Sc (also $\frac{7}{2}^-$ to $\frac{7}{2}^-$) to show a similar $\log ft$. However, the observed value is 8.5—the decay is slower by a factor of $10^5$! The transition of the 27th neutron to the 21st proton is thus a more complicated process, and the other six neutrons in the $f_{7/2}$ shell must have a significant influence on the decay. (Some general features of these many-particle states were discussed in Section 5.3.) Table 9.4 summarizes the observed $\frac{7}{2}^-$ to $\frac{7}{2}^-$ $\beta$ decays of the $f_{7/2}$ shell nuclei ($20 \leq N, Z \leq 28$). You can see that decays in which the odd particle is not required to change its state ($Z - N' = N - Z' = 0$) have $\log ft$ values in the superallowed category (about 3.5); as the value of $Z - N'$ increases, the change of state is correspondingly greater and the $\log ft$ increases, on the average by about one unit (a factor of 10 in the half-life) for each step in $Z - N'$.

## REFERENCES FOR ADDITIONAL READING

Other references on "classical" $\beta$ decay include Chapter 13 of J. M. Blatt and V. F. Weisskopf, *Theoretical Nuclear Physics* (New York: Wiley, 1952); Chapter 17 of R. D. Evans, *The Atomic Nucleus* (New York: McGraw-Hill, 1955); and I. Kaplan, *Nuclear Physics* (Reading, MA: Addison-Wesley, 1955).

More comprehensive reference works, ranging from elementary to advanced material, include H. F. Schopper, *Weak Interactions and Nuclear Beta Decay* (Amsterdam: North-Holland, 1966); C. S. Wu and S. A. Moszkowski, *Beta*

*Decay* (New York: Wiley-Interscience, 1966); E. J. Konopinski, *The Theory of Beta Decay* (Oxford: Oxford University Press, 1966); M. Morita, *Beta Decay and Muon Capture* (Reading, MA: Benjamin, 1973). An introductory work which includes reprints of some of the classic papers on β decay is C. Strachan, *The Theory of Beta Decay* (Oxford: Pergamon, 1969).

Beta spectroscopy is the subject of C. S. Wu, in *Nuclear Spectroscopy*, part A, edited by F. Ajzenberg-Selove (New York: Academic, 1959), Section I.E.1. A survey of measurements of the shapes of β spectra can be found in H. Daniel, *Rev. Mod. Phys.* **40**, 659 (1968). Many details of β decay are discussed in Chapters 22–24 of *Alpha-, Beta- and Gamma-Ray Spectroscopy*, edited by K. Siegbahn (Amsterdam: North-Holland, 1965).

A general elementary work on neutrino physics is G. M. Lewis, *Neutrinos* (London: Wykeham, 1970). A useful review of the early experimental work can be found in J. S. Allen, *The Neutrino* (Princeton, NJ: Princeton University Press, 1958). A survey of the literature about neutrinos is given in the resource letter of L. M. Lederman, *Am. J. Phys.* **38**, 129 (1970). For a discussion of the solar neutrino experiment, see the article by John Bahcall in the July 1969 issue of *Scientific American*.

Reviews of β-delayed nucleon emission can be found in J. Cerny and J. C. Hardy, *Ann. Rev. Nucl. Sci.* **27**, 333 (1977), and in J. C. Hardy, *Nuclear Spectroscopy and Reactions*, edited by J. Cerny (New York: Academic, 1974), Chapter 8B.

A popular-level treatment of parity is Martin Gardner, *The Ambidextrous Universe* (New York: Scribner, 1979).

Double-beta decay is reviewed in D. Bryman and C. Picciotto, *Rev. Mod. Phys.* **50**, 11 (1978).

## PROBLEMS

1. Compute the $Q$ values for the following $\beta^-$ decays: (a) $^{65}$Ni $\rightarrow$ $^{65}$Cu; (b) $^{11}$Be $\rightarrow$ $^{11}$B; (c) $^{193}$Os $\rightarrow$ $^{193}$Ir.

2. Compute the $Q$ values for the following $\beta^+$ and $\varepsilon$ decays: (a) $^{10}$C $\rightarrow$ $^{10}$B; (b) $^{152}$Eu $\rightarrow$ $^{152}$Sm; (c) $^{89}$Zr $\rightarrow$ $^{89}$Y.

3. $^{196}$Au can decay by $\beta^-$, $\beta^+$, and $\varepsilon$. Find the $Q$ values for the three decay modes.

4. The maximum kinetic energy of the positron spectrum emitted in the decay $^{11}$C $\rightarrow$ $^{11}$B is 1.983 ± 0.003 MeV. Use this information and the known mass of $^{11}$B to calculate the mass of $^{11}$C.

5. In the decay of $^6$He to $^6$Li, the maximum β kinetic energy is 3.510 ± 0.004 MeV. Find the mass of $^6$He, given the mass of $^6$Li.

6. In the decay of $^{47}$Ca to $^{47}$Sc, what energy is given to the neutrino when the electron has a kinetic energy of 1.100 MeV?

7. The β decay of $^{191}$Os leads only to an excited state of $^{191}$Ir at 171 keV. Compute the maximum kinetic energy of the β spectrum.

8. (a) If the β-decay energy is large compared with $m_e c^2$, find a simplified form of Equation 9.25 and show that the average value of $T_e$ (not the value

of $T_e$ where $N(T_e)$ has its maximum) is equal to $Q/2$. (b) In the case of $\beta$-decay energies that are small compared with $m_e c^2$, show that the average value of $T_e$ is $Q/3$.

9. Supply the missing component(s) in the following processes:
    (a) $\bar{\nu} + {}^3\text{He} \rightarrow$
    (b) ${}^6\text{He} \rightarrow {}^6\text{Li} + e^- +$
    (c) $e^- + {}^8\text{B} \rightarrow$
    (d) $\nu + {}^{12}\text{C} \rightarrow$
    (e) ${}^{40}\text{K} \rightarrow \nu +$
    (f) ${}^{40}\text{K} \rightarrow \bar{\nu} +$

10. What is the kinetic energy given to the proton in the decay of the neutron when (a) the electron has negligibly small kinetic energy; (b) the neutrino has negligibly small energy?

11. One of the processes that is most likely responsible for the production of neutrinos in the sun is the electron-capture decay of ${}^7\text{Be}$. Compute the energy of the emitted neutrino and the kinetic energy of the ${}^7\text{Li}$ nucleus.

12. Defining the $Q$ value as $(m_i - m_f)c^2$, compute the range of neutrino energies in the solar fusion reaction $p + p \rightarrow d + e^+ + \nu$. Assume the initial protons to have negligible kinetic energies.

13. (a) For neutrino capture reactions $\nu + {}^A\text{X} \rightarrow e^- + {}^A\text{X}'$, show that the $Q$ value, defined as in the case of decays as $Q = (m_i - m_f)c^2$, is just $[m({}^A\text{X}) - m({}^A\text{X}')]c^2$ using atomic masses. (b) Neglecting the small kinetic energy given to the final nucleus (to conserve momentum), this $Q$ value is equal to the minimum energy the neutrino must have to cause the reaction. Compute the minimum neutrino energy necessary for capture by ${}^{37}\text{Cl}$, by ${}^{71}\text{Ga}$, and by ${}^{115}\text{In}$. (c) In the Davis experiment (Section 9.6), ${}^{37}\text{Cl}$ is used to detect $\nu$ from solar fusion; ${}^{71}\text{Ga}$ and ${}^{115}\text{In}$ have also been proposed as solar neutrino detectors. Comment on the use of these detectors to observe neutrinos from the basic fusion reaction $p + p \rightarrow d + e^+ + \nu$ (see Problem 12) and from the decay of ${}^7\text{Be}$ (see Problem 11).

14. Classify the following decays according to degree of forbiddenness:
    (a) ${}^{89}\text{Sr} \left(\frac{5}{2}^+\right) \rightarrow {}^{89}\text{Y} \left(\frac{1}{2}^-\right)$
    (b) ${}^{36}\text{Cl} \left(2^+\right) \rightarrow {}^{36}\text{Ar} \left(0^+\right)$
    (c) ${}^{26}\text{Al} \left(5^+\right) \rightarrow {}^{26}\text{Mg*} \left(2^+\right)$
    (d) ${}^{26}\text{Si} \left(0^+\right) \rightarrow {}^{26}\text{Al*} \left(0^+\right) \rightarrow {}^{26}\text{Mg} \left(0^+\right)$
    (e) ${}^{97}\text{Zr} \left(\frac{1}{2}^+\right) \rightarrow {}^{97}\text{Nb*}\left(\frac{1}{2}^-\right)$

15. Show that the slope of the electron energy spectrum for allowed decays is zero near $T_e = Q$ if $m_\nu = 0$ but becomes infinite if $m_\nu \neq 0$.

16. Electron-capture decays can originate with any atomic shell $K, L, \dots$. For a wide range of nuclei, the L-capture probability is about 11% of the K-capture probability. Justify this ratio with an estimate based on the probability to locate an orbital electron near the nucleus. For this rough estimate, ignore any effects of electron screening.

17. (a) Consider a $0^+ \rightarrow 0^+$ $\beta^-$ decay. Using the helicity, Equation 9.38, of the emitted $e^-$ and $\nu$, deduce whether the $e^-$ and $\nu$ tend to be emitted parallel

or antiparallel to one another. (b) Repeat for a $1^+ \rightarrow 0^+$ $\beta^-$ decay. (c) What are the implications of these results for the recoil of the nucleus? (d) Would any of your conclusions differ in the case of $\beta^+$ decay?

18. $^{20}$Na decays to an excited state of $^{20}$Ne through the emission of positrons of maximum kinetic energy 5.55 MeV. The excited state decays by $\alpha$ emission to the ground state of $^{16}$O. Compute the energy of the emitted $\alpha$.

19. Following the decay of $^{17}$Ne, a highly excited state in $^{17}$F emits a 10.597 MeV proton in decaying to the ground state of $^{16}$O. What is the maximum energy of the positrons emitted in the decay to the $^{17}$F excited state?

20. A certain $\beta$-decay process has three components, with maximum energies 0.672, 0.536, and 0.256 MeV. The first component has two coincident $\gamma$ rays: 0.468 and 0.316 MeV, which are also coincident with each other. The second component has coincident $\gamma$'s of 0.604, 0.308, 0.136, 0.468, 0.612, 0.296, and 0.316 MeV. The third $\beta$ component is in coincidence with all of the above, plus 0.885, 0.589, 0.416, and 0.280 MeV. Use this information to construct a decay scheme and find the mass difference between the nuclear ground states.

21. The decay of $^{198}$Au to $^{198}$Pt by electron capture has not been observed, even though the very similar decay of $^{196}$Au to $^{196}$Pt by electron capture proceeds strongly. Examine the spectroscopic features of these decays and explain why the $^{198}$Au electron-capture decay is not observed. (Use the *Table of Isotopes* or a similar spectroscopic reference.)

22. From collections of nuclear spectroscopic data, find and tabulate $ft$ values for $\frac{3}{2}^+ \rightarrow \frac{1}{2}^+$ allowed decays in the region of $N$ or $Z = 14$ to 20 ($d_{3/2}$ and $s_{1/2}$ shells). Also tabulate the allowed $\frac{3}{2}^-$ to $\frac{1}{2}^-$ decays for $N$ or $Z = 2$ to 8 ($p_{3/2}$ and $p_{1/2}$ shells). Discuss any systematic differences between the two sets of values.

23. Using systematic collections of nuclear data (such as the *Table of Isotopes* or the *Nuclear Data Sheets*), tabulate the available information on $0^+ \rightarrow 0^+$ $\beta$ transitions between $f_{7/2}$ nuclei ($20 \leq Z,N \leq 28$). Discuss the coupling of the odd proton and odd neutron, and explain the observed $ft$ values.

24. Tabulate the available information on $g_{9/2} \rightarrow g_{7/2}$ positron decays of odd-mass nuclei; $g_{9/2}$ protons are generally found in the range $40 \leq Z \leq 50$, and $g_{7/2}$ neutrons are usually between $N = 50$ and $N = 66$. Try to account for the $ft$ values. (*Note:* The GT decay is sometimes called a "spin-flip" process.)

25. There are many $\beta$-decaying odd-$Z$, odd-$N$ nuclei with $2^-$ spin-parity assignments. These can decay to the $0^+$ ground state or the $2^+$ first excited state of the neighboring even-$Z$, even-$N$ nucleus. (a) Use a general nuclear spectroscopy reference (*Table of Isotopes* or the *Nuclear Data Sheets*) to tabulate the $ft$ values for the $2^+$ and $0^+$ final states from as many of these decay processes as you can find. (b) The $2^- \rightarrow 0^+$ decay is a first-forbidden process, in which the $\beta$ decay must carry 2 units of total angular momentum, while in the $2^- \rightarrow 2^+$ decay it can carry 0, 1, or 2 units of angular momentum. Use your compilations of $ft$ values to make some general conclusions on the relative probability of the $\beta$ decay carrying 2 units of

angular momentum. (c) To examine whether there might be an explanation for this effect in terms of the $0^+$ and $2^+$ nuclear wave functions, make a similar tabulation of decays from $1^-$ states, that is, of $1^- \to 0^+$ and $1^- \to 2^+$ decays. Both these first-forbidden decays carry one unit of total angular momentum. (Why?) Do you observe a systematic difference in $ft$ values between $0^+$ and $2^+$ final states? What do you conclude about the probable effect of the final nuclear state on the $\beta$ decays from $2^-$ initial states?

26. There are several examples of allowed $\beta$ decays that have larger than average $ft$ values, which can be explained with reference to the nuclear structure. Consider, for example, the following cases: (a) $^{65}\text{Ni} \to {}^{65}\text{Cu}$ and $^{65}\text{Zn} \to {}^{65}\text{Cu}$, in which the ground state-ground state decays are both $\frac{5}{2}^-$ to $\frac{3}{2}^-$ Gamow-Teller decays, but the $ft$ values are 1–2 orders of magnitude larger than for allowed decays to other low-lying states; (b) $^{115}\text{Te} \to {}^{115}\text{Sb}$ and $^{115}\text{Sb} \to {}^{115}\text{Sn*}$; in the $^{115}\text{Te}$ decay, the $\frac{7}{2}^+ \to \frac{5}{2}^+$ transition to the $^{115}\text{Sb}$ ground state is not seen, and in the $^{115}\text{Sb}$ decay, a $\frac{7}{2}^+$ low-lying excited state is populated only weakly, with an $ft$ value again 1–2 orders of magnitude larger than values for neighboring excited states. Find the shell model identification of these states and thus explain why the allowed decay mode is inhibited. Use the *Table of Isotopes* to find other examples of inhibited decays with the same shell-model assignments.

# 10

# GAMMA DECAY

Most $\alpha$ and $\beta$ decays, and in fact most nuclear reactions as well, leave the final nucleus in an excited state. These excited states decay rapidly to the ground state through the emission of one or more $\gamma$ rays, which are photons of electromagnetic radiation like X rays or visible light. Gamma rays have energies typically in the range of 0.1 to 10 MeV, characteristic of the energy difference between nuclear states, and thus corresponding wavelengths between $10^4$ and 100 fm. These wavelengths are far shorter than those of the other types of electromagnetic radiations that we normally encounter; visible light, for example, has wavelengths $10^6$ times longer than $\gamma$ rays.

The detail and richness of our knowledge of nuclear spectroscopy depends on what we know of the excited states, and so studies of $\gamma$-ray emission have become the standard technique of nuclear spectroscopy. Other factors that contribute to the popularity and utility of this method include the relative ease of observing $\gamma$ rays (negligible absorption and scattering in air, for instance, contrary to the behavior of $\alpha$ and $\beta$ radiations) and the accuracy with which their energies (and thus by deduction the energies of the excited states) can be measured. Furthermore, studying $\gamma$ emission and its competing process, internal conversion, allows us to deduce the spins and parities of the excited states.

## 10.1  ENERGETICS OF $\gamma$ DECAY

Let's consider the decay of a nucleus of mass $M$ at rest, from an initial excited state $E_i$ to a final state $E_f$. To conserve linear momentum, the final nucleus will not be at rest but must have a recoil momentum $p_R$ and corresponding recoil kinetic energy $T_R$, which we assume to be nonrelativistic ($T_R = p_R^2/2M$). Conservation of total energy and momentum give

$$E_i = E_f + E_\gamma + T_R$$
$$0 = \boldsymbol{p}_R + \boldsymbol{p}_\gamma \tag{10.1}$$

It follows that $p_R = p_\gamma$; the nucleus recoils with a momentum equal and opposite to that of the $\gamma$ ray. Defining $\Delta E = E_i - E_f$ and using the relativistic relationship $E_\gamma = cp_\gamma$,

$$\Delta E = E_\gamma + \frac{E_\gamma^2}{2Mc^2} \tag{10.2}$$

which has the solution

$$E_\gamma = Mc^2 \left[ -1 \pm \left( 1 + 2\frac{\Delta E}{Mc^2} \right)^{1/2} \right] \qquad (10.3)$$

The energy differences $\Delta E$ are typically of the order of MeV, while the rest energies $Mc^2$ are of order $A \times 10^3$ MeV, where $A$ is the mass number. Thus $\Delta E \ll Mc^2$ and to a precision of the order of $10^{-4}$ to $10^{-5}$ we keep only the first three terms in the expansion of the square root:

$$E_\gamma \cong \Delta E - \frac{(\Delta E)^2}{2Mc^2} \qquad (10.4)$$

which also follows directly from Equation 10.2 with the approximation $\Delta E \cong E_\gamma$.

The actual $\gamma$-ray energy is thus diminished somewhat from the maximum available decay energy $\Delta E$. This recoil correction to the energy is generally negligible, amounting to a $10^{-5}$ correction that is usually far smaller than the experimental uncertainty with which we can measure energies. There is one circumstance in which the recoil plays an important role; this case, known as the Mössbauer effect, is discussed in Section 10.9. Except for this case, we will in the remainder of this chapter assume $E_\gamma = \Delta E$.

For low-energy $\gamma$ rays, the recoil energy is less than 1 eV and has a negligible effect. High-energy $\gamma$ rays (such as the 5–10-MeV radiations emitted following neutron capture) give recoils in the range of 100 eV, which may be sufficient to drive the recoiling atom from its position in a solid lattice. Effects of this sort are known as *radiation damage* and have an important place in the study of solids.

## 10.2 CLASSICAL ELECTROMAGNETIC RADIATION

As you will recall from your study of modern physics, electromagnetic radiation can be treated either as a classical wave phenomenon or else as a quantum phenomenon. The type of treatment we use is determined by the kind of physical effect we are trying to describe. For analyzing radiations from individual atoms and nuclei the quantum description is most appropriate, but we can more easily understand the quantum calculations of electromagnetic radiation if we first review the classical description.

Static (i.e., constant in time) distributions of charges and currents give static electric and magnetic fields. In Section 3.5, we discussed how these fields can be analyzed in terms of the *multipole moments* of the charge distribution—dipole moment, quadrupole moment, and so on. These multipole moments give characteristic fields, and we can conveniently study the dipole field, quadrupole field, and so on.

If the charge and current distributions vary with time, particularly if they vary sinusoidally with circular frequency $\omega$, a *radiation field* is produced. The radiation field (which is studied at a distance from the source that is large compared with the size of the source) can be analyzed, like the static field, in terms of its multipole character. As an example, we consider the lowest multipole order, the dipole field.

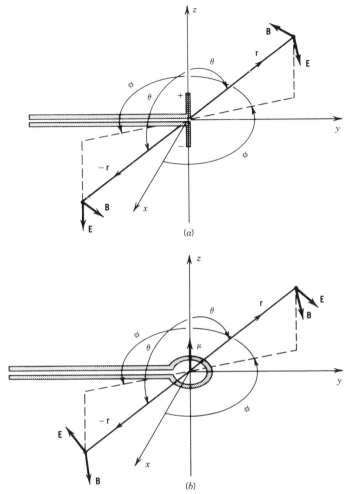

**Figure 10.1** Electric and magnetic fields from (a) an electric dipole and (b) a magnetic dipole. In each case the dipole moment is along the z axis. The vectors show the radiation fields **E** and **B** at a particular instant of time. The wires along the negative y axes should be imagined as connected to a current source of frequency ω, and to be twisted so as to make no contribution themselves to the radiation. Also shown are the behaviors of **E** and **B** under the spatial reflection r → −r; note the differences between the two cases.

A static *electric* dipole consists of equal and opposite charges $+q$ and $-q$ separated by a fixed distance $z$; the electric dipole moment is then $d = qz$. A static *magnetic* dipole can be represented as a circular current loop of current $i$ enclosing area $A$; the magnetic dipole moment is $\mu = iA$. We can produce electromagnetic radiation fields by varying the dipole moments; for example, we can allow the charges to oscillate along the $z$ axis, so that $d(t) = qz \cos \omega t$, thereby producing an electric dipole radiation field. Similarly, we could vary the current so that $\mu(t) = iA \cos \omega t$. Figure 10.1 shows the radiation fields produced in these two cases. The alternating electric dipole, Figure 10.1a, can be regarded as a linear current element, for which the magnetic field lines form circles concentric with the $z$ axis. The magnetic field vector $B$ is tangent to the circles, and the electric field direction must be chosen so that $E \times B$ is in the direction of propagation of the radiation. The magnetic dipole, Figure 10.1$b$, has the magnetic field lines that we often associate with a bar magnet.

There are three characteristics of the dipole radiation field that are important for us to consider:

1.  The power radiated into a small element of area, in a direction at an angle $\theta$ with respect to the $z$ axis, varies as $\sin^2 \theta$. The average radiated power can be calculated based on wave theory or on quantum theory, and by the correspondence principle the two calculations must agree when we extend the quantum result to the classical limit. This characteristic $\sin^2 \theta$ dependence of dipole radiation must therefore be a characteristic result of the quantum calculation as well. Higher order multipoles, such as quadrupole radiation, have a different angular distribution. In fact, as we shall see, measuring the angular distribution of the radiation is a convenient way to determine which multipoles are present in the radiation.

2.  Electric and magnetic dipole fields have opposite parity. Consider the effect of the transformation $r \rightarrow -r$. The magnetic field of the electric dipole clearly changes sign; thus $B(r) = -B(-r)$. For the magnetic dipole, on the other hand, there is no change of sign, so $B(r) = B(-r)$. Thus the electric and magnetic dipoles, which give identical angular distributions, differ in the parity of the radiation fields. Electric dipole radiation has odd parity, while magnetic dipole radiation has even parity.

3.  The average radiated power (energy emitted per unit time) is

$$P = \frac{1}{12\pi\epsilon_0} \frac{\omega^4}{c^3} d^2 \qquad (10.5)$$

for electric dipoles, and

$$P = \frac{1}{12\pi\epsilon_0} \frac{\omega^4}{c^5} \mu^2 \qquad (10.6)$$

for magnetic dipoles. Here $d$ and $\mu$ represent the *amplitudes* of the time-varying dipole moments.

Without entering into a detailed discussion of electromagnetic theory, we can extend these properties of dipole radiation to multipole radiation in general. We first define the index $L$ of the radiation so that $2^L$ is the multipole order ($L = 1$ for dipole, $L = 2$ for quadrupole, and so on). With $E$ for electric and $M$ for magnetic, we can generalize the above three properties of dipole radiation.

1.  The angular distribution of $2^L$-pole radiation, relative to a properly chosen direction, is governed by the *Legendre polynomial* $P_{2L}(\cos \theta)$. The most common cases are dipole, for which $P_2 = \frac{1}{2}(3 \cos^2 \theta - 1)$, and quadrupole, with $P_4 = \frac{1}{8}(35 \cos^4 \theta - 30 \cos^2 \theta + 3)$.

2. The parity of the radiation field is

$$
\pi(ML) = (-1)^{L+1}
$$
$$
\pi(EL) = (-1)^{L} \qquad (10.7)
$$

Notice that electric and magnetic multipoles of the same order always have opposite parity.

3. The radiated power is, using $\sigma = E$ or $M$ to represent electric or magnetic radiation,

$$
P(\sigma L) = \frac{2(L+1)c}{\epsilon_0 L[(2L+1)!!]^2} \left(\frac{\omega}{c}\right)^{2L+2} [m(\sigma L)]^2 \qquad (10.8)
$$

where $m(\sigma L)$ is the amplitude of the (time-varying) electric or magnetic multipole moment, and where the double factorial $(2L+1)!!$ indicates $(2L+1) \cdot (2L-1) \cdots 3 \cdot 1$. The generalized multipole moment $m(\sigma L)$ differs, for $L = 1$, from the electric dipole moment $d$ and magnetic dipole moment $\mu$ through some relatively unimportant numerical factors of order unity. From now on, we shall deal only with the generalized moments in our discussion of $\gamma$ radiation.

## 10.3 TRANSITION TO QUANTUM MECHANICS

To carry the classical theory into the quantum domain, all we must do is quantize the sources of the radiation field, the classical multipole moments. In Equation 10.8, it is necessary to replace the *multipole moments* by appropriate *multipole operators* that change the nucleus from its initial state $\psi_i$ to the final state $\psi_f$. As we have discussed for $\alpha$ and $\beta$ radiation, the decay probability is governed by the *matrix element* of the *multipole operator*

$$
m_{fi}(\sigma L) = \int \psi_f^* m(\sigma L) \psi_i \, dv \qquad (10.9)
$$

The integral is carried out over the volume of the nucleus. We shall not discuss the form of the operator $m(\sigma L)$, except to say that its function is to change the nuclear state from $\psi_i$ to $\psi_f$ while simultaneously creating a photon of the proper energy, parity, and multipole order.

If we regard Equation 10.8 as the energy radiated per unit time in the form of photons, each of which has energy $\hbar\omega$, then the probability per unit time for photon emission (that is, the decay constant) is

$$
\lambda(\sigma L) - \frac{P(\sigma L)}{\hbar\omega} = \frac{2(L+1)}{\epsilon_0 \hbar L[(2L+1)!!]^2} \left(\frac{\omega}{c}\right)^{2L+1} [m_{fi}(\sigma L)]^2 \quad (10.10)
$$

This expression for the decay constant can be carried no further until we evaluate the matrix element $m_{fi}(\sigma L)$, which requires knowledge of the initial and final wave functions. We can simplify the calculation and make some corresponding estimates of the $\gamma$-ray emission probabilities if we assume the transition is due to a single proton that changes from one shell-model state to another. In the case of electric transitions, the multipole operator includes a term of the form $er^L Y_{LM}(\theta, \phi)$, which reduces to $ez$ for $L = 1$ (dipole) radiation as expected, and to $e(3z^2 - r^2)$ for $L = 2$ (quadrupole) radiation, analogous to the calculation of the static quadrupole moment in Equation 3.36. If we take the radial parts of the nuclear wave functions $\psi_i$ and $\psi_f$ to be constant up to the nuclear radius $R$ and zero for $r > R$, then the radial part of the transition probability is of the form

$$\int_0^R r^2 r^L \, dr \bigg/ \int_0^R r^2 \, dr = \frac{3}{L+3} R^L \tag{10.11}$$

where the integral in the denominator is included for normalization and the $r^2$ factor comes from the volume element. Including this factor in the matrix element, and replacing the angular integrals by unity, which is also a reasonable estimate, the $EL$ transition probability is estimated to be

$$\lambda(EL) \cong \frac{8\pi(L+1)}{L[(2L+1)!!]^2} \frac{e^2}{4\pi\epsilon_0 \hbar c} \left(\frac{E}{\hbar c}\right)^{2L+1} \left(\frac{3}{L+3}\right)^2 cR^{2L} \tag{10.12}$$

With $R = R_0 A^{1/3}$, we can make the following estimates for some of the lower multipole orders

$$\lambda(E1) = 1.0 \times 10^{14} A^{2/3} E^3$$

$$\lambda(E2) = 7.3 \times 10^7 A^{4/3} E^5$$

$$\lambda(E3) = 34 A^2 E^7 \tag{10.13}$$

$$\lambda(E4) = 1.1 \times 10^{-5} A^{8/3} E^9$$

where $\lambda$ is in $s^{-1}$ and $E$ is in MeV.

For magnetic transitions, the radial integral includes the term $r^{L-1}$, and the same assumption as above about the constancy of the nuclear wave function gives the factor $3R^{L-1}/(L+2)$. The magnetic operator also includes a factor that depends on the nuclear magnetic moment of the proton. The result for the $ML$ transition probability is

$$\lambda(ML) \cong \frac{8\pi(L+1)}{L[(2L+1)!!]^2} \left(\mu_p - \frac{1}{L+1}\right)^2 \left(\frac{\hbar}{m_p c}\right)^2 \left(\frac{e^2}{4\pi\epsilon_0 \hbar c}\right)$$

$$\times \left(\frac{E}{\hbar c}\right)^{2L+1} \left(\frac{3}{L+2}\right)^2 cR^{2L-2} \tag{10.14}$$

where again several angular momentum factors of order unity have been neglected. It is customary to replace the factor $[\mu_p - 1/(L+1)]^2$ by 10, which gives the following estimates for the lower multiple orders*:

$$\lambda(M1) = 5.6 \times 10^{13} E^3$$

$$\lambda(M2) = 3.5 \times 10^7 A^{2/3} E^5$$

$$\lambda(M3) = 16 A^{4/3} E^7 \tag{10.15}$$

$$\lambda(M4) = 4.5 \times 10^{-6} A^2 E^9$$

---

*The numerical coefficients in Equation 10.15 differ slightly from those sometimes found in the literature. The difference arises because the factor $3/(L+2)$ in Equation 10.14 is often replaced by $3/(L+3)$, to make Equation 10.14 look more like Equation 10.12. We choose to maintain the form of Equation 10.14, so that to agree with some $ML$ Weisskopf estimates in the literature, Equations 10.15 must be multiplied by $(L+2)^2/(L+3)^2$.

These estimates for the transition rates are known as *Weisskopf estimates* and are not meant to be true theoretical calculations to be compared with measured transition rates. Instead, they provide us with reasonable relative comparisons of the transition rates. For example, if the observed decay rate of a certain $\gamma$ transition is many orders of magnitude smaller than the Weisskopf estimate, we might suspect that a poor match-up of initial and final wave functions is slowing the transition. Similarly, if the transition rate were much greater than the Weisskopf estimate, we might guess that more than one single nucleon is responsible for the transition.

Based on the Weisskopf estimates, we can immediately draw two conclusions about transition probabilities: (1) The lower multipolarities are dominant—increasing the multipole order by one unit reduces the transition probability by a factor of about $10^{-5}$. A similar effect occurs in atoms, in which the most common transitions are dipole. (2) For a given multipole order, electric radiation is more likely than magnetic radiation by about two orders of magnitude in medium and heavy nuclei. We shall see in Section 10.7 how these expectations agree with observations.

## 10.4 ANGULAR MOMENTUM AND PARITY SELECTION RULES

A classical electromagnetic field produced by oscillating charges and currents transmits not only energy but angular momentum as well. If, for example, we surround the charges and currents with a large spherical absorbing shell, the shell can be made to rotate by the absorbed radiation. The rate at which angular momentum is radiated is proportional to the rate at which energy is radiated.

When we go over to the quantum limit, we can preserve the proportionality if each photon carries a definite angular momentum. The multipole operator of order $L$ includes the factor $Y_{LM}(\theta, \phi)$, which is associated with an angular momentum $L$. We therefore conclude that a multipole of order $L$ transfers an angular momentum of $L\hbar$ per photon.

Let's consider a $\gamma$ transition from an initial excited state of angular momentum $I_i$ and parity $\pi_i$ to a final state $I_f$ and $\pi_f$. For the moment we will assume $I_i \neq I_f$. Conservation of angular momentum requires that the total initial angular momentum be equal to the total final angular momentum. In vector terms,

$$I_i = L + I_f$$

Since $I_i$, $L$, and $I_f$ must form a closed vector triangle, the possible values of $L$ are restricted. The largest possible value of $L$ is $I_i + I_f$ and the smallest possible value is $|I_i - I_f|$. For example, if $I_i = \frac{3}{2}$ and $I_f = \frac{5}{2}$, the possible values of $L$ are 1, 2, 3, and 4; the radiation field in this case would consist of a mixture of dipole, quadrupole, octupole ($L = 3$) and hexadecapole ($L = 4$) radiation.

Whether the emitted radiation is of the electric or magnetic type is determined by the relative parity of the initial and final levels. If there is no change in parity ($\Delta\pi$ = no), then the radiation field must have even parity; if the parity changes in the transition ($\Delta\pi$ = yes), then the radiation field must have odd parity. As shown by Equation 10.7, electric and magnetic multipoles differ in their parities. Electric transitions have even parity if $L$ = even, while magnetic transitions have even parity if $L$ = odd. Therefore a $\Delta\pi$ = no transition would consist of even electric multipoles and odd magnetic multipoles. A $\Delta\pi$ = yes transition, on the other hand, would consist of odd electric and even magnetic multipoles. In our previous example ($I_i = \frac{3}{2}$ to $I_f = \frac{5}{2}$), let us assume that $\pi_i = \pi_f$, so that $\Delta\pi$ = no. We have already concluded that $L$ = 1, 2, 3, or 4. The $L$ = 1 radiation must be of magnetic character (odd magnetic and even electric multipoles for $\Delta\pi$ = no), the $L$ = 2 radiation of electric character, and so on. The radiation field must thus be $M1$, $E2$, $M3$, and $E4$ radiation. If our two states had $\pi_i = -\pi_f$ ($\Delta\pi$ = yes) then the radiation field would be $E1$, $M2$, $E3$, and $M4$ radiation.

We therefore have the following angular momentum and parity *selection rules*:

$$|I_i - I_f| \le L \le I_i + I_f \qquad \text{(no } L = 0)$$

$$\begin{aligned} \Delta\pi = \text{no:} \qquad &\text{even electric, odd magnetic} \\ \Delta\pi = \text{yes:} \qquad &\text{odd electric, even magnetic} \end{aligned} \qquad (10.16)$$

The exception to the angular momentum selection rule occurs when $I_i = I_f$ because *there are no monopole ($L = 0$) transitions* in which a single photon is emitted. Classically, the monopole moment is just the electric charge, which does not vary with time. (A spherical charge distribution of radius $R$ gives only a pure $1/r^2$ Coulomb field for $r > R$. Even if we allow the sphere to undergo radial oscillations, the Coulomb field for $r > R$ is unaffected and no radiation is produced.) For transitions in which $I_i = I_f$, the lowest possible $\gamma$-ray multipole order is dipole ($L = 1$).

The case in which either $I_i$ or $I_f$ is zero is particularly simple, for then only a pure multipole transition is emitted. For example, the first excited $2^+$ ($I_i = 2$, $\pi_i$ = even) state in even-$Z$, even-$N$ nuclei decays to the $0^+$ ground state through the emission of a pure electric quadrupole ($E2$) transition. The above selection rules give immediately $L = 2$ and electric radiation for $\Delta\pi$ = no.

For $I_i = I_f = 0$, the selection rules would give only $L = 0$, which as we have already discussed is not permitted for radiative transitions. A few even-even nuclei have $0^+$ first excited states, which are forbidden to decay to the $0^+$ ground state by $\gamma$ emission. These states decay instead through *internal conversion*, which we discuss in Section 10.6. In this process the excitation energy is emitted by ejecting an *orbital* electron, the wave function of which penetrates the nuclear volume and samples the monopole distribution at $r < R$, where the potential *does* fluctuate.

Usually the spins $I_i$ and $I_f$ have values for which the selection rules permit several multipoles to be emitted. The single-particle (Weisskopf) estimates permit us to make some general predictions about which multipole is most likely to be emitted. Let us consider the previous example of an $I_i = \frac{3}{2}^+$ to $I_f = \frac{5}{2}^+$ transition ($M1$, $E2$, $M3$, $E4$). We assume a medium-weight nucleus ($A$ = 125, so $A^{2/3}$ = 25) and a transition energy $E$ = 1 MeV. The estimates (Equations 10.13 and 10.15) give emission probabilities in the ratio

$$\lambda(M1):\lambda(E2):\lambda(M3):\lambda(E4) = 1:1.4\times10^{-3}:2.1\times10^{-10}:1.3\times10^{-13}$$

You can see that the lower multipoles ($M1$ and $E2$) are far more likely than the higher ones. In practice we could regard this transition as being composed of $M1$ radiation with possibly a small mixture of $E2$. If the transition were $\Delta\pi$ = yes, the multipoles would be *E*1, *M*2, *E*3, *M*4 with the ratios

$$\lambda(E1):\lambda(M2):\lambda(E3):\lambda(M4) = 1:2.3 \times 10^{-7}:2.1 \times 10^{-10}:2.1 \times 10^{-17}$$

Here only the *E*1 is expected to contribute to the transition.

There are rather general expectations, based on the single-particle estimates:

1.  The lowest permitted multipole usually dominates.
2.  Electric multipole emission is more probable than the same magnetic multipole emission by a factor of order $10^2$ for medium and heavy nuclei. (Of course, the selection rules prohibit *EL* and *ML* from competing in the same radiation field.)
3.  Emission of multipole $L + 1$ is less probable than emission of multipole $L$ by a factor of the order of about $10^{-5}$.
4.  Combining 2 and 3, we have the following (here $L' = L + 1$)

$$\frac{\lambda(EL')}{\lambda(ML)} = \frac{\lambda(EL')}{\lambda(EL)} \cdot \frac{\lambda(EL)}{\lambda(ML)} \approx 10^{-5} \times 10^2 \approx 10^{-3}$$

$$\frac{\lambda(ML')}{\lambda(EL)} = \frac{\lambda(ML')}{\lambda(ML)} \cdot \frac{\lambda(ML)}{\lambda(EL)} \approx 10^{-5} \times 10^{-2} \approx 10^{-7}$$

You can thus see why *M*2 competes with *E*1 far less effectively than *E*2 competes with *M*1. Keep in mind, however, that these are only estimates based on some very crude approximations. The properties of specific nuclear states can modify these estimates by many orders of magnitude; for example, we often find cases in which $\lambda(E2) > \lambda(M1)$, especially in transitions between vibrational or rotational collective states.

## 10.5   ANGULAR DISTRIBUTION AND POLARIZATION MEASUREMENTS

In this section we consider the experimental techniques that help us to distinguish one multipole from another. Measuring the energy of a γ ray emitted in a certain transition gives us no information on the multipole character, and even if we know $I_i$ and $I_f$, all we can do is restrict the range of possible *L* values, not determine how much of each is present. (In fact, more frequently the reverse process is used—we might know $I_f$ and restrict the range of $I_i$ by measuring *L*.) Even measuring the lifetime is of limited usefulness because of the many assumptions made in obtaining the Weisskopf estimates. To determine the multipole order of the γ radiation, we must measure the angular distribution of the radiation, and to distinguish electric from magnetic radiations, it is necessary to do additional measurements, such as to measure the polarization of the radiation.

By way of illustration, let us consider a dipole transition from $I_i = 1$ to $I_f = 0$. The initial state includes three sublevels with $m_i = +1, 0, -1$; the final state has only one sublevel, with $m_f = 0$. The angular distribution generally depends on the values of $m_i$ and $m_f$. For example, in the case $m_i = 0$ to $m_f = 0$, the γ

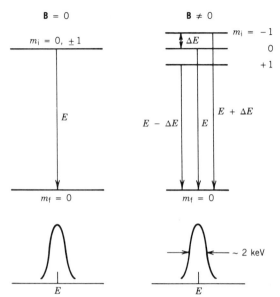

**Figure 10.2** The nuclear Zeeman effect. In a magnetic field $B$, the $2I_i + 1$ sublevels of the state $I_i$ are split into equally spaced states differing in energy by $\Delta E = \mu B / I_i$; for the case illustrated, $I_i = 1$, $2I_i + 1 = 3$, $I_f = 0$, and $\Delta E = \mu B$. Three transitions satisfy the dipole selection rule $\Delta m = 0, \pm 1$. The observed $\gamma$ emission lines are shown below each diagram; because the energy resolution is much greater than the splitting $\Delta E$, we cannot resolve the individual components.

emission probability varies as $\sin^2 \theta$ (where the angle is defined with respect to the $z$ axis that we use to measure the components of $I_i$). This is in fact the quantum analog of the case of radiation from the classical dipole we considered in Section 10.2. The transitions from $m_i = \pm 1$ to $m_f = 0$ have angular distributions that vary as $\frac{1}{2}(1 + \cos^2 \theta)$.

If we could pick out one of the three initial $m$ states and measure the angular distribution of only that component of the transition, we would observe the characteristic angular distribution. The simplest scheme to do this would be to place the nuclei in a very strong magnetic field, so that the interaction of the magnetic moment $\mu$ of level $I_i$ would give a splitting of that level depending on the relative orientations of $I_i$ and the field $B$. (This is exactly analogous to the Zeeman effect in atoms.) Figure 10.2 shows a representation of this situation. Before the field is turned on, there is one transition of energy $E$. When the field is present, the splitting of the levels gives three transitions of energy $E$, $E + \Delta E$, and $E - \Delta E$, where $\Delta E = \mu B$. If we could pick out only the component with energy $E + \Delta E$, for example, we would see the $\frac{1}{2}(1 + \cos^2 \theta)$ distribution, relative to the direction of the field. We can estimate the magnitude of $\Delta E$—for a typical magnetic moment of 1 nuclear magneton in a large field of 10 T, $\Delta E \simeq 10^{-6}$ eV. This small value of $\Delta E$ is far below the energy differences we can resolve with $\gamma$-ray detectors, which typically cannot separate or resolve transitions within 2 keV of each other. Thus what we actually observe is a mixture of all possible $m_i \rightarrow m_f$ ($+1 \rightarrow 0$, $0 \rightarrow 0$, $-1 \rightarrow 0$). If we let $W(\theta)$ represent the observed angular distribution, then

$$W(\theta) = \sum_{m_i} p(m_i) W_{m_i \to m_f}(\theta)$$

where $p(m_i)$ is the *population* of the initial state, the fraction of the nuclei that occupies each sublevel.

Under normal circumstances all the populations are equal, $p(+1) = p(0) = p(-1) = \frac{1}{3}$, so that

$$W(\theta) \propto \frac{1}{3}\left[\frac{1}{2}(1 + \cos^2\theta)\right] + \frac{1}{3}(\sin^2\theta) + \frac{1}{3}\left[\frac{1}{2}(1 + \cos^2\theta)\right] = \text{constant}$$

that is, the angular distribution disappears and the radiation intensity is independent of direction.

There are two methods that can be used to create unequal populations $p(m_i)$ resulting in nonconstant $W(\theta)$. In the first method, we place the nuclei in a strong magnetic field, as we described previously, but at the same time we cool them to very low temperature, so low that the populations are made unequal by the Boltzmann distribution, $p(m_i) \propto e^{-m_i(\Delta E/kT)}$. To have unequal populations,

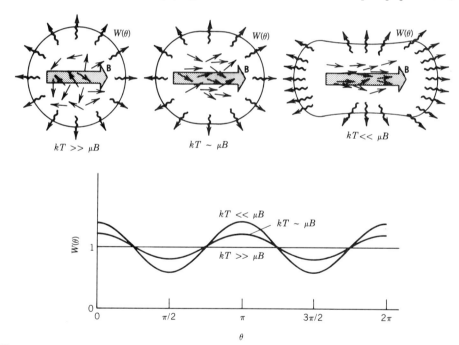

**Figure 10.3** Angular distributions of nuclei with spins oriented at low temperatures. At top left is shown the distribution of radiation expected at high temperature; the magnetic field has essentially no effect in orienting the nuclear spins because of the thermal motion. At intermediate temperature (top center), the spins begin to align with the field, and the radiation distribution becomes nonuniform. At very low temperature, the spins are essentially completely aligned with the field. Measuring the angular distribution for dipole radiation would give the results shown at the bottom.

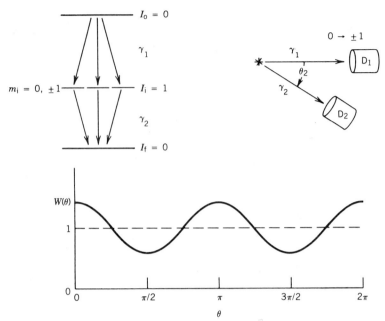

**Figure 10.4**   Angular correlation measurements. In a cascade of two radiations, here assumed to be $0 \rightarrow 1 \rightarrow 0$, the angular distribution of $\gamma_2$ is measured relative to the direction of $\gamma_1$. A typical result for two dipole transitions is shown at the bottom.

the exponential must be different from one, which means $\Delta E$ must be of the order of $kT$. (At high temperature, say room temperature, $kT \simeq 1/40$ eV, and $\Delta E \ll kT$, with the previous estimate of $10^{-6}$ eV for $\Delta E$.) To have $\Delta E \sim kT$, we must cool the nuclei to $T \sim 0.01$ K. This is accomplished using continuously operating refrigerators, called helium dilution refrigerators, and this method, called low-temperature *nuclear orientation*, has become a powerful technique for determining $\gamma$-ray multipole characters and for inferring nuclear spin assignments. Figure 10.3 shows representative angular distributions for dipole radiation. Note that in this method we still cannot distinguish one component of the transition from another; we merely create a situation in which the various components contribute to the mixture with unequal weights.

The second method consists of creating an unequal mixture of populations $p(m_i)$ by observing a previous radiation. Let us assume for simplicity that the level $I_i$ is populated by a transition from a state of spin $I_0 = 0$, so that there is a *cascade* $0 \rightarrow 1 \rightarrow 0$ of two radiations $\gamma_1$ and $\gamma_2$, as shown in Figure 10.4. Let's observe the first radiation in a certain direction, which we use to define the $z$ axis; the second radiation is observed in a direction that makes an angle $\theta_2$ with respect to the axis. With respect to the $z$ axis, the first radiation has the same angular distribution we discussed above; for $m_0 = 0$ to $m_i = 0$, the distribution is proportional to $\sin^2 \theta_1$ and for $m_0 = 0$ to $m_i = \pm 1$, to $\frac{1}{2}(1 + \cos^2 \theta_1)$. Since we *define* the $z$ axis by the direction of $\gamma_1$, it follows that $\theta_1 = 0$, and so the $0 \rightarrow 0$ transition cannot be emitted in that direction. That is, the nuclei for which $\gamma_2$ is observed following $\gamma_1$ must have a population of $p(m_i) = 0$ for $m_i = 0$. Thus the angular distribution of $\gamma_2$ relative to $\gamma_1$ is

$$W(\theta) \propto \tfrac{1}{2}\left[\tfrac{1}{2}(1 + \cos^2\theta)\right] + 0(\sin^2\theta) + \tfrac{1}{2}\left[\tfrac{1}{2}(1 + \cos^2\theta)\right] \propto 1 + \cos^2\theta$$

This type of experiment is called an *angular correlation*; again, we do not observe the individual components of $\gamma_2$, but we create an unequal *m*-state population distribution of the state $I_i$.

We have considered these examples of angular distribution measurements for the simplest case of pure dipole radiation. In general, the angular distribution or correlation of multipole radiation is of the form of a polynomial in even powers of $\cos\theta$:

$$W(\theta) = 1 + \sum_{k=1}^{L} a_{2k}\cos^{2k}\theta \tag{10.17}$$

where the coefficients $a_{2k}$ depend on $I_i$, $I_f$, and $L$, and also on whether we are doing a low-temperature angular distribution or an angular correlation experiment. For example, for the angular correlation $4^+ \overset{\gamma_1}{\to} 2^+ \overset{\gamma_2}{\to} 0^+$, where $\gamma_1$ and $\gamma_2$ are $E2$ radiation ($\gamma_2$ is pure $E2$ by the selection rules (Equation 10.16); $\gamma_1$ has a negligible mixture of $M3$ and higher multipoles), $a_2 = \tfrac{1}{8}$ and $a_4 = \tfrac{1}{24}$, while for

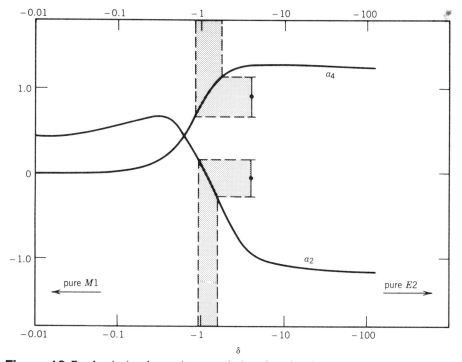

**Figure 10.5** Analysis of angular correlation data for the ratio of $E2$ to $M1$ matrix elements in a transition. The vertical error bars show the ranges of the experimentally determined $a_2$ and $a_4$, each of which gives a corresponding value for $\delta$. The $a_2$ and $a_4$ curves are derived from theory for this $2 \to 2 \to 0$ cascade in $^{110}$Cd. Data from K. S. Krane and R. M. Steffen, *Phys. Rev. C* **2**, 724 (1970).

$0^+ \to 2^+ \to 0^+$ the angular correlation is $a_2 = -3$, $a_4 = 4$. Returning to the original goal of this discussion, let us consider the angular correlation for $2^+ \xrightarrow{\gamma_1} 2^+ \xrightarrow{\gamma_2} 0^+$, where $\gamma_1$ is a mixture of $M1$ and $E2$ (neglecting the higher multipoles). The coefficients $a_2$ and $a_4$ depend on the relative amounts of $M1$ and $E2$ radiation; Figure 10.5 shows $a_2$ and $a_4$ as they vary with the parameter $\delta$, which essentially is $m_{\mathrm{fi}}(E2)/m_{\mathrm{fi}}(M1)$, where $m_{\mathrm{fi}}(\sigma L)$ is the transition matrix element defined by Equation 10.9. The fraction of $E2$ radiation is $\delta^2/(1 + \delta^2)$ and the fraction of $M1$ radiation is $1/(1 + \delta^2)$. As an example, the case of the 818–658-keV $2^+ \to 2^+ \to 0^+$ cascade in $^{110}$Cd has the measured values $a_2 = -0.06 \pm 0.22$, $a_4 = 0.89 \pm 0.24$. As shown in Figure 10.5, the deduced ratio of the multipole matrix elements is $\delta = -1.2 \pm 0.2$, corresponding to the 818-keV radiation being 59% $E2$ and 41% $M1$. This accurate knowledge of multipole character is extremely important in evaluating nuclear models and deducing partial lifetimes, as we discuss in Section 10.7. Thus angular distribution and correlation measurements have an extremely important role in nuclear spectroscopy.

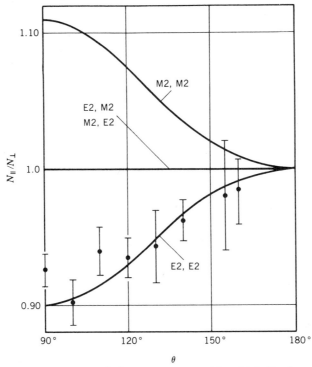

**Figure 10.6** An angular correlation measurement in which the linear polarization of the radiation is measured. The angle $\theta$ refers, as in Figures 10.4 and 7.43, to the angle between the two radiations. Data shown are for two transitions in the decay of $^{46}$Sc, obtained by F. Metzger and M. Deutsch, *Phys. Rev.* **78**, 551 (1950). Theoretical curves are drawn for various combinations of $E2$ and $M2$ radiations. The results indicate convincingly that both transitions must be of $E2$ character, consistent with the presently known $4^+ \to 2^+ \to 0^+$ level scheme.

To determine whether the radiation is electric or magnetic in character, additional measurements are necessary for in the angular distribution $a_2$ and $a_4$ are identical for $E$ and $M$ radiation. In Figure 10.1, you can see that the $E$ vector of the radiation field is parallel to the dipole axis for electric radiation but normal to it for magnetic radiation. The same characteristic carries into the complete quantum description, and we can distinguish $E$ from $M$ radiation by determining the directional relationship between the axis of the emitting nucleus, the direction of the emitted radiation, and its $E$ field. The plane formed by the radiation propagation direction $r$ and the field $E$ is called the *plane of polarization*. (Given $r$ and $E$, we can deduce $B$ because the radiation propagates in the direction $E \times B$. Choosing $r$ and $E$ to define the polarization is merely a convention and has no intrinsic significance.) As in the previous measurement, we must begin with an unequal distribution of $m$ states. (In the classical case, this would be equivalent to knowing the direction of the axis of the dipole since it is possible that a given direction for $E$ can correspond to an electric dipole in the $z$ direction or a magnetic dipole in the $y$ direction.)

This type of measurement is called a *linear polarization distribution* and is usually accomplished by taking advantage of the polarization dependence of Compton scattering (see Section 7.9 and Figure 7.43). Figure 10.6 shows an example of an angular correlation in which the linear polarization of $\gamma_2$ is observed. As before, the observation of the previous radiation $\gamma_1$ in effect provides the unequal $m$-state distribution, and we measure the linear polarization of $\gamma_2$ by the intensity of the Compton-scattered photons as a function of $\phi$.

## 10.6 INTERNAL CONVERSION

Internal conversion is an electromagnetic process that competes with $\gamma$ emission. In this case the electromagnetic multipole fields of the nucleus do not result in the emission of a photon; instead, the fields interact with the atomic electrons and cause one of the electrons to be emitted from the atom. In contrast to $\beta$ decay, the electron is not created in the decay process but rather is a previously existing electron in an atomic orbit. For this reason internal conversion decay rates can be altered slightly by changing the chemical environment of the atom, thus changing somewhat the atomic orbits. Keep in mind, however, that this is *not* a two-step process in which a photon is first emitted by the nucleus and then knocks loose an orbiting electron by a process analogous to the photoelectric effect; such a process would have a negligibly small probability to occur.

The transition energy $\Delta E$ appears in this case as the kinetic energy $T_e$ of the emitted electron, less the binding energy $B$ that must be supplied to knock the electron loose from its atomic shell:

$$T_e = \Delta E - B \qquad (10.18)$$

As we did in our discussion of nuclear binding energy, we take $B$ to be a positive number. The energy of a bound state is of course negative, and we regard the binding energy as that which we must supply to go from that state up to zero energy. Because the electron binding energy varies with the atomic orbital, for a

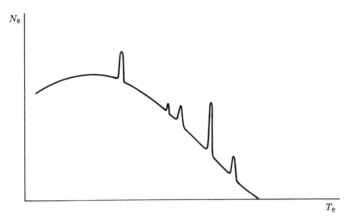

**Figure 10.7** A typical electron spectrum such as might be emitted from a radioactive nucleus. A few discrete conversion electron peaks ride on the continuous background from $\beta$ decay.

given transition $\Delta E$ there will be internal conversion electrons emitted with differing energies. The observed electron spectrum from a source with a single $\gamma$ emission thus consists of a number of individual components; these are discrete components, however, and not at all continuous like the electrons emitted in $\beta$ decay. Most radioactive sources will emit both $\beta$-decay and internal conversion electrons, and it is relatively easy to pick out the discrete conversion electron peaks riding on the continuous $\beta$ spectrum (Figure 10.7).

Equation 10.18 suggests that the internal conversion process has a threshold energy equal to the electron binding energy in a particular shell; as a result, the conversion electrons are labeled according to the electronic shell from which they come: K, L, M, and so on, corresponding to principal atomic quantum numbers $n = 1, 2, 3, \ldots$ . Furthermore, if we observe at very high resolution, we can even see the substructure corresponding to the individual electrons in the shell. For example, the L ($n = 2$) shell has the atomic orbitals $2s_{1/2}$, $2p_{1/2}$, and $2p_{3/2}$; electrons originating from these shells are called, respectively, $L_I$, $L_{II}$, and $L_{III}$ conversion electrons.

Following the conversion process, the atom is left with a vacancy in one of the electronic shells. This vacancy is filled very rapidly by electrons from higher shells, and thus we observe characteristic X-ray emission accompanying the conversion electrons. For this reason, when we study the $\gamma$ emission from a radioactive source we usually find X rays near the low-energy end of the spectrum.

As an illustration of the calculation of electron energies, consider the $\beta$ decay of $^{203}$Hg to $^{203}$Tl, following which a single $\gamma$ ray of energy 279.190 keV is emitted. To calculate the energies of the conversion electrons, we must look up the electron binding energies in the *daughter* Tl because it is from that atom that the electron emission takes place. (We will assume that the atomic shells have enough time to settle down between the $\beta$ emission and the subsequent $\gamma$ or conversion electron emission; this may not necessarily be true and will depend on the chemical environment and on the lifetime of the excited state.) Electron binding energies are conveniently tabulated in Appendix III of the *Table of*

*Isotopes.* For Tl, we find the following:

$$B(\mathrm{K}) = 85.529 \text{ keV}$$
$$B(\mathrm{L_I}) = 15.347 \text{ keV}$$
$$B(\mathrm{L_{II}}) = 14.698 \text{ keV}$$
$$B(\mathrm{L_{III}}) = 12.657 \text{ keV}$$
$$B(\mathrm{M_I}) = 3.704 \text{ keV}$$

and so on through the M, N, and O shells. We therefore expect to find conversion electrons emitted with the following energies:

$$T_e(\mathrm{K}) = 279.190 - 85.529 = 193.661 \text{ keV}$$
$$T_e(\mathrm{L_I}) = 279.190 - 15.347 = 263.843 \text{ keV}$$
$$T_e(\mathrm{L_{II}}) = 279.190 - 14.698 = 264.492 \text{ keV}$$
$$T_e(\mathrm{L_{III}}) = 279.190 - 12.657 = 266.533 \text{ keV}$$
$$T_e(\mathrm{M_I}) = 279.190 - 3.704 = 275.486 \text{ keV}$$

Figure 10.8 shows the electron spectrum for $^{203}$Hg. You can see the continuous $\beta$ spectrum as well as the electron lines at the energies we have calculated.

One feature that is immediately apparent is the varying intensities of the conversion electrons from the decay. This variation, as we shall see, depends on the multipole character of the radiation field; in fact, measuring the relative probabilities of conversion electron emission is one of the primary ways to determine multipole characters.

In some cases, internal conversion is heavily favored over $\gamma$ emission; in others it may be completely negligible compared with $\gamma$ emission. As a general rule, it is necessary to correct for internal conversion when calculating the probability for $\gamma$ emission. That is, if we know the half-life of a particular nuclear level, then the total decay probability $\lambda_t$ (equal to $0.693/t_{1/2}$) has two components, one ($\lambda_\gamma$) arising from $\gamma$ emission and another ($\lambda_e$) arising from internal conversion:

$$\lambda_t = \lambda_\gamma + \lambda_e \tag{10.19}$$

The level decays more rapidly through the combined process than it would if we considered $\gamma$ emission alone. It is (as we shall see) convenient to define the *internal conversion coefficient* $\alpha$ as

$$\alpha = \frac{\lambda_e}{\lambda_\gamma} \tag{10.20}$$

That is, $\alpha$ gives the probability of electron emission relative to $\gamma$ emission and ranges from very small ($\approx 0$) to very large. The total decay probability then becomes

$$\lambda_t = \lambda_\gamma(1 + \alpha) \tag{10.21}$$

We let $\alpha$ represent the *total* internal conversion coefficient and define *partial* coefficients representing the individual atomic shells:

$$\lambda_t = \lambda_\gamma + \lambda_{e,\mathrm{K}} + \lambda_{e,\mathrm{L}} + \lambda_{e,\mathrm{M}} + \cdots$$
$$= \lambda_\gamma(1 + \alpha_\mathrm{K} + \alpha_\mathrm{L} + \alpha_\mathrm{M} + \cdots) \tag{10.22}$$

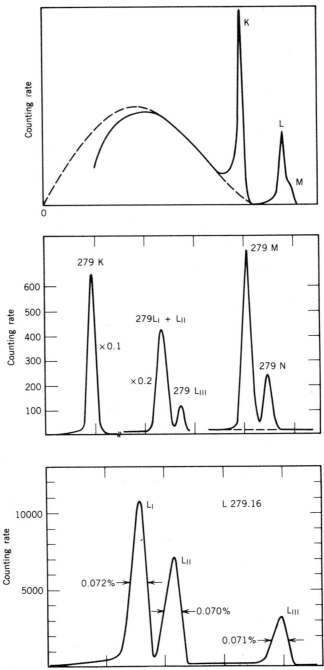

**Figure 10.8** Electron spectrum from the decay of $^{203}$Hg. At top, the continuous β spectrum can be seen, along with the K and unresolved L and M conversion lines. In the middle is shown the conversion spectrum at higher resolution; the L and M lines are now well separated, and even L$_{III}$ is resolved. At yet higher resolution (bottom) L$_I$ and L$_{II}$ are clearly separated. *Sources*: (top) A. H. Wapstra et al., *Physica* **20**, 169 (1954); (middle) Z. Sujkowski, *Ark. Fys.* **20**, 243 (1961); (bottom) C. J. Herrlander and R. L. Graham, *Nucl. Phys.* **58**, 544 (1964).

and thus

$$\alpha = \alpha_K + \alpha_L + \alpha_M + \cdots \tag{10.23}$$

Of course, considering the subshells, we could break these down further, such as

$$\alpha_L = \alpha_{L_I} + \alpha_{L_{II}} + \alpha_{L_{III}} \tag{10.24}$$

and similarly for other shells.

The calculation of internal conversion coefficients is a difficult process, beyond the level of the present text. Let's instead try to justify some of the general results and indicate the way in which the calculation differs from the similar calculation for $\gamma$ emission. Because the process is electromagnetic in origin, the matrix element that governs the process is quite similar to that of Equation 10.9 with two exceptions: the initial state includes a bound electron, so that $\psi_i = \psi_{i,N}\psi_{i,e}$ where N indicates the nuclear wave function and e indicates the electron wave function. Similarly, $\psi_f = \psi_{f,N}\psi_{f,e}$ where in this case $\psi_{f,e}$ is the free-particle wave function $e^{-i\mathbf{k}\cdot\mathbf{r}_e}$. To a very good approximation, the atomic wave function varies relatively little over the nucleus, and we can replace $\psi_{i,e}(\mathbf{r}_e)$ with its value at $r_e = 0$. The important detail, however, is that all of the specifically nuclear information is contained in $\psi_{i,N}$ and $\psi_{f,N}$, and that the same electromagnetic multipole operator $m(\sigma L)$ governs both $\gamma$ emission and internal conversion. The *nuclear* part of the matrix element of Equation 10.9 is therefore identical for both processes:

$$\lambda_\gamma(\sigma L) \propto |m_{fi}(\sigma L)|^2$$

$$\lambda_e(\sigma L) \propto |m_{fi}(\sigma L)|^2 \tag{10.25}$$

and thus the internal conversion coefficient $\alpha$, the ratio of $\lambda_e$ and $\lambda_\gamma$, is *independent* of the details of nuclear structure. The coefficient $\alpha$ will depend on the atomic number of the atom in which the process occurs, and on the energy of the transition and its multipolarity (hence, indirectly on the nuclear structure). We can therefore calculate and display general tables or graphs of $\alpha$ for different $Z$, $T_e$, and $L$.

We are oversimplifying here just a bit because the electron wave function $\psi_{i,e}$ does penetrate the nucleus and does sample the specific nuclear wave function, but it has a very slight, usually negligible, effect on the conversion coefficient.

A nonrelativistic calculation gives the following instructive results for electric ($E$) and magnetic ($M$) multipoles:

$$\alpha(EL) \cong \frac{Z^3}{n^3}\left(\frac{L}{L+1}\right)\left(\frac{e^2}{4\pi\epsilon_0 \hbar c}\right)^4 \left(\frac{2m_e c^2}{E}\right)^{L+5/2} \tag{10.26}$$

$$\alpha(ML) \cong \frac{Z^3}{n^3}\left(\frac{e^2}{4\pi\epsilon_0 \hbar c}\right)^4 \left(\frac{2m_e c^2}{E}\right)^{L+3/2} \tag{10.27}$$

In these expressions, $Z$ is the atomic number of the atom in which the conversion takes place (the daughter, in the case of transitions following $\beta$ decay) and $n$ is the principal quantum number of the bound electron wave function; the factor $(Z/n)^3$ comes from the term $|\psi_{i,e}(0)|^2$ that appears in the conversion rate (the hydrogenic wave functions of Table 2.5 show the factor $(Z/n)^{3/2}$ in the normalization constant). The dimensionless factor $(e^2/4\pi\epsilon_0 \hbar c)$ is the fine structure constant with a value close to $\frac{1}{137}$.

These expressions for the conversion coefficients are only approximate, for the electron must be treated relativistically (transition energies are typically 0.5–1 MeV, so it is *not* true that $T_e \ll m_e c^2$) and the "point" nucleus Coulomb wave functions such as those of Table 2.5 do not properly take into account the important effects that occur when the electron penetrates the nucleus. Tabulations of conversion coefficients based on more rigorous calculations are listed at the end of the chapter. The approximate expressions do, however, illustrate a number of features of the conversion coefficients:

1. They increase as $Z^3$, and so the conversion process is more important for heavy nuclei than for light nuclei. For example, the 1.27-MeV $E2$ transition in $^{22}_{10}$Ne has $\alpha_K = 6.8 \times 10^{-6}$ and the 1.22-MeV $E2$ transition in $^{182}_{74}$W has $\alpha_K = 2.5 \times 10^{-3}$; their ratio is very nearly equal to $(10/74)^3$, as expected.

2. The conversion coefficient decreases rapidly with increasing transition energy. (In contrast, the probability for $\gamma$ emission increases rapidly with energy.) For example, in $^{56}$Co there are three $M1$ transitions, with energies 158 keV ($\alpha_K = 0.011$), 270 keV ($\alpha_K = 0.0034$), and 812 keV ($\alpha_K = 0.00025$). These decrease approximately as $E^{-2.5}$, as expected based on Equation 10.27.

3. The conversion coefficients increase rapidly as the multipole order increases; in fact, for the higher $L$ values, conversion electron emission may be far more probable than $\gamma$ emission. For example, in $^{99}$Tc there is an $M1$ transition of 141 keV with $\alpha_K = 0.10$, while an $M4$ transition of 143 keV has $\alpha_K = 30$. Based on Equation 10.27 we would expect the ratio $\alpha_K(143)/\alpha_K(141)$ to be about $(2m_e c^2/E)^3$, or about 370, which is quite consistent with the observed ratio (about 300).

4. The conversion coefficients for higher atomic shells ($n > 1$) decrease like $1/n^3$. Thus for a given transition, we might roughly expect $\alpha_K/\alpha_L \simeq 8$. Using the correct electronic wave functions will cause this estimate to vary considerably, but many experimental values of $\alpha_K/\alpha_L$ do fall in the range of 3–6, so even in this case our estimate serves us well.

We therefore expect relatively large K-shell conversion coefficients for low-energy, high-multipolarity transitions in heavy nuclei, with smaller values in other cases (higher atomic shells, higher transition energy, lighter nuclei, lower multipoles).

While these estimates give us reasonable qualitative values, for quantitative comparisons with experimental results, we must do detailed computations of the conversion coefficients, using the proper atomic wave functions. Figure 10.9 shows some results of such calculations. Notice that the coefficients differ considerably for $EL$ and $ML$ transitions; thus measurement of $\alpha$ allows us to determine the relative parities of nuclear states.

There is one other application for which internal conversion is an essential tool —the observation of $E0$ transitions, which are forbidden to go by electromagnetic radiation because the nuclear monopole moment (that is, its charge) cannot radiate to points external to the nucleus. The $E0$ transition is particularly important in decays from $0^+$ initial states to $0^+$ final states, which cannot occur

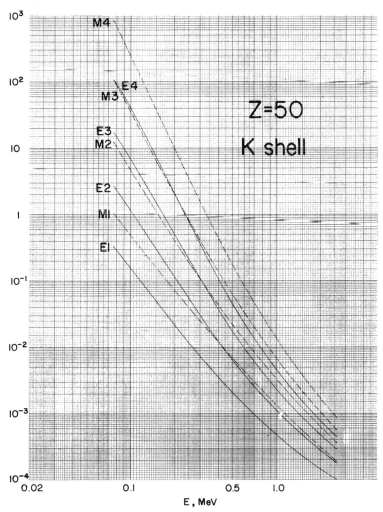

**Figure 10.9**  K-shell internal conversion coefficients for $Z = 50$. From *Table of Isotopes*, edited by C. M. Lederer and V. S. Shirley (New York: Wiley, 1978).

by any other direct process. We can regard the nucleus in this case as a spherically symmetric ball of charge; the only possible motion is a pulsation, which does not alter the electric field at points external to the sphere and thus produces no radiation. Electronic orbits that do not vanish near $r = 0$ (that is, s states) can sample the varying potential *within* the pulsating nucleus, and so a transfer of energy to the electron is possible.

Because no $\gamma$ rays are emitted, it is not possible to define a conversion coefficient ($\alpha$ is infinite for $\lambda_\gamma = 0$). We can illustrate a particular case in which the decay occurs by the $^{72}$Ge level scheme shown in Figure 10.10. The excited $0^+$ state decays to the ground state by $E0$ conversion with a half-life of 0.42 $\mu$s. The nearby $2^+$ state can decay by $\gamma$ emission much more rapidly; the internal conversion coefficient for that decay is only $4.9 \times 10^{-4}$. Comparison of these rates will involve evaluating the nuclear matrix elements (because we cannot take

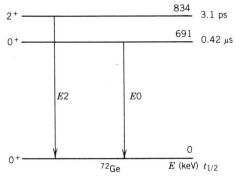

**Figure 10.10** Energy levels in $^{72}$Ge.

the ratio $\lambda_e/\lambda_\gamma$ for the $E0$ transition, its matrix elements do not cancel), but may reveal to us something about the internal structure of the excited $0^+$ state.

## 10.7 LIFETIMES FOR γ EMISSION

In Chapter 7, we discussed techniques for measuring the half-lives of excited states. One primary usefulness of these experimental values is for comparing with theoretical values calculated on the basis of different models of the nucleus. If we compare various calculated values with the experimental one, we can often draw some conclusions about the structure of the nucleus.

Before we do this, we must first evaluate the *partial* decay rate for γ emission, as we did at the end of Section 6.5. Let's consider the example of the decay shown in Figure 10.11. The half-life of the 1317-keV level has been measured to be 8.7 ps. Its total decay constant is therefore

$$\lambda_t = \frac{0.693}{t_{1/2}} = \frac{0.693}{8.7 \times 10^{-12}\ \text{s}} = 8.0 \times 10^{10}\ \text{s}^{-1}$$

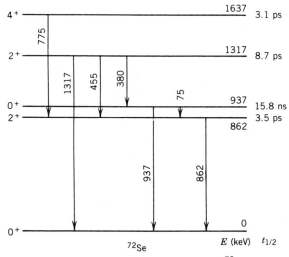

**Figure 10.11** Energy levels in $^{72}$Se.

This decay rate is just the sum of the decay rates of the three transitions that depopulate the state:

$$\lambda_t = \lambda_{t,1317} + \lambda_{t,455} + \lambda_{t,380}$$
$$= \lambda_{\gamma,1317}(1 + \alpha_{1317}) + \lambda_{\gamma,455}(1 + \alpha_{455}) + \lambda_{\gamma,380}(1 + \alpha_{380})$$

The conversion coefficients can be found in standard reference works (they have not been measured in this case) and are sufficiently small ($< 0.01$) that they can be neglected (compared with 1) to the precision of this calculation. Thus

$$\lambda_t = \lambda_{\gamma,1317} + \lambda_{\gamma,455} + \lambda_{\gamma,380}$$

The relative intensities of the three $\gamma$ rays have been measured to be

$$\lambda_{\gamma,1317} : \lambda_{\gamma,455} : \lambda_{\gamma,380} = 51 : 39 : 10$$

and so the partial decay rates for the three $\gamma$ rays are

$$\lambda_{\gamma,1317} = 0.51(8.0 \times 10^{10} \text{ s}^{-1}) = 4.1 \times 10^{10} \text{ s}^{-1}$$
$$\lambda_{\gamma,455} = 0.39(8.0 \times 10^{10} \text{ s}^{-1}) = 3.1 \times 10^{10} \text{ s}^{-1}$$
$$\lambda_{\gamma,380} = 0.10(8.0 \times 10^{10} \text{ s}^{-1}) = 0.80 \times 10^{10} \text{ s}^{-1}$$

It is these *partial rates for $\gamma$ emission* that can be compared with calculated values, such as the Weisskopf estimates of Equation 10.13. Let us calculate the expected values of $\lambda(E2)$:

$$\lambda_{E2,1317} = 8.7 \times 10^{10} \text{ s}^{-1}$$
$$\lambda_{E2,455} = 4.3 \times 10^{8} \text{ s}^{-1}$$
$$\lambda_{E2,380} = 1.7 \times 10^{8} \text{ s}^{-1}$$

The case of the 937-keV level ($t_{1/2} = 15.8$ ns) can be handled similarly:

$$\lambda_t = \frac{0.693}{15.8 \text{ ns}} = 4.39 \times 10^{7} \text{ s}^{-1}$$
$$\lambda_t = \lambda_{t,937} + \lambda_{t,75}$$
$$= \lambda_{e,937} + \lambda_{\gamma,75}(1 + \alpha_{75})$$

because the 937-keV transition is of the $0 \rightarrow 0$, $E0$ type we discussed in the previous section. The total conversion coefficient of the 75-keV transition is about 2.4 (from tables or graphs). Experimentally, it is known that $\lambda_{\gamma,75} : \lambda_{e,937} = 73 : 27$, and thus

$$\lambda_{e,937} = 4.3 \times 10^{6} \text{ s}^{-1}$$
$$\lambda_{\gamma,75} = 1.16 \times 10^{\prime} \text{ s}^{-1}$$

Finally, for the 862-keV transition, we find

$$\lambda_{\gamma,862} = 2.0 \times 10^{11} \text{ s}^{-1}$$

From the Weisskopf estimates, we would compute

$$\lambda_{E2,75} = 5.2 \times 10^{4} \text{ s}^{-1}$$
$$\lambda_{E2,862} = 1.0 \times 10^{10} \text{ s}^{-1}$$

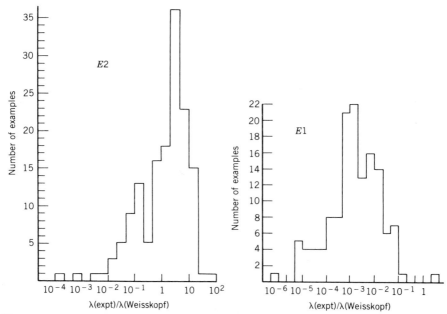

**Figure 10.12** Systematics of $E2$ and $E1$ transitions. The figures show the number of cases versus the ratio between the observed decay rate and the value calculated from the Weisskopf formulas. From S. J. Skorka et al., *Nucl. Data* **2**, 347 (1966).

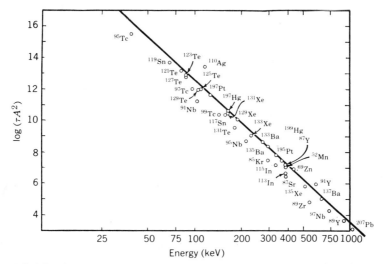

**Figure 10.13** Systematics of $M4$ transitions. The data are plotted in terms of the mean life (the reciprocal of the decay constant $\lambda$). The straight line is determined from Equation 10.15. Note especially the good agreement between the data points and the expected $E^{-9}$ dependence. From M. Goldhaber and A. W. Sunyar, *Phys. Rev.* **83**, 906 (1951).

One feature that generally emerges from these calculations is that the measured transition rates are frequently at least an order of magnitude larger than the Weisskopf estimates for $E2$ transitions. This is strong evidence for the *collective* aspects of nuclear structure discussed in Chapter 5—the Weisskopf estimates are based on the assumption that the transition arises from the motion of a *single nucleon*, and the fact that these are too small indicates that *many* nucleons must be taking part in the transition. Figure 10.12 summarizes similar results for many $E2$ transitions, and you can see that this enhancement or acceleration of the single-particle $E2$ rate is quite a common feature. No such effect occurs for $E1$ transitions, which are generally slower than the single-particle rates. On the other hand, consider Figure 10.13 which shows the systematic behavior of $M4$ transitions. Here the agreement between theory and experiment is excellent.

## 10.8 GAMMA-RAY SPECTROSCOPY

The study of the $\gamma$ radiations emitted by radioactive sources is one of the primary means to learn about the structure of excited nuclear states. Gamma-ray detection is relatively easy to accomplish, and can be done at high resolution (transitions a mere 2 keV apart can be cleanly separated by good $\gamma$ detectors) and with high precision (uncertainties of a few eV in typical cases and more than an order of magnitude better in the best cases). Knowledge of the locations and properties of the excited states is essential for the evaluation of calculations based on any nuclear model, and $\gamma$-ray spectroscopy is the most direct, precise, and often the easiest way to obtain that information.

Let's consider how the "ideal" $\gamma$-ray experiments might proceed to provide us with the information we need about nuclear excited states:

1. A spectrum of the $\gamma$ rays shows us the energies and intensities of the transitions.
2. Coincidence measurements give us clues about how these transitions might be arranged among the excited states.
3. Measuring internal conversion coefficients can give clues about the character of the radiation and the relative spins and parities of the initial and final states. Additional clues may come from angular distribution and correlation measurements.
4. Absolute transition probabilities can be found by determining the half-lives of the levels.

As a first example, we consider the decay of $^{108m}$Ag ($t_{1/2} = 127$ y). Figure 10.14 shows the conversion electron and $\gamma$-ray spectra, and Table 10.1 shows the deduced relative $\gamma$ and electron intensities.

The first excited state of $^{108}$Pd is known from many nuclear reaction studies to be at an energy of 434 keV and to be a $2^+$ state, as in nearly all even-$Z$, even-$N$ nuclei. We therefore recognize the 434-keV transition shown in Table 10.1 as representing the decay of this state, and we assume the transition to be a pure $E2$, $2^+ \rightarrow 0^+$ transition. From conversion coefficient tables, we can determine that the theoretical conversion coefficients are $\alpha_K(434) = 7.89 \times 10^{-3}$, $\alpha_{L+M+\cdots} = 1.0 \times 10^{-3}$. Given the intensities from Table 10.1, the remaining conversion

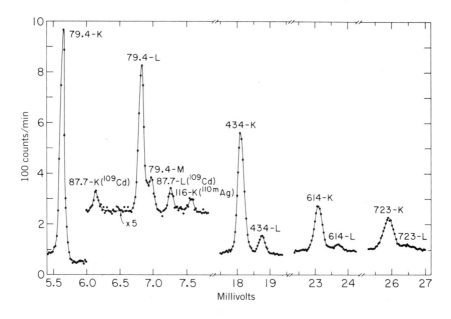

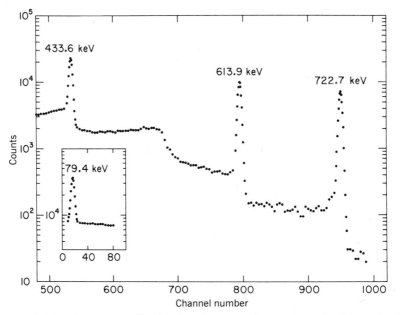

**Figure 10.14**   Gamma-ray (bottom) and conversion electron (top) spectra following the $^{108m}$Ag decay. The $\gamma$ spectrum was obtained with a Ge(Li) detector. The electron spectrum, obtained with a magnetic spectrometer, shows the high resolution necessary to separate the K and L lines. From O. C. Kistner and A. W. Sunyar, *Phys. Rev.* **143**, 918 (1966).

**Table 10.1** Conversion Electron and $\gamma$-Ray Intensities of Transitions Following the Decay of $^{108m}$Ag

| Transition Energy (keV) | Relative $\gamma$ Intensity | Relative Electron Intensity | Conversion Coefficient (units of $10^{-3}$) | |
|---|---|---|---|---|
| | | | Experimental | Theoretical |
| 79.2 | $7.3 \pm 0.8$ | $204 \pm 10$ (K) | $220 \pm 26$ | $270\ (E1), 710\ (M1), 2400\ (E2)$ |
| | | $25 \pm 2$ (L) | $27 \pm 4$ | $33\ (E1), 88\ (M1), 777\ (E2)$ |
| 434.0 | $\equiv 100$ | $\equiv 100$ (K) | $\equiv 7.89$ | $7.89\ (E2)$ |
| | | $14.8 \pm 2.3$ (L + $\cdots$) | $1.17 \pm 0.18$ | $1.02\ (E2)$ |
| 614.4 | $103 \pm 3$ | $37 \pm 3$ (K) | $2.83 \pm 0.24$ | $1.03\ (E1), 3.01\ (M1), 2.92\ (E2)$ |
| | | $5.1 \pm 1.6$ (L + $\cdots$) | $0.39 \pm 0.12$ | $0.12\ (E1), 0.35\ (M1), 0.36\ (E2)$ |
| 632.9 | $0.16 \pm 0.02$ | | | |
| 723.0 | $102 \pm 3$ | $25.0 \pm 1.2$ (K) | $1.93 \pm 0.11$ | $0.72\ (E1), 2.06\ (M1), 1.91\ (E2)$ |
| | | $4.6 \pm 0.8$ (L + $\cdots$) | $0.35 \pm 0.06$ | $0.08\ (E1), 0.24\ (M1), 0.23\ (E2)$ |

*Source*: Experimental data from O. C. Kistner and A. W. Sunyar, *Phys. Rev.* **143**, 918 (1966).

coefficients can be computed in the following way:

$$\alpha_K(434) = \frac{I_K(434)}{I_\gamma(434)}$$

$$\alpha_K(614) = \frac{I_K(614)}{I_\gamma(614)}$$

where $I$ represents the tabulated intensities.

Since we have *relative*, rather than *absolute*, intensities, we form the ratio

$$\frac{\alpha_K(614)}{\alpha_K(434)} = \frac{I_K(614)}{I_K(434)} \cdot \frac{I_\gamma(434)}{I_\gamma(614)}$$

$$\alpha_K(614) = \alpha_K(434) \cdot \frac{I_K(614)}{I_K(434)} \cdot \frac{I_\gamma(434)}{I_\gamma(614)}$$

$$= 7.89 \times 10^{-3} \cdot \frac{37 \pm 3}{100} \cdot \frac{100}{103 \pm 3}$$

$$= (2.83 \pm 0.24) \times 10^{-3}$$

From a similar procedure, the remaining conversion coefficients can be calculated, as listed in Table 10.1, and from tabulated values, we can find the theoretical values also listed in the table.

The 614- and 723-keV transitions are either $M1$ or $E2$ character, but from the conversion coefficients we cannot decide which is correct because it happens (for this atomic number and $\gamma$-ray energy range) the $M1$ and $E2$ values are nearly equal.

Further information can be obtained from coincidence experiments, illustrated in Figure 10.15. The low-energy end of the electron spectrum shows another

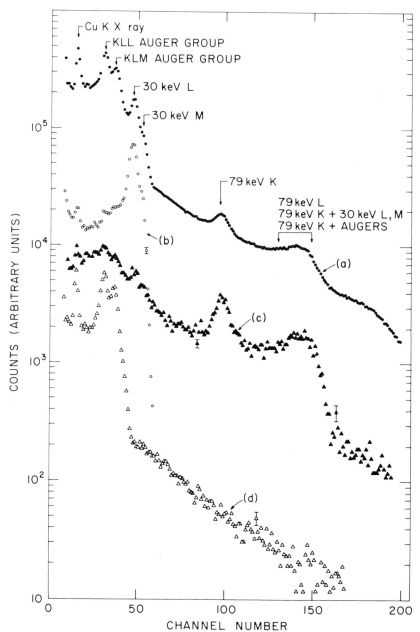

**Figure 10.15** Coincidence spectra in the $^{108m}$Ag decay. (*a*) Singles spectrum for comparison. (*b*) Spectrum in coincidence with 79-keV γ ray. (*c*) Spectrum in coincidence with K X rays. (*d*) Spectrum in coincidence with 434-, 614-, and 723-keV γ's. From O. C. Kistner and A. W. Sunyar, *Phys. Rev.* **143**, 918 (1966).

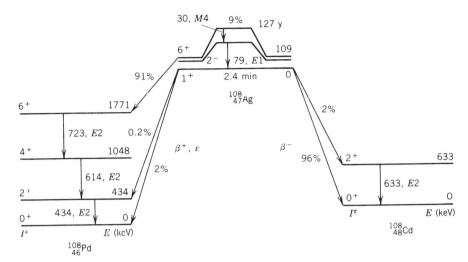

**Figure 10.16** Decay scheme of $^{108}$Ag. The 127-y isomer decays 91% by $\beta$ decay and 9% through $\gamma$ rays. The branching ratios of the $\beta$ decay of the 2.4-min ground state are also shown.

transition, of energy 30.4 keV; its $\gamma$ intensity is too low to be observed, but from the relative subshell intensities $L_I/(L_{II} + L_{III})$ it is deduced to be of $M4$ character. Neither this transition nor the 79.2-keV transition are in coincidence with the 434-, 614-, or 723-keV transitions. This strongly suggests that the 30.4- and 79.2-keV transitions take place in $^{108}$Ag following the decay of the isomer. Furthermore, the coincidence relationship among the 434-, 614-, and 723-keV transitions suggests that they follow one another in cascade.

The resulting decay scheme deduced on the basis of these (and other) data is shown in Figure 10.16. The spins of the 1048- and 1771-keV levels are deduced from angular correlation and linear polarization measurements, and the transitions are then of $E2$ multipolarity. The 633-keV transition is deduced to take place in $^{108}$Cd, and must follow the decay of the $^{108}$Ag ground state. (Otherwise, the $\beta$ decay would be $6^+ \rightarrow 2^+$, a fourth-forbidden process not likely to be observed.)

The 127-y isomeric state decays 91% by an allowed $\beta$ decay to the $6^+$ state in $^{108}$Pd, which then decays through the cascade of three transitions. Because there is negligible $\beta$ feeding of the $4^+$ or $2^+$ levels (the $2^+$ level is fed by only 0.2% of 9% of the decays), the three transitions should all have the same total ($\gamma$ plus electron) intensity, as can be confirmed from the data of Table 10.1. The $^{108}$Ag ground state decays through four different allowed $\beta$ branches, with a half-life of 2.4 min.

As a second example, we consider the decay of isomeric $^{180m}$Hf. A $\gamma$-ray spectrum is shown in Figure 10.17, and the deduced $\gamma$ intensities and conversion coefficients are shown in Table 10.2. The conversion coefficients of the 93-, 215-, 332-, and 443-keV transitions strongly suggest $E2$ multipolarity, while for 57- and 501-keV, they suggest $E1$ and $E3$, respectively.

Since $^{180}$Hf is an even-$Z$, even-$N$ nucleus, we expect the first excited state to be a $2^+$ state, and such a state has been identified at 93 keV from a variety of experiments. In this region, we expect the sequence of $0^+, 2^+, 4^+, 6^+, \ldots$ rota-

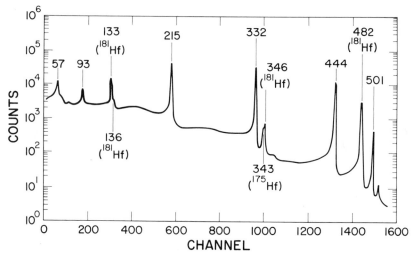

**Figure 10.17** Gamma ray spectrum from the decay of $^{180m}$Hf. The radioactive sample was made by neutron activation, which also produced other isotopes of Hf; γ rays from these other isotopes are labeled.

tional states with characteristic $I(I + 1)$ spacing, and so we expect to find a $4^+$ state at an energy of $\frac{10}{3}$ of the $2^+$ energy, or 310 keV, with a $4^+$–$2^+$ transition of energy $310 - 93 = 217$ keV. This is presumably the 215-keV $E2$ transition seen in the γ spectrum. The $6^+$ state is expected at 651 keV, and the $6^+$–$4^+$ transition should be of energy $651 - 310 = 341$ keV, which we identify with the observed 332-keV transition. The $8^+$ energy is expected to be 1116 keV, and the $8^+$–$6^+$ transition (1116 − 651 = 465 kcV) we guess to be the observed 443 keV. To place the 57- and 501-keV transitions, we notice the energy sum $57.5 + 443.2 = 500.7$, which strongly suggests that a level 57.5 keV above the $8^+$ level emits both the 57.5- and 500.7-keV transitions, with the 500.7-keV transition proceeding directly to the $6^+$ state in parallel with the cascade $57.5 + 443.2$ keV. To check this assumption, we examine the total intensities of the transitions, which we find, according to Equation 10.21, by $I_T = I_\gamma(1 + \alpha)$, where $\alpha$ is the *total* conversion coefficient (the sum of the individual coefficients). From the data of Table 10.2, the following relative intensities can be computed:

$$I_T(57.5) = 87.1 \pm 2.9 \qquad I_T(332.3) = 105.3 \pm 1.1$$

$$I_T(93.3) = 108.1 \pm 4.8 \qquad I_T(443.2) = 89.7 \pm 1.6$$

$$I_T(215.3) = 103.4 \pm 1.0 \qquad I_T(500.7) = 16.2 \pm 0.4$$

Note that, within the uncertainties, $I_T(93.3) = I_T(215.3) = I_T(332.3)$, as expected for transitions emitted in a cascade, one following the other. The 443.2 intensity is smaller, suggesting an alternative branch to the $6^+$ state other than through the 443.2-keV, $8^+$–$6^+$ transition, consistent with the assumption that the 500.7-keV transition goes to the $6^+$ state. Finally, note that $I_T(57.5) = I_T(443.2)$, suggesting they are in cascade, and that $I_T(57.5) + I_T(500.7) = I_T(332.3)$, suggesting the two branches that feed the $6^+$ level, one by the 57.5-keV transition and another by the 500.7-keV transition.

**Table 10.2**  Gamma Decays of $^{180m}$Hf

| Transition Energy (keV) | Relative γ Intensity | Conversion Coefficient | |
|---|---|---|---|
| | | Experiment (atomic shell) | Theory (multipole) |
| 57.5 | 51.3 ± 1.2 | 0.430 ± 0.029 (L) | 0.452 ($E1$) |
| | | 0.083 ± 0.007 (M) | 0.105 ($E1$) |
| | | 0.023 ± 0.004 (N + ··· ) | 0.031 ($E1$) |
| 93.3 | 17.6 ± 0.4 | ≡ 1.10 (K) | 1.10 ($E2$) |
| | | 3.13 ± 0.19 (L) | 2.72 ($E2$) |
| | | 0.91 ± 0.11 (M + ··· ) | 0.85 ($E2$) |
| 215.3 | 86.2 ± 0.8 | 0.123 ± 0.009 (K) | 0.14 ($E2$) |
| | | 0.077 ± 0.011 (L) | 0.071 ($E2$) |
| 332.3 | ≡ 100.0 ± 1.0 | 0.038 ± 0.003 (K) | 0.042 ($E2$) |
| | | 0.015 ± 0.002 (L) | 0.013 ($E2$) |
| 443.2 | 87.7 ± 1.6 | 0.0189 ± 0.0017 (K) | 0.020 ($E2$) |
| | | 0.0044 ± 0.0007 (L) | 0.005 ($E2$) |
| 500.7 | 15.4 ± 0.4 | 0.037 ± 0.012 (K) | 0.124 ($M2$), 0.038 ($E3$) |
| | | 0.016 ± 0.005 (L) | 0.062 ($M2$), 0.016 ($E3$) |

*Source*: Data for this table came from various sources compiled in the *Nuclear Data Sheets* **15**, 559 (1975).

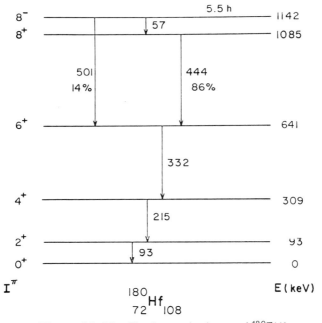

**Figure 10.18**   The isomeric decay of $^{180m}$Hf.

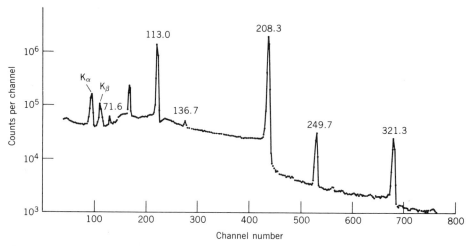

**Figure 10.19** Gamma-ray spectrum from the decay of $^{177}$Lu. The peak at about 90 keV is from an impurity activity produced in this neutron-activated sample. (Data from the author's laboratory.)

The parity of the level that emits the 57.5- and 500.7-keV transitions must be negative (because of their $E1$ and $E3$ characters, respectively), and its spin must be 7, 8, or 9 to permit decay to $8^+$ via the $E1$ transition. We can eliminate the $7^-$ possibility because it would allow the 500.7-keV transition to $6^+$ to be of $E1$ character, and $9^-$ can be eliminated by angular correlation experiments, which show the 500.7-keV transition to have a small ($\approx 3\%$) $M2$ component, which the conversion coefficients are not sensitive enough to reveal.

The resulting decay scheme is shown in Figure 10.18. Of course, many additional measurements are required to confirm that our guesses based on the $\gamma$

**Table 10.3** $\gamma$ Transitions in the Decay of $^{177}$Lu

| Transition Energy (keV) | Relative $\gamma$ Intensity | Conversion Coefficient | |
|---|---|---|---|
| | | Experiment (atomic shell) | Theory (multipole) |
| 71.6 | $2.4 \pm 0.1$ | $0.90 \pm 0.11$ (K) | $0.72$ ($E1$), $92.0$ ($M2$) |
| | | $0.087 \pm 0.010$ (L$_{\mathrm{I}}$) | $0.071$ ($E1$), $24.3$ ($M2$) |
| 113.0 | $\equiv 100$ | $0.78 \pm 0.05$ (K) | $2.4$ ($M1$), $0.74$ ($E2$) |
| | | $0.091 \pm 0.011$ (L$_{\mathrm{I}}$) | $0.345$ ($M1$), $0.076$ ($E2$) |
| 136.7 | $0.92 \pm 0.06$ | $0.49 \pm 0.08$ (K) | $1.45$ ($M1$), $0.48$ ($E2$) |
| | | $0.08 \pm 0.02$ (L$_{\mathrm{I}}$) | $0.20$ ($M1$), $0.05$ ($E2$) |
| 208.3 | $164 \pm 10$ | $0.046 \pm 0.004$ (K) | $0.046$ ($E1$), $2.05$ ($M2$) |
| | | $0.0063 \pm 0.0006$ (L$_{\mathrm{I}}$) | $0.0053$ ($E1$), $0.38$ ($M2$) |
| 249.7 | $3.0 \pm 0.2$ | $0.101 \pm 0.009$ (K) | $0.26$ ($M1$), $0.094$ ($E2$) |
| 321.3 | $3.6 \pm 0.2$ | $0.102 \pm 0.013$ (K) | $0.0155$ ($E1$), $0.49$ ($M2$) |
| | | $0.017 \pm 0.002$ (L$_{\mathrm{I}}$) | $0.0018$ ($E1$), $0.081$ ($M2$) |

*Sources*: Gamma-ray data from A. J. Haverfield et al., *Nucl. Phys. A* **94**, 337 (1967). Electron conversion data from A. P. Agnihotry et al., *Phys. Rev. C* **9**, 336 (1974).

decays are in fact correct. The 5.5-h half-life is one of the longest known for γ-emitting isomers.

For our final example, we consider the 6.7-day decay of $^{177}$Lu. The γ-ray spectrum is shown in Figure 10.19 and the γ and conversion data are listed in Table 10.3. The β decays of $^{177}$Lu to $^{177}$Hf were discussed in Section 9.10; there are four decays, leading to the ground state and to excited states at 113, 250, and 321 keV. Transitions from each of these states to the ground state can be seen among those listed in Table 10.3, and we therefore assume the excited states to be at energies of 113.0, 249.7, and 321.3 keV. (These γ energies are determined to about ±0.1 keV, while the β energies are determined only to ±2-3 keV.) We also find in the transitions of Table 10.3 all possible transitions connecting the excited states:

$$321.3 - 249.7 \text{ keV} = 71.6 \text{ keV}$$

$$321.3 - 113.0 \text{ keV} = 208.3 \text{ keV}$$

$$249.7 - 113.0 \text{ keV} = 136.7 \text{ keV}$$

The ground state and first two excited states of $^{177}$Hf are known to be the lowest states of a rotational band with spins $\frac{7}{2}^-, \frac{9}{2}^-, \frac{11}{2}^-, \ldots$. We would expect the 113.0-keV transition to be $\frac{9}{2}^- \rightarrow \frac{7}{2}^-$ and thus of $M1 + E2$ character, the 136.7-keV transition to be $\frac{11}{2}^- \rightarrow \frac{9}{2}^-$ and likewise $M1 + E2$, and the 249.7-keV transition to be $\frac{11}{2}^- \rightarrow \frac{7}{2}^-$ $E2$. These expectations are consistent with the conversion coefficients. The 321.3 level decays to all three lower levels through $E1 + M2$ transitions, and thus only a $\frac{9}{2}^+$ assignment is possible.

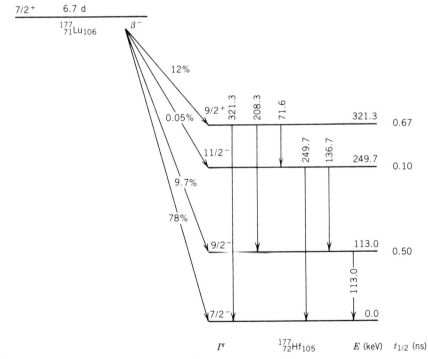

**Figure 10.20** Decay scheme of $^{177}$Lu to $^{177}$Hf. The β branching intensities are deduced indirectly from the γ-ray intensities and differ from those of Figure 9.29.

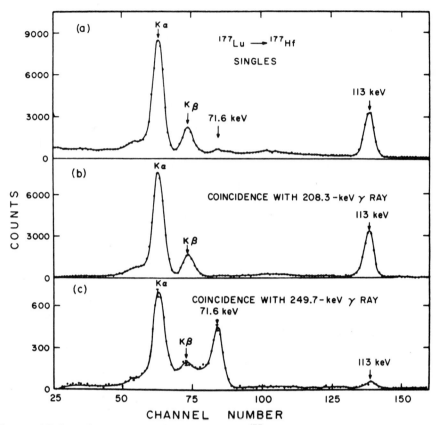

**Figure 10.21** Coincidence spectra from the $^{177}$Lu decay. Note the strong coincidence between 208 and 113; also note how the 71.6-keV peak, which is barely observable in the singles spectrum at the top, appears prominently in coincidence with 249.7 keV. (The small apparent 113.0-keV peak in the 249.7 keV coincidence spectrum is an artifact.) Data from A. P. Agnihotry et al., *Phys. Rev. C* **9**, 336 (1974).

The postulated decay scheme is shown in Figure 10.20. To verify our assumptions about the placement of the transitions, we can do γ-ray coincidence measurements, an example of which is shown in Figure 10.21. The coincidence spectrum shows quite plainly the 113.0–208.3 and 71.6–249.7 coincidence relationships.

As a final note on this decay, we point out that the L-subshell ratios can determine the relative $M1$ and $E2$ components of the 113.0-keV transition. Figure 10.22 shows the dependence of the $L_I/L_{III}$ and $L_I/L_{II}$ ratios on the $E2$ component. These very precise data show that the transition is 94–95% $E2$ and thus only 5–6% $M1$. In this case, the nuclear wave functions enhance the $E2$ transition probability to such an extent that $E2$ dominates over $M1$.

In this section we have given some examples of decay schemes that can be elucidated through γ-ray and conversion electron spectroscopy. It is of course impossible to isolate any one technique from all others in determining the properties of nuclear states, but from the examples discussed here you should

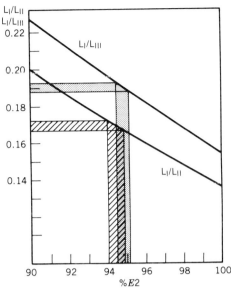

**Figure 10.22** Determination of the $E2$ fraction of the 113.0-keV transition in $^{177}$Hf from L-subshell data. The horizontal bars give the ranges of the experimental L-subshell ratios, and the vertical bars give the corresponding $E2$ fractions. The transition is deduced to be 94–95% $E2$ (and thus 5–6% $M1$). Data from S. Högberg et al., Z. Phys. **254**, 89 (1972).

appreciate the detailed and precise information that can be obtained from these methods.

## 10.9 NUCLEAR RESONANCE FLUORESCENCE AND THE MÖSSBAUER EFFECT

The inverse process of $\gamma$-ray emission is $\gamma$-ray absorption—a nucleus in its ground state absorbs a photon of energy $E_\gamma$ and jumps to an excited state at an energy $\Delta E$ above the ground state. The relationship between $E_\gamma$ and $\Delta E$ follows from a procedure similar to that used to obtain Equation 10.4:

$$\Delta E = E_\gamma - \frac{E_\gamma^2}{2Mc^2} \tag{10.28}$$

if we assume the absorbing nucleus to be initially at rest. The difference between $\Delta E$ and $E_\gamma$ comes about because of the recoil of the nucleus after absorbing the photon.

Let us assume we have a source of $\gamma$ radiation of continuously variable energy. The cross section for resonant absorption of a photon is

$$\sigma_0 = 2\pi \left(\frac{\hbar c}{E_\gamma}\right)^2 \frac{2I_e + 1}{2I_g + 1}\frac{1}{1 + \alpha} \tag{10.29}$$

where $\alpha$ is the total conversion coefficient and $I_g$ and $I_e$ are the spin quantum

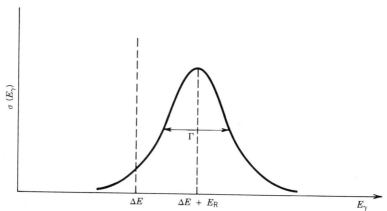

**Figure 10.23** The resonant absorption cross section of Equation 10.30. The recoil energy $E_R$ shifts the resonance slightly from the value $E_\gamma = \Delta E$ expected in the absence of recoil.

numbers of the ground and excited states. Because the energy of the excited state is not "sharp," the absorption will take place even when the $\gamma$ energy differs somewhat from the resonant value. As we discussed in Chapter 6, any state that has a mean life $\tau$ has a width $\Gamma = \hbar/\tau$, and measurement of the energy of the state gives a distribution of the form of Equation 6.20 and Figure 6.3. If we were to pass a beam of photons through a collection of bare nuclei (so as to eliminate scattering and absorption processes due to atomic electrons), then the resonant absorption cross section is

$$\sigma(E_\gamma) = \sigma_0 \frac{(\Gamma/2)^2}{\left[E_\gamma - (\Delta E + E_R)\right]^2 + (\Gamma/2)^2} \qquad (10.30)$$

where $E_R$ is the recoil correction $E_\gamma^2/2Mc^2$. This distribution is plotted in Figure 10.23. For typical nuclear states of mean lives ns to ps, the widths will be in the range of $10^{-6}$ to $10^{-3}$ eV.

Figure 10.24 shows a schematic view of the resonant absorption experiment. As $E_\gamma$ is varied, the resonance curve of Figure 10.23 is traced. At energies $E_\gamma$ far

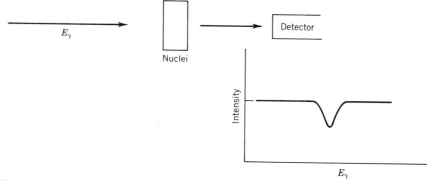

**Figure 10.24** Schematic of an experiment to observe resonant absorption by nuclei.

from the resonance, the nuclei are transparent to the radiation, and no absorption occurs. At the resonance, the transmitted intensity reaches a minimum value.

In practice, we would be unlikely to observe the *natural linewidth* $\Gamma$. A primary additional contributor to the observed linewidth is the *Doppler broadening* $\Delta$, which arises because the nuclei are not at rest (as we assumed) but in fact are in thermal motion at any temperature $T$. The photons as emitted or absorbed in the lab frame appear Doppler shifted with energies $E'_\gamma = E_\gamma(1 \pm v_x/c)$, where $v_x$ is the velocity component along the photon direction. If the motion of the nuclei is represented by the usual Maxwell velocity distribution $e^{-[(1/2)mv_x^2]/kT}$ there will be a distribution of energies of the form $e^{-(mc^2/2kT)(1 - E'_\gamma/E_\gamma)^2}$. This gives a Gaussian distribution of width

$$\Delta = 2\sqrt{\ln 2}\, E_\gamma \sqrt{\frac{2kT}{Mc^2}} \qquad (10.31)$$

At room temperature, $kT \simeq 0.025$ eV, and for a 100-keV transition in a medium-weight nucleus $\Delta \simeq 0.1$ eV, which dominates the natural linewidth for most nuclear transitions. Even cooling to low temperature (for instance, to 4 K in thermal contact with a reservoir of liquid helium) reduces the width by only an order of magnitude to 0.01 eV. The width observed in experiments such as that of Figure 10.24 will be a combination of the natural linewidth plus additional contributions including Doppler broadening.

Tunable sources of photons of the sort needed for the resonance experiment do not exist. (The best one can do is the continuous electromagnetic spectrum from bremsstrahlung or synchrotron radiation produced by charged-particle accelerators capable of reaching relativistic energies.) In our laboratories, we must therefore make do with ordinary sources of $\gamma$ radiation that emit only at discrete energies. However, to do the resonant absorption, we must find a radioactive source that emits a $\gamma$ ray of an energy within at most 0.1 eV of the desired resonant energy $\Delta E + E_R$. It is of course extremely unlikely to find such radiation, and with the proper multipole character besides. It therefore makes sense for us to try to use a source in which the $\gamma$ radiation is emitted in the same *downward* transition that we are trying to excite *upward* by resonant absorption. Consider, for example, the decay of $^{198}$Au; following the $\beta$ decay to $^{198}$Hg, a single strong $\gamma$-ray of energy 412 keV is emitted. If we now allow that $\gamma$ ray to fall upon a target of stable $^{198}$Hg nuclei, there is a possibility of absorption and excitation from the ground state to the 412-keV excited state. The mean life of the 412-keV state is 32 ps, corresponding to a width of $2 \times 10^{-5}$ eV. The recoil energy $E_R$ is $E_\gamma^2/2Mc^2 = 0.46$ eV, and it is important to note that the recoil affects both the *emitted* and *absorbed* transitions. That is, the emitted radiation has energy $\Delta E - E_R$, while for absorption we must supply an energy of $\Delta E + E_R$. The situation is indicated in Figure 10.25, in which we have assumed the lines to have the room-temperature Doppler width of 0.36 eV. As you can see, there is minimal overlap between the emission and absorption lines, and thus little probability of resonant excitation.

(Contrast this with the case of atomic radiation. Optical transitions have energies of a few eV; the recoil correction in Hg would be $2.7 \times 10^{-12}$ eV and there would be almost complete overlap between the profile of the source and

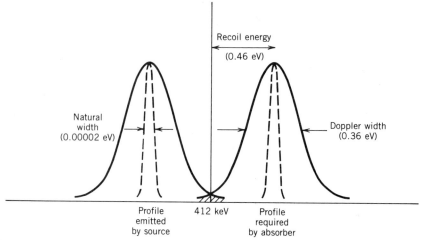

**Figure 10.25** The emitted radiation is shifted downward in energy by $E_R$, while the absorption requires an energy shifted upward by $E_R$. Because of the Doppler (thermal) broadening, there is a small overlap (shaded region) between the emission and absorption lines. The natural width has been greatly exaggerated on the energy scale of the diagram.

absorber transitions. Atomic resonant absorption experiments are thus relatively easy to perform.)

There are several techniques for overcoming the energy difference $2E_R$ between the source and absorber transitions. The first consists of raising the temperature, thereby increasing the Doppler broadening and the overlap of the profiles. A second method is to move the source toward the absorber at high speed to Doppler shift the emitted energy by an amount $2E_R$. Since the Doppler-shifted energy is $E_\gamma' = E_\gamma(1 + v/c)$, the speed required is

$$v = c\frac{2E_R}{E_\gamma} \tag{10.32}$$

which gives $2.2 \times 10^{-6}c$, or 670 m/s. Experiments of this type are usually done by attaching the source to the tip of a rotor in a centrifuge spinning at $10^4$–$10^5$ revolutions per minute. Figure 10.26 shows examples of results from the thermal-broadening and centrifuge techniques.

The most successful and useful technique for defeating the recoil problem is called the Mössbauer effect. In 1958, Rudolf Mössbauer performed a resonant absorption experiment using a source of $^{191}$Ir ($E_\gamma = 129$ keV; $E_R = 0.047$ eV). The emitting and absorbing nuclei were bound in a crystal lattice. Typical binding energies of an atom in a lattice are 1–10 eV, and thus there is not enough recoil energy available for the atom to leave its lattice site. The effect is somewhat like the difference between hitting a single brick with a baseball bat and hitting one brick in a solid brick wall—the entire solid lattice absorbs the recoil momentum. The mass that appears in the expression for the recoil energy becomes the mass of the entire solid, rather than the mass of one atom. In addition, a certain fraction of the atoms in a lattice (determined from statistical considerations) is in the vibrational ground state of thermal motion and thus

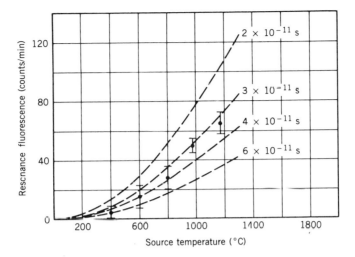

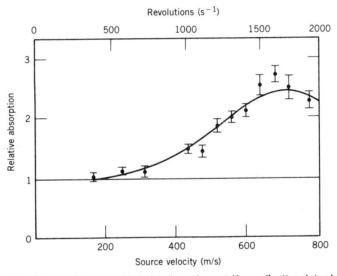

**Figure 10.26** Thermal broadening (top) and centrifuge (bottom) techniques for observing nuclear resonance in $^{198}$Hg. The data points at top show increasing resonant absorption as the temperature is increased (broadening the lines in Figure 10.25 and increasing the overlap). The dashed lines show the expected behavior for different excited-state lifetimes (that is, for different natural widths). From these data the lifetime is determined to be about 30 ps. The graph at bottom shows the result of Doppler shifting the radiation emitted by the source, by attaching it to the tip of a rotor. In this case, the emission line in Figure 10.25 is being moved to higher energy until it overlaps the absorption line. As estimated, this occurs at about 670 m/s. Thermal data from F. R. Metzger and W. B. Todd, *Phys. Rev.* **95**, 853 (1954); rotor data from W. G. Davey and P. B. Moon, *Proc. Phys. Soc. London A* **66**, 956 (1953).

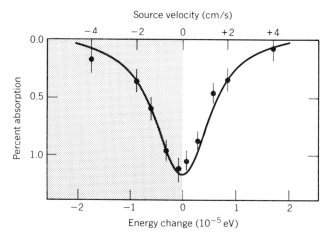

**Figure 10.27** Mössbauer effect using 129-keV γ ray from $^{191}$Ir. Because (1) the natural linewidth is obtained, and (2) the recoil is eliminated, there is essentially complete overlap between source and absorber. Doppler-shifting the source energy by an amount slightly greater that the natural linewidth ($10^{-5}$ eV) is sufficient to destroy the resonance. From original data by R. Mössbauer, *Z. Naturforsch. A* **14**, 211 (1959).

shows very little thermal Doppler broadening. The result is very narrow and overlapping emission and absorption lines, each characterized by the natural linewidth ($3 \times 10^{-6}$ eV in $^{191}$Ir). To demonstrate this phenomenon, we have only to move the source and absorber relative to one another at low speed; if the speed is such that the Doppler shift is greater than the natural linewidth, the resonance will be destroyed. For a total linewidth of $6 \times 10^{-6}$ eV (because both source and absorber have the natural width), the necessary speed is about $5 \times 10^{-11}c$, or about 15 mm/s, a considerable improvement over the 700 m/s needed for the centrifuge experiment! Figure 10.27 shows the resulting resonance, first obtained by Mössbauer in 1958.

What is remarkable about the Mössbauer effect is its extreme precision for the measurement of relative energies. For instance, suppose we modified the environment of the source or absorber nuclei in such a way that the energy difference between the initial and final nuclear states shifted by a very small amount $\delta E$. Using the Mössbauer effect, we should be able to measure this shift, as long is it is of the same order as the width of the resonance. (If the shift is too small compared with the width of the resonance, it is very difficult to measure.) In the case of $^{191}$Ir, where the observed width is about $10^{-5}$ eV, this would amount to measuring a change in energy of $10^{-5}$ eV out of a gamma-ray energy of $10^5$ eV, or an effect of one part in $10^{10}$. In $^{57}$Fe, which is more frequently used for the Mössbauer effect, the observed (natural) linewidth is of order $10^{-8}$ eV, and in this case experimental effects of the order of one part in $10^{12}$ can be measured!

Although we will not go into the details of the theory of the Mössbauer effect, it is worthwhile to consider briefly one other aspect of the resonance, its depth, which is determined by the fraction of the nuclei in the lattice that emits (or absorbs) with no recoil. The calculation of the recoil-free fraction $f$ depends on properties of the solid more detailed than the simple question of whether or not the recoil energy exceeds the lattice binding energy. Solids can absorb energy in many ways other than by removing atoms from their lattice sites. At low energies and temperatures, the primary way is through lattice vibrations, called *phonons*. (Propagation of these phonons through a lattice is responsible for familiar properties such as mechanical and acoustical waves.) These vibrations occur at a spectrum of frequencies, from zero up to a maximum, $\omega_{\max}$. The energy corresponding to the highest vibrational frequency is usually expressed in terms of the corresponding temperature, called the *Debye temperature*, $\theta_D$, defined so that $\hbar\omega_{\max} = k\theta_D$, where $k$ is the Boltzmann constant. For typical materials, $\hbar\omega_{\max} \sim 0.01$ eV and $\theta_D \sim 1000$ K. The recoilless fraction is

$$ f = \exp\left[ -\frac{\langle x^2 \rangle}{(\lambda/2\pi)^2} \right] \tag{10.33} $$

where $\langle x^2 \rangle$ is the mean-square vibrational amplitude of the emitting nucleus and $\lambda$ is the wavelength of the $\gamma$ ray. Using the Bose-Einstein distribution function for the vibrational phonon spectrum permits the calculation of the mean-square amplitude, and gives for the recoilless fraction

$$ f = \exp\left\{ -\frac{6E_R}{k\theta_D}\left[ \frac{1}{4} + \left(\frac{T}{\theta_D}\right)^2 \int_0^{\theta_D/T} \frac{x\,dx}{e^x - 1} \right] \right\} \tag{10.34} $$

At low temperatures $T \ll \theta_D$, the last term in the exponent is negligible. Values of $\theta_D$ do not vary greatly among metals ($\theta_D \sim 400$ K for Fe and 300 K for Ir), so the recoil energy $E_R$ is essential in determining the recoilless fraction. For the 14.4 keV transition of $^{57}$Fe, $E_R = 0.002$ eV and $f = 0.92$, while for Ir, $f \approx 0.1$. (The second term in the exponent of Equation 10.34 is always negative, and thus this term will work to make $f$ smaller than these low-temperature estimates.) Because recoilless processes are needed in both the source and absorber, the overall recoilless fraction is determined by the product of the factors in the source and absorber. It is therefore not surprising that Mössbauer's original experiment with Ir showed an effect of only 1%, while Fe shows a much larger effect.

The Mössbauer effect has found applications in an enormous variety of areas. Its primary usefulness is in those applications in which we must determine the properties of the physical or chemical environment of a nucleus, but an important exploitation of the extreme precision of the method was the determination of the change in energy of photons falling in the gravitational field of the Earth, called the *gravitational red shift*. One of the cornerstones of Einstein's General Theory of Relativity is the Principle of Equivalence, according to which the effects of a local uniform gravitational field cannot be distinguished from those of a uniformly accelerated reference frame. If we were to observe the emission and absorption of radiation in an accelerated reference frame, in which $H$ is the distance between the source and absorber, then in the time $H/c$ necessary for radiation to travel from the source to the absorber, the absorber would acquire a velocity $gH/c$, where $g$ is the acceleration, chosen to be numerically equal (and in the opposite direction) to the gravitational acceleration of the uniform field. The radiation photons are therefore Doppler shifted,

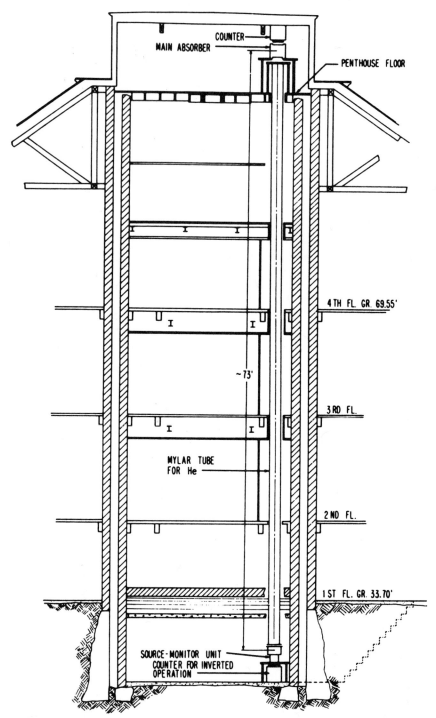

**Figure 10.28** Schematic view of the gravitational red shift experiment in the Harvard tower. To reduce absorption in air, the 14.4-keV photons traveled in He gas in a Mylar tube. In the configuration shown, the source is at the bottom and the absorber and counter are at the top. From R. V. Pound, in *Mössbauer Spectroscopy II*, edited by U. Gonser (Berlin: Springer-Verlag, 1981), p. 31.

according to

$$\frac{\Delta E}{E} = \frac{\Delta v}{c} = \frac{gH}{c^2} \tag{10.35}$$

This amounts to about $1 \times 10^{-16}$ per meter in the Earth's gravitational field.

In the original experiment of Pound and Rebka, *Phys. Rev. Lett.* **4**, 337 (1960), $^{57}$Fe was used (from a 1-Ci source of $^{57}$Co), and the 14.4-keV photons were allowed to travel 22.5 m up the tower of the Jefferson Physical Laboratory at Harvard (Figure 10.28). The expected effect was of order $2 \times 10^{-15}$, which required heroic efforts even for $^{57}$Fe with a sensitivity ($\Gamma/E_\gamma$) of roughly $3 \times 10^{-13}$. To observe the small shift (about $10^{-2}$ of the width of the resonance), Pound and Rebka concentrated on the portions of the sides of the resonance curve with the largest slope. To reduce systematic effects, it was necessary to monitor with great precision the temperature of the source and absorber (temperature differences between source and absorber would cause unequal Doppler broadenings, which would simulate a shift in the peak) and periodically to interchange source and absorber to allow the photons to travel in the opposite direction. After four months of experiments, the result was $\Delta E/E = (4.902 \pm 0.041) \times 10^{-15}$, compared with the expected value $4.905 \times 10^{-15}$ for the 45-m round trip. This careful experiment represents one of the most precise tests of the General Theory of Relativity, and it would not have been possible without the great sensitivity provided by the Mössbauer effect. [For a review of this experiment, see R. V. Pound, in *Mössbauer Spectroscopy II*, edited by U. Gonser (Berlin: Springer-Verlag, 1981), p. 31.]

As mentioned above, perhaps the primary application of the Mössbauer effect has been in studying the interaction of nuclei with their physical and chemical environments. The interaction of nuclear electromagnetic moments with the fields

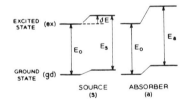

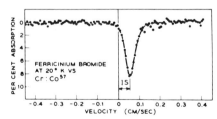

**Figure 10.29**  The isomer shift. In different materials, the ground and excited states show different shifts from the overlap of electronic wave functions with the nucleus. The effect on the resonance is to shift it away from zero relative velocity. From G. K. Wertheim, *Mössbauer Effect: Principles and Applications* (New York: Academic, 1964).

of the environment is usually called *hyperfine interactions*, and will be discussed in more detail in Chapter 16. For the present, we give a few examples of the application of Mössbauer spectroscopy to their study.

In the first case, we measure simply the effect of the penetration of atomic wave functions into the nuclear volume. This quantity, called $\Delta E$ in Equation 3.12, represents the difference in energy between electronic levels calculated for a "point" nucleus and for a uniformly charged spherical nucleus of radius $R$. Although our goal in Chapter 3 was to calculate the effect on the atomic levels, a bit of thought should convince you that nuclear levels should be shifted by an equal but opposite energy because the observed total (atomic + nuclear) energy cannot change under the influence of internal forces. If we let $E_0$ represent the photon energy in the absence of this effect ($E_0 = E_e - E_g$, where e is the excited state and g is the ground state), then the observed energy is

$$E = (E_e + \Delta E_e) - (E_g + \Delta E_g)$$
$$= E_0 + \Delta E_e - \Delta E_g$$

(10.36)

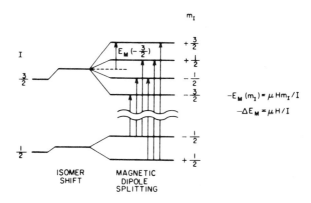

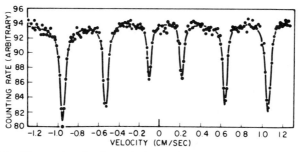

**Figure 10.30** Magnetic dipole splitting of nuclear levels observed with the Mössbauer effect. At top right are shown the nuclear *m* states split by a magnetic field. If the ground state and excited state have different nuclear magnetic dipole moments, the energy splittings $\Delta E_M$ will be different, as shown; in the case illustrated, the moments also have opposite signs. For dipole transitions, only $\Delta m_I = 0$ or $\pm 1$ can occur, so 6 individual components are seen. From G. K. Wertheim, *Mössbauer Effect: Principles and Applications* (New York: Academic, 1964).

because the ground and excited states will have different nuclear wave functions and thus different radii. If the source and absorber in a Mössbauer experiment had the same chemical environment, the resonance would not be affected, but if the source and absorber are different, then the transition energies are slightly different. In this case, one of the peaks in Figure 10.25 would be shifted somewhat, relative to where it appears when the source and absorber are similar. The effect on the Mössbauer spectrum is to shift the center of the resonance away from zero velocity. This effect is called the *isomer shift* (or sometimes *chemical shift*) and is illustrated in Figure 10.29. You can see that this is a small effect, of the order of one part in $10^{12}$.

In another kind of hyperfine coupling, we study the splitting of the nuclear levels in a magnetic field. In atomic physics, the effect of a magnetic field on spectral lines is called the *Zeeman effect* and corresponds to the removal of the $m$ degeneracy of a level of angular momentum $I$ in a magnetic field—the field splits the level into $2I + 1$ equally spaced sublevels. Atomic wavelengths are typically shifted by one part in $10^4$ by the Zeeman effect; nuclear magnetic moments are only $10^{-3}$ of atomic magnetic moments, and nuclear transition energies are $10^5$ of atomic energies, and so the resulting nuclear effect is one part in $10^{12}$. Figure

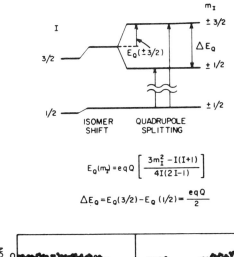

$$E_Q(m_I) = eqQ \left[ \frac{3m_I^2 - I(I+1)}{4I(2I-1)} \right]$$

$$\Delta E_Q = E_Q(3/2) - E_Q(1/2) = \frac{eqQ}{2}$$

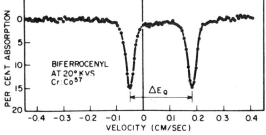

**Figure 10.31** Electric quadrupole hyperfine splitting. In the expression for the splitting energy, $Q$ represents the nuclear electric quadrupole moment and $q$ is the electric field gradient (sometimes called $V_{zz}$ or $\partial^2 V / \partial z^2$). In this case only two lines appear; in addition there is an isomer shift that moves the center of the Mössbauer spectrum away from zero velocity. From G. K. Wertheim, *Mössbauer Effect: Principles and Applications* (New York: Academic, 1964).

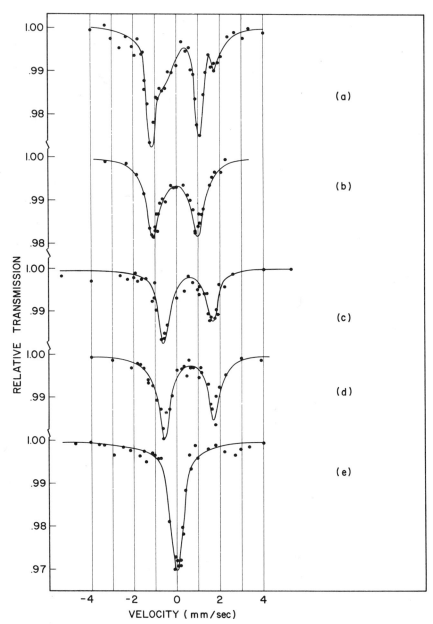

**Figure 10.32** Mössbauer effect experiments showing the differing chemical environments of Fe in hemoglobin. The source for these experiments was $^{57}$Co (which decays to $^{57}$Fe) in Pt; absorbers were (a) rat red cells at 4 K, (b) oxygenated rat hemoglobin at 77 K, (c) human hemoglobin in $CO_2$ at 77 K, (d) human hemoglobin in $N_2$ at 77 K, and (e) human hemoglobin in CO at 77 K. Note the different isomer shifts and electric quadrupole splittings. From U. Gonser et al., *Science* **143**, 680 (1964). Copyright © 1964, AAAS.

10.30 shows an example of this for $^{57}$Fe. An isomer shift is present, in addition to the magnetic hyperfine structure. The magnetic moments of the ground and excited states are unequal, and in fact the Mössbauer spectrum can give the magnetic moment of the excited state to great precision, when (as in $^{57}$Fe) the ground state is stable and its moment is preciscly known. It is also possible, if the moment is known, to deduce the size of the magnetic field, the determination of which can give important clues about atomic structure.

Finally, the nuclear quadrupole moment can interact with an electric field gradient to give an electric quadrupole splitting. This splitting is proportional to $m^2$ and thus does not distinguish between $+m$ and $-m$. Furthermore, it vanishes when $I = \frac{1}{2}$. Figure 10.31 shows an example in the case of $^{57}$Fe. Note that 2 lines appear in the case of electric quadrupole interactions in $^{57}$Fe (whereas 6 appear in the magnetic dipole case). For other isotopes having different ground-state and excited-state spins, the number of lines will be different

These same studies can be extended to materials with a variety of applications. For example, the protein hemoglobin gives blood its red color and is responsible for binding to the oxygen in the bloodstream. Hemoglobin is rich in Fe and can therefore be used in Mössbauer experiments. Figure 10.32 shows some typical results. Oxygenated blood shows a quadrupole-split Mössbauer spectrum, but with an isomer shift slightly different from that of deoxygenated blood. Venous blood shows a mixture of the two types. Blood exposed to CO shows neither quadrupole splitting nor isomer shift. Oxygenated hemoglobin shows magnetic dipole splittings when placed in a strong magnetic field; deoxygenated hemo-globin does not. All of the above refer to the ferrous ion ($Fe^{2+}$); ferric ions ($Fe^{3+}$), the presence of which may indicate certain blood diseases, give very different Mössbauer spectra.

## REFERENCES FOR ADDITIONAL READING

A comprehensive work on electromagnetic multipole radiation is M. E. Rose, *Multipole Fields* (New York: Wiley, 1955); Chapter 12 of J. M. Blatt and V. F. Weisskopf, *Theoretical Nuclear Physics* (New York: Wiley, 1952) also discusses classical multipole radiation. The quantum mechanics of the electromagnetic radiation field is considered in detail in Sections 3.1–3.9 of M. G. Bowler, *Nuclear Physics* (Oxford: Pergamon, 1973). Also of interest are Chapters 15 (multipole radiation) and 16 (internal conversion) of *Alpha-, Beta- and Gamma-Ray Spectroscopy*, edited by K. Siegbahn (Amsterdam: North-Holland, 1965).

Many theoretical and experimental aspects of $\gamma$ emission are reviewed in *The Electromagnetic Interaction in Nuclear Spectroscopy*, edited by W. D. Hamilton (Amsterdam: North-Holland, 1975).

Tabulations of theoretical conversion coefficients can be found in R. S. Hager and E. C. Seltzer, *Nuclear Data Tables A* **4**, 1 (1968); F. Rösel, H. M. Fries, K. Alder, and H. C. Pauli, *Atomic Data and Nuclear Data Tables* **21**, 91 (1978); L. A. Sliv and I. M. Band, in *Alpha-, Beta- and Gamma-Ray Spectroscopy*, edited by K. Siegbahn (Amsterdam: North-Holland, 1965), Appendix 5.

The Mössbauer effect is reviewed in G. K. Weitheim, *Mössbauer Effect: Principles and Applications* (New York: Academic, 1964), and T. C. Gibb,

*Principles of Mössbauer Spectroscopy* (New York: Halsted, 1976). Collections of recent papers on the application of the Mössbauer effect are *Applications of Mössbauer Spectroscopy*, Vols. I and II, edited by Richard L. Cohen (New York: Academic, 1976 and 1980), and *Mössbauer Spectroscopy*, edited by U. Gonser (Berlin: Springer-Verlag, 1981).

## PROBLEMS

1. Each of the following nuclei emits a photon in a $\gamma$ transition between an excited state and the ground state. Given the energy of the photon, find the energy of the excited state and comment on the relationship between the nuclear recoil energy and the experimental uncertainty in the photon energy: (a) $320.08419 \pm 0.00042$ keV in $^{51}$V; (b) $1475.786 \pm 0.005$ keV in $^{110}$Cd; (c) $1274.545 \pm 0.017$ keV in $^{22}$Ne; (d) $3451.152 \pm 0.047$ keV in $^{56}$Fe; (e) $884.54174 \pm 0.00074$ keV in $^{192}$Ir.

2. Following the decay of $^{198}$Au, three $\gamma$'s are observed to be emitted from states in $^{198}$Hg; their energies (in keV) are $\gamma_1$, $411.80441 \pm 0.00015$; $\gamma_2$, $675.88743 \pm 0.00069$; and $\gamma_3$, $1087.69033 \pm 0.00074$. It is suggested that there are two excited states $E_1$ and $E_2$ in $^{198}$Hg that are populated in the decay, and that the $\gamma$'s correspond respectively to the transitions $E_1 \rightarrow E_0$, $E_2 \rightarrow E_1$, and $E_2 \rightarrow E_0$ (where $E_0$ represents the ground state). If this hypothesis were correct, we would expect $E_{\gamma_1} + E_{\gamma_2} = E_{\gamma_3}$, which is almost but not quite true according to the experimental uncertainties. Show how the proper inclusion of the nuclear recoil resolves the discrepancy.

3. The calculation of the emission probability for electric quadrupole radiation involves a term of the form of Equation 3.36, with the proper labeling of initial and final wave functions. From such an integral, verify the parity selection rule for electric quadrupole transitions.

4. (a) For a light nucleus ($A \approx 10$), compute the ratio of the emission probabilities for quadrupole and dipole radiation according to the Weisskopf estimates. Consider all possible choices for the parities of the initial and final states. (b) Repeat for a heavy nucleus ($A \approx 200$).

5. In a nucleus described by the rotational model (see Figure 5.22), the second excited state is always $4^+$. This state decays by $E2$ radiation to the $2^+$ state. Justify this observation by calculating, using the Weisskopf estimates, the ratio between the $E2$ decay probability and (a) the octupole ($L = 3$) and hexadecapole ($L = 4$) decays to the $2^+$ state and (b) the hexadecapole decay to the ground state. (*Note:* These are collective rotational states, for which the Weisskopf estimates should not be taken too seriously.)

6. For the following $\gamma$ transitions, give all permitted multipoles and indicate which multipole might be the most intense in the emitted radiation.
   (a) $\frac{9}{2}^- \rightarrow \frac{7}{2}^+$     (d) $4^+ \rightarrow 2^+$
   (b) $\frac{1}{2}^- \rightarrow \frac{7}{2}^-$     (e) $\frac{11}{2}^- \rightarrow \frac{3}{2}^+$
   (c) $1^- \rightarrow 2^+$     (f) $3^+ \rightarrow 3^+$

7. A certain decay process leads to final states in an even-$Z$, even-$N$ nucleus and gives only three $\gamma$ rays of energies 100, 200, and 300 keV, which are

found to be (respectively) of $E1$, $E2$, and $E3$ multipolarity. Construct two different possible level schemes for this nucleus (consistent with known systematics of nuclear structure) and label the states with their most likely spin-parity assignments. Suggest experiments that might distinguish between your proposed level schemes.

**8.** A nucleus has the following sequence of states beginning with the ground state: $\frac{3}{2}^{+}$, $\frac{7}{2}^{+}$, $\frac{5}{2}^{+}$, $\frac{1}{2}^{-}$, and $\frac{3}{2}^{-}$. Draw a level scheme showing the intense $\gamma$ transitions likely to be emitted and indicate their multipole assignment.

**9.** The isomeric $2^{+}$ state of $^{60}$Co at 58.6 keV decays to the $5^{+}$ ground state. Internal conversion competes with $\gamma$ emission; the observed internal conversion coefficients are $\alpha_{K} = 41$, $\alpha_{L} = 7$, $\alpha_{M} = 1$. (a) Compute the expected half-life of the $2^{+}$ state if the transition multipolarity is assumed to be $M3$, and compare with the observed half-life of 10.5 min. (b) If the transition also contained a small component of $E4$ radiation, how would your estimate for the half-life be affected? (c) The $2^{+}$ state also decays by direct $\beta$ emission to $^{60}$Ni. The maximum $\beta$ energy is 1.55 MeV and the log $ft$ is 7.2. The $2^{+}$ state decays 0.25% by $\beta$ emission and 99.75% by $\gamma$ emission and internal conversion. What is the effect on the calculated half-life of including the $\beta$ emission?

**10.** An even-$Z$, even-$N$ nucleus has the following sequence of levels above its $0^{+}$ ground state: $2^{+}$ (89 keV), $4^{+}$ (288 keV), $6^{+}$ (585 keV), $0^{+}$ (1050 keV), $2^{+}$ (1129 keV). (a) Draw an energy level diagram and show all reasonably probable $\gamma$ transitions and their dominant multipole assignments. (b) By considering also internal conversion, what additional transitions and multipoles would appear?

**11.** (a) Pick half a dozen or so typical examples of $2^{+} \rightarrow 0^{+}$ transitions from the first excited states of "rotational" nuclei, $150 < A < 190$. (Use standard reference works for nuclear spectroscopic data.) Compute the ratio between the observed $\gamma$ decay rate and the corresponding Weisskopf estimate. Be sure to correct the measured lifetimes for internal conversion if necessary. (b) Repeat for "vibrational" nuclei, $60 < A < 150$, excluding cases at closed shells. (c) Draw any apparent conclusion about the difference between "rotational" and "vibrational" $\gamma$ transitions.

**12.** The *Table of Isotopes* shows multipole assignments of $\gamma$ transitions and lifetimes of excited states. By searching through the data given there, prepare a graph similar to Figure 10.13 showing the lifetimes for $M2$ transitions. Verify the dependence on the transition energy. (Be sure to use the *partial* lifetime in the cases in which a level can decay through several transitions.) Among the cases you should consider are $^{39}$Ar, $^{73}$As, $^{117}$Eu, $^{165}$Ho, $^{181}$Ta, and $^{182}$W, but you should find many other instances as well.

**13.** Among the nuclei in which there are known $E4$ transitions are $^{44}$Sc, $^{52}$Mn, $^{86}$Rb, $^{93}$Mo, $^{114}$In, and $^{202}$Pb. Look up the partial lifetimes for these transitions from collections of spectroscopic data, and compare with values calculated from the Weisskopf estimates.

**14.** The isotope $^{113}$Cd captures a very low-energy neutron, leading to an excited state of $^{114}$Cd, which emits a $\gamma$ ray leading directly to the $^{114}$Cd ground state. (a) Find the energy of the $\gamma$ ray, neglecting the nuclear recoil. (b) Calculate the kinetic energy of the recoiling $^{114}$Cd.

15. In Section 5.2, the states of the vibrational model for even-$Z$, even-$N$ nuclei up to the $0^+$, $2^+$, $3^+$, $4^+$, $6^+$ three-phonon multiplet were discussed. The model also gives selection rules for $\gamma$ emission: the phonon number must change by exactly one unit, and only $E2$ transitions are permitted. Draw a vibrational level scheme showing all permitted $\gamma$ transitions starting with the three-phonon multiplet (use Figure 5.19 as a basis).

16. A certain decay scheme shows the following $\gamma$ energies (in keV): 32.7, 42.1, 74.8, 84.0, 126.1, and 158.8. Coincidence studies reveal two features of the decay: only one of the $\gamma$'s has none of the others in coincidence with it, and none of the $\gamma$'s is in coincidence with more than three of the others. The $\gamma$'s are preceded by a $\beta$ decay that populates only one level. From this information suggest a possible level scheme. (*Note*: There are two different arrangements of the $\gamma$'s that are consistent with the information given.)

17. In a study of the conversion electrons emitted in a decay process, the following electron energies were measured (in keV): 207.40, 204.64, 193.36, 157.57, 154.81, 143.53, 125.10, 75.27, 49.03, 46.27, 34.99. The electron binding energies are known to be 83.10 keV (K shell), 14.84 keV (L shell), 3.56 keV (M shell), and 0.80 keV (N shell). What is the minimum number of $\gamma$'s that can produce the observed electron groups, and what are the $\gamma$ energies?

18. Based on the information given in Figure 10.18 and Table 10.2, find all partial lifetimes for $\gamma$ and electron emission for the $8^-$ level of $^{180}$Hf.

19. For each of the following Mössbauer-effect transitions compute the natural width, the Doppler width at room temperature, the Doppler width at liquid helium temperature (4 K), and the nuclear recoil energy: (a) 73 keV, 6.3 ns in $^{193}$Ir; (b) 14.4 keV, 98 ns in $^{57}$Fe; (c) 6.2 keV, 6.8 $\mu$s in $^{181}$Ta; (d) 23.9 keV, 17.8 ns in $^{119}$Sn; (e) 95 keV, 22 ps in $^{165}$Ho. Half-lives are given.

20. The absorption of the 27.8 keV magnetic dipole Mössbauer transition in $^{129}$I takes the nucleus from its $\frac{7}{2}^+$ ground state to a $\frac{5}{2}^+$ excited state. The values of the magnetic dipole and electric quadrupole moments are: $\mu(\frac{7}{2}) = +2.6$ $\mu_N$, $\mu(\frac{5}{2}) = +2.8\ \mu_N$, $Q(\frac{7}{2}) = -0.55$ b, $Q(\frac{5}{2}) = -0.68$ b. Make a sketch of the $m$-state splittings (similar to Figures 10.30 and 10.31) for the magnetic dipole and electric quadrupole cases and show the number of components in the Mössbauer spectrum. *Hint*: The magnetic field ($B$ or $H$) and the electric field gradient ($q$ or $V_{zz}$) can be regarded as positive.

21. In most of our work in nuclear physics, we regard the decay constant $\lambda$ as a true constant for a given nuclear species. However, in this chapter and in the previous chapter you have studied two processes in which the nuclear decay rate could be sensitive to the chemical state of the atom. Discuss these two processes and explain how the atomic state might influence the nuclear decay rate. [For a discussion and some examples of cases in which this can occur, see the review by G. T. Emery, *Ann. Rev. Nucl. Sci.* **22**, 165 (1972).]

# UNIT III
## NUCLEAR
## REACTIONS

# 11

# NUCLEAR REACTIONS

If energetic particles from a reactor or accelerator (or even from a radioactive source) are allowed to fall upon bulk matter, there is the possibility of a nuclear reaction taking place. The first such nuclear reactions were done in Rutherford's laboratory, using $\alpha$ particles from a radioactive source. In some of these early experiments, the $\alpha$ particles merely rebounded elastically from the target nuclei; this phenomenon, known ever since as Rutherford scattering, gave us the first evidence for the existence of atomic nuclei. In other experiments, Rutherford was able to observe a change or transmutation of nuclear species, as in this reaction done in 1919:

$$\alpha + {}^{14}\text{N} \rightarrow {}^{17}\text{O} + \text{p}$$

The first particle accelerator capable of inducing nuclear reactions was built by Cockcroft and Walton, who in 1930 observed the reaction

$$\text{p} + {}^{7}\text{Li} \rightarrow {}^{4}\text{He} + \alpha$$

In this chapter we discuss various types of nuclear reactions and their properties. In most cases, we deal with light projectiles, usually $A \leq 4$, incident on heavy targets; there is, however, much current interest in reactions induced by accelerating heavy ions (usually $A \leq 40$, but even beams as heavy as uranium are considered). We also deal only with reactions that are classified as "low energy," that is, of the order of 10 MeV per nucleon or less. In the range of 100 MeV– 1 GeV, called "medium energy," meson production can take place, and protons and neutrons can transform into each other. At "high energy," all sorts of exotic particles can be produced, and we can even rearrange the quarks that are the constituents of nucleons. We discuss these latter types of reactions in Chapters 17 and 18.

## 11.1 TYPES OF REACTIONS AND CONSERVATION LAWS

A typical nuclear reaction is written

$$\text{a} + \text{X} \rightarrow \text{Y} + \text{b}$$

where a is the accelerated projectile, X is the target (usually stationary in the laboratory), and Y and b are the reaction products. Usually, Y will be a heavy product that stops in the target and is not directly observed, while b is a light particle that can be detected and measured. Generally, a and b will be nucleons or light nuclei, but occasionally b will be a γ ray, in which case the reaction is called *radiative capture*. (If a is a γ ray, the reaction is called the *nuclear photoeffect*.)

An alternative and compact way of indicating the same reaction is

$$X(a, b)Y$$

which is convenient because it gives us a natural way to refer to a general class of reactions with common properties, for example (α, n) or (n, γ) reactions.

We classify reactions in many ways. If the incident and outgoing particles are the same (and correspondingly X and Y are the same nucleus), it is a *scattering* process, *elastic* if Y and b are in their ground states and *inelastic* if Y or b is in an excited state (from which it will generally decay quickly by γ emission). Sometimes a and b are the same particle, but the reaction causes yet another nucleon to be ejected separately (so that there are three particles in the final state); this is called a *knockout* reaction. In a *transfer reaction*, one or two nucleons are transferred between projectile and target, such as an incoming deuteron turning into an outgoing proton or neutron, thereby adding one nucleon to the target X to form Y. Reactions can also be classified by the mechanism that governs the process. In *direct reactions* (of which transfer reactions are an important subgroup), only very few nucleons take part in the reaction, with the remaining nucleons of the target serving as passive spectators. Such reactions might insert or remove a single nucleon from a shell-model state and might therefore serve as ways to explore the shell structure of nuclei. Many excited states of Y can be reached in these reactions. The other extreme is the *compound nucleus* mechanism, in which the incoming and target nuclei merge briefly for a complete sharing of energy before the outgoing nucleon is ejected, somewhat like evaporation of a molecule from a hot liquid. Between these two extremes are the *resonance reactions*, in which the incoming particle forms a "quasibound" state before the outgoing particle is ejected.

## Observables

We have at our disposal techniques for measuring the energies of the outgoing particles to high precision (perhaps 10 keV resolution with a magnetic spectrometer). We can determine the direction of emission of the outgoing particle, and observe its angular distribution (usually relative to the axis of the original beam) by counting the number emitted at various angles. The *differential cross section* is obtained from the probability to observe particle b with a certain energy and at a certain angle $(\theta, \phi)$ with respect to the beam axis. Integrating the differential cross section over all angles, we get the total cross section for particle b to be emitted at a certain energy (which is also sometimes called a differential cross section). We can also integrate over all energies of b to get the absolute total cross section, which is in effect the probability to form nucleus Y in the reaction.

This quantity is of interest in, for instance, neutron activation or radioisotope production.

By doing *polarization* experiments, we can deduce the spin orientation of the product nucleus Y or perhaps the spin dependence of the reaction cross section. For these experiments we may need an incident beam of polarized particles, a target of polarized nuclei, and a spectrometer for analyzing the polarization of the outgoing particle b.

We can simultaneously observe the γ radiations or conversion electrons from the decay of excited states of Y. This is usually done in coincidence with the particle b to help us decide which excited states the radiations come from. We can also observe the angular distribution of the γ radiations, as an aid in interpreting the properties of the excited states, especially in deducing their spin-parity assignments.

## Conservation Laws

In analyzing nuclear reactions, we apply the same conservation laws we applied in studying radioactive decays. *Conservation of total energy and linear momentum* can be used to relate the unknown but perhaps measurable energies of the products to the known and controllable energy of the projectile. We can thus use the measured energy of b to deduce the excitation energy of states of Y or the mass difference between X and Y. *Conservation of proton and neutron number* is a result of the low energy of the process, in which no meson formation or quark rearrangement take place. (The weak interaction is also negligible on the time scale of nuclear reactions, about $10^{-16}$ to $10^{-22}$ s.) At higher energies we still conserve total nucleon (or, as we discuss in Chapter 18, baryon) number, but at low energy we conserve *separately* proton number *and* neutron number. *Conservation of angular momentum* enables us to relate the spin assignments of the reacting particles and the orbital angular momentum carried by the outgoing particle, which can be deduced by measuring its angular distribution. We can thus deduce the spin assignments of nuclear states. *Conservation of parity* also applies; the net parity *before* the reaction must equal the net parity *after* the reaction. If we know the orbital angular momentum of the outgoing particle, we can use the $(-1)^{\ell}$ rule and the other known parities in the reaction to deduce unknown parities of excited states. In Section 11.3 we discuss yet another quantity that is conserved in nuclear reactions.

## 11.2 ENERGETICS OF NUCLEAR REACTIONS

Conservation of total relativistic energy in our basic reaction gives

$$m_X c^2 + T_X + m_a c^2 + T_a = m_Y c^2 + T_Y + m_b c^2 + T_b \qquad (11.1)$$

where the $T$'s are kinetic energies (for which we can use the nonrelativistic approximation $\frac{1}{2}mv^2$ at low energy) and the $m$'s are rest masses. We define the *reaction Q value*, in analogy with radioactive decay $Q$ values, as the initial mass energy minus the final mass energy:

$$Q = (m_{\text{initial}} - m_{\text{final}})c^2$$
$$= (m_X + m_a - m_Y - m_b)c^2 \qquad (11.2)$$

which is the same as the excess kinetic energy of the *final* products:

$$Q = T_{\text{final}} - T_{\text{initial}}$$
$$= T_Y + T_b - T_X - T_a \qquad (11.3)$$

The $Q$ value may be positive, negative, or zero. If $Q > 0$ ($m_{\text{initial}} > m_{\text{final}}$ or $T_{\text{final}} > T_{\text{initial}}$) the reaction is said to be *exoergic* or *exothermic*; in this case nuclear mass or binding energy is released as kinetic energy of the final products. When $Q < 0$ ($m_{\text{initial}} < m_{\text{final}}$ or $T_{\text{final}} < T_{\text{initial}}$) the reaction is *endoergic* or *endothermic*, and initial kinetic energy is converted into nuclear mass or binding energy. The changes in mass and energy must of course be related by the familiar expression from special relativity, $\Delta E = \Delta mc^2$—any change in the kinetic energy of the system of reacting particles must be balanced by an equal change in its rest energy.

Equations 11.1–11.3 are valid in any frame of reference in which we choose to work. Let's apply them first to the laboratory reference frame, in which the target nuclei are considered to be at rest (room-temperature thermal energies are negligible on the MeV scale of nuclear reactions). If we define a reaction plane by the direction of the incident beam and one of the outgoing particles, then conserving the component of momentum perpendicular to that plane shows immediately that the motion of the second outgoing particle must lie in the plane as well. Figure 11.1 shows the basic geometry in the reaction plane. Conserving linear momentum along and perpendicular to the beam direction gives

$$p_a = p_b \cos\theta + p_Y \cos\xi \qquad (11.4a)$$
$$0 = p_b \sin\theta - p_Y \sin\xi \qquad (11.4b)$$

Regarding $Q$ as a known quantity and $T_a$ (and therefore $p_a$) as a parameter that we control, Equations 11.3 and 11.4a, b represent three equations in four unknowns ($\theta$, $\xi$, $T_b$, and $T_Y$) which have no unique solution. If, as is usually the case, we do not observe particle Y, we can eliminate $\xi$ and $T_Y$ from the equations

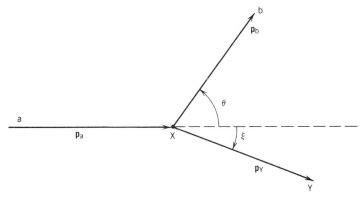

**Figure 11.1** Basic reaction geometry for a + X → b + Y.

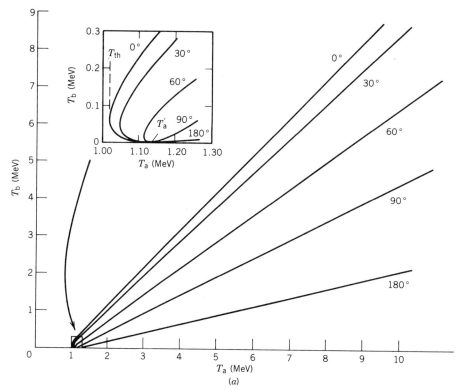

**Figure 11.2** (a) $T_a$ vs $T_b$ for the reaction $^3$H(p, n)$^3$He. The inset shows the region of double-valued behavior near 1.0 MeV.

to find a relationship between $T_b$ and $\theta$:

$$T_b^{1/2} = \frac{(m_a m_b T_a)^{1/2}\cos\theta \pm \{ m_a m_b T_a \cos^2\theta + (m_Y + m_b)[m_Y Q + (m_Y - m_a)T_a]\}^{1/2}}{m_Y + m_b}$$

(11.5)

This expression is plotted in Figure 11.2a for the reaction $^3$H(p, n)$^3$He, for which $Q = -763.75$ keV. Except for a very small energy region between 1.019 and 1.147 MeV, there is a one-to-one correspondence (for a given $T_a$) between $T_b$ and $\theta$. That is, keeping the incident energy fixed, choosing a value of $\theta$ to observe the outgoing particles automatically then selects their energy.

Several other features of Figure 11.2 are apparent, which you should be able to show explicitly from Equation 11.5:

1. There is an absolute minimum value of $T_a$ below which the reaction is not possible. This occurs only for $Q < 0$ and is called the *threshold energy* $T_{th}$:

$$T_{th} = (-Q)\frac{m_Y + m_b}{m_Y + m_b - m_a}$$

(11.6)

The threshold condition always occurs for $\theta = 0°$ (and therefore $\xi = 0°$)—the products Y and b move in a common direction (but still as separate nuclei).

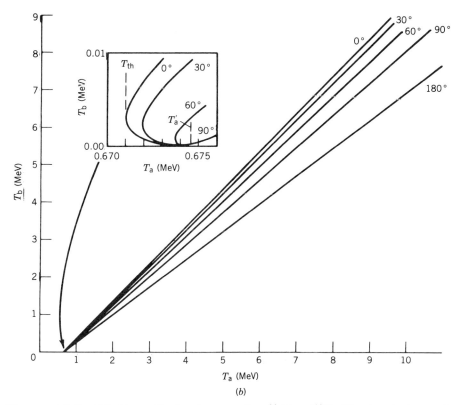

**Figure 11.2** (b) $T_a$ vs $T_b$ for the reaction $^{14}C(p, n)^{14}N$. The inset shows the double-valued region.

No energy is "wasted" in giving them momentum transverse to the beam direction. If $Q > 0$, there is no threshold condition and the reaction will "go" even for very small energies, although we may have to contend with Coulomb barriers not considered here and which will tend to keep a and X outside the range of each other's nuclear force.

2.  The double-valued situation occurs for incident energies between $T_{th}$ and the upper limit

$$T_a' = (-Q)\frac{m_Y}{m_Y - m_a} \tag{11.7}$$

This also occurs only for $Q < 0$, and is important only for reactions involving nuclei of comparable masses. Using Equations 11.6 and 11.7 we can approximate this range as

$$T_a' - T_{th} \cong T_{th}\frac{m_a m_b}{m_Y(m_Y - m_a)}\left(1 - \frac{m_b}{m_Y} + \cdots\right) \tag{11.8}$$

and you can see that if a and b have mass numbers of 4 or less and if Y is a medium or heavy nucleus, then the range $(T_a' - T_{th})$ becomes much smaller

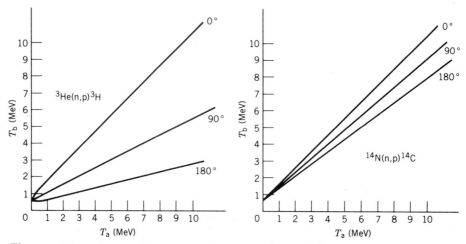

**Figure 11.3** $T_a$ vs $T_b$ for the reactions $^3$He(n, p)$^3$H and $^{14}$N(n, p)$^{14}$C. No double-valued behavior occurs.

than 1% of the threshold energy. Figure 11.2b shows the double-valued region for the reaction $^{14}$C(p, n)$^{14}$N.

3.  There is also a maximum angle $\theta_m$ at which this double-valued behavior occurs, the value for which is determined for $T_a$ in the permitted range by the vanishing of the argument in the square root of Equation 11.5:

$$\cos^2 \theta_m = -\frac{(m_Y + m_b)[m_Y Q + (m_Y - m_a)T_a]}{m_a m_b T_a} \quad (11.9)$$

When $T_a = T_a'$, the double-valued behavior occurs between $\theta = 0°$ and $\theta_m = 90°$; near $T_a = T_{th}$ it occurs only near $\theta_m = 0°$.

4.  Reactions with $Q > 0$ have neither a threshold nor a double-valued behavior, as you can see by reversing the reactions shown in Figures 11.2a and 11.2b, $^3$He(n, p)$^3$H and $^{14}$N(n, p)$^{14}$C, for which we can in each case make the single transformation $-Q \rightarrow +Q$. Figure 11.3 shows the $T_b$ vs $T_a$ graphs for these cases. The reactions occur down to $T_a \rightarrow 0$ (no threshold), and the curves are single-valued for all $\theta$ and $T_a$.

If, for a given $\theta$ and $T_a$, we measure $T_b$, then we can determine the $Q$ value of the reaction and deduce the mass relationships among the constituents. If we know $m_a$, $m_b$, and $m_X$, we then have a way of determining the mass of Y. Solving Equation 11.5 for $Q$, we obtain

$$Q = T_b\left(1 + \frac{m_b}{m_Y}\right) - T_a\left(1 - \frac{m_a}{m_Y}\right) - 2\left(\frac{m_a}{m_Y}\frac{m_b}{m_Y}T_a T_b\right)^{1/2}\cos\theta \quad (11.10)$$

This procedure is not strictly valid, for $m_Y$ also appears on the right side of the equation, but it is usually of sufficient accuracy to replace the masses with the integer mass numbers, especially if we measure at 90° where the last term vanishes.

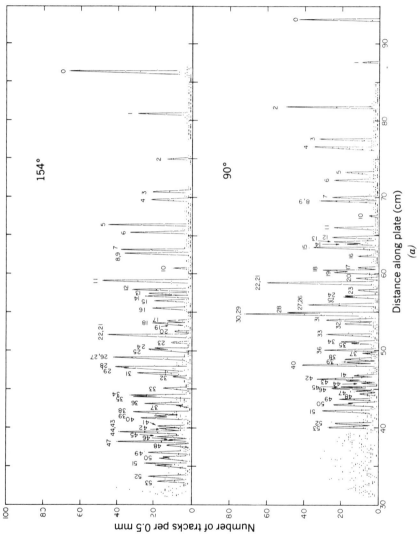

**Figure 11.4** (*a*) Spectrum of protons from the reaction $^{58}$Ni($^4$He, p)$^{61}$Cu. The highest energy proton group populates the ground state of $^{61}$Cu, while the remaining groups lead to excited states (numbered 1, 2, 3, . . .). The spectra taken at angles of 90 and 154° show a very dramatic angular dependence; note especially the change in the cross section for groups 1 and 2 at the two angles. (*b*) γ rays following the reaction. (*c*) Deduced partial level scheme of $^{61}$Cu. Data from E. J. Hoffman, D. G. Sarantites, and N.-H. Lu, *Nucl. Phys. A* **173**, 146 (1971).

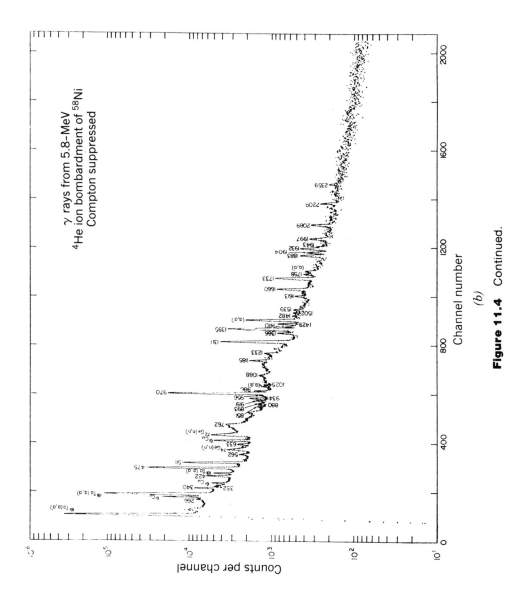

**Figure 11.4** Continued.

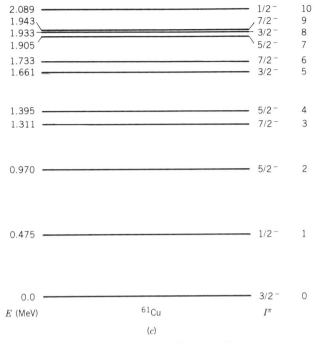

| E (MeV) | | $^{61}$Cu | $I^{\pi}$ | |
|---|---|---|---|---|
| 2.089 | | | 1/2$^-$ | 10 |
| 1.943 | | | 7/2$^-$ | 9 |
| 1.933 | | | 3/2$^-$ | 8 |
| 1.905 | | | 5/2$^-$ | 7 |
| 1.733 | | | 7/2$^-$ | 6 |
| 1.661 | | | 3/2$^-$ | 5 |
| 1.395 | | | 5/2$^-$ | 4 |
| 1.311 | | | 7/2$^-$ | 3 |
| 0.970 | | | 5/2$^-$ | 2 |
| 0.475 | | | 1/2$^-$ | 1 |
| 0.0 | | | 3/2$^-$ | 0 |

(c)

**Figure 11.4**   Continued.

As an example of the application of this technique, we consider the reaction $^{26}$Mg($^{7}$Li, $^{8}$B)$^{25}$Ne. The nucleus $^{26}$Mg already has a neutron excess, and the removal of two additional protons in the reaction results in the final nucleus $^{25}$Ne with a large excess of neutrons. Data reported by Wilcox et al., *Phys. Rev. Lett.* **30**, 866 (1973), show a $^{8}$B peak about 55.8 MeV observed at a lab angle of 10° when the incident $^{7}$Li beam energy is 78.9 MeV. Using Equation 11.10 with mass numbers instead of masses gives $Q = -22.27$ MeV, which gives 24.99790 u for the mass of $^{25}$Ne. Iterating the calculation a second time with the actual masses instead of the mass numbers does not change the result even at this level of precision.

If the reaction reaches excited states of Y, the $Q$-value equation should include the mass energy of the excited state.

$$Q_{ex} = (m_X + m_a - m_Y^* - m_b)c^2$$

$$= Q_0 - E_{ex} \tag{11.11}$$

where $Q_0$ is the $Q$ value corresponding to the ground state of Y, and where we have used $m_Y^* c^2 = m_Y c^2 + E_{ex}$ as the mass energy of the excited state ($E_{ex}$ is the excitation energy above the ground state). The largest observed value of $T_b$ is normally for reactions leading to the ground state, and we can thus use Equation 11.10 to find $Q_0$. Successively smaller values of $T_b$ correspond to higher excited states, and by measuring $T_b$ we can deduce $Q_{ex}$ and the excitation energy $E_{ex}$.

Figure 11.4 shows an example of this kind of measurement. The peaks in the figure serve to determine $T_b$, from which the following $Q$ values and excited-state energies are obtained (energy uncertainties are about $\pm 0.005$ MeV):

| Peak | $Q$ (MeV) | $E_{ex}$ (MeV) |
|------|-----------|----------------|
| 0 | 3.152 | 0.0 |
| 1 | 3.631 | 0.479 |
| 2 | 4.122 | 0.970 |
| 3 | 4.464 | 1.312 |
| 4 | 4.547 | 1.395 |
| 5 | 4.810 | 1.658 |
| 6 | 4.884 | 1.732 |
| 7 | 5.061 | 1.919 |
| 8, 9 | 5.090 | 1.938 |
| 10 | 5.240 | 2.088 |

leading to the excited states shown in the figure. The spectrum of the γ rays emitted following the reaction is also shown in the figure, and transitions can be seen corresponding to each of the deduced values of $E_{ex}$ and therefore interpreted as direct transitions from the excited state to the ground state. Finally, angular distribution studies following the reaction can be used to deduce the spin-parity assignments of the excited states, leading to the level scheme shown in the figure. Notice how the various bits of data complement and supplement one another in building up the level scheme; from the γ rays alone, for example, we cannot tell which transitions connect the ground state with an excited state and therefore what the energies are of the excited states. The proton spectrum, however, gives us the excited-state energies directly, and turning to the γ-ray energies, which can be measured with greater precision, we can obtain more precise values for the energies of the states.

## 11.3 ISOSPIN

The interactions of a nucleon with its surroundings (other nucleons, for instance) in most cases do not depend on whether the nucleon has spin components $m_s = +\frac{1}{2}$ or $m_s = -\frac{1}{2}$ relative to an arbitrarily chosen $z$ axis. That is, there is no need to distinguish in the formalism of nuclear physics between a "spin-up" nucleon and a "spin-down" nucleon. The multiplicity of spin orientations (two, for a single nucleon) may enter into the equations, for example in the statistics of the interaction, but the actual value of the projection does not appear. The exception to this situation comes about when a magnetic field is applied; the magnetic interaction of a nucleon depends on its spin component relative to the direction of the external field.

The charge independence of nuclear forces means that in most instances we do not need to distinguish in the formalism between neutrons and protons, and this leads us to group them together as members of a common family, the nucleons. The formalism for nuclear interactions may depend on the multiplicity of nucleon states (two) but it is independent of whether the nucleons are protons or neutrons. The exception, of course, is the electromagnetic interaction, which can distinguish between protons and neutrons; with respect to the strong nuclear force alone, the symmetry between neutrons and protons remains valid.

This two-state degeneracy leads naturally to a formalism analogous to that of the magnetic interaction of a spin-$\frac{1}{2}$ particle. The neutron and proton are treated as two different states of a single particle, the nucleon. The nucleon is assigned a fictitious spin vector, called the *isospin*.* The two degenerate nuclear states of the nucleon in the absence of electromagnetic fields, like the two degenerate spin states of a nucleon in the absence of a magnetic field, are then "isospin-up," which we arbitrarily assign to the proton, and "isospin-down," the neutron.[†] That is, for a nucleon with isospin quantum number $t = \frac{1}{2}$, a proton has $m_t = +\frac{1}{2}$ and a neutron has $m_t = -\frac{1}{2}$. These projections are measured with respect to an arbitrary axis called the "3-axis" in a coordinate system whose axes are labeled 1, 2, and 3, in order to distinguish it from the laboratory $z$ axis of the $x, y, z$ coordinate system. The isospin obeys the usual rules for angular momentum vectors; thus we use an isospin vector $t$ of length $\sqrt{t(t+1)}\,\hbar$ and with 3-axis projections $t_3 = m_t \hbar$.

For a system of several nucleons, the isospin follows coupling rules identical with the rules of ordinary angular momentum vectors. A two-nucleon system, for example, can have total isospin $T$ of 0 or 1, corresponding (semiclassically) to the antiparallel or parallel orientations of the two isospin-$\frac{1}{2}$ vectors. The 3-axis component of the total isospin vector, $T_3$, is the sum of the 3-axis components of the individual nucleons, and thus for any nucleus,

$$T_3 = \tfrac{1}{2}(Z - N) \tag{11.12}$$

expressed in units of $\hbar$ which will not be shown explicitly.

For a given nucleus, $T_3$ is determined by the number of neutrons and protons. For any value of $T_3$, the total isospin quantum number $T$ can take any value at least as great as $|T_3|$. Two related questions that immediately follow are: Can we assign the quantum number $T$ to individual nuclear states? Is such an assignment useful, for example, in predicting decay or reaction probabilities?

We consider as an example the two-nucleon system, which can have $T$ of 0 or 1. There are thus four possible 3-axis components: $T_3 = +1$ (two protons), $T_3 = -1$ (two neutrons), and two combinations with $T_3 = 0$ (one proton and one neutron). The first two states must have $T = 1$, while the latter two can have $T = 0$ and $T = 1$. If the nuclear interaction is perfectly charge independent (and if we "turn off" the electromagnetic interaction), then the three 3-axis projections of $T = 1$ $(+1, 0, -1)$ must have the same energy, while the single $T = 0$ state may have a different energy. In fact, we know that the isospin triplet (which is the $I = 0$ singlet of ordinary spin) is unbound, as discussed in Chapter 4.

---

*Isospin is often called *isotopic spin* or *isobaric spin*, the former because the value of its projection, equal to $\frac{1}{2}(Z - N)$, distinguishes between isotopes and the latter because the isospin quantum number is valid to label isobaric multiplets. The name "isospin" avoids the controversy and is now the generally accepted term.

[†]Originally, nuclear physicists defined the neutron as the isospin-up member of the nucleon family. Particle physicists also use isospin to label the different charge states of strongly interacting particles, but they stress the connection with electric charge by choosing isospin-up for the proton. This choice has now been accepted by nuclear physicists as well.

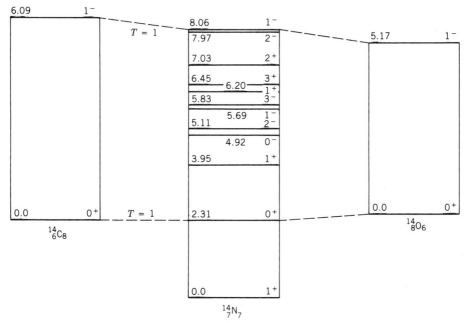

**Figure 11.5** Lower energy levels of $A = 14$ isobars. The ground states of $^{14}$C and $^{14}$O have been shifted relative to $^{14}$N by the neutron-proton mass difference and the Coulomb energy; the respective shifts are 2.36 and 2.44 MeV. Energy levels in $^{14}$C and $^{14}$O have $T = 1$; levels in $^{14}$N have $T = 0$ except the $T = 1$ levels at 2.31 and 8.06 MeV. Based on data compiled by F. Ajzenberg-Selove, *Nucl. Phys. A* **449**, 53 (1986).

A clearer example of the isospin assignments can be found in the nuclei of the $A = 14$ system. Figure 11.5 shows the states of $^{14}$C ($T_3 = -1$), $^{14}$N ($T_3 = 0$), and $^{14}$O ($T_3 = +1$). For $^{14}$N, we know that any integer $T$ can give a 3-axis component of 0, and the possible values of $T$ therefore range from 0 up to a maximum of $A/2$ or 7. The tendency toward nuclear symmetry (reflected in the symmetry term of the semiempirical mass formula) implies that the lowest states will most likely have $T = |T_3|$, that is, the smallest possible value of $T$. This will certainly apply to the ground state, but the excited states must be assigned on the basis of reaction or decay studies or symmetry arguments. In Figure 11.5, the energies have been adjusted so that the neutron-proton mass difference (an electromagnetic effect) and the Coulomb energy of the nucleus have been removed. The energies of the states should then be due only to the nuclear force. Note that the energies of the $0^+$ states in the three nuclei are nearly identical; these are the states of the $T = 1$ triplet. Similar agreement is obtained for the $1^-$ triplet.

Such speculations regarding the $T$ assignments can be verified through decay and reaction studies. For example, angular momentum coupling theory leads to selection rules for $E1$ transitions: $\Delta T$ must be 0 or $\pm 1$, except that transitions from $T = 0$ to $T = 0$ are forbidden and $\Delta T = 0$ transitions are forbidden in nuclei with $T_3 = 0$. To test these rules, we examine the half-lives for the $1^-$ to $0^+$ $E1$ transitions in $^{14}$O, $^{14}$C, and $^{14}$N. The measured half-lives of the analogous states are, respectively, $1.2 \times 10^{-4}$, $< 7$, and 27 fs. The $^{14}$N transition, which is a

$\Delta T = 0$ $E1$ transition in a $T_3 = 0$ nucleus, is forbidden by the isospin selection rule and is indeed strongly inhibited, as its longer half-life indicates. (The Weisskopf estimate for the half-life is about $7 \times 10^{-3}$ fs.)

Consider also the decay of the $1^-$, $T = 0$ level at 5.69 MeV in $^{14}$N. The $E1$ decay to the $1^+$, $T = 0$ ground state should be inhibited by the selection rule, while the $E1$ decay to the $0^+$, $T = 1$ level at 2.31 MeV is permitted. The higher energy transition ought to be greater in intensity by about a factor of 5, owing to the $E^3$ dependence of the $E1$ transition probability, yet the lower energy transition is observed to have about twice the intensity. The effect of the isospin selection rule is a reduction in the expected relative intensity of the 5.69-MeV $E1$ transition by about an order of magnitude.

Similar selection rules operate in $\beta$ decay. The Fermi matrix element is forbidden unless $\Delta T = 0$, which is the case in the mirror decays listed in the top half of Table 9.3. The nonmirror decays are those with $\Delta T = 1$, and the Fermi contribution to the transition is reduced by several orders of magnitude by the violation of the isospin selection rule. The $0^+$ to $0^+$ decays, which on the basis of ordinary angular momentum alone should be pure Fermi decays of the superallowed category as in Table 9.2, are inhibited by three orders of magnitude if $\Delta T \neq 0$; the log $ft$ values rise from about 3.5 for the $\Delta T = 0$ decays permitted by the isospin selection rule to 7 or larger for the $\Delta T \neq 0$ isospin-forbidden decays.

Nuclear reactions also show effects of isospin. Because the nuclear force does not distinguish between protons and neutrons, the isospin must be absolutely conserved in all nuclear reactions. The 3-axis component is automatically conserved when the numbers of protons and neutrons remain constant, but it is also true that the total isospin quantum number $T$ remains invariant in reactions. Consider the reaction $^{16}$O $+ ^2$H $\rightarrow ^{14}$N $+ ^4$He, leading to states in $^{14}$N. All four reacting particles have $T = 0$ ground states; thus $T$ is conserved if the product particles remain in their ground states. Excitation of $^4$He is unlikely in low-energy reactions, for its first excited state is above 20 MeV, and thus it is expected that only $T = 0$ excited states in $^{14}$N can be reached in the reaction; the 2.31-MeV, $T = 1$ state should not be populated. Any small population observed for that state must arise from isospin impurities in the reacting particles. The cross section to reach the 2.31-MeV state is observed to be about 2 orders of magnitude smaller than the cross sections to reach the neighboring $T = 0$ states, showing the effectiveness of the isospin selection rule. In the similar reaction $^{12}$C$(\alpha, d)^{14}$N the cross section for the 2.31-MeV state is 3 orders of magnitude smaller than the isospin-allowed cross sections, and in $^{10}$B$(^6$Li, d$)^{14}$N and $^{12}$C$(^6$Li, $\alpha)^{14}$N it is at least two orders of magnitude smaller. By way of contrast, in $^{10}$B$(^7$Li, $^3$H$)^{14}$N, the $T = 1$ level is populated with a strength comparable to that of the neighboring $T = 0$ level; the isospin selection rule does not inhibit the probability to reach the $T = 1$ level. (The initial nuclei have a total $T$ of $\frac{1}{2}$; the $\frac{1}{2}$ isospin of $^3$H can couple to either $T = 0$ or $T = 1$ in $^{14}$N to give a resultant of $\frac{1}{2}$.)

The members of an isospin multiplet, as for example pairs of mirror nuclei or a set of the three states connected by the dashed lines in Figure 11.5, are called *isobaric analog states*, a term which was previously introduced in the discussion of $\beta$ decay in Section 9.8. The analog states in neighboring nuclei have identical nucleon wave functions, except for the change in the number of protons and neutrons. In the $^{14}$C and $^{14}$O ground states, the nucleons are strongly coupled pairwise (with two coupled proton holes in $^{14}$C and two coupled neutron holes in $^{14}$O), and the 2.31-MeV analog state in $^{14}$N must have a similar wave function, with the odd proton hole and neutron hole strongly paired.

Because analog states are obtained by exchanging a proton for a neutron, they tend to be strongly populated in $\beta$ decay (see Figure 9.17) and in (p, n) or (n, p) reactions. In medium and heavy nuclei, placing a proton into a state formerly occupied by a neutron involves a large energy transfer, because with $N > Z$ the newly placed neutron occupies a considerably higher shell-model state than the former proton. Analog states may appear in medium and heavy nuclei at energies of 10 MeV and above, and thus they generally do not contribute to low-energy reaction and decay studies.

## 11.4 REACTION CROSS SECTIONS

In Chapter 4 we considered the nature of cross sections and the application to nucleon-nucleon scattering. In this section we give some more general definitions of various measurable quantities that are loosely grouped under the heading "cross section."

Roughly speaking, the cross section is a measure of the relative probability for the reaction to occur. If we have a detector placed to record particle b emitted in a direction ($\theta$, $\phi$) with respect to the beam direction, the detector defines a small solid angle $d\Omega$ at the target nucleus (Figure 11.6). Let the current of incident particles be $I_a$ particles per unit time, and let the target show to the beam $N$ target nuclei per unit area. If the outgoing particles appear at a rate $R_b$, then the reaction cross section is

$$\sigma = \frac{R_b}{I_a N} \tag{11.13}$$

Defined in this way, $\sigma$ has the dimension of area per nucleus, but it may be very much larger or smaller than the geometrical area of the disc of the target nucleus

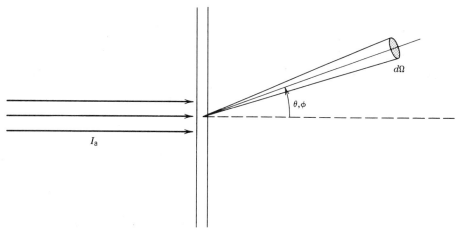

**Figure 11.6** Reaction geometry showing incident beam, target, and outgoing beam going into solid angle $d\Omega$ at $\theta$, $\phi$.

seen by the incoming beam. For a typical nucleus of radius $R = 6$ fm, the geometrical area $\pi R^2$ is about 100 fm$^2 = 1$ b; for neutron capture by $^{135}$Xe, the cross section is about $10^6$ b, while for other much more improbable reactions the cross section may be measured in millibarns or microbarns. You should think of $\sigma$ as a quantity which has the dimension of an area, but which is proportional to the reaction probability.

Our detector occupies only a small solid angle $d\Omega$ and therefore does not observe *all* of the outgoing particles; only a small fraction $dR_b$ are actually counted, and therefore only a fraction of the cross section $d\sigma$ will be deduced. Moreover, the outgoing particles will not in general be emitted uniformly in all directions, but will have an angular distribution that will depend on $\theta$ and possibly also on $\phi$. If we let this angular distribution function be arbitrarily represented by $r(\theta, \phi)$, then $dR_b = r(\theta, \phi)\, d\Omega/4\pi$. (The $4\pi$ is introduced to make $d\Omega/4\pi$ a pure fraction.) Then

$$\frac{d\sigma}{d\Omega} = \frac{r(\theta, \phi)}{4\pi I_a N} \tag{11.14}$$

The quantity $d\sigma/d\Omega$ is called the *differential cross section*, and its measurement gives us important information on the angular distribution of the reaction products. In the literature, it is often called $\sigma(\theta, \phi)$ or $\sigma(\theta)$ or sometimes (unfortunately) just "cross section." (If you see a graph of "cross section" vs $\theta$, you should know that what is intended is differential cross section.) Because solid angle is measured in steradians (the surface of a sphere subtends a solid angle of $4\pi$ steradians at its center), units of differential cross section are barns/steradian. The reaction cross section $\sigma$ can be found by integrating $d\sigma/d\Omega$ over all angles; with $d\Omega = \sin\theta\, d\theta\, d\phi$ we have*

$$\sigma = \int \frac{d\sigma}{d\Omega}\, d\Omega = \int_0^\pi \sin\theta\, d\theta \int_0^{2\pi} d\phi\, \frac{d\sigma}{d\Omega} \tag{11.15}$$

Notice that if $d\sigma/d\Omega$ is constant (independent of angle), the integral gives $\sigma = 4\pi(d\sigma/d\Omega)$. This justifies the insertion of the constant $4\pi$ into Equation 11.14, for now $r(\theta, \phi)$ reduces to the constant $R_b$ and Equation 11.14 agrees with Equation 11.13.

In many nuclear physics applications, we are not concerned simply with the probability to find particle b emitted at a certain angle; we also want to find it with a certain energy, corresponding to a particular energy of the residual nucleus Y We therefore must modify the definition of cross section to give the probability to observe b in the angular range of $d\Omega$ and in the energy range $dE_b$. This gives the so-called *doubly differential cross section* $d^2\sigma/dE_b\, d\Omega$. In the literature, this additional energy dependence is often not explicitly stated; usually the cross sections are plotted as $d\sigma/d\Omega$ vs $\theta$ *leading to a specific final energy state*. This is in reality $d^2\sigma/dE_b\, d\Omega$, although it may not be labeled as such. For discrete states, there may be only a single level within the energy range $dE_b$, and the

---

*An element of area on the surface of a sphere is $r^2\, d\Omega$ or $r^2 \sin\theta\, d\theta\, d\phi$ in spherical coordinates. Hence $d\Omega = \sin\theta\, d\theta\, d\phi$.

**Table 11.1** Reaction Cross Sections

| Cross Sections | Symbol | Technique | Possible Application |
|---|---|---|---|
| Total | $\sigma_t$ | Attenuation of beam | Shielding |
| Reaction | $\sigma$ | Integrate over all angles and all energies of b (all excited states of Y) | Production of radioisotope Y in a nuclear reaction |
| Differential (Angular) | $d\sigma/d\Omega$ | Observe b at $(\theta, \phi)$ but integrate over all energies | Formation of beam of b particles in a certain direction (or recoil of Y in a certain direction) |
| Differential (Energy) | $d\sigma/dE$ | Don't observe b, but observe excitation of Y by subsequent $\gamma$ emission | Study of decay of excited states of Y |
| Doubly differential | $d^2\sigma/dE_b\, d\Omega$ | Observe b at $(\theta, \phi)$ at a specific energy | Information on excited states of Y by angular distribution of b |

distinction becomes unimportant. If, on the other hand, we do not observe the direction of particle b (by surrounding the target area with $4\pi$ solid angle of detectors, or by not observing b at all), then we measure yet another differential cross section $d\sigma/dE$, where now $E$ may represent an excitation energy of Y.

There is still another cross section that may be of interest, the *total cross section* $\sigma_t$. Here, for a specific incident particle a, we add the reaction cross sections $\sigma$ for all possible different outgoing particles b, no matter what their direction or energy. Such a determination would tell us the probability for an incident particle to have *any reaction at all* with the target and thus be removed from the beam of incident particles. This can be deduced directly by measuring the loss in intensity of a collimated beam in passing through a certain thickness of the target material.

When we discuss a specific reaction then, the exact meaning of the term cross section will depend on exactly what we measure. Table 11.1 summarizes these different measurements, how they might be accomplished, and the application to which the result might be put. For example, if we wish to produce a radioactive isotope as the residual nucleus Y, we have absolutely no interest in the direction of emission of particle b, nor in the excited states of Y that may be populated, for they will quickly decay by $\gamma$ emission to the ground state of Y. The literature often does not discriminate carefully among these definitions, and often they are called merely "cross section." It is almost always obvious in context which cross section is meant, and therefore not strictly necessary to distinguish carefully among them.

## 11.5  EXPERIMENTAL TECHNIQUES

A typical nuclear reaction study requires a beam of particles, a target, and a detection system. Beams of charged particles are produced by a variety of different types of accelerators (see Chapter 15), and neutron beams are available from nuclear reactors and as secondary beams from charged-particle accelerators. To do precision spectroscopy of the outgoing particle b and the residual nucleus Y, the beam must satisfy several criteria:

1. It must be *highly collimated and focused*, so that we have a precise reference direction to determine $\theta$ and $\phi$ for angular distribution measurements.

2. It must have a *sharply defined energy*; otherwise, in trying to observe a specific excited state by finding $Q_{ex}$ and $E_{ex}$ from Equation 11.5, we might find that variations in $T_a$ would give two or more different $E_{ex}$ for the same $T_b$.

3. It must be of *high intensity*, so that we can gather the necessary statistics for precise experiments.

4. If we wish to do timing measurements (such as to measure the lifetimes of excited states of Y), the beam must be *sharply pulsed* to provide a reference signal for the formation of the state, and the pulses must be separated in time by at least the time resolution of the measuring apparatus and preferably by a time of the order of the one we are trying to measure.

5. Under ideal circumstances, the accelerator beam should be *easily select-able*—we should be able to change the incident energy $T_a$ or even the type of incident particle in a reasonable time. The stringent tuning requirements of modern large accelerators and the demands that high currents put on ion sources make this requirement hard to meet in practice. Accelerator beam time is often scheduled far in advance (6 months to a year is common), so that experiments with common beam requirements can be grouped together, thus minimizing the beam tuning time.

6. The intensity of the incident beam should be nearly *constant and easily measurable*, for we must know it to determine the cross section. If we move a detector from one position to another, we must know if the change in the observed rate of detection of particle b comes from the angular dependence of the differential cross section or merely from a change in the incident beam intensity.

7. The beam may be *polarized* (that is, the spins of the incident particles all aligned in a certain direction) or *unpolarized*, according to the desire of the experimenters.

8. The beam must be transported to the target through a *high-vacuum system* so as to prevent beam degradation and production of unwanted products by collisions with air molecules.

Types of targets vary widely, according to the goals of the experiment. If we want to measure the yield of a reaction (that is, $\sigma$ or $\sigma_t$), perhaps through observation of the attenuation of the beam or the decay of radioisotope Y, then we may choose a thick, solid target. Such a target might degrade, scatter, or even stop the outgoing particles b, which does not bother us in this kind of measurement. On the other hand, if we wish to observe b unaffected by interactions in the target, a very thin target is required. Thin metal foils are often used as targets, but for nonmetals, including compounds such as oxides, the target material is often placed on a thin backing, which does not contribute to the reaction or affect the passage of particle b. For many applications, extremely rare (and often expensive) targets of separated isotopes are used. A high-intensity, highly focused beam (typically a few mm in diameter) delivers considerable thermal power to the target (absorption of 1 $\mu$A of 10 MeV protons delivers 10 W), which is enough to burn up thin targets; therefore a way must be found to cool the target and extract the heat generated by the beam. As with the beam, it should be relatively easy to change targets so that valuable beam time is not wasted. For some applications, it may be desirable to polarize the spins of the target nuclei.

The detectors may consist of some (or all) of the following: particle detectors or detector telescopes to determine the energy and type of the outgoing particles, magnetic spectrometers for good energy resolution (sometimes necessary to identify close-lying excited states of Y), position-sensitive particle detectors (such as multiwire proportional counters) to do accurate angular distribution work, $\gamma$-ray detectors to observe the de-excitation of the excited states of Y (possibly in coincidence with particle b), polarimeters to measure the polarization of the particles b, and so on. Because beam time is a precious commodity at a modern accelerator facility, the emphasis is always on getting the largest amount of data in the shortest possible time. Therefore multidetector configurations are very common; many signals arrive simultaneously at the detectors and are stored by an on-line computer system for later "re-play" and analysis. (Keeping the beam and the detectors going during the experiment usually demands all the attention of the experimenters and leaves little time for data analysis!)

## 11.6   COULOMB SCATTERING

Because the nucleus has a distribution of electric charge, it can be studied by the electric (Coulomb) scattering of a beam of charged particles. This scattering may be either elastic or inelastic.

Elastic Coulomb scattering is called *Rutherford scattering* because early (1911–1913) experiments on the scattering of $\alpha$ particles in Rutherford's laboratory by Geiger and Marsden led originally to the discovery of the existence of the nucleus. The basic geometry for the scattering is shown in Figure 11.7. As is always the case for unbound orbits in a $1/r^2$ force, the scattered particle follows a hyperbolic path. (We will assume the target nucleus to be infinitely massive, so that the scattering center remains fixed.) The particle approaches the target nucleus along a straight line that would pass a distance $b$ from the nucleus in the absence of the repulsive force; this distance is called the *impact parameter*. The scattering angle is $\theta$. Very far from the nucleus, the incident particle has negligible Coulomb potential energy; its total energy is thus only the incident kinetic energy $T_a = \frac{1}{2}mv_0^2$. Its angular momentum relative to the target nucleus is $|r \times mv| = mv_0 b$ at large distances. In passing close to the target nucleus, the particle reaches a minimum separation distance $r_{min}$ (which depends on $b$), the

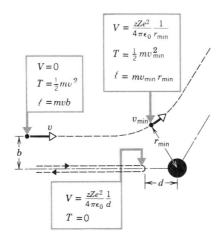

**Figure 11.7**  The trajectory of a particle undergoing Rutherford scattering, showing the closest approach to the target nucleus.

absolute minimum value of which occurs in a head-on collision ($b = 0$), in which the particle comes instantaneously to rest before reversing its motion. At this point it has exchanged its initial kinetic energy for Coulomb potential energy:

$$\tfrac{1}{2}mv_0^2 = \frac{1}{4\pi\epsilon_0}\frac{zZe^2}{d} \tag{11.16}$$

where $ze$ is the charge of the projectile and $Ze$ the target. The distance $d$ is called the *distance of closest approach*. At intermediate points in the trajectory, the energy is partly kinetic and partly potential; conservation of energy gives (for any value of the impact parameter)

$$\tfrac{1}{2}mv_0^2 = \tfrac{1}{2}mv^2 + \frac{1}{4\pi\epsilon_0}\frac{zZe^2}{r} \tag{11.17}$$

The scattering has cylindrical symmetry about the beam axis (because the Coulomb force is symmetric), and therefore the cross section is independent of the azimuthal angle $\phi$. We therefore work in a ring or annular geometry (Figure 11.8). Particles with impact parameters between $b$ and $b + db$ are scattered into the ring at angles between $\theta$ and $\theta + d\theta$. Let the target have $n$ nuclei per unit volume, and assume the target to be thin enough so that it is unlikely to have any "shadowing" of one nucleus by another. The target is considered to be a foil of thickness $x$. Then the number of nuclei per unit area is $nx$, and the fraction $df$ of the incident particles that pass through the annular ring of area $2\pi b \, db$ is

$$df = nx(2\pi b \, db) \tag{11.18}$$

The fraction $f$ with impact parameters less than $b$ is

$$f = nx\pi b^2 \tag{11.19}$$

If particles scattered with impact parameter $b$ emerge at angle $\theta$, then Equation 11.19 also gives the fraction that are scattered at angles greater than $\theta$, but to carry the discussion further we need a relationship between $b$ and $\theta$. (We are

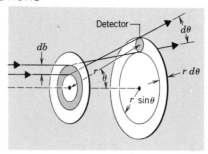

**Figure 11.8** Particles entering the ring between $b$ and $b + db$ are distributed uniformly along a ring of angular width $d\theta$. A detector is at a distance $r$ from the scattering foil.

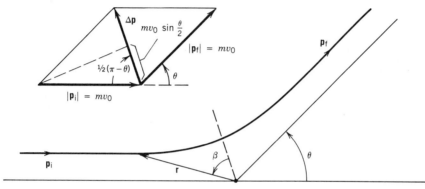

**Figure 11.9** The hyperbolic trajectory of a scattered particle. The instantaneous coordinates are $r, \beta$. The change in momentum is $\Delta\boldsymbol{p}$, in the direction of the dashed line that bisects $(\pi - \theta)$.

assuming that each incident particle is scattered only once—more about this assumption later.)

The net linear momentum of the scattered particles changes in direction only; far from the scattering, the incident and the final linear momentum are both $mv_0$. (This follows from the assumption that the target is so massive that it does not move.) The change in the momentum vector, as shown in Figure 11.9, is a vector of magnitude

$$\Delta p = 2mv_0 \sin \frac{\theta}{2} \tag{11.20}$$

in the direction of the bisector of $\pi - \theta$. According to Newton's second law in the form $F = dp/dt$, this is equal to the net impulse of the Coulomb force in that direction:

$$\Delta p = \int dp = \int F\,dt = \frac{zZe^2}{4\pi\epsilon_0} \int \frac{dt}{r^2} \cos\beta \tag{11.21}$$

where $\beta$ is the angle between the bisector and instantaneous vector $\boldsymbol{r}$ locating the particle. In the initial position far from the scattering, which we take to be time $t = 0$, the angle $\beta$ has the value $-(\pi/2 - \theta/2)$; in the final position ($t = \infty$), the angle $\beta$ is $+(\pi/2 - \theta/2)$.

The instantaneous velocity $v$ can be written in terms of radial (along $r$) and tangential components:

$$v = \frac{dr}{dt}\hat{r} + r\frac{d\beta}{dt}\hat{\beta}$$ (11.22)

where $\hat{r}$ and $\hat{\beta}$ indicate unit vectors in the radial and tangential directions, respectively. Only the tangential component contributes to the angular momentum about the nucleus:

$$\ell = |m\mathbf{r} \times \mathbf{v}| = mr^2\frac{d\beta}{dt}$$ (11.23)

Far from the nucleus, the angular momentum has the value $mv_0b$; conservation of angular momentum gives

$$mv_0b = mr^2\frac{d\beta}{dt}$$

$$\frac{dt}{r^2} = \frac{d\beta}{v_0b}$$ (11.24)

and substituting into Equation 11.21 gives

$$\Delta p = \frac{zZe^2}{4\pi\epsilon_0 v_0 b}\int_{-(\pi/2-\theta/2)}^{+(\pi/2-\theta/2)}\cos\beta\,d\beta$$

$$= \frac{zZe^2}{2\pi\epsilon_0 v_0 b}\cos\frac{\theta}{2}$$ (11.25)

Combining this result with Equation 11.20 gives the needed relationship between $b$ and $\theta$:

$$b = \frac{d}{2}\cot\frac{\theta}{2}$$ (11.26)

where $d$ is the distance of closest approach from Equation 11.16. Combining Equations 11.18 and 11.26,

$$|df| = \pi nx\frac{d^2}{4}\cot\frac{\theta}{2}\csc^2\frac{\theta}{2}\,d\theta$$ (11.27)

and the rate at which particles reach the ring, per unit solid angle, is

$$r(\theta,\phi) = \frac{I_a|df|}{d\Omega/4\pi}$$ (11.28)

where $I_a$ is the rate at which incident particles fall on the target (and hence $I_a|df|$ is the number that fall between $b$ and $b + db$). With $d\Omega = 2\pi\sin\theta\,d\theta$ for the ring geometry (that is, $\sin\theta\,d\theta\,d\phi$ integrated over $\phi$), the net result is

$$\frac{d\sigma}{d\Omega} = \left(\frac{zZe^2}{4\pi\epsilon_0}\right)^2\left(\frac{1}{4T_a}\right)^2\frac{1}{\sin^4\frac{\theta}{2}}$$ (11.29)

This is the differential cross section for Rutherford scattering, usually called the

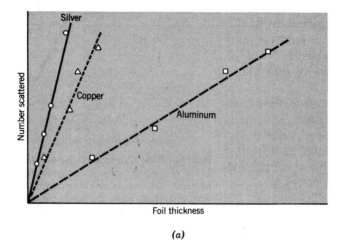

(a)

**Figure 11.10** (a) The dependence of scattering rate on foil thickness for three different scattering foils.

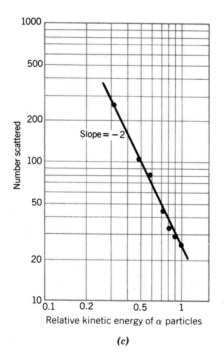

(c)

**Figure 11.10** (c) The dependence of scattering rate on the kinetic energy of the incident α particles for scattering by a single foil. Note the log-log scale; the slope of −2 shows that $\log N \propto -2 \log T$, or $N \propto T^{-2}$, as expected from the Rutherford formula.

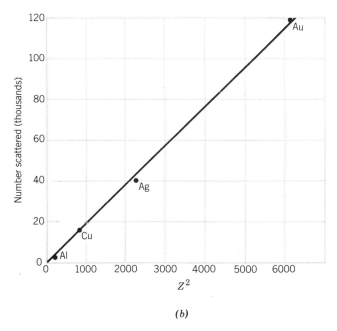

(b)

**Figure 11.10** (b) The dependence of scattering rate on the nuclear charge $Z$ for foils of different materials. The data are plotted against $Z^2$

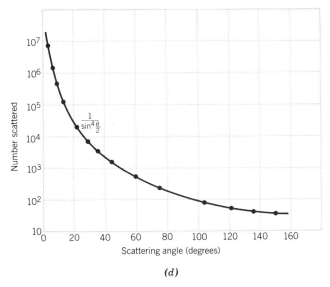

(d)

**Figure 11.10** (d) The dependence of scattering rate on the scattering angle $\theta$, using a gold foil. The $\sin^{-4}(\theta/2)$ dependence is exactly as predicted by the Rutherford formula.

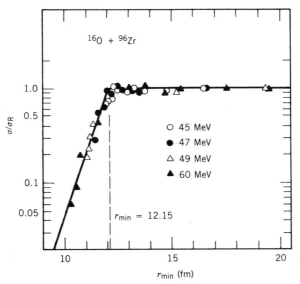

**Figure 11.11** Elastic scattering of $^{16}$O from $^{96}$Zr at several incident energies. The horizontal axis shows the minimum separation distance $r_{min}$ between projectile and target, which varies with $b$ and therefore with $\theta$. The vertical axis shows the cross section in terms of the calculated Rutherford cross section. Nuclear scattering effects appear at separations of less than 12.15 fm; this corresponds to $R_0 = 1.7$ fm, considerably greater than the mean radius of 1.25 fm, but consistent with a "skin thickness" of about 0.5 fm which allows the two nuclear distributions to overlap at these larger distances. From P. R. Christensen et al., *Nucl. Phys. A* **207**, 33 (1973).

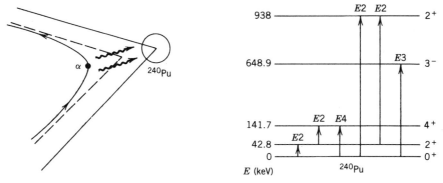

**Figure 11.12** Inelastic Coulomb scattering (Coulomb excitation). The projectile exchanges energy with the target through the Coulomb interaction (exchanged photons are shown as wavy lines) and the target $^{240}$Pu, originally in its ground state, can be driven to excited states. Several different modes of excitation are shown, including two-step processes. The spectrum of inelastically scattered $\alpha$'s shows which excited states of $^{240}$Pu have been excited. Data from C. E. Bemis, Jr., et al., *Phys. Rev. C* **8**, 1466 (1973).

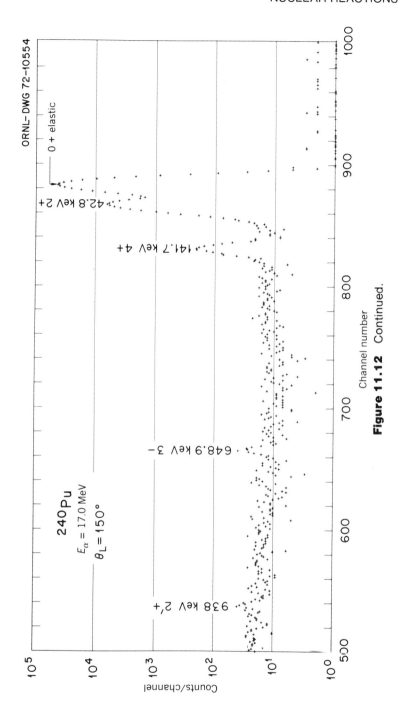

**Figure 11.12**  Continued.

*Rutherford cross section.* Note especially the $\sin^{-4}(\theta/2)$ dependence, which is characteristic.

In a difficult and painstaking series of experiments, Geiger and Marsden verified three aspects of the Rutherford formula: the dependence on $Z^2$, $T_a^{-2}$, and $\sin^{-4}(\theta/2)$. Figure 11.10 shows the excellent agreement with the predictions of the formula.

The most noteworthy aspect of the scattering experiment, and the detail that led Rutherford to his concept of the nuclear atom, is the fraction scattered at large angles, say beyond 90°. For example, we consider a gold foil of thickness $2.0 \times 10^{-4}$ cm, on which 8.0 MeV $\alpha$'s are incident. From Equation 11.26, we find $b = 14$ fm, from which Equation 11.19 gives $f = 7.5 \times 10^{-5}$. This is a large fraction to be scattered at such angles and requires a dense, compact nucleus as the scatterer.

Let's look at the situation for small angles. For the above gold foil, there are about $12 \times 10^{18}$ nuclei/cm$^2$, or of the order of 0.003 nm lateral spacing between nuclei, as seen by an incoming $\alpha$ particle. This means about $\frac{2}{3}$ of the $\alpha$ particles have an impact parameter of 0.001 nm or greater. For this impact parameter, the scattering angle is 1.6°. Thus the mean scattering angle is of order 1° or less. To appear at large angles, we must have either many scatterings, each at a small angle, or a single scattering at large angle. Of course, if there are many individual scatterings of a random nature, some will tend to increase the net scattering angle and others will tend to decrease it. To observe scattering at a total angle of about $N\theta_{\text{mean}}$, there must be about $N^2$ individual scatterings. If we observe scattering at a fixed angle $\theta$ much larger than 1° and if we vary the thickness $x$ of the scattering foil, then we expect the probability to observe scattered particles to vary as $\sqrt{x}$ for multiple scattering, while it should vary with $x$ in the case of single scattering (simply because there are linearly more chances to have single large-angle scattering as the number of nuclei increases). Figure 11.10*a* shows the variation of the number of scattered particles with $x$, and the linear behavior is quite apparent.

Our treatment of Rutherford scattering has been based entirely on classical concepts; no quantum effects are included. In particular, the uncertainty principle renders doubtful any treatment based on fixed trajectories and particle orbits. Any attempt to locate a particle with an impact parameter of arbitrarily small uncertainty would introduce a large uncertainty in the transverse momentum and thus in the scattering angle. We are not discussing the experimental difficulty of "aiming" a beam with a specific impact parameter; the range of impact parameters is automatically included in the variation with $\theta$ of $d\sigma/d\Omega$. What we discuss here is whether the assumption of a definite trajectory has introduced pathological errors into the derivation of the Rutherford cross section.

Corresponding to an uncertainty $\Delta b$ in the impact parameter will be an uncertainty $\Delta p$ in the transverse momentum of order $\hbar/\Delta b$. Our classical derivation makes sense only if $\Delta b \ll b$ and $\Delta p \ll p_{\text{transverse}}$:

$$b \, \Delta p_{\text{transverse}} \gg \Delta b \, \Delta p \gtrsim \hbar$$

$$\frac{b \, \Delta p_{\text{transverse}}}{\hbar} \gg 1 \tag{11.30}$$

We consider two extreme cases: (1) 90° scattering, for which $b = d/2 = 14$ fm and $\Delta p_{\text{transverse}} = mv_0 = 250$ MeV/$c$, where we have assumed 8-MeV $\alpha$'s incident on gold. For this case the ratio in Equation 11.30 is about 18, reasonably far from the quantum limit. (2) Small-angle scattering ($\theta \simeq 1°$), with $b = 1600$ fm and $\Delta p_{\text{transverse}} \simeq mv_0 \tan \theta \simeq 4$ MeV/$c$. The ratio is now about 32, again far from the quantum limit.

Ultimately what justifies the classical calculation is a happy accident of quantum physics: the quantum calculation of the Coulomb scattering cross section gives the same result as the classical calculation, Equation 11.29. This is a peculiarity of the $1/r^2$ force, in which the exact quantum result contains no factors of $\hbar$, and thus the "classical limit" of $\hbar \to 0$ leaves the quantum result unchanged.

As we increase the energy of the incident particle, we will eventually reach a point where the distance of closest approach decreases to the nuclear radius, and thus the projectile and target feel each other's nuclear force. The Rutherford formula, which was derived on the basis of Coulomb interactions only, fails at that point to account for the cross section, as we illustrated in Figure 3.11. (The cross section then includes Coulomb and nuclear parts, as in the case of proton-proton scattering, Equation 4.43.) The internuclear separation at which the Rutherford formula fails is then a measure of the nuclear radius, as illustrated in Figure 11.11.

Up to now we have considered only elastic Coulomb scattering. Inelastic Coulomb scattering is called *Coulomb excitation*; after the encounter the nucleus (and possibly, although not usually, the projectile) is left in an excited state, from which it decays rapidly with the emission of $\gamma$ rays. We can think of this process as the emission and absorption of virtual photons, with the most likely mode being $E2$. This process has therefore been extensively used to study the first excited $2^+$ states of even-$Z$, even-$N$ nuclei. Because the $0^+ \to 2^+$ photon absorption probability is closely related to the $2^+ \to 0^+$ photon emission probability, the Coulomb excitation probability can give a measure of the half-life of the $2^+$ state. Moreover, since the $2^+$ state lives much longer than the time it takes for the encounter between target and projectile, there is a second-order interaction between the projectile and the excited-state nuclei of the target. This can have several effects, including photon absorption causing a $2^+ \to 4^+$ upward transition and a change in the $m$-state population of the $2^+$ state from the interaction of its quadrupole moment with the electric field gradient of the moving projectile.

Figure 11.12 shows some sample results from inelastic Coulomb scattering. The reduced energy of the detected particles exactly matches the energy simultaneously observed in $\gamma$-ray emission from the excited states.

## 11.7  NUCLEAR SCATTERING

The elastic nuclear scattering of particles bears a strong resemblance to a familiar problem from optics: the diffraction of light by an opaque disk (Figures 4.3 and 11.13). In the optical case, diffraction at the sharp edge results in a series of maxima and minima; the first minimum occurs at $\theta \sim \lambda/R$, the succeeding minima are roughly (but not exactly) equally spaced, and the intervening maxima are of steadily and substantially decreasing intensity.

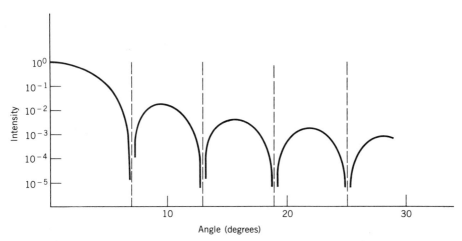

**Figure 11.13** Diffraction pattern of light incident on a circular aperture; a circular disk gives a similar pattern. The minima have intensity of zero. The curve is drawn for a wavelength equal to ten times the diameter of the aperture or disk.

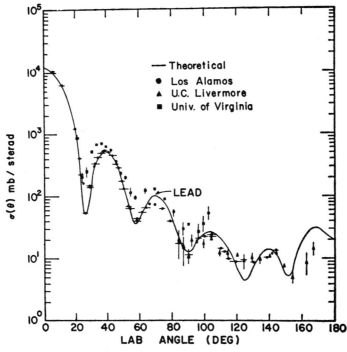

**Figure 11.14** Elastic scattering of 14-MeV neutrons from Pb. From S. Fernbach, *Rev. Mod. Phys.* **30**, 414 (1958).

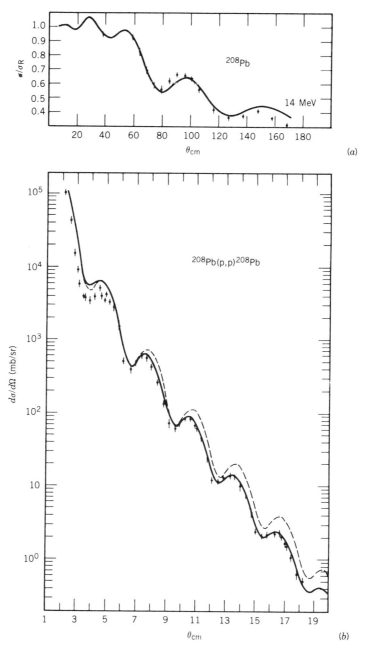

**Figure 11.15** Elastic scattering of protons from $^{208}$Pb. In (*a*), at low energy (14 MeV), the diffraction-like behavior occurs only at large angles (beyond 60°), where nuclear scattering occurs, because the closest distance between projectile and target (12.6 fm at 60° from the Rutherford formulas) agrees with the internuclear distance appropriate for nuclear interactions (11.8 fm), calculated using $R_0 = 1.7$ fm, as in Figure 11.11. Compare this figure with that for neutron scattering at the same energy, Figure 11.14. In (*b*), the incident energy is 1050 MeV and the Coulomb barrier is easily penetrated, so diffraction effects occur at small angles. (*a*) From J. S. Eck and W. J. Thompson, *Nucl. Phys. A* **237**, 83 (1975). (*b*) From G. Igo, in *High Energy Physics and Nuclear Structure — 1975,* edited by D. Nagle et al. (New York: American Institute of Physics, 1975).

A nucleus is a strongly absorbing object for nucleons, and thus the analogy with the opaque disk is quite valid. For charged particles, we must deal with the interference between nuclear and Coulomb scattering, as in Figure 4.9 and Equation 4.43. It is this effect that is responsible for the deviation of scattering cross sections from the Rutherford formula, as shown in Figure 11.11. If we wish to observe the elastic scattering of nucleons, in the form of the "diffraction-like" pattern, we must eliminate the effects of Rutherford scattering, which we can do in either of two ways. The first is to use uncharged neutrons as the scattered particle. Figure 11.14 shows an example of neutron elastic scattering. One particular difference between the nuclear scattering and optical diffraction is that the minima do not fall to zero. This is a direct result of the diffuseness of the nuclear surface—nuclei do not have sharp edges.

For charged particles, we must take two steps to reduce the effect of interference with Coulomb scattering: we work at higher energy, so that the Rutherford cross section is small and the projectile can more easily penetrate to feel the nuclear interaction, and we observe at larger angles, where again the Rutherford cross section is small and where the small impact parameter also helps to guarantee nuclear penetration. An example of nucleon elastic scattering is shown in Figure 11.15. Again, the diffraction-like effects are apparent.

One result of nucleon elastic scattering studies is the determination of the nuclear radius. Although the value may depend somewhat on the potential model used to analyze the scattering (such as the square well discussed in Chapter 4), the results are generally quite consistent with $R = R_0 A^{1/3}$ with $R_0 = 1.25$ fm as in other studies. In Section 11.9 we discuss in more detail the implication of these experiments on our knowledge of the potential.

Inelastic nuclear scattering, like inelastic Coulomb scattering, results when the target nucleus takes energy from the projectile and reaches excited states. (It is also possible for projectiles to be placed in excited states; we ignore this effect for now.) If we measure the energy distribution of scattered projectiles at a fixed angle, we observe a single elastic peak, which is the highest energy scattered projectile. Each inelastic peak corresponds to a specific excited state of the target nucleus. Figure 11.4 showed an example of inelastic nuclear scattering, and another example is discussed in Section 11.11. From the locations of the inelastic peaks, we can learn the energies of the excited states; from their relative heights we learn the relative cross sections for excitation of each state, which tells us something of the wave function of the excited state. We can also measure an angular distribution of scattered projectiles for any excited state, from which we can learn the spin and parity of the excited states.

## 11.8   SCATTERING AND REACTION CROSS SECTIONS

In this section we cover some details of reaction cross sections more thoroughly than in our previous discussion in Section 4.2. You may wish to review that discussion before proceeding.

We take the $z$ axis to be the direction of the incident beam and assume it can be represented by a plane wave $e^{ikz}$ corresponding to momentum $p = \hbar k$. The outgoing particles will be represented by spherical waves, and so the manipulations become easier if we express the incident plane wave as a superposition of spherical waves:

$$\psi_{\text{inc}} = Ae^{ikz} = A \sum_{\ell=0}^{\infty} i^{\ell}(2\ell + 1)\, j_{\ell}(kr)\, P_{\ell}(\cos\theta) \qquad (11.31)$$

where $A$ is an appropriately chosen normalization constant. The radial functions $j_{\ell}(kr)$ are *spherical Bessel functions* which were previously given in Table 2.3; they are solutions to the radial part of the Schrödinger equation, Equation 2.60, in a region far from the target where the nuclear potential vanishes. The angular functions $P_{\ell}(\cos\theta)$ are Legendre polynomials:

$$P_0(\cos\theta) = 1$$
$$P_1(\cos\theta) = \cos\theta \qquad (11.32)$$
$$P_2(\cos\theta) = \tfrac{1}{2}(3\cos^2\theta - 1)$$

This expansion of the incident (and eventually, the scattered) wave is called the *partial wave expansion*, with each partial wave corresponding to a specific angular momentum $\ell$. Such a procedure is valid if the nuclear potential is assumed to be central. What makes the method useful is that it is often sufficient to consider the effect of the nuclear potential on at most only a few of the lowest partial waves (such as $\ell = 0$ or s-wave nucleon-nucleon scattering discussed in Chapter 4). If the particles of momentum $p$ interact with impact parameter $b$, then the (semiclassical) relative angular momentum will be

$$\ell\hbar = pb$$

or

$$b = \ell\frac{\hbar}{p} = \ell\frac{\lambda}{2\pi} = \ell\lambdabar \qquad (11.33)$$

where $\lambdabar = \lambda/2\pi$ is called the *reduced* de Broglie wavelength. Incidentally, $\lambdabar = k^{-1}$.

According to quantum mechanics, $\ell$ can only be defined in integer units, and thus the semiclassical estimate should be revised somewhat. That is, particles with (semiclassical) angular momenta between $0\hbar$ and $1\hbar$ will interact through impact parameters between 0 and $\lambdabar$, and thus effectively over an area (cross section) of at most $\pi\lambdabar^2$. With $\hbar \leq \ell \leq 2\hbar$, the cross section is a ring of inner radius $\lambdabar$ and outer radius $2\lambdabar$, and thus of area $3\pi\lambdabar^2$. We can thus divide the interaction area into a number of zones, each corresponding to a specific angular momentum $\ell$ and each having area $\pi[(\ell + 1)\lambdabar]^2 - \pi(\ell\lambdabar)^2 = (2\ell + 1)\pi\lambdabar^2$. We can estimate the maximum impact parameter for nuclear scattering to be about $R = R_1 + R_2$ (the sum of the radii of the incident and target nuclei), and thus the maximum $\ell$ value likely to occur is $R/\lambdabar$, and the total cross section is correspondingly

$$\sigma = \sum_{\ell=0}^{R/\lambdabar} (2\ell + 1)\pi\lambdabar^2 = \pi(R + \lambdabar)^2 \qquad (11.34)$$

This is a reasonable estimate, for it includes not only an interaction distance $R$, but it allows the incident particle's wave nature to spread over a distance of the order of $\lambdabar$, making the effective interaction radius $(R + \lambdabar)$. We will see later how the exact calculation modifies this estimate.

When the wave is far from the nucleus, the $j_\ell(kr)$ have the following convenient expansion:

$$j_\ell(kr) \cong \frac{\sin\left(kr - \frac{1}{2}\ell\pi\right)}{kr} \qquad (kr \gg \ell)$$

$$= i\,\frac{e^{-i(kr - \ell\pi/2)} - e^{+i(kr - \ell\pi/2)}}{2kr} \qquad (11.35)$$

so that

$$\psi_{\text{inc}} = \frac{A}{2kr} \sum_{\ell=0}^{\infty} i^{\ell+1}(2\ell + 1)\left[e^{-i(kr - \ell\pi/2)} - e^{+i(kr - \ell\pi/2)}\right]P_\ell(\cos\theta) \quad (11.36)$$

The first term in brackets, involving $e^{-ikr}$, represents an incoming spherical wave converging on the target, while the second term, in $e^{+ikr}$, represents an outgoing spherical wave emerging from the target nucleus. The superposition of these two spherical waves, of course, gives the plane wave.

The scattering can affect only the outgoing wave, and can affect it in either of two ways: through a change in phase (as in the phase shift discussed in Chapter 4), and through a change in amplitude. The change in amplitude suggests that there may be fewer particles coming out than there were going in, which may appear to be a loss in the net number of particles. However, keep in mind that the wave function represents only those particles of momentum $\hbar k$. If there is inelastic scattering (or some other nuclear reaction), the energy (or even the identity) of the outgoing particle may change. It is therefore not surprising that there may be fewer particles in the $e^{ikr}$ term following inelastic scattering. It has become customary to refer to a specific set of conditions (exclusive of direction of travel) of the outgoing particle and residual nucleus as a reaction *channel*. The reaction may thus proceed through the elastic channel or through any one of many inelastic channels. Some channels may be *closed* to the reacting particles, if there is not enough energy or angular momentum to permit a specific final configuration to be reached.

We account for the changes in the $\ell$th outgoing partial wave by introducing the complex coefficient $\eta_\ell$ into the outgoing ($e^{ikr}$) term of Equation 11.36:

$$\psi = \frac{A}{2kr} \sum_{\ell=0}^{\infty} i^{\ell+1}(2\ell + 1)\left[e^{-i(kr - \ell\pi/2)} - \eta_\ell e^{+i(kr - \ell\pi/2)}\right]P_\ell(\cos\theta) \quad (11.37)$$

This wave represents a superposition of the incident and scattered waves: $\psi = \psi_{\text{inc}} + \psi_{\text{sc}}$, exactly as in Equation 4.23. To find the scattered wave itself, we subtract Equation 11.37 from Equation 11.36:

$$\psi_{\text{sc}} = \frac{A}{2kr} \sum_{\ell=0}^{\infty} i^{\ell+1}(2\ell + 1)(1 - \eta_\ell)e^{i(kr - \ell\pi/2)}P_\ell(\cos\theta)$$

$$= \frac{A}{2k}\frac{e^{ikr}}{r} \sum_{\ell=0}^{\infty} (2\ell + 1)i(1 - \eta_\ell)P_\ell(\cos\theta) \qquad (11.38)$$

Because we have accounted for only those parts of $\psi_{sc}$ with wave number $k$ identical with the incident wave, this represents only *elastic scattering*. As we did in Equation 4.24 we now find the scattered current density:

$$j_{sc} = \frac{\hbar}{2mi}\left(\psi_{sc}^* \frac{\partial \psi_{sc}}{\partial r} - \frac{\partial \psi_{sc}^*}{\partial r}\psi_{sc}\right) \tag{11.39}$$

$$= |A|^2 \frac{\hbar}{4mkr^2}\left|\sum_{\ell=0}^{\infty}(2\ell+1)i(1-\eta_\ell)P_\ell(\cos\theta)\right|^2 \tag{11.40}$$

The incident current is identical with Equation 4.26:

$$j_{inc} - \frac{\hbar k}{m}|A|^2 \tag{11.41}$$

and by analogy with Equation 4.27, the differential cross section is

$$\frac{d\sigma}{d\Omega} = \frac{1}{4k^2}\left|\sum_{\ell=0}^{\infty}(2\ell+1)i(1-\eta_\ell)P_\ell(\cos\theta)\right|^2 \tag{11.42}$$

To find the total cross section, we require the integral of the Legendre polynomials:

$$\int P_\ell(\cos\theta)P_{\ell'}(\cos\theta)\sin\theta\, d\theta\, d\phi = \frac{4\pi}{2\ell+1} \qquad \text{if } \ell=\ell'$$

$$= 0 \qquad \text{if } \ell \neq \ell' \tag{11.43}$$

Thus

$$\sigma_{sc} = \sum_{\ell=0}^{\infty}\pi\lambda^2(2\ell+1)|1-\eta_\ell|^2 \tag{11.44}$$

If elastic scattering were the only process that could occur, then $|\eta_\ell| = 1$ and it is conventional to write $\eta_\ell = e^{2i\delta_\ell}$ where $\delta_\ell$ is the *phase shift* of the $\ell$th partial wave. For this case, $|1-\eta_\ell|^2 = 4\sin^2\delta_\ell$ and

$$\sigma_{sc} = \sum_{\ell=0}^{\infty}4\pi\lambda^2(2\ell+1)\sin^2\delta_\ell \tag{11.45}$$

which reduces directly to Equation 4.30 for $\ell = 0$.

If there are other processes in addition to elastic scattering (inelastic scattering or other reactions) then Equation 11.45 is not valid, because $|\eta_\ell| < 1$. We group all of these processes together under the term *reaction cross section* $\sigma_r$, where we take "reaction" to mean all nuclear processes *except* elastic scattering. To find this cross section, we must examine Equation 11.37 to find the rate at which particles are "disappearing" from the channel with wave number $k$. That is, we find the difference between the incoming current and the outgoing current using, respectively, the first and second terms of Equation 11.37:

$$|j_{\text{in}}| - |j_{\text{out}}| = \frac{|A|^2 \hbar}{4mkr^2}\left\{\left|\sum_{\ell=0}^{\infty}(2\ell+1)i^{\ell+1}e^{i\ell\pi/2}P_\ell(\cos\theta)\right|^2\right.$$

$$\left. -\left|\sum_{\ell=0}^{\infty}(2\ell+1)i^{\ell+1}\eta_\ell e^{-i\ell\pi/2}P_\ell(\cos\theta)\right|^2\right\} \quad (11.46)$$

and the reaction cross section becomes

$$\sigma_r = \sum_{\ell=0}^{\infty}\pi\lambda^2(2\ell+1)\left(1-|\eta_\ell|^2\right) \quad (11.47)$$

The *total* cross section, including all processes, is

$$\sigma_t = \sigma_{\text{sc}} + \sigma_r$$

$$= \sum_{\ell=0}^{\infty}2\pi\lambda^2(2\ell+1)(1-\text{Re }\eta_\ell) \quad (11.48)$$

You should note the following details about these results:

1.  It is possible to have elastic scattering in the absence of other processes; that is, if $|\eta_\ell| = 1$, then Equation 11.47 vanishes. It is *not* possible, however, to have reactions without also having elastic scattering; that is, any choice of $\eta_\ell$ for which $\sigma_r \neq 0$ for a given partial wave automatically gives $\sigma_{\text{sc}} \neq 0$ for that partial wave. We can understand this with reference to the diffraction model of scattering we considered in Section 11.7. If particles are removed from the incident beam, creating a "shadow" behind the target nucleus, incident particles will be diffracted into the shadow.

2.  For a "black disk" absorber, as in Equation 11.34, in which all partial waves are completely absorbed up to $\ell = R/\lambda$ ($\eta_\ell = 0$ for complete absorption) and unaffected for $\ell > R/\lambda$ ($\eta_\ell = 1$), then

$$\sigma_{\text{sc}} = \pi(R+\lambda)^2 \quad (11.49)$$

and

$$\sigma_r = \pi(R+\lambda)^2 \quad (11.50)$$

so that

$$\sigma_t = 2\pi(R+\lambda)^2 \quad (11.51)$$

The total cross section is *twice* the geometrical area! The explanation for this nonclassical effect can also be found in the "shadow" region—the target nucleus cannot simply absorb and throw a sharp shadow. It must also diffract into the shadow region.

The program for using these results to study nuclear structure is similar to that of Chapter 4 for nucleon–nucleon scattering. We can guess at a form for the nuclear potential, solve the Schrödinger equation inside the interaction region

$0 \leq r \leq R$, and match boundary conditions at the surface. In this way we should be able to calculate $\eta_\ell$ and, by comparison with experimental values of $\sigma_{sc}$ and $\sigma_r$, evaluate whether our chosen form for the potential is reasonable. In practice this is very difficult for all but the elastic channel because all of the inelastic and reaction channels are coupled together leading to a complicated system of coupled equations. We discuss one particular technique, the optical model for elastic scattering, in Section 11.9.

## 11.9 THE OPTICAL MODEL

A simple model used to account in a general way for elastic scattering in the presence of absorptive effects is the *optical model*, so called because the calculation resembles that of light incident on a somewhat opaque glass sphere. (The model is also called the "cloudy crystal ball model.")

In this model, we represent the scattering in terms of a complex potential $U(r)$:

$$U(r) = V(r) + iW(r) \tag{11.52}$$

where the real functions $V$ and $W$ are selected to give the potential its proper radial dependence. The real part, $V(r)$, is responsible for the elastic scattering; it describes the ordinary nuclear interaction between target and projectile and may therefore be very similar to a shell-model potential. The imaginary part, $W(r)$, is responsible for the absorption. We can demonstrate this by considering a square-well form for $U(r)$:

$$
\begin{aligned}
U(r) &= -V_0 - iW_0 \qquad r < R \\
&= 0 \qquad\qquad\quad\; r > R
\end{aligned}
\tag{11.53}
$$

The outgoing scattered wave we take to be in the form of $e^{ikr}/r$, with $k = \sqrt{2m(E + V_0 + iW_0)/\hbar^2}$, which follows from solving the Schrödinger equation in the usual way for this potential. The wave number $k$ is thus complex: $k = k_r + ik_i$, where $k_r$ and $k_i$ are the real and imaginary parts, respectively. The wave function behaves like $e^{ik_r r} \cdot e^{-k_i r}/r$, and the radial probability density is proportional to $e^{-2k_i r}$. The wave is therefore exponentially attenuated as it passes through the nucleus. (Choosing $W_0 > 0$ in Equation 11.53 gives a loss in intensity, rather than a gain.) If we assume that the absorption is relatively weak (that is, $W_0$ is small compared with $E + V_0$), then we can use the binomial theorem to expand the expression for $k$:

$$k \cong \sqrt{\frac{2m(E + V_0)}{\hbar^2}} + \frac{iW_0}{2}\sqrt{\frac{2m}{\hbar^2}\left(\frac{1}{E + V_0}\right)} \tag{11.54}$$

The usual shell-model potential has a depth $V_0$ of the order of 40 MeV, and we can take $E = 10$ MeV for a typical low-energy projectile. The distance over which the intensity is attenuated by $e^{-1}$ (a sort of mean free path) is

$$d = \frac{1}{2k_i} = \frac{1}{W_0}\sqrt{\frac{\hbar^2(E + V_0)}{2m}} \tag{11.55}$$

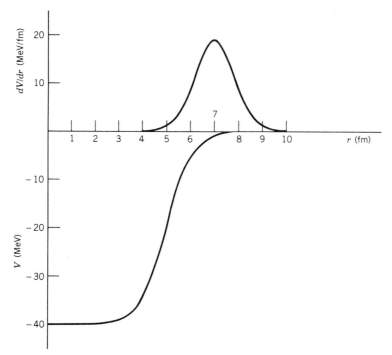

**Figure 11.16**  The optical model functions $V(r)$ and $W(r) = dV/dr$. Typical parameters chosen are $V_0 = 40$ MeV, $R = 1.25A^{1/3}$, $a = 0.523$ fm, and $A = 64$.

If this distance is to be at most of the order of the nuclear radius (say, 3 fm) then $W_0 \simeq 11$ MeV. Thus for the usual case, in which absorption is relatively weaker than elastic scattering, we estimate $|V| \sim 40$ MeV, $|W| \sim 10$ MeV.

The procedure for applying the optical model might be as follows: First, we must choose a form for the potential. The square-well form is often adequate (with $R \simeq 1.4A^{1/3}$, a bit larger than usual to account for the diffuse nuclear surface), but a more detailed form is often chosen:

$$V(r) = \frac{-V_0}{1 + e^{(r-R)/a}} \tag{11.56}$$

exactly as in the case of the shell model, Equation 5.1. The constants $V_0$, $R$, and $a$ are adjusted to give the best fits with the scattering data. The absorptive part $W(r)$ at low energies must have a very different form. Because of the exclusion principle, the tightly bound nucleons in the nuclear interior cannot participate in absorbing incident nucleons. Only the "valence" nucleons near the surface can absorb the relatively low energy carried by the incident particle. The function $W(r)$ is thus often chosen as proportional to $dV/dr$, which has the proper shape of being large only near the surface, as shown in Figure 11.16. (At higher energy, where the inner nucleons can also participate in absorption, $W(r)$ may look more like $V(r)$.) A *spin-orbit* term is also included in modern optical potentials. It is also peaked near the surface, because the spin density of the inner nucleons vanishes. Finally, a Coulomb term must be included if the incident particle is

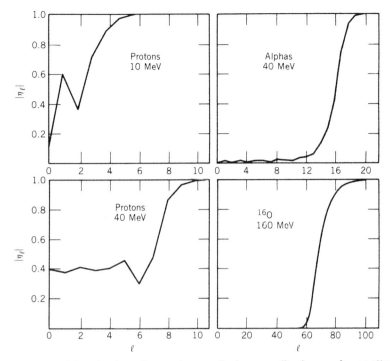

**Figure 11.17** Magnitudes of complex scattering amplitudes $\eta_\ell$ for scattering of various projectiles on a target of $^{58}$Ni. The approach of $|\eta_\ell|$ to 1 at higher energy corresponds to $\sigma_r \to 0$, so that few particles are absorbed and only elastic scattering takes place. From G. R. Satchler, *Introduction to Nuclear Reactions* (Wiley: New York, 1980).

charged. For the chosen potential, the Schrödinger equation can be solved, and equating boundary conditions at $r = R$, as we did in Chapter 4 for the nucleon-nucleon problem, gives the complex scattering amplitudes $\eta_\ell$, which can be used to compare calculated cross sections with experiment. Figure 11.17 shows examples of some $\eta_\ell$ values.

The complete optical model fits to scattering data often are very impressive. Figure 11.18 shows an example of several fits to elastic scattering cross sections and polarizations.

The optical model is useful only in discussing average behavior in reactions such as scattering. Many of the interesting features of the microscopic structure of nuclei are accounted for indirectly only in this average way. The calculation using the optical model, as described in this section, does not deal with where the absorbed particles actually go; they simply disappear from the elastic channel. In fact, the many interactions between the nucleons of the target and projectile are so complicated that representing them by a single potential is itself a significant approximation. Nevertheless, the optical model is successful in accounting for elastic and inelastic scattering and leads us to an understanding of the interactions of nuclei.

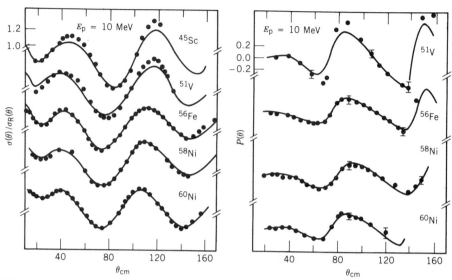

**Figure 11.18** Optical-model fits to differential cross sections (at left, shown as a ratio to the Rutherford cross section) and polarizations, for 10-MeV protons scattered elastically from various targets. The solid lines are the fits to the data using the best set of optical-model parameters. From F. D. Becchetti, Jr., and G. W. Greenlees, *Phys. Rev.* **182**, 1190 (1969).

## 11.10 COMPOUND-NUCLEUS REACTIONS

Suppose an incident particle enters a target nucleus with an impact parameter small compared with the nuclear radius. It then will have a high probability of interacting with one of the nucleons of the target, possibly through a simple scattering. The recoiling struck nucleon and the incident particle (now with less energy) can each make successive collisions with other nucleons, and after several such interactions, the incident energy is shared among many of the nucleons of the combined system of projectile + target. The average increase in energy of any single nucleon is not enough to free it from the nucleus, but as many more-or-less random collisions occur, there is a statistical distribution in energies and a small probability for a single nucleon to gain a large enough share of the energy to escape, much as molecules evaporate from a hot liquid.

Such reactions have a definite intermediate state, after the absorption of the incident particle but before the emission of the outgoing particle (or particles). This intermediate state is called the *compound nucleus*. Symbolically then the reaction a + X → Y + b becomes

$$a + X \rightarrow C^* \rightarrow Y + b$$

where C* indicates the compound nucleus.

As might be assumed from seeing the reaction written in this form, we can consider a reaction that proceeds through the compound nucleus to be a two-step process: the formation and then the subsequent decay of the compound nucleus. A given compound nucleus may decay in a variety of different ways, and essential to the compound-nucleus model of nuclear reactions is the assumption that *the relative probability for decay into any specific set of final products is independent of the means of formation of the compound nucleus.* The decay probability depends only on the total energy given to the system; in effect, the compound nucleus "forgets" the process of formation and decays governed primarily by statistical rules.

Let's consider a specific example. The compound nucleus $^{64}$Zn* can be formed through several reaction processes, including p $+ ^{63}$Cu and $\alpha + ^{60}$Ni. It can also decay in a variety of ways, including $^{63}$Zn + n, $^{62}$Zn + 2n, and $^{62}$Cu + p + n. That is

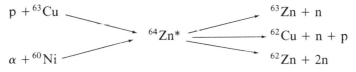

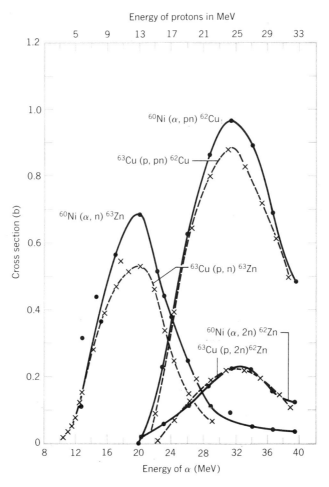

**Figure 11.19** Cross sections for different reactions leading to the compound nucleus $^{64}$Zn show very similar characteristics, consistent with the basic assumptions of the compound nucleus model. From S. N. Goshal, *Phys. Rev.* **80**, 939 (1950).

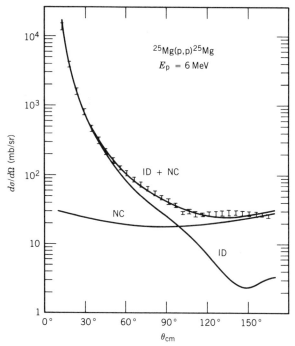

**Figure 11.20** The curve marked NC shows the contribution from compound-nucleus formation to the cross section of the reaction $^{25}$Mg(p, p)$^{25}$Mg. The curve marked ID shows the contribution from direct reactions. Note that the direct part has a strong angular dependence, while the compound-nucleus part shows little angular dependence. From A. Gallmann et al., *Nucl. Phys.* **88**, 654 (1966).

If this model were correct, we would expect for example that the relative cross sections for $^{63}$Cu(p, n)$^{63}$Zn and $^{60}$Ni($\alpha$, n)$^{63}$Zn would be the same at incident energies that give the same excitation energy to $^{64}$Zn*. Figure 11.19 shows the cross sections for the three final states, with the energy scales for the incident protons and $\alpha$'s shifted so that they correspond to a common excitation of the compound nucleus. The agreement between the three pairs of cross sections is remarkably good, showing that indeed, the decay of $^{64}$Zn* into any specific final state is nearly independent of how it was originally formed.

The compound-nucleus model works best for low incident energies (10–20 MeV), where the incident projectile has a small chance of escaping from the nucleus with its identity and most of its energy intact. It also works best for medium-weight and heavy nuclei, where the nuclear interior is large enough to absorb the incident energy.

Another characteristic of compound-nucleus reactions is the angular distribution of the products. Because of the random interactions among the nucleons, we expect the outgoing particle to be emitted with a nearly isotropic angular distribution (that is, the same in all directions). This expectation is quite consistent with experiment, as shown in Figure 11.20. In cases in which a heavy ion is the incident particle, large amounts of angular momentum can be transferred to the compound nucleus, and to extract that angular momentum the

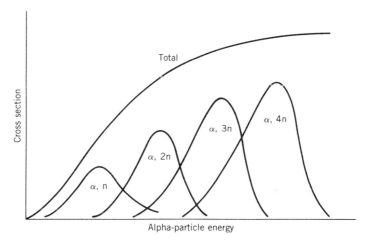

**Figure 11.21**  At higher incident energies, it is more likely that additional neutrons will "evaporate" from the compound nucleus.

emitted particles tend to be emitted at right angles to the angular momentum, and thus preferentially at 0 and 180°. With light projectiles, this effect is negligible.

The "evaporation" analogy mentioned previously is really quite appropriate. In fact, the more energy we give to the compound nucleus, the more particles are likely to evaporate. For each final state, the cross section has the Gaussian-like shape shown in Figure 11.19. Figure 11.21 shows the cross sections for ($\alpha$, $x$n) reactions, where $x = 1, 2, 3, \ldots$ . For each reaction, the cross section increases to a maximum and then decreases as the higher energy makes it more likely for an additional neutron to be emitted.

## 11.11  DIRECT REACTIONS

At the opposite extreme from compound-nucleus reactions are *direct* reactions, in which the incident particle interacts primarily at the surface of the target nucleus; such reactions are also called *peripheral* processes. As the energy of the incident particle is increased, its de Broglie wavelength decreases, until it becomes more likely to interact with a nucleon-sized object than with a nucleus-sized object. A 1-MeV incident nucleon has a de Broglie wavelength of about 4 fm, and thus does not "see" individual nucleons; it is more likely to interact through a compound-nucleus reaction A 20-MeV nucleon has a de Broglie wavelength of about 1 fm and therefore may be able to participate in direct processes. Direct processes are most likely to involve one nucleon or very few valence nucleons near the surface of the target nucleus.

Of course, it may be possible to have direct and compound-nucleus processes both contribute to a given reaction. How can we distinguish their contributions or decide which may be more important? There are two principal differences that can be observed experimentally: (1) Direct processes occur very rapidly, in a time of the order of $10^{-22}$ s, while compound-nuclear processes typically take much longer, perhaps $10^{-16}$ to $10^{-18}$ s. This additional time is necessary for the distribution and reconcentration of the incident energy. There are ingenious experimental techniques for distinguishing between these two incredibly short intervals of time. (2) The angular distributions of the outgoing particles in direct reactions tend to be more sharply peaked than in the case of compound-nuclear reactions.

Inelastic scattering could proceed either through a direct process or a compound nucleus, largely depending on the energy of the incident particle. The *deuteron stripping reaction* (d, n), which is an example of a *transfer reaction* in which a single proton is transferred from projectile to target, may also go by either mechanism. Another deuteron stripping reaction (d, p) may be more likely to go by a direct process, for the "evaporation" of protons from the compound nucleus is inhibited by the Coulomb barrier. The ($\alpha$, n) reaction is less likely to be a direct process, for it would involve a single transfer of three nucleons into valence states of the target, a highly improbable process.

One particularly important application of single-particle transfer reactions, especially (d, p) and (d, n), is the study of low-lying shell-model excited states. Several such states may be populated in a given reaction; we can choose a particular excited state from the energy of the outgoing nucleon. Once we have done so, we would like to determine just which shell-model state it is. For this we need the angular distribution of the emitted particles, which often give the spin and parity of the state that is populated in a particular reaction. Angular distributions therefore are of critical importance in studies of transfer reactions. (*Pickup reactions*, for example (p, d), in which the projectile takes a nucleon from the target, also give information on single-particle states.)

Let's consider in somewhat more detail the angular momentum transfer in a deuteron stripping reaction. In the geometry of Figure 11.22, an incident particle with momentum $\boldsymbol{p}_a$ gives an outgoing particle with momentum $\boldsymbol{p}_b$, while the residual nucleus (target nucleus plus transferred nucleon) must recoil with momentum $\boldsymbol{p} = \boldsymbol{p}_a - \boldsymbol{p}_b$. In a direct process, we may assume that the transferred nucleon instantaneously has the recoil momentum and that it must be placed in an orbit with orbital angular momentum $\ell = Rp$, assuming that the interaction

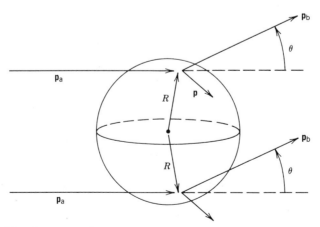

**Figure 11.22** Geometry for direct reactions occurring primarily on the nuclear surface.

takes place at the surface of the nucleus. The momentum vectors are related by the law of cosines:

$$p^2 = p_a^2 + p_b^2 - 2p_a p_b \cos\theta$$

$$= (p_a - p_b)^2 + 2p_a p_b(1 - \cos\theta) \qquad (11.57)$$

Given the energies of the incident and outgoing particles, we then have a direct relationship between $\ell$ and $\theta$—particles emerging at a given angle should correspond to a specific angular momentum of the orbiting particle.

Consider a specific example, the $(d, p)$ reaction on $^{90}$Zr leading to single neutron shell-model states in $^{91}$Zr. The $Q$ value is about 5 MeV, so an incident deuteron at 5 MeV gives a proton at about 10 MeV, less any excitation in $^{91}$Zr. Since at these energies $p_a \simeq p_b \simeq 140$ MeV/$c$, Equation 11.57 gives

$$\ell = \left[ \frac{2c^2 p_a p_b (2\sin^2\theta/2)}{\hbar^2 c^2/R^2} \right]^{1/2} \cong 8\sin\frac{\theta}{2}$$

For each angular momentum transfer, we expect to find outgoing protons at the following angles: $\ell = 0, 0°$; $\ell = 1, 14°$; $\ell = 2, 29°$; $\ell = 3, 44°$.

This simple semiclassical estimate will be changed by the intrinsic spins of the particles, which we neglected. There will also be interference between scatterings that occur on opposite sides of the nucleus, as shown in Figure 11.22. These interferences result in maxima and minima in the angular distributions.

Figure 11.23 shows the result of studies of $(d, p)$ reactions on $^{90}$Zr. You can see several low-lying states in the proton spectrum, and from their angular distributions we can assign them to specific spins and parities in $^{91}$Zr. Notice the appearance of maxima and minima in the angular distribution. The angular momentum transfer, as usual, also gives us the change in parity of the reactions, $\ell = $ even for no change in parity and $\ell = $ odd for a change in parity. If we are studying shell-model states in odd-$A$ nuclei by single-particle transfer reactions such as $(d, p)$, we will use an even-$Z$, even-$N$ nucleus as target, and so the initial spin and parity are $0^+$. If the orbital angular momentum transferred is $\ell$, then the final nuclear state reached will be $\ell \pm \frac{1}{2}$, allowing for the contribution of the spin of the transferred nucleon. For $\ell = 2$, for instance, we can reach states of $j = \frac{3}{2}$ or $\frac{5}{2}$, both with even parity.

The complete theory of direct reactions is far too detailed for this text, but we can sketch the outline of the calculation as an exercise in applications of the principles of quantum mechanics. The transition amplitude for the system to go from the initial state $(X + a)$ to the final state $(Y + b)$ is governed by the usual quantum mechanical matrix element:

$$M = \int \psi_Y^* \psi_b^* V \psi_X \psi_a \, dv \qquad (11.58)$$

The interaction $V$ must be a very complicated function of many nuclear coordinates. A simplifying assumption is the plane-wave *Born approximation*, in which $\psi_a$ and $\psi_b$ are treated as plane waves. Expanding the resulting exponential $e^{i p \cdot r/\hbar}$ using a spherical wave expansion of the form of Equation 11.31 and making the simplifying assumption that the interaction takes place on the nuclear

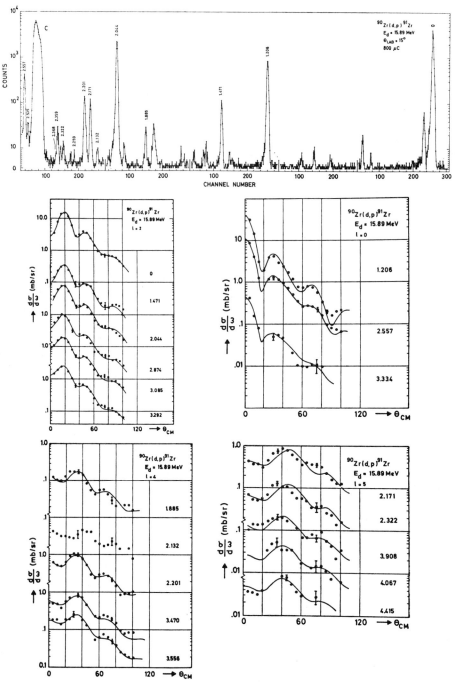

**Figure 11.23** (top) Proton spectrum from $^{90}$Zr(d, p)$^{91}$Zr. Peaks are identified with the final states in $^{91}$Zr populated. The large peak at the left is from a carbon impurity. (bottom) Angular distributions fitted to determine the $\ell$ value. Note that the location of the first maximum shifts to larger angles with increasing $\ell$, as predicted by Equation 11.57. See Figure 11.24 for the deduced excited states. Data from H. P. Blok et al., *Nucl. Phys. A* **273**, 142 (1976).

surface, so the integral is evaluated only at $r = R$, the matrix element is proportional to $j_\ell(kR)$ where $k = p/\hbar$ contains the explicit angular dependence through Equation 11.57. The cross section then depends on $[j_\ell(kR)]^2$, which gives results of the form of Figure 11.23.

Taking this calculation one step further, we use the optical model to account for the fact that the incoming and outgoing plane waves are changed (or distorted) by the nucleus. This gives the *distorted-wave Born approximation*, or DWBA. We can even put in explicit shell-model wave functions for the final state, and ultimately we find a differential cross section for the reaction. Because there are no "pure" shell-model states, the calculated cross section may describe many different final states. Each will have a differential cross section whose shape can be accurately calculated based on this model, but the amplitude of the cross section for any particular state depends on the fraction of the pure shell-model state included in the wave function for that state. The measured cross section is thus reduced from the calculated shell-model single-particle value by a number

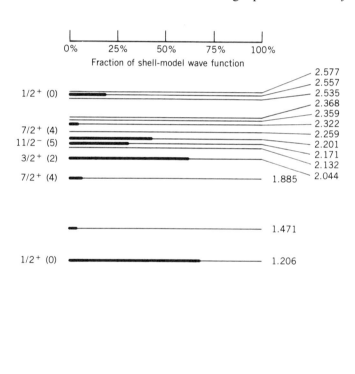

**Figure 11.24** Deduced level scheme for $^{91}$Zr. Each $\ell$ value (except zero) deduced from the angular distributions of Figure 11.23 leads to a definite parity assignment but to two possible $I$ values, $\ell \pm \frac{1}{2}$. Which one is correct must be determined from other experiments. The fraction of the single-particle strengths represented by each level is indicated by the length of the shading; thus the ground state is nearly pure $d_{5/2}$ shell-model state.

between 0 and 1 called the *spectroscopic factor* $S$:

$$\left(\frac{d\sigma}{d\Omega}\right)_{\text{meas}} = S\left(\frac{d\sigma}{d\Omega}\right)_{\text{calc}} \tag{11.59}$$

A pure shell-model state would have $S = 1$. In practice we often find the shell-model wave function to be distributed over many states. Figure 11.24 shows examples of the spectroscopic factors measured for $^{91}$Zr.

## 11.12 RESONANCE REACTIONS

The compound-nucleus model of nuclear reactions treats the unbound nuclear states as if they formed a structureless continuum. That is, there may be discrete nuclear states, but there are so many of them and they are so close together that they form a continuous spectrum. Each of these supposed discrete states is unstable against decay and therefore has a certain *width*; when the states are so numerous that their spacing is much less than the widths of the individual states, the compound-nucleus continuum results.

The bound states studied by direct reactions are at the opposite end of the scale. Because they are stable against particle emission, their mean lives are much longer (for example, characteristic of $\gamma$ decay) and their corresponding widths are much smaller. A state with a lifetime of 1 ps, for instance, has a width of about $10^{-3}$ eV, far smaller than the typical spacing of bound states. We are therefore justified in treating these as discrete states with definite wavefunctions.

Between these two extremes is the *resonance* region—discrete levels in the compound-nucleus region. These levels have a high probability of formation (large cross sections), and their widths are very small because at low incident energy, where these resonances are most likely to occur, the quasibound state that is formed usually has only two modes of decay available to it—re-ejecting the incident particle, as in elastic or inelastic scattering, or $\gamma$ emission.

To obtain a qualitative understanding of the formation of resonances, we represent the nuclear potential seen by the captured particle as a square well. The oscillatory wave functions inside and outside the well must be matched smoothly, as we did in Figure 4.7a for nucleon-nucleon scattering. Figure 11.25 shows several examples of how this might occur. Depending on the phase of the wave function inside the nucleus, the smooth matching can result in substantial variations between the relative amplitudes of the wave functions inside and outside the nucleus. In case ($a$), the incident particle has relatively little probability to penetrate the nucleus and form a quasibound state; in case ($c$), there is a very high probability to penetrate. As we vary the energy of the incident particle, we vary the relative phase of the inner and outer wave functions; the location of the matching point *and* the relative amplitudes vary accordingly. Only for certain incident energies do we achieve the conditions shown in part ($c$) of Figure 11.25. These are the energies of the *resonances* in the cross section.

In a single, isolated resonance of energy $E_{\text{R}}$ and width $\Gamma$, the energy profile of the cross section in the vicinity of the resonance will have the character of the energy distribution of any decaying state of lifetime $\tau = \hbar/\Gamma$; see, for example, Equation 6.20 or Figure 6.3. The resonance will occur where the total cross

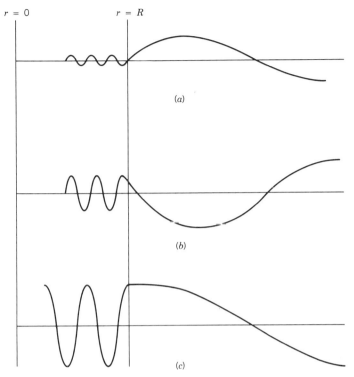

r = 0          r = R

(a)

(b)

(c)

**Figure 11.25** (a) Far from resonance, the exterior and interior wave functions match badly, and little penetration of the nucleus occurs. (b) As the match improves, there is a higher probability to penetrate. (c) At resonance the amplitudes match exactly, the incident particle penetrates easily, and the cross section rises to a maximum.

section has a maximum; from Equation 11.48, assuming only one partial wave $\ell$ is important for the resonant state, there will be a scattering resonance where $\eta_\ell = -1$, corresponding to a phase shift $\delta_\ell = \pi/2$.

The shape of the resonance can be obtained by expanding the phase shift about the value $\delta_\ell = \pi/2$. Better convergence of the Taylor series expansion is obtained if we expand the cotangent of $\delta_\ell$:

$$\cot \delta_\ell(E) = \cot \delta_\ell(E_R) + (E - E_R)\left(\frac{\partial \cot \delta_\ell}{\partial E}\right)_{E=E_R}$$

$$+ \tfrac{1}{2}(E - E_R)^2\left(\frac{\partial^2 \cot \delta_\ell}{\partial E^2}\right)_{E=E_R} + \cdots \qquad (11.60)$$

in which

$$\left(\frac{\partial \cot \delta_\ell}{\partial E}\right)_{E=E_R} = -\left(\frac{\partial \delta_\ell}{\partial E}\right)_{E=E_R} \qquad (11.61)$$

Defining the width $\Gamma$ as

$$\Gamma = 2\left(\frac{\partial \delta_\ell}{\partial E}\right)^{-1}_{E=E_R} \qquad (11.62)$$

then it can be shown that the second-order term vanishes, and thus (neglecting higher-order terms)

$$\cot \delta_\ell = -\frac{(E - E_R)}{\Gamma/2} \tag{11.63}$$

Because $\Gamma$ is the full width of the resonance, the cross section should fall to half of the central value at $E - E_R = \pm\Gamma/2$. From Equation 11.63, this occurs when $\cot \delta_\ell = \pm 1$, or $\delta_\ell = \pi/4, 3\pi/4$ (compared with $\delta_\ell = \pi/2$ at the center of the resonance). The cross section depends on $\sin^2 \delta_\ell$, which does indeed fall to half the central value at $\delta_\ell = \pi/4$ and $3\pi/4$. The width defined by Equation 11.62 is thus entirely consistent with the width shown in Figure 6.3.

From Equation 11.63, we find

$$\sin \delta_\ell = \frac{\Gamma/2}{\left[ (E - E_R)^2 + \Gamma^2/4 \right]^{1/2}} \tag{11.64}$$

and the scattering cross section becomes, using Equation 11.45

$$\sigma_{sc} = \frac{\pi}{k^2} (2\ell + 1) \frac{\Gamma^2}{(E - E_R)^2 + \Gamma^2/4} \tag{11.65}$$

This result can be generalized in two ways. In the first place, we can account for the effect of reacting particles with spin. If $s_a$ and $s_X$ are the spins of the incident and target particles, and if $I$ is the total angular momentum of the resonance,

$$I = s_a + s_X + \ell \tag{11.66}$$

then the factor $(2\ell + 1)$ in Equation 11.65 should be replaced by the more general statistical factor

$$g = \frac{2I + 1}{(2s_a + 1)(2s_X + 1)} \tag{11.67}$$

Note that $g$ reduces to $(2\ell + 1)$ for spinless particles.

The second change we must make is to allow for partial entrance and exit widths. If the resonance has many ways to decay, then the total width $\Gamma$ is the sum of all the partial widths $\Gamma_i$

$$\Gamma = \sum_i \Gamma_i \tag{11.68}$$

The $\Gamma^2$ factor in the denominator of Equation 11.65 is related to the decay width of the resonant state and therefore to its lifetime: $\Gamma = \hbar/\tau$. The observation of only a single entrance or exit channel does not affect this factor, for the resonance always decays with the same lifetime $\tau$. In the analogous situation in

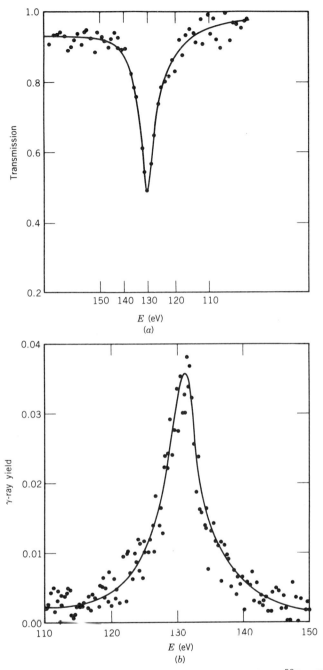

**Figure 11.26**  130-eV neutron resonance in scattering from $^{59}$Co. Part (*a*) shows the intensity of neutrons transmitted through a target of $^{59}$Co; at the resonance there is the highest probability for a reaction and the intensity of the transmitted beam drops to a minimum. In (*b*), the γ-ray yield is shown for neutron radiative capture by $^{59}$Co. Here the yield of γ rays is maximum where the reaction has the largest probability. From J. E. Lynn, *The Theory of Neutron Resonance Reactions* (Oxford: Clarendon, 1968).

radioactive decay, the activity decays with time according to the total decay constant, even though we might observe only a single branch with a very different partial decay constant. The $\Gamma^2$ factor in the numerator, on the other hand, is directly related to the formation of the resonance and to its probability to decay into a particular exit channel. In the case of elastic scattering, for which Equation 11.65 was derived, the entrance and exit channels are identical. That is, for the reaction a + X → a + X, we should use the partial widths $\Gamma_{aX}$ of the entrance and exit channels:

$$\sigma = \frac{\pi}{k^2} g \frac{\left(\Gamma_{aX}\right)^2}{\left(E - E_R\right)^2 + \Gamma^2/4} \tag{11.69}$$

Similarly, for the reaction a + X → b + Y, a different exit width must be used:

$$\sigma = \frac{\pi}{k^2} g \frac{\Gamma_{aX}\Gamma_{bY}}{\left(E - E_R\right)^2 + \Gamma^2/4} \tag{11.70}$$

Equations 11.69 and 11.70 are examples of the *Breit-Wigner formula* for the shape of a single, isolated resonance. Figure 11.26 shows such a resonance with the Breit-Wigner shape. The cross section for resonant absorption of γ radiation has a similar shape, as given by Equations 10.29 and 10.30.

Many elastic scattering resonances have shapes slightly different from that suggested by the Breit-Wigner formula. This originates with another contribution to the reaction amplitude from direct scattering of the incident particle by the nuclear potential, without forming the resonant state. This alternative process is called *potential scattering* or *shape-elastic scattering*. Potential scattering and

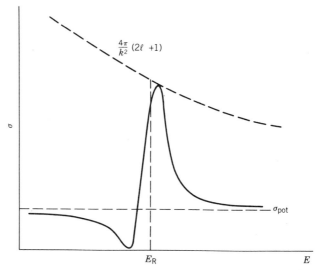

**Figure 11.27** Interference between resonance and potential scattering produces resonances with this characteristic shape.

resonant scattering both contribute to the elastic scattering amplitude, and interference between the two processes causes variation in the cross section. Interference can cause the combined cross section to be smaller than it would be for either process alone. It is therefore not correct simply to add the cross sections for the two processes. We can account for the two processes by writing

$$\eta_\ell = e^{2i(\delta_{\ell R} + \delta_{\ell P})} \qquad (11.71)$$

where $\delta_{\ell R}$ is the resonant phase shift, as in Equations 11.63 or 11.64, and $\delta_{\ell P}$ is an additional contribution to the phase shift from potential scattering. From

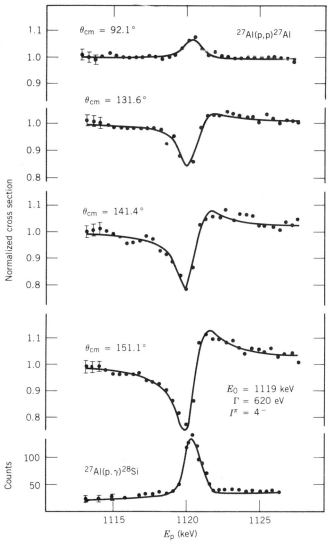

**Figure 11.28** Resonances in the reaction $^{27}Al(p, p)^{27}Al$. The resonances occur in the nucleus $^{28}Si$. Note that the $(p, \gamma)$ yield shows a resonance at the same energy. From A. Tveter, *Nucl. Phys. A* **185**, 433 (1972).

Equation 11.44 we find the cross section

$$\sigma_{sc} = \frac{\pi}{k^2}(2\ell + 1)\left|e^{-2i\delta_{\ell P}} - 1 + \frac{i\Gamma}{(E - E_R) + i\Gamma/2}\right|^2 \qquad (11.72)$$

Far from the resonance, $(E - E_R) \gg \Gamma/2$ and the potential scattering term dominates:

$$\sigma \cong \sigma_{pot} = \frac{4\pi}{k^2}(2\ell + 1)\sin^2\delta_{\ell P} \qquad (11.73)$$

At $E = E_R$, the resonant term dominates and

$$\sigma \cong \sigma_{res} = \frac{4\pi}{k^2}(2\ell + 1) \qquad (11.74)$$

Near the resonance there is interference between the two terms, which produces the characteristic shape shown in Figure 11.27. According to this model, we

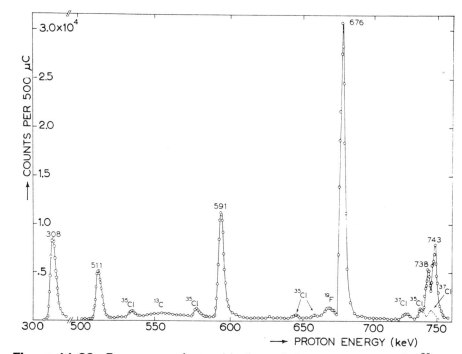

**Figure 11.29** Resonances observed in the radiative proton capture by $^{23}$Na. In this case, the total yield of $\gamma$ rays in the energy range 3–13 MeV was measured as a function of the incident proton energy. The Cl peaks appear because the target used was NaCl. From P. W. M. Glaudemans and P. M. Endt, *Nucl. Phys.* **30**, 30 (1962).

expect an interference "dip" on the low-$E$ side of the resonance. The resonance height should decrease roughly as $k^{-2}$ (that is, as $E^{-1}$) with increasing incident energy, and the nonresonant "background" from potential scattering should remain roughly constant. Figure 11.28 shows scattering cross sections with the resonant structure clearly visible. The expectations of the resonance model are clearly fulfilled.

Radiative capture reactions also show a resonant structure. Figure 11.29 shows examples of (p, $\gamma$) reactions. Note that this is not a $\gamma$ spectrum in the conventional sense—the horizontal axis shows the incident proton energy, not the emitted $\gamma$ energy.

Resonances observed in neutron scattering are discussed in more detail in Chapter 12.

## 11.13  HEAVY-ION REACTIONS

From the point of view of nuclear reactions, a heavy ion is defined to be any projectile with $A > 4$. Accelerators devoted to the study of heavy-ion reactions can produce beams of ions up to $^{238}$U, at typical energies of the order of 1–10 MeV per nucleon, although much higher energies are also possible.

The variety of processes than can occur in heavy-ion reactions is indicated schematically in Figure 11.30. At large impact parameters, Coulomb effects dominate, and Rutherford scattering or Coulomb excitation may occur. When the nuclear densities of the target and projectile just begin to overlap, nuclear reactions can occur, and at small overlap ordinary elastic or inelastic scattering and few-nucleon transfer through direct reactions may occur, as discussed previously in this chapter.

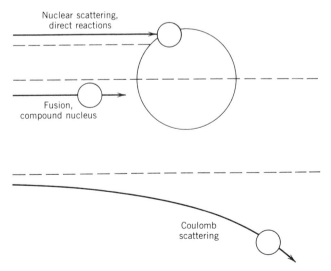

**Figure 11.30**  Processes in heavy-ion scattering depend on the impact parameter, when energies are large enough to penetrate the Coulomb barrier.

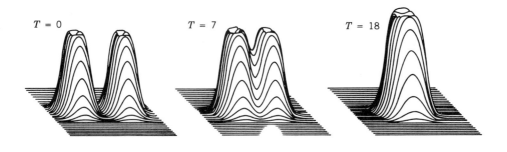

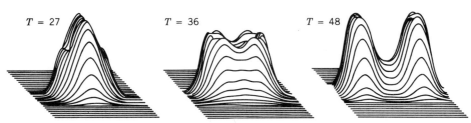

**Figure 11.31** Two $^{12}$C nuclei colliding, shown at various times (each unit of time is $3.3 \times 10^{-24}$ s, and the area shown represents 1 fm$^2$). The vertical scale shows the nuclear density, which reaches a peak at $T = 11$. Note the internal oscillations that occur before the compound system breaks apart. The energy of the incident projectile was about 700 MeV. (Courtesy Ronald Y. Cusson and Joachim Maruhn, Oak Ridge National Laboratory; from *Scientific American*, p. 59, December 1978.)

At small impact parameters, new and unusual features emerge in these reactions. If the impact parameter is small enough that the nuclei can overlap completely, a compound nucleus, representing complete fusion of the two nuclei, can form as an intermediate state. However, to overcome the repulsive Coulomb barrier, the incident ion must be quite energetic, and thus the compound nucleus is formed with a considerable excitation energy. This compound nucleus may be an unusual state of nuclear matter that cannot be achieved in reactions with light nuclei. Because of the large incident energy, the compound nucleus may achieve a density or a "temperature" (that is, a mean internal kinetic energy per nucleon) beyond what can be achieved in reactions with light ions. The analysis of these compound states and their decay modes thus represent a challenge for nuclear theory—can we extrapolate from an equation of state for "ordinary" nuclear matter to one for "extraordinary" nuclear matter? Figure 11.31 shows an attempt to calculate the intermediate states through which the highly excited compound nucleus $^{24}$Mg progresses in the course of the reaction $^{12}$C + $^{12}$C.

Once the excited compound state is formed, there are many channels available for its decay. It can split more-or-less in half, either through the original entrance channel ($^{24}$Mg $\rightarrow$ $^{12}$C + $^{12}$C) or through a closely related channel ($^{24}$Mg $\rightarrow$ $^{15}$O

$+ {}^9$Be). For heavier nuclei, the study of the fission mode provides a check on theories derived from the study of the more familiar cases of transuranic fission, described in Chapter 13.

A more probable means of decay of the compound nucleus is through particle emission, for fission is inhibited by a substantial Coulomb barrier. Emission of charged particles (protons or $\alpha$'s) is also inhibited by a Coulomb barrier. In reactions with heavy nuclei the compound nucleus is extremely proton rich, but the preferred decay mode is still neutron emission; this remains so, even for heavy nuclei with a proton excess of 10–20 or more. It is thus possible to study nuclei far from stability on the proton-rich side through (HI, $x$n) reactions, where HI indicates any heavy ion and $x$ may be in the range 5–10.

A particular application of these reactions is in the search for stable or nearly stable nuclei of *superheavy elements*. The transuranic atoms that have been studied through the neutron capture–$\beta$-decay technique move up the atomic number scale in single steps, but the technique loses its applicability for the nuclei around $Z = 104$ or 105, where the half-lives become very short ($\sim$ seconds) for decay by spontaneous fission. As $Z$ increases, the spontaneous fission half-life should continue to decrease (because the Coulomb energy, which makes the nucleus more unstable to fission, increases like $Z^2$), until we approach the region of the next closed shell or "magic number" for protons, which has been calculated to be $Z = 114$ (rather than 126, as is already known for neutrons).

It is possible to search for superheavy nuclei directly, by bombarding the heaviest possible quasistable targets (${}^{249}_{98}$Cf, with $t_{1/2} = 351$ y) with beams such as ${}^{32}$S or ${}^{40}$Ca, in the hope of producing stable products around $Z = 114$, $N = 164$ following few-nucleon emission from the compound state. Another possibility would be to produce a highly unstable, extremely heavy compound state in a reaction such as ${}^{238}$U $+ {}^{238}$U, in the hope that one of the fission decay channels would have a high probability of producing a stable superheavy nucleus. To date no success has been reported from either of these approaches, but the effort continues.

Another unique feature of heavy-ion reactions is the transfer of large amounts of angular momentum to the compound nucleus. For example, in the reaction

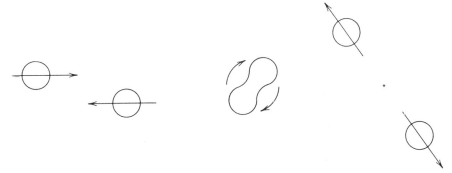

**Figure 11.32**   In nuclear molecule formation, there is not quite complete fusion of the two particles; they retain a "memory" of their previous character and break apart accordingly. The internal energy of the system can show rotational and vibrational structures, just like an ordinary molecule.

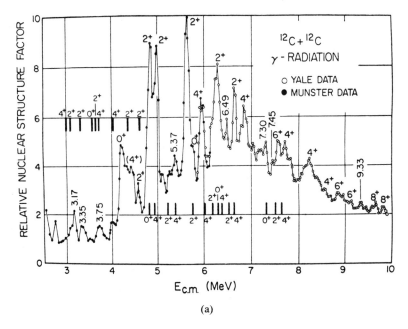

(a)

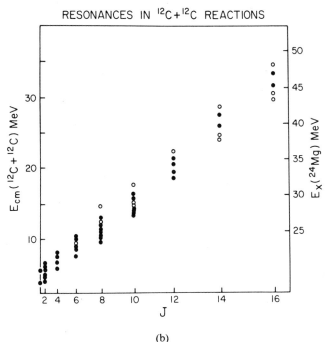

(b)

**Figure 11.33** (a) The $^{12}C + ^{12}C$ molecular states. The vertical scale shows the cross section with the "uninteresting" Coulomb penetrability factor removed, leaving the nuclear structure factor. The resonances are labeled with the spin-parity assignments which can be grouped into rotational sequences $0^+, 2^+, 4^+, 6^+, \ldots$ (b) Plotting the internal excitation energies of the resonances against $l(l + 1)$ reveals that the states do indeed form rotational sequences. From T. M. Cormier, *Ann. Rev. Nucl. Particle Sci.* **32**, 271 (1982).

$^{40}$Ca + $^{197}$Au, the Coulomb barrier is about 200 MeV. If we use 200 MeV incident $^{40}$Ca, a grazing collision will provide about 140$\hbar$ of angular momentum to the system. Even at collisions with smaller impact parameters, it would not be unusual to transfer an angular momentum of $\ell \gtrsim 40\hbar$ to the compound system. At such rotational velocities, the nuclear force may not be able to provide the necessary centripetal acceleration, and the compound system may be completely unstable and therefore unable to form. In such a case, a new type of system is possible, called a *nuclear molecule*. Figure 11.32 illustrates the process schematically. The two nuclei do not form a compound system, corresponding to complete sharing of the incident energy. Instead a system analogous to a diatomic molecule is formed, exists for a short time, and then breaks apart in the same configuration as the incident particles. Because the decay occurs into the original particles, the combined system retains a considerable "memory" of its formation, contrary to the basic assumption of the compound-nucleus model. Evidence for such molecular states comes from observing the rotational and vibrational excitations that correspond closely with those observed in ordinary molecules. Figure 11.33 shows an example of the states observed in the $^{12}$C + $^{12}$C nuclear molecule. Resonances in the cross section correspond to the rotational and vibrational states permitted in the molecular system.

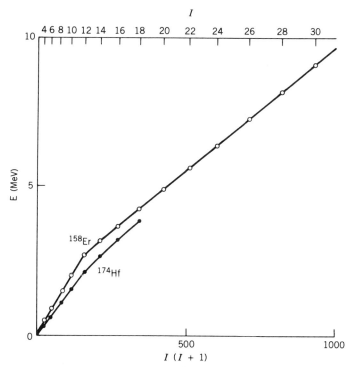

**Figure 11.34** Rotational energies of $^{158}$Er and $^{174}$Hf. Neither case shows the expected linear dependence of $E$ on $I(I + 1)$, but in $^{174}$Hf the deviation is relatively gradual, while in $^{158}$Er there appears to be a sudden change in slope (and therefore in moment of inertia) in the neighborhood of $I = 12 - 14$.

In heavier nuclei, highly excited states with $\ell \gtrsim 40$–$50\hbar$ can be populated in the compound system. The emission of a few neutrons from the excited system can change the angular momentum only little (a 5-MeV neutron carries at most only $\ell \sim 3\hbar$), and following the neutron emission, excited bound states in the final nucleus can be formed with angular momentum quantum numbers of 40 or so. Assuming the product nucleus to be of the deformed even-$Z$, even-$N$ variety, the excited states will show a rotational spectrum of the type illustrated in Figure 5.22. The rotational energies are given by Equation 5.17:

$$E = \frac{\hbar^2}{2\mathscr{I}} I(I + 1) \tag{11.75}$$

and the states should cascade down toward the ground state through a sequence of $E2$ $\gamma$ transitions as in Figure 10.18. The observation of these cascade $E2$ transitions provides a way to study these excited states. In particular, we can study whether the assumption of a fixed, constant moment of inertia $\mathscr{I}$ remains valid at such high excitations. One way to test this assumption is to plot the energies of the states against $I(I + 1)$ and to see if the slope remains constant, as predicted by Equation 11.75. Figure 11.34 is an example of such a plot, and there appears to be some deviation from the expected linear behavior.

There is a more instructive way to plot the data on the rotational structure. From Equation 11.75, the energy of a transition from state $I$ to the next lower state $I - 2$ is

$$E(I) - E(I - 2) = \frac{\hbar^2}{2\mathscr{I}} (4I - 2) \tag{11.76}$$

The transition energies should increase linearly with $I$; Figure 11.35 shows that this is true for the lower transitions, but becomes less valid as we go to larger $I$, and in fact the behavior changes completely at about $I = 16$, but then seems to restore itself as we go to higher states.

Let's assume that the moment of inertia is not constant, but increases gradually as we go to more rapidly rotating states; this effect, known classically as "centrifugal stretching," would not occur for a rigid rotor but would occur for a fluid. Because rotating nuclei have moments of inertia somewhere between that of a rigid rotor and of a fluid, as described in Equations 5.18 and 5.19, it is not surprising that centrifugal stretching occurs. Representing the rotational energy in terms of the rotational frequency

$$E = \tfrac{1}{2} \mathscr{I} \omega^2 \tag{11.77}$$

we can then assume $\mathscr{I}$ varies either with increasing angular momentum,

$$\mathscr{I} = \mathscr{I}_0 + kI(I + 1) \tag{11.78}$$

or with increasing rotational frequency,

$$\mathscr{I} = \mathscr{I}_0 + k'\omega^2 \tag{11.79}$$

where $k$ and $k'$ are appropriate proportionality constants. From Equation 11.76,

$$\frac{2\mathscr{I}}{\hbar^2} = \frac{4I - 2}{E(I) - E(I - 2)} \tag{11.80}$$

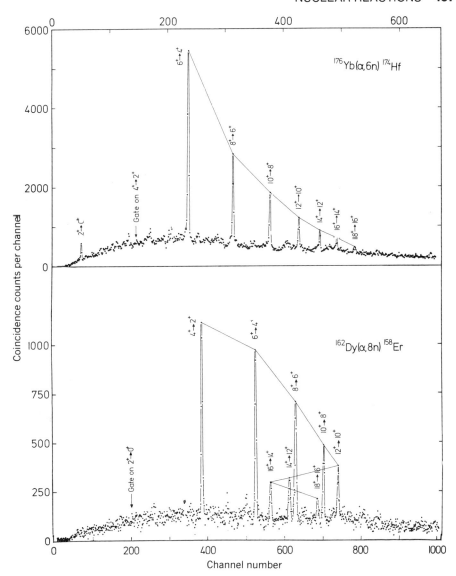

**Figure 11.35** γ-ray spectra of transitions between rotational states in $^{158}$Er and $^{174}$Hf. For a perfect rotor, the γ-ray energies should increase monotonically with $I$. This is so for $^{174}$Hf, but for $^{158}$Er the energy begins to *decrease* with $I$ in the range $I = 12-16$, and after $I = 16$ the energy again begins to increase. From R. M. Lieder and H. Ryde, in *Advances in Nuclear Physics*, Vol. 10, edited by M. Baranger and E. Vogt (New York: Plenum, 1978).

and plotting $\mathcal{I}$, measured in these units, against $\omega^2$, we ought to see either a constant $\mathcal{I}_0$ if no stretching occurs, or a linear behavior. Figure 11.36 shows an example of such a relationship. There appears to be a gradual increase in $\mathcal{I}$ among the lower angular momentum states, then a radical change in behavior around $I = 16$, and then a return to the gradual stretching. This effect is known as *backbending*, and occurs because the rotational energy exceeds the energy

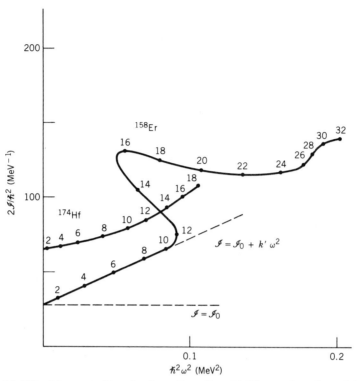

**Figure 11.36**  Moment of inertia, from Equation 11.80, as a function of $\hbar^2\omega^2$, from the semiclassical formula $\frac{1}{4}[E(I) - E(I - 2)]^2$. Note the gradual increase in the moment of inertia for the lower states in both $^{158}$Er and $^{174}$Hf, and note also the backbending in $^{158}$Er.

needed to break a pair of coupled nucleons. When that occurs (at an energy corresponding to $I = 16$), the unpaired nucleons go into different orbits and change the nuclear moment of inertia. The situation then remains stable until about $I = 30$, where another pair is broken and another change in moment of inertia occurs.

The study of the properties of nuclei at high angular momentum is another example of an unusual state of nuclear matter accessible only through heavy-ion reactions.

A final example of the nuclear structure studies that can be done through heavy-ion reactions is the $\alpha$-particle transfer reaction, such as $(^{16}\text{O}, {}^{12}\text{C})$. In our discussion of $\alpha$ decay in Chapter 8, we alluded to the "preformation" of the $\alpha$ particle inside the nucleus. Because the $\alpha$ particle is such a stable structure, we can consider the nucleons in a nucleus to have a high probability of occasionally forming an $\alpha$ particle, even in nuclei that do not $\alpha$ decay. This leads to the $\alpha$-*cluster* model of nuclei, in which we look for nuclear structure characteristic of such clusters. States populated in $(^{16}\text{O}, {}^{12}\text{C})$ reactions, in which four nucleons are simultaneously transferred to the target nucleus, might be analyzed in terms of the transfer of an $\alpha$ cluster from $^{16}$O to the target. Figure 11.37 illustrates the cross sections for the formation of states in $^{20}$Ne through $\alpha$-transfer reactions.

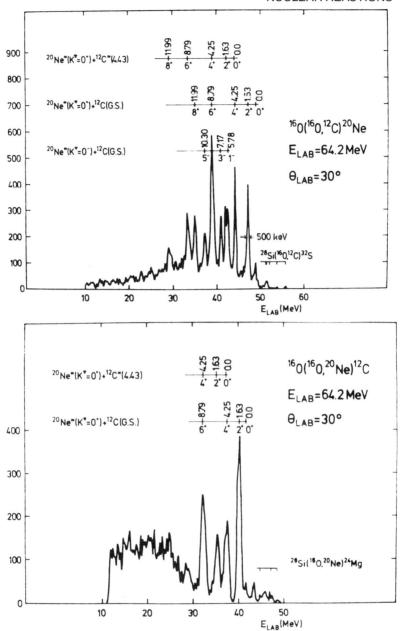

**Figure 11.37** α-particle transfer reactions leading to $^{20}$Ne. At top is the observed $^{12}$C spectrum and at bottom is the $^{20}$Ne spectrum. Individual excited states in $^{20}$Ne are labeled, with some peaks in the spectrum assigned to reactions in which $^{12}$C is left in its first excited state at 4.43 MeV. On the following page are shown the excited states of $^{20}$Ne. Notice the selectivity of the reaction in populating certain states and not others; the dashed lines show states that are not populated in the reaction. Because the $^{16}$O projectile and target are doubly closed-shell nuclei ($Z = 8, N = 8$), the observed states in $^{20}$Ne correspond to the addition of an α particle to a doubly magic core; that is, the four valence nucleons are coupled to a resultant spin of zero, but can carry a net orbital angular momentum. Only a small subset of the $^{20}$Ne states will have this character. Data from H. H. Rossner et al., *Nucl. Phys. A* **218**, 606 (1974).

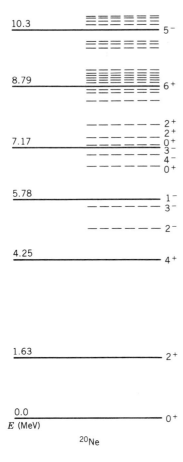

Figure 11.37 Continued.

## REFERENCES FOR ADDITIONAL READING

A more complete description of nuclear reactions, at about the same level as this chapter, is by G. R. Satchler, *Introduction to Nuclear Reactions* (New York: Wiley, 1980). A reference at a similar level is I. E. McCarthy, *Nuclear Reactions* (Oxford: Pergamon, 1966). A more advanced text is P. E. Hodgson, *Nuclear Reactions and Nuclear Structure* (Oxford: Clarendon, 1971).

The most comprehensive recent review of all aspects of nuclear reactions is the four-part collection, *Nuclear Spectroscopy and Reactions*, edited by Joseph Cerny (New York: Academic, 1974). It includes chapters on direct reactions, resonance reactions, scattering, Coulomb excitation, and heavy-ion reactions, as well as material on experimental techniques.

Comprehensive summaries of the role of isospin in nuclear decays and reactions can be found in *Isospin in Nuclear Physics*, edited by D. H. Wilkinson (Amsterdam: North-Holland, 1969). Wilkinson's introduction is highly recommended as a review of the historical development of isospin and of nuclear theory.

Effects at high rotational velocities are discussed by R. M. Lieder and H. Ryde, in *Advances in Nuclear Physics*, Vol. 10, edited by M. Baranger and E. Vogt (New

York: Plenum, 1978); see also the brief review by R. M. Diamond and F. S. Stephens, *Nature* **310**, 457 (1984).

## PROBLEMS

1. Complete the following reactions:

$$^{27}\text{Al} + \text{p} \rightarrow \quad + \text{n}$$

$$^{32}\text{S} + \alpha \rightarrow \quad + \gamma$$

$$^{197}\text{Au} + {}^{12}\text{C} \rightarrow {}^{206}\text{At} +$$

$$^{116}\text{Sn} + \quad \rightarrow {}^{117}\text{Sn} + \text{p}$$

2. (a) Solve Equations 11.3 and 11.4 for $\cos\theta$. (b) Determine the relationship between $\cos\theta$ and $p_b$ for elastic scattering. (c) Show that there is a maximum value of $\theta$ only when $m_a > m_Y$. (d) Find the maximum angle at which $\alpha$ particles appear after elastic scattering from hydrogen and from deuterium.

3. It is desired to study the first excited state of $^{16}\text{O}$, which is at an energy of 6.049 MeV. (a) Using the $(\alpha, \text{n})$ reaction on a target of $^{13}\text{C}$, what is the minimum energy of incident alphas which will populate the excited state? (b) In what direction will the resulting neutrons travel? (c) If it is desired to detect the neutrons at 90° to the incident beam, what is the minimum $\alpha$ energy that can result in the excited state being populated?

4. (a) In Coulomb scattering of 7.50-MeV protons by a target of $^7\text{Li}$, what is the energy of the elastically scattered protons at 90°? (b) What is the energy of the inelastically scattered protons at 90° when the $^7\text{Li}$ is left in its first excited state (0.477 MeV)?

5. The (n, p) reaction can be regarded as equivalent to $\beta^+$ decay in that the same initial and final nuclei are involved. Derive a general expression relating the $Q$ value of the (n, p) reaction to the maximum energy release in $\beta^+$ decay. Find several examples to verify your derived relationship.

6. The $Q$ value for the reaction $^9\text{Be}\,(\text{p}, \text{d})^8\text{Be}$ is $559.5 \pm 0.4$ keV. Use this value along with the accurately known masses of $^9\text{Be}$, $^2\text{H}$, and $^1\text{H}$ to find the mass of $^8\text{Be}$.

7. (a) Calculate the $Q$ value of the reaction $\text{p} + {}^4\text{He} \rightarrow {}^2\text{H} + {}^3\text{He}$. (b) What is the threshold energy for protons incident on He? For $\alpha$'s incident on hydrogen?

8. For the reaction $^2\text{H} + {}^2\text{H} \rightarrow {}^3\text{He} + \text{n}$, plot the energy of the outgoing neutron as a function of angle for $^2\text{H}$ incident on $^2\text{H}$ at rest. Use incident energies of 0.00, 2.50, and 5.00 MeV.

9. Compute the $Q$ values for the reactions (a) $^6\text{Li} + \text{p} \rightarrow {}^3\text{He} + {}^4\text{He}$; (b) $^{59}\text{Co} + \text{p} \rightarrow \text{n} + {}^{59}\text{Ni}$; (c) $^{40}\text{Ca} + \alpha \rightarrow \text{n} + {}^{43}\text{Ti}$.

10. For the following endoergic reactions, find the $Q$ value and the threshold kinetic energy, assuming in each case that the lighter particle is incident on the heavier particle at rest: (a) $^7\text{Li} + \text{p} \rightarrow {}^7\text{Be} + \text{n}$; (b) $^{12}\text{C} + \text{p} \rightarrow \text{n} + {}^{12}\text{N}$; (c) $^{35}\text{Cl} + \alpha \rightarrow \text{n} + {}^{38}\text{K}$.

11. At threshold, the product particles Y + b move at the same velocity. Use momentum conservation to derive a relationship between $T_a$ and $T_b$ at threshold, and then substitute your expression into Equation 11.5 to obtain the threshold condition (11.6).

12. It is desired to study the low-lying excited states of $^{35}$Cl (1.219, 1.763, 2.646, 2.694, 3.003, 3.163 MeV) through the $^{32}$S$(\alpha, p)$ reaction. (a) With incident $\alpha$ particles of 5.000 MeV, which of these excited states can be reached? (b) Again with 5.000-MeV incident $\alpha$'s, find the proton energies observed at 0, 45, and 90°.

13. In the reaction $^{7}$Li + p $\rightarrow ^{4}$He + $^{4}$He (18.6 MeV protons incident on a lithium target) the differential cross section (in the center-of-mass system) reaches a maximum of about 1.67 barns/steradian at a center-of-mass angle of 75°. (a) Sketch the reaction kinematics in the laboratory system, labeling all momenta, directions, and energies. (b) Assuming a target thickness of 1.0 mg/cm$^2$ and a beam of protons of current 1.0 $\mu$A spread over an area of 1 cm$^2$, find the number of $\alpha$ particles per second in the above geometry that would strike a detector of area 0.5 cm$^2$ located 12.0 cm from the target.

14. The radioactive isotope $^{15}$O, which has important medical applications (see Chapter 20), can be produced in the reaction $^{12}$C$(\alpha, n)$. (a) The cross section reaches a peak when the laboratory energy of the incident $\alpha$ particles is 14.6 MeV. What is the excitation energy of the compound nuclear state? (b) The reaction cross section at the above incident energy is 25 mb. Assuming a carbon target of 0.10 mg/cm$^2$ and a current of 20 nA of $\alpha$'s, compute the $^{15}$O activity that results after 4.0 min of irradiation.

15. In a Coulomb excitation experiment, $\alpha$ particles are inelastically scattered from $^{160}$Dy nuclei. (a) If the incident energy is 5.600 MeV, what is the energy of the elastically scattered $\alpha$'s observed at $\theta = 150°$? (b) States in $^{160}$Dy are known at $2^+(0.087$ MeV), $4^+(0.284$ MeV), and $2^+(0.966$ MeV). Considering only the $E2$ excitation mode, find the energies of the inelastically scattered $\alpha$'s observed at 150°.

16. What should be the incident energy of a beam of protons to be Coulomb scattered by gold nuclei, if it is desired that the minimum distance between projectile and target should correspond to the two nuclei just touching at their surfaces?

17. Alpha particles of energy 8.0 MeV are incident at a rate of $3.0 \times 10^7$ per second on a gold foil of thickness $4.0 \times 10^{-6}$ m. A detector in the form of an annular ring is placed 3.0 cm from the scattering foil and concentric with the beam direction; the annulus has an inner radius of 0.50 cm and an outer radius of 0.70 cm. What is the rate at which scattered particles strike the detector?

18. Alpha particles of energy 6.50 MeV are Coulomb scattered by a gold foil. (a) What is the impact parameter when the scattered particles are observed at 90°? (b) Again for scattering at 90°, find the smallest distance between the $\alpha$ particles and the nucleus, and also find the kinetic and potential energies of the $\alpha$ particle at that distance.

**19.** Protons of energy 4.00 MeV are Coulomb scattered by a silver foil of thickness $4.0 \times 10^{-6}$ m. What fraction of the incident protons is scattered at angles (a) beyond 90°? (b) Less than 10°? (c) Between 5 and 10°?

**20.** Derive Equations 11.49–11.51 for "black disk" scattering.

**21.** Give the compound nucleus resulting from protons bombarding an aluminum target, and give at least five different ways for the compound nucleus to decay.

**22.** For the states of $^{61}$Cu populated in the $(\alpha, p)$ reaction, Figure 11.4, find the $\ell$ transfer for each of the states.

**23.** In the $(d, p)$ reaction leading to states in $^{91}$Zr, as shown in Figures 11.23 and 11.24, discuss the possible final angular momentum states if the reaction could proceed by a compound-nucleus mechanism. As an example, consider whether it still is possible to associate a final $I^{\pi} = \frac{7}{2}^{+}$ state uniquely with $\ell = 4$. Discuss other final states as well.

**24.** The low-lying levels of $^{43}$Sc were illustrated in Figure 5.12. It is desired to populate the states up to the $\frac{7}{2}^{-}$ excited state with the $(d, n)$ reaction. Estimate the most likely angle for the outgoing neutrons for each excited state. (Try to estimate the excited-state energies from the figure.)

**25.** The $(d, p)$ reaction on $^{49}$Ti $(\frac{7}{2}^{-}$ ground state) populates the "collective" $0^{+}$, $2^{+}$, and $4^{+}$ states at 0.000, 1.555, and 2.675 MeV (respectively) in $^{50}$Ti. What are the angular momentum values transferred in the direct reaction?

**26.** The $(^{3}\text{He}, p)$ reaction on an even-$Z$, even-$N$ target leads to certain final states identified with the transfer of either $\ell = 0$, 2, or 4. (a) For each choice, list the possible spin-parity assignments in the final nucleus. (b) In some cases, the analysis suggests that certain states are populated by a mixture of $\ell = 0$ and $\ell = 2$, while others are populated by a mixture of $\ell = 2$ and $\ell = 4$. Is it possible to make a unique determination of the final spin in either of these cases?

**27.** The $(d, p)$ reaction on $^{52}$Cr leads to the $\frac{3}{2}^{-}$ ground state of $^{53}$Cr. How would the analysis of the angular momentum transfer in this reaction differ between an analysis in terms of direct reactions and one in terms of compound-nucleus reactions?

# 12

# NEUTRON PHYSICS

As the uncharged member of the nucleon pair, the neutron plays a fundamental role in the study of nuclear forces. Unaffected by the Coulomb barrier, neutrons of even very low energy (eV or less) can penetrate the nucleus and initiate nuclear reactions. In contrast, part of our lack of understanding of processes in the interior of stars results from the difficulty of studying proton-induced reactions at energies as low as keV. On the other hand, the lack of Coulomb interaction presents some experimental problems when using neutrons as a nuclear probe: energy selection and focusing of an incident neutron beam are difficult, and neutrons do not produce primary ionization events in detectors (neutrons passing through matter have negligible interactions with the atomic electrons).

The first experimental observation of the neutron occurred in 1930, when Bothe and Becker bombarded beryllium with $\alpha$ particles (from radioactive decay) and obtained a very penetrating but nonionizing radiation, which they assumed to be a high-energy $\gamma$ ray. Soon thereafter, Curie and Joliot noticed that when this radiation fell on paraffin, an energetic proton was emitted. From the range of these protons, they determined their energy to be 5.3 MeV. If the radiation under study were indeed $\gamma$'s, protons could be knocked loose from paraffin by a Compton-like collision; from the Compton-scattering formula, they computed that the energy of this "$\gamma$ radiation" would be at least 52 MeV to release such protons. An emitted $\gamma$ of such an energy seemed extremely unlikely. In 1932, Chadwick provided the correct explanation, identifying the unknown radiation as a neutral (therefore penetrating and nonionizing) particle with a mass nearly the same as that of the proton. Thus in a head-on collision, a 5.3-MeV neutron could transfer its energy entirely to the struck proton. Chadwick did additional recoil experiments with neutrons and confirmed his hypothesis, and he is generally credited with being the discoverer of the neutron.

The free neutron is unstable against $\beta$ decay, with a half-life of 10.6 min. In nuclei, the bound neutron may be much longer-lived (even stable) or much shorter-lived. Despite the instability of free neutrons, their properties are measured to high precision, particularly the magnetic dipole moment, $\mu = -1.91304184 \pm 0.00000088 \ \mu_N$, and the neutron-proton mass difference, $m_n - m_p = 1.29340 \pm 0.00003$ MeV.

Basic research with neutrons goes back almost to the earliest days of nuclear physics, and it continues to be a vital and exciting research field today. For example, interference effects with neutron beams have permitted some basic aspects of quantum mechanics to be demonstrated for the first time. The *electric dipole moment* of the neutron should vanish if the neutron were an elementary particle or even a composite particle in which the binding forces were symmetric with respect to the parity and time-reversal operations. Many careful and detailed experiments have been done, and all indicate a vanishing electric dipole moment, but the limit has been pushed so low ($10^{-25}$ e · cm) that it is almost possible to distinguish among certain competing theories for the interactions among the elementary particles. The so-called Grand Unified Theories that attempt to unify the strong (nuclear), electromagnetic, and weak ($\beta$-decay) interactions predict that the conservation of nucleon number (actually baryon number) can break down, and that a neutron could convert into its antiparticle, the antineutron, and then back again to a neutron. No evidence has yet been seen for this effect either, but current research is trying to improve the limits on our knowledge of the neutron-antineutron conversion frequency.

## 12.1   NEUTRON SOURCES

Beams of neutrons can be produced from a variety of nuclear reactions. We cannot accelerate neutrons as we can charged particles, but we can start with high-energy neutrons and reduce their energy through collisions with atoms of various materials. This process of slowing is called "moderating" the neutrons. The resulting neutrons can have very low energies, which by convention are given the following designations:

| | |
|---|---|
| Thermal | $E \simeq 0.025$ eV |
| Epithermal | $E \sim 1$ eV |
| Slow | $E \sim 1$ keV |
| Fast | $E = 100$ keV–10 MeV |

**α-Beryllium Sources**   The reaction responsible for the discovery of the neutron can be used to produce a source of neutrons suitable for use in the laboratory. The stable isotope of beryllium, $^9$Be, has a relatively loosely bound neutron (1.7 MeV binding energy). If a typical $\alpha$ particle from a radioactive decay (5–6 MeV) strikes a $^9$Be nucleus, a neutron can be released:

$$^4\text{He} + {}^9\text{Be} \rightarrow {}^{12}\text{C} + \text{n}$$

The $Q$ value for this reaction is 5.7 MeV. If we mix together a long-lived $\alpha$-emitting material, such as $^{226}$Ra, and $^9$Be, there will be a constant rate of neutron production. From $^{226}$Ra and its daughters there are $\alpha$'s emitted with energies from about 5 to nearly 8 MeV, and thus we find neutrons with an energy spectrum up to 13 MeV. The neutrons are not monoenergetic because of (1) the many $\alpha$ groups, (2) the slowing of $\alpha$'s that will occur by collision in any solid material, (3) the various directions of emission that can occur for the neutrons relative to the $\alpha$'s (whose direction we do not know), and (4) the possibility that $^{12}$C is left in an excited state. The most probable neutron energy is about 5 MeV, and the neutron production rate is about $10^7$ neutrons per second for each Ci of $^{226}$Ra. A typical neutron spectrum is shown in Figure 12.1.

Because of the high $\gamma$ emission of $^{226}$Ra and its daughters, the radium-beryllium neutron source has largely been replaced with sources using $^{210}$Po (138 d),

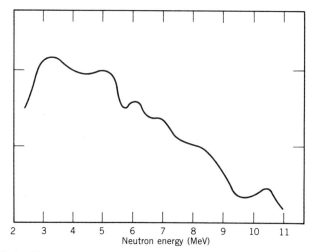

**Figure 12.1** Neutron energy spectrum from a Ra-Be source, measured with a proton recoil counter. Several neutron groups are present; they result from reactions induced by $\alpha$'s with differing energies and in which the $^{12}$C is left either in the ground state or the 4.43- or 7.6-MeV excited states.

$^{238}$Pu (86 y), and $^{241}$Am (458 y). These sources produce about $2-3 \times 10^6$ neutrons per second per Ci of $\alpha$ activity.

**Photoneutron Sources** In a process similar to the $(\alpha, n)$ sources discussed above, we can use the $(\gamma, n)$ reaction to produce neutrons. The advantage of photoneutron production is that we can make the neutrons more nearly monoenergetic, particularly if the photon source is nearly monoenergetic. For example, $^{24}$Na emits a $\gamma$ of 2.76 MeV, absorption of which would be sufficient to overcome the neutron binding energy of $^9$Be:

$$\gamma + {}^9\text{Be} \rightarrow {}^8\text{Be} + n$$

The yield is acceptable ($2 \times 10^6$ neutrons/s per Ci of $^{24}$Na), but the half-life is short (15 h). The neutron energy is about 0.8 MeV. A longer-lived isotope $^{124}$Sb (60 d) emits a strong $\gamma$ whose energy just exceeds the $^9$Be neutron binding energy; the emitted neutron has a much lower energy, about 24 keV.

**Spontaneous Fission** A common source of neutrons is the spontaneous fission of isotopes such as $^{252}$Cf (2.65 y). Neutrons are produced directly in the fission process, at a rate of about 4 per fission. The fission occurs in only about 3% of the decays ($\alpha$ decay accounts for the rest), and the neutron production rate is $2.3 \times 10^{12}$ neutrons/s per gram of $^{252}$Cf or $4.3 \times 10^9$ n/s per Ci of $^{252}$Cf. The neutron energies are characteristic of fission—a continuous distribution with an average energy of 1–3 MeV.

**Nuclear Reactions** There are of course many nuclear reactions that produce neutrons. These require an accelerator to produce a beam of particles to initiate the reaction, and thus they are not as convenient as the radioactive-decay type of sources discussed previously. However, by carefully selecting the incident energy and the angle at which we observe the emitted neutron, we can obtain a

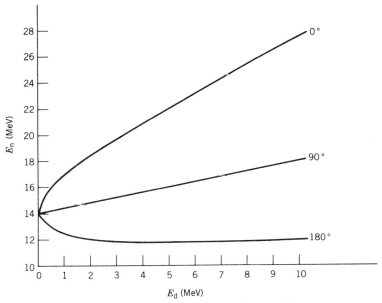

**Figure 12.2** Neutrons emitted in the $^3$H(d, n)$^4$He reaction.

reasonably monoenergetic beam of almost any desired energy. Some reactions that might be used are

$$^3\text{H} + \text{d} \ \rightarrow {}^4\text{He} + \text{n} \qquad Q = +17.6 \text{ MeV}$$

$$^9\text{Be} + {}^4\text{He} \rightarrow {}^{12}\text{C} + \text{n} \qquad Q = +5.7 \text{ MeV}$$

$$^7\text{Li} + \text{p} \ \rightarrow {}^7\text{Be} + \text{n} \qquad Q = -1.6 \text{ MeV}$$

$$^2\text{H} + \text{d} \ \rightarrow {}^3\text{He} + \text{n} \qquad Q = +3.3 \text{ MeV}$$

Figure 12.2 illustrates the dependence of the neutron energy in the first of these reactions on the incident energy and on the direction of the outgoing neutron.

**Reactor Sources** The neutron flux near the core of a nuclear fission reactor can be quite high—typically $10^{14}$ neutrons/cm$^2$/s. The energy spectrum extends to 5–7 MeV but peaks at 1–2 MeV. These neutrons are generally reduced to thermal energies within the reactor, but there are also fast neutrons present in the core. Cutting a small hole in the shielding of the reactor vessel permits a beam of neutrons to be extracted into the laboratory for experiments. The high neutron fluxes from a reactor are particularly useful for production of radioisotopes by neutron capture, as in neutron activation analysis.

## 12.2 ABSORPTION AND MODERATION OF NEUTRONS

As a beam of neutrons travels through bulk matter, the intensity will decrease as neutrons are removed from the beam by nuclear reactions. For fast neutrons, many reactions such as (n, p), (n, $\alpha$), or (n, 2n) are possible, but for slow or thermal neutrons the primary cause of their disappearance is capture, in the form of the (n, γ) reaction. Often the cross sections for these capture reactions are dominated by one or more resonances, where the cross section becomes very large. Off resonance, the cross section decreases with increasing velocity like $v^{-1}$; thus as the neutrons slow down (become moderated) due to elastic and inelastic scattering processes, absorption becomes more probable. Neutrons with initial energy in the 1-MeV range would undergo many scattering processes until their energies were reduced to the eV range, where they would have a high probability of resonant or nonresonant absorption.

In crossing a thickness $dx$ of material, the neutrons will encounter $n\,dx$ atoms per unit surface area of the beam or the material, where $n$ is the number of atoms per unit volume of the material. If $\sigma_t$ is the total cross section (including scattering processes, which will tend to divert neutrons from the beam), then the loss in intensity $I$ is

$$dI = -I\sigma_t n\,dx \tag{12.1}$$

and the intensity decreases with absorber thickness according to an exponential relationship:

$$I = I_0 e^{-\sigma_t n x} \tag{12.2}$$

Keep in mind that this expression refers only to monoenergetic neutrons—the original intensity of neutrons of a certain energy decreases according to Equation 12.2. Of course, we may at the same time be creating neutrons of lower energy (by scattering, for example), which may have a very different cross section, but this effect is not included in Equation 12.2. We therefore cannot use it reliably to calculate the decrease in the *total number* of neutrons, only the change in intensity of those with a given initial energy.

Let's consider an elastic collision between a neutron of initial energy $E$ and velocity $v$ with a target atom of mass $A$ initially at rest. Elementary application of the laws of conservation of energy and linear momentum gives the ratio between the final neutron energy $E'$ and the initial energy:

$$\frac{E'}{E} = \frac{A^2 + 1 + 2A\cos\theta}{(A + 1)^2} \tag{12.3}$$

where $\theta$ is the scattering angle in the center-of-mass system (but $E$ and $E'$ are measured in the laboratory system). For no scattering ($\theta = 0$), Equation 12.3 gives $E'/E = 1$, as it should. The maximum energy loss occurs for a head-on collision ($\theta = 180°$):

$$\left(\frac{E'}{E}\right)_{min} = \left(\frac{A - 1}{A + 1}\right)^2 \tag{12.4}$$

Notice that for $A = 1$ (scattering from hydrogen), the neutron gives all its energy to the struck proton.

For neutron energies of about 10 MeV and below, the scattering is mostly s wave and thus (in the center-of-mass system) largely independent of $\theta$. The values of $E'/E$ are uniformly distributed between $E'/E = 1$ and the minimum value given by Equation 12.4, as shown in Figure 12.3a.

Because each neutron will scatter many times, we must repeatedly calculate the energy loss. In the case of the second scattering, the incident neutrons are no

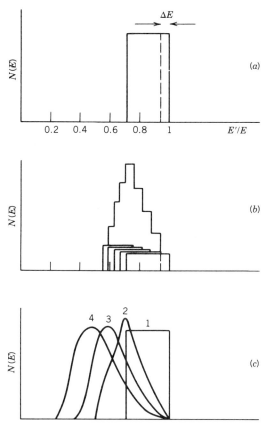

**Figure 12.3** (*a*) A monoenergetic neutron of energy *E* gives, after a single s-wave scattering from $^{12}$C, a flat distribution of laboratory energies *E'* from $0.72E$ to *E*. (*b*) Dividing the scattered distribution into five narrow, nearly monoenergetic distributions of width $\Delta E$, we get after a second scattering the five flat distributions shown, whose sum is the peaked distribution. (*c*) An exact calculation of the energy distribution after 1, 2, 3, and 4 scatterings.

longer monoenergetic but rather are distributed as in Figure 12.3*a*. We can approximate this effect by considering each interval of width $\Delta E$ to be a new generation of approximately monoenergetic neutrons giving the result shown in Figure 12.3*b*. Continuing this process, we obtain the succeeding "generations" of energy distributions shown in Figure 12.3*c*.

To make the calculations more quantitative, we define the parameter $\xi$ to represent the average value of $\log(E/E')$ after a single collision:

$$\xi = \left[ \log \frac{E}{E'} \right]_{\text{av}} \tag{12.5}$$

$$= \frac{\int \log \left[ \dfrac{(A+1)^2}{A^2 + 1 + 2A \cos \theta} \right] d\Omega}{\int d\Omega} \tag{12.6}$$

**Table 12.1** Moderating Properties of Various Nuclei

| Nucleus | $\xi$ | $n$ (for thermalization) |
|---------|-------|--------------------------|
| $^1\text{H}$ | 1.00 | 18 |
| $^2\text{H}$ | 0.725 | 25 |
| $^4\text{He}$ | 0.425 | 43 |
| $^{12}\text{C}$ | 0.158 | 110 |
| $^{238}\text{U}$ | 0.0084 | 2200 |

where $d\Omega$ is the element of solid angle in the center-of-mass system. Here again we assume the scattering to be isotropic. Carrying out the integration gives

$$\xi = 1 + \frac{(A-1)^2}{2A} \log \frac{A-1}{A+1} \tag{12.7}$$

The average value of log $E'$ is decreased after each collision by an amount $\xi$, and after $n$ collisions, the average value of log $E'$ is log $E'_n$:

$$\log E'_n = \log E - n\xi \tag{12.8}$$

which follows directly from Equation 12.5.

Table 12.1 shows values of $\xi$ for some commonly used moderators. If our goal is to reduce the average neutron energy from that which is typical for neutrons emitted in fission ($E \sim 2$ MeV) to that which is characteristic of thermal motion ($E'_n \sim 0.025$ eV), the number of generations of collisions is shown in Table 12.1.

The previous calculation has assumed the atoms from which the neutrons scatter to be at rest. This is certainly a good approximation for MeV neutrons, but as thermal energies are approached, we find the thermal motion of the atoms of the moderator to be comparable to the speeds of the neutrons. The scattering in this case is better analyzed using statistical mechanics, and we can simply assume that after a sufficient time the neutrons will reach thermal equilibrium with the moderator at a temperature $T$. In this case, the neutrons are described by a Maxwellian speed distribution:

$$f(v)\, dv = 4\pi n \left(\frac{m}{2\pi kT}\right)^{3/2} v^2\, e^{-mv^2/2kT}\, dv \tag{12.9}$$

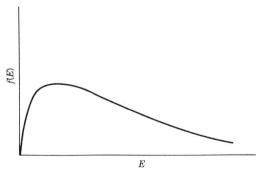

**Figure 12.4** Maxwellian energy distribution, a representation of the neutron energy spectrum after many scatterings.

where $f(v)\,dv$ gives the fraction of neutrons with speeds between $v$ and $v + dv$. Here $m$ is the neutron mass and $n$ is the total number of neutrons per unit volume. Rewriting this in terms of energy gives

$$f(E)\,dE = \frac{2\pi n}{(\pi kT)^{3/2}} E^{1/2} e^{-E/kT}\,dE \tag{12.10}$$

which is illustrated in Figure 12.4 and looks similar to Figure 12.3c, showing the thermalizing effect of even a few generations of collisions.

## 12.3 NEUTRON DETECTORS

Because neutrons produce no direct ionization events, neutron detectors must be based on detecting the secondary events produced by nuclear reactions, such as $(n, p)$, $(n, \alpha)$, $(n, \gamma)$, or $(n, \text{fission})$, or by nuclear scattering from light charged particles, which are then detected.

For slow and thermal neutrons, detectors based on the $(n, p)$ and $(n, \alpha)$ reactions provide a direct means for observing neutrons from the signal left by the energetic p or $\alpha$ resulting from the reaction. The isotope $^{10}$B is commonly used, by producing an ionization chamber or a proportional counter filled with $BF_3$ gas or lined with boron metal or a boron compound. The reaction is

$$^{10}\text{B} + \text{n} \rightarrow {}^7\text{Li}^* + \alpha$$

where the $^7$Li is preferentially left in an excited state with energy 0.48 MeV. (Natural boron consists of about 20% of the isotope $^{10}$B, so materials enriched in $^{10}$B increase the efficiency of the detector.) For thermal neutrons, the cross section is about 3840 b, a very large value, and the cross section follows the $1/v$ law up to about 100 keV, so the dependence of cross section on incident energy is featureless (no resonances are present) and predictable (Figure 12.5).

There is also another advantage of the $1/v$ dependence of the cross section. Suppose we are observing a collimated beam of neutrons or an isotropic flux (perhaps near the core of a reactor) that has a velocity distribution of $n(v)\,dv$ neutrons per unit volume with speeds between $v$ and $v + dv$. The flux passing through the detector will be $n(v)v\,dv$, and if the counter contains $N$ boron nuclei each with cross section $\sigma$, the probability per second of an interaction (or counting rate, if we are able to detect and count every interaction) is

$$dR = N\sigma n(v)v\,dv \tag{12.11}$$

for neutrons of speeds between $v$ and $v + dv$. For neutrons of all speeds, the total counting rate is

$$R = \int N\sigma n(v)v\,dv \tag{12.12}$$

$$= NC \int n(v)\,dv \tag{12.13}$$

where the last step assumes that $\sigma \propto v^{-1}$, so that the product $\sigma v$ is the constant $C$. The integral then gives the total number of neutrons per unit volume $n$, and

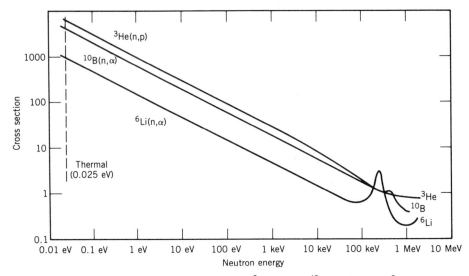

**Figure 12.5** Neutron cross sections for $^3$He(n, p), $^{10}$B(n, $\alpha$), and $^6$Li(n,$\alpha$). The cross section shows the $1/v$ behavior for $E < 1$ keV, but begins to show resonances above 100 keV.

the counting rate is

$$R = NCn \tag{12.14}$$

That is, $R$ is directly proportional to the neutron density for *any distribution of velocities* as long as we can neglect any contributions to the neutron flux outside the $1/v$ region of cross section.

The $Q$ value for the reaction leading to the $^7$Li excited state is 2.31 MeV, and for incident neutrons of kinetic energy small compared with this value, momentum conservation requires the sharing of energy between $^7$Li and $\alpha$ so that the $\alpha$ is given a kinetic energy of 1.47 MeV. The kinetic energy of the incident neutron, if it is in the eV or even keV range, will not substantially modify this value. Unless either particle strikes the wall, we will detect simultaneously the $^7$Li ($T = 0.84$ MeV) as well, and the neutron then leaves as its signature a 2.31-MeV energy loss in the counter. Because we cannot measure MeV energies to eV or keV precision in a proportional counter, we cannot use such a device to measure such low neutron energies.

Other similar devices are based on $^6$Li (n, $\alpha$), with $Q = 4.78$ MeV and $\sigma = 940$ b for thermal neutrons, and $^3$He (n, p), with $Q = 0.765$ MeV and $\sigma = 5330$ b. A comparison of the neutron cross sections for these reactions is shown in Figure 12.5.

Another way of measuring neutron intensities is by exposing to the neutrons a material which becomes radioactive after neutron capture and which has a known capture cross section for neutrons of a particular energy. If we remove the material from the neutron flux and measure the induced radioactivity (using a $\gamma$-ray detector, for example), we can determine the neutron flux.

Among the earliest devices used for determining neutron energies were mechanical devices, for example the velocity selector, a rotating shutter made of a highly absorbing material, such as Cd for thermal neutrons (Figure 12.6). This

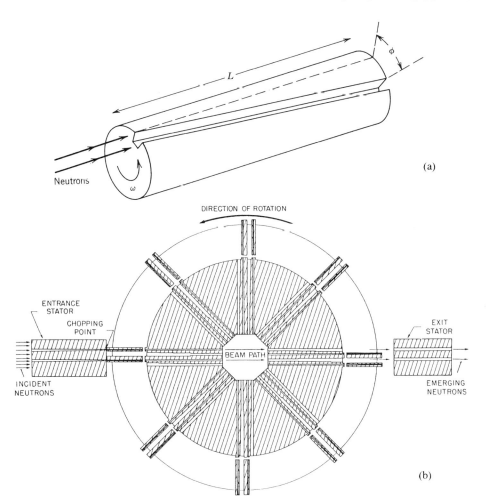

**Figure 12.6** (*a*) Neutron velocity selector, consisting of a rotating cylinder with one or more helical slots cut into its surface. The cylinder is made of a material, such as cadmium, with a high absorption for neutrons. The selector will pass neutrons of velocity *v* that travel the length *L* of the cylinder in the time that it takes it to rotate through the angle $\phi$; that is, $t = L/v = \phi/\omega$, so that $v = L\omega/\phi$. Changing the angular speed $\omega$ permits selection of the neutron velocity. (*b*) Rotating shutter or "chopper" for producing pulses of neutrons. A continuous stream of neutrons enters from the left and a pulse of neutrons emerges at right if the rotor slits line up with the entrance slits. The rotor is made of stainless steel with phenolic slits. From R. E. Chrien, in *Neutron Sources for Basic Physics and Applications*, edited by S. Cierjacks (Oxford: Pergamon, 1983).

device is practical only for velocities in the thermal region, but it can be used to select neutrons from the continuous velocity distribution such as is produced by a reactor.

Another way of measuring velocities is through a variant of the time-of-flight technique (see Section 7.9 and Figure 7.39). If we have neutrons in a short pulse, we can time their travel over a distance of several meters (thermal neutrons have

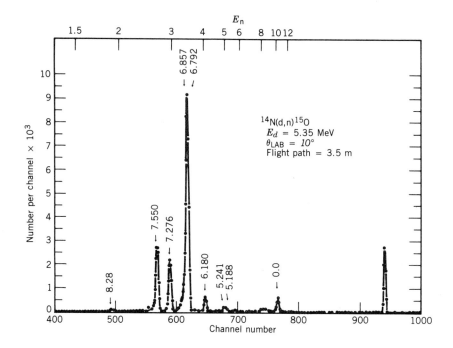

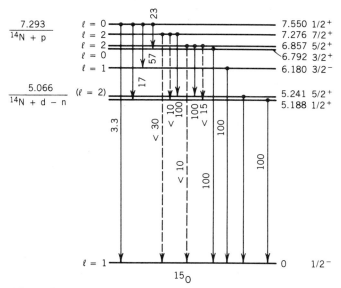

**Figure 12.7** Time-of-flight spectrum of neutrons emitted in the reaction $^{14}$N(d, n)$^{15}$O. Timing was done with reference to a pulsed beam of deuterons. The peak at far right comes from $\gamma$ rays, which of course travel at the highest possible speed. The neutron peaks correspond to the ground and excited states of $^{15}$O, as shown. The $\ell$ values come from measurement of the angular distribution of the cross section, or $d\sigma/d\Omega$. From R. C. Ritter, E. Sheldon, and M. Strang, *Nucl. Phys.* *A* **140**, 609 (1970).

$v = 2200$ m/s, and the time of travel is thus an easily measurable $10^{-3}$ s). For higher energies, longer flight paths of order 100 m and increased sensitivity of short-timing techniques can give accurate velocity measurements for neutrons up to MeV energies.

The initial pulse of neutrons for a timing measurement can be provided by a "chopper" of the kind shown in Figure 12.6, or else by a pulsed charged-particle accelerator, in which the neutrons are produced through reactions such as those listed in Section 12.1. If the initial pulse includes a wide range of velocities, the start-stop technique using a time-to-amplitude converter can display the energy spectrum of the neutrons, as in Figure 12.7.

Very precise energy determinations can be done in the thermal region using crystal diffraction. Thermal neutrons have a de Broglie wavelength of about 0.1 nm, about the same as the spacing between atoms in a crystal lattice. If a beam of thermal neutrons is incident on a crystal, the wave nature of the beam will be revealed through a set of interference maxima that occur at angles determined by the Bragg condition:

$$n\lambda = 2d \sin \theta \qquad (12.15)$$

where $d$ is the lattice spacing, $n$ the order of the interference maximum, and $\theta$ the angle the incident and reflected beams make with the surface of the crystal. This technique, which is used frequently to study the crystalline properties or atomic spacing of materials, is discussed in Section 12.6.

For measurement of the energies of fast neutrons, the most common method is to use the recoil following elastic scattering between the neutron and a light target (H, $^2$H, $^3$He, $^4$He, etc.). This elastic scattering was previously discussed in Section 12.2 in connection with neutron moderation, and we can use the results derived in that section. In the discussion following Equation 12.4, we showed that the scattered neutron has a continuous range of energies from $E' = E$ down to the minimum value given by Equation 12.4. The struck nucleus has a corresponding recoil energy

$$E_R = E - E' \qquad (12.16)$$

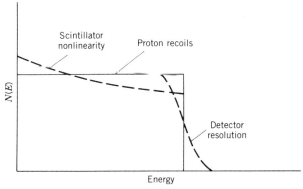

**Figure 12.8** An ideal spectrum of proton recoils (from monoenergetic incident neutrons) can be distorted by detector resolution and scintillator nonlinearity.

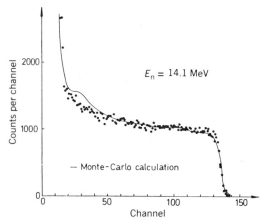

**Figure 12.9** Spectrum of monoenergetic 14-MeV neutrons observed in an organic scintillator. From M. Bormann et al., *Nucl. Instrum. Methods* **88**, 245 (1970).

ranging from zero up to a maximum

$$(E_R)_{max} = E - (E')_{min}$$

$$= E \frac{4A}{(A + 1)^2} \tag{12.17}$$

For hydrogen, $(E_R)_{max} = E$, while for $^3$He, $(E_R)_{max} = 0.75E$. The response of an ideal proton recoil detector to monoenergetic neutrons is shown in Figure 12.8.

The proton recoil signal is normally observed by using a scintillating material that is rich in hydrogen, such as plastic or an organic liquid. The scintillator therefore serves as both the proton target for the neutron and the detector for the recoiling proton. Taking into account the energy resolution of the scintillator and geometrical effects, the observed energy spectrum for monoenergetic neutrons looks like the continuous distribution shown in Figure 12.9. If the incident neutrons have several distinct energy components, the unfolding of the superimposed spectra of recoils may be difficult.

The efficiency of proton-recoil scintillation detectors for MeV neutrons can be of the order of 50%.

## 12.4 NEUTRON REACTIONS AND CROSS SECTIONS

The formalism for analyzing nuclear reactions has already been discussed in Chapter 11. In this section we give some examples of applications to neutron-induced reactions and show the specific aspects of nuclear structure that can be probed.

Let's first consider the $1/v$ dependence of the low-energy neutron cross section. We can obtain this result using two very different approaches. In Section 11.8, we obtained an estimate for the reaction cross section, Equation 11.50, of

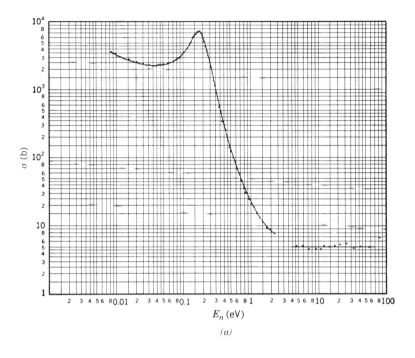

$(a)$

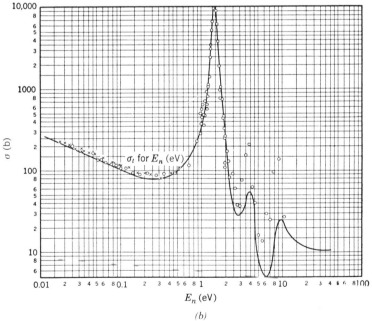

$(b)$

**Figure 12.10**   Resonances in neutron total cross sections. ($a$) A single isolated resonance superimposed on a $1/v$ background in Cd. ($b$) Several close-lying resonances in In. ($c$) Many sharp peaks in the resonance region in $^{238}$U. ($d$) In $^{32}$S, the relative heights of the resonance peaks are indications of the spin of the resonances. *Sources*: ($a$), ($b$) H. H. Goldsmith et al., *Rev. Mod. Phys.* **19**, 259 (1947); ($c$) D. J. Hughes and R. B. Schwartz, Brookhaven National Laboratory Report BNL 325 (1958); ($d$) J. M. Blatt and V. F. Weisskopf, *Theoretical Nuclear Physics* (New York: Wiley, 1952).

$\sigma_r = \pi(R + \lambda)^2$ based on a total absorption model. A primary modification to this estimate would include the reflection of the incident neutron wave function at the nuclear surface—how likely is it that the incident particle will penetrate to the region of nuclear potential, where it can be absorbed?

The transmission probability for a rectangular potential barrier was calculated in Section 2.3, and including this factor the cross section is estimated as

$$\sigma = \pi(R + \lambda)^2 \frac{4kK}{(k + K)^2} \tag{12.18}$$

where $K = \sqrt{2m(E + V_0)/\hbar^2}$ for a barrier of depth $-V_0$, and $k = \sqrt{2mE/\hbar^2}$.

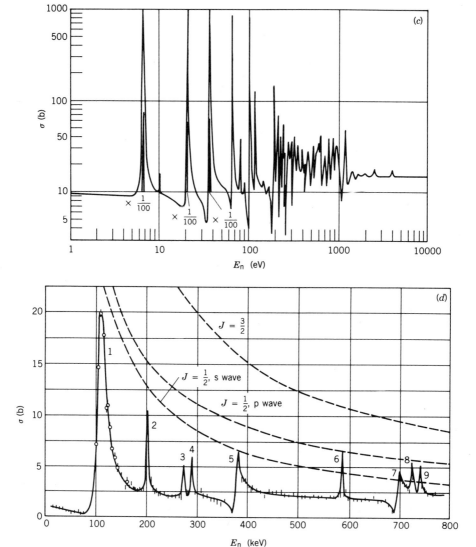

**Figure 12.10** Continued.

For low-energy neutrons, $E \ll V_0$ and $k \ll K$; also $\lambda = k^{-1} \gg R$, so

$$\sigma \cong \frac{4\pi}{kK} \qquad (12.19)$$

and since $k = p/\hbar = mv/\hbar$, we have the $1/v$ dependence of the cross section.

A similar result can be obtained from quite another approach, using the single-level resonance formula, Equation 11.70. Following neutron capture, the primary decay mechanism is $\gamma$ emission, the probability for which is virtually independent of any small variation in the resonance or incident energy. We can therefore take $\Gamma$ as independent of the neutron energy. The neutron width $\Gamma_n$, which refers to the entrance channel, is dependent on the density of final states $dn/dE$ available to the captured neutron, which according to Equation 9.15 is proportional to the velocity of the neutron. (This is somewhat similar to $\alpha$ decay, in which the decay probability includes a factor proportional to $v$ that originates from considering the frequency with which the $\alpha$ particle presents itself to the nuclear barrier in preparation for decay.) Far from the resonance, $E \ll E_R$ and

$$\sigma \cong \frac{\pi}{k^2} \frac{\Gamma_n \Gamma}{E_R^2 + \Gamma^2/4} \propto \frac{1}{v} \qquad (12.20)$$

when $\Gamma_n \propto v$.

As indicated by the cross sections plotted in Figure 12.5, the $1/v$ law is followed quite accurately for reactions far from the resonance region.

In the resonance region there is no exact theory for predicting the location of the resonances; the structure may be dominated by a single, isolated resonance (as in Cd, Figure 12.10*a*), or a complex structure (as in In and U, Figures 12.10*b*

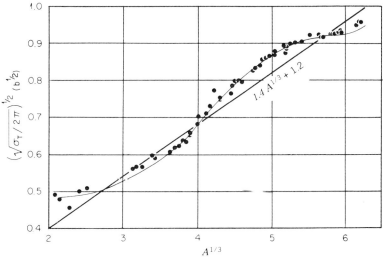

**Figure 12.11**  Total cross section for 14-MeV neutrons, plotted against $A^{1/3}$. The straight line represents $R + \lambda = 1.4A^{1/3} + 1.2$ fm, and it fits the data reasonably well. The curve through the data points is an improved calculation using the optical model. From G. R. Satchler, *Introduction to Nuclear Reactions* (Halstead: New York, 1980).

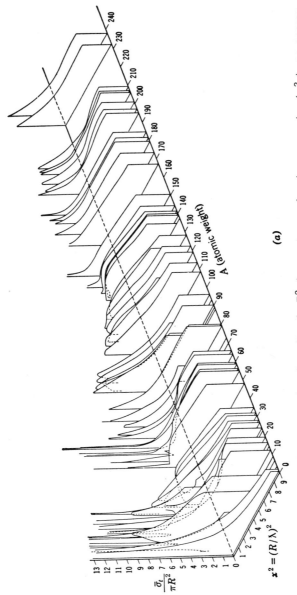

**Figure 12.12** The total cross section, in units of $\pi R^2$, for neutrons of various energies ($x^2$ is an energy parameter in dimensionless form). The top curves (*a*) represent experimental results, and the bottom curves (*b*) are the results of calculations using the optical model.

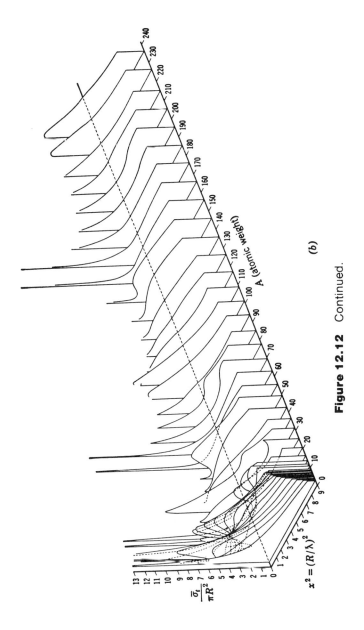

**Figure 12.12** Continued.

and *c*). In some cases, the intensity of a resonance can be used to determine its spin. For instance, in the resonance structure of $^{32}$S (which is an even-even, spin zero nucleus), s-wave capture leads to a resonance with total spin $I = \frac{1}{2}$ (from the intrinsic spin of the neutron). P-wave capture can give $I = \frac{1}{2}$ or $\frac{3}{2}$, depending on whether the $\ell$ and the $s$ of the neutron are antiparallel or parallel. The relationship between the cross section at resonance and $\ell$ is slightly more complicated than the $(2\ell + 1)$ dependence predicted in Equation 11.47, but the general effect of the increase in cross sections is apparent.

As we move to higher incident energies, the microscopic properties become less important because the observed structure often represents an average over many overlapping resonances. An indication of the applicability of an average property is obtained from the geometrical estimate for the *total* cross section, Equation 11.51. Figure 12.11 shows the quantity $(\sigma_t/2\pi)^{1/2}$ plotted to show $R + \lambda$, where $R = R_0 A^{1/3}$ and $\lambda = 1.2$ fm for 14-MeV neutrons. The radius $R_0$ is taken as 1.4 fm, somewhat larger than the usual estimate of 1.25 fm because it takes into account the skin thickness of the nuclear interaction.

The variations of the total cross section with energy and with $A$ are indicated by Figure 12.12. We can use the optical model (Section 11.9) to calculate the neutron total cross section for these nuclei, with the results also shown in Figure 12.12. The agreement with the observed gross structure is reasonably good.

## 12.5 NEUTRON CAPTURE

Figure 12.13 shows some of the processes that can occur following neutron capture. Although it is of course possible to re-emit the neutron, for heavy nuclei and very low energy incident neutrons this mode of decay of the compound or resonant state is suppressed, and $\gamma$ emission is the most probable decay process.

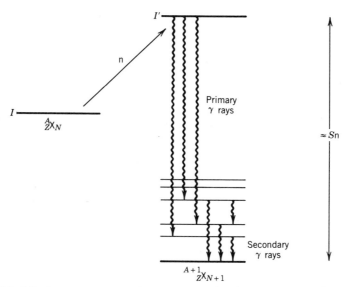

**Figure 12.13** Low-energy neutron capture leads to a state $I'$, which emits primary $\gamma$ rays followed by secondary $\gamma$ rays.

(Charged-particle emission is inhibited by the Coulomb barrier and unlikely to occur in any but the very lightest nuclei.) The excitation energy $E_x$ of $A'$ is just $S_n + E_n$, the neutron separation energy plus the energy of the incident neutron. For low-energy neutrons, $E_x$ is typically 5–10 MeV.

Neutron capture reactions can be used to determine the energy and spin-parity assignments of the capturing states. Let's assume the original nucleus had spin $I$ and parity $\pi$ (+ or −). The spin $I'$ of the capturing state is determined by the neutron orbital angular momentum $\ell$ and spin angular momentum $s$ added to that of the target nucleus:

$$I' = I + \ell + s \tag{12.21}$$

and the parities are related by

$$\pi' = \pi(-1)^\ell \tag{12.22}$$

For incident neutrons in the thermal range of energies, only s-wave capture will occur, for which $I' = I \pm \frac{1}{2}$ (excepting $I = 0$, in which case $I' = \frac{1}{2}$ only) and $\pi' = \pi$.

The capture state can usually be treated as a compound-nuclear state, that is, as many unresolved individual levels having $I' = I + \frac{1}{2}$ and $I' = I - \frac{1}{2}$. Occasionally, the capture may be dominated by one or a few resonances, and in certain selected cases it is possible to populate only only the $I' = I + \frac{1}{2}$ or $I' = I - \frac{1}{2}$ resonances by aligning the spins of the target nuclei either parallel or antiparallel to the neutron spins.

In the $\gamma$ decay of the capture state, all excited states of $A'$ can be populated to some extent, limited only by the selection rules for $\gamma$ radiation. For dipole radiation, the maximum change in spin is one unit, and thus the primary radiation (that which comes directly from the capture state) reaches states with spins from $I' + 1$ to $I' - 1$. Taking into account the relationship between $I'$ and $I$, the excited states populated by the primary radiation will have spins ranging from $I - \frac{3}{2}$ to $I + \frac{3}{2}$ (except for $I \leq 1$). For dipole radiation, the emission probability is proportional to $E_\gamma^3$, as in Equation 10.12. It is therefore most likely to observe the highest energy transitions, that is, those that populate the lowest excited states. The $\gamma$ spectrum therefore shows two principal components: the *primary* component, consisting of direct 5–10-MeV radiations from the capture state to lower excited states, usually below 2 MeV, and the *secondary* component, consisting of low-energy radiations between the low-lying excited states.

Electric dipole primary radiation is roughly 100 times more likely than magnetic dipole, and so it is most likely to populate lower excited states whose parity is opposite to that of the capture state. However, magnetic dipole radiation is often present, as is radiation of multipole order higher than dipole; these radiations are usually far less intense than the electric dipole radiation.

One of the useful features in studying neutron capture $\gamma$ rays is that the primary radiation is almost completely unselective in populating the lower excited states—no strong selection rules based on nuclear structure will forbid primary decays to any particular excited state. Thus if, for instance, the capture state has $I' = 4$, we expect dipole radiation to populate directly *all* spin 3, 4, and 5 states. This is in contrast to $\alpha$ or $\beta$ decay, in which the emission process is governed by selection rules which may forbid decays to states that, considering

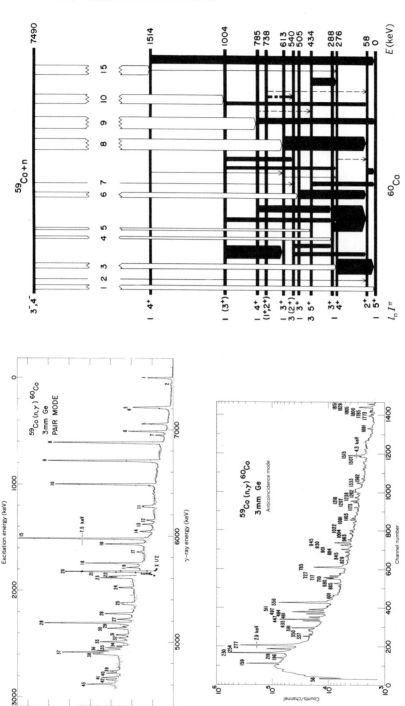

**Figure 12.14** Gamma rays following thermal neutron capture by ⁵⁹Co. (Top left) Primary transitions from the capture state, shown in white on the level scheme at right. (Bottom left) Secondary transitions between the low-lying excited states, shown in black at right. Because ⁵⁹Co has spin $\frac{7}{2}$ and negative parity, the capture state must be a combination of spin-parity 3⁻ and 4⁻. Electric dipole primary transitions will populate only low-lying 2⁺, 3⁺, 4⁺, and 5⁺ states, as shown. No primary transition leading to the 1⁺ state at 738 keV is expected, and none is observed. Because the capture state is mostly 4⁻, the 2⁺ states are only weakly populated (the thickness of the lines at right is proportional to the transition intensity); note transitions number 2 and 7. Data from E. B. Shera and D. W. Hafemeister, *Phys. Rev.* **150**, 894 (1966).

angular momentum couplings only, we might expect to be populated. Neutron capture can therefore serve as a rather *complete* means of spectroscopy of excited states—if we can only unravel what must be a very complicated spectrum of radiations, we should be able to learn a great deal about the number and location of excited states.

Figure 12.14 shows an example of the primary and secondary $\gamma$ radiations emitted following neutron capture, including the detailed structure of the excited states that can be deduced.

Another application of the (n, $\gamma$) reaction occurs when the ground state (or a long-lived isomeric state) of $A'$ is itself radioactive. We therefore accumulate activity of $A'$ (usually without bothering to observe the decay $\gamma$'s from the capture state). The activity builds up according to Equation 6.24, which can be expressed in the following useful form:

$$\mathscr{A} = 0.602 \frac{m}{A} \sigma \frac{\phi}{3.7 \times 10^{10}} (1 - e^{-\lambda t}) \qquad (12.23)$$

where $\mathscr{A}$ is the activity in Curies, $m/A$ is the dimensionless ratio between the mass of the target and its atomic mass, $\sigma$ is the capture cross section (usually for thermal neutrons) in units of barns, $\phi$ is the neutron flux in neutrons/cm$^2$/s, and $t$ is the duration of the neutron bombardment.

This technique has a variety of applications: with a known cross section, measuring $\mathscr{A}$ gives the neutron flux, and thus we have a measure of neutron intensity. If $\phi$ is known, on the other hand, we can determine unknown cross sections. The most common application, however, is to use cases with known $\phi$ and $\sigma$ to do qualitative analysis to determine $m$. After exposing an unknown sample to neutrons, we can observe many different radiations from the radioactive decays of those isotopes that can be produced by neutron capture. Careful measurement of the $\gamma$-ray spectrum permits the determination of which isotopes are present and in what quantities, and we can thus deduce the original quantities present in the irradiated sample. This technique is called *neutron activation analysis* and it has important applications in a variety of areas, including environmental pollution research, archeology, and forensic science. We discuss this application of nuclear techniques in Chapter 20.

## 12.6  INTERFERENCE AND DIFFRACTION WITH NEUTRONS

Beams of neutrons provide an excellent way to observe effects depending on the wave behavior of material particles. In Chapter 11 we saw examples of the diffraction of nucleons in scattering from nuclei, in cases in which the wavelength of the incident particle was comparable to the nuclear size. By moving to other ranges of wavelengths, however, we can observe effects such as single and double slit or thin film interference, which are more frequently demonstrated only with optical radiations.

In Figure 12.15 are shown results from single-slit and double-slit interference experiments. The interference effects are very apparent and remind us of the analogous results using light waves.

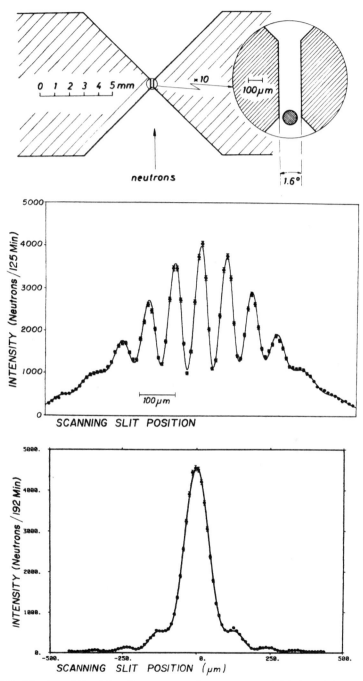

**Figure 12.15** (Top) Double-slit apparatus for neutrons. A highly absorbing boron wire is mounted in the gap between two pieces of neutron-absorbing glass to form a double slit. (Middle) Double-slit interference pattern. The dots are the experimental points and the curve is calculated from the Schrödinger equation for a neutron wavelength of 1.845 nm. (Bottom) A single-slit interference pattern. All figures from A. Zeilinger, R. Gaehler, C. G. Shull, and W. Treimer, in *Neutron Scattering—1981*, edited by J. Faber (New York: American Institute of Physics, 1982), p. 93.

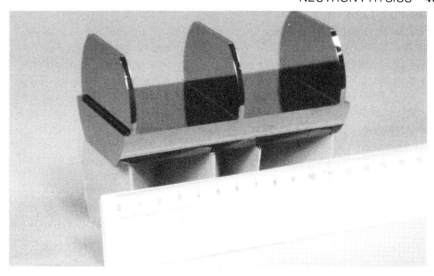

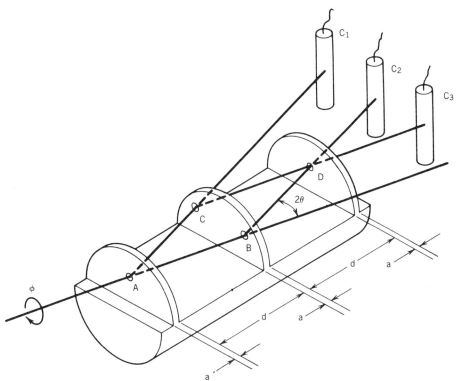

**Figure 12.16** (Top) Photograph of neutron interferometer, carved from a single block of silicon 5 cm in diameter and 9 cm long. The three slabs are attached to the base as they were in the original crystal, and therefore the atoms in the slabs are spatially coherent. Photograph courtesy of Professor Samuel A. Werner. (Bottom) Schematic diagram of interferometer. Counter $C_1$ counts the non-interfering beam, while $C_2$ and $C_3$ count the two interfering beams which re-combine at D. From S. A. Werner, J.-L. Staudenmann, R. Colella, and A. W. Overhauser, in *Neutron Interferometry*, edited by U. Bonse and H. Rauch (Oxford: Clarendon Press, 1979), p. 209.

The most ingenious demonstrations of neutron interference have been done with the interferometer shown in Figure 12.16. The device consists of three identical "beam splitters," each of which passes an undeviated transmitted beam and a diffracted beam deflected by an angle $2\theta$. (We discuss neutron diffraction later in this section.) To obtain identical diffractions at the three beam splitters and to guarantee precise alignment of the three planes, the entire apparatus was carved from a large single crystal of silicon, of length 9 cm and diameter 5 cm. The function of the third slab is to recombine the two interfering beams $BD$ and $CD$ into outgoing beams that contain the interference information as a net intensity variation. (The variations in the two beams are complementary to one another and contain the same information.) Without the recombination done by the third crystal, the interference would occur only at the single point $D$ where the interfering beams cross. Figure 12.17 shows the interference effects observed with this device.

The neutron interferometer has been used in two remarkable experiments which demonstrate aspects of quantum behavior that have not been accessible to testing by any other means. In the first of these, the interferometer was rotated about the horizontal axis $AB$ in Figure 12.16. At a certain value of the rotation angle, one of the beams, $ABD$ for instance, will lie at a greater height than the other ($ACD$) and will therefore experience a different gravitational potential, $mgy$. The differing gravitational potential between the two beams causes a relative phase shift that changes the way they interfere when they are recombined. As the angle $\phi$ (which measures the rotation about the axis) is varied, the intensity in the recombined beam will vary accordingly. Figure 12.18 shows the resulting variation in intenity as $\phi$ is varied. This experiment, done 50 years after de Broglie's work and after the introduction of the Schrödinger equation, is the first to demonstrate that the Schrödinger equation and the laws of quantum mechanics apply to gravitational fields. The experimental result shows great sensitivity to physical parameters—it was possible to deduce the sag in the crystal caused by bending under its own weight (about 2 nm) and also to observe the Coriolis effect on the two neutron paths due to the rotation of the Earth.

Another unusual and fundamental interferometry experiment with neutrons is done by causing one of the neutron paths ($AB$, for example) to pass through a magnetic field. In the region of the magnetic field, the magnetic moment of the neutron will precess about the field direction, and this precession causes a change in the phase of the neutron wave function, which is again observed as a variation in the intensity of the recombined beam.

Quantum mechanics predicts an unusual (and previously untested) result for spin-$\frac{1}{2}$ particles—the phase of the wave function upon rotation by an angle $\alpha$ is $e^{is\alpha}$, where $s$ is the spin in units of $\hbar$. When $\alpha$ is $2\pi$ (360°), the phase is $e^{i\pi}$ or $-1$. Thus a complete 360° rotation of the system causes a change in phase of the wave function! This is certainly a nonclassical effect, for in classical physics a 360° rotation leaves all equations unchanged. However, the change in phase of the interfering beams will cause the interference to go from, for instance, a maximum (corresponding to $|\psi|^2 = |\psi_{ACD} + \psi_{ABD}|^2$) to a minimum ($|\psi|^2 = |\psi_{ACD} - \psi_{ABD}|^2$). The neutron interference experiments not only observed the effects of the spin precession of the neutrons, they demonstrated for the first time that a 720° rotation, not a 360° rotation, is required to bring the system back to

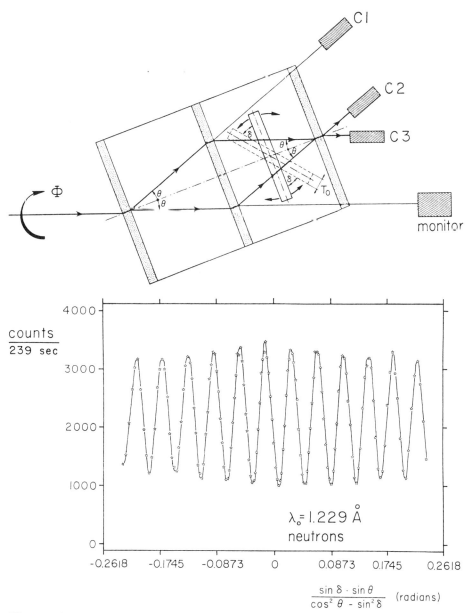

**Figure 12.17**  A silicon slab placed between the second and third diffracting planes is free to rotate (angle $\delta$ in the figure at top). As It does, the two neutron beams must pass through different thicknesses of silicon, which causes a phase shift and a change in intensity when they are recombined. The lower part of the figure shows the observed intensity variation, measured in counter $C_3$, as $\delta$ is varied. From S. Werner, J.-L. Staudenmann, R. Colella, and A. W. Overhauser, in *Neutron Interferometry*, edited by U. Bonse and H. Rauch (Oxford: Clarendon, 1979), p. 209.

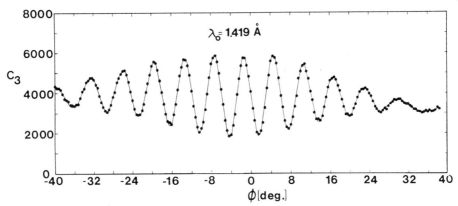

**Figure 12.18** Neutron interference in a gravitational field. As the interferometer rotates by the angle $\phi$ about a horizontal axis, as in Figure 12.17, the two interfering beams change gravitational potentials, which causes a phase difference and a corresponding intensity variation in the recombined beams, as shown. Notice that a rotation of 3–4° changes the counting rate in detector $C_3$ by a factor of 3. From S. Werner, J.-L. Staudenmann, R. Colella, and A. W. Overhauser, in *Neutron Interferometry*, edited by U. Bonse and H. Rauch (Oxford: Clarendon, 1979), p. 209.

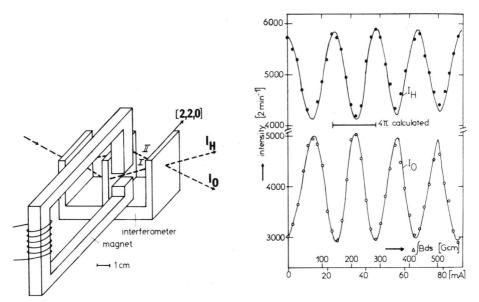

**Figure 12.19** Neutron interferometer used to demonstrate rotation of neutron spins in a magnetic field. Unpolarized neutrons of $\lambda = 1.8$ nm are incident on the interferometer; one of the interfering beams passes through the magnetic field, the strength of which may be varied. The neutron spins in branch I precess relative to those in branch II. The data show that a phase change of $4\pi$ is necessary to obtain constructive interference. From H. Rauch et al., *Phys. Lett. A* **54**, 425 (1975).

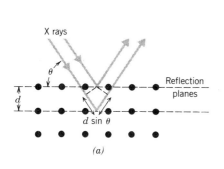

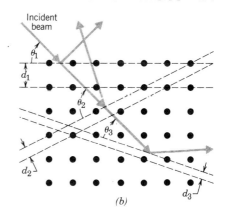

**Figure 12.20** (*a*) An incident beam can be reflected from two adjacent parallel rows of atoms in a lattice. If the path difference, $2d\sin\theta$, is an integral number of wavelengths, there will be constructive interference. (*b*) There are many possible ways to choose the set of parallel reflecting lattice planes. Corresponding to each will be a particular value of $\theta$ and of $d$.

its original configuration. Figure 12.19 shows the variation in the interfering intensity as the magnetic field is varied.

Probably the most frequent application of neutron interference is in the diffraction of neutrons by crystals. The diffraction can be used to provide a source or a detector of monoenergetic neutrons, or it can be used to study crystal structures of materials. Figure 12.20 illustrates the basic application of the Bragg law, Equation 12.15, to the diffraction by a single crystal. For neutrons whose wavelengths are about the same order as the crystalline atomic spacing $d$, interference results in an apparent "reflection" from planes drawn through atoms of the lattice. Typical atomic spacings in solids are of order 0.1 nm, and a de Broglie wavelength of 0.1 nm corresponds to thermal neutron energies.

The choice of a particular set of reflection planes is arbitrary; Figure 12.20 also shows several other choices, corresponding to different angles $\theta$ and different spacings $d$. In three dimensions, we can specify the choice of planes with a set of indices that give essentially the number of fundamental lattice spacings along the three coordinate axes. Figure 12.21 shows some examples of different lattice planes in three dimensions.

The intensity at any particular reflection may be a complicated function of crystal properties and neutron scattering amplitudes (which may involve coherent sums over different isotopes that may be present in the crystal). Figure 12.22 is an example of a scattering spectrum taken for neutrons of a particular wavelength. The scattering sample in this case was a powder consisting of many microcrystals, so that all possible orientations are simultaneously observed.

An alternative approach would be to begin with a beam of neutrons containing a mixture of energies, such as might be obtained from a reactor. If the beam falls on a scattering crystal, at a certain angle (as in Figure 12.23) we will observe monoenergetic neutrons that satisfy the Bragg condition (assuming that the same angle does not happen to satisfy the Bragg condition for a different wavelength and set of crystal planes). If we change the angle $\theta$ slightly, the same Bragg condition will be satisfied for a slightly different wavelength or energy. We thus

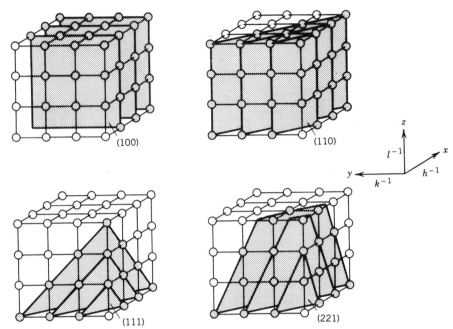

**Figure 12.21** Planes of atoms in three-dimensional cubic lattice (this is a three-dimensional analog of Figure 12.20*b*). The indices (*hkl*) give integers in the ratio of the *inverse* of the coordinates of the plane with respect to the three directional axes *x, y, z*. Thus $x = 1$, $y = \infty$, $z = \infty$ (upper left) reduces to (100); $x = 1$, $y = 1$, $z = 2$ (lower right) becomes $(11\frac{1}{2})$ and then (221) by changing to integers in the same ratio.

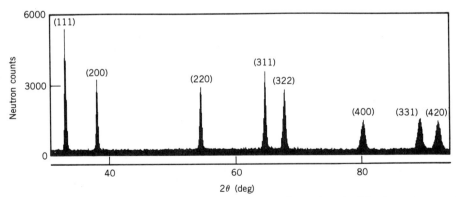

**Figure 12.22** Diffraction pattern of nickel powder using incident neutrons of wavelength 0.114 nm. The scattering angle $2\theta$ is plotted, and the peaks are labeled with the indices of the lattice planes, as in Figure 12.21. From G. E. Bacon, *Neutron Diffraction* (Oxford: Clarendon, 1975).

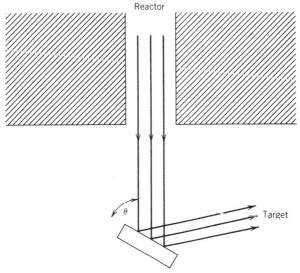

Reactor

$\theta$

Target

**Figure 12.23** Monochromator for producing a beam of monoenergetic neutrons. A collimated beam of reactor neutrons with a broad spectrum of wavelengths is Bragg reflected from a single crystal. For a particular value of $\theta$, there will be an interference maximum for a certain wavelength, and thus by varying $\theta$ we can choose the wavelength. As in Figure 12.20b, there may be other Bragg reflected peaks at other angles, which are not shown.

have a *neutron monochromator*, a source of neutrons of any particular energy, with a spread in energy determined by the angular spread of the beam.

Powder diffraction patterns such as shown in Figure 12.22 can be used to deduce the type of crystal structure, as in the cubic lattice characteristic of nickel. Similar studies are possible for more complex systems, including the relatively disordered structures of liquids and glasses, and the exceedingly complicated structures of biological molecules.

The primary mechanism for neutron scattering is the nuclear scattering that we have discussed previously in this chapter and in Chapter 11. An alternative mechanism is *magnetic* scattering, in which the neutron magnetic moment scatters from the electronic magnetic moment. If it is possible to separate nuclear from magnetic scattering (as, for instance, by using polarized neutrons), it is possible to deduce magnetic structures, which can be more complicated and more interesting than the physical arrangement of atoms in the lattice. Figure 12.24 shows the neutron diffraction pattern of MnO above and below the Néel temperature, which is for antiferromagnets what the Curie temperature is for ferromagnets. Below the Néel temperature, MnO is antiferromagnetic, in which there are alternating layers with opposite spin orientations. (In a ferromagnetic structure, all spins would be parallel.) Figure 12.25 shows a representation of the structure. The basic distance needed for the structure to repeat itself (called the *unit cell* dimension) is twice what it would be if we were above the Néel temperature where the magnetic structure repeats in nonalternating fashion. There is a factor of 2 difference in the basic distance $d$ that appears in the Bragg

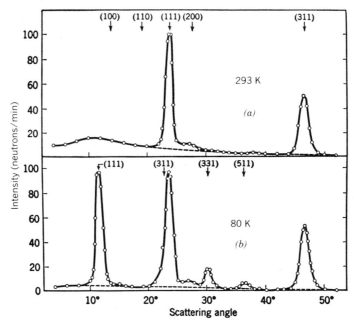

**Figure 12.24**   The neutron diffraction pattern of MnO (*a*) above and (*b*) below the antiferromagnetic transition temperature. Notice the factor of 2 difference in the location of the 111 peak. From C. G. Shull and J. S. Smart, *Phys. Rev.* **76**, 1256 (1949).

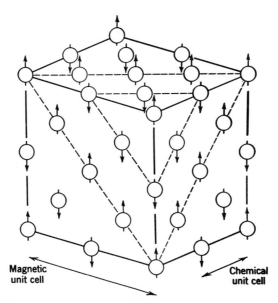

**Figure 12.25**   The crystal structure of MnO. In the antiferromagnetic structure, the lattice repeats over a distance of 2*a* (where *a* is the spacing of the Mn atoms) while at high temperature the spins are all aligned and the lattice repeats over a distance of *a*.

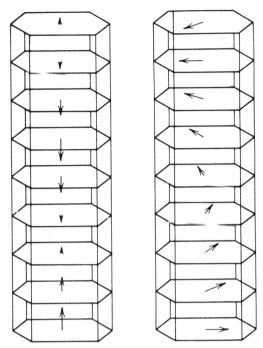

**Figure 12.26** Two examples of rare-earth magnetic structures studied by neutron diffraction. The basic crystal structure is that of stacked planes of hexagonal arrays of atoms. The arrows show the effective magnetic dipole moments. The structure at left is characteristic of erbium metal in the temperature range 52–85 K, and that at right describes terbium in the range of 220–230 K. Below 220 K, the terbium moment does not rotate from one plane to the next, but points in a fixed direction. Below 20 K, erbium shows a moment that has a rotating component in the hexagonal plane and a perpendicular component of fixed magnitude.

equation, and the result is that the peaks for magnetic scattering appear at values of $\theta$ such that $\sin \theta$ is half what it would be for nuclear scattering.

The rare-earth metals have rather similar crystal structures of the hexagonal type, but very different magnetic properties that also change with temperature as in the case of MnO. In some cases the electronic magnetic moment is perpendicular to the hexagonal plane, but may be modulated in unusual ways from one layer to the next, as for example several layers pointing up and then several down, or else sinusoidally. In other structures, the electronic moment lies in the hexagonal plane, but it rotates through a certain angle from one layer to the next. These magnetic structures, illustrated in Figure 12.26, can be studied only through neutron diffraction.

## REFERENCES FOR ADDITIONAL READING

A nontechnical introduction to neutron physics is D. J. Hughes, *The Neutron Story* (New York: Doubleday, 1959). Modern research with extremely low-energy neutrons is described in "Ultracold Neutrons" by R. Golub, W. Mampe, J. M. Pendlebury, and P. Ageron in the June 1979 issue of *Scientific American*.

For general surveys of neutron physics, see L. F. Curtiss, *Introduction to Neutron Physics* (Princeton, NJ: Van Nostrand, 1959); D. J. Hughes, *Pile Neutron Research* (Cambridge: Addison-Wesley, 1953); D. J. Hughes, *Neutron Optics* (New York: Interscience, 1954).

For more advanced and detailed references, see I. I. Gurevich and L. V. Tarasov, *Low-Energy Neutron Physics* (Amsterdam: North-Holland, 1968); J. E. Lynn, *The Theory of Neutron Resonance Reactions* (Oxford: Clarendon, 1968); *Fast Neutron Physics*, edited by J. B. Marion and J. L. Fowler (New York: Interscience, 1960); C. G. Windsor, *Pulsed Neutron Scattering* (New York: Halsted, 1981).

A comprehensive survey of neutron diffraction is G. E. Bacon, *Neutron Diffraction*, 3rd ed. (Oxford: Clarendon, 1975).

Reports on recent work with neutrons can be found in *Neutron Sources for Basic Physics and Applications*, edited by S. Cierjacks (Oxford: Pergamon, 1983); *Neutron Interferometry*, edited by U. Bonse and H. Rauch (Oxford: Clarendon, 1979); *Neutron Scattering—1981*, edited by John Faber, Jr. (New York: American Institute of Physics, 1982).

## PROBLEMS

1. In the reaction $\alpha + {}^9\text{Be} \rightarrow {}^{12}\text{C} + \text{n}$, find the maximum and minimum neutron energies when the incident $\alpha$ energy is (a) 7.0 MeV; (b) 1.0 MeV.

2. If the neutrons in Problem 1 are emitted uniformly in space, sketch their energy spectrum, that is, the relative number emitted at each energy from the minimum to the maximum.

3. A 5.00-MeV proton beam is available to produce neutrons from the reaction ${}^7\text{Li}(\text{p}, \text{n}){}^7\text{Be}$. A neutron beam of 1.75 MeV is needed for a particular experiment. (a) In what direction relative to the incident beam should we look to find 1.75-MeV neutrons? (b) If we extract the neutrons through a 1-cm aperture located 10 cm from the reaction target, what will be the energy spread of the neutron beam?

4. Neutrons were first observed through the "knock-on" protons produced in direct head-on scattering. If protons are observed at an energy of 5.3 MeV, compute the incident energy if the incident particle is (a) a photon, or (b) a neutron.

5. A neutron of kinetic energy $E$ scatters elastically from a target atom at rest. Find the maximum energy lost by the neutron if the target is (a) ${}^2\text{H}$; (b) ${}^{12}\text{C}$; (c) ${}^{238}\text{U}$.

6. For neutron moderation by protons, make a sketch showing the average energy of a beam of initial 2.0-MeV neutrons as a function of time, assuming the mean time between collisions is $t$. How long does it take for the neutrons to lose 50% of their energy? How long for 90%?

7. A point source of fast neutrons (a radium-beryllium source, for example) is placed at the center of a tank of water. Make a sketch showing the density of thermal neutrons as a function of the distance from the source. Your sketch should indicate general features only; don't try to be too specific. The

mean free path of neutrons in water is a few centimeters for fast neutrons to a few tenths of a centimeter for slow neutrons.

**8.** Derive Equation 12.7.

**9.** Find the number of collisions necessary to reduce the energy of a 1 MeV neutron to 1 eV in scattering by (a) hydrogen; (b) carbon; (c) beryllium; (d) iron.

**10.** In scattering between incident neutrons and target protons at rest, show that the angle between the two scattered particles is always $90°$.

**11.** The intensity of a source of thermal neutrons is to be measured by counting the induced radioactivity in a thin foil of indium metal exposed to the neutrons. The foil is $3.0 \times 3.0$ mm in area and $1.0$ $\mu$m in thickness. The activation of $^{115}$In to $^{116}$In ($t_{1/2} = 54$ min) takes place with a cross section of 160 b for thermal neutrons. The foil is irradiated for 1 min, but after the conclusion of the irradiation, the counting of the decays from the foil cannot be started for 30 min. The efficiency of the detecting system is only $2.4 \times 10^{-4}$, and in 1 h of counting $4.85 \times 10^{4}$ counts are accumulated. What is the thermal neutron flux?

**12.** In low-energy neutron capture by $^{55}$Mn, the following primary $\gamma$ transitions are seen (energies in MeV): 7.2703, 7.2438, 7.1597, 7.0578, 6.9287. (a) Deduce the corresponding energies of the $^{56}$Mn excited states. (b) $^{55}$Mn has $I^{\pi} = \frac{5}{2}^{-}$. What are the possible $I^{\pi}$ assignments of the $^{56}$Mn excited states populated through primary transitions?

**13.** (a) $^{143}$Nd has a $\frac{7}{2}^{-}$ ground state. Following thermal neutron capture, would you expect to see a strong primary transition from the capture state to the $^{144}$Nd ground state? (b) Would you expect the same in the case of capture by $^{119}$Sn ($\frac{1}{2}^{+}$)?

**14.** A beam of neutrons of energy 50 keV is incident on a slab of graphite of thickness 2.5 cm. The transmitted beam has intensity of 26% of the original beam. What value does this indicate for the scattering phase shift? (It is a good approximation to neglect all but the s-wave phase shift. Why?)

**15.** A beam of neutrons of wavelength $\lambda$ is incident in the $x$ direction on a crystal represented by the cubic lattice of Figure 12.21. The spacing between neighboring atoms along the coordinate axes is $a$. (a) Find the spacing $d$ between the lattice planes for the 4 cases in terms of $a$. (b) Find the direction in space of the scattered neutrons in each case for the first-order interference.

# 13

# NUCLEAR FISSION

The development of nuclear physics occurred very rapidly in the 1930s. Following the discovery of the neutron by Chadwick in 1932, it was a natural next step to study the effects of exposing various nuclei to neutrons (in effect, carrying on the work begun a few years earlier with charged-particle beams from the first accelerators). Enrico Fermi and his co-workers in Italy exposed many elements to neutrons and studied the induced radioactivity following neutron capture. They discovered that many nuclei decay by $\beta^-$ emission following neutron capture, as the nuclei try to compensate for the excess neutron by converting a neutron to a proton. The result is a residual nucleus of atomic number one unit higher. (Fermi received the Nobel Prize in 1938 for this work.) The next natural step was to use this technique to increase the atomic number to produce *transuranic* elements, those beyond the heaviest naturally occurring element (uranium) found in nature. Indeed, the technique of irradiating uranium with neutrons did reveal $\beta^-$ activities that were at first identified with new transuranic elements, but efforts to separate these elements chemically to study their properties produced confusing results. In particular, the induced activity appeared to show a chemical behavior similar to barium; consequently it was at first assumed to be radium, which appears just below barium in the periodic table and thus has atomic structure and chemical properties very similar to those of barium. However, radium is produced from uranium through the $(n, 2\alpha)$ reaction, which must be very improbable. In 1939, using careful radiochemical techniques, Hahn and Strassmann showed that in fact the induced activity was barium itself and not merely something chemically similar to it. Further work revealed many other intermediate-mass nuclei formed in the bombardment of uranium by neutrons, and experiments using ionization chambers showed that the energy released following neutron capture was very large, of the order of 100 MeV and certainly far greater than any previously observed $\alpha$-decay energy. With this evidence, Meitner and Frisch in 1939 proposed that the uranium nuclei following neutron capture are highly unstable and split nearly in half or *fission* (a term borrowed from the biologists' description of cell division).

Fission results primarily from the competition between the nuclear and Coulomb forces in heavy nuclei. The total nuclear binding energy increases roughly in proportion to $A$, while the Coulomb repulsion energy of the protons increases faster, like $Z^2$. If we regard the emission of a heavy fragment as a decay process similar to $\alpha$ decay, then we can regard heavy nuclei as residing very close to the top of the potential well in Figure 8.3, where the Coulomb barrier is very thin and easily penetrable. Fission can thus occur spontaneously as a natural decay process, or it can be induced through the absorption of a relatively low-energy particle, such as a neutron or a photon (producing excited states or compound-nuclear states that are high enough in energy to surmount or more easily penetrate the barrier).

Although any nucleus will fission if we provide enough excitation energy, as a practical matter the process is important only for heavy nuclei (thorium and beyond). The applicability of fission for obtaining large total energy releases was realized soon after its discovery, for another characteristic of the process is that every neutron-induced fission event produces, in addition to the two heavy fragments, several neutrons which can themselves induce new fission events. This *chain reaction* of fissions can occur very rapidly and without control, as in a fission explosive, or slowly and under careful control, as in a fission reactor. Because of these spectacular and awesome applications, nuclear fission plays a prominent role in many technical processes and in political decisions as well.

## 13.1 WHY NUCLEI FISSION

The energetic preference for nuclei to fission can be understood immediately from the binding energy per nucleon, Figure 3.16. A heavy nucleus in the uranium region has a binding energy of about 7.6 MeV/nucleon. If $^{238}$U were to divide into two equal fragments with $A \simeq 119$, their binding energy per nucleon would be about 8.5 MeV. Going to a more tightly bound system means that energy must be released; that is, the energy changes from bound $^{238}_{92}$U at $-238 \times 7.6 = -1809$ MeV to two bound $^{119}_{46}$Pd nuclei at $-2 \times 119 \times 8.5 = -2033$ MeV. To conserve energy, the final state must include an extra 214 MeV, which can appear in a variety of forms (neutrons, $\beta$ and $\gamma$ emissions from the fragments) but which appears primarily ($\sim 80\%$) as kinetic energy of the fragments as Coulomb repulsion drives them apart. In calculating decay probabilities, there is a term that depends on the energy release—the more energy is released, the more ways there are for the decay products to share the energy, the greater the number of final states to decay into, and the higher the decay probability. With such a large energy release, fission ought to be a decay means that is readily available for these nuclei as they "climb the curve of binding energy."

While the fission decay mode does indeed exist, it is not nearly so probable as our discussion might indicate—it does not compete successfully with the spontaneous $\alpha$ decay of $^{238}$U ($t_{1/2} = 4.5 \times 10^9$ y, while the partial half-life for fission is about $10^{16}$ y), and it does not become an important decay process until we get to nuclei of mass 250 and above. What inhibits the fission process is the Coulomb barrier, which also inhibits the analogous $\alpha$-decay process. If we divide $^{238}$U into two identical fragments that are just touching at their surfaces (separation $= R_1$

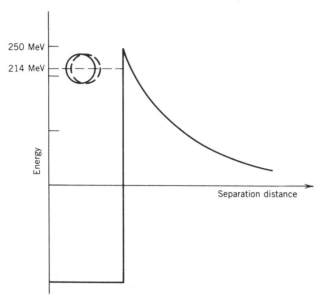

**Figure 13.1** Inside its nuclear potential well, $^{238}$U may perhaps exist instantaneously as two fragments of $^{119}$Pd, but the Coulomb barrier prevents them from separating.

$+ R_2$ where $R_1 = R_2 = 1.25(119)^{1/3} = 6.1$ fm), the Coulomb barrier is

$$V = \frac{1}{4\pi\epsilon_0} \frac{Z_1 Z_2 e^2}{R}$$

$$= (1.44 \text{ MeV} \cdot \text{fm})\frac{(46)^2}{12.2 \text{ fm}} = 250 \text{ MeV}$$

If we regard the zero of our energy scale to be the two fragments at rest separated by an infinite distance, then we can represent this system by Figure 13.1. Inside the region of nuclear potential, $^{238}$U can exist as two $^{119}$Pd nuclei because of the enormous number of final states accessible with a 214-MeV energy release. However, the Coulomb barrier prevents the fragments from separating, and the decay probability is small because the barrier cannot be penetrated.

This very crude calculation may indicate why fission is inhibited from readily occurring, but it should not be taken too seriously because the numbers we have used (250 MeV for the barrier height and 214 MeV for the energy release) are only estimates and can easily be modified by 10 to 20%. For example, the assumption that $^{238}$U splits into two identical fragments may not be very realistic. If the two fragments have masses and atomic numbers in roughly a $2:1$ ratio, such as $^{79}_{30}$Zn and $^{159}_{62}$Sm, the Coulomb barrier height is reduced from 250 to 221 MeV. The release of a few neutrons will change the mass numbers of the final fragments and can produce more nearly stable and tightly bound fragments (nuclei such as $^{119}$Pd, $^{79}$Zn, and $^{159}$Sm have a large neutron excess and are unlikely to be formed in fission). Also, the Coulomb barrier calculation based on a sharp edge at $R = R_1 + R_2$ is quite unlikely to be strictly correct.

NUCLEAR FISSION **481**

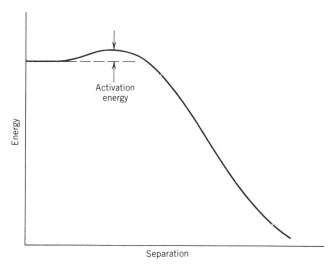

**Figure 13.2** A smooth potential barrier opposing the spontaneous fission of $^{238}$U. To surmount the fission barrier, we must supply an amount of energy equal to the activation energy.

What is certainly true, though, is that the height of the Coulomb barrier is *roughly* equal to the energy released in fission of heavy nuclei, and there are certain to be *some* nuclei for which the energy release puts the two fragments just below the Coulomb barrier, giving them a reasonably good chance to penetrate. These are the *spontaneously fissioning* nuclei, for which fission competes success-fully with other radioactive decay processes. There may be others whose sep-arated state would put them *above* the barrier and that would (if formed) instantly spontaneously fission. Such nuclei of course do not exist in nature; calculations suggest that the barrier against fission is zero at about $A = 300$. Still other nuclei may be far enough below the barrier that spontaneous fission is not observed, but absorption of a relatively small amount of energy, such as from a low-energy neutron or photon, forms an intermediate state (perhaps a compound nuclear state) that is at or above the barrier so that *induced fission* occurs readily, that is, it competes successfully with other modes of decay of the compound nucleus. If the intermediate state is below the barrier, fission is inhibited and other decay modes, including re-emission of the absorbed particle, may dominate. Subthreshold fission may have important implications for nuclear structure for there are often resonances that can enhance the fission probability, as we discuss in Section 13.4. The ability of a nucleus to undergo induced fission will depend critically on the energy of the intermediate system; for some nuclei, absorption of thermal neutrons may be sufficient to push them over the barrier, while for others, fast (MeV) neutrons may be required. Figure 13.2 shows a more realistic representation of the fission barrier for heavy nuclei.

A more detailed calculation of the energy needed to induce fission is shown in Figure 13.3, which gives essentially the height of the fission barrier above the ground state (usually called the *activation energy*). This calculation is based on the liquid-drop model, which treats only average nuclear properties; the inclusion of more sophisticated effects based on the shell model modifies the calculation

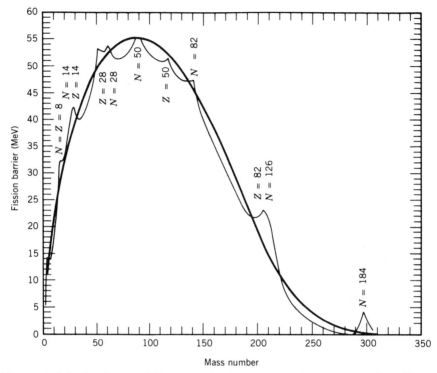

**Figure 13.3** Variation of fission activation energy with mass number. The dark curve is based on the liquid-drop model, calculated only for the most stable isotope at each mass number, and the light curve shows the effect of including shell structure. Note the typical 5-MeV energies around uranium, the vanishing energy around mass 280 (making these nuclei extremely unstable to spontaneous fission), and the stability around mass 300 from the expected neutron shell closure. From W. D. Myers and W. J. Swiatecki, *Nucl. Phys.* **81**, 1 (1966).

somewhat and suggests that certain super-heavy nuclei around $A = 300$ may be more stable against fission. Other consequences of shell structure are discussed in Section 13.4.

An instructive approach to understanding fission can be obtained from the semiempirical mass formula, Equation 3.28. Let us consider the effect on the binding energy of an initially spherical nucleus that we gradually stretch. The stretching can be done to keep the volume constant, but the surface and Coulomb terms, which were calculated originally for spherical nuclei, will certainly be affected by the stretching. We represent the stretched nucleus as an ellipsoid of revolution, which has a volume of $\frac{4}{3}\pi ab^2$, where $a$ is the semimajor axis and $b$ is the semiminor axis. The deviation of the ellipsoid from a sphere of radius $R$ is given in terms of a distortion parameter $\varepsilon$ as

$$a = R(1 + \varepsilon)$$
$$b = R(1 + \varepsilon)^{-1/2}$$

(13.1)

where $\varepsilon$ is the eccentricity of the ellipse and is related to the deformation parameter $\beta$ defined in Equation 5.15 as $\varepsilon = \beta\sqrt{5/4\pi}$. Note that $R^3 =$

$ab^2$, which keeps the volume constant as we increase the distortion. As a sphere is stretched to distort it into an ellipsoid, its surface area increases as $S = 4\pi R^2(1 + \frac{2}{5}\varepsilon^2 + \cdots)$; the surface energy term in the semiempirical mass formula increases accordingly. The Coulomb energy term can be shown to be modified by the factor $(1 - \frac{1}{5}\varepsilon^2 + \cdots)$, and thus the *difference* in energy (actually the decrease in binding energy) between a spherical nucleus and an ellipsoid of the same volume is

$$\begin{aligned}
\Delta E &= B(\varepsilon) - B(\varepsilon = 0) \\
&= -a_s A^{2/3}\left(1 + \tfrac{2}{5}\varepsilon^2 + \cdots\right) - a_c Z^2 A^{-1/3}\left(1 - \tfrac{1}{5}\varepsilon^2 + \cdots\right) \\
&\quad + a_s A^{2/3} + a_c Z^2 A^{-1/3} \\
&\cong \left(-\tfrac{2}{5}a_s A^{2/3} + \tfrac{1}{5}a_c Z^2 A^{-1/3}\right)\varepsilon^2
\end{aligned} \qquad (13.2)$$

If the second term is larger than the first, the energy difference is positive—we *gain* energy through the stretching, and the more the nucleus is stretched, the more energy is gained. Such a nucleus is unstable against this stretching and will readily undergo fission. Therefore, for spontaneous fission, the condition is

$$\tfrac{1}{5}a_c Z^2 A^{-1/3} > \tfrac{2}{5}a_s A^{2/3}$$

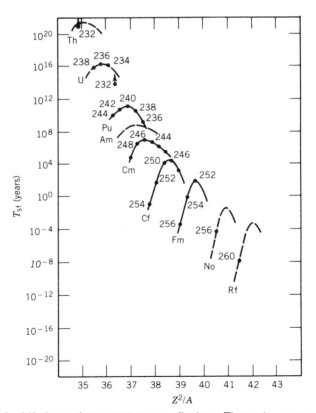

**Figure 13.4** Lifetimes for spontaneous fission. There is a general trend of decreasing lifetime with increasing $Z^2/A$. From V. M. Strutinsky and H. C. Pauli, in *Physics and Chemistry of Fission* (Vienna: IAEA, 1969), p. 155.

**Figure 13.5** Representation of nuclear shapes in fission.

and using the values of $a_s$ and $a_c$ gives

$$\frac{Z^2}{A} > 47 \tag{13.3}$$

This estimate must be modified somewhat to account for quantum mechanical barrier penetration, which permits spontaneous fission even when the deformation energy is negative. Furthermore, the nuclei in the region around uranium have permanent equilibrium deformation—the equilibrium shape is ellipsoidal rather than spherical. Nevertheless, the parameter $Z^2/A$ does serve as a rough indicator of the ability of a nucleus to fission spontaneously, as shown in Figure 13.4. The larger the value of $Z^2/A$, the shorter is the half-life for spontaneous fission. Extrapolating the spontaneous fission half-lives gives about $10^{-20}$ s for $Z^2/A \simeq 47$; thus the fission would occur "instantly" on the nuclear time scale for nuclei beyond the critical value of $Z^2/A$. No such nuclei are known, however, but a nucleus with $A = 300$ and $Z/A = 0.4$ has $Z^2/A = 48$, so this estimate is consistent with the zero activation energy for $A = 300$ as shown in Figure 13.3.

From a physical standpoint, what occurs during stretching can be described by the sequence of drawings shown in Figure 13.5. As the distortion becomes more extreme, the center becomes "pinched off" and the nucleus fissions.

## 13.2 CHARACTERISTICS OF FISSION

### Mass Distribution of Fragments

A typical neutron-induced fission reaction is

$$^{235}\text{U} + \text{n} \rightarrow {}^{93}\text{Rb} + {}^{141}\text{Cs} + 2\text{n}$$

which is possible for incident neutrons of thermal energies. The fission products are not determined uniquely—there is a distribution of masses of the two fission products of the form shown in Figure 13.6. The distribution must be symmetric about the center—for every heavy fragment, there must be a corresponding light fragment, but notice that fission into equal or nearly equal fragments ($A_1 \approx A_2$) is less probable by a factor of about 600 relative to the maximum yield of fragments with $A_1 \simeq 95$ and $A_2 \simeq 140$. Surprisingly, a convincing explanation for this mass distribution, which is characteristic of low-energy fission processes, has not been found. In contrast, fissions induced by very energetic particles show mass distributions that favor equal-mass fragments.

### Number of Emitted Neutrons

The fission fragments in the vicinity of $A = 95$ and $A = 140$ must share 92 protons. If they do so in rough proportion to their masses, the nuclei formed will be ${}^{95}_{37}\text{Rb}_{58}$ and ${}^{140}_{55}\text{Cs}_{85}$. These nuclei are extremely rich in neutrons—the most

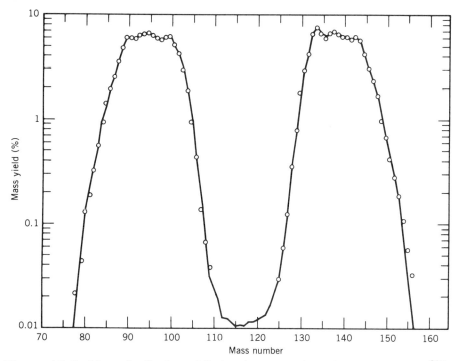

**Figure 13.6**  Mass distribution of fission fragments from thermal fission of $^{235}$U. Note the symmetry of the heavy and light distributions, even in the small variations near the maxima. From G. J. DiIorio, *Direct Physical Measurement of Mass Yields in Thermal Fission of Uranium 235* (New York: Garland, 1979).

stable nuclei in this mass region have $Z/A \simeq 0.41$, while for these fission products, $Z/A = 0.39$. The stable $A = 95$ isobar has $Z = 42$, and the stable $A = 140$ isobar has $Z = 58$. The fission fragments shed this neutron excess through emission of one or more neutrons *at the instant of fission* (within $10^{-16}$ s). These neutrons are known as *prompt neutrons*, and the number of prompt neutrons emitted in a given fission event will vary with the nature of the two fragments (and with the energy of the incident particle in the case of induced fission). The average number of prompt neutrons, called $\nu$, is characteristic of the particular fission process; for thermal neutron-induced fission, the experimentally observed values of $\nu$ are 2.48 for $^{233}$U, 2.42 for $^{235}$U, and 2.86 for $^{239}$Pu. The distribution about the mean shows the statistical behavior expected for an evaporation process, as illustrated by Figure 13.7. The Gaussian distribution that governs the neutron emission is remarkably independent of the fissioning nucleus, as all nuclei seem to follow the same behavior.

In addition to the prompt neutrons, *delayed neutrons* are often emitted in fission. These neutrons are emitted following the $\beta$ decay of the fission fragments and are examples of the $\beta$-delayed nucleon emission discussed in Section 9.8. The delay times are typically quite short, usually of the order of seconds. Figure 13.8 shows a typical case. Following the 6-s $\beta$ decay of $^{93}$Rb, $^{93}$Sr is left in a highly excited state, the energy of which happens to exceed the neutron separation

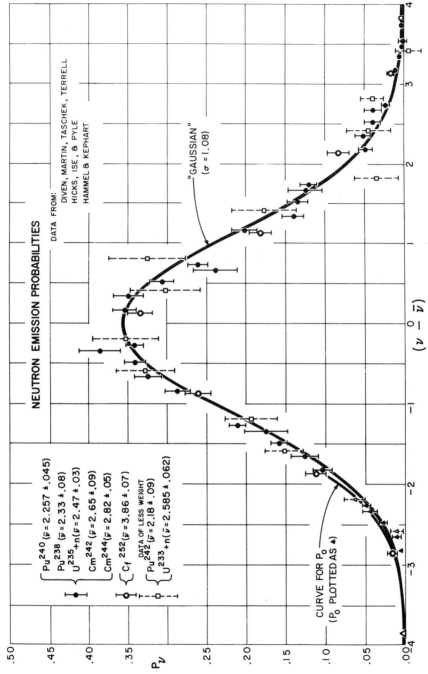

**Figure 13.7** Distribution of fission neutrons. Even though the average number of neutrons $\bar{\nu}$ changes with the fissioning nucleus, the distribution about the average is independent of the original nucleus. From J. Terrell, in *Physics and Chemistry of Fission*, Vol. 2 (Vienna: IAEA, 1965), p. 3.

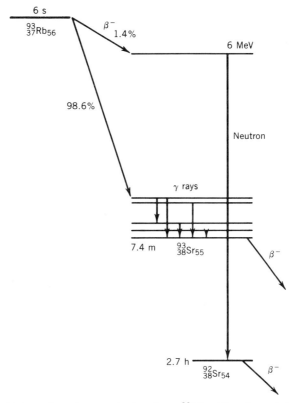

**Figure 13.8** Delayed neutron emission from $^{93}$Rb. After the original $\beta$ decay, the excited state of $^{93}$Sr has enough energy to decay by neutron emission to $^{92}$Sr. The neutrons are delayed relative to the prompt fission neutrons by a time characteristic of the mean lifetime of $^{93}$Rb.

energy. The state therefore can decay by neutron emission in competition with $\gamma$ emission; in this case, the neutron branch occurs with an intensity of 1.4%.

The total intensity of delayed neutrons amounts to about 1 per 100 fissions, but these neutrons are essential for the control of nuclear reactors. No mechanical system could respond rapidly enough to prevent statistical variations in the prompt neutrons from causing the reactor to run out of control, but it is indeed possible to achieve control using the delayed neutrons.

## Radioactive Decay Processes

The initial fission products are highly radioactive and decay toward stable isobars by emitting many $\beta$ and $\gamma$ radiations (which contribute ultimately to the total energy release in fission). Some sample decay chains are

$$^{93}\text{Rb} \xrightarrow{6\text{ s}} {}^{93}\text{Sr} \xrightarrow{7\text{ min}} {}^{93}\text{Y} \xrightarrow{10\text{ h}} {}^{93}\text{Zr} \xrightarrow{10^6\text{ y}} {}^{93}\text{Nb}$$

$$^{141}\text{Cs} \xrightarrow{25\text{ s}} {}^{141}\text{Ba} \xrightarrow{18\text{ min}} {}^{141}\text{La} \xrightarrow{4\text{ h}} {}^{141}\text{Ce} \xrightarrow{33\text{ d}} {}^{141}\text{Pr}$$

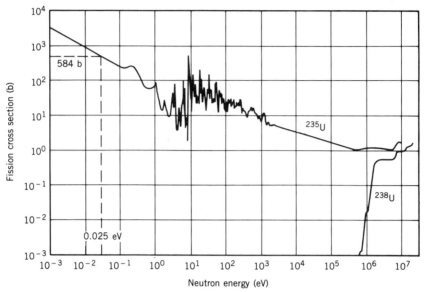

**Figure 13.9** Cross sections for neutron-induced fission of $^{235}$U and $^{238}$U.

These radioactive products are the waste products of nuclear reactors. Many decay very quickly, but others have long half-lives, especially near the stable members of the series.

## Fission Cross Sections

Figure 13.9 shows the cross sections for neutron-induced fission of $^{235}$U and $^{238}$U. The $^{235}$U cross section shows many features in common with the neutron cross sections we considered in Chapter 11. The thermal region shows the usual $1/v$ dependence of the cross section, and numerous resonances dominate the structure in the region 1–100 eV. For $^{235}$U, the thermal cross section for fission (584 b) dominates over scattering (9 b) and radiative capture (97 b). Note also that the thermal cross section is 3 orders of magnitude larger than the cross section for fast neutrons. If we wish to use the MeV neutrons emitted in fission to induce new fission events, the neutrons must first be moderated to thermal energies to increase the cross section. For $^{238}$U, there is no fission at all in the thermal region; only for fast neutron energies will fission occur. This extreme difference in behavior, as we see in the next section, results from the relationship between the excitation energy of the compound system and the activation energy needed to overcome the barrier.

## 13.3 ENERGY IN FISSION

In this section we consider in more detail the energy involved in fission. When $^{235}$U captures a neutron to form the compound state $^{236}$U*, the *excitation energy* is

$$E_{ex} = \left[ m\left(^{236}\text{U}^*\right) - m\left(^{236}\text{U}\right) \right] c^2$$

The energy of the compound state can be found directly from the mass energies of $^{235}$U and n, if we assume the neutron's kinetic energy is so small (i.e., in the thermal region) that it is negligible.

$$m\left(^{236}U^*\right) = m\left(^{235}U\right) + m_n$$

$$= (235.043924 \text{ u} + 1.008665 \text{ u})$$

$$= 236.052589 \text{ u}$$

$$E_{ex} = (236.052589 \text{ u} - 236.045563 \text{ u})\, 931.502 \text{ MeV/u}$$

$$= 6.5 \text{ MeV}$$

The activation energy (energy needed to overcome the fission barrier, as in Figure 13.2) for $^{236}$U is calculated also to be 6.2 MeV. Thus, the energy needed to excite $^{236}$U into a fissionable state (the activation energy) is exceeded by the energy we get by adding a neutron to $^{235}$U. This means that $^{235}$U can be fissioned with zero-energy neutrons, consistent with its large observed fission cross section in the thermal region.

A similar calculation for $^{238}$U + n → $^{239}$U* gives $E_{ex}$ = 4.8 MeV, which is far smaller than the calculated activation energy of $^{239}$U, 6.6 MeV. (Table 13.1 and Figure 13.10 give calculated activation energies for nuclei in this region.) Neutrons of at least MeV energy are therefore required for fission of $^{238}$U, which is consistent with the observed threshold for neutron-induced fission of $^{238}$U, as indicated in Figure 13.9.

The primary explanation for the extreme differences in the fissionability of $^{235}$U and $^{238}$U thus lies in the difference between their excitation energies, respectively 6.5 and 4.8 MeV. This difference in turn can be understood in terms

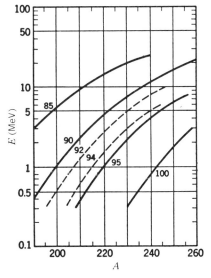

**Figure 13.10**  Activation energies of heavy nuclei. This graph shows the explicit *A* and *Z* dependence (which is not shown in Figure 13.3), calculated using the liquid-drop model. From S. Frankel and N. Metropolis, *Phys. Rev.* **72**, 914 (1947).

**Table 13.1**  Fission Cross Sections for Thermal Neutrons

| Nuclide | Cross Section (b) | $A + 1$ Activation Energy (MeV) |
|---|---|---|
| $^{227}_{90}\text{Th}_{137}$ | $200 \pm 20$ | |
| $^{228}_{90}\text{Th}_{138}$ | $< 0.3$ | |
| $^{229}_{90}\text{Th}_{139}$ | $30 \pm 3$ | 8.3 |
| $^{230}_{90}\text{Th}_{140}$ | $< 0.001$ | 8.3 |
| $^{230}_{91}\text{Pa}_{139}$ | $1500 \pm 300$ | 7.6 |
| $^{231}_{91}\text{Pa}_{140}$ | $0.019 \pm 0.003$ | 7.6 |
| $^{232}_{91}\text{Pa}_{141}$ | $700 \pm 100$ | 7.2 |
| $^{233}_{91}\text{Pa}_{142}$ | $< 0.1$ | 7.1 |
| $^{231}_{92}\text{U}_{139}$ | $300 \pm 300$ | 6.8 |
| $^{232}_{92}\text{U}_{140}$ | $76 \pm 4$ | 6.9 |
| $^{233}_{92}\text{U}_{141}$ | $530 \pm 5$ | 6.5 |
| $^{234}_{92}\text{U}_{142}$ | $< 0.005$ | 6.5 |
| $^{235}_{92}\text{U}_{143}$ | $584 \pm 1$ | 6.2 |
| $^{238}_{92}\text{U}_{146}$ | $(2.7 \pm 0.3) \times 10^{-6}$ | 6.6 |
| $^{234}_{93}\text{Np}_{141}$ | $1000 \pm 400$ | 5.9 |
| $^{236}_{93}\text{Np}_{143}$ | $3000 \pm 600$ | 5.9 |
| $^{237}_{93}\text{Np}_{144}$ | $0.020 \pm 0.005$ | 6.2 |
| $^{238}_{93}\text{Np}_{145}$ | $17 \pm 1$ | 6.0 |
| $^{239}_{93}\text{Np}_{146}$ | $< 0.001$ | 6.3 |
| $^{238}_{94}\text{Pu}_{144}$ | $17 \pm 1$ | 6.2 |
| $^{239}_{94}\text{Pu}_{145}$ | $742 \pm 3$ | 6.0 |
| $^{240}_{94}\text{Pu}_{146}$ | $< 0.08$ | 6.3 |
| $^{241}_{94}\text{Pu}_{147}$ | $1010 \pm 10$ | 6.0 |
| $^{242}_{94}\text{Pu}_{148}$ | $< 0.2$ | 6.2 |
| $^{241}_{95}\text{Am}_{146}$ | $3.24 \pm 0.15$ | 6.5 |
| $^{242}_{95}\text{Am}_{147}$ | $2100 \pm 200$ | 6.2 |
| $^{243}_{95}\text{Am}_{148}$ | $< 0.08$ | 6.3 |
| $^{244}_{95}\text{Am}_{149}$ | $2200 \pm 300$ | 6.0 |
| $^{243}_{96}\text{Cm}_{147}$ | $610 + 30$ | 6.1 |
| $^{244}_{96}\text{Cm}_{148}$ | $1.0 \pm 0.5$ | 6.3 |
| $^{245}_{96}\text{Cm}_{149}$ | $2000 \pm 200$ | 5.9 |
| $^{246}_{96}\text{Cm}_{150}$ | $0.2 \pm 0.1$ | 6.0 |

of the pairing energy term $\delta$ of the semiempirical mass formula, Equation 3.28. Figure 13.11 illustrates the effect of the pairing term. The binding energy of $^{236}\text{U}$ is increased (that is, its ground-state energy is lowered) by an amount $\delta$, which is roughly 0.56 MeV; the excitation energy is correspondingly increased by $\delta$ over what it would be in the absence of pairing. In the case of $^{238}\text{U}$, the energy of the ground state *before capture* is lowered by $\delta$, and as a result the energy of the

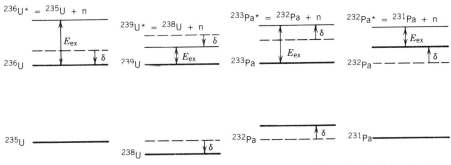

**Figure 13.11** Effect of pairing on excitation energies. The dashed levels show the nuclear energies in the absence of pairing, which are then raised or lowered by the amount $\delta$ when the effect of pairing is included.

capture state is correspondingly lower. The excitation energy is therefore *reduced* by $\delta$ relative to its value without the pairing force term. The difference in excitation energy between $^{235}$U + n and $^{238}$U + n is therefore about $2\delta$ or 1.1 MeV, which nicely accounts for most of the observed difference.

Similar considerations based on the pairing force lead us to expect the following additional results, as indicated in Figure 13.11—an increase in the excitation energy for neutron-induced fission in $^{232}$Pa and a decrease in $^{231}$Pa. We therefore expect that $^{232}$Pa will fission more easily with thermal neutrons than will $^{231}$Pa. Table 13.1 shows the thermal neutron cross sections for a variety of nuclei in this region. You can see quite clearly that neutron capture by *any* odd-$N$ nucleus will lead to a more stable (paired) even-$N$ configuration, with a corresponding increase in $E_{ex}$ by $\delta$. Capture by even-$N$ nuclei will lead to reductions in $E_{ex}$ by the same amount. We therefore expect that odd-$N$ nuclei will in general have much larger thermal neutron cross sections than even-$N$ nuclei, and the data listed in Table 13.1 are consistent with this expectation. The large fission cross sections of $^{235}$U and $^{239}$Pu owe much to the pairing force!

We now turn to a consideration of the energy released in fission and where it goes. Let us begin by considering a specific set of final products:

$$^{235}\text{U} + \text{n} \rightarrow {}^{236}\text{U}^* \rightarrow {}^{93}\text{Rb} + {}^{141}\text{Cs} + 2\text{n}$$

Using the masses of $^{93}$Rb (92.92172 u) and $^{141}$Cs (140.91949 u), the $Q$ value is computed to be 181 MeV. Other final products (possibly with a different number of neutrons emitted) will give energy releases of roughly the same magnitude, and it is quite reasonable to take 200 MeV as an average figure for the energy released in the fission of $^{235}$U. (This value is consistent with the estimate of 0.9 MeV per nucleon introduced earlier, and also with the Coulomb repulsion energy of the two fragments separated roughly by $R = R_1 + R_2$.) The Coulomb repulsion will drive the two fragments apart, converting the potential energy to kinetic, and giving them of order 200 MeV of kinetic energy in the process. Figure 13.12 shows the observed energy distribution of the two fission fragments. There are two highly probable energies, at about 66 and 98 MeV. These correspond to the heavy and the light fragment, respectively. Because the neutrons carry very little momentum, conservation of momentum requires that the two fragments have nearly equal (and opposite) momenta: $m_1 v_1 = m_2 v_2$, from which it is easy to see that the ratio between the kinetic energies should be the inverse of the ratio of the

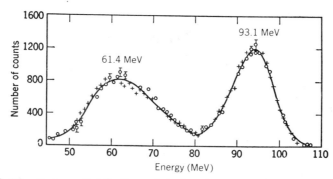

**Figure 13.12** Energy distribution of fission fragments from thermal fission of $^{235}$U. These data were taken with an ion chamber, which subsequent work showed to be slightly miscalibrated for fission fragments; the deduced energies are about 5 MeV too low. From J. L. Fowler and L. Rosen, *Phys. Rev.* **72**, 926 (1947).

masses:

$$\frac{\frac{1}{2}m_1v_1^2}{\frac{1}{2}m_2v_2^2} = \frac{m_2}{m_1} \tag{13.4}$$

The ratio 66 MeV/98 MeV = 0.67 is entirely consistent with the ratio of the most probable masses obtained from Figure 13.6, $(140/95)^{-1} = 0.68$. This distribution of energies thus appears to be characteristic of the distribution of masses, and suggests that on the average 165 MeV or roughly 80% of the fission energy appears as kinetic energy of the fragments.

Figure 13.13 shows the distribution of prompt neutron energies following fission of $^{235}$U. The mean energy is about 2 MeV, and with an average of about 2.5 neutrons per fission, the average energy carried by the neutrons in fission is about 5 MeV. This also suggests that the average neutron carries a momentum only 2% that of either of the fragments, thus justifying neglecting the neutron momentum in the estimate of Equation 13.4.

Other energy releases in fission appear as follows:

prompt γ rays, 8 MeV

β decays from radioactive fragments, 19 MeV

γ decays from radioactive fragments, 7 MeV

These are only estimates and will obviously depend on the exact nature of the fragments and their decays; within 1–2 MeV, they are characteristic of most decays. The prompt γ's are emitted essentially at the instant of fission (later than the prompt neutrons, perhaps, but still within $10^{-14}$ s.) The β and γ radiations are emitted according to the decay schemes of the chain of heavy and light fission fragments. The β energy is, on the average, about 30–40% given to β particles with the remainder ($\sim$ 12 MeV) going to neutrinos. The neutrino energy is lost and contributes neither to the energy recovery or heating of reactor fuel nor to the dangers posed by radioactive waste products.

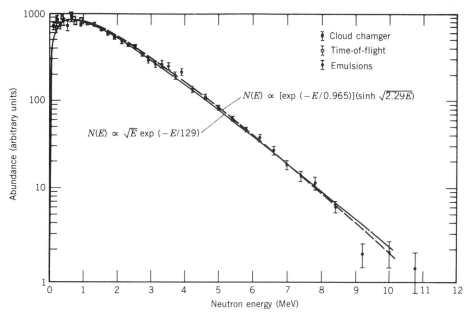

**Figure 13.13** Energy spectrum of neutrons emitted in the thermal-neutron fission of $^{235}$U. From R. B. Leachman, in *Proceedings of the International Conference on the Peaceful Uses of Atomic Energy*, Vol. 2 (New York: United Nations, 1956), p. 193.

## 13.4 FISSION AND NUCLEAR STRUCTURE

Fission is generally treated as a collective phenomenon according to the liquid-drop model, and the analogy with a charged drop of liquid is not only helpful analytically, but also provides a useful mental image of the process. It is therefore perhaps surprising to learn that shell effects play an important and in many cases a decisive role in determining the outcome of fission. As a clue to the importance of shell structure we consider in somewhat more detail the asymmetric mass distribution of the fragments (Figure 13.6). Figure 13.14 shows the mass distributions for $^{236}$U and for several other fissioning nuclei of heavier mass. These distributions reveal an unexpected feature—the mass distributions for the heavy fragments overlap quite well, while the lighter fragment shows a large variation. Comparing $^{236}$U with $^{256}$Fm, we note that $Z$, $N$, and $A$ all increase by about 8.5%, and if the liquid-drop model of fission were a complete description of the process, we should expect both the heavy and light fragment distributions to shift by about 8.5% between $^{236}$U and $^{256}$Fm; that is, the average masses should go from about 95 and 140 in $^{236}$U to about 103 and 152 in $^{256}$Fm. Instead, the observed average masses in $^{256}$Fm are about 114 and 141. The 20 additional mass units in $^{256}$Fm nearly all find themselves given to the lighter fragment.

A more dramatic indication of this effect is illustrated in Figure 13.15, which shows the average masses of the light and heavy fragments over a mass range from 228 to 256. The average mass of the heavy fragment stays nearly constant at about 140, while the average mass of the lighter fragment increases linearly as $A$ increases. Throughout this entire range, the added nucleons all go to the lighter

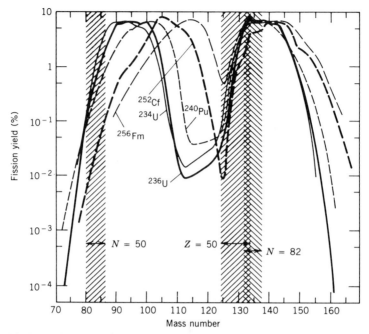

**Figure 13.14** Mass distribution of fission fragments from thermal-neutron fission of $^{233}$U, $^{235}$U, and $^{239}$Pu, along with spontaneous fission of $^{252}$Cf and $^{256}$Fm. Shaded areas show approximate locations of closed-shell nuclei. From A. C. Wahl, in *Physics and Chemistry of Fission*, Vol. I (Vienna: IAEA, 1965), p. 317. Data for $^{256}$Fm from K. F. Flynn et al., *Phys. Rev. C* **5**, 1725 (1972).

fragment, while in a liquid-drop fission we would expect the average masses to scale roughly with the mass of the drop.

The explanation for this unusual observed behavior lies with the shell model. Figure 13.14 shows the regions in which we would expect to find fission fragments with shell-model "magic numbers" of protons or neutrons. That is, for $Z = 50$ we have a stable nucleus with $Z/A = 0.4$ (and thus with $A = 125$) and neutron-rich fission products ranging down to a minimum of about $Z/A \simeq 0.38$ (corresponding to $A = 132$ and thus about 7 neutrons from stability). Just at the lower edge of the heavy fragment mass distribution there is the doubly magic nucleus $^{132}_{50}$Sn$_{82}$. This exceptionally stable configuration determines the low edge of the mass distribution of the heavier fragment. No such effect occurs for the lighter fragment, and indeed the light fragment mass distribution has practically no overlap even with singly magic nuclei and is thus unaffected by shell closures.

The most drastic effect of shell structure occurs in the fission barrier itself. As we begin to stretch the nucleus (characterized by an eccentricity parameter $\varepsilon$), the energy increases like $\varepsilon^2$ as in Equation 13.2, giving an approximate parabolic dependence. The single-particle states in the now deformed nucleus vary with deformation as in Figure 5.29. Notice that for some states in Figure 5.29, the energy increases with deformation, while for some others it decreases. If the valence nucleons are in a state that happens to have a positive slope, the net increase in energy with deformation will be a bit faster than the parabola, for the single-particle energy also increases with $\varepsilon$. At some point, however, as $\varepsilon$

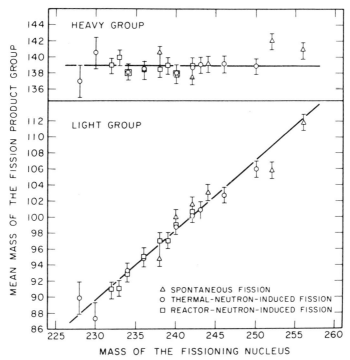

**Figure 13.15** Dependence of average masses of heavy and light groups of fission fragments on mass of fissioning nucleus. From K. F. Flynn et al., *Phys. Rev. C* **5**, 1725 (1972).

increases in Figure 5.29 we encounter a crossing with a state of negative slope. The valence particle, choosing the state of lowest energy, now follows the new state, and the net change of energy with $\varepsilon$ is a bit *below* the parabola. It remains so until a new crossing with a state whose energy increases with $\varepsilon$, so that the total energy is once again above the parabola. This oscillation due to the changing behavior of the valence particles with $\varepsilon$ is shown in Figure 13.16. At the point where fission begins to occur, the form of the single barrier that was shown in Figure 13.2 becomes modified and the energy dependence introduced by the single-particle shell structure results in a fission barrier with two humps. The net effect is that to have a high probability to fission, we no longer need to excite the nucleus so it is close to the top of the barrier. If we excite it above the bottom of the well between the two humps, penetration of the two thinner barriers becomes more probable and fission can occur.

The introduction of the double-humped barrier was necessary following the 1962 discovery of *fission (or shape) isomers*—isomeric excited states with unusually short half-lives for spontaneous fission. Several dozens of the fission isomers are known today. They are typically at excitations of 2–3 MeV (and thus far below the barrier height of 6–7 MeV), but their half-lives for spontaneous fission are in the range of $10^{-6}$–$10^{-9}$ s. It was postulated that these isomers were actually states in the intermediate potential well and could decay either by fission through a relatively thin barrier, or by $\gamma$ emission back to the ground state. Figure 13.17 illustrates the situation. The ordinary ground state is usually one of

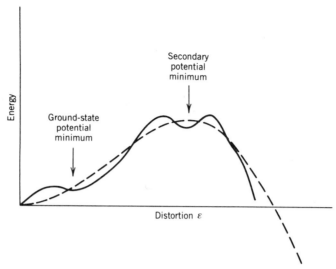

**Figure 13.16**  The dashed line shows the nuclear energy in the absence of shell effects; it is a much magnified section of Figure 13.2 and is approximately parabolic for small distortions. The solid line oscillates above and below the dashed line owing to the shell structure; the result is a distorted ground-state potential minumum and a more distorted secondary potential minimum.

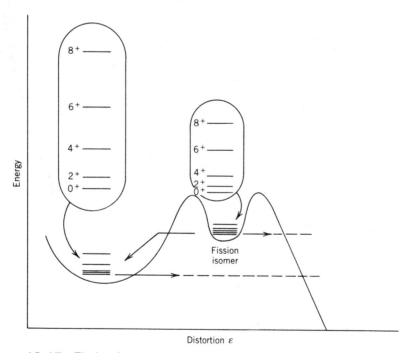

**Figure 13.17**  Fission isomers are states in the secondary potential well. They have a higher probability to fission compared with the ordinary ground state because they must penetrate a much thinner potential barrier. The rotational excited states show that the second well corresponds to a larger deformation and therefore to a larger moment of inertia.

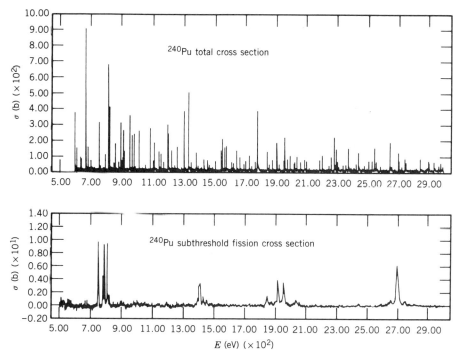

**Figure 13.18** Comparison of fission and total cross sections of $^{240}$Pu. Note the grouping of the fission resonances. From V. M. Strutinsky and H. C. Pauli, in *Physics and Chemistry of Fission* (Vienna: IAEA, 1969), p. 155.

modest stable deformation, and the well corresponds to states of a much larger deformation. It is easy to see why these states have spontaneous fission lifetimes short compared with the ground states.

This explanation of the fission isomers was confirmed by measurements of the rotational spectra of the excited states in the second potential well. The rotational spectrum was presented in Equation 5.17. As the deformation increases, the moment of inertia becomes larger and the rotational states become closer together in energy. Figure 13.17 shows two sequences of rotational states, one corresponding to about twice the moment of inertia of the other. These states have been identified from the observation of the $E2$ $\gamma$ transitions between them.

Another influence of the second potential well is on the structure of the resonances in the fission cross section. As in Figure 13.9, there are many individual fission resonances in the eV–keV region. We can regard these resonances as originating from excited states in the first potential well. In *any* nucleus, at an excitation of 6 MeV following neutron capture, the level density is very high, and there are many states separated in energy by an average spacing of the order of eV. Not all of these excited compound states following neutron capture are likely to fission. Figure 13.18 compares the fission cross sections with the total neutron cross section for $^{240}$Pu. You can see that the fission picks out a few of the resonances that are likely to fission. The other resonances decay through other processes, perhaps by $\gamma$ or neutron emission.

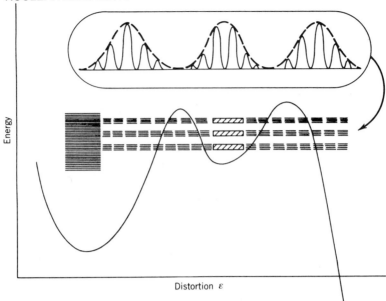

**Figure 13.19** There are many closely spaced states in the first well and a few broad, widely separated states in the second well. Fission resonances occur where states in the first well match in energy (and in spin-parity) with states in the second well. If we reach these selected states in the first well, we will observe them to fission with high probability. The resultant fission resonant structure is shown.

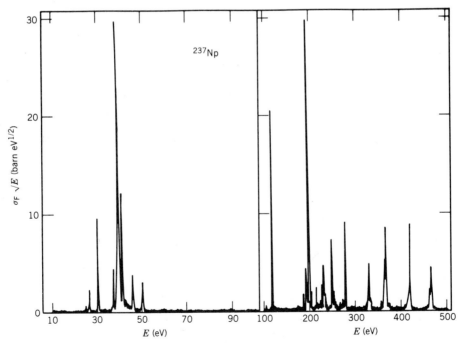

**Figure 13.20** Fission resonances in $^{237}$Np. The broad peaks in the low-resolution spectrum at right are the many unresolved states corresponding to the broad states of the second well. A high-resolution view of a resonance around 40 eV is shown at left, which reveals the narrow states characteristic of the first well. From D. Paya et al., *Physics and Chemistry of Fission* (Vienna: IAEA, 1969), p. 307.

A closer inspection of Figure 13.18 shows a more interesting effect—the fission resonances appear to cluster in well-separated groups. This effect occurs because the second well is not quite as deep as the first. The density of states of any nucleus depends on the excitation energy above the ground state—the higher we go above the ground state, the closer together are the states. States in the second well at the same energy as those in the first well are on the average further apart—perhaps the average spacing is 100–1000 eV in the second well and 1–10 eV in the first well. Another difference is that states in the second well have a higher probability to fission (because they must penetrate only one barrier) and thus a greater width than states in the first well. Figure 13.19 gives a schematic view of the relationship between the states in the two wells, showing how the fissioning states are selected through their overlap in energy between narrow, closely spaced states in the first well, and broader, more widely spaced states in the second well. The effect on the cross section is also shown and gives rise to the resonance structure shown in Figure 13.18.

The actual neutron fission resonances involve relatively complicated couplings of wave functions representing states in the first and second wells. Because angular momentum and parity are always good quantum numbers, we can combine the states only if they have the same angular momentum. Each of the broad states in the second well has a definite spin-parity assignment, and it will

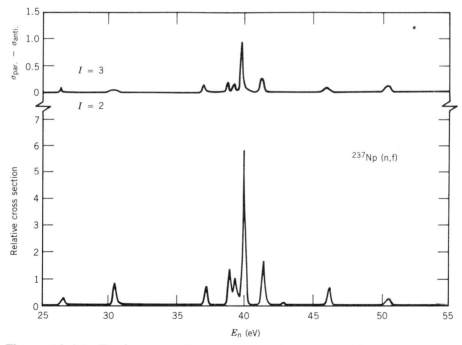

**Figure 13.21** The bottom portion shows the 40-eV group of fission resonances of $^{237}$Np, as in Figure 13.20. The top portion shows the difference in cross section between neutron and $^{237}$Np spins parallel and antiparallel, which should be positive for $I = 3$ and negative for $I = 2$. Clearly the entire group has $I = 3$, consistent with the common origin of the states with the broad resonance of the second well. From G. A. Keyworth et al., *Phys. Rev. C* **8**, 2352 (1973).

couple only with overlapping states from the first well of the same spin-parity assignment. Thus not all of the states of the first well within the energy width of a state of the second well will show these enhanced resonant fission cross sections; only those with the proper spin-parity assignment. If we excite the states in the first well by capture of s-wave neutrons, the only possible spin assignments of the levels in the first well are $I + \frac{1}{2}$ and $I - \frac{1}{2}$, where $I$ is the spin of the original target nucleus. We therefore expect that the groupings consist of many resonances with identical spin-parity assignments within each group (but possibly differing from one group to another). Verification of this hypothesis comes from experiments done using fission induced by polarized neutrons incident on polarized targets. For example, consider the case of $^{237}$Np ($I = \frac{5}{2}$). The capture states can be $I = 2$ or 3. If we do the experiment with the spins of the incident neutrons parallel to those of the target, only $I = 3$ states can be populated, and we ought to be able to pick out the $I = 3$ resonances. Figure 13.20 shows the fission cross section for $^{237}$Np. The broad unresolved peaks are the states of the second well, each containing many individual states. A high-resolution view of the resonance near 40 eV is shown in Figure 13.21, and the cross section

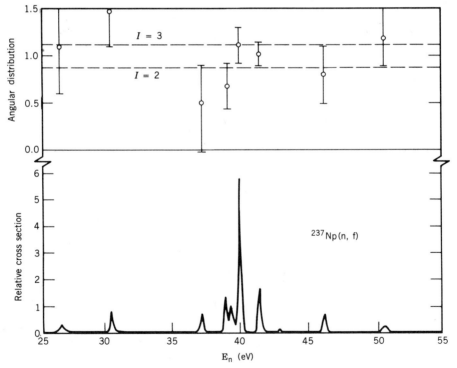

**Figure 13.22** As in Figure 13.21, the bottom portion shows the region of the 40-eV group of fission resonances of $^{237}$Np. The top shows the angular distribution coefficient of the fission fragments from aligned $^{237}$Np. The results are generally consistent with $I = 3$, but do not discriminate between $I = 2$ and $I = 3$ as well as the data of Figure 13.21. From R. Kuiken, N. J. Pattenden, and H. Postma, *Nucl. Phys. A* **196**, 389 (1972).

measured with neutron and target spins parallel shows every peak of the multiplet, indicating they all have $I = 3$. The 120-eV broad resonance is also observed to be $I = 3$, while the groups at about 230 and 375 eV are $I = 2$.

An alternative approach to this same end is to study the angular distribution of the fragments produced following unpolarized neutron capture on nuclei whose spins are aligned. Figure 13.22 shows a sample set of results for the 40-eV group of resonances in $^{237}$Np. Although the discrimination between $I = 2$ and $I = 3$ is not as absolute as in the case of the studies with polarized neutrons, the data do indicate similar conclusions regarding $I = 3$ for this group of resonances.

## 13.5 CONTROLLED FISSION REACTIONS

Let's consider an infinitely large mass of uranium, which we will for the present assume to be of natural isotopic composition (0.72% $^{235}$U, 99.28% $^{238}$U). A single fission event will produce, on the average, about 2.5 neutrons. Each of these "second-generation" neutrons is capable of producing yet another fission event producing still more neutrons, and so on. This process is the *chain reaction*. Each fission event releases about 200 MeV in the form of kinetic energy of heavy fragments (that is, heat) and radiation.

It is convenient to define the *neutron reproduction factor $k_\infty$* (for an infinite medium, that is, ignoring loss of neutrons through leakage at the surface). The reproduction factor gives the net change in the number of thermal neutrons from one generation to the next; on the average, each thermal neutron produces $k_\infty$ new thermal neutrons. For a chain reaction to continue, we must have $k_\infty \geq 1$. Although we have an average of 2.5 neutrons emitted per fission, these are *fast neutrons*, for which the fission cross section is small. It is advantageous to *moderate* these neutrons to thermal velocities because of the large thermal cross section (about 580 b) resulting from the $1/v$ law. In the process, many neutrons can become absorbed or otherwise lost to the chain reaction, and the 2.5 fast neutrons per fission can easily become $< 1$ thermal neutron, effectively halting the reaction.

As we discussed in Section 12.2, neutrons lose energy in elastic collisions with nuclei, and a popular choice for a moderator is carbon in the form of graphite blocks. (The best choice for a moderator is the lightest nucleus, to which the neutron transfers the largest possible energy in an elastic collision. Carbon is a reasonable choice because it is available as a solid, thus with a high density of scattering atoms, and is inexpensive and easy to handle.) A lattice of blocks of uranium alternating with graphite is called a chain-reacting *pile*, and the first such pile was constructed by Fermi and his collaborators in a squash court at the University of Chicago in 1942. If the reproduction factor $k$ (for a finite pile) is exactly 1.0, the pile is said to be *critical*; for $k < 1$, the pile is *subcritical* and for $k > 1$ it is *supercritical*. To maintain a steady release of energy, we would like for the pile to be exactly critical.

To calculate the reproduction factor $k_\infty$, we must follow the fate of a collection of thermal neutrons from one generation to the next. Let's assume we have $N$ thermal neutrons in the present generation. Even though each fission produces on the average $\nu$ neutrons, we will *not* have $\nu N$ fast fission neutrons immediately available, for not every one of the original collection will cause a fission event. Some will be absorbed through other processes, most notably $(n, \gamma)$ reactions in both $^{235}$U and $^{238}$U. We define $\eta$ as the mean number of fission neutrons produced per original thermal neutron. It is clear that $\eta < \nu$, for some of the original thermal neutrons do not cause fissions. If we let $\sigma_f$ represent the fission cross section and $\sigma_a$ the cross section for other absorptive processes (both of these cross sections are evaluated for thermal neutrons), then the relative probability for a neutron to cause fission is $\sigma_f/(\sigma_f + \sigma_a)$ and

$$\eta = \nu \frac{\sigma_f}{\sigma_f + \sigma_a} \tag{13.5}$$

For $^{235}$U, $\sigma_f = 584$ b and $\sigma_a = 97$ b, so that $\eta = 2.08$ fast neutrons are produced per thermal neutron. $^{238}$U is not fissionable with thermal neutrons, and so $\sigma_f = 0$, but $\sigma_a = 2.75$ b. For a natural mixture of $^{235}$U and $^{238}$U, the effective fission and absorption cross sections are

$$\sigma_f = \frac{0.72}{100} \sigma_f(235) + \frac{99.28}{100} \sigma_f(238)$$
$$= 4.20 \text{ b}$$

$$\sigma_a = \frac{0.72}{100} \sigma_a(235) + \frac{99.28}{100} \sigma_a(238)$$
$$= 3.43 \text{ b}$$

and the effective value of $\eta$ becomes 1.33. This is already very close to 1.0, so we must minimize other ways that neutrons can be lost, in order to obtain a critical reactor. If we use enriched uranium, in which the fraction of $^{235}$U is raised to 3%, the effective value of $\eta$ becomes 1.84, considerably further from the critical value and allowing more neutrons to be lost by other means and still maintain the criticality condition.

At this point the $N$ thermal neutrons have been partly absorbed and the remainder have caused fissions, and we now have $\eta N$ fast neutrons which must be reduced to thermal energies. As these fast neutrons journey through the chain-reacting pile, some of them may encounter $^{238}$U nuclei which have a small cross section (about 1 b) for fission by fast neutrons. This causes a small increase in the number of neutrons, so we introduce a new factor $\varepsilon$, the *fast fission factor*; the number of fast neutrons is now $\eta \varepsilon N$. The value of $\varepsilon$ is about 1.03 for natural uranium.

Moderation of the neutrons is accomplished by intermixing the reactor fuel with a light moderator, such as carbon, usually in the form of graphite. As indicated in Table 12.1, it takes about 100 collisions with carbon for MeV neutrons to be thermalized. In the process, they must pass through the region from 10–100 eV, in which $^{238}$U has many capture resonances (Figure 13.23) with cross sections in the range of 1000 b (greater than the $^{235}$U fission cross section). If we are to have any thermal neutrons at all, we must find a way for the neutrons to avoid resonance capture. If the uranium and the graphite are intimately mixed, such as in the form of fine powders, it will be almost impossible for neutrons to avoid resonant capture in $^{238}$U. In this kind of mixture, a neutron may scatter

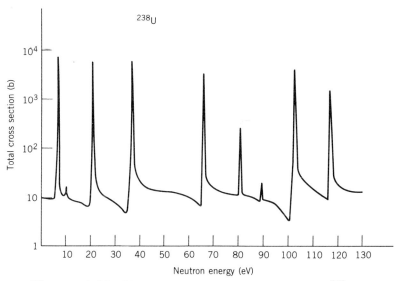

**Figure 13.23** Neutron-capture resonance region of $^{238}$U.

very few times from carbon before encountering a $^{238}$U nucleus, and therefore will very likely pass near $^{238}$U when its energy is in the critical region. If we make the lumps of graphite larger, we will eventually reach a configuration in which the neutrons can be completely thermalized through many scatterings without leaving the graphite and encountering any $^{238}$U. In this way it is possible to avoid the dangerous resonance region. The average distance needed by a fission neutron to slow to thermal energies in graphite is about 19 cm. If we therefore construct the pile as a matrix of uranium fuel elements separated by about 19 cm of graphite, we minimize the neutron losses due to resonant capture. Of course, it will still be possible for some neutrons to wander too close to the surface of the graphite and enter the uranium before they are fully thermalized, and thus resonant capture cannot be completely eliminated. We account for this effect by including a reduction factor $p$, the *resonance escape probability*, in the number of neutrons remaining after thermalization, now $\eta \varepsilon p N$. A typical value of $p$ might be 0.9.

Once the neutrons are successfully thermalized, we must immediately get them into the uranium, so the graphite lumps mustn't be made too large. In any case, there is a probability of capture of thermal neutrons by graphite and by any structural components of the reactor (such as the material used to encapsulate the fuel elements). One reason for choosing carbon as a moderator is that it has a very small thermal cross section (0.0034 b), but there is a lot of it present. The *thermal utilization factor $f$* gives the fraction of the thermal neutrons that are actually available to $^{235}$U and $^{238}$U. This factor may also typically be about 0.9.

The number of neutrons which finally survive capture by the moderator and other materials is $\eta \varepsilon p f N$, and whether this is greater or smaller than the original number $N$ determines the criticality of the reactor. The reproduction factor is

$$k_{\infty} = \eta \varepsilon p f \qquad (13.6)$$

which is known for obvious reasons as the *four-factor formula*. Figure 13.24 shows the processes that can occur to neutrons during a reactor cycle.

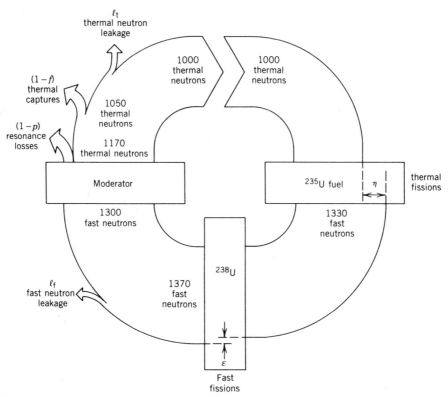

$\ell_t$
thermal neutron
leakage

$(1-f)$
thermal
captures

$(1-p)$
resonance
losses

1000
thermal
neutrons

1000
thermal
neutrons

1050
thermal
neutrons

1170
thermal neutrons

Moderator

$^{235}$U fuel $\quad \eta \quad$ thermal fissions

1300
fast neutrons

1330
fast
neutrons

$^{238}$U

$\ell_f$
fast neutron
leakage

1370
fast
neutrons

$\varepsilon$

Fast
fissions

**Figure 13.24** Schematic representation of processes occurring during a single generation of neutrons. The cycle has been drawn for a reproduction factor $k$ of exactly 1.000.

The actual design of the pile will be a compromise of attempts to optimize the three factors that depend on the geometry ($\varepsilon$, $p$, $f$). For instance, large lumps of U tend to reduce $p$ because resonant absorption occurs primarily on the surface; that is, a neutron of energy 10–100 eV that enters a uranium lump is unlikely to get very far before absorption. The uranium in the center of the lump never sees such neutrons, and that central mass of uranium can be treated from the perspective of resonant absorption as if it were not present. The larger the lumps of uranium, the more effective is the surface in shielding the central uranium from neutron absorption. On the other hand, if the lumps become too large, the same effect occurs for the thermal neutrons that cause fission—more fission occurs near the surface and the density of thermal neutrons decreases toward the center of the lump of uranium.

Using the four factors estimated for the natural uranium and graphite pile, the reproduction factor is estimated to be $k_\infty = 1.11$. This estimate is still not appropriate for an actual pile, for we have ignored leakage of neutrons at the surface. This leakage must be considered both for fast neutrons and for thermalized neutrons. If $\ell_f$ and $\ell_t$ are the fractions of each that are lost, the complete formula for the reproduction factor is

$$k = \eta \varepsilon p f (1 - \ell_f)(1 - \ell_t) \tag{13.7}$$

The larger the pile, the smaller is the surface area to volume ratio and the smaller is the fraction of neutrons that leak. If $\ell_f$ and $\ell_t$ are small, then $k_\infty - k \approx k(\ell_f + \ell_t)$. We expect that the total leakage $(\ell_f + \ell_t)$ decreases as the surface area increases. Furthermore, the leakage must increase with the distance a neutron is able to travel before absorption. This distance is called the *migration length M* and it includes two contributions: the diffusion length $L_d$ for thermal neutrons, the distance a thermal neutron can travel on the average before absorption, and the slowing distance $L_s$ over which a fast neutron slows to thermal energies:

$$M = \left( L_d^2 + L_s^2 \right)^{1/2} \tag{13.8}$$

For graphite, $L_s = 18.7$ cm and $L_d - 50.8$ cm.

If the pile has dimension $R$ (radius, if it is spherical, or length of a side if a cube), it is reasonable to suppose that $(k_\infty - k) \propto R^{-2}$ and also that $k_\infty - k$ depends on $M$; if these are the only physical parameters involved, then from a dimensional analysis we expect

$$k_\infty - k \propto \frac{M^2}{R^2} \tag{13.9}$$

and there will be a critical size $R_c$ corresponding to $k = 1$

$$R_c \propto \frac{M}{\sqrt{k_\infty - 1}} \tag{13.10}$$

The proportionality constant needed to make Equation 13.10 an equality depends on the geometry but is of order unity. For a spherical arrangement,

$$R_c = \frac{\pi M}{\sqrt{k_\infty - 1}} \tag{13.11}$$

and our estimates for the natural uranium-graphite reactor give $R_c = 5$ m. A spherical lattice of radius 5 m consisting of uranium and graphite blocks should "go critical." The size can be decreased somewhat by surrounding the pile with a material that reflects escaping neutrons back into the pile.

Before leaving this brief introduction to reactor theory, let's consider the time constants involved in neutron multiplication. The neutrons are characterized by a time constant $\tau$, which includes the time necessary to moderate (about $10^{-6}$ s) and a diffusion time at thermal energies before absorption (about $10^{-3}$ s). If the reproduction factor is $k$, and if there are $N$ neutrons at time $t$, then there are, on the average, $kN$ neutrons at time $t + \tau$, $k^2N$ at time $t + 2\tau$, and so on. In a short time interval $dt$, the increase is

$$dN = (kN - N)\frac{dt}{\tau} \tag{13.12}$$

from which

$$N(t) = N_0 e^{(k-1)t/\tau} \tag{13.13}$$

If $k = 1$, then $N = $ constant; this would be the desired operating mode of a reactor. If $k < 1$, the number of neutrons decays exponentially. If $k > 1$, the number of neutrons grows exponentially with time, with a time constant characterized by $\tau/(k - 1)$. Even if the reactor is just slightly supercritical ($k = 1.01$), the time constant is of order 0.1 s. It would be dangerous to operate a reactor in which the number of neutrons could increase by a factor of $e^{10}$ ($= 22,000$) in 1 s. In practice, control of the number of neutrons is achieved by inserting into the pile a material such as cadmium, which is highly absorptive of thermal neutrons; the cadmium *control rods* are under mechanical control and can be gradually removed from the pile or rapidly inserted. If the pile is designed so that it is just slightly subcritical for prompt neutrons, the small number of delayed neutrons can be used to achieve critical operation, and since the delayed-neutron decay time constants are fairly long (seconds to minutes) the control rods can be manipulated to achieve a constant reaction rate.

## 13.6 FISSION REACTORS

Nuclear reactor engineering is a complex discipline and we cannot do it justice with a short summary. Nevertheless, there are enough general properties and categories of reactors that it is instructive to try to list them and describe some of them.

All reactors consist of the same essential elements: the *fuel*, or fissile material; a *moderator* to thermalize the neutrons (which may not be present in a reactor using fast neutrons); a *reflector* surrounding the *core* (fuel elements plus moderators) to reduce neutron leakage and thereby reduce the critical size of a reactor; a *containment vessel* to prevent the escape of radioactive fission products, some of which are gases; *shielding* to prevent neutrons and $\gamma$ rays from causing biological harm to operating personnel; a *coolant* to remove heat from the core; a *control system* allowing the operator to control the power level and to keep it constant during normal operation; and various *emergency systems* designed to prevent runaway operation in the event of a failure of the control or coolant systems.

**Types of reactors** The first and perhaps most basic classification is the use for which the reactor is intended, and we can loosely define three categories: power generation, research, and conversion. *Power reactors* are nothing more than devices for extracting the kinetic energy of the fission fragments as heat and converting the heat energy to electrical energy, perhaps by boiling water and driving a turbine with the resulting steam. Design considerations for power reactors therefore involve as much attention to thermodynamic details of the efficiency of heat engines as to nuclear engineering. The fuel assembly represents a relatively small fraction of the cost of a power reactor; most of the expenses are associated with the shielding and containment vessel and with the electrical generating equipment. It is therefore economically advantageous to build large power reactors; 10 reactors each producing 100 MW are far more costly than a single 1000-MW reactor. The corresponding thermal wastes are, however, considerable; it must be emphasized that these thermal wastes have nothing to do with the specific nuclear nature of the power source—*any* 1000 MW heat engine is, according to the Second Law of Thermodynamics, going to discharge a great deal of heat in a relatively confined area. (In addition, the nuclear power plant operates at somewhat lower thermodynamic efficiency than a fossil-fuel plant, and therefore discharges more heat.)

*Research reactors* are generally designed to produce neutrons for research in areas such as nuclear or solid-state physics. These generally operate at low power levels, in the range of 1–10 MW. Principal design features of research reactors may be large neutron flux ($10^{13}$ n/cm$^2$/s is typical, but specialized reactors may exceed that by 1–2 orders of magnitude in steady operation and by considerably more in pulsed operation), ease of access to the neutrons (many research reactors have a beam pipe leading from the region of the core to a research facility outside the shielding, for example, a neutron diffraction apparatus to study crystal structures), and quality of neutron spectrum (a *thermal column* is a block of graphite near the core of sufficient thickness to eliminate fast neutrons and allow small samples placed within it to be exposed to neutrons with a relatively pure thermal spectrum).

*Converters* are reactors designed to have high efficiency for converting material that is not fissionable with thermal neutrons to material that is. The specific conversions that are generally used are $^{238}$U to $^{239}$Pu and $^{232}$Th to $^{233}$U. In both cases, the conversion involves the capture of a neutron followed by two $\beta$ decays:

$$^{238}\text{U} + \text{n} \rightarrow {}^{239}\text{U} \xrightarrow{\text{23 min}} {}^{239}\text{Np} + \beta^- + \bar{\nu}$$

$$\downarrow 2.3 \text{ d}$$
$$\xrightarrow{\phantom{xx}} {}^{239}\text{Pu} + \beta^- + \bar{\nu}$$

$$^{232}\text{Th} + \text{n} \rightarrow {}^{233}\text{Th} \xrightarrow{\text{22 min}} {}^{233}\text{Pa} + \beta^- + \bar{\nu}$$

$$\downarrow 27 \text{ d}$$
$$\xrightarrow{\phantom{xx}} {}^{233}\text{U} + \beta^- + \bar{\nu}$$

Isotopes such as $^{238}$U and $^{232}$Th, which can be converted to thermally fissile material, are called *fertile*. In principle it is possible to design a reactor in which the value of $\eta$ were at least 2.0. With two neutrons emitted per neutron absorbed in the fuel, one neutron could go to sustaining the chain reactions and the other to the fertile material. Assuming other losses could be minimized, if $\eta > 2$ then a reactor could produce more fissile material than it consumed. Such a reactor is called a *breeder*. For thermal neutrons, $^{239}$Pu has $\eta = 2.1$ and it is unlikely that losses of neutrons (resonance capture, leakage, etc.) could be kept low enough to permit breeding. For fast (MeV) neutrons, $\eta = 3$ for $^{239}$Pu and breeding becomes a strong possibility. Much research has therefore gone into the design and construction of fast breeder reactors. One advantage of this method of production of fissile material is that, following a sufficient time of breeding, the newly made fissile material can be separated from the fertile material by chemical means, which are considerably easier than the methods that must be used to separate fissile and nonfissile isotopes such as $^{235}$U and $^{238}$U from one another.

**Neutron Energy**   It is possible to design reactors to operate with thermal, intermediate, or fast neutrons. In the last section, the neutron budget for thermal reactors was discussed. One advantage of the intermediate-energy (eV to keV) reactor is that it requires considerably less volume than a thermal reactor, and these reactors were originally developed for use in propulsion, as for example in submarines. Fast reactors require no moderator at all (and consequently cannot use a moderating material such as water as coolant circulating through the fuel). Because of the lower cross sections for fast neutrons, fast reactors require of the order of 10–100 times the fuel of thermal reactors providing the same power; however, because of the lack of a moderator, the core of a fast reactor occupies a much smaller volume than the core of a thermal reactor.

**Type of Fuel**   The most commonly used fuels are natural uranium (0.72% $^{235}$U), enriched uranium ($> 0.72\%$ $^{235}$U), $^{239}$Pu, and $^{233}$U. The latter two are obtained from chemical processing of fertile material from converter or breeder reactors. Enriched uranium, which is the most commonly used fuel in power reactors, is produced in quantity at great expense from processes sensitive to the small mass difference between $^{235}$U and $^{238}$U. One such process is *gaseous diffusion*, in which uranium hexafluoride gas ($UF_6$) is forced through a porous barrier. The diffusion coefficient of a gas is inversely proportional to the square root of its mass, and the lighter isotope diffuses slightly more rapidly owing to its higher average speed in a mixture of $^{235}$U and $^{238}$U at thermal equilibrium. (The mean kinetic energies will be equal, so the molecule with the lighter mass has the higher speed.) The enrichment after a single passage through the barrier is very small, of the order of 0.4% *relative* (that is, the original 0.72% of $^{235}$U becomes $1.004 \times 0.72\%$), a very small increase, and the process must be repeated *thousands* of times before highly enriched (nearly pure) $^{235}$U can be obtained. Highly enriched $^{235}$U and $^{239}$Pu are bomb-grade fissile materials, while uranium at 2–3% enrichment, used in certain types of reactors, is not.

**Moderator**   The ideal moderator would (1) be cheap and abundant, (2) be chemically stable, (3) have a mass of nearly one (to absorb the maximum amount of energy in a collision with a neutron), (4) be a liquid or a solid so its density will be high, and (5) have a minimal neutron capture cross section. Carbon, in the form of graphite, satisfies (1), (2), (4), and (5), and we must compensate for a relatively small loss in neutron energy per collision by increasing the amount of moderator. Ordinary water satisfies (1), (2), (3), and (4), but the protons in water have a very high cross section to capture a neutron (n + p $\rightarrow$ d + $\gamma$). Heavy water ($D_2O$) has a very small neutron capture cross section, but captures that do occur lead to radioactive tritium, a particularly nasty product for biological systems. It is also quite rare and relatively expensive to separate, although it is not nearly so difficult to separate deuterium from ordinary hydrogen as it is one isotope of uranium from another; the mass ratio of 2 : 1 leads to quite dramatic effects. Because of the small capture cross section, heavy-water moderated reactors can use natural uranium as fuel; the extra neutron absorption in light-water moderated reactors requires enriched uranium. Beryllium ($Z = 4$) and BeO are also used as moderators, but are difficult to work with and dangerous poisons.

**Assembly**   Reactors are usually classified as *heterogeneous* when the moderator and fuel are lumped, or *homogeneous* when they are mixed together. Homogeneous reactors are more tractable to mathematical analysis then heterogeneous reactors, for which the calculation of the thermal utilization factor $f$ and the resonance escape probability $p$ can be difficult. A homogeneous natural uranium-graphite mixture cannot go critical, but a heterogeneous arrangement can.

**Coolant** The coolant is an essential element of the reactor for without it the heat generated would melt the core (called "meltdown"). In the design of power

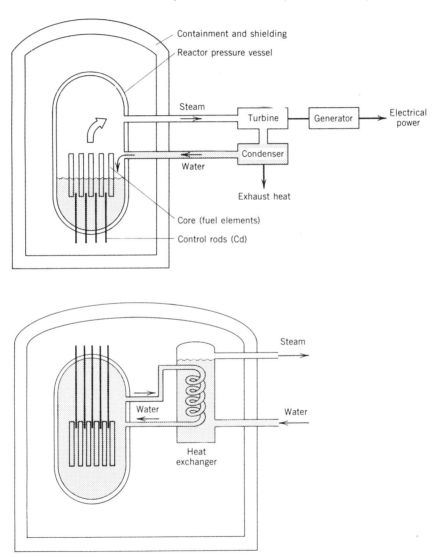

**Figure 13.25** Schematic diagram of boiling-water (top) and pressurized-water (bottom) reactors. The core consists of a number of rods containing pellets of uranium oxide in a metal (zirconium alloy or stainless steel) housing. Control rods of cadmium can be inserted into the core to absorb neutrons and keep the power level stable. The boiling-water reactor is shown driving electrical generating equipment. Many details, including the important emergency core cooling system, are omitted.

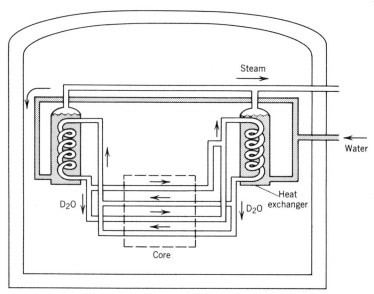

**Figure 13.26** Schematic diagram of Canadian deuterium-uranium (CANDU) reactor. A closed system of $D_2O$ circulates through the core and then exchanges heat with ordinary water, the steam from which can drive a turbine. The $D_2O$ is under pressure to keep it in the liquid state.

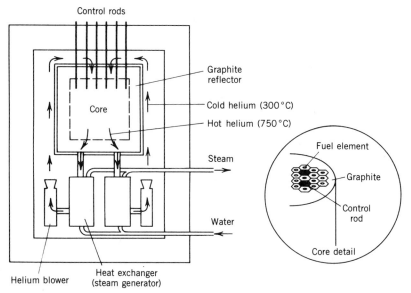

**Figure 13.27** Schematic diagram of gas-cooled reactor. Helium gas flows through the core to extract the heat; the helium is then used to produce steam. A detail of the core is shown at right. The fuel elements are in the form of hexagonal rods containing the fissionable material, graphite moderator, and a channel for gas flow. The core is surrounded with a graphite reflector.

reactors, an essential feature is the ability of the coolant to transfer heat efficiently. Coolant materials can be gases (air, $CO_2$, helium), water or other liquids, or even liquid metals, which have a large heat capacity. Because steam has a small heat capacity, reactors that use water as a coolant may keep it under high pressure (of order 100 atmospheres) allowing it to remain as a liquid well above its normal boiling point. These are called *pressurized-water reactors*. Liquid sodium has been studied as a coolant in fast breeder reactors, which have a high density of fuel, thus requiring efficient heat transfers from a relatively small volume, and which cannot permit the moderating effect of water as coolant. Although liquid sodium is extremely corrosive and becomes radioactive through its large capture cross section, its high boiling temperature allows it to remain a liquid under ordinary pressure and eliminates the need for a high-pressure system.

Based on these design considerations, we can now examine some of the common reactor systems. Figure 13.25 illustrates the boiling-water and pressurized-water reactors, which might be used for power generation. These designs use ordinary (light) water as both moderator and coolant. The boiling-water reactor circulates water through the core (as moderator) and then into the generating apparatus as steam. A strong pressure vessel is needed to contain the steam. Even though pure water does not become radioactive (except for a tiny fraction of $^{18}O$ which has a very small neutron capture cross section), impurities even at the level of parts per million will become highly radioactive in the large neutron flux at the reactor core, and the design calls for the circulation of this radioactive steam beyond the containment vessel. The pressurized-water reactor

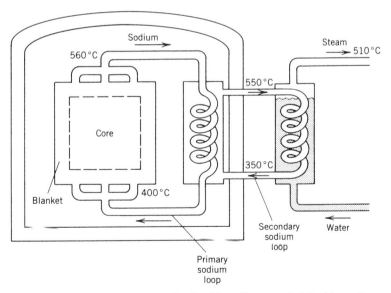

**Figure 13.28**  Schematic diagram of a liquid-sodium-cooled fast breeder reactor. The core may consist of $^{235}U$ and $^{239}Pu$, while the blanket contains the fertile $^{238}U$ that will breed into fissionable material. An intermediate heat exchanger is necessary to keep sodium and water (which react explosively) from simultaneously being present in the reactor core, and also to keep water away from the highly radioactive sodium in the primary loop.

eliminates this potential danger by exchanging heat between the high-pressure water that circulates through the core and a separate steam line to drive the electrical generators. As mentioned previously, reactors moderated with light water must use enriched uranium; enrichments of 2–3% are typical for these reactors. Because of the relative availability of enriched uranium in the United States, many reactors have been built with these basic designs using light water.

Canada, on the other hand, has a supply of natural uranium and the ability to produce heavy water, and so the Canadian power reactors are moderated with $D_2O$ (and use either $D_2O$ or $H_2O$ as coolant). Figure 13.26 shows the basic design of a heavy-water moderated reactor.

Graphite-moderated reactors are of the lumped (heterogeneous) arrangement and can use natural or enriched uranium as fuel. Most designs are gas cooled, as shown in Figure 13.27. Fuel assemblies are constructed to permit easy gas flow for good heat transfer. Most of the power reactors in Great Britain are graphite-moderated, gas-cooled reactors.

An example of a fast reactor is the liquid-sodium fast breeder reactor, illustrated in Figure 13.28.

## 13.7 RADIOACTIVE FISSION PRODUCTS

The two nuclear fragments produced in fission have an excess of neutrons and therefore undergo negative $\beta$ decay to reach the stable isobar in the mass chain of each. These radioactive fission products and their decays are important for several reasons: (1) A small portion (10–15%) of the fission energy is released through these $\beta$ and $\gamma$ emissions. After the reactor is shut down (that is, after the control rods are inserted and fission is halted), the heat generated by these decay processes continues and decreases on a time scale characteristic of the radioactive decay half-lives of the products. (2) As each chain approaches stability, two effects occur—the decay energy decreases (see Figure 3.18) and the half-life increases. The residue of very long-lived ($t_{1/2} \sim$ years) decaying products constitutes the major share of the hazardous waste product of nuclear reactors. (3) Certain of the fission products that accumulate in the fuel elements during reactor operation can interfere with the fission process because of their large neutron-capture cross sections. (4) Many of the radioactive products have important research applications, and extraction from spent fuel elements is the only economic means of producing some of these isotopes. We will discuss each of these four topics in turn.

As was estimated in Section 13.3, the total $\beta$-decay energy from the fragment decays is about 19 MeV. This of course represents several sequential decays, perhaps ranging in energy from 6–7 MeV down to 1 MeV as stability is approached. In each decaying nucleus, there will be many excited states that can be populated; however, the $\beta$-emission probability, Equation 9.28, favors higher energy decays, and therefore the most probable decays are those carrying the largest energies and populating the lowest excited states of the daughter nucleus. It is for this reason that the average $\gamma$-decay energy from the fission products is only about 7 MeV, much smaller than the $\beta$ energy.

The $\beta$-decay energy is shared by the electron and the antineutrino. The antineutrino escapes from the fuel element and contributes neither to the decay heating nor to the radiation. The mean energy given to the electron is about 0.3–0.4 of the decay energy, and thus on the average only 6–8 MeV of the 19 MeV appears as heat.

To calculate the actual energy that appears as $\beta$ and $\gamma$ decays, we must know exactly which isotopes are produced, and we must also know all of the $\beta$ and $\gamma$ branching intensities and decay probabilities. Much of this information is not available, especially for the shorter-lived fission products at the beginning of the decay chains. Instead, we take an average figure, based on estimates of the mass differences and half-lives:

$$\beta(t) = 1.26t^{-1.2} \text{ MeV/s} \qquad (13.14)$$

$$\gamma(t) = 1.40t^{-1.2} \text{ MeV/s} \qquad (13.15)$$

where $\beta$ and $\gamma$ give the power per fission deposited as $\beta$ and $\gamma$ decays, respectively, as functions of the time for $t$ between 1 and $10^6$ s.

As a reactor is operated, fission products are continuously produced while others decay. When the reactor is shut down, the energy of the still-decaying products will depend on how long the reactor has operated and at what power level. If the reactor operates from time 0 to time $T$, the power emitted as $\beta$ and $\gamma$ decays at time $t$ after the shutdown is

$$P(t) = 4.10 \times 10^{11}\left[t^{-0.2} - (t + T)^{-0.2}\right] \text{ MeV/s} \qquad (13.16)$$

per watt of original operating level. (That is, if the reactor operated before shutdown at $10^3$ MW, multiply this expression by $10^9$.) Figure 13.29 shows the

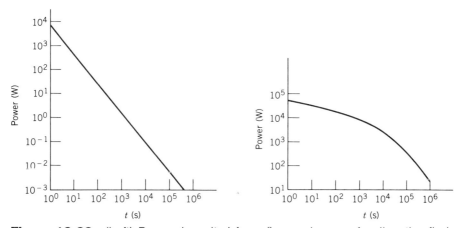

**Figure 13.29** (Left) Power deposited from $\beta$ or $\gamma$ decays of radioactive fission products per second of reactor operation for a 1-MW reactor ($3.1 \times 10^{16}$ fissions/s). (Right) Total radioactive decay heat from a 1-MW reactor at time $t$ following shutdown, after the reactor has been operating for $T = 8$ h. Note that the decay power remains at or above 1% of the reactor power for a time of order 1 h after the shutdown. For power reactors, scale both figures upward by a factor of 1000.

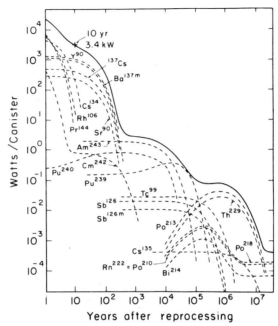

**Figure 13.30** Radioactive power produced from decays of fission products and actinides. This amount of power would result from the waste products of a 1000-MW power reactor operating for about one month. The solid curve is the sum of the contributions of the individual isotopes. From B. L. Cohen, *Rev. Mod. Phys.* **49**, 1 (1977).

decay of the $\beta$ and $\gamma$ activities and the decay power level after shutdown, called *after-heat*, of a research reactor operating at 1 MW.

When we consider the longer-lived activities that might characterize spent fuel elements, we are interested in a time scale of the order of years. Figure 13.30 shows the activities of the long-lived fission products and actinides. After reprocessing, these activities must be isolated from biological systems for times of the order of $10^5$–$10^6$ y. Several schemes have been suggested for achieving this isolation, but no system has yet been adopted that can guarantee isolation over this time scale. Any leakage into groundwater or the food chain would be expected to result in an increase in deaths from cancer. It is, however, interesting to note that over the long time periods involved, the burn-up of uranium through fission actually *reduces* the natural background and the consequent risk of cancer. This natural background radiation is released into the environment in the burning of coal, and it is released in particulate form, which poses a special hazard to the lungs. It is estimated that a coal-burning power plant operating at 1000 MW annually releases into the atmosphere 23 kg of uranium and 46 kg of thorium, along with their radioactive decay products, most notably radon. The radiation exposure to the general population from coal-burning power plants is several times that of nuclear power reactors. We know far more about the risks of exposure to radioactive fission products than we know about the effects of this radioactive release from coal-burning power plants.

Prominent among the fission products that can actually modify reactor operation is $^{135}$Xe, which can be formed directly in $^{235}$U fission (with a yield of 0.2%)

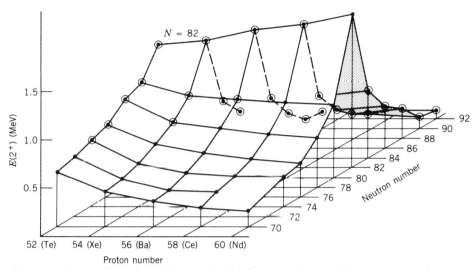

**Figure 13.31** Energies of first excited $2^+$ states of even-$Z$, even-$N$ nuclei in the mass-140 region. The closed shell at $N = 82$ is apparent from the systematic peak in the $2^+$ energy, and the rapid decrease to low energies beyond the peak is characteristic of nuclei with nonspherical equilibrium shapes (deformed nuclei). The circled points are those on the neutron-rich side studied from the decay of fission fragments.

or as a result of the decay of $^{135}$I (which has a higher yield, about 6.4%). The mass-135 decay chain is

$$^{135}\text{Sb} \xrightarrow{\text{1.7 s}} {}^{135}\text{Te} \xrightarrow{\text{19 s}} {}^{135}\text{I} \xrightarrow{\text{6.6 h}} {}^{135}\text{Xe} \xrightarrow{\text{9.1 h}} {}^{135}\text{Cs} \xrightarrow{\text{3 My}} {}^{135}\text{Ba}$$

The Sb and Te activities decay in a time short compared with the effects we are considering here, so we will consider the $^{135}$I as the primary activity in the sequence. In a time scale of the order of hours, there is a build-up of $^{135}$Xe, which has a thermal neutron capture cross section of $2.7 \times 10^6$ b. This large cross section tends to reduce the number of thermal neutrons available for fission, and there must be a corresponding correction in the control system if the reactor power level is to remain constant. This effect of $^{135}$Xe (and other isotopes) is called *fission product poisoning*. (See Problems 14 and 15.)

Fission products have many applications in research. The most apparent is the study of the nuclear spectroscopy in the decays of the fission products. For example, the interesting region around mass 140–150 has been carefully studied from the $\gamma$ decays of separated fragments. To do careful and precise spectroscopy, it is important to deal with only a single decay, and so methods have been developed for rapid chemical and mass separation of fission products, so that the short-lived fragments can also be studied. Figure 13.31 shows an example of what can be learned from these studies. As neutrons are added beyond the neutron shell closure at $N = 82$, nuclei begin to acquire a stable deformation, as indicated by the decrease of the energy of the first excited $2^+$ state. To study the effect of adding additional neutrons, it is necessary to go to more neutron-rich nuclei, for which the short-lived fission products are essential.

Fission products with medical applications include iodine isotopes $^{131}$I and $^{132}$I, which are used in studies of the human thyroid gland. The thyroid takes in iodine from the blood plasma and uses the iodine to manufacture thyroxine, the hormone that regulates metabolism. The isotope $^{131}$I ($t_{1/2} = 8.0$ d) is obtained from fission products and is administered to the patient in an oral dose of several $\mu$Ci. A $\gamma$-ray detector placed at the patient's neck can follow the build-up of radioactive iodine in the thyroid. In this way, it is possible to diagnose overactive and underactive thyroid function. The isotope $^{132}$I ($t_{1/2} = 2.3$ h) is also used, and because of its shorter half-life, the radiation dose absorbed by the patient is much smaller. The short half-life would ordinarily result in difficulties in extracting and purifying the isotope before too much of it had decayed; however, $^{132}$I is the daughter of 78-h $^{132}$Te, and it is possible to "milk" $^{132}$I from a source of $^{132}$Te, for instance, by treating it with a solvent that dissolves I but not Te.

Much larger doses ($\sim$ mCi) of radioactive iodine are used to reduce the functioning of an overactive thyroid, and still larger doses ($\sim$ Ci) can be used to treat cancers of the thyroid.

Iodine is convenient to obtain from fission products because of its volatility—heating the spent fuel converts the iodine to gas and makes it possible to extract it without resorting to dangerous chemical procedures. Other gaseous fission products include Xe, which can be administered to patients by inhalation and used to measure lung function.

## 13.8 A NATURAL FISSION REACTOR

Fission of course occurs as a natural process, through the spontaneous decay of certain heavy nuclei. Many of these heavy nuclei have decay half-lives that are short compared with the time since their formation (most likely in supernova explosions), and so they do not now exist in nature, but we can infer their former presence by the fission products they leave behind. A particular example is $^{244}$Pu, with $t_{1/2} = 81 \times 10^6$ y. Assuming that the solar system formed primarily from the debris of a supernova explosion that occurred at least $4.5 \times 10^9$ y ago, nearly all of the $^{244}$Pu that might have been formed would long since have decayed away, leaving behind the products of its fission. One popular place to search for such records of the early universe is in meteorites, which are unaffected by the geological evolution and possible redistribution of minerals that have taken place on Earth. Indeed, such meteorites show an excess isotopic abundance of $^{136}$Xe, which is assumed to be due to the spontaneous fission of $^{244}$Pu.

Extrapolating back over geological time is possible because there are few naturally occurring phenomena that can alter relative isotopic abundances over time scales of the order of $10^9$ years. The isotopic abundances we observe today are fixed by those that resulted from the supernova explosion (see Chapter 19 for a discussion of the formation of the elements), with modifications from the radioactive decay of unstable isotopes. Furthermore, we expect that these processes are common throughout the solar system, including in solid extraterrestrial minerals that we observe as meteorites. In the case of Xe, which forms no chemical compounds under normal conditions, the primordial isotopes were released during the era when the Earth was molten and they have been thoroughly mixed in the atmosphere. In meteorites, this has not occurred, and often the Xe produced by primordial fissions can be released from microscopic inclusions within the mineral and analyzed by mass spectroscopy to reveal isotopic abundances different from terrestrial Xe.

Owing to the need for exact control of $^{235}$U enrichments, its natural abundance has been carefully and precisely measured in minerals from a variety of sources. The accepted value is $0.00720 \pm 0.00001$ for the fraction of $^{235}$U in natural uranium. The small limit of uncertainty not only represents the precision of the measured values, but is also indicative of the variation in samples from very different locations where uranium is mined—the western United States, Canada, Australia, and Africa; even the rocks returned from the moon by the Apollo missions show the same value.

Of course, we expect this ratio to vary over geological time because both $^{235}$U and $^{238}$U are radioactive, with $^{235}$U having a shorter half-life ($7.0 \times 10^8$ y) than $^{238}$U ($4.5 \times 10^9$ y). Because of the shorter half-life, $^{235}$U decays more rapidly than $^{238}$U, and in times past $^{235}$U must have had a greater relative abundance. If $t = 0$ represents the present, and if $N_5$ and $N_8$ represent, respectively, the number of 235 and 238 nuclei and $\lambda_5$ and $\lambda_8$ their decay constants, then the present abundance is

$$r(0) = \frac{N_5(0)}{N_5(0) + N_8(0)} = 0.00720 \qquad (13.17)$$

and previous abundances were

$$r(t) = \frac{N_5(0)\, e^{-\lambda_5 t}}{N_5(0)\, e^{-\lambda_5 t} + N_8(0)\, e^{-\lambda_8 t}} \qquad (13.18)$$

where $t$ is negative. Figure 13.32 is a plot of this ratio. Notice that about $2 \times 10^9$ years ago, the $^{235}$U fraction was of order 3%; as we know from our previous discussion of water-moderated reactors, with $^{235}$U enrichments of 3% the neutron absorption in natural water as moderator does not prevent the construction of a critical assembly. Going further back, we might postulate that the extreme conditions in a supernova (where the heavy elements may have been produced) would not be expected to discriminate significantly between $^{235}$U and $^{238}$U (recall the similarity of the fast neutron cross sections, Figure 13.9). Thus the extrapolation back to about $r = 0.5$ should give an estimate of the time of formation of these elements; Figure 13.32 shows this to be about $6 \times 10^9$ y, consistent with values obtained from other sources for the age of the solar system. See Section 19.6 for a more rigorous discussion of the $^{235}$U/$^{238}$U ratio and its use in dating the galaxy.

Because of the extreme precision and uniformity of the measured abundance of $^{235}$U, even small anomalies are particularly apparent. In 1972, a sample of uranium, mined at Oklo in what is now the Republic of Gabon on the west coast of Africa, was analyzed by the French Atomic Energy Commission and showed a $^{235}$U abundance of only 0.00717, about 3 standard deviations below the accepted value. This small deviation was enough to excite their curiosity, and analysis of additional samples showed even smaller $^{235}$U abundances, as low as 0.00440. Since the only known process that could lead to a reduction in $^{235}$U concentra-

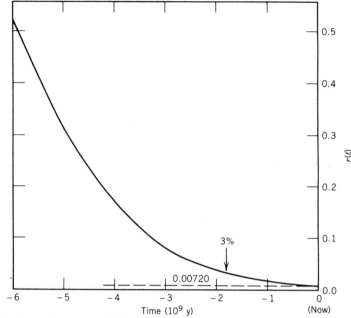

**Figure 13.32** Fraction of $^{235}$U in natural uranium. The present fraction is 0.0072, and $2 \times 10^9$ y ago it was about 3%.

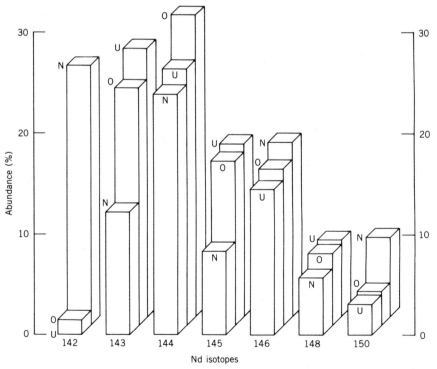

**Figure 13.33** Abundances of neodymium isotopes. N indicates the abundance of natural Nd, U indicates abundances in $^{235}$U fission, and O indicates the observed abundances in the Oklo ore samples. The Oklo abundances are in much better agreement with those of $^{235}$U fission than with those of natural Nd.

tion is fission by low-energy neutrons, the French workers theorized that a natural nuclear reactor operated in the Oklo site about $2 \times 10^9$ y ago, when the $^{235}U$ abundance was high enough ($\sim 3\%$) to permit the operation of a reactor moderated by groundwater. Of course, the reactor could have operated even before $2 \times 10^9$ y ago; however, the formation of deposits of uranium such as in the Oklo region requires the transport of uranyl ions ($UO_2^{2+}$) in groundwater, and it is now believed that before the evolution of oxygen-producing bacteria about $2 \times 10^9$ y ago there was insufficient oxygen in the water to have formed the uranium deposit. Thus the need for highly oxygenated water and for $^{235}U$ enrichments of 3% place fairly rigid limits on the age of the reactor.

Estimating the total size of the uranium deposit and its deficiency of $^{235}U$, it has been calculated that about *5 tons* of $^{235}U$ were fissioned; with 200 MeV released per fission, this amounts to a total energy release of $2 \times 10^{30}$ MeV, or about $10^8$ MW·h. A contemporary power reactor, by comparison, may operate at about $10^3$ MW. However, it is unlikely that the Oklo reactor operated at this power level for the boiling of the water would have removed the moderator and terminated the reactor until groundwater could collect again. The average power level was probably more like 0.01 MW, suggesting that the reactor may have operated for $10^{10}$ h, or $10^6$ y!

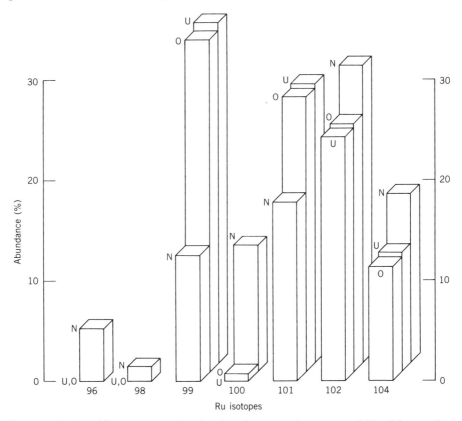

**Figure 13.34** Abundances of ruthenium isotopes from natural Ru (N), uranium fission (U), and Oklo deposits (O). As in the case of the Nd isotopes, the Oklo abundances agree with those of fission and disagree with those of natural Ru.

Confirmation of the remarkable hypothesis of the natural reactor was found in the observation of the abundances of fission products in the Oklo minerals. Figure 13.33 shows the relative abundances of Nd isotopes in natural Nd, in the Oklo deposits, and in residues of $^{235}$U fission products. The disagreement of the Oklo Nd abundances with natural Nd and the agreement with the fission distribution supports the identification of the natural reactor. Figure 13.34 shows similar relationships for the abundances of the Ru isotopes. The $^{100}$Ru abundance differs slightly from that of the ordinary fission of $^{235}$U. This small excess is believed due to neutron capture by (radioactive) fission product $^{99}$Tc, from the chain

$$^{99}\text{Tc} + \text{n} \rightarrow {}^{100}\text{Tc} \xrightarrow{\beta} {}^{100}\text{Ru}$$
$$\xLeftarrow{\beta} {}^{99}\text{Ru} + \text{n}$$

Of the two possible paths leading to $^{100}$Ru, the upper one has the larger cross section, and the long half-life of $^{99}$Tc ($2.1 \times 10^5$ y) ensures that it will be present long enough to capture a neutron. The observed excess of $^{100}$Ru can be used to estimate the duration of the Oklo reactor, and the values obtained are about $10^6$ y, in agreement with the estimate based on the $^{235}$U depletion.

One final interesting feature of the Oklo reactor should be mentioned—the fission products are still in place in the reactor zone and have migrated very little. Despite climatic changes, no substantial movement of these fission products has taken place over the past $2 \times 10^9$ y. This suggests that there may be merit in the present schemes for burying waste products of power reactors in geologically stable formations.

## 13.9  FISSION EXPLOSIVES

If the exponential increase in energy release of a supercritical assembly of $^{235}$U or $^{239}$Pu is permitted to continue without control, a highly unstable situation will rapidly occur. The energy released in the fissionable material must be dissipated, and in doing so will often blow apart the fissionable fuel, thereby rendering it subcritical. In a reactor core, the $^{235}$U is neither sufficiently enriched nor sufficiently concentrated to result in an explosion, even if it were allowed to go supercritical. To make a nuclear explosive, it is necessary to collect subcritical pieces into a supercritical assembly, and to do it so quickly that the exponential fission energy release produces the desired blast effects before the inertia of the supercritical mass is overcome and the fuel itself is blown apart into a subcritical state.

Two basic designs were employed to construct fission-based explosives. The first is the gun type, illustrated in Figure 13.35. In this case the mass of pure $^{235}$U is cut into a sphere with a plug removed from its center. The plug is then fired rapidly into the center of the sphere, and the assembly goes supercritical. The bomb that was dropped on Hiroshima, Japan, in 1945 was of this type. The energy release was about $10^{14}$ J, or equivalent to a conventional bomb containing

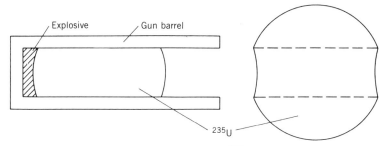

**Figure 13.35** The gun-type design of the $^{235}$U explosive. Not shown are the tamper, which surrounds the critical mass and reflects escaping neutrons back into the explosive, and the initiator, a source of neutrons that guarantees that a neutron will be present to begin the chain reaction at the instant the critical mass is achieved.

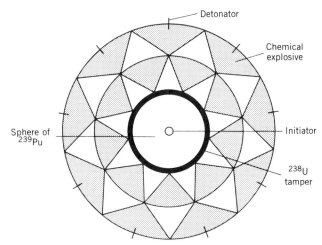

**Figure 13.36** The $^{239}$Pu implosion bomb. Detonation of the conventional explosive compresses the $^{239}$Pu core into a supercritical state. An initiator at the center provides neutrons to begin the chain reaction. The $^{238}$U tamper reflects neutrons back into the core (and can also provide additional neutrons through fissions induced by escaping fast neutrons).

about 20 kilotons of TNT. (The energy release from one ton of TNT is about $4 \times 10^9$ J.)

A second design is the implosion type, illustrated in Figure 13.36. Here a solid spherical subcritical mass of fissionable material is surrounded by a spherical shell of conventional explosives. When the conventional explosives are detonated in exact synchronization, a spherical shock wave compresses the fissionable material into a supercritical state. The first nuclear explosive, tested near Alamogordo, New Mexico, in 1945 and the one later dropped on Nagasaki, Japan, were of this type. The yields of these weapons were, like the $^{235}$U bomb, in the range of 20 kilotons of TNT equivalent.

Although many of the details of the construction of fission explosives are classified, based on the known physical dimensions (see Figure 13.37), we can estimate that the fissile material occupies a sphere of perhaps 10 cm diameter,

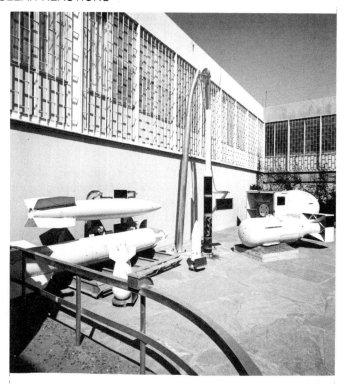

**Figure 13.37** The large casing on the right side is from "Fat Man," the plutonium implosion-type weapon. The smaller one in front of it is "Little Boy," a $^{235}$U gun-barrel type. The smaller casings on the left are from kiloton-range tactical weapons, and the upper casing at left is from a thermonuclear explosive. Photo courtesy Los Alamos National Laboratory.

and thus has a mass of roughly 10 kg. (The fast-neutron fission cross sections for $^{235}$U and $^{239}$Pu are in the range 1–2 b, where the mean free path of a neutron would be of order 10 cm. The size is thus determined by the distance over which a neutron has a high probability of interacting.) An energy release of $10^{14}$ J (20 kilotons of TNT equivalent) would require about $3 \times 10^{24}$ fission events (at 200 MeV released per fission), which corresponds to the complete fissioning of the atoms in about 1 kg of fissile material. Thus only 10% of the fuel in the explosive actually fissions before the energy released reduces the density of the fissile materials to the subcritical region.

The effects of nuclear explosives fall into several categories: the blast itself (a rapidly spreading shock wave), heat radiation (sometimes called a "fireball"), direct nuclear radiation (mostly neutrons and $\gamma$ rays from fission), and indirect nuclear radiation (from the radioactive decays of fission products).

The direct blast damage from a nuclear explosion can be considered as a rapidly expanding spherical wavefront carrying a sudden increase and a following decrease in air pressure. The energy density in this spherical wavefront decreases like $1/r^2$ from a purely geometrical effect, but even at a distance of 1 km from a 20-kiloton explosion, the increase in pressure is of order one atmosphere, sufficient to destroy brick buildings. At 2 km, the pressure increase would be only about 0.25 atmosphere, or perhaps 3 psi, which is sufficient to destroy wood-frame buildings and to send debris flying at speeds of the order 100 miles/h.

The heat radiation also decreases like $1/r^2$ as the heat wave spreads, but it also decreases by an additional exponential factor from absorption and scattering by the atmosphere. At a distance of 2 km from a 20-kiloton explosion, the heat wave (which arrives within about 2 s following the explosion) is still sufficient to produce third-degree burns on exposed skin and to ignite flammable materials such as wood and cloth. An indirect effect of the many fires produced by this radiation is the "firestorm," in which the heat rising from the fires produces ground-level winds of 50–100 miles/h which increase the intensity of the fires and help to spread them. (This is not a specifically nuclear effect; incendiary bombing raids with conventional bombs produce the same effect, but a nuclear explosion is capable of producing a firestorm from a single bomb.)

Direct nuclear radiations (neutrons and $\gamma$'s) also decrease both like $1/r^2$ and exponentially, and the distance over which lethal doses of radiation would be received is roughly the same as the distance for severe blast damage or fatal burns (of the order of 1 km for a 20-kiloton explosion) in unprotected population. Even the smaller doses received at greater distances can have severe long-term effects, including increased incidences of leukemia, cancer, and genetic defects.

The long-lived radioactive products of fission are generally vaporized in the explosion and fall to the ground as radioactive fallout. Some of this material can be carried, as a vaporized cloud, high into the atmosphere where it is spread by prevailing high-altitude winds. This "cloud" of radioactivity may circulate in the upper atmosphere for a year or more as it gradually falls back to the ground. Many short-lived products decay during this time, and the major isotopes remaining after this period are $^{90}$Sr and $^{137}$Cs. The isotope $^{90}$Sr is particularly worrisome, for it is chemically similar to calcium and can concentrate in the bones, where its decays can produce bone cancer.

If the bomb is exploded low enough above the ground that material from the surface of the Earth is vaporized by the fireball, then this additional vaporized material can mix with the cloud of vaporized fission products and condense into particles as the fireball spreads and cools. These particles fall back to the ground relatively rapidly (over several hours) and over a range determined by the low-altitude local winds. Population downwind from the blast, even tens or hundreds of miles, can be exposed to serious, perhaps even lethal, doses of radiation from the decay of shorter-lived fission products. (Hence, we can distinguish two basic ways to detonate a nuclear explosive—a high-altitude "air burst," the goal of which is blast damage over a relatively large area and which might be used as a means of destruction of cities, and a low altitude "ground burst" which concentrates the blast damage over a smaller area, such as might be desired in an attack against blast-hardened underground missile silos, but which produces extreme levels of local fallout in the process.)

As awesome and frightening as these effects of nuclear explosions may be, it is even more frightening that within a decade after the development of fission weapons came the testing of fusion-based thermonuclear weapons, with yields 1000 times as great. Even though many of these effects we have discussed scale as the square root or cube root of the yield, the resulting complete destruction of a population center of 10-mile radius (e.g., most of metropolitan New York City or Los Angeles) from a single explosion is truly staggering. The principles, yields, and effects of thermonuclear weapons are considered in Section 14.5.

## REFERENCES FOR ADDITIONAL READING

A comprehensive and advanced work on the theory of nuclear fission is R. Vandenbosch and J. R. Huizenga, *Nuclear Fission* (New York: Academic, 1973); a similar work is L. Wilets, *Theories of Nuclear Fission* (Oxford: Clarendon, 1964). The early paper on the application of the liquid-drop model to fission by N. Bohr and J. Wheeler, *Phys. Rev.* **56**, 426 (1939) is essential reading. Also of historical interest is H. G. Graetzer and D. L. Anderson, *The Discovery of Nuclear Fission: A Documentary History* (New York: Van Nostrand, 1971).

For general reviews of fission, see also I. Halpern, *Ann. Rev. Nucl. Sci.* **9**, 245 (1959) and J. S. Fraser and J. C. D. Milton, *Ann. Rev. Nucl. Sci.* **16**, 379 (1966).

Experimental data on fission studies, especially cross sections, have been collected in a series of conference reports under the title *Physics and Chemistry of Fission* and published serially by the International Atomic Energy Agency (Vienna). A similar series of Proceedings of International Conferences on the Peaceful Uses of Atomic Energy has been published by the United Nations. •

A popular-level review of nuclear energy and its effects is D. R. Inglis, *Nuclear Energy—Its Physics and Its Social Challenge* (Reading, MA: Addison-Wesley, 1973).

There are many texts and monographs on reactor physics and design, including: I. R. Cameron, *Nuclear Fission Reactors* (New York: Plenum, 1982); J. R. Lamarsh, *Introduction to Nuclear Reactor Theory* (Reading, MA: Addison-Wesley, 1966); J. J. Duderstadt and L. J. Hamilton, *Nuclear Reactor Analysis* (New York: Wiley, 1976); L. Massimo, *Physics of High-Temperature Reactors* (Oxford: Pergamon, 1976).

Breeder reactors are reviewed by W. Häfele, D. Faude, E. A. Fischer, and H. J. Lane, *Ann. Rev. Nucl. Sci.* **20**, 393 (1970), and A. M. Perry and A. M. Weinberg, *Ann. Rev. Nucl. Sci.* **22**, 317 (1972).

For a survey of the problems of management of radioactive wastes, see Bernard L. Cohen, *Rev. Mod. Phys.* **49**, 1 (1977) and *Am. J. Phys.* **54**, 38 (1986); a review of this topic by a study group of the American Physical Society was published in *Rev. Mod. Phys.* **50**, S1 (1978). The radiation released from coal-burning power plants is discussed by J. P. McBride et al., *Science* **202**, 1045 (1978).

Many technical details of reactor safety and radiation protection are reviewed in *Nuclear Power and Its Environmental Effects*, by S. Glasstone and W. H. Jordan, published by the American Nuclear Society in 1980. The American Association of Physics Teachers has published six papers given at a 1982 symposium on nuclear power; see *The Status of the Nuclear Enterprise*, edited by Morris W. Firebaugh.

The Oklo reactor is reviewed in the conference proceedings, *The Oklo Phenomenon*, published by the International Atomic Energy Agency (Vienna) in 1975. Other references to Oklo are P. K. Kuroda, *The Origin of the Chemical Elements and the Oklo Phenomenon* (Berlin: Springer-Verlag, 1982) and M. Maurette, *Ann. Rev. Nucl. Sci.* **26**, 319 (1976). For a popular-level account, see G. A. Cowan, "A Natural Fission Reactor," in the July 1976 issue of *Scientific American*.

Fission-based nuclear explosives are now mainly of historical interest (references for fusion explosives, including the effects of nuclear weapons, will be given in the next chapter). The now-declassified summary of the wartime work on the Manhattan Project can be found in H. D. Smyth, *Atomic Energy for Military Purposes* (Washington, DC: U.S. Government Printing Office, 1946). More details of the work at Los Alamos are described in L. Lamont, *Day of Trinity* (New York: Atheneum, 1965). A similar work, describing a more personal view of bomb design, is John McPhee, *The Curve of Binding Energy* (New York: Farrar, Strauss and Giroux, 1974). A recent historical work from the British perspective is A. McKay, *The Making of the Atomic Age* (Oxford: Oxford University Press, 1984).

## PROBLEMS

1. Compute the energy release ($Q$ value) of: (a) $^{235}U + n \rightarrow {}^{90}Kr + {}^{144}Ba + ?$; (b) $^{239}Pu + \gamma \rightarrow {}^{92}Sr + ? + 3n$; (c) $^{252}Cf \rightarrow {}^{106}Nb + ? + 4n$.

2. In the fission of $^{236}U$ into two fragments $A_1$ and $A_2 = 236 - A_1$, plot the Coulomb repulsion energy of the two fragments if they are formed just touching at their surfaces. Consider all values of $A_1$ from 1 to 235, and assume each fragment has the same $Z/A$ ratio as $^{236}U$.

3. $^{252}Cf$ has a half-life of 2.64 y and a spontaneous fission branch of 3.09%. How many fission neutrons per second are emitted per milligram of $^{252}Cf$? (See Figure 13.7.)

4. The isotope $^{254}Cf$ decays almost exclusively by spontaneous fission, with a half-life of 60.5 days. The energy released is about 225 MeV per fission. (a) Calculate the total fission power produced by 1.0 $\mu$g of $^{254}Cf$. (b) Assuming Cf to behave like an ordinary metal, what would be the rise in temperature of the Cf sample per minute from the fission heating?

5. Estimate the neutron energy needed to produce fission of $^{208}Pb$. Is it likely that such neutrons would be released in the fission, making possible a self-sustaining reaction?

6. Given that the activation energy of $^{236}U$ is 6.2 MeV, what is the minimum-energy $\alpha$ particle that can produce fission following bombardment of a $^{232}Th$ target?

7. Compare the excitation and activation energies for thermal neutron-induced fission of (a) $^{239}Pu$; (b) $^{231}Pa$; (c) $^{237}Np$; (d) $^{238}Np$.

8. Which of the following heavy nuclei would you expect to have large thermal cross sections? (Use arguments based on the pairing-energy discussion given in Section 13.3.) (a) $^{251}Cf$; (b) $^{253}Es$; (c) $^{255}Fm$; (d) $^{250}Bk$.

9. Use Figure 13.7 to estimate the fraction of neutron-induced fissions of $^{235}U$ that produce no prompt neutrons.

10. Discuss the formation of $^{244}Pu$ in the core of a reactor operating with enriched uranium as a fuel. (Information on cross sections and decays can

be found in the *Table of Isotopes.*) This isotope is one of the longest-lived remnants of supernova explosions, in which it is also formed through neutron captures.

11. In $^{239}$Pu, the thermal fission cross section is 742 b, while the cross section for other (nonfission) absorptive processes is 267 b. Each fission produces, on the average, 2.86 fast neutrons. What is the mean number of fission neutrons produced by $^{239}$Pu per thermal neutron?

12. In calculating the heat deposited in a reactor core by the radioactive decays of the fission products, is it necessary to know the details of $\beta$ and $\gamma$ spectroscopy of each decay process? That is, given the $Q$ value of each decay, we know the sum of the total $\beta$ and $\gamma$ energies emitted in each decay process. Is this knowledge sufficient for calculating the decay heat? Give some examples to justify your arguments.

13. In a water-moderated uranium reactor, what will be the effect on the fission rate of an increase in the temperature of the core? Consider especially the possible qualitative effects of temperature on $p$ and $f$.

14. Consider the fission product poisoning by $^{135}$I and $^{135}$Xe. Let I be produced directly in fission with yield $\gamma_I = 6.4\%$ per fission, and let it be depleted by its decay (to $^{135}$Xe) and by neutron capture (to $^{136}$I). Let Xe be produced directly in fission ($\gamma_{Xe} = 0.23\%$) and indirectly in $^{135}$I decay, and let Xe be depleted by its decay and by neutron capture. (a) Write differential equations for the change $dn$ in the number density of I and of Xe resulting from these processes. Your results should be in terms of the fission cross section $\sigma_f$, the neutron flux $\phi$, the number density $n_U$ of fissionable U, the decay constants $\lambda_I$ and $\lambda_{Xe}$, the capture cross sections $\sigma_I$ (7 b) and $\sigma_{Xe}$ (2.7 $\times$ $10^6$ b), and the yields $\gamma_I$ and $\gamma_{Xe}$. (b) Show that the neutron capture contribution can be neglected for I but not for Xe. (c) Subject to the initial conditions $n_I(0) = 0$ and $n_{Xe}(0) = 0$ and assuming $n_U$ and $\phi$ to be constant, solve the differential equations to obtain $n_I(t)$ and $n_{Xe}(t)$. Show that the equilibrium values at large $t$ are $n_I = \gamma_I \sigma_f n_U \phi / \lambda_I$ and $n_{Xe} = (\gamma_I + \gamma_{Xe}) \sigma_f n_U \phi / (\lambda_{Xe} + \sigma_{Xe} \phi)$. (d) After reaching equilibrium, the reactor is shut down at $t = t_0$. In terms of $t' = t - t_0$, find $n_I(t')$ and $n_{Xe}(t')$; sketch their behavior from $t = 0$ to $t \gg t_0$. What problems might arise in starting the reactor again?

15. Another example of fission product poisoning (see previous problem) occurs in the mass-149 decay chain, $^{149}$Nd $\rightarrow$ $^{149}$Pm $\rightarrow$ $^{149}$Sm. The thermal cross sections are 1400 b for $^{149}$Pm and 42,000 b for $^{149}$Sm. Calculate and plot the number density of Sm nuclei as a function of time in a reactor that operates at a constant power level (keeping $\phi$ constant) for several weeks, following which it is turned off. (*Hint:* Nd decays sufficiently rapidly on this time scale that you may assume the Pm to be formed only directly in fission with yield $\gamma_{Pm} = 1.13\%$.)

16. (a) Given a collection of $n$ thermal neutrons in a natural uranium reactor, how many $^{235}$U atoms (of $N$ total fuel atoms) disappear following the absorption of these neutrons? (b) How many $^{239}$Pu atoms appear? Consider the production of $^{239}$U from absorption of the original thermal neutrons and also of the secondary fast fission neutrons, taking $p = 0.9$ and $\varepsilon = 1.0$.

(c) The ratio between the numbers found in (b) and (a) is called the *conversion ratio* and represents a measure of whether a converter reactor will function as a breeder, that is, produce more fuel than it consumes. Evaluate the conversion ratio and comment. (d) Repeat the calculation for a reactor fueled with a mixture of 1% fissile $^{233}$U ($\sigma_f = 530$ b, $\sigma_a = 47.7$ b, $\nu = 2.49$) and 99% fertile $^{232}$Th ($\sigma_f = 0$, $\sigma_a = 7.4$ b).

17. Suppose you were given the assignment of calculating the time dependence of radioactive decay heat produced by fission products after a reactor is shut down. List the information you would need, and discuss whether the needed input data could be measured or estimated. Carefully specify any assumptions that would need to be made. Be as realistic as possible in your approach to this problem—your resources of time and money needed to do experiments are not infinite.

18. The thermal utilization factor $f$ depends on the ratio of the total number of captures in the fuel to the number in the fuel plus everything else (moderator, structural components, etc.). Suppose we have a homogeneous, graphite-moderated reactor containing 20 times as much graphite (by weight) as uranium; the uranium is enriched to 2% in $^{235}$U. Compute the thermal utilization factor. How would the calculation be different in a heterogeneous (lumped) arrangement?

# 14

# NUCLEAR FUSION

Figure 3.16 suggests an alternative to fission for extracting energy from the nucleus—we can climb the binding energy curve toward the more stable nuclei by beginning with the very light nuclei, rather than the very heavy nuclei as in fission. That is, if we combine two light nuclei into a nucleus below $A = 56$, energy is released. This process is called *nuclear fusion* because two light nuclei are fused into a heavier one.

As an energy source, fusion has several obvious advantages over fission: the light nuclei are plentiful and easy to obtain, and the end products of fusion are usually light, stable nuclei rather than heavy radioactive ones. There is one considerable disadvantage, however: before light nuclei can be combined, their mutual Coulomb repulsion must be overcome. Fission induced by neutrons has no Coulomb barrier and thus very low-energy incident particles can be used; indeed, the cross section for $^{235}U$ increases as we reduce the neutron energy. On the other hand, cross sections for reactions induced by charged particles tend to decrease with decreasing energy.

Consider the fission of two $^{20}Ne$ nuclei to form $^{40}Ca$. The $Q$ value is about 20.7 MeV, or about 0.5 MeV/nucleon, comparable to the energy released in fission. However, before the nuclear forces of the two $^{20}Ne$ nuclei can interact, we must move them close enough together so that their nuclear distributions begin to overlap. At the point where their surfaces are just touching, the Coulomb repulsion is 21.2 MeV. If we were to perform a nuclear reaction in which two $^{20}Ne$ were brought together with a total of 21.2 MeV of kinetic energy, the final energy of the system would be 41.9 MeV, representing the initial 21.2 MeV of kinetic energy *plus* the 20.7 MeV released in the reaction (the $Q$ value). The energy gain is therefore a factor of two—21 MeV goes in and 42 MeV comes out.

Accelerating $^{20}Ne$ to 21.2 MeV against a $^{20}Ne$ target is certainly no problem, but heavy-ion accelerators ($^{20}Ne$ is called a "heavy ion" by accelerator physicists, for whom "light ion" means only H or He) produce beams in the nanoampere to microampere range. At a current of $10^{-6}$ A, even if every particle in the beam were to react (an unlikely prospect—scattering is several orders of magnitude more likely than fusion) the power output would be about 2 W, hardly enough even to light the accelerator laboratory!

An alternative approach would be to heat a container of neon gas until the thermal energy is large enough to have a high probability of two nuclei approaching one another and colliding with 21.2 MeV of energy. Because the thermal energy is used to overcome the Coulomb barrier that inhibits fusion, this process is called *thermonuclear* fusion. For the mean kinetic energy per molecule of a gas ($\frac{3}{2}kT$) to be $\frac{1}{2}$ of 21.2 MeV, we must have $kT = 7$ MeV. At room temperature, $kT = 0.025$ eV, and so this process requires a temperature about $3 \times 10^8$ times room temperature, or about $10^{11}$ K!

Despite these various drawbacks, fusion energy is at present a subject of vigorous and intensive research to perfect techniques for heating fusible nuclei and for increasing their density so the number of reactions becomes large enough that the energy output is comparable to that of a fission reactor ($10^9$ W). Of course, fusion also powers the sun and other stars and is therefore ultimately responsible for the evolution of life on Earth. Understanding fusion is critical for understanding the end products of stellar reactions, when the thermonuclear fuel is mostly exhausted and a star may pass through a nova or supernova stage, ending as a chunk of cosmic ash or as a neutron star or black hole. Most horribly, thermonuclear weapons are the curse of our civilization and a constant threat to our very existence.

In this chapter, we cover the basic physics of fusion processes and direct applications to solar fusion, controlled fusion reactors, and thermonuclear weapons. In Chapter 19, we consider further applications in nuclear astrophysics.

## 14.1  BASIC FUSION PROCESSES

As we discussed in the introduction, fusion is not at all a "natural" process on Earth (as fission is) because of the substantial limitations imposed by the Coulomb barrier. Once we overcome the barrier, fusion becomes very likely, as the two overlapping nuclei quickly reach a state of minimum energy. The basic fusion processes are thus considerably simpler to understand and explain than fission processes. The most elementary fusion reaction, $p + p \rightarrow {}^2$He, is not possible, owing to the instability of $^2$He (but an alternative process analogous to $\beta$ decay and leading to $^2$H is a primary first step in solar fusion; see Section 14.3). Another elementary reaction is

$$^2\text{H} + {}^2\text{H} \rightarrow {}^4\text{He} + \gamma$$

where the $\gamma$ is essential for energy balance, since $^4$He has no excited states. The energy release ($Q$ value) is 23.8 MeV, which happens to be greater than both the neutron and proton separation energies of $^4$He. More likely reactions are thus

$$^2\text{H} + {}^2\text{H} \rightarrow {}^3\text{He} + \text{n} \qquad (Q = 3.3 \text{ MeV})$$

$$^2\text{H} + {}^2\text{H} \rightarrow {}^3\text{H} + \text{p} \qquad (Q = 4.0 \text{ MeV})$$

These are called *deuterium-deuterium or D-D reactions*.

Of course, the more stable the end product formed, the greater is the energy release in the reaction. A reaction that forms $^4$He would be likely to show a particularly large energy release:

$$^{2}\text{H} + ^{3}\text{H} \rightarrow ^{4}\text{He} + \text{n} \qquad (Q = 17.6 \text{ MeV})$$

This reaction is called the *deuterium-tritium or D-T* reaction. If the incident particles have negligibly small kinetic energies, the $^{4}\text{He}$ and n share 17.6 MeV consistent with linear momentum conservation, and a monoenergetic neutron with energy 14.1 MeV emerges. This reaction often serves as a source of fast neutrons. Because of the large energy release (and because the Coulomb barrier is no higher than in the D-D reactions), the D-T reaction has been selected for use in controlled fusion reactors. A disadvantage is that most of the energy is given to the neutron, in which form it is not particularly easy to extract. In fission, very little of the energy release is given to the neutrons, and the kinetic energy of the fission fragments is easy to extract.

The fusion of four protons ultimately to form $^{4}\text{He}$ (in several steps) is responsible for the thermonuclear energy released in stars similar to the sun. The next step, once the hydrogen fuel has been used up, is helium fusion. The simplest reaction, $^{4}\text{He} + ^{4}\text{He} \rightarrow ^{8}\text{Be}$, is not observed because $^{8}\text{Be}$ breaks up into two $^{4}\text{He}$ again almost as fast as it is formed ($10^{-16}$ s). Instead, a more complicated process takes place:

$$3\,^{4}\text{He} \rightarrow ^{12}\text{C}$$

The probability of bringing together three particles at one point is negligibly small. Instead, the process (in stars) first produces a small equilibrium concentration of $^{8}\text{Be}$, and capture of a third $\alpha$ particle by the $^{8}\text{Be}$ happens to occur through a resonance in $^{12}\text{C}$ where the cross section becomes large enough to give a reasonable probability for capture before the $^{8}\text{Be}$ splits apart. The larger Coulomb barrier of the helium reactions relative to hydrogen reactions means that helium burning occurs only in hotter (and older) stars. At still higher temperatures, other reactions occur that can produce energy from $^{12}\text{C}$ and heavier products (up to $^{56}\text{Fe}$). In Section 14.3 we consider the hydrogen-burning processes in the sun, and in Chapter 19 we consider other aspects of nuclear reactions in stars.

## 14.2 CHARACTERISTICS OF FUSION

**Energy Release** The calculation of the energy release in fusion is much more direct than that for fission. It is necessary simply to calculate the $Q$ value of the appropriate reaction. For most applications of fusion, from controlled fusion reactors to solar processes, the reacting particles have energies in the range of 1–10 keV. The initial kinetic energies are thus small compared with the $Q$ values of several MeV. The energy released, and the final total energy of the product particles, will then be equal to the $Q$ value:

$$\tfrac{1}{2}m_{\text{b}}v_{\text{b}}^{2} + \tfrac{1}{2}m_{\text{Y}}v_{\text{Y}}^{2} \simeq Q \qquad (14.1)$$

for product particles b and Y. Again neglecting the initial motions, the final momenta are equal and opposite:

$$m_{\text{b}}v_{\text{b}} \simeq m_{\text{Y}}v_{\text{Y}} \qquad (14.2)$$

and thus

$$\tfrac{1}{2}m_b v_b^2 \simeq \frac{Q}{1 + m_b/m_Y} \tag{14.3}$$

$$\tfrac{1}{2}m_Y v_Y^2 \simeq \frac{Q}{1 + m_Y/m_b} \tag{14.4}$$

from which we can calculate the distribution in energy for the elementary fusion reactions given in the previous sections.

One consequence of this energy sharing is immediately apparent—the lighter product particle takes the larger share of the energy. The ratio of the kinetic energies is, from Equation 14.2, just

$$\frac{\tfrac{1}{2}m_b v_b^2}{\tfrac{1}{2}m_Y v_Y^2} = \frac{m_Y}{m_b} \tag{14.5}$$

Thus in the case of the D-T reaction, the product neutron has 80% of the energy. In the D-D reactions, the product proton or neutron has 75% of the available energy.

**Coulomb Barrier**  If $R_a$ and $R_X$ are the radii of the reacting particles, the Coulomb barrier is

$$V_c = \frac{e^2}{4\pi\epsilon_0} \frac{Z_a Z_X}{R_a + R_X} \tag{14.6}$$

when the particles just touch at their surfaces. The effect of the Coulomb barrier on the fusion reaction is very similar to the effect of the Coulomb barrier on $\alpha$ decay. The product $Z_a Z_X$ will ultimately appear in an exponential barrier penetration probability, and so the fusion cross section is extremely sensitive to the Coulomb barrier. The fusion probability therefore decreases rapidly with $Z_a Z_X$, and the barrier is lowest for the hydrogen isotopes. For the D-T reaction, $V_c = 0.4$ MeV, but even though the barrier is lowest, it is still far above the typical incident particle energy of 1–10 keV. Of course, it is not necessary for the particles to be above the barrier; just as in $\alpha$ decay, it is the barrier penetration probability that determines the outcome.

**Cross Section**  The fusion cross section can be adapted from the basic expression for nuclear reaction cross sections, Equation 11.70. For particles reacting at thermal energies, it is likely that the reaction will occur far from any resonance, and thus the energy dependence of the cross section comes mainly from two terms: the $k^{-2}$ factor (which gives a $v^{-2}$ dependence), and the partial reaction probability, which for two charged particles includes a barrier penetration factor of the form $e^{-2G}$ as in the case of $\alpha$ decay, Equation 8.15. Thus

$$\sigma \propto \frac{1}{v^2} e^{-2G} \tag{14.7}$$

where $G$ is essentially given by Equation 8.17, substituting for $Q$ the center-of-mass energy $E$ of the reacting particles. Since $E \ll B$, we can approximate

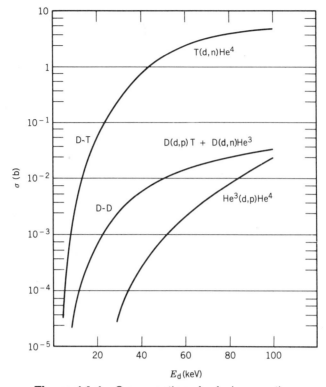

**Figure 14.1** Cross sections for fusion reactions.

Equation 8.17 as

$$G \simeq \frac{e^2}{4\pi\epsilon_0} \frac{\pi Z_a Z_X}{\hbar v} \tag{14.8}$$

In these expressions $v$ represents the relative velocity of the reacting particles. The proportionality factor needed in Equation 14.7 will involve nuclear matrix elements and statistical factors depending on the spins of the particles, but the entire energy dependence is accounted for by the factors included in Equation 14.7. Figure 14.1 shows a plot of the resulting expression.

**Reaction Rate** As discussed in Sections 6.3 and 12.3, the rate for a nuclear reaction depends on the product $\sigma v$. For neutron-induced reactions outside the resonance region, $\sigma \propto 1/v$ and so $\sigma v = $ constant. For fusion reactions, this is not the case. Moreover, in thermonuclear fusion there will be a distribution of particle speeds described by the usual Maxwell-Boltzmann velocity distribution:

$$n(v) \propto e^{-mv^2/2kT} \tag{14.9}$$

where $n(v)v^2\, dv$ gives the relative probability to find a particle with speed between $v$ and $v + dv$ in a collection of particles in thermal equilibrium at

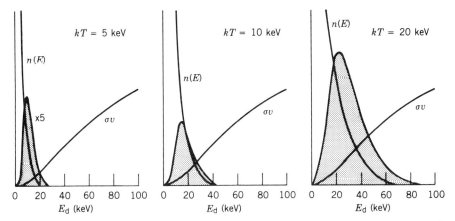

**Figure 14.2**  Folding of the product $\sigma v$ with a Maxwell-Boltzmann energy distribution at temperatures corresponding to $kT = 5$, 10, and 20 keV. The curve $n(E)$ shows the Maxwell-Boltzmann distribution, proportional to $E^{1/2} \exp(-E/kT)$; the curve $n(E)$ falls to zero at low energies, which is not shown in the graphs. The shaded area shows the product. Note the great increase in shaded area with $kT$, corresponding to an increase of $\langle \sigma v \rangle$ as is shown in Figure 14.3. The deuteron energy $E_d$ is one-half of the total center-of-mass reaction energy $E$.

temperature $T$. In such a collection of nuclei undergoing thermonuclear fusion, it is appropriate to calculate $\sigma v$ averaged over all speeds or energies:

$$\langle \sigma v \rangle \propto \int_0^\infty \frac{1}{v} e^{-2G} e^{-mv^2/2kT} v^2 \, dv \tag{14.10}$$

or

$$\langle \sigma v \rangle \propto \int_0^\infty e^{-2G} e^{-E/kT} \, dE \tag{14.11}$$

Figure 14.2 shows an example of the folding of the product $\sigma v$ with the Maxwell-Boltzmann distribution function for various temperatures. At low temperature, there is little overlap between $n(E)$ and $\sigma v$, and the average is small. At very high $T$, the area of the Maxwell-Boltzmann distribution becomes small and again the average value of $\sigma v$ is small. At intermediate temperatures, $\langle \sigma v \rangle$ rises to a maximum. Figure 14.3 shows $\langle \sigma v \rangle$ for several fusion reactions as a function of the temperature. At extremely high temperatures ($T \sim 10^{10}$ K, corresponding to MeV energies), the D-T reaction may become less favorable than others, but in the temperature region that is likely to be achievable in a thermonuclear fusion reactor (1–10 keV or $T \sim 10^7$–$10^8$ K) the D-T reaction is clearly favored.

The simple theory used in this section to evaluate $\langle \sigma v \rangle$ would be appropriate only for the D-D reaction. For reactions involving two different nuclei (D-T, for example), we should take better account of the velocities of the different species. The cross section and reaction rate should involve a relative velocity, $\sigma(v_{rel})$ and $\langle \sigma(v_{rel}) v_{rel} \rangle$, and the average should be done over the Maxwell-Boltzmann distribution of both species. Even though the specific details may be more complicated, the general conclusions regarding fusion reaction rates remain valid.

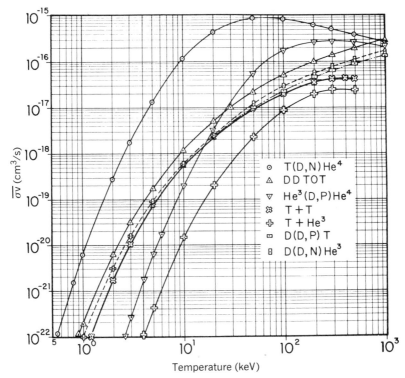

**Figure 14.3**   Values of $\sigma v$ averaged over a Maxwell-Boltzmann energy distribution for various fusion reactions. From D. Keefe, *Ann. Rev. Nucl. Particle Sci.* **32**, 391 (1982).

## 14.3   SOLAR FUSION

Before discussing how we achieve thermonuclear fusion on Earth, let's consider the sun, which we can regard as an extremely successful prototype of a self-sustaining thermonuclear reactor. As far as we can tell from the fossil records on Earth, the sun's output has been nearly constant over a time scale of more than $10^9$ years.

The basic process in the sun (and most other stars) is the fusion of hydrogen into helium. Hydrogen is by far the most abundant material in the universe—more than 90% of the atoms in the universe are hydrogen, and all but less than 1% of the remainder are helium. (This helium was formed during the early stages of the evolution of the universe and not as a result of later stellar processes.) All reactions in any fusion cycle must be two-body reactions (two particles in the initial state) because the simultaneous collision of three particles is too improbable an event to be significant. The first step in the fusion process must be the combination of two protons to form the only stable two-nucleon system:

$$^1H + {}^1H \rightarrow {}^2H + e^+ + \nu \qquad (Q = 1.44 \text{ MeV})$$

The $\nu$ in the final state signals a weak interaction process, which must occur to turn a proton into a neutron (not enough energy is available to create a $\pi$ meson and to have $p \rightarrow n + \pi^+$). The cross sections for weak interaction processes are very small; for the formation of deuterium, the cross section is calculated to be of the order of $10^{-33}$ b at keV energies and $10^{-23}$ b at MeV energies. The central temperature of the sun is about $15 \times 10^6$ K, corresponding to a mean proton energy of about 1 keV, but to calculate the reaction rate it is necessary to find $\langle \sigma v \rangle$ averaged over all energies, and the easy penetration of the Coulomb barrier for MeV particles in the high-energy tail of the Maxwell-Boltzmann distribution compensates somewhat for the low intensity in the tail. The reaction rate is nevertheless very small, and even at the high densities at the core of the sun (about 125 g/cm$^3$, or $7.5 \times 10^{25}$ protons/cm$^3$) the reaction rate is only about $5 \times 10^{-18}$/s per proton. What keeps the sun radiating is the enormous number of reacting protons, of the order of $10^{56}$, so that the total reaction rate is of the order of $10^{38}$/s. This step in the solar fusion cycle is often called the "bottleneck" because it is the slowest and least probable step.

Following deuteron formation, it becomes very likely for the following reaction to occur:

$$^2\text{H} + {}^1\text{H} \rightarrow {}^3\text{He} + \gamma \qquad (Q = 5.49 \text{ MeV})$$

It is very unlikely at this point to observe D-D reactions because of the small number of deuterons present—only one deuteron is formed for every $\sim 10^{18}$ protons, and thus it is about $10^{18}$ times more likely that a deuteron will react with a proton than with another deuteron. Deuterons are thus "cooked" to $^3$He nearly as rapidly as they are formed.

Reactions of $^3$He with protons are not possible:

$$^3\text{He} + {}^1\text{H} \rightarrow {}^4\text{Li} \rightarrow {}^3\text{He} + {}^1\text{H}$$

The isotope $^4$Li does not exist as a bound system, and it breaks up as soon as it is formed. It is also unlikely for $^3$He to react with $^2$H because the density of $^2$H is very low and because the $^2$H is converted to $^3$He very rapidly. The fate of a $^3$He is thus to wander until it finds another $^3$He:

$$^3\text{He} + {}^3\text{He} \rightarrow {}^4\text{He} + 2\,{}^1\text{H} + \gamma \qquad (Q = 12.86 \text{ MeV})$$

The complete process is indicated schematically in Figure 14.4, and is known as the *proton-proton cycle*. The net reaction is the conversion of four protons to helium:

$$4\,{}^1\text{H} \rightarrow {}^4\text{He} + 2e^+ + 2\nu$$

To find the total $Q$ value, we must keep in mind that we have been discussing reactions with bare nuclear particles. Let's add four electrons to each side of this process, giving four neutral H atoms on the left, and a neutral He atom on the right with two additional electrons to annihilate the positrons. The net process is thus $4\,{}^1\text{H} \rightarrow {}^4\text{He}$, with a $Q$ value of 26.7 MeV. The energy converted to solar radiation per cycle is slightly less, for the neutrinos emerge directly from the core without contributing to the heating of the *photosphere*, the outer region of the sun where the energy released in nuclear reactions is converted to light.

An alternative fate for the $^3$He is to encounter an $\alpha$ particle:

$$^3\text{He} + {}^4\text{He} \rightarrow {}^7\text{Be} + \gamma$$

followed either by

$$^7\text{Be} + e^- \rightarrow {}^7\text{Li} + \nu$$

and

$$^7\text{Li} + p \rightarrow 2\,{}^4\text{He}$$

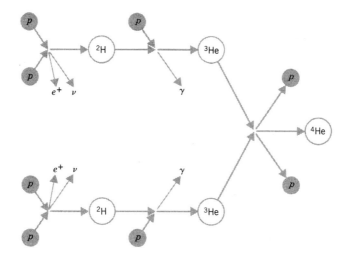

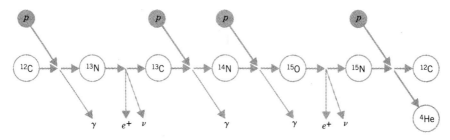

**Figure 14.4** (Top) Sequence of processes in proton-proton chain of fusion reactions. (Bottom) Carbon (CNO) cycle of fusion reactions.

or perhaps by the sequence

$$^{7}\text{Be} + \text{p} \rightarrow {}^{8}\text{B} + \gamma$$

$$^{8}\text{B} \rightarrow {}^{8}\text{Be} + e^{+} + \nu$$

$$^{8}\text{Be} \rightarrow 2\ {}^{4}\text{He}$$

The net reaction and the net $Q$ value are the same for these three possible paths. Which one is actually chosen depends on the composition of the star and on its temperature. In the case of the sun, we can test for these three alternatives by observing the neutrinos. In the first case, we get a continuous distribution of neutrinos with a maximum energy of 0.42 MeV. In the second case, the two-body $^{7}\text{Be}$ electron capture gives a monoenergetic neutrino of energy 0.862 MeV, while the $^{8}\text{B}$ decay gives a continuous neutrino distribution with an endpoint of 14 MeV. To observe these neutrinos, Davis has devised a sensitive experiment which was previously described in Section 9.6. After many years of experiments, Davis has observed only a small fraction (one-third or less) of the expected number of neutrinos from the sun.

Neutrino observation gives us in effect a window to the deep solar interior, for the visible light that reaches us from the sun is characteristic of its surface and represents $\gamma$ rays from reactions in the core that are scattered many thousands of times on their journey to the surface. It also takes perhaps millions of years for the radiation to reach the surface, so the light we see today results from solar processes that occurred millions of years ago. The neutrinos, on the other hand, come to us directly from the core at the speed of light. It is therefore very important to try to understand why the results of the Davis experiment disagree with theory and to devise other experiments that are sensitive to other components of the neutrino spectrum.

If in addition to hydrogen and helium there are heavier elements present in the interior of a star, a different series of fusion reactions can occur. One such series is the *carbon or CNO cycle*:

$$^{12}\text{C} + {}^1\text{H} \rightarrow {}^{13}\text{N} + \gamma$$

$$^{13}\text{N} \rightarrow {}^{13}\text{C} + e^+ + \nu$$

$$^{13}\text{C} + {}^1\text{H} \rightarrow {}^{14}\text{N} + \gamma$$

$$^{14}\text{N} + {}^1\text{H} \rightarrow {}^{15}\text{O} + \gamma$$

$$^{15}\text{O} \rightarrow {}^{15}\text{N} + e^+ + \nu$$

$$^{15}\text{N} + {}^1\text{H} \rightarrow {}^{12}\text{C} + {}^4\text{He}$$

In this case the $^{12}\text{C}$ is neither created nor destroyed, but acts as a catalyst to aid in the fusion process. The net process is $4\ ^1\text{H} \rightarrow {}^4\text{He} + 2e^+ + 2\nu$, exactly as in the proton-proton cycle, and the $Q$ value is the same. The carbon cycle can proceed more rapidly than the proton-proton cycle because it does not have a process analogous to the deuterium bottleneck. However, the Coulomb barrier is 6 or 7 times higher for proton reactions with carbon and nitrogen than for proton-proton reactions. The carbon cycle will thus be dominant at higher temperatures, where the additional thermal energy is needed to increase the probability to penetrate the Coulomb barrier (Figure 14.5).

The mean solar radiation reaching the Earth is about $1.4 \times 10^3$ W/m$^2$, which (if distributed uniformly in space) means the sun's total output is about $4 \times 10^{26}$ W. Each fusion reaction produces about 25 MeV, and thus there must be about $10^{38}$ fusion reactions per second, consuming $4 \times 10^{38}$ protons per second. At this rate, the sun can be expected to continue to burn its hydrogen fuel for another $10^{10}$ y.

Once a star has exhausted its hydrogen fuel, helium fusion reactions can take place, with $3\ ^4\text{He} \rightarrow {}^{12}\text{C}$ at the higher temperature needed to penetrate the Coulomb barrier. Other reactions involving fusion of light nuclei and $\alpha$-particle capture can continue to release energy, until the process ends near $^{56}\text{Fe}$, beyond which there is no energy gain in combining nuclei. This relatively simple recipe not only helps to explain some of the many categories of stars we observe, but it also gives us a means to understand the relative abundances of various atomic species (light even-$Z$ atoms, made through successive $\alpha$ captures on $^{12}\text{C}$, are far

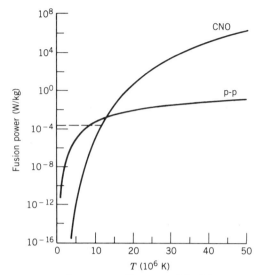

**Figure 14.5** Power generation per unit mass of fuel for proton-proton and CNO processes. The dashed line indicates the sun's power of about $2 \times 10^{-4}$ W / kg.

more abundant then neighboring odd-$Z$ atoms; nearly everything above Fe is less abundant than almost everything below Fe). We consider further details of nuclear astrophysics, including the evolution of stars and the production of the elements, in Chapter 19.

## 14.4 CONTROLLED FUSION REACTORS

The essence of controlling fusion reactions and extracting usable energy is to heat a thermonuclear fuel to temperatures of the order of $10^8$ K (mean particle kinetic energies of 10 keV) while simultaneously maintaining a high enough density for a long enough time that the rate of fusion reactions will be large enough to generate the desired power. At these temperatures, the atoms must become ionized (for hydrogen, only 13.6 eV is needed to strip the electron), and so the fuel is a hot mixture of clouds of positive ions and negative electrons but is overall electrically neutral. Such a situation is called a *plasma*, and the dynamical equations that govern plasma behavior are beyond the scope of this text. The electrostatic properties of a plasma determine a length scale called the *Debye length*,

$$L_{\mathrm{D}} = \left( \frac{4\pi\epsilon_0}{e^2} \frac{kT}{4\pi n} \right)^{1/2} \tag{14.12}$$

where $n$ is the mean ion or electron density (which will of course be subject to considerable local fluctuations on a microscopic scale). Using solid densities ($10^{28}$ m$^{-3}$) as a rough scale factor, the Debye length for a 10-keV plasma is of the order of $10^{-8}$ m and the number of particles in a volume of the plasma of dimension of one Debye length is about $10^4$. For a more rarified plasma, say of density only $10^{22}/$m$^3$, the Debye length is $10^{-5}$ m and the number of particles in a volume of dimension one Debye length is $10^7$. In either of these two extreme cases, there are two basic properties: the physical size of the reacting plasma is far larger than the Debye length in dimension, and there are many particles in any characteristic volume. It is these two properties that permit the use of plasma equations to describe the hot thermonuclear fuel.

A major problem is obviously to confine the plasma—for the hot fuel to exchange energy with the walls of its container would simultaneously cool the fuel and melt the container. There are at present two schemes under investigation for confining the thermonuclear fuel: *magnetic confinement* and *inertial confinement*. In magnetic confinement, the plasma is confined by a carefully designed magnetic field. In inertial confinement, a solid pellet is suddenly heated and compressed by being struck simultaneously from many directions with intense beams of photons or particles.

Confinement of a plasma is of course not absolute—there will be many ways for a plasma to lose energy. The primary mechanism is *bremsstrahlung*, in which the Coulomb scattering of two particles produces an acceleration which in turn gives rise to the emission of radiation. The largest accelerations are suffered by the lightest particles (electrons), but because the electrons and ions are approximately in thermal equilibrium, any loss by the electrons is felt also by the ions, which are then less energetic and less successful in penetrating the Coulomb barrier. The power radiated by an electron experiencing an acceleration $a$ is

$$P = \frac{e^2 a^2}{6\pi\epsilon_0 c^3} \qquad (14.13)$$

If the electron is at a distance $r$ from an ion of charge $Z$, the acceleration is

$$a = \frac{F}{m_e} = \frac{Ze^2}{4\pi\epsilon_0 m_e r^2} \qquad (14.14)$$

If $\tau$ is the characteristic time during which the ion and electron interact, then the number of ions encountered at a distance $r$ will be (see Figure 14.6) $(n)(v_e\tau)(2\pi r\,dr)$, where $n$ is the density of positive ions. Thus

$$dP = \frac{e^2 n}{6\pi\epsilon_0 c^3} \frac{Z^2 e^4 v_e \tau (2\pi r\,dr)}{(4\pi\epsilon_0)^2 m_e^2 r^4} \qquad (14.15)$$

where we let $dP$ be the contribution to the total power from electrons scattered at impact parameters between $r$ and $r + dr$. The characteristic interaction time $\tau$ can be estimated as $r/v_e$ and thus

$$dP = \frac{4\pi e^6 Z^2 n}{3(4\pi\epsilon_0)^3 m_e^2 c^3} \frac{dr}{r^2} \qquad (14.16)$$

Integrating from $r_{min}$ to $r_{max}$ gives the total power radiated by a single electron, and multiplying the result by the density of electrons $n_e$ gives the power per unit volume radiated by the plasma. We can take $r_{max} \sim \infty$ and for $r_{min}$ it is tempting to try the distance of closest approach, which for 10 keV electrons turns out to be $144Z$ fm. If we calculate the quantum mechanical uncertainty in the electron's position, taking $\Delta p \sim p \approx 100$ keV/$c$, then $\Delta x$ is of the order of 2000 fm. We therefore cannot specify $r_{min}$ as precisely as $144Z$ fm, and we should take as a

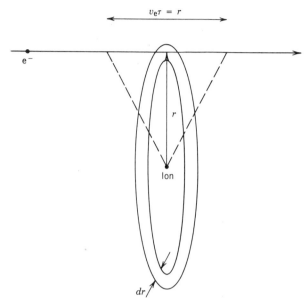

**Figure 14.6** Geometry for estimating bremsstrahlung losses. The interaction time $\tau$ is approximately $r/v_e$ from the portion of the path where the Coulomb force has the greatest effect. All electrons that pass through the ring of radius $r$ and thickness $dr$ suffer similar losses.

better estimate $r_{min} = \hbar/m_e v_e$. Carrying out the integration, the power per unit volume radiated in bremsstrahlung becomes

$$P_{br} = \frac{4\pi nn_e Z^2 e^6 v_e}{3(4\pi\epsilon_0)^3 m_e c^3 \hbar} \tag{14.17}$$

Continuing the estimate, we put for $v_e$ the velocity corresponding to the mean kinetic energy of the Maxwell-Boltzmann distribution, $v_e \simeq \sqrt{3kT/m_e}$. Evaluating all the numerical coefficients, the final estimate is

$$P_{br} = 0.5 \times 10^{-36} Z^2 nn_e (kT)^{1/2} \ \text{W/m}^3 \tag{14.18}$$

where $kT$ is in keV. The reaction rate for fusion reactions is $n_1 n_2 \langle \sigma v \rangle$, where $n_1$ and $n_2$ are the densities of the two kinds of fusing ions; if there is only one kind of ion, as in D-D fusion, the product $n_1 n_2$ should be replaced with $\frac{1}{2}n^2$, where the factor of $\frac{1}{2}$ corrects for double counting of reactions. Comparing Equation 14.18 with the fusion rates in Figure 14.7, we see that there is a temperature at which the fusion output will exceed the bremsstrahlung loss, which is of the order of 4 keV for D-T and 40 keV for D-D reactions. This suggests the superiority of the choice of D-T for fuel. Note also that the bremsstrahlung losses increase as $Z^2$; therefore fusion reactions using nuclei other than hydrogen have substantially greater bremsstrahlung losses as well as generally smaller reaction rates in the keV region (because of the Coulomb barrier).

We therefore choose to operate our fusion reactor at a temperature where the power gain from fusion exceeds the bremsstrahlung losses. Other radiation losses, including synchrotron radiation from charged particles orbiting about magnetic

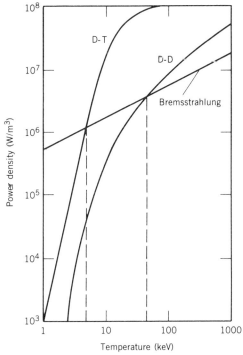

**Figure 14.7** Comparison of bremsstrahlung losses with power outputs of D-D and D-T reactions, assuming an ion density of $10^{21}/m^3$. The vertical lines show the temperatures above which fusion power exceeds bremsstrahlung losses.

field lines, can also be neglected. The fusion reactor will have a net energy gain if the energy released in fusion reactions exceeds the radiation losses *and* the original energy investment in heating the plasma to the operating temperature. If we operate at a temperature above 4 keV, even at 10 keV, the D-T fusion gain is greater than the radiation loss, and we can neglect the loss in energy to radiation. The energy released per unit volume from fusion reactions in the plasma is

$$E_f = \tfrac{1}{4}n^2\langle \sigma v \rangle Q\tau \qquad (14.19)$$

where we assume that the densities of D and T are each equal to $\tfrac{1}{2}n$ (so that the total $n$ is equal to $n_e$), $Q$ is the energy released per reaction (17.6 MeV for D-T), and $\tau$ is the length of time the plasma is confined so that reactions can occur. Note that Equation 14.19 is of the form: reactions per unit time (reaction rate) per unit volume times energy release per reaction times the time during which reactions take place.

The thermal energy per unit volume needed to raise the ions and electrons to temperature $T$ is $\tfrac{3}{2}nkT$ (for the ions) and $\tfrac{3}{2}n_e kT$ (for the electrons). With $n = n_e$, the thermal energy is

$$E_{th} = 3nkT \qquad (14.20)$$

Let us review the sequence. We supply an energy of $E_{th}$ to heat the plasma, then if we are able to confine it for a time $\tau$ we can extract fusion energy $E_f$. The

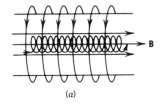

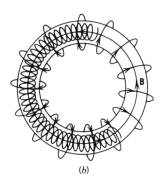

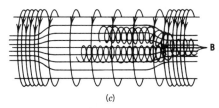

**Figure 14.8** (a) Confinement by uniform axial magnetic field. The field **B** is established by the large current-carrying coils. The particles spiral about **B**. (b) In a toroidal geometry, the particles follow the magnetic field lines as they spiral, but there is a gradual drift toward the outer wall. (c) In a magnetic mirror, the particles again follow the field lines, but are reflected from the high-field region.

reactor shows a net energy gain if

$$E_f > E_{th}$$

$$\tfrac{1}{4} n^2 \langle \sigma v \rangle Q \tau > 3nkT \tag{14.21}$$

or

$$n\tau > \frac{12kT}{\langle \sigma v \rangle Q} \tag{14.22}$$

For an operating temperature of 10 keV for the D-T reaction, $\langle \sigma v \rangle \sim 10^{-22}$ m$^3$/s and thus $n\tau > 10^{20}$ s/m$^3$. This estimate of the minimum necessary product of ion density and confinement time is called the *Lawson criterion* and represents the goal of reactor designers. It will, of course, be different for different operating temperatures and for different fuels. For D-D reactions, Figure 14.7 shows that 10 keV is too low an operating temperature (bremsstrahlung losses are too great). Operating at $kT = 100$ keV gives $\langle \sigma v \rangle \simeq 0.5 \times 10^{-22}$ and thus $n\tau > 10^{22}$ s/m$^3$. We therefore need a hundredfold increase in the density of ions or the confinement time or some combination of both to gain energy in a D-D reactor.

Let's turn now to an examination of the basic reactor types and see how close they come to meeting the Lawson criterion. The simplest magnetic confinement is a uniform magnetic field—charged particles spiral about the field direction, as indicated in Figure 14.8. This is sufficient to confine the particle in only two directions. To prevent the loss of particles along the axis, there are two solutions shown: we can form a torus, thus keeping the spiral in a ring, or we can form a high density of magnetic field lines which reflects the particles back into the

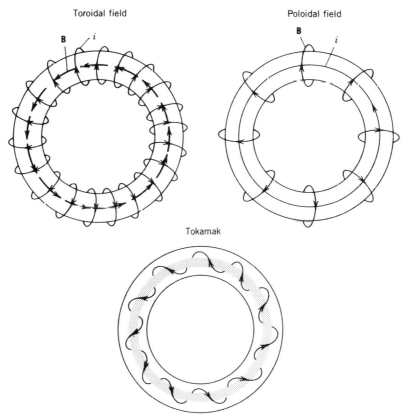

**Figure 14.9** Principle of tokamak method of magnetic confinement. A toroidal field is produced by a winding of coils, and a poloidal field is produced by an axial current; these two fields are combined in the tokamak design. (The current-carrying coils are not shown.) The resulting field lines form a helix, through which the ions can travel in closed orbits.

low-field region and is hence known as a *magnetic mirror*. In any real toroidal winding, the field is weaker at larger radii, and thus as a particle spirals it sees a region of lower field which lets the spiral radius become larger and lets the particle approach the outer wall. To reduce this effect, a magnetic field component along the surface of the toroid is introduced; this is called a *poloidal* field. It can be achieved using a set of external coils, as illustrated in Figure 14.9, or by passing a current along the axis of the toroid through the plasma itself. The current serves the dual purpose of heating the plasma and confining the particles. The basic design is called a *tokamak* after the Russian acronym for the device.

The tokamak is at present one of the two most promising candidates for the basic design of a fusion power reactor. In actual tokamak developmental facilities, such as the one illustrated in Figure 14.10, the poloidal field is provided not by a set of external coils (which are used in a different device, known as a stellarator), but by a current induced in the plasma itself by a set of external windings that function in essence as the primary of a transformer. The ohmic heating resulting from the current (which is of the order of a few MA) also helps

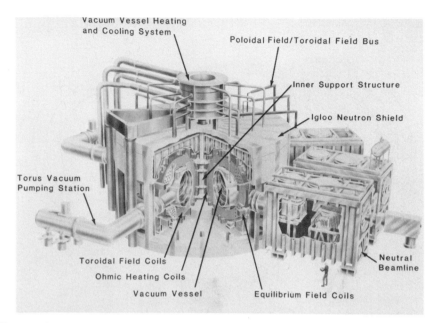

**Figure 14.10** The Tokamak Fusion Test Reactor (TFTR). The radius of the vacuum chamber is about 2 m. The ohmic heating coils produce a plasma current of $2.5 \times 10^6$ A. Neutral-beam heating of about 10 MW contributes to the plasma. Drawing courtesy Princeton Plasma Physics Laboratory.

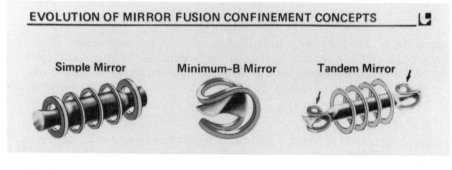

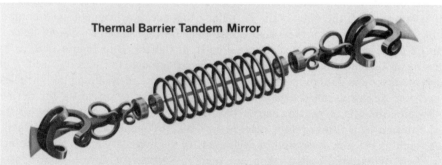

**Figure 14.11** Basic concepts of linear mirror devices. The bottom drawing shows the combination of devices that would be used in a design such as that shown in Figure 14.12. Courtesy Lawrence Livermore National Laboratory, University of California, and U.S. Department of Energy.

to heat the plasma. The use of transformers requires the tokamak to operate in a pulsed mode, which is a fundamental limitation on the operation of the device; current pulses in the present generation of tokamaks have a duration of the order of 1 s.

Additional heating must be provided to raise the plasma to temperatures of the order of 10–100 keV. Two methods under investigation are radiofrequency (rf) heating and neutral beam injection (NBI). The rf waves radiated into the plasma drive the electrons and induce toroidal currents that heat the plasma. In the NBI technique, a beam of H or D ions is accelerated to energies of 10–100 keV and then neutralized (for example, by charge-exchange reactions in passing through a cell of neutral H or D atoms). The neutral atoms can pass undeviated through the magnetic fields of the tokamak and into the plasma, where they rapidly lose energy to the plasma through Coulomb scattering from ions and electrons.

Auxiliary heating systems of tens of MW will be required to achieve ignition of the plasma, after which the 3.5-MeV $\alpha$'s resulting from D-T fusion will provide the necessary heat to sustain the reactions. The charged $\alpha$ particles are confined to the plasma by the magnetic fields and eventually lose energy to the plasma through collisions.

Magnetic mirror machines generally try to make a three-dimensional "magnetic well" to confine the plasma because the simple linear mirror device is not effective in confining the plasma if it tries to bulge toward the walls in the low-field region. A so-called "minimum-B" configuration consists of a single coil shaped like the seams on a baseball or tennis ball as indicated in Figure 14.11. A combination of the simple mirror principle and the minimum-B coils to trap particles in tandem (called a *tandem mirror* device) is illustrated in Figure 14.11.

A design for a tandem mirror facility is shown in Figure 14.12. The actual operating parameters are similar to those of the tokamak. Auxiliary heating by rf waves and NBI are essential to raise the temperature of the plasma, as in the tokamak design.

Inertial confinement fusion takes a very different approach. A tiny pellet containing deuterium and tritium is suddenly struck with an intense laser pulse that both heats the pellet and compresses it to high density. The goal of this technique is to achieve densities and temperatures that are high enough that fusion can occur before the pellet simply expands and blows apart. In the design for a power plant based on this process, it is anticipated that many (10–100) pellets per second would be used, and the cycle of fuel injection, compression, ignition, and power stroke is somewhat similar in principle to the cycle of the internal combustion engine.

To make a crude estimate of the requirements for such a device, let's note that the time necessary for the compressed pellet to blow apart will be determined by the speed of propagation of mechanical waves in the medium, which will be of the same order as the mean thermal speed of the particles in the medium. (For an ordinary solid at room temperature, the mean thermal velocities are of the order of $10^3$ m/s; the speed of mechanical waves, sound for instance, is of the same order.) At $kT \sim 10$ keV, the mean thermal speed is about $10^6$ m/s. If we consider a pellet compressed to a diameter of 0.1–1 mm, it could be expected to blow apart in about $10^{-9}$–$10^{-10}$ s. Applying Lawson's criterion for a D-T mixture, if the confinement time were as short as $10^{-9}$–$10^{-10}$ s, we would need a

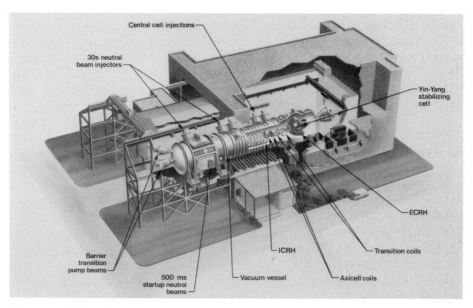

**Figure 14.12** Design of Magnetic Fusion Test Facility (MFTF-B). The cutaway portion shows the mirror devices drawn in Figure 14.11. Courtesy Lawrence Livermore National Laboratory, University of California, and U.S. Department of Energy.

density of at least $10^{29}$–$10^{30}$/m$^3$, which is two orders of magnitude greater than ordinary liquid or solid densities for hydrogen.

To heat a spherical pellet of diameter 1 mm to a mean thermal energy of 10 keV per particle, the total thermal energy that must be supplied is about

$$E_{th} \sim \tfrac{4}{3}\pi(0.5 \text{ mm})^3 \times 10^{29} \text{ m}^{-3} \times 10^4 \text{ eV} \approx 10^5 \text{ J}$$

That is, we must supply an energy of the order of $10^5$ J in about $10^{-9}$ s for a net power of $10^{14}$ W! Our estimate for the power that must be supplied is surely too low, for a large portion of the energy supplied to the pellet is sure to be absorbed in "boiling off" of surface particles, so we probably should increase by another order of magnitude the estimate of the total energy we must supply actually to heat and compress the core of the pellet. Furthermore, lasers are notoriously inefficient devices for converting electrical energy into radiation; 10% is probably the best that can be hoped for, but 1% is more typical. Therefore, the electrical power needed for the lasers may approach $10^{17}$ W. Fortunately, that power need be provided only for short intervals of time, but nevertheless the magnitude of $10^{17}$ W is a staggering number—compare it for instance with the entire electrical generating capacity of the United States, which is of the order of $10^{12}$ W!

To run an inertial confinement reactor at a net energy gain, it is apparent that we must exceed considerably the Lawson criterion. We should hope to achieve compressions perhaps approaching 1000 times ordinary density, and to operate well above 10 keV thermal energy per particle.

The sequence of processes in laser-driven fusion might be as follows: A pellet is injected into the machine and is simultaneously struck from many directions by an intense laser pulse. The outer layer of the solid pellet is immediately vaporized and forms a plasma, which continues to absorb the laser radiation. The plasma itself is unconfined and it rapidly "blows off" or *ablates*, which (by Newton's third law) drives a compressional shock wave back into the remaining core of the pellet; this shock wave compresses and heats the core to the point at which thermonuclear ignition can occur at the highest density region near the center. The $\alpha$ particles resulting from the fusion rapidly lose their energy in collisions with ions in the dense fuel. This contributes additional heating, and the thermonuclear burn propagates outward, finally blowing the pellet apart and ending the reaction.

The requirements on lasers imposed by the break-even energy condition are severe, and certainly beyond the capability of present lasers. Particularly serious is the low efficiency (1–10%) for converting electrical energy into radiation. Therefore, alternative approaches to inertial confinement fusion are being explored using beams of charged particles instead of lasers. The particles must deposit their energy in a distance which is of the order of the radius of the pellet, thus contributing the maximum amount of the heating and shock wave. Taking a pellet of about 100 times solid density with a diameter of 0.1–1 mm, the range for the particles should be about 0.1 g/cm². From Figure 7.2 for protons in dense matter (take Pb, for example), a particle energy of 5 MeV is required. Adopting the previous estimate of $10^5$ J as the thermal energy needed to heat the fuel, we

**Table 14.1**  Magnetic Confinement Fusion Devices

| Size[a] (m) | $B_{max}$ (T) | Device | Location | Type | Year | $n\tau$ ($10^{20}$ s/m³) | $kT$ (keV) |
|---|---|---|---|---|---|---|---|
| 1.3 | 3.5 | PLT (Princeton Large Torus) | Princeton (USA) | Tokamak | 1978 | 0.4 | 10 |
| 2.5 | 5.2 | TFTR (Tokamak Fusion Test Reactor) | Princeton (USA) | Tokamak | 1984 | 0.15 | 6 |
| 2.8 | 2.8 | JET (Joint European Torus) | England | Tokamak | 1983 | 0.5 | 5 |
| 0.64 | 12 | Alcator-C | MIT (USA) | Tokamak | 1983 | 0.8 | 1.7 |
| 3.0 | 5.0 | JT-60 | Japan | Tokamak | 1986 | 0.2–0.6 | 5–10 |
| 2.4 | 3.5 | T-20 | USSR | Tokamak | 1986 | 1.0 | 7–10 |
| 5 | 0.1 | TMX (Tandem Mirror Experiment) | Livermore (USA) | Tandem mirror | 1980 | 0.01 | 10 |
| 16 | 1.0 | MFTF-B (Mirror Fusion Test Facility) | Livermore (USA) | Tandem mirror | 1988 | 0.1 | 15 |

[a] For tokamaks, the size entry gives the major radius of the toroid; for the linear mirrors, the size indicates the length of the central chamber.

would require

$$10^5 \text{ J} \times \frac{1}{1.6 \times 10^{-19} \text{ J/eV}} \times \frac{1 \text{ proton}}{5 \times 10^6 \text{ eV}} \simeq 10^{17} \text{ proton}$$

in a pulse of $10^{-9}$ s duration; that is, we would need a current of the order of $20 \times 10^6$ A! Using electrons instead of protons, Figures 7.3 and 7.4 indicate an energy of about 0.5 MeV, and to obtain $10^{14}$ W we would need a current of $200 \times 10^6$ A. If we use beams of heavy ions (U, for instance), we can use Equation 7.6 to estimate that a range of 0.1 g/cm$^2$ for U would be equivalent to a range of protons of about 3 g/cm$^2$. Figure 7.2 does not show ranges as high as 3 g/cm$^2$, but we can extrapolate the line for heavy matter to obtain an energy of about 35 MeV for protons to have such a range, and since Equation 7.6 is based on a comparison of ranges of different particles with the same velocity, the energy of the U would be about $240 \times 35 = 8$ GeV. The necessary current would be about $10^4$ A. Although the energies we have estimated for the beams of electrons, light ions (protons) and heavy ions (U) are all quite reasonable and well within present accelerator technology, the currents are far beyond present accelerators, which are limited to mA. Research therefore is now in progress to solve the beam transport and focusing problems that will arise at these high currents, and it is hoped that charged-particle beams will offer a reasonable alternative to lasers.

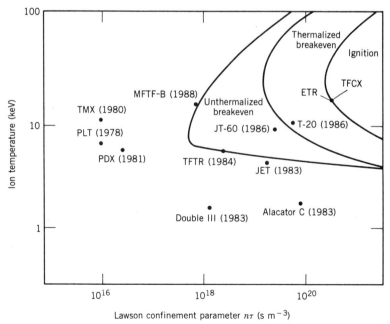

**Figure 14.13** Approach to the Lawson criterion of various past and future magnetic confinement fusion devices. Some of these are listed in Table 14.1. The region of thermalized breakeven corresponds to exceeding the Lawson criterion; ignition indicates the region in which heat from fusion products will sustain the reaction. ETR (Engineering Test Reactor) and TFCX (Toroidal Fusion Core Experiment) are planned or proposed facilities.

As of this writing, there is a great deal of research and development activity throughout the world on the various aspects of controlled fusion we have briefly outlined. The entire area is extremely fast-moving, and new discoveries and achievements are rapidly announced. Rather than try to project into a very uncertain future, let's instead look at the steady progress that has been made up to now at approaching the Lawson criterion or providing the necessary ignition power. Table 14.1 lists some of the magnetic confinement devices that have been built and tested and some that are expected to become operational during the

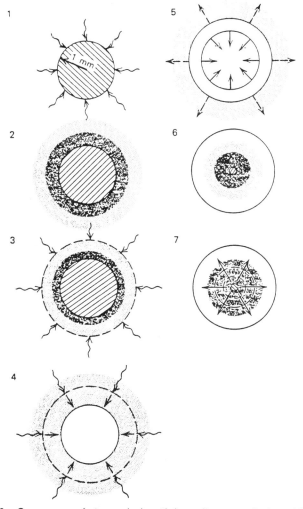

**Figure 14.14** Sequence of stages in inertial confinement fusion: (1) irradiation of fuel pellet by lasers; (2) formation of plasma atmosphere; (3) additional laser beam absorption in atmosphere; (4) ablation and resulting imploding shock wave; (5) shock wave compressing core; (6) ignition of core; (7) burn propagating outward. From J. J. Duderstadt and G. A. Moses, *Inertial Confinement Fusion* (New York: Wiley, 1982).

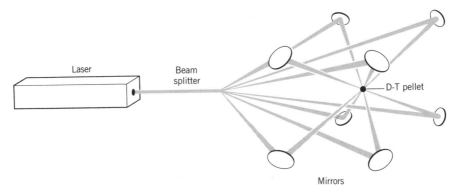

**Figure 14.15** Irradiation of D-T pellet by multiple laser beams.

next several years. Figure 14.13 shows the steady march toward satisfying the Lawson criterion that has been made in the past decade.

By comparison, research in inertial confinement techniques has not made nearly as much progress, although the physics is perhaps better understood than in the case of magnetic confinement. The major technological difficulty has been in achieving the necessary laser power. Figure 14.14 shows the sequence of events that occur in the pellet, and Figure 14.15 shows a schematic diagram of how the pellet might be exposed to the laser radiation.

**Figure 14.16** The Nova inertial confinement fusion facility of the Lawrence Livermore Laboratory. The photo shows the target chamber and 5 of the 10 laser beam tubes. Courtesy Lawrence Livermore National Laboratory, University of California, and U.S. Department of Energy.

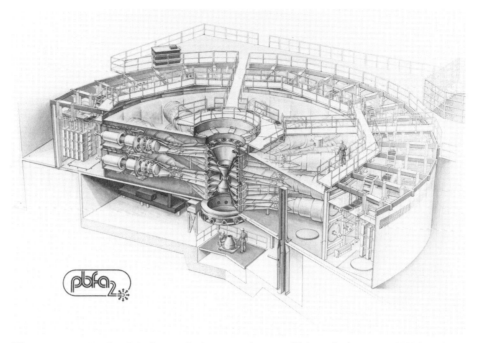

**Figure 14.17** Particle beam fusion accelerator. Thirty-six beams of lithium ions are focused on a D-T target, in a pulse of $10^{14}$ W of duration 20 ns. Because the efficiency of absorption of the beam's energy is nearly 100%, this device may be able to achieve ignition of the D-T fuel. Drawing courtesy Sandia National Laboratory.

At the Lawrence Livermore Laboratory in California, the Nova laser (Figure 14.16) provides 10 beams to strike the target. The laser itself is based on Nd, in which the primary lasing transition is 1060 nm, in the infrared region. Absorption of the laser light is enhanced by going to shorter wavelengths, so the lasing medium incorporates a crystal of potassium dihydrogen phosphate (KDP), which acts as a nonlinear optical medium and produces higher harmonics $(\lambda/2, \lambda/3, \lambda/4, \ldots)$ of the incident light. There is a particularly high efficiency for converting to the first harmonic at 530 nm. The Nova system is expected to produce 100-kJ nanosecond pulses, for a power of order 100 TW.

Valuable basic research has been and will be done with the Nd laser systems, but they are not suitable for practical fusion devices because the Nd glass must cool for about one hour after a laser pulse (thus repetition rates of 10–100/s are impossible) and the lasers themselves are quite inefficient, about 0.1%. Research is presently underway to study the use of $CO_2$ gas lasers, which permit high repetition rates and high laser efficiency ($\sim 10\%$). A disadvantage is the long wavelength (10.6 $\mu$m), where absorption by the pellet's plasma surface layer to create ablation is less efficient.

A particle-beam test device is illustrated in Figure 14.17. Thirty-six individual light-ion (or electron) accelerators are arranged around the pellet injection region. The latest development of this device will focus beams of 100 TW on the pellet, with about 4 MJ of ion energy in a 35-ns pulse. The time necessary for charging

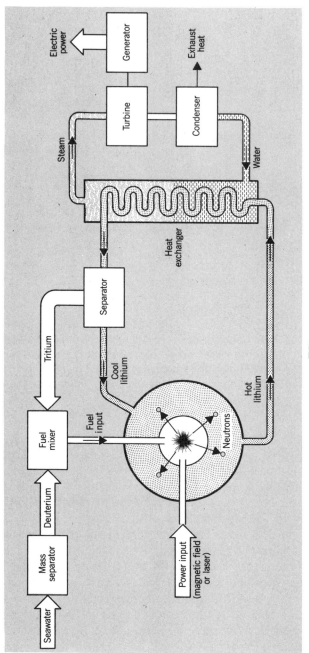

**Figure 14.18** Design for a fusion power plant.

the enormous capacitative energy storage devices may limit the repetition rate of this system.

This introduction to controlled fusion devices is by no means complete but it represents a realistic summary of the state of the art as of late 1986. Research is taking place simultaneously on many fronts, and while many magnetic confinement and inertial confinement devices are well within the last order of magnitude (or even factor of 2) of exceeding the Lawson criterion, there is as yet no single leading candidate for the basic design of a controlled fusion reactor. Despite the many technological problems, progress in the last decade has been considerable, and it is not overly optimistic to extrapolate to the break-even point within the next decade or two.

A proposed design for a fusion power reactor is shown in Figure 14.18. The D-T reaction gives a neutron and an $\alpha$ particle, which deposits its energy within the fuel and contributes to the heating of the D-T mixture. The neutron carries 80% of the energy (14.1 MeV) and can be captured by a blanket of lithium through the reactions $^6\text{Li} + \text{n} \rightarrow {}^4\text{He} + {}^3\text{H}$ or $^7\text{Li} + \text{n} \rightarrow {}^4\text{He} + {}^3\text{H} + \text{n}$. The energetic $^4\text{He}$ and $^3\text{H}$ then deposit their energy in the lithium as heat. The hot liquid lithium then drives a steam generator which is used to run a turbine. The $^3\text{H}$ is extracted to produce fuel for the reactor.

## 14.5 THERMONUCLEAR WEAPONS

Once the first thermonuclear explosives were detonated by the United States and the USSR in the early 1950s, the fission explosives in the strategic arsenals of both nations were soon replaced with thermonuclear weapons of explosive energy 2–3 orders of magnitude greater than the early fission weapons. The kiloton-range explosives that devastated Hiroshima and Nagasaki have given way to megaton explosives. It gives but small comfort to note that the effective destructive capability of a weapon increases only as the square or cube root of its explosive energy.

Most of the details of the construction of thermonuclear weapons are classified, but enough is publicly known that we may make some general observations about their operation. As a starting point, all of the previous discussions regarding the ignition temperature of fusable fuel remain valid, and the only rapid and mobile source capable of achieving such temperatures is a nuclear explosion. Thus a fusion weapon includes a fission explosive as an initiator. The radiation from the fission explosion is responsible for heating and compressing the thermonuclear fuel.

The first thermonuclear explosive used a liquified fuel mixture, which required a cumbersome refrigeration apparatus to achieve and maintain the low temperatures necessary to liquify hydrogen. Present-day weapons use solid lithium deuteride as fuel, made with separated isotope $^6\text{Li}$. The neutrons released from the primary fission explosion (and the subsequent fusions) convert the $^6\text{Li}$ into tritium:

$$^6\text{Li} + \text{n} \rightarrow {}^3\text{H} + {}^4\text{He} \qquad (Q = 4.78 \text{ MeV})$$

Even for low-energy neutrons, the tritium carries enough energy (2.7 MeV) to penetrate easily the D-T Coulomb barrier and initiate the fusion reactions.

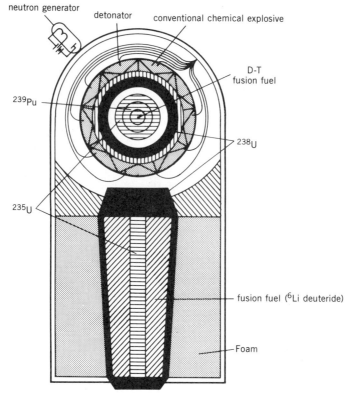

**Figure 14.19** Schematic diagram of thermonuclear explosive. The detonation of the chemical explosive compresses a $^{238}$U shell about $^{235}$U and $^{239}$Pu, driving them into criticality and initiating a fission explosion. The fission reaction is "boosted" by a small amount of D-T fuel at the center of the sphere; D-T fusion provides additional neutrons for the fission explosion. The X rays and $\gamma$ rays from the fission explosion vaporize the polystyrene foam which compresses the $^{238}$U tamper about the main fusion fuel and also heats the fusion fuel to its ignition temperature, beginning the thermonuclear reaction. The fast neutrons released in the fusion reactions cause fissions in $^{235}$U and $^{238}$U, increasing the total yield of the weapon. The diagram and description are from an article by Howard Morland published in the November 1979 issue of *The Progressive*; a more complete version can be found in H. Morland, *The Secret That Exploded* (New York: Random House, 1981).

Somewhat similar to the inertial confinement technique of controlled fusion, the heat, neutrons, and $\alpha$ particles contribute to sustaining the reaction until the expansion of the fuel terminates the reaction.

The fast neutrons released in the fusion can be used to add additional energy to the explosive by surrounding the fusion fuel with a casing of $^{238}$U, which fissions with fast neutrons. The operation and energy release in a thermonuclear weapon are thus dependent on a fission-fusion-fission cycle. About half the yield of a typical strategic weapon may come from the final fission processes.

A highly schematic diagram of a thermonuclear weapon is shown in Figure 14.19 and from the photograph of an actual weapon housing shown in Figure 14.20 it is not difficult to infer the relative size and placement of the components.

**Figure 14.20** The casing of a thermonuclear weapon. From H. Morland, *The Secret That Exploded* (New York: Random House, 1981).

The destructive effects of a 1–10-megaton thermonuclear weapon can be anticipated from the discussion of fission weapons in Chapter 13. The radius within which there would be virtually complete destruction by blast effects and firestorm varies roughly as the cube root of the weapon yield, and is thus an order of magnitude beyond the 1 km distance characteristic of a typical fission weapon. Over a radius of perhaps 10 km, a single weapon produces essentially complete destruction.

Reducing the fission yields (by eliminating the $^{238}$U casing) eliminates most of the radioactive fission products from the debris of the explosion and creates a relatively "clean" explosion, that is, free of the long-term effects of fallout and surface radioactivity associated with fission products. Because the yield of an explosion is the primary goal of a strategic weapon, it is believed that most of the strategic weapons in the U.S. arsenal are "dirty" weapons of the fission-fusion-fission variety.

There are also relatively low-yield battlefield nuclear explosives, of the tactical category. Among these is the enhanced radiation or neutron weapon. By eliminating the $^{238}$U casing, these small weapons (in the kiloton range) produce an intense burst of neutrons, the object of which is to deliver a lethal dose of radiation to an army advancing in armored vehicles, such as tanks. It is estimated that a 1-kiloton neutron bomb will subject personnel shielded by armor plate to a radiation dose of about $10^3$ rads over a radius of 1 km. Such a dose will cause death within days. The enhanced radiation weapons were designed to penetrate armor and thus halt an advancing army, particularly one that is invading in the defender's home territory (in which case minimizing blast damage is highly desirable; the use of megaton explosives would destroy the country in order to save it).

The effects of fusion weapons on population and structures are similar to those of fission weapons, discussed in Section 13.9, with appropriate scale factors introduced to account for differences in explosive yields. One additional effect not considered in Chapter 13 is the electromagnetic pulse (EMP). The prompt $\gamma$ rays and X rays released in the explosion interact with air molecules (through Compton scattering and ionization) to create a large current of negative electrons flowing outward from the point of the explosion. These electrons are accelerated by the Earth's magnetic field and give rise to a traveling electromagnetic wave in the form of a pulse. An explosion several hundred kilometers above the center of the United States would be within the line of sight of the entire United States and would expose the country to electric fields of the order of $10^4$ V/m for a 1-megaton blast. Such a pulse could be destructive of electrical power networks and communications grids necessary for decisive action in time of war.

The scenario of an all-out nuclear war is thus particularly horrible—perhaps 1000 or more strategic warheads in the megaton range explode within a few minutes over the entire United States. Perhaps 50% of the U.S. population would die in such an attack, from the combination of blast, heat, fire, and radiation. Blast damage and firestorms would lay waste to most cities, and EMP effects would destroy electrical power and communication facilities. The remaining population would have to contend with the long-term effects of fallout and with the disease and starvation that are likely to follow from the general destruction. Recent calculations indicate that the dust and smoke created from a general exchange of nuclear weapons will circulate throughout the atmosphere, blocking enough sunlight that agriculture would become impossible and major climate changes would bring about a perpetual "nuclear winter." Surely all petty disagreements among nations must become insignificant when measured against such outcomes of a general nuclear war.

The present strategic arsenal of the United States includes 1000 Minuteman land-based intercontinental ballistic missiles (ICBMs), approximately half of which carry a single warhead (1.5 megatons) and half of which carry three multiple independently targeted re-entry vehicles (MIRVs) with a total yield of 0.5–1 megaton; 33 Poseidon and Trident submarines, each with 16 or 24 missiles having 8 or 10 MIRV warheads of about 50–100 kilotons each; and 332 B-52 bombers, each carrying at least four warheads of 1 megaton each. The total number of strategic warheads in the U.S. arsenal is 2152 in ICBMs (1572 megatons), 4960 in submarine-launched ballistic missiles (344 megatons), and 2698 in bombers, some equipped with air-launched missiles (1621 megatons); the totals are 9810 warheads (3537 megatons). Equivalent totals for the Soviet Union are 7741 warheads (6618 megatons). An additional 10,000–20,000 small (kiloton range) tactical weapons are also held by each side. The total megatonnage is thus of order 10,000!

Let's attempt to put this staggering figure in perspective. The total explosive energy is of the order of $10^{10}$ tons of TNT; the population of the Earth is of order $5 \times 10^9$, and thus we are each allotted a share of about 2 tons of TNT, roughly a cubic meter. Every person on Earth is thus living precariously with his or her personal cubic meter of high explosive. It is apparent to any reasonable thinker that this silly overkill capability compromises everyone's security, and the only sensible course is a reduction in the number of weapons and control of their proliferation. Achieving this reduction is a major challenge facing both physicists and politicians in the next decade.

## REFERENCES FOR ADDITIONAL READING

Among the references for the basic physics of fusion processes are those that deal also with the technology of controlled fusion and the design of fusion reactors: S. Glasstone and R. H. Lovberg, *Controlled Thermonuclear Reactions* (Princeton, NJ: Van Nostrand, 1960); W. M. Stacey, Jr., *Fusion: An Introduction to the Physics and Technology of Magnetic Confinement Fusion* (New York: Wiley, 1984); H. Motz, *The Physics of Laser Fusion* (London: Academic, 1979); J. J. Duderstadt and G. A. Moses, *Inertial Confinement Fusion* (New York: Wiley, 1982); M. O. Hagler and M. Kristiansen, *An Introduction to Controlled Thermo-nuclear Fusion* (Lexington, MA: Heath, 1977); *Fusion*, edited by Edward Teller (New York: Academic, 1981).

For reviews of fusion reactor research, see: F. L. Ribe, *Rev. Mod. Phys.* **47**, 7 (1975); R. F. Post, *Ann. Rev. Nucl. Sci.* **20**, 509 (1970); D. Keefe, *Ann. Rev. Nucl. Part. Sci.* **32**, 391 (1982); K. I. Thomassen, *Ann. Rev. Energy* **9**, 281 (1984).

An interesting nontechnical description of the social and political background of the controlled fusion research program is Joan Lisa Bromberg, *Fusion: Science, Politics, and the Invention of a New Energy Source* (Cambridge, MA: MIT Press, 1982).

Several general popular-level articles on controlled fusion have appeared in *Scientific American*: B. Coppi and J. Rem, "The Tokamak Approach in Fusion Research" (July 1972); J. L. Emmett, J. Nuckolls, and L. Wood, "Fusion Power by Laser Implosion" (June 1974); W. C. Gough and B. J. Eastland, "The Prospects of Fusion Power" (February 1971); M. J. Lubin and A. P. Fraas, "Fusion by Laser" (June 1971); Harold P. Furth, "Progress Toward a Tokamak Fusion Reactor" (August 1979); R. W. Conn, "The Engineering of Magnetic Fusion Reactors" (October 1983); R. S. Craxton, R. L. McCrory, and John M. Soures, "Progress in Laser Fusion" (August 1986).

References to work on fusion in stars will be presented in Chapter 19.

In contrast to the considerable body of literature on the technology of fission weapons, there is little published on fusion weapons. Perhaps the best-known work, at least in terms of the publicity it generated, is: Howard Morland, *The Secret That Exploded* (New York: Random House, 1981). There is much more written about the effects of nuclear weapons and about the political issues that relate to the reduction of weapon inventories. A summary of the entire literature in this area can be found in the resource letter "Physics and the Nuclear Arms Race," by D. Schroeer and J. Dowling, *Am. J. Phys.* **50**, 786 (1982). The American Association of Physics Teachers has published a collection of reprints, including the above resource letter and Morland's original article, in *Physics and the Nuclear Arms Race: Selected Reprints* (1984). Particularly worthwhile for its technical detail (and included in the above collection) is H. L. Brode, *Ann. Rev.*

*Nucl. Sci.* **18**, 153 (1968). More recent work includes: the entire March 1983 issue of *Physics Today*, which features articles on the effects of nuclear weapons and descriptions of present arsenals; A. A. Broyles, *Am. J. Phys.* **50**, 586 (1982); D. Schroeer, *Science, Technology, and the Nuclear Arms Race* (New York: Wiley, 1984), a well-written and comprehensive (but introductory) work treating weapon technologies, delivery systems, strategies, and arms control; and *Physics, Technology and the Nuclear Arms Race*, edited by D. W. Hafemeister and D. Schroeer (New York: American Institute of Physics, 1983), a collection of articles by experts in the areas of the technology, strategy, and politics of nuclear weapons.

## PROBLEMS

1. Calculate (1) the temperature necessary to overcome the Coulomb barrier and (2) the fusion energy release in a gas of: (a) $^{16}$O; (b) $^{12}$C; (c) $^{24}$Mg; (d) $^{14}$N; (e) $^{10}$B.

2. (a) Compute the $Q$ values for the basic D-D reactions: $^2$H $+ ^2$H $\rightarrow ^3$He $+$ n and $^2$H $+ ^2$H $\rightarrow ^3$H $+$ p. (b) Assuming the incident particle kinetic energies to be negligible, find the energies of the outgoing particles in both reactions.

3. (a) Calculate the ratio of the D-D fusion cross sections at 100 and 20 keV from Equation 14.7 and show that it agrees with Figure 14.1. (b) Do the same for the D-T cross section.

4. (a) Estimate the shaded areas in Figure 14.2 and show that they are consistent with the D-D curve of Figure 14.3. (b) Extend the calculation upward to 100 keV and downward to 1 keV; again compare with Figure 14.3. (Duplicate the 5 and 20 keV calculations as well, so that your calculation can be normalized against that of Figure 14.2.) Remember that for D-D reactions, $E = 2E_D$; also remember that $n(E)$ includes a $T^{-3/2}$ dependence that must be included before comparisons can be made of results at different temperatures.

5. By calculating the energy released in each of the six reactions of the carbon cycle, show that the total energy release is 26.7 MeV.

6. In analogy with the carbon cycle, fusion of hydrogen to helium can occur with $^{20}$Ne as a catalyst. Suggest a sequence of reactions, similar to the carbon cycle, in which this could occur. Calculate the energy release from each reaction and evaluate the total energy release. (*Hint*: Two successive proton captures must occur in the carbon cycle, because $^{14}$N is stable and does not $\beta$ decay. This will not necessarily happen in the neon-based cycle.)

7. Calculate the threshold for neutrino capture by $^{37}$Cl. Which of the reactions of the proton-proton cycle can produce neutrinos above this threshold? Does the carbon cycle produce detectable neutrinos?

8. Making (and justifying) any necessary assumptions about conversion efficiencies, estimate the number of D-T pellets that must be irradiated each second to operate a 1000-MW generating plant.

# 15

# ACCELERATORS

The purpose of an accelerator of charged particles is to direct against a target a beam of a specific kind of particles of a chosen energy. There are many varieties of methods for accomplishing this task, all using various arrangements of electric and magnetic fields, and in this chapter we review the general features of some of the more common types of accelerators.

As an electronic device, the accelerator shares many features in common with an ordinary television picture tube. Both require a source of charged particles (electrons from a hot filament or ionized atoms from an ion source), an electric field to accelerate the particles ($10^4$ V in a TV tube and perhaps $10^7$ V in some accelerators), focusing elements to counteract the natural tendency of the beam to diverge, deflectors to aim the beam in the desired direction, a target of selected material for the beam to strike, and a chamber to house all the components in high vacuum to prevent the beam from scattering in collisions with molecules in the air.

The design of accelerators varies greatly with the purpose for which they will be used. In some cases high energy is desired, and in other cases high intensity. Accelerators for electrons (which become relativistic at reasonably low voltages and which have long ranges in matter) differ greatly from those for heavy ions (which are generally nonrelativistic and have extremely short ranges in matter). As a practical problem, the short range of heavy ions means that the entire accelerator must be one continuous vacuum for heavy ions cannot cleanly penetrate even the thinnest "window" that would separate one vacuum chamber from another.

We can somewhat loosely classify accelerators as low, medium, or high energy. Low-energy accelerators are used to produce beams in the 10–100-MeV range, often for reaction or scattering studies to elucidate the structure of specific final states, perhaps even individual excited states. Such accelerators should have accurate energy selection and should have reasonably high currents because the ultimate precision of many experiments is limited by counting statistics. The heating of targets by intense beams can be considerable, and often targets must be cooled to prevent the heat from destroying the target.

Medium-energy accelerators operate in the range of roughly 100–1000 MeV. At these energies, collisions of nucleons with nuclei can release $\pi$ mesons, and so these accelerators are often used to study the role of meson exchange in the

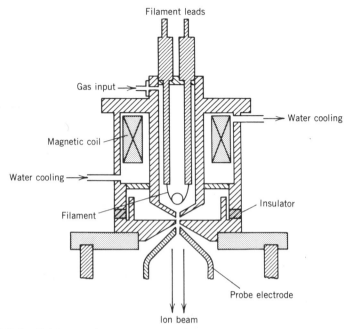

Filament leads

Gas input →

Water cooling

Magnetic coil

Water cooling →

Filament

Insulator

Probe electrode

Ion beam

**Figure 15.1** Design of basic ion source. Neutral gas atoms enter near the top and are ionized in the vicinity of a hot filament which provides an electron discharge. The magnetic field is present to focus the ions and also to concentrate the electrons leaving the filament to increase their ionizing efficiency. The beam is extracted with a high-voltage probe electrode.

nuclear force. In only a few cases are these accelerators able to resolve individual final excited states.

High-energy accelerators produce beams of 1 GeV (1000 MeV) and above. Their purpose is less to investigate nuclear structure than to produce new varieties of particles and study their properties. Here the highest possible center-of-mass energy is the primary goal, and under design at present is a new generation of machines capable of reaching beyond the level of TeV (1000 GeV or $10^6$ MeV).

The details of accelerator technology are not discussed in this chapter; instead, we present the general categories of accelerators and discuss the strengths and weaknesses of each. Before doing so, let's consider some of the equipment that is a necessary part of any accelerator facility. First comes the *ion source*, from which originates the beam of ions or electrons that is to be accelerated. In its basic operation (for ions), a gas is ionized, usually by subjecting it to an electric discharge, and the positively charged ions are extracted by acceleration toward a negative electrode at a potential of order 10 kV (see Figure 15.1). For some applications, it is desired to have a beam of *negative* ions, that is, a neutral atom that accepts an additional electron. If we pass a beam of positive ions through a neutral gas having a relatively loosely bound electron (alkali atoms, for example), there is a high probability of the positive ions capturing electrons to become negative ions. This capture is enhanced because, at an energy of order 10 keV, the ions are moving at about the same speed as the orbital electrons of the alkali vapor. Perhaps 1% of the positive ions become negatively charged, but the remaining positively charged ions can be swept out of the beam by an electric or magnetic field, leaving a relatively pure beam of negative ions.

The *beam transport* (or *beam optics*) system consists of a number of electric or magnetic devices that focus the beam and bend or deflect it along the desired path. In analogy with optics, focusing devices are often called lenses, but they consist of magnetic fields, rather than glass. Figure 15.2 shows an example of a quadrupole lens, which creates field components in the $x$ and $y$ directions of the form $B_x = by$ and $B_y = bx$, where $b$ is a constant. The beam axis is the $z$ direction, along which there is no field. The components of the Lorentz force $F = q(v \times B)$ are

$$F_x = -qv_z B_y = -qv_z bx = -kx$$
$$F_y = qv_z B_x = qv_z by = ky$$

(15.1)

Let's suppose that $b > 0$. Particles with $x$ displacement feel a restoring force that pushes them toward the $x$ axis and thus focuses the beam along the $x$ direction. In the $y$ direction, the effect is to defocus the beam. It may not appear that we gain anything from such an arrangement, but if we place two quadrupole lenses in series, with the second rotated 90° relative to the first, then along each axis ($x$ and $y$) there is both a focusing and a defocusing effect, and it can be shown that the net effect is to focus the beam. An optical analogy is shown in Figure 15.2.

Bending magnets can (like prisms in optics) change the direction of a beam and also analyze it into its components (because the radius of curvature of the path of a charged particle in a magnetic field depends on the momentum). Bending magnets can be very useful in analyzing a beam of reaction products, such as to form a secondary beam of a specific type of particle from among the debris of the reaction. Isotope separators, analogous to the mass spectrometers discussed in Chapter 3, can also be used to pick out a specific isotopic species from among the reaction products.

Targets for the accelerated beams are as varied as the uses to which the accelerator is put. For doing careful nuclear spectroscopy, such as studying specific excited states and their cross sections, it is usually desirable to have as little disturbance as possible of the incoming and outgoing beams; thus very thin targets (of order 10 μm) are used. On the other hand, if one wants to stop completely a beam of high-energy protons to create secondary particles, then thick targets (tens of cm) must be used. Both thin and thick targets must often be cooled to extract the heat deposited by the incident beam.

Finally, essential parts of any accelerator facility are the detecting and analyzing devices used to record the identity, energy, time, and direction of the reaction products. Detectors of low-energy particles and radiations were discussed in Chapter 7; devices similar in principle are used by particle physicists to study the often elusive and rare but always energetic products of high-energy reactions.

With this brief background discussion of the source, shaping, bending and analyzing of accelerator beams, let's now turn to the study of the techniques used to boost the particles to the desired energy.

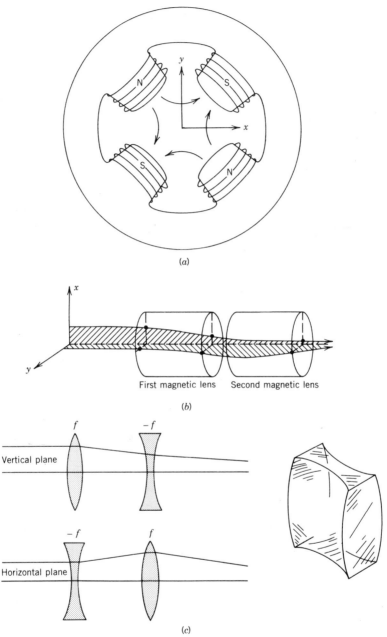

**Figure 15.2** (*a*) Cross section of quadrupole lens along beam direction. The magnetic equipotentials have the shape of hyperbolae. (*b*) Two lenses in series give a net focusing effect. (*c*) An optical analogy, using a lens that is converging in one direction and diverging in another. Two such lenses in series give a net focusing effect along both directions. Part (*c*) derived from H. Enge, *Introduction to Nuclear Physics* (Reading, MA: Addison-Wesley, 1966).

## 15.1 ELECTROSTATIC ACCELERATORS

The simplest way to accelerate a charged particle is to "drop" it through a constant potential difference $V$; if the particle has a charge $q$, it acquires a kinetic energy of $qV$. The largest potential difference that can be maintained under accelerator conditions is about $10^7$ V, and thus ions will acquire an energy in the range of 10 MeV per unit of charge. This is just the energy we need for many studies of nuclear structure, and so this type of accelerator has found wide use in nuclear physics laboratories throughout the world.

The technology of electrostatic accelerators consists entirely in establishing and maintaining a high-voltage terminal to accelerate the charged particles from the ion source. The earliest development of this type for nuclear physics applications was in 1932 by Cockcroft and Walton, who built a device that was eventually to reach a potential of 800 kV. Figure 15.3 illustrates the basic operating principle, in which capacitors are charged in parallel to a common potential, but then discharged in series; the switching between the series and parallel connections is accomplished through rectifiers.

Let the secondary voltage from the transformer be $V(t) = V_0 \sin \omega t$, where $V_0$ may be of order 100 kV. The charging of the capacitors is done through a sufficiently large load that the $RC$ time constants are large compared with the time $\omega^{-1}$ that characterizes the variation of the transformer voltage. We examine the circuit after a very long time, when the capacitors have become charged. Capacitor $C_1$ is charged to voltage $V_0$, and thus the voltage at point A varies sinusoidally between 0 and $2V_0$. With the forward conducting of rectifier $R_2$,

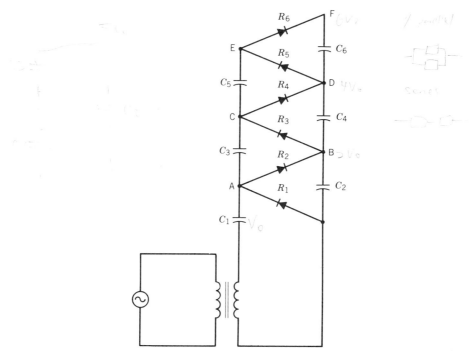

**Figure 15.3** Cockcroft-Walton high-voltage generator. For secondary transformer voltage amplitude $V_0$, the dc voltages are $V_B = 2V_0$, $V_D = 4V_0$, $V_F = 6V_0$.

point B eventually reaches a constant potential of $2V_0$; this same dc potential of $2V_0$ is imposed on point C by rectifier $R_3$, and thus the ac voltage at point C varies between $2V_0$ and $4V_0$. The rectifier $R_4$ then fixes the potential at point D to the constant value of $4V_0$, as capacitor $C_4$ charges to a voltage of $2V_0$. This chain can be continued to higher potentials, limited only by the ability of the high-voltage terminal to hold its potential without sparking to the surroundings.

In practice, there is a loss of voltage due to current carried through the load, and each cycle of the applied voltage $V(t)$ restores the lost charge in the steady state. During the charging cycle of $V(t)$ the rectifiers are all conducting, and the capacitors are effectively in parallel. During the discharging cycle of $V(t)$, the rectifiers are nonconducting and look like open circuits, in which case the capacitors are effectively in series. As a result of this charging and discharging cycle, the terminal voltage is not constant, but has a small ripple that depends directly on the external load resistance and on the period $\omega^{-1}$ of the charging voltage. The ripple also increases in geometric proportion with the number of steps in the chain.

This technique of voltage multiplication was used by Cockcroft and Walton to perform the first nuclear disintegration using artificially accelerated particles:

$$p + {}^7Li \rightarrow {}^4He + {}^4He$$

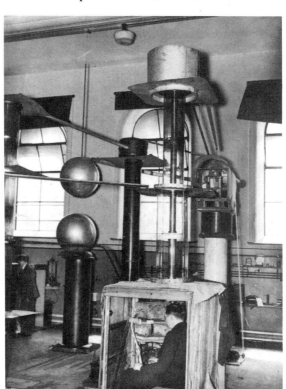

**Figure 15.4** The Cockcroft-Walton accelerator which was used for the first nuclear reaction experiment in 1932. Courtesy of the University of Cambridge, Cavendish Laboratory.

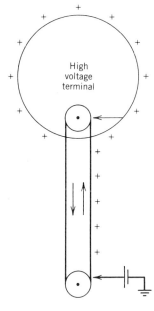

**Figure 15.5** Basic operation of Van de Graaff generator.

Because of its simplicity of design, the Cockcroft-Walton accelerator has retained more than just historical interest; it is in use today to provide sources of neutrons (for instance, $^2H + ^2H \rightarrow n + ^3He$ can be done successfully at a few hundred keV) and also as an injector of particles, especially protons, for higher energy accelerators. Figure 15.4 shows the actual accelerating facilities.

The most common type of electrostatic accelerator in use today in nuclear physics laboratories is based on the Van de Graaff generator, shown schematically in Figure 15.5. The basic principle of operation is a familiar one from elementary electrostatics: when a charged inner conductor and a hollow outer conducting shell are placed in electrical contact, all of the charge from the inner conductor will flow to the outer one, no matter how much charge resides there already or how high its potential. The resulting potential on the outer conductor is determined by its capacitance with respect to the grounded surroundings, $V = Q/C$, and in principle the potential increases without limit as we add more and more charge $Q$. In practice, a limit is imposed by the electrical breakdown (sparking) of the insulating column that supports the outer conductor or of the surrounding atmosphere (Figure 15.6).

The charge is transferred to the terminal by a continuously moving belt, originally made of insulating material such as silk. Charge is sprayed onto the belt by a corona discharge at the base of the device illustrated in Figure 15.5; a large potential difference ($+20$ kV) at the corona points ionizes the air and repels the positive ions, where they are intercepted by the moving belt. A complementary set of corona points near the upper pulley extracts the charge and transfers it to the high-voltage terminal. An ion source is located inside the terminal, and ions "fall to ground" through the potential difference $V$.

To reduce breakdown and sparking, the generator is enclosed in a pressure tank containing perhaps 10–20 atmospheres of an insulating gas which inhibits breakdown; $SF_6$ is a particularly stable gas that is in common use today. An

**Figure 15.6** An early Van de Graaff high-voltage generator. The hollow metal sphere is 15 feet in diameter. The high voltage is limited to about 2.7 million volts by sparking through the air. Courtesy of High Voltage Engineering Co.

evacuated accelerating tube guides the ions from the source to the target, which is at ground potential.

The Van de Graaff accelerator provides one enormous advantage over the Cockcroft-Walton accelerator—the terminal voltage on a Van de Graaff is extremely stable and lacks the ac ripple of the Cockcroft-Walton. Terminal voltages are constant to within $\pm 0.1\%$ ($\pm 1$–$10$ keV) which is extremely important when it is desired to measure reaction cross sections leading to specific excited states.

A disadvantage of the Van de Graaff accelerator is its low current output ($\mu$A) compared with the Cockcroft-Walton (mA). Nevertheless, currents in the $\mu$A range are quite sufficient for nuclear reaction experiments (indeed, higher currents cannot be tolerated by many targets), and as a result the Van de Graaff became the workhorse of low-energy nuclear structure physics in the 1960s and several dozen facilities were constructed at U.S. universities and research labora-

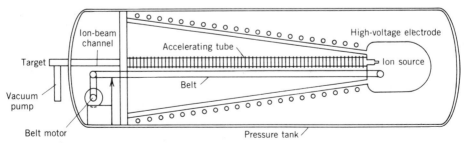

**Figure 15.7** Diagram of Van de Graaff accelerator. The ion source is inside the high-voltage terminal, and both are contained in a pressure tank to inhibit sparking.

**Figure 15.8**  A Van de Graaff accelerator, showing the pressure tank at right, the emerging beam line, and a bending magnet to direct the beam to the experimental area. Courtesy Purdue University.

tories. A commercial device manufactured by High-Voltage Engineering Corporation is illustrated in Figures 15.7 and 15.8.

Perhaps the weakest link in an accelerator facility, requiring the greatest attention by the experimenters, is the ion source. Discharge filaments may burn out and require replacement, and changing the type of accelerated ion often requires changing all or part of the ion source. Having the ion source located inside the high-voltage terminal therefore creates annoying problems for the experimenter—gaining access requires discharging the terminal, pumping out and storing the insulating gas, opening the pressure tank (while maintaining absolute standards of cleanliness to prevent dust from entering the tank and creating discharges when the voltage is restored), and then reversing the procedure to reassemble the system. The entire process requires the accelerator to be shut down for many hours. An alternative design that eliminates this problem (and gains energy for the beam in the process) is the *tandem* Van de Graaff accelerator, shown schematically in Figure 15.9. A beam of *negative* ions (the production of which is discussed in the introduction to this chapter) is accelerated from ground potential toward the high-voltage terminal in the center of the pressure tank. There they enter a foil or gas stripper, which removes $n + 1$ electrons, resulting in an ion carrying a net positive charge of $ne$. Leaving the terminal, these positive ions are accelerated away from the positive high voltage, and the net result is an ion with a kinetic energy of $(n + 1)eV$; typically, $V$ is of the order of 10 million volts for many tandem installations. The need to produce negative ions limits somewhat the current that can be obtained from the ion source, but tandem accelerators are still able to produce currents of many microamperes.

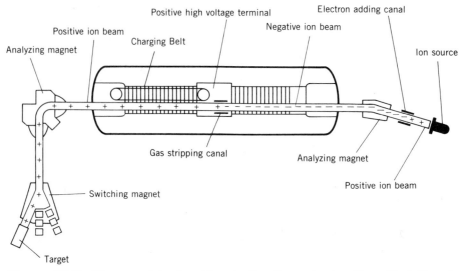

**Figure 15.9** Diagram of tandem Van de Graaff accelerator. From R. J. Van de Graaff, *Nucl. Instrum. Methods* **8**, 195 (1960).

Recent advances in accelerator technology have enabled tandem Van de Graaff accelerators to achieve terminal voltages in excess of 20 million volts. New accelerators using these high voltages have been built primarily to study heavy-ion reactions. One such machine is shown in Figure 15.10. The 25-MV "folded" tandem of the Holifield Heavy Ion Research Facility at the Oak Ridge National Laboratory uses a "pelletron" chain of metal cylinders linked by insulating spacers. Pure $SF_6$ at 7 atmospheres is used as the insulating gas. (Handling more than 2000 m$^3$ of this insulating gas is itself a major engineering problem. The gas is compressed and stored as a liquid when the pressure tank must be emptied.) The pelletron system is superior to the older rubber or fabric belts; the metal chain has a greater lifetime and reliability, and it carries a more uniform distribution of charge, thus reducing voltage fluctuations on the terminal.

Modern accelerator facilities must provide beams for a variety of users. Figure 15.10 shows the complex facilities at the Oak Ridge laboratory. Steering magnets can provide beams at any of the experimental stations. A representative sample of the possible beam energies and currents is as follows:

| Ion | Tandem Only | | Tandem + Cyclotron | |
|---|---|---|---|---|
| | Energy (MeV) | Current[a] (pnA) | Energy (MeV) | Current[a] (pnA) |
| $^9$Be | 17 | 9 | 158 | 2.3 |
| $^{16}$O | 39 | 18 | 404 | 1.9 |
| $^{58}$Ni | 142 | 29 | 889 | 0.4 |
| $^{116}$Cd | 112 | 5 | 494 | 0.2 |

[a]Particle-nanoamps (pnA) multiplied by electric charge per particle gives the conventional beam current.

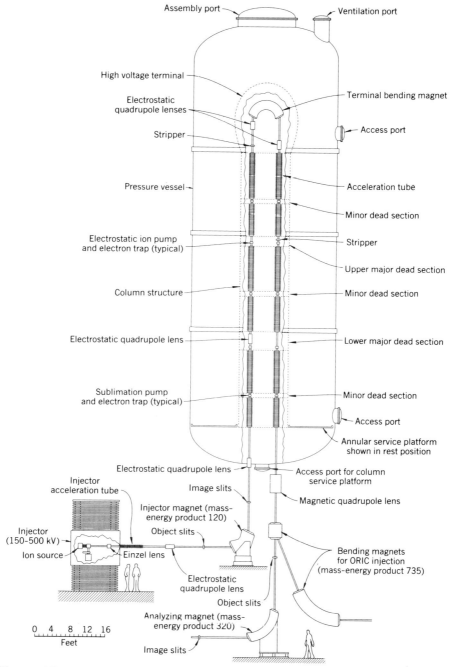

**Figure 15.10**  (*a*) The 25-MV "folded" tandem at Oak Ridge National Laboratory. The tandem serves as injector to a cyclotron (Figure 15.16). Figure courtesy of Holifield Heavy Ion Research Facility, Oak Ridge National Laboratory.

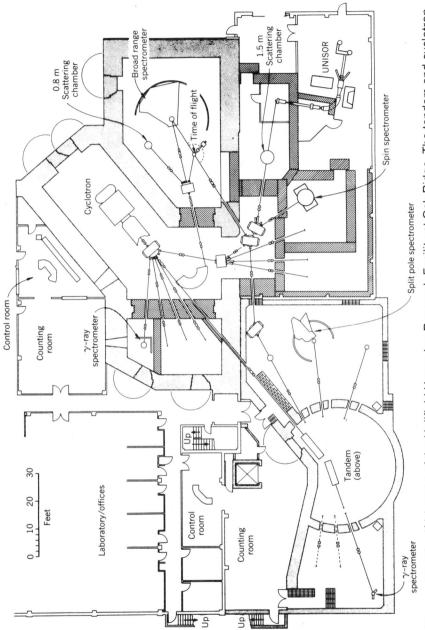

**Figure 15.10** (b) Layout of the Holifield Heavy-Ion Reserach Facility at Oak Ridge. The tandem and cyclotron are shown, along with the many experimental facilities. Figure courtesy of Holifield Heavy-Ion Research Facility, Oak Ridge National Laboratory.

## 15.2 CYCLOTRON ACCELERATORS

An alternative to the single-stage electrostatic accelerators is the circular device, in which a beam of particles makes many (perhaps hundreds) of cycles through the device, receiving a small voltage increment in each orbit until the particle energy reaches the MeV range. The earliest and simplest of these accelerators is the cyclotron, also called the magnetic resonance accelerator.

The cyclotron is illustrated schematically in Figure 15.11. The beam is bent into a circular path by a magnetic field, and the particles orbit inside two semicircular metal chambers called "dees" because of their shape. The dees are connected to a source of alternating voltage. When the particles are inside the dees, they feel no electric field and follow a circular path under the influence of the magnetic field. In the gap between the dees, however, the particles feel an accelerating voltage and gain a small energy each cycle.

The essential design idea of the cyclotron was conceived by Ernest Lawrence at the University of California at Berkeley in 1929. The critical feature is that the time it takes for a particle to travel one semicircular path is independent of the radius of the path—as particles spiral to larger radii, they also gain energy and move at greater speed, and the gain in path length is exactly compensated by the increased speed. If the half-period of the ac voltage on the dees is set equal to the semicircular orbital time, then the field alternates in exact synchronization with the passage of particles through the gap, and the particle sees an accelerating voltage each time it crosses the gap.

The Lorentz force in the circular orbit, $qvB$, provides the necessary centripetal acceleration to maintain the circular motion at an instantaneous radius $r$, and thus

$$F = qvB = \frac{mv^2}{r} \qquad (15.2)$$

and the time necessary for a semicircular orbit is

$$t = \frac{\pi r}{v} = \frac{m\pi}{qB} \qquad (15.3)$$

The frequency of the ac voltage is

$$\nu = \frac{1}{2t} = \frac{qB}{2\pi m} \qquad (15.4)$$

which is often called the *cyclotron frequency* or *cyclotron resonance frequency* for a particle of charge $q$ and mass $m$ moving in a uniform field $B$. Equation 15.4 shows that $\nu$ and $B$ are intimately linked—for a given field strength, the frequency can only have a certain value for resonance.

The velocity increases gradually as the particle spirals outward, and the greatest velocity occurs at the largest radius $R$, according to Equation 15.2:

$$v_{\text{max}} = \frac{qBR}{m} \qquad (15.5)$$

leading to a maximum kinetic energy

$$T = \tfrac{1}{2}mv_{\text{max}}^2 = \frac{q^2B^2R^2}{2m} \qquad (15.6)$$

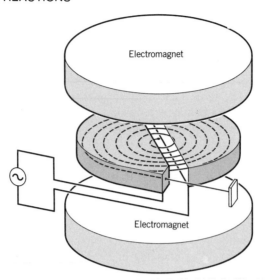

**Figure 15.11** Simplified diagram of a cyclotron accelerator. The beam spirals outward from the center, accelerated each time it crosses the gap between the dees, and is eventually extracted and directed against a target.

Equation 15.6 shows that it is advantageous to build cyclotrons with large fields and large radii. Notice also that the amplitude of the ac voltage between the dees does not appear in any of these expressions; a larger voltage means that the particle gets a larger "kick" with each orbit, but it makes a smaller number of orbits and emerges with the same energy as it would with a smaller voltage.

The first practical cyclotron for accelerating particles was built by Lawrence and M. Stanley Livingston at Berkeley in 1931. The dees had a 12.5-cm radius and the cyclotron was able to produce protons of energy 1.2 MeV in a field of about 1.3 T (13 kG); the corresponding frequency is about 20 MHz. Within a few years, the radius had been extended to about 35 cm and the basic particle energy to 10 MeV protons, 5 MeV deuterons, and 10 MeV $\alpha$'s. By the end of the 1930s, radii of 75 cm had been achieved, extending the range to 40 MeV $\alpha$'s and protons, and 20 MeV deuterons. Figure 15.12 shows the 75-cm cyclotron of the Argonne National Laboratory.

Currents in these cyclotrons are typically in the range of tens of microamperes, more than sufficient intensity for detailed studies of nuclear reactions. As a result, from the 1930s until the commercial availability of large Van de Graaff accelerators in the 1960s, the cyclotron was the most commonly used accelerator for nuclear structure studies with nuclear reactions.

As the beam in a cyclotron travels outward toward the edge of the machine, the magnetic field lines are diverted somewhat from the true vertical (Figure 15.13). There are two effects of this fringing field, one beneficial and one harmful. The curvature of the field lines gives a net force component toward the median plane, which tends to provide focusing and to counteract the tendency of the beam to diverge. At the same time, however, the field loses its uniformity and the resonance condition (Equation 15.4) can no longer be maintained if the frequency is held constant.

**Figure 15.12**  A cyclotron accelerator. The large chambers top and bottom hold the magnets, and the beam is visible as it emerges from the machine owing to its ionization of air molecules. Courtesy Argonne National Laboratory.

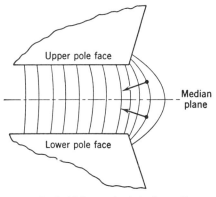

**Figure 15.13**  The magnetic field lines deviate from the vertical near the edge of a cyclotron; the resultant force on an ion (arrows) gives a vertical focusing effect.

A more serious difficulty comes from the relativistic behavior of the accelerated particles. Replacing the momentum $mv$ in Equation 15.2 with the relativistic value $\gamma mv$ where $\gamma = (1 - v^2/c^2)^{-1/2}$, we see from Equation 15.4 that to maintain the resonance condition as $v$ increases we must also increase $B$, and so the field should be larger at the larger radii. This can be accomplished by "shimming" the field, which would then show magnetic lines curving inward (opposite to those of Figure 15.13) and would have a corresponding and undesirable defocusing effect. In the basic design of the fixed-field, fixed-frequency cyclotron, there is no acceptable way to compensate for the relativistic effect, and this provides an ultimate limit on the size of such machines. For protons, an energy of about 40 MeV is the maximum that can be achieved, corresponding to $\gamma = 1.04$.

To overcome this problem, one solution is to vary the frequency, resulting in a frequency-modulated cyclotron, called a *synchrocyclotron*. To understand the operation of the synchrocyclotron, we must first discuss the concept of *phase stability* of cyclotron orbits. It is obvious that in a variable-frequency cyclotron, a continuous beam is not possible, for the time to travel the semicircular orbits will no longer be constant and equal to the half-period (which is now variable). Thus the particles travel through the cyclotron in bunches, and the frequency is swept from its maximum value (when the bunch is near the center, the particles are only slightly accelerated, and the relativistic increase in mass is slight) to its minimum value (when the bunch is ready to exit the cyclotron, the maximum energy is attained, and the mass has its largest value). Particles in the bunches will arrive at the gap between the dees at different times. Phase stability provides a sort of time-focusing effect; those particles that arrive early are delayed somewhat and on the next cycle are closer to the center of the bunch, while those that arrive late are advanced and likewise pushed closer to the center.

To see how this occurs, imagine a particle circulating at the center of a bunch and arriving at the gap at the instant the accelerating voltage passes through zero (Figure 15.14). Such a particle would circulate forever in this stable orbit, called a *synchronous* orbit. Now suppose another particle in the bunch arrives a bit earlier, at point *b* of Figure 15.14. This particle sees an accelerating potential, which will increase the energy and radius of the orbit, but the mass will increase, thereby decreasing the orbital frequency, as in Equation 15.4. Because its frequency is lower, it arrives at the next gap crossing later in phase, that is, closer to the center of the bunch. Similarly, a particle arriving originally later than the center of the bunch will be decelerated, and the decrease in mass increases the angular frequency and pushes the particles closer to the center of the bunch at the next gap crossing. (This is an often confusing property of circular orbits—a gain in energy leads to a decrease in angular speed. The same effect occurs, for very different reasons, with satellite orbits in a gravitational field; if the rear thrusters are burned, so as to increase the energy of the satellite, it moves to a larger orbital radius and the net result is a decrease in the orbital angular velocity. Under certain circumstances, the proper procedure to use in overtaking an object orbiting ahead of a spacecraft is to slow down, not speed up!) Particles in a bunch therefore may perform oscillations with respect to the synchronous orbit, moving first ahead, then closer to, and then perhaps behind the particles in the

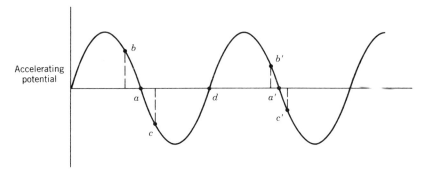

**Figure 15.14**  Phase stability in a synchrocyclotron. Particles arriving at a gap a bit early (point *b*) are accelerated and at the next gap crossing (point *b'*) are bunched closer to the particles in the synchronous orbit (points *a, a'*). The same effect occurs for particles arriving a bit late (*c, c'*). Point *d* is an unstable point, and particles just ahead of or behind *d* appear at the next crossing further ahead or behind.

synchronous orbit. In the synchrocyclotron, as the frequency is slowly decreased, the radius of the synchronous orbit will increase and with it the energy will increase. With each passage of the gap, the decreasing frequency causes particles to appear at the "early" position with respect to the synchronous orbit; these particles are both accelerated and bunched by the phase-stability effect.

Because individual accelerations tend to be small, many more orbits are required in a synchrocyclotron than in an ordinary cyclotron. We can make a rough estimate of the number of orbits required by comparing the time that it takes for the particles to travel from the center to the edge of the field with the time needed for a single orbit:

$$\text{number of orbits} = \frac{\text{total time}}{\text{time/orbit}} \sim \frac{\text{orbital frequency}}{\text{modulation frequency}}$$

The modulation frequency is the frequency with which the cyclotron frequency is swept from its maximum to its minimum value and is representative of the time it takes the ions to travel from the center to the edge. For a typical synchrocyclotron, such as the 184-inch at Berkeley, the cyclotron frequency range is from 36 to 18 MHz and the modulation rate is 64 Hz; thus the number of orbits is of order $10^5$. The energy of extracted protons is 740 MeV, and the field strength is about 2.3 T, corresponding to a frequency of about 20 MHz at the largest orbital radius (where the mass is 1.8 times the rest mass). The Berkeley synchrocyclotron, first operated in 1946, is the highest energy synchrocyclotron and provides a mean proton current of order 0.1 $\mu$A (but of course the current is pulsed, not continuous). Other comparable synchrocyclotrons have operated at Dubna in the USSR and at the European Center for Nuclear Research (CERN) in Geneva.

An alternative solution to extending cyclotrons to higher energies is to increase the magnetic field with increasing orbital radius to compensate for the increasing relativistic mass of the orbiting particles. However, as we discussed previously, this has the undesirable effect of defocusing the beam owing to the curvature of the field lines (opposite to those shown in Figure 15.13). Focusing can be restored

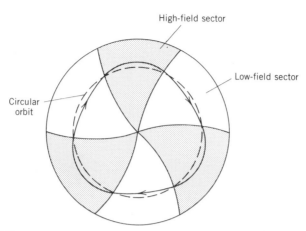

**Figure 15.15** High-field and low-field regions in an AVF cyclotron. Particles perform radial oscillations about the circular path shown.

if the magnetic field is divided into sectors of alternating high and low field (Figure 15.15). Such a cyclotron is called a sector-focusing or AVF (azimuthally varying field) cyclotron. The stable orbits in an AVF cyclotron are not circles; as shown in Figure 15.15, the particles perform radial oscillations about the circular orbit. At the boundaries between the high-field and low-field sectors, there is an azimuthal component to the field, and the Lorentz force $F = qv \times B$ gives a

**Figure 15.16** Model of the Oak Ridge Isochronous Cyclotron, showing the shape of the pole tips that give the high-field regions. A single dee is visible in the center of the machine, which has been opened for the illustration. Photo courtesy Oak Ridge National Laboratory.

vertical force that tends to keep the beam in focus in the midplane. This focusing effect must be designed to be strong enough to overcome the defocusing effect of the radially increasing field.

The major advantage of AVF cyclotrons over synchrocyclotrons is the continuous beam and thus the larger possible beam currents (of order 100 $\mu$A). One of the earliest AVF cyclotrons, the Oak Ridge Isochronous Cyclotron (ORIC), was first operated in 1961 and was capable of producing protons up to 70 MeV at currents of 100 $\mu$A. Figure 15.16 shows a representative view of ORIC, which is today used for work with heavy ions (and for which the 25-MV tandem serves as injector). The diameter is about 2 m, and the field strength is in the range of 1.5–1.7 T (15–17 kG). In the original design, ions from protons up to heavy atoms (Kr, for example in the $+12$ charge state) could be accelerated to about 100 MeV total energy. Depending on the mass of the ion, the cyclotron frequency could be varied from 22.5 MHz (for 75-MeV protons) to 3.7 MHz (for 145-MeV $^{84}$Kr). The main magnetic field is produced by three sectors of iron, whose shape is designed to provide an increase in the field of 8% at the largest radius.

A larger AVF cyclotron facility is at TRIUMF (Tri-Universities Meson Physics Facility) in Vancouver, Canada. This accelerator, shown in Figure 15.17, is designed for research into nuclear reactions with protons of up to 520 MeV. At these energies, $\pi$ mesons are produced in great quantities by proton-induced reactions, and so the accelerator provides secondary beams of $\pi$ mesons (and their decay product, muons). The high intensities permit careful experiments to be done on reactions and scatterings using these secondary beams incident on various targets. An unusual feature of this machine is its acceleration of negative $H^-$ ions, which permits convenient extraction of the beam (often a problem in cyclotrons) after stripping the two electrons (by passing the beam through a thin foil); the resulting beam of positive ions curves in the opposite direction in the magnetic field and exits the cyclotron. As shown in Figure 15.17, two stripper foils give two extracted beams, one dedicated to proton-induced reactions and a second used for $\pi$ meson production. Another cyclotron facility devoted to meson production is the 590-MeV AVF cyclotron at the Swiss Institute for Nuclear Research (SIN) near Zurich. Research into meson-nucleus interactions using such facilities is discussed in Chapter 17.

Another area of current interest for which AVF cyclotrons are used is the study of reactions of heavy ions, in which overcoming the Coulomb barrier is a significant problem. In fact, the interesting physics occurs well above the Coulomb barrier, and it is desirable to accelerate ions to 10–100 MeV per mass unit. Accelerating carbon ($Z = 6$, $A = 12$) to even 120 MeV requires a cyclotron that can accelerate protons to 40 MeV, which is beyond the range of a conventional cyclotron. The accelerating capability of cyclotrons for ions of mass $A$ times the nucleon mass $m$ and charge $q = ze$ is measured in terms of

$$K = AT/z^2$$
$$= e^2 B^2 R^2 / 2m$$

(15.7)

according to Equation 15.6. That is, $K$ depends only on the design parameters $B$ and $R$, and corresponds to the energy (in MeV) to which protons ($A = 1$, $z = 1$) would be accelerated. Heavier ions are accelerated to energies of $Kz^2/A$. Table

**Figure 15.17** The 520-MeV TRIUMF AVF cyclotron. The six sectors of the magnet can be seen in the top photo, taken during the assembly of the accelerator. At bottom is shown the layout of the accelerator building, with two extracted beams going to facilities dedicated to experiments with protons (left) and mesons (right). Courtesy of TRIUMF, Vancouver, Canada.

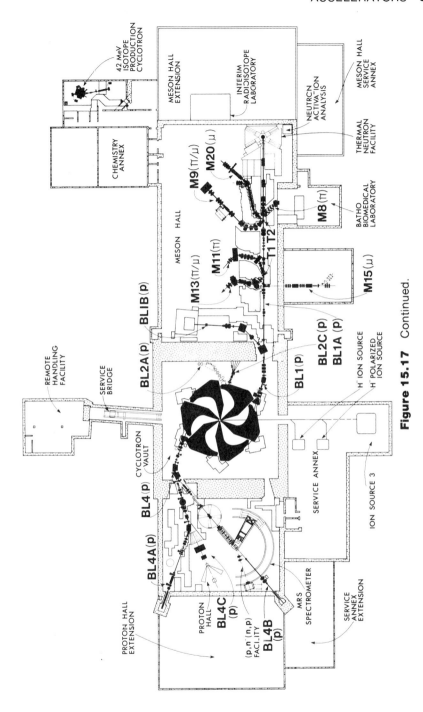

**Figure 15.17** Continued.

**Table 15.1**  Heavy-Ion Facilities using AVF Cyclotrons

| Facility[a] | Location | Injector[b] | Main Cyclotron[b] |
|---|---|---|---|
| HHIRF | Oak Ridge, Tenn. | 25-MeV Tandem | $K = 90$ |
| NSCL | East Lansing, Mich. | $K = 50$ | $K = 500$ |
| GANIL | Caen, France | $K = 25$ | $K = 400 + K = 400$ |
| JINR | Dubna, USSR | $K = 156$ | $K = 250$ |
| CYCLONE | Louvain, Belgium | — | $K = 110$ |
| | Chalk River, Canada | 13-MeV Tandem | $K = 520$ |

[a] HHIRF, Holifield Heavy Ion Research Facility; NSCL, National Superconducting Cyclotron Laboratory; GANIL, Grand Accélérateur National d'Ions Lourds; JINR, Joint Institute for Nuclear Research; CYCLONE, CYCLotron isochrONE.
[b] $K$ is in units of MeV.

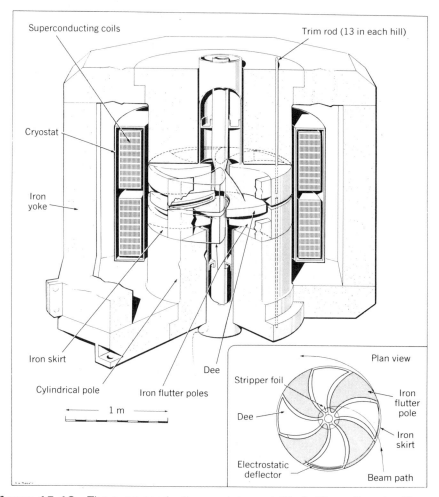

**Figure 15.18**  The superconducting cyclotron at Chalk River, Canada. The superconducting coils establish larger magnetic fields than conventional magnets and permit a more compact design for the cyclotron. Courtesy of Chalk River Nuclear Laboratories.

15.1 lists several of the heavy-ion facilities now in operation using AVF cyclotrons. Most of the listed facilities use one accelerator (a tandem Van de Graaff or another cyclotron) as injector for the AVF cyclotron. The GANIL facility uses two $K = 400$ cyclotrons to make a three-stage acceleration facility. The NSCL at Michigan State University uses superconducting magnets to raise the magnetic field to about 5 T, a factor of 2–3 greater than the capability of conventional magnets, and does so in a smaller area and at less cost. Figure 15.18 shows an example of the design of a superconducting AVF cyclotron. Typical beams from such facilities might be ions of energies from several hundred MeV for the lighter ions (C, O) to 5–10 MeV per nucleon for heavy ions ($A > 150$) and beam currents from $\mu$A for the lighter ions to tens or hundreds of nA for the heaviest ions.

## 15.3 SYNCHROTRONS

Extending the cyclotron or synchrocyclotron to higher energy means building machines of larger radii. Because the magnet is the principal factor in the cost of a cyclotron, we expect that costs of building larger cyclotrons will scale roughly as the cube of the energy. Extending the present generation of ~ 500 MeV cyclotrons (costing of order $10^8$) to the high-energy regime (even as low as 5 GeV, quite insufficient for the study of fundamental phenomena of recent or current interest) quickly puts the cost into the realm of the U.S. gross national product! The solution to this dilemma is the *synchrotron* accelerator, in which both the magnetic field strength *and* the resonant frequency are varied.

Figure 15.19 shows the simplest design for a synchrotron. The essential feature that keeps the costs reasonable as the energy is extended is that particles orbit at a very nearly constant radius at high energies. The magnetic field therefore need be applied only at the circumference, not throughout the entire circular volume, as in an ordinary cyclotron. An annular magnet, as shown in Figure 15.19, accomplishes the task. Particles follow a circular path and are accelerated by a resonant electric field as they cross a gap during each orbit. As the energy increases, the frequency of the ac voltage across the gap must increase to maintain the resonance; simultaneously, the magnetic field must increase to keep the radius constant. (Here we vary the field in time, not spatially as in the AVF cyclotron.) In a magnetic field of strength $B$, a particle of charge $e$ moves in a circular arc of radius $r$ at momentum $p = erB$. The total relativistic energy of the particle is

$$E = \sqrt{p^2c^2 + m^2c^4}$$
$$= \sqrt{e^2r^2B^2c^2 + m^2c^4} \tag{15.8}$$

The basic cyclotron condition, Equation 15.4, can then be written as

$$\nu = \frac{eBc^2}{2\pi\sqrt{e^2r^2B^2c^2 + m^2c^4}} \tag{15.9}$$

For a given $r$, Equation 15.9 gives the relationship between $B$ and $\nu$ necessary to maintain the synchronization.

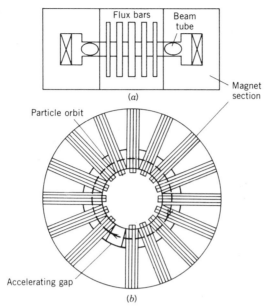

**Figure 15.19** Vertical and horizontal cross-sectional views of electron synchrotron. Several magnets bend the beam into a circle, and an electric field accelerates the particles once in each orbit. If the particle energy is high, the orbital radius remains nearly constant even as the energy increases.

Because the accelerator uses a varying frequency and magnetic field, it must operate in a pulsed, rather than a continuous, mode. This returns us to problems of the stability of the beam pulses in time and in space. The stability of the beam pulses in time comes about through phase stability, exactly as discussed previously in the case of the synchrocyclotron. Spatial stability is not independent from phase stability in the case of the synchrotron, because a particle receiving a slightly greater "kick" in the gap will move to larger $r$. It must see a larger $B$ field at this radius or it will diverge from the pulse. However, as can be inferred from the converse of Figure 15.13, if the field increases with $r$, there will be vertical defocusing. The solution to the focusing problem, generally called *strong focusing* or *alternating gradient* (*AG*) *focusing*, is very similar to the focusing provided by the quadrupole lens illustrated in Figure 15.2. The magnets are arranged in sectors, as illustrated in Figure 15.20, with alternately increasing and decreasing radial gradients. The field at the center of the beam tube has the same value in all sectors, but in one set it decreases with $r$ and in the neighboring set it increases. The variations of the field with $r$ are quite dramatic: $B_z \propto r^{+n}$ and $B_z \propto r^{-n}$ in the alternate sectors, with $n \sim 300$.

In a synchrotron for accelerating electrons, relativistic speeds are quickly reached, where $cp \gg mc^2$, and the orbital frequency is roughly constant, as can be seen from Equation 15.9. The first machines in the late 1940s and early 1950s reached energies in the range of several hundred MeV. Application of AG focusing permitted higher energies to be reached, and the Cambridge Electron Accelerator operated from 1962–1968 at an energy of 6 GeV. The orbit radius was only 36 m and the maximum magnetic field was 0.76 T. A linear accelerator

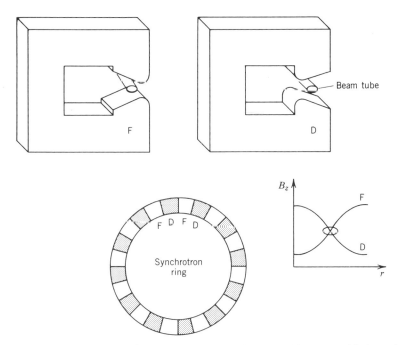

**Figure 15.20** Strong or AG focusing in a synchrotron. Magnets with focusing (F) and defocusing (D) fields alternate around the ring of the accelerator. The magnets are shown at top, and the radial dependence of the field is shown at right.

served as injector and produced an initial energy of 25 MeV, already relativistic for electrons. The resonant frequency was 476 MHz, applied at 16 cavities around the ring. A total of 48 magnet sectors provided the AG focusing. The construction cost of the facility was a relatively modest $12 million. Another similar machine at Hamburg, Germany, is the Deutscher Elektronen Synchrotron (DESY), originally designed and built for 7 GeV in 1965 and currently operating at 35 GeV as a colliding-beam accelerator (see Section 15.5).

Protons do not reach completely relativistic speeds until they have energies of several GeV. Therefore, the resonant frequency must be varied to keep the orbital radius constant. The first proton synchrotron was the "cosmotron" at Brookhaven National Laboratory, completed in 1952 and designed to produce protons at an energy of 3 GeV (Figure 15.21). The mean radius of the orbit was about 10 m, and the maximum magnetic field strength was 1.4 T. Injection energy was 3.5 MeV, and the ac oscillator frequency varied from 0.37 to 4 MHz during the acceleration, in which process the protons completed about $3 \times 10^6$ orbits. About $5 \times 10^{10}$ particles per second were produced. A competing proton synchrotron was built at the same time at the Lawrence Radiation Laboratory at Berkeley, with a slightly larger radius (18 m) and field strength (1.6 T). The Berkeley machine was called the "bevatron" (BeV, meaning billion electron volts, at that time signified the $10^9$-eV unit now called GeV); it was completed in 1954 with a design energy of 6.4 GeV. This energy was chosen to exceed the threshold for production of antiprotons $\bar{p}$ by the reaction

$$p + p \rightarrow p + p + p + \bar{p}$$

**Figure 15.21** The 3-GeV proton synchrotron at Brookhaven, called the cosmotron. Courtesy Brookhaven National Laboratory.

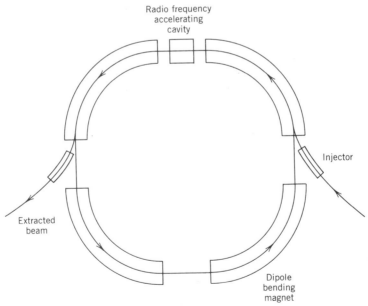

**Figure 15.22** Basic structure of the proton synchrotron, with four dipole sections to bend the beam and one rf cavity to provide the acceleration.

accomplished by accelerating protons against a hydrogen target. The discovery of the antiproton in 1956 at the bevatron earned the 1959 Nobel Prize in physics for its discoverers, Owen Chamberlain and Emilio Segrè. Figure 15.22 shows the basic design of the cosmotron and bevatron, with four magnet sections and one accelerating gap. Several other conventional proton synchrotrons were constructed during the later 1950s in the United States, England, France, and the USSR; these machines produced protons in the 1–10-GeV range.

During this same period (1950s), design studies were underway to apply the AG principle to proton synchrotrons, and by 1960 two machines were in operation, the alternating gradient synchrotron (AGS) at Brookhaven, and the CERN proton synchrotron (CPS). (CERN, the European Center for Nuclear Research, is a joint effort of many countries in western Europe, who realized that the trend toward larger and more expensive nuclear accelerators would quickly

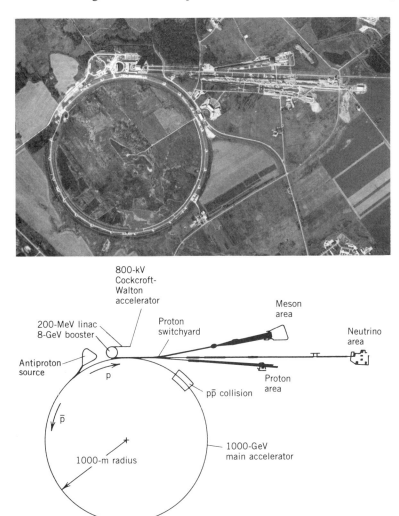

**Figure 15.23**  Layout of Fermilab proton synchrotron. Courtesy Fermi National Accelerator Laboratory.

put them beyond the economic reach of most individual nations. The CPS, constructed at Geneva, Switzerland, was the first major accelerator at the CERN facility, which has become one of the world's most active centers of high-energy physics research.) The AGS reached an energy of 33 GeV, following injection at 50 MeV, producing about $3 \times 10^{11}$ protons per second. The higher energy corresponds to a much larger radius (128 m), compared for example with the 3-GeV cosmotron (10 m). The strong focusing was provided by about 100 gradient reversals around the orbit, with field variations of $r^{\pm 300}$, and the effect of the focusing is evident in the reduced size of the beam aperture—8 cm height $\times$ 18 cm width in the AGS compared with 22 $\times$ 91 cm in the cosmotron. The CERN design group collaborated closely with the Brookhaven group, and as a result the CPS has a similar design and similar capability, with a maximum energy of 28 GeV. In the late 1960s, a 70-GeV AG proton synchrotron was built at Serpukhov in the USSR.

In the quest for ever higher energies to study the production and interactions of elementary particles, the proton synchrotron continues to be the primary accelerator. A major advance in design has been the separation of the bending and focusing functions, so that dipole magnets bend the beam and quadrupole magnets do the focusing. At the Fermi National Accelerator Laboratory (Fermilab or FNAL) in Batavia, Illinois, is a 500-GeV separated-function proton synchrotron with an orbital radius of 1000 m (Figure 15.23). The accelerator began operation in 1972 at an energy of 200 GeV. Injection at 8 GeV is provided by a sequence of three accelerators, a 0.8-MeV Cockcroft-Walton, followed by a 200-MeV drift-tube linear accelerator (see Section 15.4), and then an 8-GeV "booster" synchrotron. The peak field in the bending magnets is 1.4 T, and the resonant frequency in the ac cavities is 53 MHz. The accelerator produces pulses of roughly 1-s width every 12 s, and the magnet power required (36 MW) is enough to power a small city and necessitates special techniques for taking electrical power during the beam pulses and restoring it to the commercial power network during the off-cycle.

The SPS (super proton synchrotron) at CERN, illustrated in Figure 15.24, is a similar machine, producing protons of energy 400 GeV.

A major improvement in the Fermilab capability has been the addition of superconducting magnets, the larger field from which (about 4 T) permits a near doubling of the beam energy to about 1000 GeV or 1 TeV, from which the name "tevatron" is derived. Experimental discoveries in particle physics made at Fermilab and SPS are discussed in Chapter 18.

The development of cyclical proton accelerators proceeded in generations: conventional cyclotrons (10–100 MeV), AVF cyclotrons and synchrocyclotrons (100 MeV–1 GeV), conventional synchrotrons (1–100 GeV), and AG synchrotrons (100 Gev–1 TeV). Other than cost of the magnets and size of the ring, there is no limit on the energy that can be obtained from a proton synchrotron, and at the present time discussions are underway regarding the design and construction of the next generation, which is expected to reach 20 TeV. The use of superconducting magnets permits some economies, but even so the size estimates for the diameter range from 30 to 60 km (18 to 36 miles), depending on the strength of the field of the superconducting magnets; this should be compared with the 2-km diameter of the main ring at Fermilab. The estimated basic cost of the facility

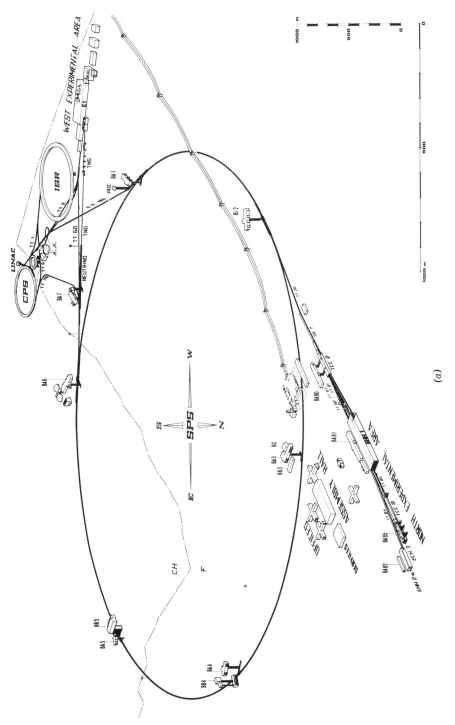

(a)

**Figure 15.24** (a) Diagram of 400-GeV CERN super proton synchrotron. The 26-GeV proton synchrotron (PS) serves as injector for the main ring, which has a diameter of 2.2 km and straddles the border between France (F) and Switzerland (CH) at a mean depth of 40 m. Courtesy of CERN.

**Figure 15.24** (*b*) The underground accelerator tunnel of the CERN SPS. The long sections are dipole magnets which bend the beam; in the distance can be seen a quadrupole magnet (the fourth magnet from the left, slightly larger than the dipoles). Courtesy CERN.

(perhaps $5 billion in 1985 dollars) also scales linearly upward from the original cost of Fermilab (about $200 million in 1972 dollars).

## 15.4 LINEAR ACCELERATORS

In a linear accelerator (often called a "linac") particles receive many individual accelerations by an ac voltage, as in a cyclotron, the difference being that they travel in a straight line in a linac. This immediately eliminates the large costs of cyclotron magnets and the defocusing effects associated with magnetic fields. The basic design of a linac is illustrated in Figure 15.25. The beam travels through a series of hollow tubular electrodes connected alternately to opposite poles of the ac voltage source. Particles are accelerated as they cross the gap between the electrodes. Upon entering the interior of an electrode, the particle drifts in a field-free region (hence the name "drift tube" given to the electrodes) for a time equal to half the period of the ac voltage. In this way the polarity of the voltage is reversed during the time the particle is within the drift tube, and it is then accelerated as it crosses the next gap.

The operation of such an accelerator is dependent on the condition that the entrance of the particles into each gap be in resonance with the electric field across the gap. If $t/2$ is the half-period of the ac voltage, then the length of the

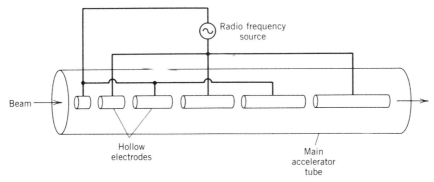

**Figure 15.25**  Basic design of linear accelerator (often called Sloan-Lawrence type). Acceleration occurs in the gaps between the hollow tubular electrodes.

$n$th drift tube for particles of speed $v_n$ must be

$$L_n = v_n t/2 \qquad \frac{L_n}{V_n} = \frac{t}{2} \tag{15.10}$$

For nonrelativistic particles of charge $e$, after passing through $n$ gaps of voltage difference $V_0$ the kinetic energy is

$$T_n = neV_0 = \tfrac{1}{2}mv_n^2 \qquad neV_0 = \tfrac{1}{2}m\left(\frac{2L_n}{t}\right)^2 \tag{15.11}$$

so that

$$= 2m\frac{L_n^2}{t^2}$$

$$L_n = \left(\frac{neV_0}{2m}\right)^{1/2} t \tag{15.12}$$

The drift-tube lengths must therefore increase as $n^{1/2}$. For relativistic particles, where $v \simeq c$, the drift-tube lengths are roughly constant.

Upon crossing the gap, particles will experience a slight radial focusing, which can be understood with reference to Figure 15.26. In the left half of the gap (region $ab$), the lines of force of the electric field focus off-axis particles toward the axis, while in region $bc$, there is a defocusing effect. However, the acceleration of the particles means they move more slowly, and thus spend more time, in region $ab$ so that the focusing effect exceeds slightly the defocusing effect. This slight focusing (which would occur for static fields) is altered by the time-varying

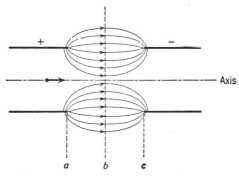

**Figure 15.26**  The electric field in the gap between two drift tubes.

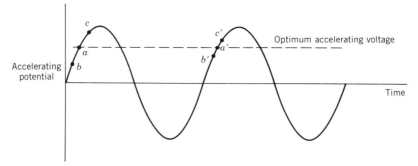

**Figure 15.27** Phase stability in a linear accelerator. Particles arriving at one gap at point *a*, when the accelerating voltage has its optimum value, cross the next gap at point *a'*, when the voltage has the same value. Particles arriving early (point *b*) are accelerated less and therefore are relatively delayed, reaching the next gap less early (*b'*). Particles arriving late (*c*) receive greater acceleration and arrive at the next gap less late (*c'*).

nature of the field, the effect of which we discuss with reference to our previous consideration of phase stability.

Phase stability in a linac is achieved when the bunch arrives on the increasing phase of the ac cycle—see Figure 15.27 and compare with Figure 15.14. Consider a bunch of particles arriving at the gap. Because the voltage is rising, particles arriving early (at the front of the bunch) do not experience the optimum voltage; they are accelerated somewhat less than the particles arriving later and they take longer to cross the drift tube. These "early" particles are thus delayed and arrive toward the center or even the end of the bunch at the next gap. Similarly, particles that reach the gap near the end of a bunch experience a larger voltage and a greater acceleration, which pushes them toward the beginning of the next bunch. For each bunch there is an optimum voltage for perfect resonance, about which the particles may oscillate from one gap to the next, but the net result is a phase stability that keeps the bunches together.

Let's now return to the radial focusing shown in Figure 15.26. Because phase stability demands that particles cross the gap as the voltage increases, the voltage will be larger while the particles are in the defocusing region *bc* of the gap. This defocusing effect exceeds the weak radial focusing, and the overall effect is a radial defocusing, which must be corrected by placing quadrupole lenses within the drift tubes.

The basic accelerator described so far can be considered to be a cavity (or a series of cavities) in which a resonant electromagnetic standing wave is present. For high energies and high currents, it is more efficient to use a traveling wave, in which we imagine the particles to travel the length of the accelerator riding the crest of a traveling wave, just as a surfboard rides the crest of an ocean wave. Power must be fed into the length of the accelerator at regular intervals to maintain the traveling wave because resistive losses are high. For this reason, linacs are operated in a pulsed mode, rather than with a continuous beam; in the pulsed mode the power need be supplied for only a small fraction (a few percent) of the time. The problem facing accelerator designers is to construct a cavity in which the phase velocity of the traveling wave exactly matches the velocity of a

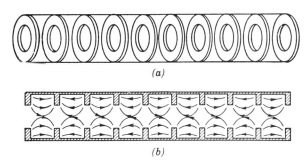

**Figure 15.28**  (*a*) A "disk-loaded" circular wave guide. (*b*) The resulting electric field.

particle as it is accelerated along the cavity. This is done using the "disk-loaded" configuration illustrated in Figure 15.28. The dimensions of the disks determine the phase velocity of the wave.

Three notable examples of linacs in the United States are worth describing. The first is the proton accelerator of the Los Alamos Meson Physics Facility (LAMPF). The accelerator consists of three stages: Cockcroft-Walton injectors, which provide beams of 0.75-MeV protons, H⁻, and polarized protons; a drift-tube section containing 165 individual drift tubes, operating at a frequency of 200 MHz and accelerating the beam to 100 MeV; and a traveling wave section, consisting of coupled cavities with a resonant frequency of 800 MHz, that accelerates the beam to its final energy of 800 MeV. The accelerator is designed to provide high intensity (1 mA, an almost unimaginable current for a particle accelerator in comparison with the more typical $\mu$A or nA). The accelerator was designed to study the nucleon-nucleon interaction in the "intermediate energy" range 200–800 MeV and to provide a high-intensity source of $\pi$ mesons for studying meson-nucleon interactions. Figure 15.29 shows the facility, which first became operational in 1972.

An example of an electron linac is the Stanford Linear Accelerator Center (SLAC). It is a traveling wave accelerator, operating at a frequency of 2856 MHz, in which electrons are accelerated to 30 GeV down its 2-mile length. The basic design is the disk-loaded waveguide (as in Figure 15.28) and it produces an average electron current of 30 $\mu$A with a pulse rate of 360 Hz and a pulse width of 1.7 $\mu$s.

Figure 15.30 shows a view of SLAC, located on the campus of Stanford University. Since its first beams were produced in 1967, the accelerator has been an essential component of the U.S. program in high-energy physics. (The present accelerator's predecessor, a linac capable of producing 1.2-GeV electrons, was important in acquiring knowledge of basic nuclear properties; the electron scattering data shown in Figure 3.1 provided the first detailed determinations of nuclear radii.) Among the important recent discoveries made at SLAC were the $J/\psi$-meson and the $\tau$ lepton (see Chapter 18). Electromagnetic radiation, primarily X rays, produced when the electron beam is diverted from the linac to a circular ring has been used as a probe of bulk and surface properties of solids. The addition of a colliding-beam storage ring has greatly extended the capabilities of the accelerator, as we discuss in Section 15.5.

**Figure 15.29** The half-mile-long proton linac at the Los Alamos Meson Physics Facility. The accelerator itself is underground. The top photo shows the access buildings and the experimental hall, and the bottom photo shows the actual traveling-wave section of the accelerator. Courtesy Los Alamos National Laboratory.

**Figure 15.30** The two-mile long, 32-GeV linear electron accelerator at Stanford. Electrons begin near the top of the photo and are accelerated toward the target areas near the bottom. The buildings near the circular road in the lower part of the figure show the experimental areas associated with the underground positron– electron collider called PEP, a storage ring 800 m in diameter. The dashed lines show the planned SLAC Linear Collider (SLC), which will produce 100-GeV positron–electron collisions. Photo courtesy Stanford Linear Accelerator Center, operated by Stanford University for the U.S. Department of Energy.

A third type of linac is designed for the acceleration of heavy ions; the SuperHILAC (Heavy Ion Linear Accelerator) at the Lawrence Berkeley Laboratory uses a drift-tube design with a Cockcroft-Walton injector to accelerate heavy ions (up to uranium) to about 9 MeV per nucleon. A similar facility is located at Darmstadt, Germany. The Berkeley SuperHILAC can also be used as the injector for a synchrotron, which then accelerates the ions to about 2.5 GeV per nucleon. The importance of the study of heavy-ion reactions was discussed in Section 11.13.

## 15.5 COLLIDING-BEAM ACCELERATORS

In the quest for higher energies to study the production of new and exotic species of particles, the goal of the accelerator designer is to convert as much of the incident kinetic energy as possible into mass energy of the new particles. Let's suppose we use a beam of protons against a hydrogen target to produce one or more product particles we denote collectively as X:

$$p + p \rightarrow p + p + X$$

(Other outcomes may also occur, in which a particle other than a proton is produced in the final state; however, certain conservation laws, discussed in Chapter 18, require the presence of two nucleon-like particles, which we assume for the present discussion to be protons.)

The threshold laboratory kinetic energy for particle production can be written as

$$T_{th} = (-Q) \frac{\text{total mass of initial and final particles}}{2 \times \text{mass of target particle}} \qquad (15.13)$$

where $Q$ has the usual meaning of $(m_i - m_f)c^2$. For the proton-proton reaction above

$$T_{th} = (m_X c^2) \frac{4m_p + m_X}{2m_p} \qquad (15.14)$$

$$= m_X c^2 \left( 2 + \frac{1}{2} \frac{m_X}{m_p} \right) \qquad (15.15)$$

In the laboratory frame of reference, $T_{th}$ must be greater than twice the rest energy of the particles we hope to produce. This somewhat unhappy situation comes about because of the need to preserve the motion of the center of mass before and after the collision—in a reaction between a moving particle and a fixed target, the final particles move with the same total linear momentum as the incident particle. Some of the energy that might otherwise go into production of X is "wasted" in conserving momentum.

If particle X is a $\pi^0$ ($mc^2 = 135$ MeV), then $T_{th} = 280$ MeV. Of the initial energy supplied by the accelerator, 48% goes into producing the new particle and 52% goes to the center-of-mass motion. If X is $p + \bar{p}$, as in the antiproton production reaction, $T_{th} = 5.63$ GeV and $m_X c^2/T_{th} = \frac{1}{3}$; only 33% of the energy goes to produce new particles. As the rest energy of X increases, the fraction of the initial energy going into particle production decreases still further. If X is one of the newly discovered W or Z particles (which carry the weak interaction in the same way that the $\pi$ carries the strong interaction) with a rest energy around 90 GeV, then $T_{th} = 4500$ GeV. In this case $m_X c^2/T_{th} = \frac{1}{50}$ and thus 98% of the initial energy is lost to the experimenter for particle production. As you can see from Equation 15.15, for the production of particles much heavier than the reacting particles, $T_{th}$ increases like $(m_X c^2)^2$. Each factor of 10 increase in the rest mass of the particles to be produced necessitates an increase by a factor of 100 in the design energy of the accelerator (and in its cost).

A resolution of this unfortunate fact of accelerator design is the *colliding-beam* accelerator, in which we bring together two beams of equal energy moving in opposite directions. In effect, we do the collision in the center-of-mass frame. The threshold energy for each beam is only $m_X c^2/2$; thus we could produce a 90-GeV particle by bringing together two beams of only 45 GeV, rather than by bringing a 4500-GeV beam against a fixed target. This enormous relaxation in the

(a)

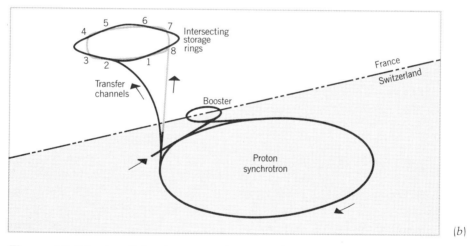

(b)

**Figure 15.31** A colliding-beam accelerator at CERN. An aerial view is shown in (a); the large circle in the foreground shows the location of the 28-GeV proton synchrotron, and the circular area in the background shows the 300-m rings in which beams of protons circulate in opposite directions. A schematic diagram is shown in (b); the protons collide at eight points around the rings. One of the collision regions is shown in (c). Photos courtesy CERN.

(c)

**Figure 15.31** Continued.

energy needed comes at the expense of the experimental difficulty of forcing beams of particles to collide. The density of particles in a beam bunch is quite low, and the ratio of the reaction rates of a fixed-target accelerator and a colliding-beam accelerator is roughly the ratio of the number of particles per cm$^2$ of the target material (often liquid hydrogen) and of the beam. This ratio is so large (a typical beam bunch may have of order $10^{11}$ particles per cm$^2$, compared with a liquid hydrogen target of about $10^{25}$ hydrogens/cm$^2$) that the reaction rate would be negligibly small for a colliding-beam accelerator.

To counteract this small reaction rate, the reactions are made to occur by first passing each beam through a *storage ring*. In a storage ring, many accelerator pulses can be kept circulating for times of the order of 1 day. At the same time, the beams are focused to occupy a far smaller area than they did upon leaving the accelerator. The net effect of the increase in the collision rate of the bunches and of the increased beam density is that the reaction rate is reduced relative to a fixed-target accelerator by a factor of the order of only $10^4$–$10^6$, which enables many experiments to be performed with colliding-beam facilities.

An example of an early proton-proton colliding-beam accelerator is the CERN Intersecting Storage Rings (ISR) shown in Figure 15.31. Beams of 28-GeV protons from the CERN proton synchrotron are directed into two storage rings, where the beams orbit in opposite directions and are made to collide at eight locations around the rings. The 56-GeV center-of-mass energy is equivalent to a beam energy of 1700 GeV in a fixed-target accelerator. After its first operation in 1972, the CERN ISR was for many years in effect the highest-energy proton accelerator for the production of new particles.

The conservation laws mentioned previously restrict the possible outcomes of collisions. The nature of the particles produced is much less restricted if we can arrange collisions between a particle and its antiparticle, such as proton–antiproton ($p\bar{p}$) or electron–positron ($e^+e^-$) collisions. To accomplish $p\bar{p}$ collisions, CERN has discontinued operation of the ISR and operated the SPS in conjunction with proton and antiproton storage rings, each at an energy of 320 GeV. The resulting facility is called Sp$\bar{p}$S. Because the number of antiprotons is extremely small, it is critical to focus the antiprotons to the smallest possible volume. Random collisions of a thermal nature tend to defocus the beam, and reducing these relative motions is equivalent to "cooling" the beam. To accomplish this, Simon van der Meer of CERN developed the process of stochastic cooling, for which he was honored in 1984 as the co-recipient of the Nobel Prize in Physics. In van der Meer's process, a sensor detects the antiproton beam profile and sends a signal (at the speed of light) along a chord of the storage ring to a "kicker" that imposes electromagnetic fields designed exactly to reverse the defocusing motions detected by the sensor. Even though the circulating beam is traveling at speeds very close to $c$, the signal arrives at the kicker just ahead of the beam pulse but in sufficient time to establish the necessary focusing fields. Using the Sp$\bar{p}$S, experimenters at CERN were able to observe the W and Z particles that carry the weak interactions. Carlo Rubbia shared the 1984 Nobel Prize with van der Meer for this discovery.

The Tevatron at Fermilab has recently been modified to operate as a $p\bar{p}$ collider with each beam having an energy of 1 TeV. The newest generation of proton synchrotron presently under design is a 20-TeV pp collider, tentatively called the Superconducting Super Collider (SSC). The 40-GeV center-of-mass energy will not only permit a great extension of the realm of particle physics toward the most fundamental level, but will also permit duplication of the conditions that existed in the early universe, just $10^{-16}$ s after the Big Bang, permitting important cosmological questions to be studied as well.

**Table 15.2**  Some Colliding-Beam Accelerators

| Name | Type | Energy per Beam (GeV) | Years of Operation |
|------|------|------|------|
| CERN ISR | pp | 28 | 1972–1984 |
| CERN Sp$\bar{p}$S | $p\bar{p}$ | 320 | 1983– |
| FNAL Tevatron | $p\bar{p}$ | 1,000 | 1985– |
| SSC | pp | 20,000 | 1995 (?) |
| SLAC SPEAR | $e^+e^-$ | 4 | 1972–1985 |
| DESY DORIS | $e^+e^-$ | 5 | 1974–1985 |
| SLAC PEP | $e^+e^-$ | 18 | 1979– |
| DESY PETRA | $e^+e^-$ | 19 (to 35) | 1978– |
| CERN LEP | $e^+e^-$ | 85 | 1989 (?) |
| SLAC SLC | $e^+e^-$ | 50 | 1987 |

Similar developments have taken place at electron accelerators to produce storage rings for electrons and positrons. At SLAC, the SPEAR facility extracted electrons and positrons (from electron collisions with a target) at about 4 GeV. That facility has now been replaced with one called PEP, in which the colliding beams of electrons and positrons each have an energy of 18 GeV. At DESY, a facility (called DORIS) consisting of colliding 5-GeV $e^+$ and $e^-$ beams was replaced by PETRA, in which the beam energies were raised first to 19 GeV and now 35 GeV. These and other proposed colliding-beam accelerators are listed in Table 15.2. There are also proposals under development for heavy-ion colliders with beams of 100 GeV, which will produce extremely "hot" compound nuclei that approximate the conditions inside supernovas or neutron stars.

## REFERENCES FOR ADDITIONAL READING

A nontechnical review of accelerators can be found in R. Gourian, *Particles and Accelerators* (New York: McGraw-Hill, 1967). See also the historical reviews by one of the pioneers of accelerator development, M. S. Livingston: "Early History of Particle Accelerators," *Adv. Electron. Electron Phys.* **50**, 1 (1980); *Particle Accelerators: A Brief History* (Cambridge, MA: Harvard University Press, 1969); and *The Development of High-Energy Accelerators* (New York: Dover, 1966). This last reference is a collection of original papers with commentary by Livingston.

More complete and advanced texts on this topic are: M. S. Livingston and J. P. Blewett, *Particle Accelerators* (New York: McGraw-Hill, 1962); J. J. Livingood, *Principles of Cyclic Particle Accelerators* (Princeton, NJ: Van Nostrand, 1961); and E. Persico, E. Ferrari, and S. E. Segrè, *Principles of Particle Accelerators* (New York: Benjamin, 1968).

Chapter 3 of J. B. A. England, *Techniques in Nuclear Structure Physics* (New York: Wiley, 1974) contains a review of accelerator properties, including more details on the Cockcroft-Walton design. Characteristics of typical accelerators are reviewed by M. H. Blewett, *Ann. Rev. Nucl. Sci.* **17**, 427 (1967). A series of articles on different types of accelerators can be found in part A of the collection *Nuclear Spectroscopy and Reactions*, edited by J. Cerny (New York: Academic, 1974). More specific information on heavy-ion accelerators is detailed in H. A. Grunder and F. B. Selph, *Ann. Rev. Nucl. Sci.* **27**, 353 (1977).

A general introduction to tandem accelerators can be found in the article by P. H. Rose and A. B. Wittkower in the August 1970 issue of *Scientific American*, and more technical details on tandems are contained in the review by S. J. Skorka, *Nucl. Instrum. Methods* **146**, 67 (1977).

An interesting pictorial history of the CERN SPS is M. Goldsmith and E. Shaw, *Europe's Giant Accelerator* (London: Taylor and Francis, 1977). The details of Fermilab are reviewed by J. R. Sanford, *Ann. Rev. Nucl. Sci.* **26**, 151 (1976) and by R. R. Wilson, "The Batavia Accelerator," in the February 1974 *Scientific American*. New developments in accelerators are discussed by R. R. Wilson in "The Next Generation of Particle Accelerators" in the January 1980 *Scientific American*. The proposed superconducting supercollider is discussed by J. D. Jackson, M. Tigner, and S. Wojcicki in the March 1986 *Scientific American*.

## PROBLEMS

1.  For the original Berkeley cyclotron ($R = 12.5$ cm, $B = 1.3$ T) compute the maximum proton energy (in MeV) and the corresponding frequency of the varying voltage.

2.  Assuming a magnetic field of 1.4 T, compute the maximum energy of protons, deuterons, and $\alpha$'s that can be obtained from a cyclotron of 75 cm radius.

3.  A Van de Graaff accelerator uses a spherical terminal with a radius of 2 m, which is charged to $2.0 \times 10^6$ V. (a) What is the total charge on the terminal? (b) If the charging belt can carry 0.1 mA, how long will it take to charge the terminal?

4.  The original design of the Berkeley 184-inch synchrocyclotron gave 350 MeV protons using a magnetic field of about 1.5 T. (a) At what radius should the protons be extracted? (b) What is the necessary range of cyclotron frequencies? (c) How long does it take to accelerate a particle? (d) What is the maximum pulse rate?

5.  Estimate the deflection angle that could be obtained in a circular accelerator, such as a synchrotron, from a 2-m-long dipole magnet with a field of 4.0 T, if a beam of protons is accelerated to 1 GeV. How many such magnets would be needed to complete the ring?

6.  In the SLAC electron linac, electrons are accelerated to 30 GeV. (a) What is the difference, in m/s, between the electron's speed and the speed of light? (b) What would be the energy of a proton moving at the same speed?

7.  In the drift-tube portion of the LAMPF accelerator, protons are accelerated from 0.75 to 100 MeV. The ac voltage has a frequency of 200 MHz. Find the lengths of the first and last drift tubes.

8.  Discuss the similarities and differences between phase stability in a linear accelerator and in a synchrocyclotron.

9.  Because accelerated charged particles radiate energy, a beam traveling a circular path must radiate. The energy loss per cycle is $\Delta E = (4\pi/3)(e^2/4\pi\epsilon_0 R)(E/mc^2)^4$, where $E$ is the total relativistic energy of the particle and $R$ the radius of its orbit. Discuss the radiation losses in cyclotrons, synchrocyclotrons, and electron and proton synchrotrons. Use parameters characteristic of the accelerators discussed in this chapter.

# UNIT IV
## EXTENSIONS
## AND
## APPLICATIONS

# 16

# NUCLEAR SPIN AND MOMENTS

In analyzing and interpreting nuclear level schemes in this text, we have repeatedly used the spin quantum number $I$ to label the individual levels. In Chapter 5 we discussed how important it is to have an established set of these spin quantum numbers to compare the observed level scheme with the predictions of a particular nuclear model. The measurement of nuclear spin assignments is one of the goals of experimental nuclear physics, and in this chapter we explore some of the techniques that are used to obtain this information.

Nuclear magnetic dipole and electric quadrupole moments have a similar importance in helping us to interpret nuclear structure. We have already discussed in Chapter 4 the clues to the deuteron structure deduced from its moments. In Chapter 5 we have seen the systematic behavior of shell-model magnetic moments, and we have also seen how the unusually large quadrupole moments of certain nuclei suggest a new feature of nuclear structure, the stable deformation.

The experimental techniques that are responsible for the determination of these spins and moments span a considerable range, from those involving nuclear radiations (angular distributions and correlations, Mössbauer effect), to those involving atomic and molecular beams (the Stern-Gerlach experiment, for instance) and radiations in the optical, microwave, and radio regions of the spectrum. In this chapter we introduce and review many of these techniques and give examples of their applications.

## 16.1 NUCLEAR SPIN

Each nuclear state is assigned a unique "spin" quantum number $I$, representing the total angular momentum (orbital plus intrinsic) of all the nucleons in the nucleus. The vector $I$ can be considered the sum of the orbital and intrinsic contributions to the angular momentum:

$$I = \sum_{i=1}^{A} (\ell_i + s_i) \tag{16.1}$$

$$= L + S \tag{16.2a}$$

$$= \sum_{i=1}^{A} j_i \tag{16.2b}$$

where the decomposition according to either Equation 16.2$a$ or 16.2$b$ is largely a matter of convenience. The *quantum number I* has the usual connection with the *vector I*:

$$|I| = \sqrt{I(I+1)}\, \hbar \tag{16.3}$$

$$I_z = m_I \hbar \qquad (m_I = I, I-1, \ldots, -I+1, -I) \tag{16.4}$$

Equation 16.1 represents what could in principle be a very complicated coupling of many vectors to a single resultant, and it may not be apparent why we can neglect this internal structure and treat the nucleus as if it were an elementary particle with a single spin quantum number, representing the intrinsic angular momentum of the "particle." This is possible only because the interactions to which we subject the nucleus, such as static electromagnetic fields, are not sufficiently strong to change the internal structure or break the coupling of nucleons that is responsible for Equation 16.1.

For the electronic motion in atoms, we can similarly define the total electronic angular momentum:

$$J = \sum_{i=1}^{Z} \left( \ell_i^{(e)} + s_i^{(e)} \right) \tag{16.5}$$

where the $\ell$ and $s$ vectors now refer to the electronic states. In analogy with the nuclear case, we can often (but not always) treat the electrons as if they were represented by a single angular momentum $J$.

Finally, there are cases in which it is most appropriate to deal with the total nuclear plus electronic angular momentum, usually called $F$:

$$F = I + J \tag{16.6}$$

The vectors $J$ and $F$ obey all the usual quantum rules for angular momentum, as in Equations 16.3 and 16.4.

The quantum numbers $I$ and $J$ may be either integral or half-integral as the number of nucleons or electrons is even or odd:

| A | Z | I | J | F |
|---|---|---|---|---|
| Even | Even | Integer | Integer | Integer |
| Odd | Even | Half-integer | Integer | Half-integer |
| Even | Odd | Integer | Half-integer | Half-integer |
| Odd | Odd | Half-integer | Half-integer | Integer |

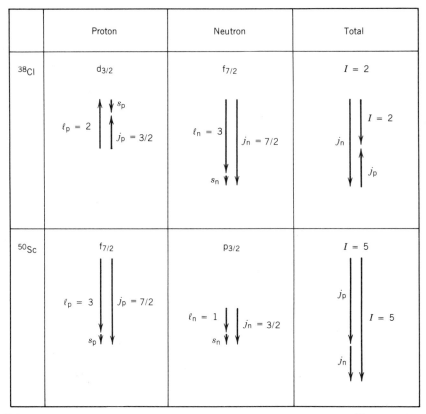

**Figure 16.1** Proton-neutron angular momentum couplings in $^{38}$Cl and $^{50}$Sc.

For nuclear ground states, there are several rules for determining the spins:

1.  All even-$Z$, even-$N$ nuclei have $I = 0$. This results from the strong tendency of nucleons to couple pairwise to zero spin.

2.  In odd-$A$ nuclei, the net spin is almost always determined by the $j$ of the last odd particle, with the remaining $A - 1$ nucleons (having even numbers of protons and neutrons) pairing to zero spin as above.

3.  In odd-$Z$, odd-$N$ nuclei, the spin is determined by the vector coupling of the $j$ of the odd proton and neutron, $I = j_p + j_n$, and thus any of several values are possible. To determine which of these possible couplings will be the ground state, we use the empirical rule that the ground state is usually the coupling with the neutron and proton intrinsic spins $s_p$ and $s_n$ parallel. As an example, consider $^{38}$Cl, which consists of a $d_{3/2}$ proton coupled to an $f_{7/2}$ neutron. For the proton, $\ell_p = 2$ and thus $s_p$ is opposite to $j_p$. For the neutron, $\ell_n = 3$ and $s_n$ is parallel to $j_n$. Arranging the coupling so that $s_p$ and $s_n$ are parallel, as in Figure 16.1, we get $I = |j_p - j_n|$ or $I = 2$, which is in fact the ground-state spin of $^{38}$Cl. (The first excited state is $I = 5$, corresponding to $I = j_p + j_n$.) On the other hand, consider $^{50}$Sc, resulting from an $f_{7/2}$ proton coupled to a $p_{3/2}$ neutron. Here making $s_p$ and $s_n$

parallel also makes $j_p$ and $j_n$ parallel, and thus $I = j_p + j_n = 5$, in agreement with observation. (The state $I = |j_p - j_n| = 2$ is a low excited state of $^{50}$Sc.) Other couplings with $I$ between $j_p + j_n$ and $|j_p - j_n|$ may be found among other low-lying excited states.

## 16.2 NUCLEAR MOMENTS

### Magnetic Dipole Moments

Classically, the magnetic dipole moment $\mu$ arises from the motion of charged particles, and we can regard $\mu$ as a means to characterize a distribution of currents whose effect on the surroundings (that is, on other moving charges) we call "magnetic." When we go over to the quantum limit we find a similar relationship, with one distinctly nonclassical addition: the intrinsic angular momentum (spin) contributes to the magnetic moment also.

Let's briefly review the classical electromagnetism that leads to magnetic dipole moments. We consider some currents distributed over a sample that occupies a certain volume in space (Figure 16.2). The distribution of currents is specified by the *current density* $j(r')$. The vector $r'$ locates a specific point of the sample relative to the origin; the *vector function* $j(r')$ then gives the magnitude and direction of the electric current per unit volume $dv'$ at that point. The recipe for calculating the magnetic field $B$ resulting from the currents is straightforward: first calculate the vector potential $A(r)$ at the observation point $r$ by integrating (summing) over all the currents in the sample:

$$A(r) = \frac{\mu_0}{4\pi} \int \frac{j(r')\, dv'}{|r - r'|} \tag{16.7}$$

and then the magnetic field follows directly from $B(r) = \nabla \times A(r)$. Following some mathematical manipulations, which can be found in standard texts on

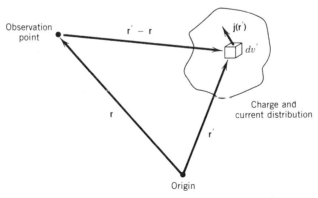

**Figure 16.2** The current element $j(r')\, dv'$ gives a contribution to the vector potential at the observation point. The total potential is found from the integral over the entire current distribution.

electromagnetism, we can rewrite the vector potential as

$$A(r) = \frac{\mu_0}{4\pi}\left\{ \frac{1}{r}\int j(r')\, dv' + \frac{1}{r^3}\int j(r')(r \cdot r')\, dv' + \cdots \right\} \qquad (16.8)$$

which can be written

$$A(r) = \frac{\mu_0}{4\pi}\frac{\mu \times r}{r^3} + \cdots \qquad (16.9)$$

where

$$\mu = \frac{1}{2}\int r' \times j(r')\, dv' \qquad (16.10)$$

The leading nonvanishing term is characterized by the *magnetic dipole moment* $\mu$ of the current distribution. What we have done, in effect, is a multipole expansion of the current distribution; the lowest-order term (dipole) is likely to be the most important. The argument of the integral for $\mu$ includes the charge density and the vector product $r' \times v'$, which in the case of a particle with mass $m$ is just $\ell/m$, where $\ell$ is the angular momentum. Going over to the quantum limit, the charge density is $e|\psi(r')|^2$, and it is entirely consistent with our previous experience with quantum mechanics to write this as

$$\mu = \frac{e}{2m}\int \psi^*(r')\ell\psi(r')\, dv' \qquad (16.11)$$

If the wave function corresponds to a state of definite $\ell_z$, then only the $z$ component of the integral is nonvanishing, and

$$\mu_z = \frac{e}{2m}\int \psi^*(r')\ell_z\psi(r')\, dv' \qquad (16.12)$$

$$\mu_z = \frac{e\hbar}{2m}m_\ell \qquad (16.13)$$

with $\ell_z = m_\ell\hbar$.

What we observe in an experiment as *the* magnetic moment is *defined* to be the value of $\mu_z$ corresponding to the maximum possible value of the $z$ component of the angular momentum. The quantum number $m_\ell$ has a maximum value of $+\ell$, and thus the magnetic moment $\mu$ is

$$\mu = \frac{e\hbar}{2m}\ell \qquad (16.14)$$

The quantity $e\hbar/2m$ has the dimensions of a magnetic moment ($\ell$ is a dimensionless quantum number) and is called a *magneton*. Putting in the proton mass for $m$, we get a *nuclear magneton* $\mu_N$:

$$\mu_N = \frac{e\hbar}{2m_p} = 3.15245 \times 10^{-8}\text{ eV/T}$$

and using the electron mass gives the *Bohr magneton* $\mu_B$:

$$\mu_B = \frac{e\hbar}{2m_e} = 5.78838 \times 10^{-5}\text{ eV/T}$$

Considering the intrinsic spin, which has no classical analog, we make a simple extension of Equation 16.14:

$$\mu = (g_\ell\ell + g_s s)\mu_N/\hbar \qquad (16.15)$$

where the *g factors* $g_\ell$ and $g_s$ account for the orbital and intrinsic contributions to μ. Their values can be adjusted as needed for individual particles: $g_\ell = 1$ for protons and $g_s$ must be measured for "free protons" in which $\ell$ does not contribute to μ. As we discuss later in this chapter, $g_s$ is measured to be 5.5856912 for protons. For neutrons, which are uncharged, we can set $g_\ell = 0$, and $g_s$ is measured to be $-3.8260837$.

In real nuclei, we must make a modification to allow for the effects of all the nucleons:

$$\mu = \sum_{i=1}^{A} [g_{\ell,i}\ell_i + g_{s,i}s_i]\mu_N/\hbar \tag{16.16}$$

which is similar to Equation 16.1 for $I$.

There is no single theory that allows us to evaluate Equation 16.16 to calculate μ because the interactions between the nucleons are strong and the relative spin orientations are not sufficiently well known. In certain cases, we can make simplifying assumptions, based on nuclear models. For example, in the independent particle shell model, we couple $A - 1$ nucleons pairwise to zero-spin combinations that do not contribute to μ. For the remaining odd nucleon, the shell-model theory gives the coupling of $\ell$ and $s$ to form $I$, which permits μ to be calculated, as we did in Section 5.1. In many other cases, we cannot ignore the effect of the "core" nucleons, and we assign them a "collective" g factor usually designated $g_R$, so that

$$\mu = \left[ g_R I_c + \sum_i (g_{\ell,i}\ell_i + g_{s,i}s_i) \right] \mu_N/\hbar \tag{16.17}$$

where $I_c$ refers to the core and the sum is carried out over a few nucleons outside the core. If we consider "pure" collective states, with no odd nucleons, the collective model gives $g_R = Z/A$, the ratio of the nuclear charge to its mass. Figure 5.16a showed that this was a good approximation for $2^+$ states of many even-*Z*, even-*N* nuclei.

### Electric Quadrupole Moments

We now consider the distribution of charges, rather than currents, within the nucleus. From an external point, the electric potential $V(r)$ appears to be

$$V(r) = \frac{1}{4\pi\epsilon_0} \int \frac{\rho(r')\,dv'}{|r - r'|} \tag{16.18}$$

which is analogous to the expression (16.7) for the magnetic vector potential. Classically, we can assign to a charge distribution a monopole (Coulomb) field, which is proportional to the total charge. If we construct a charge distribution in which the total charge vanishes, we can easily study the next highest multipole, the dipole field, the standard for which is charges of $\pm q$ located at, respectively, $z = +a/2$ and $z = -a/2$. In general, any charge distribution that lacks spherical symmetry will have a dipole field, possibly in addition to the monopole field. (One way to distinguish the two contributions to the total field is that the monopole electric field varies as $r^{-2}$ while the dipole field varies as $r^{-3}$.) Just as adding equal and opposite charges at different locations gives a dipole field, adding equal and opposite dipoles causes the vanishing of the dipole field and gives the next higher multipole, the quadrupole field. For example, we could add the dipole with charges $-q$ at the origin and $+q$ at $z = a$ to the opposite dipole with charges $-q$ at the origin and $+q$ at $z = -a$. The characteristic dependence of the electric quadrupole field is as $r^{-4}$.

Expanding the factor of $|r - r'|$ in Equation 16.18 gives immediately the mathematical details of the multipole expansion of the electric field:

$$|r - r'|^{-1} = r^{-1}\left[1 + \frac{r'^2}{r^2} - 2\frac{r'}{r}\cos\theta\right]^{-1/2} \tag{16.19}$$

$$\cong \frac{1}{r}\left\{1 - \frac{1}{2}\left(\frac{r'^2}{r^2} - 2\frac{r'}{r}\cos\theta\right) + \frac{3}{8}\left(\frac{r'^2}{r^2} - 2\frac{r'}{r}\cos\theta\right)^2 + \cdots\right\} \tag{16.20}$$

where $\theta$ is the angle between $r$ and $r'$, and where we have assumed $r \gg r'$. (That is, the observation point is far from the nucleus. For the interaction with atomic electrons, which dominates the hyperfine structure, this is a good approximation.) Thus

$$V(r) = \frac{1}{4\pi\epsilon_0}\left[\frac{1}{r}\int\rho(r')\,dv' + \frac{1}{r^2}\int\rho(r')r'\cos\theta\,dv'\right.$$

$$\left. + \frac{1}{r^3}\int\rho(r')r'^2\tfrac{1}{2}(3\cos^2\theta - 1)\,dv' + \cdots\right] \tag{16.21}$$

The integral in the first term gives the total charge $Ze$, which from the point of view of nuclear structure is uninteresting. The second term vanishes for nuclei under ordinary circumstances because nuclear states are, to a very good approximation of the order of one part in $10^7$, states of definite parity. Going over to the quantum limit and replacing $\rho(r')$ by $\psi^*(r')\psi(r')$, the integral vanishes because the integrand is an odd function of the coordinates. (Simplify the geometry somewhat by choosing the origin at the center of the nuclear charge distribution, and let $r$ define the $z$ axis. Then $r'\cos\theta$ is $z'$, and under the parity operation $z' \to -z'$ while $|\psi(r')|^2 = |\psi(-r')|^2$. The integrand is therefore odd and the integral vanishes.) The first "interesting" term in the multipole expansion is the quadrupole term, and we define *the* nuclear quadrupole moment as

$$eQ = \int\rho(r')r'^2(3\cos^2\theta' - 1)\,dv' \tag{16.22}$$

where, as in the case of the magnetic dipole moment, we refer to a specific choice of reference axis—we measure $\theta'$ from the axis corresponding to the maximum projection of the nuclear spin.

The nuclear quadrupole moment tells us whether nuclei are spherical (for which $Q = 0$) or nonspherical. If $Q > 0$, the nuclei are *prolate* deformed—in the expression (16.22), the quantity $r'^2(3\cos^2\theta' - 1) = 3z'^2 - r'^2$ is on the average positive. That is, there is more of the nuclear charge density along the $z'$ axis than within the average radius. Figure 16.3a illustrates that case. If $3z'^2 - r'^2$ is negative, the $z'$ axis contains less of the nuclear charge density and there is a

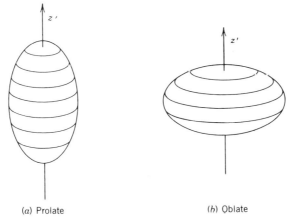

(a) Prolate                              (b) Oblate

**Figure 16.3**   Prolate and oblate charge distributions.

corresponding flattening. In this case $Q < 0$ and the deformation is *oblate* (Figure 16.3b).

The energy of interaction of the nuclear charge distribution with an *externally supplied* (perhaps from the atomic electrons) potential $V_{ext}$ is

$$E = \int \rho(r') V_{ext}(r') \, dv' \tag{16.23}$$

again integrated over the nuclear volume. (Consider how this reduces to the familiar expression for a point charge in an external field when $V_{ext}$ = constant.) If we expand $V_{ext}$ in a Taylor series about the center of the nucleus, then there is a constant term depending on $V_{ext}(0)$, which is of no interest, a dipole term which involves integrals such as

$$\int \rho(r') z' \left( \frac{\partial V_{ext}}{\partial z'} \right)_{z'=0} dv'$$

which vanishes by the same parity argument presented above, and a nonvanishing quadrupole term, proportional to integrals of the form

$$\int \rho(r') z'^2 \left( \frac{\partial^2 V_{ext}}{\partial z'^2} \right)_{z'=0} dv'$$

In all there are nine possible terms (involving $x'^2$, $x'y'$, etc.). If the external field has cylindrical symmetry (as in many cases of interest for atoms), then we can reduce the electric quadrupole contribution to the energy to the following form:

$$E_Q = \tfrac{1}{4}(eQ')\left(\tfrac{3}{2}\cos^2\theta - \tfrac{1}{2}\right)\left( \frac{\partial^2 V_{ext}}{\partial z^2} \right)_{z=0} \tag{16.24}$$

where $\theta$ is now the angle between the symmetry axis (now the $z$ axis) of $V_{ext}$ and the nuclear symmetry axis. The quadrupole moment $Q'$ is calculated with respect to the $z$ axis (the symmetry direction of $V_{ext}$), while $Q$ of Equation 16.22 is calculated with respect to the nuclear symmetry direction $z'$. In evaluating Equation 16.24, we must take into account the directional relationships among these different reference systems. The nuclear angular momentum has component $I_z$ relative to the chosen $z$ axis, and thus

$$\cos\theta = \frac{I_z}{|I|} = \frac{m_I}{\sqrt{I(I+1)}} \qquad (16.25)$$

Evaluating the expression $eQ'(\frac{3}{2}\cos^2\theta - \frac{1}{2})$, with $Q$ defined as always with respect to the axis of maximum projection of $I_z$, the result is

$$E_Q = \frac{1}{4}eQ\,\frac{3m_I^2 - I(I+1)}{I(2I-1)}\left(\frac{\partial^2 V_{\text{ext}}}{\partial z^2}\right)_{z=0} \qquad (16.26)$$

In Section 16.3 we consider the case in which the angle $\theta$ is determined by the relationship between the nuclear spin $I$ and atomic spin $J$.


## 16.3  HYPERFINE STRUCTURE

Hyperfine structure was originally taken to include those atomic effects (much smaller than the fine structure) that arise from the coupling between the electronic and nuclear angular momenta. It is thus an "internal" effect in atoms, and we cannot switch it off or modify it except by changing the nuclear or electronic structure (going to excited states, for instance). These effects were first studied by optical spectroscopists, who observed them as small perturbations in the structure of spectral lines. Modern techniques using lasers have extended these measurements to unprecedented levels of precision.

In recent years, hyperfine structure has come to include all effects that originate with the coupling of nuclear spins and moments with their environment, including the atomic electrons. The environment is often under the direct control of the experimenter, who can alter the hyperfine structure by, for example, changing an externally applied magnetic field. In this section we adopt this broad interpretation of hyperfine interactions.

Atomic states are labeled using the spectroscopic notation $n^{2S+1}L_J$, in which $L$ is indicated by the usual designation S, P, D, F, ... corresponding to $L = 0, 1, 2, 3, \ldots$ . For atomic states with only a single electron, such as the alkali atoms, the atomic spectroscopic notation is similar to the conventional notation used to designate individual electron states. Thus the sodium ground state, with its $3s_{1/2}$ electron, would be represented as $3^2S_{1/2}$. The principal quantum number $n$ is often not indicated.

We will use $I$ to represent the total nuclear angular momentum (the nuclear spin). Similarly, $J$ will represent the total (intrinsic plus orbital) electronic angular momentum. In ideal hydrogenic atoms, the electron moves in the nuclear Coulomb potential in quantum states of well defined orbital angular momentum $L$. Including the electron spin gives a second label $S$. In principle, it should not matter whether we label the electronic states of this ideal atom by the set of quantum numbers $L$, $m_L$, $S$, $m_S$ or the set $L$, $S$, $J$, $m_J$. However, the *spin-orbit* interaction, which produces the *fine structure* of electronic levels, couples $L$ and $S$ in such a way that $m_L$ and $m_S$ are no longer well-defined, and the coupling

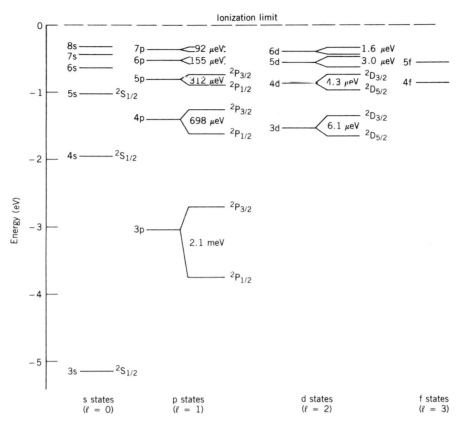

**Figure 16.4** Energy levels in atomic sodium. To the left of each level is indicated its electronic configuration; to the right is shown the spectroscopic notation for the atomic state. The fine-structure doublets are labeled with the total angular momentum $j = \ell \pm \frac{1}{2}$. The relative spacings of the doublets are consistent only within a particular $\ell$ value; that is, the P-state and D-state splittings are not to scale, but the decrease of splittings within the P states is to scale.

$J = L + S$, must be used to identify the real states. This spin-orbit interaction can be most easily understood in a semiclassical picture in the reference frame of the electron, in which the nucleus produces a current loop, which gives rise to a magnetic field at the location of the electron; this magnetic field interacts with the spin magnetic moment $\mu_s$ of the electron to give the spin-orbit contribution to the fine structure. Figure 16.4 shows the energy levels of sodium with and without the spin-orbit effect. (Of course, we cannot "turn off" the effect—it is always present.) The interaction energy between the magnetic field of the apparent motion (which is proportional to $L$) and the spin magnetic moment (which is proportional to $S$) is

$$E = -\mu_S \cdot B = f(r)L \cdot S \qquad (16.27)$$

where $f(r)$ is some function of the coordinates. Elementary quantum theory permits us to calculate the effect of this term on the energy levels, but to do so,

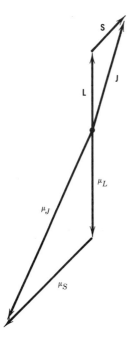

**Figure 16.5** Vector model for **J** and $\mu_J$. Note that $\mu_L$ is antiparallel with **L** and $\mu_S$ with **S**, but not $\mu_J$ with **J**.

we must evaluate the product $\boldsymbol{L} \cdot \boldsymbol{S}$. For this we use a standard trick:

$$\boldsymbol{J} = \boldsymbol{L} + \boldsymbol{S} \tag{16.28}$$

$$(\boldsymbol{J})^2 = (\boldsymbol{L})^2 + 2\boldsymbol{L} \cdot \boldsymbol{S} + (\boldsymbol{S})^2 \tag{16.29}$$

$$\boldsymbol{L} \cdot \boldsymbol{S} = \tfrac{1}{2}\left[(\boldsymbol{J})^2 - (\boldsymbol{L})^2 - (\boldsymbol{S})^2\right] \tag{16.30}$$

$$\langle \boldsymbol{L} \cdot \boldsymbol{S} \rangle = \tfrac{1}{2}\hbar^2\left[J(J+1) - L(L+1) - S(S+1)\right] \tag{16.31}$$

where the last step follows from replacing each squared angular momentum by its quantum mechanical expectation value.

Associated with this coupling is a magnetic moment, denoted by $\mu_J$:

$$\mu_J = \mu_L + \mu_S \tag{16.32}$$

where $\mu_L$ and $\mu_S$ are the magnetic moments associated with the orbital and spin motions:

$$\mu_L = -g_L L \mu_B/\hbar \qquad \mu_S = -g_S S \mu_B/\hbar \tag{16.33}$$

where $\mu_B$ is the Bohr magneton. The $g$ factors have the values $g_L = 1$ and $g_S = 2.00232$ (or, to a good approximation, $g_S = 2$). We cannot write similarly $\mu_J = -g_J J \mu_B/\hbar$ because $\mu_J$ and $\boldsymbol{J}$ are not in general parallel—see Figure 16.5. Instead, we redefine the value of the observable $\mu_J$ as that component of $\mu_J$ along the direction of $\boldsymbol{J}$, in which case

$$\mu_J = -g_J J \mu_B \tag{16.34}$$

where now $J$ is the quantum number, not the vector $\boldsymbol{J}$. In this case $g_J$ is given by the Landé formula

$$g_J = g_L \frac{J(J+1) + L(L+1) - S(S+1)}{2J(J+1)}$$

$$+ g_S \frac{J(J+1) - L(L+1) + S(S+1)}{2J(J+1)} \tag{16.35}$$

The reason for considering this spin-orbit coupling (which involves no nuclear properties) in such detail, is that the calculation of the magnetic dipole *hyperfine* coupling (which does involve the nucleus) is exactly analogous. The motion of the electrons produces a magnetic field $\boldsymbol{B}_e$ at the nucleus, which interacts with the nuclear magnetic moment:

$$E_M = -\boldsymbol{\mu}_I \cdot \boldsymbol{B}_e \tag{16.36}$$

If we now take $\boldsymbol{B}_e \propto \boldsymbol{J}$ and $\mu_I = g_I I \mu_N$, then we obtain the magnetic hyperfine interaction

$$E_M = A\boldsymbol{I} \cdot \boldsymbol{J}/\hbar^2 \tag{16.37}$$

where $A$ may involve many atomic or nuclear properties. Using the same procedure as above to evaluate the product $\boldsymbol{I} \cdot \boldsymbol{J}$, we first define a new total (electronic plus nuclear) angular momentum vector,

$$\boldsymbol{F} = \boldsymbol{I} + \boldsymbol{J} \tag{16.38}$$

The vector coupling of $\boldsymbol{I}$ and $\boldsymbol{J}$ to $\boldsymbol{F}$ allows the vector $\boldsymbol{F}$ to range from $\boldsymbol{I} + \boldsymbol{J}$ to $|\boldsymbol{I} - \boldsymbol{J}|$. The quantum number $F$ can take a similar range, from $I + J$ to $|I - J|$ in steps of 1. If $I > J$, the possible $F$ values are $I + J$, $I + J - 1$, $I + J - 2, \ldots,$ $I - J + 1$, $I - J$. There are thus $2J + 1$ different possible $F$ values. If $J > I$, a similar argument shows that the number will be $2I + 1$. The combined atomic and nuclear angular momentum states thus consist of a multiplet of levels whose number is the minimum of $2I + 1$ or $2J + 1$.

Continuing in the calculation of $\boldsymbol{I} \cdot \boldsymbol{J}$ as we did above for $\boldsymbol{L} \cdot \boldsymbol{S}$,

$$(\boldsymbol{F})^2 = (\boldsymbol{I})^2 + 2\boldsymbol{I} \cdot \boldsymbol{J} + (\boldsymbol{J})^2 \tag{16.39}$$

$$\boldsymbol{I} \cdot \boldsymbol{J} = \tfrac{1}{2}\left[ (\boldsymbol{F})^2 - (\boldsymbol{I})^2 - (\boldsymbol{J})^2 \right] \tag{16.40}$$

$$\langle \boldsymbol{I} \cdot \boldsymbol{J} \rangle = \tfrac{1}{2}\hbar^2 \left[ F(F+1) - I(I+1) - J(J+1) \right] \tag{16.41}$$

Let's consider, for example, the case of sodium, which we will assume to consist entirely of $^{23}$Na ($I = \tfrac{3}{2}$). The $s_{1/2}$ electron gives the ground-state configuration of $^2S_{1/2}$ ($J = 1/2$), and so the possible values of $F$ are, according to the usual vector-coupling rules of quantum mechanics, $F = 2$ and $F = 1$. The lowest atomic excited states of Na result from the electronic 3p states and are the spin-orbit doublet $^2P_{1/2}$ and $^2P_{3/2}$; the transitions from these states to the ground state are the well-known yellow "D lines" at $\lambda = 589.0$ and $589.6$ nm. The fine-structure splitting $E(^2P_{3/2}) - E(^2P_{1/2})$ is 0.0021 eV. The $F$ states of $^2P_{1/2}$ are $F = 2$ and $F = 1$, while for $^2P_{3/2}$ the vector coupling gives $F = 3, 2, 1, 0$. Figure 16.6*a* shows the energy levels of Na including the hyperfine splitting. (The parameter $A$ is different for each hyperfine multiplet.) Typical energy differences of hyperfine multiplets are only about $10^{-7}$ to $10^{-6}$ eV, far smaller than the fine

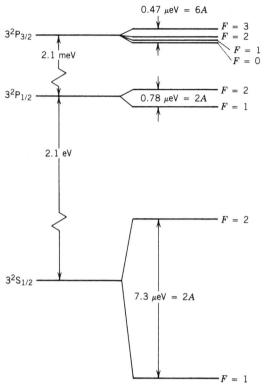

**Figure 16.6** (a) Hyperfine splitting in sodium. Each multiplet of $F$ states has a different value of the splitting parameter $A$.

structure. With modern techniques using tunable lasers, these hyperfine splittings can be measured with great precision. Figure 16.6b shows an example of a measurement of the hyperfine multiplets of the sodium D lines. The observed transitions have very nearly their natural linewidths. For the commonly observed electric dipole transitions, the selection rule $\Delta F = 0, \pm 1$ is followed.

Let's examine more closely the energy differences between the $F$ states of the $^2P_{3/2}$ multiplet of $^{23}$Na. The energy difference between adjacent $F$ states ($F + 1$ and $F$) should be, according to Equations 16.37 and 16.41,

$$\Delta E = E(F + 1) - E(F) = \frac{A}{\hbar^2}[\langle \boldsymbol{I} \cdot \boldsymbol{J} \rangle_{F+1} - \langle \boldsymbol{I} \cdot \boldsymbol{J} \rangle_F] \qquad (16.42)$$

$$= \frac{A}{2}[(F + 1)(F + 2) - F(F + 1)] \qquad (16.43)$$

$$= A(F + 1) \qquad (16.44)$$

The interval $F = 1 \rightarrow F = 0$, with $\Delta E/h = 16.4$ MHz, gives $A/h = 16.4$ MHz. For the interval $F = 2 \rightarrow F = 1$, with $\Delta E/h = 36.1$ MHz, we find $A/h = 18$ MHz, while $F = 3 \rightarrow F = 2$ gives $A/h = 20.8$ MHz. Since $A$ should involve only the atomic or nuclear state, which are identical for the members of the multiplet, we expect $A$ to be constant for a given multiplet, while these values

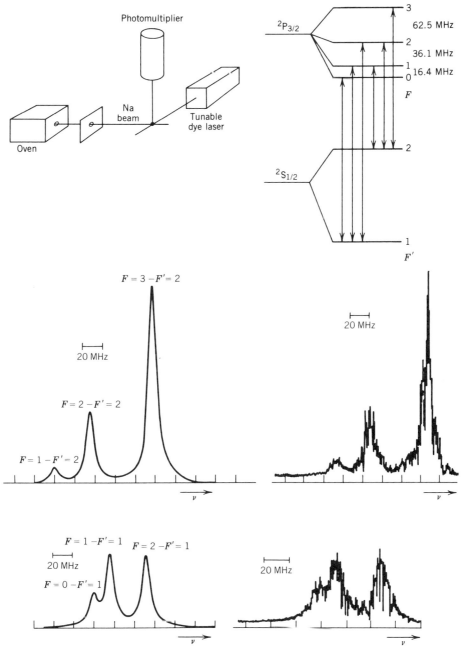

**Figure 16.6** (*b*) Hyperfine structure in sodium detected by atomic-beam fluorescence spectroscopy. As the frequency of the laser is varied, resonant absorption and re-emission will occur at frequencies corresponding to the $^2S_{1/2}$–$^2P_{3/2}$ energy difference, and the photomultiplier detects the fluorescent radiation. At top right are shown the hyperfine multiplets of the $^2S_{1/2}$ and $^2P_{3/2}$ states, along with the permitted $\Delta F = \pm 1, 0$ electric dipole transitions. The calculated and observed fluorescence spectra are shown at bottom for the two sets of transitions originating with the $F' = 1$ and $F' = 2$ members of the ground-state multiplet. The deduced $^2P_{3/2}$ hyperfine splittings are shown. Data from W. Lange et al., *Opt. Commun.* **8**, 157 (1973).

derived from the observed splittings are not at all constant. The difference arises from the *electric quadrupole* part of the hyperfine structure. Up to now we considered only the interaction of the magnetic field of the electronic motion with the magnetic dipole moment of the nucleus. We now must consider how the electric field gradient $\partial^2 V / \partial z^2$ established by the electronic motion interacts with the nuclear electric quadrupole moment $Q$. We will not derive the expression here, but merely state the result:

$$E_Q = eQ \frac{\partial^2 V}{\partial z^2} \frac{\left[ 3\langle \mathbf{I} \cdot \mathbf{J} \rangle^2 \hbar^{-4} + \frac{3}{2}\langle \mathbf{I} \cdot \mathbf{J} \rangle \hbar^{-2} - I(I+1)J(J+1) \right]}{2I(2I-1)J(2J-1)} \quad (16.45)$$

It can be shown that this vanishes for $I = \frac{1}{2}$ or for $J = \frac{1}{2}$, and thus there is no contribution to the $^{23}$Na structure in the $^2S_{1/2}$ or $^2P_{1/2}$ states. In the $^2P_{3/2}$ state, we can work out the intervals

$$\Delta E_{3 \to 2} = 3A + B \quad (16.46)$$

$$\Delta E_{2 \to 1} = 2A - B \quad (16.47)$$

$$\Delta E_{1 \to 0} = A - B \quad (16.48)$$

where $B = eQ(\partial^2 V / \partial z^2)$. This is the *electric quadrupole hyperfine parameter*, and is analogous to the magnetic dipole hyperfine parameter $A$. The values $A/h = 19.7$ MHz, $B/h = 3.3$ MHz reproduce the observed intervals fairly well.

The precise determination of the quantities $A$ and $B$, based on measuring the energies of the transitions between the $F$ states, gives us a way to determine the nuclear moments $\mu$ and $Q$. Extracting $\mu$ and $Q$ from $A$ and $B$, however, requires knowledge of the quantities $\mathbf{B}_e$ and $\partial^2 V / \partial z^2$, the magnetic field and electric field gradient due to the electronic motion. Since these are very difficult to calculate precisely, in practice what is usually done is to measure the hyperfine structure parameters in two isotopes, one of which has a known moment. Taking the ratio between the hyperfine parameters gives the ratio between the moments:

$$\frac{A_1}{A_2} = \frac{(\mu_I/I)_1}{(\mu_I/I)_2} \quad (16.49)$$

where 1 and 2 refer to the known and unknown moments, respectively. Because nuclei are not points but have finite volume, the hyperfine interaction $-\mu_I \cdot \mathbf{B}_e$ should properly be calculated as an integral over the nuclear volume, and if isotopes 1 and 2 have different nuclear radii, the evaluations of this integral over the volumes of the two nuclei may show a slight effect of the differences in volume. A small correction to the ratio $A_1/A_2$, amounting generally to less than 1%, is usually necessary. This correction, called the *hyperfine anomaly*, is dependent on the nuclear structure and can itself reveal interesting properties of the nucleus.

As a final topic, we consider the *Zeeman effect*, which results from placing the atom in an external magnetic field $\mathbf{B}_{ext}$. The complete interaction is, neglecting the electric quadrupole effect,

$$E = A\mathbf{I} \cdot \mathbf{J}/\hbar^2 - \mu_J \cdot \mathbf{B}_{ext} - \mu_I \cdot \mathbf{B}_{ext} \quad (16.50)$$

because both the nuclear and electronic magnetic moments interact with $\mathbf{B}_{ext}$. If the magnetic field is very weak, so that $\mu_J B_{ext} \ll A$, then it is appropriate to use

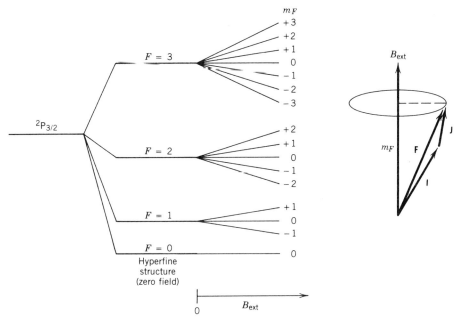

**Figure 16.7** Zeeman splitting of $^2P_{3/2}$ hyperfine states when $I = \frac{3}{2}$. I and J are coupled to **F**, which has a constant projection $m_F$ along the external field direction.

the notation of the $F$ states. We take

$$\boldsymbol{\mu}_J = -g_J \boldsymbol{J}\mu_B/\hbar \tag{16.51}$$

$$\boldsymbol{\mu}_I = g_I \boldsymbol{I}\mu_N/\hbar = g_I' \boldsymbol{I}\mu_B/\hbar \tag{16.52}$$

where $g_I' = g_I \mu_N/\mu_B$, a very small number compared with $g_J$. Then

$$E = E_F + g_F \mu_B B_{ext} m_F \tag{16.53}$$

where $E_F$ is the energy of the $F$ state in the absence of $\boldsymbol{B}_{ext}$, as in Equation 16.37, $m_F$ is the magnetic quantum number associated with $F$ ($m_F = -F, -F + 1, \dots, F - 1, F$), and $g_F$ is a combined $g$ factor calculated in analogy to the Landé formula:

$$g_F = g_J \frac{F(F + 1) + J(J + 1) - I(I + 1)}{2F(F + 1)}$$

$$g_I' \frac{F(F + 1) - J(J + 1) + I(I + 1)}{2F(F + 1)} \tag{16.54}$$

Figure 16.7 illustrates the effect of the magnetic field on some representative $F$ states. Note that the sublevels are split equally (by an amount $g_F \mu_B B_{ext}$), and that the splitting increases linearly with $B_{ext}$. This condition is true in the weak field region; typically, $A/h \sim 100$ MHz and $\mu_J \sim 1\ \mu_B$, so that $B_{ext} \ll A/\mu_J \sim (100\ \text{MHz})(h)/\mu_B \sim 10^{-2}$ T. For fields comparable to or greater than 0.01 T (100 G), this approximation is not valid.

The reason for the loss in validity of the approximation is that the interaction of $\mu_J$ with $B_{ext}$ becomes at high fields comparable to the term $A\boldsymbol{I} \cdot \boldsymbol{J}$ that is

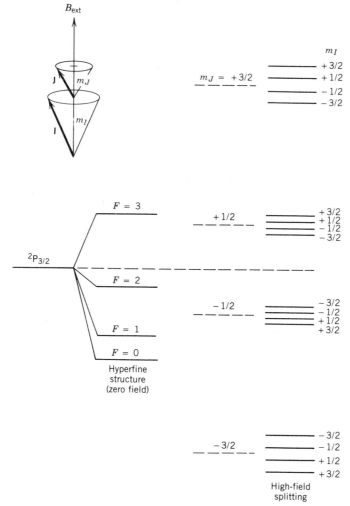

**Figure 16.8**   Splitting of states in a large magnetic field. I and J are decoupled and have their separate components $m_I$ and $m_J$ along $\mathbf{B}_{ext}$.

responsible for the coupling of $\boldsymbol{I}$ and $\boldsymbol{J}$ to $\boldsymbol{F}$. This large field breaks the coupling of $\boldsymbol{I}$ and $\boldsymbol{J}$, and it no longer is proper to consider the states in terms of $F$. Instead, the labels $I, m_I$ and $J, m_J$ become appropriate, and the energies are

$$E = Am_I m_J + g_J \mu_B B_{ext} m_J - g_I \mu_N B_{ext} m_I \qquad (16.55)$$

The Zeeman effect in this limit is called the *Paschen-Back effect*. At these high fields ($\gg 10^{-2}$ T), the second term in the energy becomes the most important. Figure 16.8 shows the energy levels in the high-field limit. You can see that the splitting of the $m_J$ states is the major effect, with each $m_J$ level broken up into $2I + 1$ $m_I$ levels, split by a much smaller amount (less than the $m_J$ splitting by $\mu_N/\mu_B \sim 10^{-3}$). The dependence on the external field is again linear.

In the intermediate-field region $B_{ext} \sim A/\mu_J$, no such simplifications are possible. The $m_F$ states of the low-field region go gradually to the $m_I, m_J$ states of the high-field region (because $m_F = m_I + m_J$), but the states become mixed; that is, the state with $m_F = 1$ will include a mixture of the states with $F = 1, 2,$ and 3, and it is not correct to try to associate a definite $F$ value with states in the intermediate field region. Figure 16.9 shows an example of the behavior of the states in the intermediate region. The smooth curves that connect to $F = 2$ and $F = 1$ at zero field indicate that those particular intermediate-field states become the appropriate pure $F$ states at zero field, not that they are always associated with that $F$ value.

Summarizing the effect of hyperfine structure, we have seen that: (1) The coupling of the nuclear and electronic total angular momentum $I + J$ to $F$ gives hyperfine states ranging from $F = I + J$ to $F = |I - J|$. If $I < J$ there are $2I + 1$ of these states (or $2J + 1$, if $J < I$). Counting these states can then give a direct measure of the nuclear spin $I$. (2) The energy splitting between the $F$ states in zero field gives the hyperfine structure parameters $A$ and $B$, which can in turn give the nuclear moments $\mu$ and $Q$. (3) The splitting of the states in an external magnetic field gives $g_F$ for weak fields or $g_I$ and $g_J$ for strong fields.

In the next section we discuss applications of these principles in the measurement of nuclear moments.

## 16.4 MEASURING NUCLEAR MOMENTS

### Proton and Neutron Moments

The magnetic dipole moment of the electron is a case of developments in theory and experiment going hand in hand to produce an understanding of a fundamental property of nature. For an ideal Dirac electron with spin $\frac{1}{2}$, we would expect $g = -2$ exactly (that is, $\mu_e$ is exactly one Bohr magneton). The measured value is very close but not exact: $\mu_e = -1.0011596567 \pm 0.0000000035 \, \mu_B$, and theory based on quantum electrodynamics is able to reproduce this small deviation from ideal Dirac behavior to within the experimental error of less than 1 part in $10^8$. This is one of the most remarkable agreements between theory and experiment in all of physics; it is a triumph in which theorists and experimentalists can share pride in the refinements of their respective crafts.

The case of the nucleons is unfortunately not so satisfying. First of all, the $g$ factors deviate considerably from the expected Dirac behavior for protons (measured 5.6 vs expected 2) and neutrons (measured $-3.8$ vs expected 0). Dirac behavior requires point particles, which the proton and neutron certainly are not. Moreover, present-day theory based on the quark model (see Chapter 18) gives agreement with experiment for the ratio $\mu_p / \mu_n$ only to about 1%, and the agreement with the individual values of $\mu_p$ and $\mu_n$ is much worse. Nevertheless, it still is necessary to obtain the best possible experimental values of $\mu_p$ and $\mu_n$, in the expectation that theory will soon catch up with experiment in helping us to understand these fundamental properties.

The difficulty of making absolute determinations of magnetic moments is that what we actually measure is an energy or a frequency that involves the product of the magnetic moment and magnetic field, $\mu B$. Magnetic fields can never be made sufficiently homogeneous over a bulk sample nor be measured to sufficient

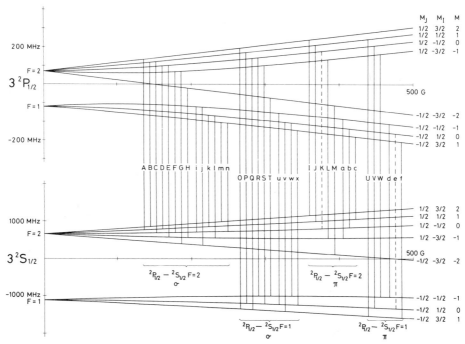

**Figure 16.9** Splitting of the $^2S_{1/2}$ and $^2P_{1/2}$ states of sodium in a magnetic field. The field increases to the right in the above figure. At low fields, the splitting is linear in the field and the states are labeled with $m_F$, as in Figure 16.7; at high fields, the states form multiplets of $m_I$ and $m_J$, as in Figure 16.8. At right are shown the $\Delta m = \pm 1$ dipole transitions for different field values up to 280 G. The transitions are labeled with letters corresponding to those in the diagram above. The measurement was made by passing a beam of sodium atoms through a static magnetic field and irradiating it with light from a tunable dye laser. Varying the laser frequency allows individual transitions to be observed. From L. Windholz, *Z. Phys.* **322**, 203 (1985).

accuracy to enable a determination of $\mu_p$ or $\mu_n$ to the order of 1 part in $10^8$. We therefore arrange to measure a ratio in which the field value cancels to first order and in which other corrections can be estimated to sufficient precision in second order. For the proton, we measure $\mu_p/\mu_e$; given the accurate measured value of $\mu_e$, it is then possible to determine $\mu_p$. For the neutron, it is $\mu_n/\mu_p$ that is measured, from which $\mu_n$ can be obtained.

The proton moment is measured by comparing proton and electron "spin-flip" resonances measured *simultaneously* in atomic hydrogen. Figure 16.10 shows the hyperfine structure in hydrogen in the "strong-field" region where $I$ and $J$ are decoupled. In an applied field of 0.35 T, there are three resonant transitions, called $\nu_{p1}$, $\nu_{p2}$, and $\nu_e$ in the figure. You can see from the quantum numbers that $\nu_{p1}$ and $\nu_{p2}$ flip the proton spin from $m_I = +\frac{1}{2}$ to $m_I = -\frac{1}{2}$ or the reverse, leaving the electron spin unchanged. On the other hand, $\nu_e$ flips the electron spin between $m_J = +\frac{1}{2}$ and $m_J = -\frac{1}{2}$, leaving the proton spin unaffected.

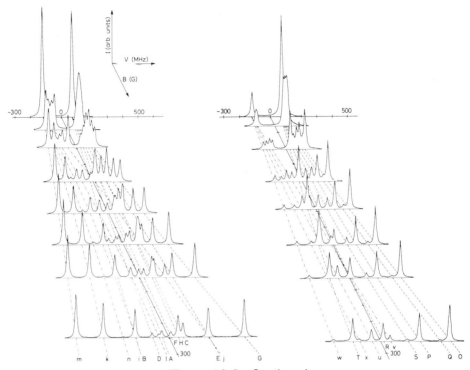

**Figure 16.9** Continued.

The hyperfine interaction is

$$E = \frac{A\mathbf{I}\cdot\mathbf{J}}{\hbar^2} - g_J\mu_B\frac{\mathbf{J}\cdot\mathbf{B}}{\hbar} - g_p\mu_B\frac{\mathbf{I}\cdot\mathbf{B}}{\hbar} \qquad (16.56)$$

where $g_J$ and $g_p$ represent the electron and proton $g$ factors, respectively. In the 1s state of atomic hydrogen, the Landé formula, Equation 16.35, shows that $g_J = g_e$. The four energy states of this interaction are

$$E_1 = \tfrac{1}{4}A(1 + 2x) - h\nu_p \qquad (16.57)$$

$$E_2 = -\tfrac{1}{4}A + \tfrac{1}{2}A(1 + x^2)^{1/2} \qquad (16.58)$$

$$E_3 = \tfrac{1}{4}A(1 - 2x) + h\nu_p \qquad (16.59)$$

$$E_4 = -\tfrac{1}{4}A - \tfrac{1}{2}A(1 + x^2)^{1/2} \qquad (16.60)$$

where $\nu_p$ is the proton nuclear magnetic resonance (NMR) frequency in the field $B$ $(\nu_p = g_p\mu_N B/h)$ and $x = (R + 1)h\nu_p/A$. The ratio $R \equiv -g_J/g_p$ is the object of the experiment. (The minus sign is present to make $R$ a positive number since $g_e$ is usually taken to be negative.) After a bit of algebra, we find

$$R = \frac{y + 1}{y - 1} \qquad (16.61)$$

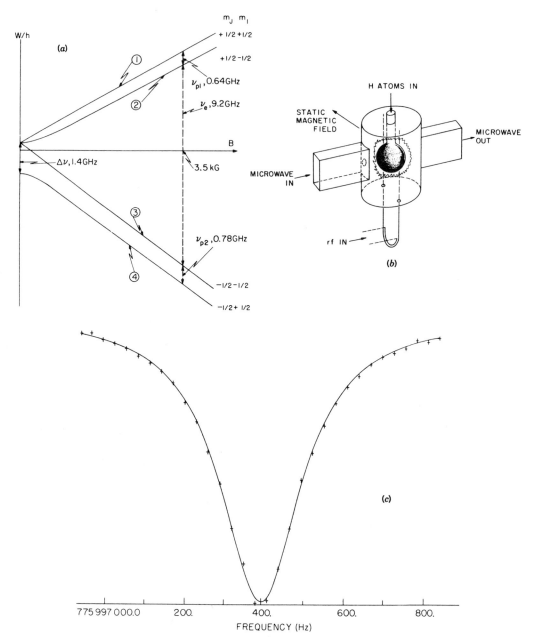

**Figure 16.10** (*a*) The energy levels of hydrogen in a static field. In the high-field region, the states are described by $m_I$ and $m_J$, and the transitions $\nu_{p1}$ and $\nu_{p2}$ flip the proton spins ($m_I = -\frac{1}{2} \leftrightarrow m_I = +\frac{1}{2}$). (*b*) A resonant cavity for observing the transitions. H atoms are kept in a storage bulb in a static field of 3.5 kG, where they experience microwave radiation at the frequency $\nu_e$ and simultaneous radio-frequency radiation at $\nu_{p1}$ or $\nu_{p2}$. Double resonance is observed by detecting the electron transition as the rf is varied, as shown in (*c*), and the frequency (after small corrections) is deduced to be $\nu_{p2} = 775{,}997{,}375.46 \pm 0.04$ Hz. From P. F. Winkler et al., *Phys. Rev. A* **5**, 83 (1972).

where

$$y = \frac{\nu_e - \nu_{p1}}{\nu_e + \nu_{p1}} \left( 1 + \frac{2A/h}{\nu_e - \nu_{p1}} \right) \qquad (16.62)$$

with a similar result based on $\nu_{p2}$ following from the substitution $\nu_{p2} = A/h - \nu_{p1}$. (The hydrogen 1s hyperfine splitting in zero field is one of the most precisely determined quantities in nature with a value $A/h = 1,420,405,751.768 \pm 0.002$ Hz.) The final result, in an experiment reported by P. F. Winkler et al., *Phys. Rev. A* **5**, 83 (1972), is $R = -g_J/g_p = 658.210706 \pm 0.000006$. There are several small corrections that must be applied, foremost among which is that relativistic and quantum electrodynamical effects modify the $g$ factors of protons and electrons in hydrogen from their "bare" values. After all corrections are put in, the resultant value is

$$\mu_p = 0.001\,521\,032\,181 \pm 0.000\,000\,000\,015\,\mu_B$$

or, in nuclear magnetons,

$$\mu_p = 2.7928456 \pm 0.0000011\,\mu_N$$

Notice that the result in units of $\mu_B$ has a precision of about 1 part in $10^8$, while the result in $\mu_N$ is much less precise. The difference lies in the relatively larger uncertainty in the value of the ratio $m_e/m_p$ needed to convert from $\mu_B$ to $\mu_N$.

The most precise recent measurement of the neutron magnetic moment was reported by G. L. Greene et al., *Phys. Rev. D* **20**, 2139 (1979). In this experiment the measured ratio was $R \equiv -\mu_n/\mu_p$, obtained using an apparatus capable of doing a resonant spin rotation of polarized protons and neutrons (not simultaneously, as in the most ideal experiment, but successively and in comparison with the same standard). Polarized particles were passed through the apparatus; the polarization was rotated by applying a signal at the Larmor frequency $\omega_L = \mu B/I\hbar$. If the spin rotates by 180° (or $3 \times 180°$, $5 \times 180°$, etc.) the polarization analyzer records a minimum in intensity. If the rotation is through a multiple of 360°, the initial and final spin directions are the same and a maximum number of particles passes through the polarization analyzer. Figure 16.11 shows the resonant Larmor precession effect on the recorded intensities of protons and neutrons. The static field was about 0.0018 T, giving Larmor frequencies in the range of 100 kHz. Neutrons were obtained from a nuclear reactor (at a mean velocity of about 180 m/s and a rate of $1.5 \times 10^5$/s), and protons in a stream of water were used. A central beam pipe could alternately be switched to carry neutrons or water, while an outer pipe always carried water. The neutron and proton resonances in the central pipe could be compared with the proton resonances in the water in the outer pipe, thus giving a common reference. Of course, small corrections must be made—for instance, the presence of water in the central pipe causes slight changes in the static field in the outer pipe, thus causing the standard reference frequency to change somewhat for protons compared to its value when the central pipe contains neutrons. Other corrections must be applied to the final result, particularly the difference between the magnetic moment of free protons and protons in water (amounting to about 2 parts in $10^5$, a small effect but large on the scale of the desired precision of 1 part in $10^8$). The net

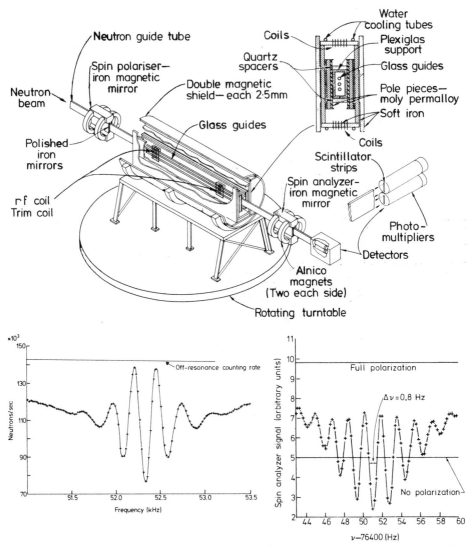

**Figure 16.11** Resonance apparatus for neutrons. Neutrons are polarized and pass through a static-field region where they can be resonated by a radiofrequency field. Protons (in water) can also be passed through the apparatus for comparison. Sample resonance curves are shown at bottom for neutrons (left) and protons (right). From G. L. Greene et al., *Phys. Rev. D* **20**, 2139 (1979).

result of Greene et al. is

$$\mu_n = -0.001\,041\,875\,64 \pm 0.000\,000\,000\,26\ \mu_B$$

or

$$\mu_n = -1.913\,041\,84 \pm 0.000\,000\,88\ \mu_N$$

As a final topic in this section, we consider the measurements of the magnetic dipole moment and electric quadrupole moment of the deuteron, $^2$H. These parameters are essential for understanding the nucleon-nucleon interaction, as we discussed in Chapter 4. The deuteron moments were measured by the molecular beam magnetic resonance technique (which is essentially the same as the atomic beam resonance method to be discussed in the next subsection) in a classic experiment by J. M. B. Kellogg, I. I. Rabi, N. F. Ramsey, and J. R. Zacharias, *Phys. Rev.* **57**, 677 (1940). A beam of $D_2$ molecules passes simultaneously through a region of static magnetic field (which gives a Zeeman splitting of the levels) and a varying radiofrequency (rf) field which causes resonances between the Zeeman states. In this resonance technique, the frequency $\nu$ of the rf field is kept constant, and the magnitude of the static field is varied until the Zeeman splitting of a particular pair of levels just equals $h\nu$. At this point there is absorption of the rf photons, and the apparatus is constructed so that a decrease in the transmitted intensity of the beam results. If the Zeeman levels are not equally split, several different resonances may be observed, each corresponding to the splitting of a particular pair of levels.

Because the two identical nuclei in $D_2$ have integral spins, the total wave function must be symmetric in the coordinates. The *molecular* ground state has rotational angular momentum $\ell = 0$, and the nuclear spin state must be symmetrical; thus the total nuclear spin must be $I_t = 0$ or 2. In the first excited rotational state $\ell = 1$, so the spin function is antisymmetric and $I_t = 1$. The source of the $D_2$ beam was kept at liquid-nitrogen temperature, so that the resulting Boltzmann population gave about 56% of the beam in the $\ell = 0$ state (with 1/6 of those having $I_t = 0$ and 5/6 having $I_t = 2$, in accordance with the relative statistical weights $2I_t + 1$), with another 33% in the $\ell = 1$, $I_t = 1$ state. The remaining 11% was in states with $\ell \geq 2$, which did not contribute to the experiment.

The resonance spectrum is shown in Figure 16.12. The deep central resonance is from the $\ell = 0$, $I_t = 2$ molecules and corresponds to the four equal-energy transitions between the five Zeeman states of the $I_t = 2$ multiplet: $m_I = +2 \rightarrow m_I = +1$, and so on. From the location of the resonance at $B = 0.201$ T in a rf

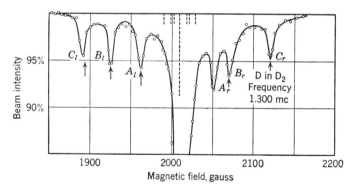

**Figure 16.12** Molecular beam resonance spectrum of $D_2$. The apparatus is of the type shown in Figure 16.14. The dashed lines would correspond to the expected locations of the transitions shown in Figure 16.13, including the spin-orbit term but not including the quadrupole interaction. The actual positions of the resonances deviate considerably and indicate the effect of the nuclear quadrupole moment. The transitions are labeled corresponding to Figure 16.13. From J. M. B. Kellogg et al., *Phys. Rev.* **57**, 677 (1940).

field of $\nu = 1.300$ MHz, we calculate as follows:

$$h\nu = \mu_D B$$

$$\mu_D = \frac{h\nu}{B} = \frac{\left(4.14 \times 10^{-15} \text{ eV} \cdot \text{s}\right)\left(1.300 \times 10^6 \text{ Hz}\right)}{0.201 \text{ T}}$$

$$= 2.68 \times 10^{-8} \text{ eV/T}$$

$$= 0.850 \ \mu_N$$

(A more precise value is today available by a direct comparison of the frequen-

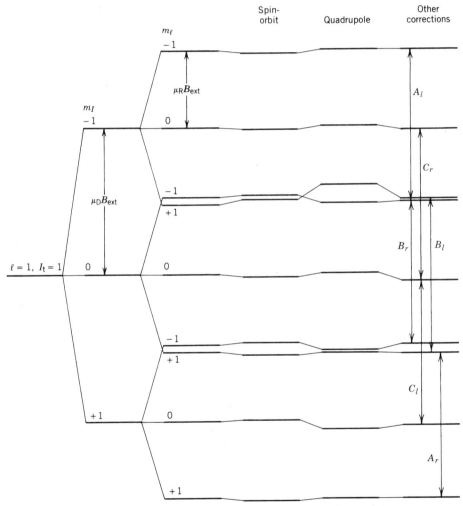

**Figure 16.13** Hyperfine structure of $D_2$ molecule. The external field gives the Zeeman splitting of the levels, arising from the nuclear moment $\mu_D$ and the rotational magnetic moment of the molecule $\mu_R$. Allowed transitons are those for which $\Delta m_\ell = 0$ and $\Delta m_I = \pm 1$. The spin-orbit and quadrupole interactions give a small correction to the levels, and other corrections, including the dipole-dipole interaction of the two nuclei, are also important. The allowed transitions are labeled to correspond with those in Figure 16.12.

cies of nuclear magnetic resonance for $^1$H and $^2$H: $\nu(^2\text{H})/\nu(^1\text{H}) = 0.15350609 \pm 0.00000002$; given the previous value for $\mu_p$, it follows that $\mu_D = 0.8574376 \pm 0.0000004$.)

The weaker satellite resonances in Figure 16.12 come from the $\ell = 1$, $I_t = 1$ states. Figure 16.13 shows the Zeeman splitting of the states. Considering only the magnetic interactions with $\mu_D$ and with the rotation of the molecule (think of the rotating molecule as a current loop that interacts with the field) there would be six possible $\Delta m_I = \pm 1$, $\Delta m_\ell = 0$ transitions, all of which would have the same energy (equal to the energy corresponding to the central resonance). Small effects, known for the $H_2$ molecule, lead to the predicted locations for the six resonances indicated by the dashed lines in Figure 16.12. The substantial differences between these predicted positions of the six lines and their observed positions are due to the quadrupole moment of the deuteron. From the spacing of the lines, with corrections for other small effects, the deduced quadrupole splitting corresponds to

$$\frac{e^2 \dfrac{\partial^2 V}{\partial z^2} Q}{\mu_D} = -70 \text{ G} = -0.0070 \text{ T}$$

and thus

$$e^2 \frac{\partial^2 V}{\partial z^2} Q = 1.89 \times 10^{-10} \text{ eV}$$

Calculating $\partial^2 V/\partial z^2$ from the molecular wave function is a difficult process, but it has been done and the resulting quadrupole moment is

$$Q = +0.00288 \pm 0.00002 \text{ b}$$

This original experiment of Kellogg et al. has remained for over forty years the best measure of the quadrupole moment of the deuteron.

## Stable Nuclear Ground States

**Atomic Beam Magnetic Resonance**   Much of the pioneering work on precise determination of nuclear magnetic dipole moments was done by I. Rabi and his colleagues, who developed the atomic beam magnetic resonance method. The apparatus is indicated schematically in Figure 16.14. A beam of atoms (or molecules) is prepared in an oven; the beam emerges with a Maxwellian speed distribution. The beam passes through an evacuated chamber containing three magnets A, C, and B. At the end of the chamber is a detecting device that records the beam current. Magnets A and B produce inhomogeneous fields (constant along the beam direction but varying at right angles to the beam direction). The fields in the A and B regions are in the same direction, but the field gradients $\partial B/\partial z$ are opposite. In region C, there is a homogeneous field $B_0$.

As in a Stern–Gerlach type of apparatus, the inhomogeneous field causes the beam to deflect, and those atoms that enter region A with the proper initial velocity will be deflected so as to pass through the slit S. Magnet B, with the opposite field gradient, refocuses the beam on the detector. For this to occur, the

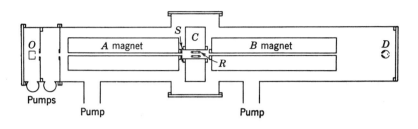

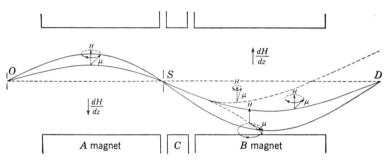

**Figure 16.14** Molecular beam resonance apparatus. The beam is produced by the oven $O$ and reaches the detector $D$ only if the beam passes through the slit $S$. A resonant radiofrequency field can change the $m$ state and therefore the orientation of the magnetic moment, in which case the molecules will miss the detector (dashed line). From I. I. Rabi et al., *Phys. Rev.* **53**, 318 (1938).

force in regions A and B must be equal in magnitude; since the force depends on the $z$ component of the atomic or nuclear angular momentum, if $J_z$ or $I_z$ is unchanged in passing through region C, then the identical field gradients give total refocusing and maximize the observed current. Because atomic magnetic moments are typically $10^3$ times larger than nuclear magnetic moments, atomic effects will ordinarily dominate in these experiments. If we wish to study nuclear moments, then we must use atoms in molecules in which the total electronic angular momentum is zero, which we assume to be the case in the following discussion.

In region C, the nuclear magnetic moments precess about $B_0$ with constant $I_z = m_I \hbar$. If we now apply perpendicular to $B_0$ an oscillatory field whose frequency $\nu$ is carefully chosen so that $h\nu$ is equal to the energy difference between the substates $m_I$ and $m_I \pm 1$, then the nuclei can absorb one of these photons and change the value of $I_z$. Nuclei that experience this absorption and change in $I_z$ will not be refocused in region B and consequently will miss the detector.

The energy of the nuclear magnetic moment $\mu = g_I I \mu_N$ in the field $B_0$ is

$$E = g_I B_0 \mu_N m_I \tag{16.63}$$

and the magnitude of the separation between the states $m_I$ and $m_I \pm 1$ is

$$\Delta E = g_I B_0 \mu_N \tag{16.64}$$

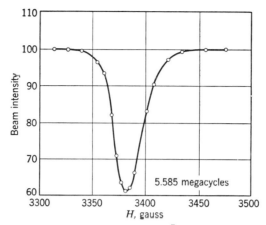

**Figure 16.15** Measurement of resonance of $^7$Li using apparatus of Figure 16.14. From I. I. Rabi et al., *Phys. Rev.* **53**, 318 (1938).

The resonance condition is thus

$$g_I B_0 \mu_N = h\nu \tag{16.65}$$

and either $\nu$ or $B_0$ can be varied until resonance is achieved, as observed by a decrease in the beam current reaching the detector.

Figure 16.15 shows early results of Rabi's group for a beam of $^7$Li. The resonance occurs for $B_0 = 0.3385$ T at a frequency of $\nu = 5.585$ MHz. From Equation 16.65 we calculate $g_I = 2.167$ and thus $\mu = 3.250\ \mu_N$ (since it is known that $I = \frac{3}{2}$).

**Nuclear Magnetic Resonance**   The process of nuclear magnetic resonance (NMR) shares some similarities with the atomic beam magnetic resonance process. The main differences are (1) the nuclei are contained in materials in solid or liquid

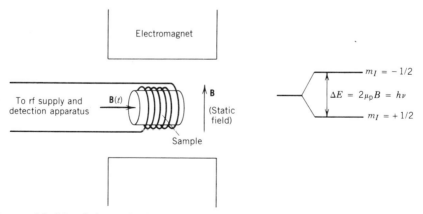

**Figure 16.16** Schematic diagram of nuclear magnetic resonance apparatus. For protons (or other spin-$\frac{1}{2}$ nuclei), the energy splitting of the two Zeeman states is shown at right; the frequency of the rf signal is varied until power is absorbed in transitions between the states.

form, and (2) the resonance is detected by observing the radiofrequency power absorbed at resonance. A typical apparatus is sketched in Figure 16.16. There is again a static field and a sinusoidally varying radiofrequency field at right angles.

For simplicity, let's consider a material, such as a hydrocarbon, that is rich in protons. The static field $B$ establishes the $z$ axis of the coordinate system, relative to which the proton spin can be parallel or antiparallel, that is $m_I = \pm \frac{1}{2}$. The energy of the parallel state is $-\mu_p B$ and that of the antiparallel state is $+\mu_p B$. The energy difference is

$$\Delta E = 2\mu_p B \tag{16.66}$$

and if we apply a sinusoidally varying signal at frequency $\nu = \Delta E/h$ (that is, if we flood the sample with photons of that frequency), the protons will absorb energy as they flip from one spin orientation to the other. Thus at

$$\nu = \frac{2\mu_p B}{h} \tag{16.67}$$

we will observe energy being absorbed as the protons flip back and forth. For a modest laboratory field of $B = 1$ T, the resonant frequency is

$$\nu = \frac{2(2.79)(5.05 \times 10^{-27} \text{ J/T})(1 \text{ T})}{6.63 \times 10^{-34} \text{ J} \cdot \text{s}}$$

$$= 42.5 \text{ MHz}$$

which is in the radiofrequency range of the spectrum.

In an actual experiment, the number of protons in either spin orientation is determined by the Boltzmann factor $e^{-\Delta E/kT}$. At room temperature, $kT \simeq 0.025$ eV, while $\Delta E = 1.8 \times 10^{-7}$ eV at $B = 1$ T. There is thus a relatively small imbalance of protons in the lower energy (parallel) state, of the order of $1 - e^{-\Delta E/kT} = 7 \times 10^{-6}$. Although one might expect that the protons in the higher state could emit photons by stimulated emission and thus replace virtually all of the photons absorbed by the lower state, this is not the case. Protons in the upper state can lose their energy and "fall back" to the lower state by transferring energy directly to the surrounding material through the process called *spin-lattice relaxation*. The spin-lattice relaxation time is typically in the range of seconds to milliseconds, depending on the nature of the host material and on the temperature.

---

**Figure 16.17** Hyperfine structure measurements of optical transitions. (Left) Magnetic hyperfine splitting in $^{181}$Ta ($I = \frac{7}{2}$). The spectra labeled A, B, and C show the results of different ionizing voltages producing different ionized states of Ta. The hyperfine splitting constants are $A = 25.50 \pm 0.15$ GHz and $4.86 \pm 0.15$ GHz for the atomic ground and excited states, respectively, from which the nuclear moment $\mu = 2.36 \pm 0.02$ $\mu_N$ is deduced. From J. Sugar and V. Kaufman, *Phys. Rev. C* **12**, 1336 (1975). (Right) Electric quadrupole splitting in Sb. The top shows the multiplet of unsplit atomic states, and the middle shows the hyperfine splittings for two of the atomic transitions, with $I = \frac{3}{2}$. At bottom is shown the observed splitting; the primed labels refer to $^{123}$Sb and unprimed labels to $^{121}$Sb. The deduced quadrupole moments are $Q(121) = -0.36 \pm 0.04$ b and $Q(123) = -0.49 \pm 0.05$ b. From B. Buchholz et al., *Z. Phys. A* **288**, 247 (1978).

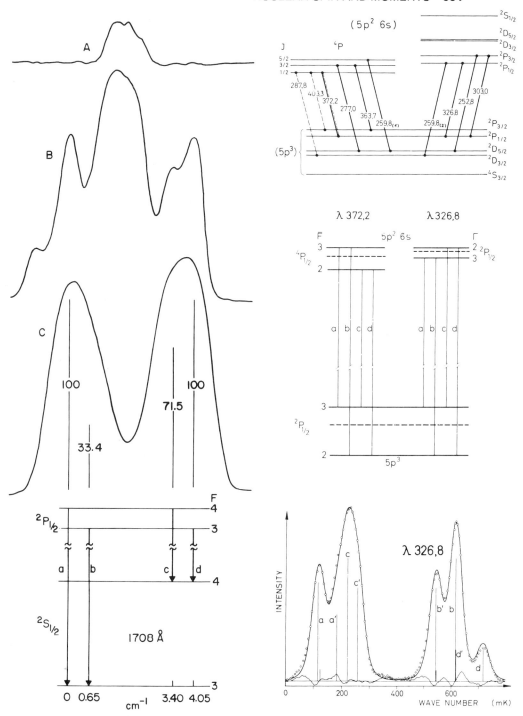

**Figure 16.17**

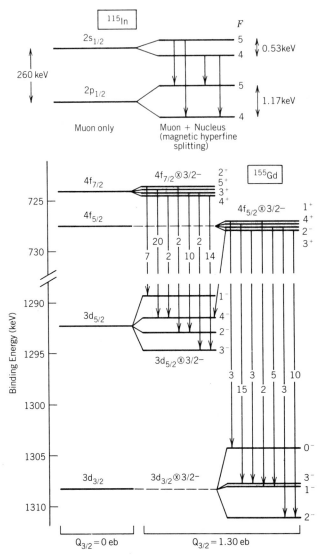

**Figure 16.18** Muonic X-ray spectra and hyperfine splitting. (Top) At top left are shown the $2s_{1/2}$ and $2p_{1/2}$ muonic (electron-like) states in $^{115}$In ($I = \frac{9}{2}$). The coupling between $J = \frac{1}{2}$ and $I = \frac{9}{2}$ gives $F = 4$ or 5 for both states, with four possible components to the transition. At top right is shown the observed spectrum; the four components, only barely resolved, are indicated by the vertical lines. The delayed γ ray comes from a nuclear reaction initiated by nuclear muon capture. Data from R. Link, *Z. Phys.* **269**, 163 (1974). (Bottom) Electric quadrupole hyperfine coupling in muonic Gd. At left are shown the hyperfine multiplets of the M X-rays (4f to 3d) resulting from coupling of the muonic spin-orbit doublets to the nuclear spin of $\frac{3}{2}$. The observed spectra are shown at right, with $^{155}$Gd and the similar $^{157}$Gd compared with spin-0 $^{154}$Gd. The vertical arrows at left indicate the individual transitions; the numbers on the arrows give their relative intensities. From the observed splittings, the nuclear quadrupole moment is deduced to be $Q = 1.30 \pm 0.02$ b. From Y. Tanaka et al., *Phys. Lett. B* **108**, 8 (1982).

Nuclear magnetic resonance can be applied to any nucleus (other than spin 0) in virtually any environment. The resonant frequency corresponds to the difference in energy between adjacent magnetic substates, as in Equation 16.66. Because we can measure resonant frequencies with great precision, magnetic moments can be precisely determined. We also use nuclei with known moments to probe the electronic environment, either through measuring the local value of $B$ or the spin-lattice relaxation time. Organic compounds are often enriched with isotopes such as $^{13}C$ or $^{17}O$ (because $^{12}C$ and $^{16}O$ have spin zero), and the NMR

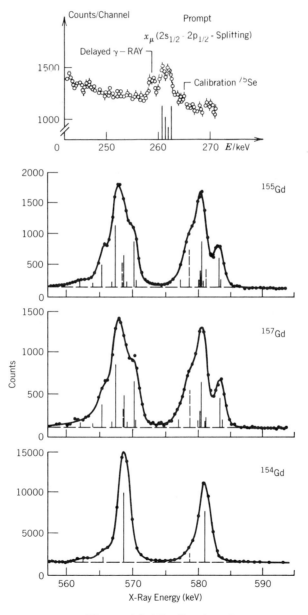

**Figure 16.18** Continued.

signal can be used to study the chemical bonding. In ferromagnets, the applied field is enhanced by a considerable factor at the nuclear site. In iron, which is magnetically saturated in applied fields of the order of 0.1 T, the field at an Fe nucleus is about 33 T. The atomic structure can have a significant effect on this field; if we prepare a dilute alloy of gold atoms in Fe, the field at the Au nuclei can be 115 T. In other ferromagnetic materials, such as the rare-earth metals, local fields of several hundred tesla are found. These so-called *impurity hyperfine fields* can be determined with great precision by NMR, and their systematic variation with impurity contributes to our knowledge of atomic and solid-state physics.

**Optical Hyperfine Structure**   If we could observe, at very high resolution, the optical transitions between the $F$ states, such as those of $^{23}$Na illustrated in Figure 16.6, we could deduce directly the hyperfine structure constant $A$. In practice, this may often be difficult, for the experimental linewidths may be the same as (or larger than) the hyperfine spacing. The "classical" method of optical spectroscopy, in which we excite the atoms thermally or by an electrical discharge and observe the emitted light with a high-resolution spectrometer, is thus of limited usefulness. More success has been obtained in recent years using tunable lasers, in which we pass a beam of the atoms through a laser beam and watch for resonance fluorescence; the atom absorbs and then re-emits light when the laser is tuned to the proper optical frequency corresponding to the difference between the $F$ states in the optical ground and excited multiplets. (The Na data of Figure 16.6 were obtained in this way.) Figure 16.17 shows some optical hyperfine spectra and the deduced nuclear moments.

One of the major difficulties of this method is the knowledge of the electronic wave functions required to go from the hyperfine parameters $A$ and $B$ to the nuclear moments $\mu$ and $Q$. For the alkali elements and certain others, the electronic wave functions are known well enough to permit extraction of $\mu$ and $Q$, but for many others, the many-electron wave functions are poorly known. One solution is to use a *muonic atom*, in which a single negative muon is captured into a set of hydrogen-like levels in an atom. Because the muon is about 200 times more massive than an electron, its orbits have only 1/200 of the radius of the corresponding electronic orbits. By the time the muon (which may be captured originally into a state of very large principal quantum number $n$) reaches about $n = 14$, it crosses the electronic 1s orbit; the inner muon orbits are thus well inside the electronic orbits and relatively unaffected by electron screening or other uncertainties arising from the electrons. We can therefore treat the muon with a (relativistic) one-electron calculation to high precision. The photons of the $2p \rightarrow 1s$ muonic transition, called $K_\alpha$ *muonic X rays*, have energies in the range of one to several MeV for medium to heavy nuclei. The muonic levels couple to the nucleus through the hyperfine interaction, exactly as in the case of electronic levels. The $s_{1/2}$ muonic level is thus split into two levels with $F = I \pm \frac{1}{2}$ where again $I$ is the nuclear angular momentum quantum number. The splitting involves the nuclear magnetic moment through the magnetic dipole hyperfine constant $A$, which has the same meaning as the electronic hyperfine constant but differs considerably in value because of the differences between the muonic and electronic magnetic moments and wave functions. In typical cases, $A$ for the muonic hyperfine splitting is of the order of $10^3$ eV, while for the electronic case it is of order of $10^{-6}$ eV. Even though the muon hyperfine splitting is larger, we cannot measure it as precisely because the best photon detectors have resolutions of a few keV in the MeV energy range and we cannot separate the muonic hyperfine components—what is observed is a broadening of the lines. Figure 16.18 shows some representative muonic X-ray spectra including hyperfine splitting. The technique is not extremely useful for measurements of magnetic dipole moments.

Electric quadrupole moments, on the other hand, can be determined with high precision because the muonic quadrupole hyperfine parameter $B$ can be 10 keV or greater, allowing the structure to be clearly resolved. The wave functions can again be obtained to high precision, and values of $Q$ can be obtained directly from $B$. Figure 16.18 shows a sample muonic spectrum split by the electric quadrupole interaction.

### Radioactive Ground States

**Optical Hyperfine Structure with Atomic Beams** One of the most exciting recent developments in the study of nuclear ground states is the ability to follow a chain of isotopes over 10–20 or more mass numbers. Because usually only few of these are stable or relatively long-lived, this requires that we have a technique capable of determining the moments of states that are very short-lived (half-lives of seconds). Moreover, to make careful comparisons of neighboring moments, we must be able to accomplish the measurement with high precision. Finally, as these short-lived species are produced in nuclear reactions, we must have a technique that can extract the measured values from observing relatively few atoms.

One basic method for these measurements uses a variation on the atomic beam magnetic resonance technique, and we consider for example the measurement of a sequence of Na isotopes. As shown in Figure 16.19, a beam containing a mixture of many neutron-rich Na isotopes is produced by bombarding a uranium target with protons. (Other reaction products are of course formed, but they do not affect the experiment and we can ignore them.) The atoms, originally in the $^2S_{1/2}$ ground state, are excited by a tunable dye laser to the $^2P_{1/2}$ state (the sodium $D_1$ line). Both states are split into hyperfine doublets with $F = I \pm \frac{1}{2}$ (but with different $A$ values). The laser excitation occurs in a weak magnetic field, where $m_F$ serves to label the substates, but then the beam travels to a high-field region, where $I$ and $J$ are decoupled and where the proper labels are $m_I$ and $m_J$. A focusing magnet is set to focus the atoms with $m_J = +\frac{1}{2}$ and to defocus those with $m_J = -\frac{1}{2}$. The focused atoms are ionized and identified using a mass spectrometer.

Let's consider the four transitions labeled a, b, c, and d in Figure 16.19. The $m_J = +\frac{1}{2}$ states come only from the $F = I + \frac{1}{2}$ level (because the passage of the beam from the low-field to the high-field region is adiabatic, so that the spin orientation is maintained). Transitions a and b depopulate the $m_J = +\frac{1}{2}$ substates by means of excitation to the $^2P_{1/2}$ levels. Of course, those levels decay quickly back down to $^2S_{1/2}$ before the atoms are collected by the mass spec-

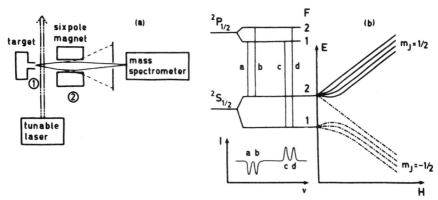

**Figure 16.19** Atomic beam resonance experiment for short-lived radioactive nuclei. The beam originates at left with an accelerator target, where the desired species is produced in a nuclear reaction. The laser excites the atoms from the $^2S_{1/2}$ ground state to the $^2P_{1/2}$ excited state; as shown at right there are four possible resonances among the states of the two hyperfine doublets. Only atoms with $m_J = +\frac{1}{2}$ are focused and counted at the mass spectrometer. As the laser frequency is varied, the upward transitions a and b decrease the counting rate of $m_J = +\frac{1}{2}$ atoms and the upward transitions c and d increase the counting rate. From G. Huber et al., *Phys. Rev. C* **18**, 2342 (1978).

trometer; some decays will populate $m_J = +\frac{1}{2}$ and some $m_J = -\frac{1}{2}$ substates of the $^2S_{1/2}$ ground state. There will, however, be a net loss of atoms in the $m_J = +\frac{1}{2}$ substate because all atoms excited by a and b transitions start with $m_J = +\frac{1}{2}$, but not all end that way. Similarly, the c and d transitions eventually contribute to *increase* the number of $m_J = +\frac{1}{2}$ atoms because none of those excited by c or d start with $m_J = +\frac{1}{2}$, but some may end that way following the decay of the $^2P_{1/2}$ level. As the laser frequency is tuned through the four resonances, decreases in the mass spectrometer output current occur at the energies of a and b, while increases occur for c and d.

Figure 16.20 shows the results of the experiment for the isotopes $^{21}$Na–$^{31}$Na. You can see an *isotope shift*—the pattern moves gradually to the right with increasing mass number and thus with increasing nuclear radius. For each isotope, we can find the hyperfine splitting of the $^2P_{1/2}$ state ($E_b - E_a$ or $E_d - E_c$) and the $^2S_{1/2}$ state ($E_c - E_a$ or $E_d - E_b$). Differences in the hyperfine constant $A$ from one isotope to another must be due to variations in the nuclear $g$ factor:

$$\frac{A_x}{A_{23}} = \frac{g_x}{g_{23}} \tag{16.68}$$

where the subscript 23 refers to the stable isotope $^{23}$Na that is used for comparison. By comparing the hyperfine splittings it is possible to determine the magnetic moments. Table 16.1 shows a summary of the measured values. Note the high precision (0.1%) and the short half-lives, down to the millisecond range. These small differences in the moments between otherwise similar states (such as the four $I = \frac{3}{2}$ states) can serve as keys to understanding small changes in nuclear structure arising from the addition of two neutrons.

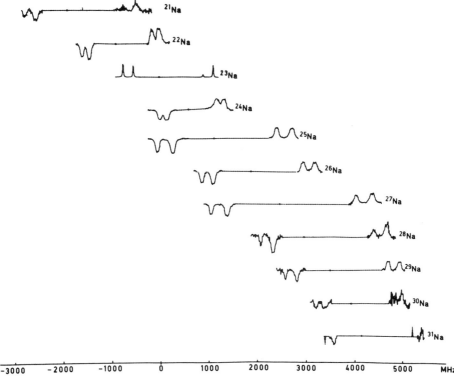

**Figure 16.20** Data obtained with the apparatus of Figure 16.19 for a variety of sodium isotopes. The changes in the realtive spacings of the doublets arise from differences between the magnetic moments of the isotopes. The gradual drift to the right is the isotope shift, which arises from the increase of the nuclear radius with increasing mass number. From G. Huber et al., *Phys. Rev. C* **18**, 2342 (1978).

Other measurements, using different techniques but having in common optical excitation with a tunable laser, have been used to observe sequences of Rb ($A = 76$ to 97), Cs ($A = 118$ to 145), and Ba ($A = 123$ to 145), and others as well.

**Low-Temperature Nuclear Orientation** In Section 10.5 the orientation of nuclear states at ultralow temperature ($T \ll 1$ K) was briefly discussed. Let's suppose we have a nucleus of spin $I$ and magnetic moment $\mu$, which we place in a magnetic field $B$. The result is a Zeeman splitting of the $2I + 1$ levels, corresponding to the different orientations between the vectors $I$ (or $\mu$) and $B$ (Usually we let $B$ define the $z$ axis.) For a typical large laboratory field of 1 T, the splitting $\Delta E$ is about

$$\Delta E = \frac{\mu_I}{I} B = g_I B \mu_N = 3 \times 10^{-8} \text{ eV}$$

for a large nuclear moment corresponding to $g_I = 1$. If we observe the decay of the nuclear state (such as by observing $\beta$ or $\gamma$ radiations emitted), we will not be able to detect the extremely small energy splitting, and if all of the $m_I$ states have

**Table 16.1** Hyperfine Parameters, Magnetic Moments, and Isotope Shifts for Sodium Isotopes

| $^{x}$Na | $T_{1/2}$ | $I$ | $A(^2S_{1/2})$ (MHz) | $\mu_I(\mu_N)$ | $A(^2P_{1/2})$ (MHz) | $IS^{23,\,x}$ (MHz) |
|---|---|---|---|---|---|---|
| $^{21}$Na | 22.5 s | $\frac{3}{2}$ | 953.233(11) | 2.38612(10) | 102.6(1.8) | $-1596.7(2.3)$ |
| $^{22}$Na | 2.60 y | 3 | 348.75(1) | 1.746(3) | 37.0(1) | $-758.5(7)$ |
| $^{23}$Na | stable | $\frac{3}{2}$ | 885.8130644(5) | 2.2175203(22) | 94.25(15) | |
| $^{24}$Na | 15.02 h | 4 | 253.185018(23) | 1.6902(5) | 28.2(2.7) | 706.4(6.2) |
| $^{25}$Na | 60.0 s | $\frac{5}{2}$ | 882.8(1.0) | 3.683(4) | 94.5(5) | 1347.2(1.3) |
| $^{26}$Na | 1.07 s | 3 | 569.4(3) | 2.851(2) | 61.0(3) | 1937.5(9) |
| $^{27}$Na | 290 ms | $\frac{5}{2}$ | 933.6(1.1) | 3.895(5) | 100.2(1.1) | 2481.3(2.0) |
| $^{28}$Na | 30.5 ms | 1 | 1453.4(2.9) | 2.426(3) | 156.0(2.7) | 2985.8(2.7) |
| $^{29}$Na | 43 ms | $\frac{3}{2}$ | 978.3(3.0) | 2.449(8) | 104.4(3.0) | 3446.2(3.8) |
| $^{30}$Na | 53 ms | 2 | 624.0(3.0) | 2.083(10) | 66.2(2.8) | 3883.5(6.0) |
| $^{31}$Na | 17 ms | $\frac{3}{2}$ | 912(15) | 2.283(38) | | 4286(16) |

From G. Huber et al., *Phys. Rev. C* **18**, 2342 (1978). Figures in parentheses indicate uncertainties.

the same population $p(m_I)$, then the radiation has a uniform (or isotropic) angular distribution, as we could show by considering the angular distribution for each possible transition $m_i \rightarrow m_f$ and adding them with the population factors $p(m_I)$ as weights (as we did in Section 10.5). At room temperature, the thermal energy is of the order of 0.025 eV, and the Boltzmann factors $e^{-E(m_I)/kT}$, which determine $p(m_I)$, are virtually identical. If we cool the nuclei down to very low temperatures, we can reduce $kT$ by a factor of $10^4$–$10^5$, and if at the same time we prepare the nuclei in a ferromagnetic environment where the local field (the so-called "hyperfine" magnetic field $B_{hf}$) may be 100 times larger than our laboratory field, the Boltzmann factors can differ substantially from one another,

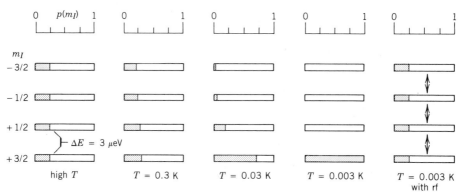

**Figure 16.21** Zeeman-split $I = \frac{3}{2}$ nuclear levels in a static magnetic field. The shaded portions indicate the relative populations according to the Boltzmann factors; at high temperature, the populations are all equal to $(2I + 1)^{-1} = 0.25$. At the lowest temperature, essentially only the lowest state is populated. A rf signal at the resonant frequency $\Delta E / h$ drives the populations back to equality, even at the lowest temperature.

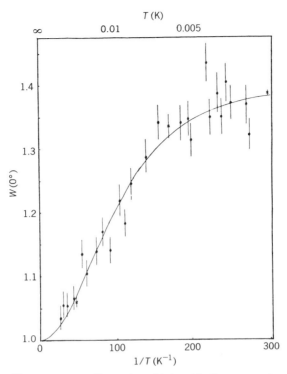

**Figure 16.22** The $\gamma$-ray counting rate, obtained in the geometry shown in Figure 10.3, for the decay of $^{191}$Pt. Note that the temperature scale at bottom is inverted; that is, temperature decreases to the right. At high temperature, $W(0°) = 1$, and as the temperature decreases, $W(0°)$ increases as more of the nuclei occupy the lowest Zeeman state as in Figure 16.21. The solid line is a fit to the data, from which the value $\mu = 0.49 \pm 0.02\ \mu_N$ is obtained. From W. M. Lattimer et al., *J. Phys. G* **7**, 1713 (1981).

and we get a very nonuniform (or anisotropic) angular distribution. Figure 16.21 illustrates the energy splittings and Boltzmann populations for a state with $I = \frac{3}{2}$.

Figure 16.22 shows the angular distribution $W(0°)$ as a function of the temperature for the decay of $^{191}$Pt prepared in ferromagnetic iron. The nuclei in this case were cooled to a temperature of 0.003 K. The variation of $W(0°)$ with $T$ can be used to deduce the magnetic moment, and the value obtained is $|\mu| = 0.49 \pm 0.02\ \mu_N$. This method is not sensitive to the sign of $\mu$; we get the same $W(0°)$ whether the $m_I = +\frac{3}{2}$ or $m_I - \frac{3}{2}$ state is the lowest.

An improvement in the measured moment can be achieved by combining this technique with NMR. If we expose the nuclei to a radiofrequency field whose frequency is carefully selected to correspond to the energy difference $\Delta E$ (that is, $\nu = \Delta E/h$), then even at the coldest temperatures, the absorption of this signal tends to equalize the $m_I$-state populations, and $W(0°)$ approaches its high-temperature value ($\approx 1$). Varying the frequency permits the resonance to be traced, as shown in Figure 16.23. The center of the resonance can be found very precisely: $\nu = 319.48 \pm 0.06$ MHz. In principle, this should enable us to deduce

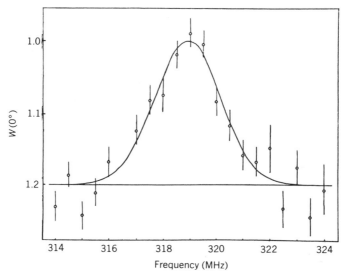

**Figure 16.23** Resonance spectrum of the counting rate $W(0°)$ in the decay of $^{191}$Pt oriented at low temperature ($T = 0.013$ K, corresponding to about $W(0°) = 1.2$ in Figure 16.22). When the resonance is reached, the populations of the nuclear $m$ states are suddenly driven toward equilibrium, as in the rightmost column of Figure 16.21, and the counting rate approaches 1.0. From W. M. Lattimer et al., *J. Phys. G* **7**, 1713 (1981).

$\mu$ to an accuracy of about 0.01%, but in practice we are limited by the precision of $B_{hf}$ (typically, 0.1–1%) and the hyperfine anomaly (0.01–0.1%).

This method of combining nuclear orientation with NMR has become very popular in recent years for determining nuclear magnetic dipole moments. It has even been extended to extremely short-lived radioactive nuclei, as in the laser experiments described above, by coupling the low-temperature orientation apparatus to an accelerator. The short-lived species are produced by nuclear reactions and are then transported to and implanted into the ferromagnetic host lattice at low temperature. Nuclei with half-lives as short as 1 min have been studied with this technique.

### Excited States

**Perturbed Angular Distributions and Correlations** Consider a nuclear state $I_i$ which has a certain mean life $\tau$. Let's suppose we measure the decay of that state as a function of time, as illustrated by the time-to-amplitude converter (TAC) spectra shown in Figures 7.32 or 7.33. Suppose we place the nuclei in a magnetic field (usually at room temperature) and detect the radiation emitted by $I_i$ in a plane perpendicular to the field. During the lifetime $\tau$, the nuclei will precess about the field direction (Figure 16.24). The radiation is emitted in a certain preferred direction with respect to the nuclear spin, and because the spin direction is varying, the counting rate will fluctuate up or down according to the particular direction the spin happens to be taking at the instant of emission. The

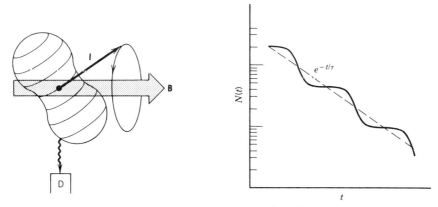

**Figure 16.24** As the nuclear spin $I$ precesses about **B** during the mean life $\tau$, the $\gamma$-ray counting rate $N(t)$ observed in detector D will fluctuate, because the radiation emission pattern carried by the precessing spin varies its orientation relative to the detector.

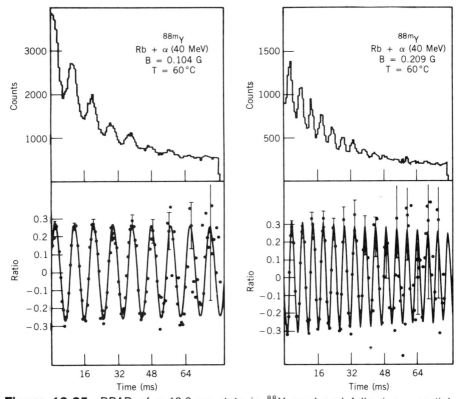

**Figure 16.25** DPAD of a 13.9-ms state in $^{88}$Y produced following $\alpha$-particle bombardment of Rb. The top part shows the raw time spectrum, as in Figure 16.24, and the bottom half shows the same data with the exponential decay factor removed. The sinusoidal fit to the data gives the Larmor frequency and thus the nuclear $g$ factor. At right, doubling the field doubles the frequency. From L. Varnell and I. Rezanka, *Nucl. Phys. A* **255**, 75 (1975).

frequency, called the Larmor frequency, is

$$\omega_L = \frac{\mu_I B}{I_i \hbar} \tag{16.69}$$

Superimposed on the exponential decay of the state is a fluctuation at the Larmor frequency, and the result would look somewhat like Figure 16.24.

If the state $I_i$ is formed following a nuclear reaction, the "start" signal for the TAC is the beam pulse from the accelerator and if the radiations (usually $\gamma$'s) are detected in singles mode, the experiment is called an *angular distribution*. If the start signal is a previous radiation (also usually $\gamma$'s) which populates the state $I_i$ following a relatively long-lived radioactive decay, the radiations are detected in coincidence and the experiment is called an *angular correlation*. In either case, the interaction of the nuclear moment of the state $I_i$ changes or *perturbs* the angular distribution or correlation that would be observed if, for instance, we turned the field off. These experiments are thus known as *perturbed angular distributions* (PAD) and *perturbed angular correlations* (PAC). If the nuclear state is long-lived enough that we can observe its exponential decay as a function of time, the experiment is called time *differential*, and hence DPAD or DPAC.

Figure 16.25 shows results of a typical DPAD experiment. The $8^+$ state in $^{88}$Y has a mean life of 13.9 ms and is produced following the bombardment of Rb by $\alpha$ particles. The beam pulse of $\alpha$'s in this experiment came every 100 ms. The

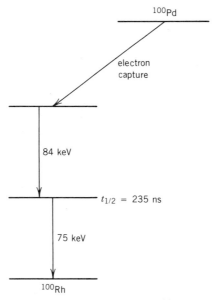

**Figure 16.26** DPAC experiment in the decay of $^{100}$Pd. At top right is shown the unperturbed time spectrum, in which the start signal comes from the 84-keV $\gamma$. In an external magnetic field of 2.22 kG, the time spectra (middle right) of two opposite field directions show the characteristic Larmor precession; the sinusoidal fit to the combined and exponentially corrected data at bottom right gives the nuclear *g* factor of the 75-keV excited state. From E. Matthias et al., *Phys. Rev.* **140**, B264 (1965).

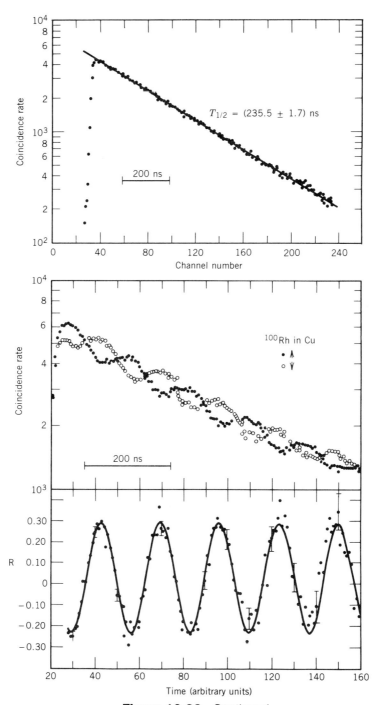

**Figure 16.26** Continued

TAC spectrum is plotted on linear scale rather than a semilog scale, but you can quite clearly see the modulation at the Larmor frequency. Removing the exponential decay factor shows the precession frequency directly, and it is apparent that doubling the field also doubles the frequency. The deduced $g$ factor is $g_I = +0.598 \pm 0.012$, indicative of the high precision that the method can yield. The calculated $g$ factor is in good agreement with the shell-model value obtained by coupling a $g_{9/2}$ proton to a $g_{9/2}$ neutron.

Figure 16.26 shows results from a DPAC experiment on an excited state of $^{100}$Rh. You can see the unperturbed time spectrum, the perturbed spectrum for two opposite field directions (giving Larmor precessions in opposite directions), and the net variation with the exponential decay factor removed. The $g$ factor in this case is deduced to be $g_I = +2.13 \pm 0.03$. Again, the $g$ factor helps us to understand the nuclear structure of the state; the calculated $g$ factor for a configuration of a $g_{9/2}$ proton coupled to a $d_{5/2}$ neutron is approximately $g_I = 2$, but in this case with 5 protons in the $g_{9/2}$ shell and 5 neutrons beyond the $N = 50$ magic number, the extreme independent particle model is probably not valid and more complicated wave functions must be used.

The time-differential method is useful for states that live long enough to display their decay as a function of time. For states with mean lives shorter than about 1 ns, this is not possible. Moreover, even using the large internal magnetic fields in ferromagnets, the largest Larmor frequencies are of order $10^9$ Hz, and thus even if we could display the time-differential pattern, it would include at

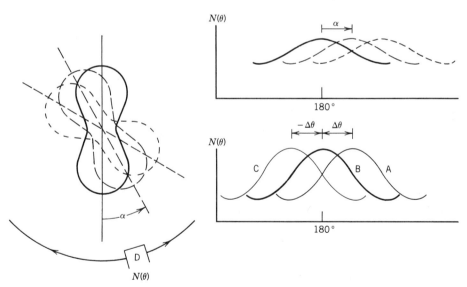

**Figure 16.27** Schematic of integral PAC or PAD experiment. The radiation pattern precesses through an angle $\alpha = \omega_L t$, which varies with the actual time $t$ the nucleus spends in the excited state that emits the radiation. At top right are shown the individual contributions to $N(\theta)$ for several such times; we cannot observe these distributions, but instead we measure the average of all such contributions over the exponential decay, which gives the distribution A at lower right, shifted by an angle $\Delta\theta$ relative to the unperturbed distribution (curve B). If the external magnetic field direction is reversed, the net shift $\Delta\theta$ reverses (curve C).

most only one cycle of the periodic variation and we would lose the high precision obtained from the multicycle fits shown in Figures 16.25 and 16.26. Because we cannot observe the variations in time, we in effect integrate (or average) over all times, and the technique is called time *integral* and labeled IPAD or IPAC (where I = integral).

The effect of the perturbation can be predicted by referring to Figure 16.27. Each nuclear spin rotates through a small angle $\alpha = \omega_L t$; the number of nuclei with a given rotation angle decreases exponentially with the size of the angle (because the number of nuclei that survive for a time $t$ decreases exponentially). The average of all these small rotations will be a net rotation angle $\Delta\theta$, as indicated in Figure 16.27. The maximum (or minimum) of the angular distribution pattern rotates away from 180° by $\Delta\theta$, which is usually a few degrees. If we reverse the direction of the magnetic field, $\Delta\theta$ likewise reverses. For small rotations, $\Delta\theta \simeq \omega_L \tau$. Figure 16.28 shows some sample results from IPAC and IPAD measurements.

Up to now we have discussed only magnetic dipole moment measurements. Electric quadrupole moments of excited states can also be measured using PAD and PAC techniques, but the behavior of a nucleus in an electric quadrupole field is more complex than a simple Larmor precession. The "wiggles" superimposed on the exponential decay even in the simplest electric quadrupole case have at least two Fourier components, and if the field lacks axial symmetry, more complex patterns can be observed. Nevertheless, the patterns can be analyzed to obtain the quadrupole moments, as illustrated in Figure 16.29.

**Mössbauer Effect**   Consider the Mössbauer spectrum that was shown in Figure 10.30, and notice that the lines are not equally spaced—the center pair seems closer together than any other pair of adjacent lines. Let's see how we can understand this effect. We will ignore the isomer shift, which merely changes the center of the pattern and has no effect on the relative spacing of the pairs. Let $E_{3/2}$ represent the energy of the upper state in the absence of magnetic splitting, and similarly $E_{1/2}$ for the ground state. In the presence of the field $B$, the levels are

$$E'_{3/2}(m_{3/2}) = E_{3/2} - g_{3/2}Bm_{3/2} \tag{16.70}$$

$$E'_{1/2}(m_{1/2}) = E_{1/2} - g_{1/2}Bm_{1/2} \tag{16.71}$$

where $g_I$ represents the $g$ factors (note that in the figure, $g_{3/2} < 0$ and $g_{1/2} > 0$, as is the case for $^{57}$Fe), $m_{3/2}$ is the magnetic quantum number of the $I = \frac{3}{2}$ state $(+\frac{3}{2}, +\frac{1}{2}, -\frac{1}{2}, -\frac{3}{2})$, and $m_{1/2}$ is the same for the ground state $(+\frac{1}{2}, -\frac{1}{2})$. Let the lines be numbered left to right, $E_1$ to $E_6$. Then

$$E_2 = E'_{3/2}\left(m_{3/2} = -\tfrac{1}{2}\right) - E'_{1/2}\left(m_{1/2} = -\tfrac{1}{2}\right)$$

$$= \left(E_{3/2} - E_{1/2}\right) + \left[g_{1/2}\left(-\tfrac{1}{2}\right) - g_{3/2}\left(-\tfrac{1}{2}\right)\right]B \tag{16.72}$$

$$E_1 = E'_{3/2}\left(m_{3/2} = -\tfrac{3}{2}\right) - E'_{1/2}\left(m_{1/2} = -\tfrac{1}{2}\right)$$

$$= \left(E_{3/2} - E_{1/2}\right) + \left[g_{1/2}\left(-\tfrac{1}{2}\right) - g_{3/2}\left(-\tfrac{3}{2}\right)\right]B \tag{16.73}$$

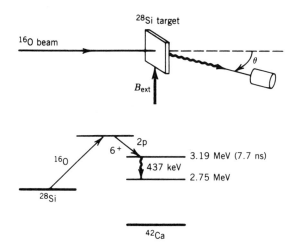

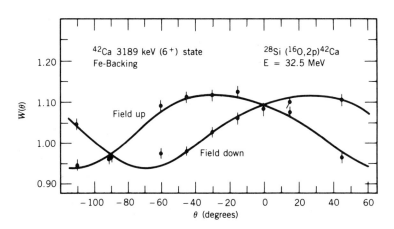

**Figure 16.28** (Above) Time-integral PAD measurement. The 7.7-ns state is formed following the nuclear reaction; the angular distribution is measured relative to the beam direction. The maximum of $W(\theta)$ would be at 0° in the absence of an external field, but it shifts right or left according to the field direction as the nuclei precess in the 3.19-MeV level. From M. Uhrmacher et al., *Phys. Lett. B* **56**, 247 (1975). (Facing page) Time-integral PAC; here $\theta$ is the correlation angle between the two $\gamma$'s. The solid curve is the expected $W(\theta)$ with no external field; the dashed curve shows the shift in an external field of 4.3 kG and gives for the 0.37-MeV level $g_l = 0.07 \pm 0.03$. From H. Frauenfelder et al., *Phys. Rev.* **93**, 1126 (1954).

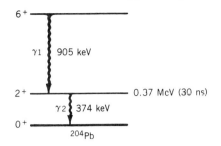

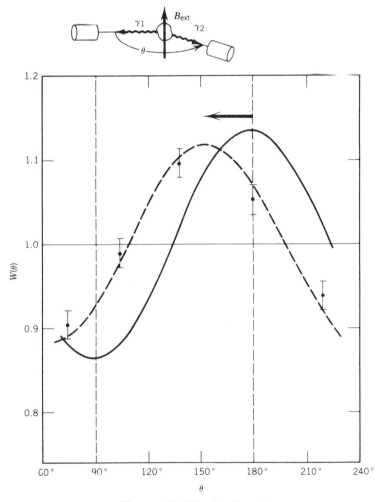

**Figure 16.28** Continued.

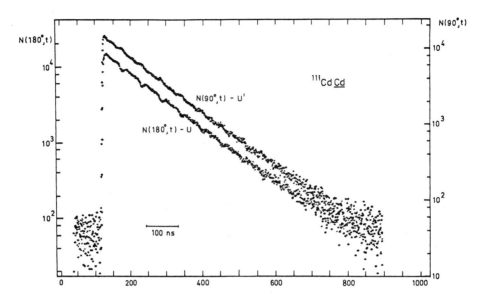

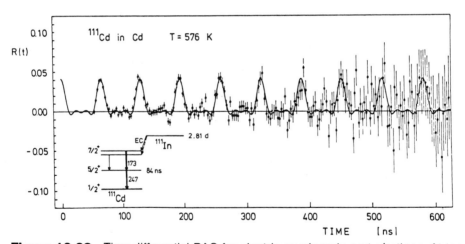

**Figure 16.29** Time-differential PAC for electric quadrupole perturbations. At top are shown the time spectra for angles of 180 and 90° between the two detectors; note that the "wiggles" do not have the sinusiodal variation as in the case of magnetic dipole perturbations. At bottom is shown the time spectrum corrected for the exponential decay; this shape is characteristic of three components, with frequencies $f$, $2f$, and $3f$. From W. Witthuhn and W. Engel, in *Hyperfine Interactions of Radioactive Nuclei*, edited by J. Christiansen (Berlin: Springer-Verlag, 1983).

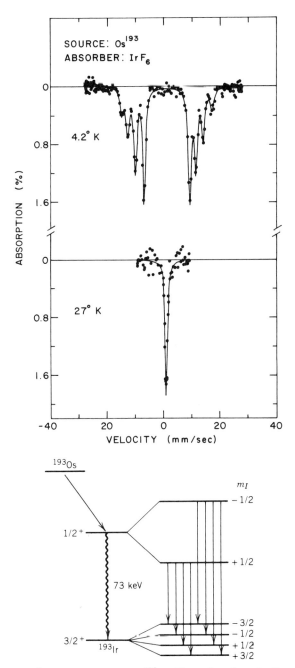

**Figure 16.30** Mössbauer effect in $^{193}$Ir. The absorber is the compound IrF$_6$, which is antiferromagnetic below 8 K. The Mössbauer velocity spectra above show the 73-keV transition from the first excited state to the ground state split into its $\Delta m$ components below 8 K and unsplit above. The eight transitions among the individual Zeeman states are shown at bottom. From the relative spacings of the inner and outer lines, the ratio of the excited-state and ground-state $g$ factors can be deduced. Data from G. J. Perlow et al., *Phys. Rev. Lett.* **23**, 680 (1969).

and the spacing between the lowest-energy pair of lines is

$$E_2 - E_1 = -g_{3/2}B \tag{16.74}$$

A similar calculation shows that $E_3 - E_2$, $E_5 - E_4$, and $E_6 - E_5$ give the same result. (This can be seen directly by looking at the arrows in the top part of this figure.) However,

$$E_3 = E'_{3/2}\left(m_{3/2} = +\tfrac{1}{2}\right) - E'_{1/2}\left(m_{1/2} = -\tfrac{1}{2}\right)$$
$$= \left(E_{3/2} - E_{1/2}\right) + \left[g_{1/2}\left(-\tfrac{1}{2}\right) - g_{3/2}\left(\tfrac{1}{2}\right)\right]B \tag{16.75}$$

$$E_4 = E'_{3/2}\left(m_{3/2} = -\tfrac{1}{2}\right) - E'_{1/2}\left(m_{1/2} = +\tfrac{1}{2}\right)$$
$$= \left(E_{3/2} - E_{1/2}\right) + \left[g_{1/2}\left(+\tfrac{1}{2}\right) - g_{3/2}\left(-\tfrac{1}{2}\right)\right]B \tag{16.76}$$

$$E_4 - E_3 = \left(g_{1/2} + g_{3/2}\right)B \tag{16.77}$$

In the case in which we do not know $B$ very precisely, we can form the ratio between the central spacing $E_4 - E_3$ and any pair of outer spacings:

$$R = \frac{E_4 - E_3}{E_2 - E_1} = -\left(1 + \frac{g_{1/2}}{g_{3/2}}\right) \tag{16.78}$$

If the ground-state $g$ factor is known, the excited-state $g$ factor can be determined. In one experiment, reported by R. S. Preston, S. S. Hanna, and J. Heberle (*Phys. Rev.* **128**, 2207 (1962)), the ratio $\mu_{3/2}/\mu_{1/2}$ was determined to be $-1.715 \pm 0.004$. The ground-state moment is known from NMR measurements to be $+0.0905 \pm 0.0001$ $\mu_N$, and thus the deduced excited-state moment is $\mu_{3/2} = -0.1552 \pm 0.0004$ $\mu_N$. Without using the ratio technique, the precision would be limited by the knowledge of $B$ to about 1%.

Figure 16.30 shows another example in the case of $^{193}$Ir. In this case, the $\tfrac{3}{2}$ and $\tfrac{1}{2}$ states are reversed, and the photon is a mixture of $M1$ and $E2$, so that $\Delta m = \pm 2$ transitions occur and the Mössbauer spectrum consists of eight (rather than six) lines. The observed spacings give $g_{1/2}/g_{3/2} = 9.045 \pm 0.039$, and given the NMR result $\mu_{3/2} = +0.1591 \pm 0.0006$ $\mu_N$, the deduced value of the excited-state moment is $\mu_{1/2} = +0.504 \pm 0.003$ $\mu_N$, after several corrections, including a fairly large hyperfine anomaly, have been applied.

## REFERENCES FOR ADDITIONAL READING

Experimental values of nuclear moments are tabulated in *Table of Isotopes*, edited by C. M. Lederer and V. S. Shirley (New York: Wiley, 1978) and in G. H. Fuller, *J. Phys. Chem. Ref. Data* **5**, 835 (1976). Both of these collections specify the methods used to determine the moments.

Reviews of atomic structure that may provide useful background material for the chapter are: J. C. Willmott, *Atomic Physics* (Chichester, England: Wiley, 1975) and G. K. Woodgate, *Elementary Atomic Structure*, 2nd edition (Oxford: Clarendon, 1980). These are both introductory works; for more advanced treatments of atomic structure, see E. U. Condon and G. H. Shortley, *The Theory of Atomic Spectra* (Cambridge: Cambridge University Press, 1967), and H. Kuhn, *Atomic Spectra*, 2nd edition (London: Longman, 1970).

The classic reference on the theory and practice of measuring nuclear moments is H. Kopfermann, *Nuclear Moments* (New York: Academic, 1958). A more advanced summary discussing many of the techniques for measuring nuclear moments is *Hyperfine Interactions*, edited by A. J. Freeman and R. B. Frankel (New York: Academic, 1967). A slightly different (and more elementary) approach to the same topic is P. J. Wheatley, *The Chemical Consequences of Nuclear Spin* (Amsterdam: North-Holland, 1970).

The study of hyperfine structure has been revolutionized by the application of lasers. Two reviews of this topic are: D. E. Murnick and M. S. Feld, *Ann. Rev. Nucl. Part. Sci.* **29**, 411 (1979) and P. Jacquinot and R. Klapisch, *Rep. Prog. Phys.* **42**, 77 (1979).

Reports from a conference devoted to this topic can be found in *Lasers in Nuclear Physics*, edited by C. E. Bemis and H. K. Carter (New York: Harwood Academic, 1982). Two brief articles by O. Redi and H. Schuessler summarizing current research can be found in the February 1981 issue of *Physics Today*.

## PROBLEMS

1.  The empirical rule for the ground-state spin assignments of odd-$Z$, odd-$N$ nuclei suggests that the odd neutron and odd proton intrinsic spins couple in a parallel configuration. Justify this empirical rule on the basis of the known properties of the nucleon–nucleon interaction.

2.  What values would you expect for the ground state spin-parities of: $^{24}$Na ($d_{3/2}$ proton + $d_{5/2}$ neutron); $^{26}$Na ($d_{5/2}$ proton + $s_{1/2}$ neutron); $^{68}$Cu ($p_{3/2}$ proton + $p_{1/2}$ neutron); $^{198}$Au ($d_{3/2}$ proton + $p_{1/2}$ neutron)?

3.  Using the vector relationships shown in Figure 16.5, derive Equation 16.35. (*Hint:* It may be helpful to review the calculation leading to Equation 5.8.)

4.  (a) Calculate the Landé $g$ factor for an atom with $S = \frac{3}{2}$, $L = 2$, $J = \frac{1}{2}$. (b) Give a physical explanation for this value by carefully drawing a vector diagram similar to Figure 16.5. (c) Find one other combination of $S$, $L$, and $J$ that gives the same $g_J$.

5.  For the case $J = \frac{5}{2}$, $I = 1$, show the magnetic hyperfine structure splitting in zero field, weak field, and strong field (similar to Figures 16.7 and 16.8, but note that in the special case $J = \frac{3}{2}$, illustrated in Figure 16.7, $g_F$ is the same for all states in the hyperfine multiplet, which will not in general be true).

6.  Rubidium has two stable isotopes, $^{85}$Rb ($I = \frac{5}{2}$) and $^{87}$Rb ($I = \frac{3}{2}$). The electronic state is that of the alkali atom (a single electron in an s state). Make a schematic sketch of the magnetic hyperfine structure of these two isotopes, and show the splitting in weak and strong magnetic fields.

7.  For the case of atomic $J = 1$ and nuclear $I = \frac{3}{2}$, draw the zero field magnetic hyperfine multiplet, showing the splitting in terms of $A$. (b) If there is also a small electric quadrupole interaction, compute the multiplet energies in terms of $A$ and $B$.

8.  The atomic structure of Cs is similar to that of Na, but the nuclear spin is $\frac{7}{2}$ in the only stable isotope, $^{133}$Cs. Sketch the hyperfine structure in zero field

of the $s_{1/2}$, $p_{1/2}$, and $p_{3/2}$ states in $^{133}$Cs and show the structure of the electric dipole transitions.

9.  In an odd-$Z$, odd-$N$ nucleus, we can regard the ground state as the coupling between an odd proton and an odd neutron, $I = j_p + j_n$. If the proton and neutron states have $g$ factors $g_p$ and $g_n$, show that the combination has $g$ factor

$$g = \tfrac{1}{2}(g_p + g_n) + \frac{(g_p - g_n)\left[ j_p(j_p + 1) - j_n(j_n + 1)\right]}{2I(I + 1)}$$

    (*Hint*: The magnetic moment of the combination must be defined, as in Equation 16.34, as the component of $g_p j_p + g_n j_n$ along $I$.)

10.  Use the result of the previous problem to evaluate the magnetic moments of the following nuclei, and compare with the experimental values: (a) $^{14}$N, $I^\pi = 1^+$, $\mu = +0.40\ \mu_N$; (b) $^{60}$Co, $I^\pi = 5^+$, $\mu = +3.8\ \mu_N$; (c) $^{84}$Rb, $I^\pi = 2^-$, $\mu = -1.3\ \mu_N$. In each case, use the shell model to find the proton and neutron single-particle states, and evaluate the single-particle proton and neutron $g$ factors with $g_s = 0.6g_s(\text{free})$.

11.  (a) Calculate the Boltzmann populations of an $I = 2$ nucleus with a magnetic moment of 3.0 $\mu_N$ in an external field of 25 T. Assume temperatures of 300 K (room temperature), 4 K (liquid helium), and 0.01 K. (b) How large a magnetic field would be required to produce a nuclear Zeeman splitting $\Delta E$ roughly equal to $kT$ at room temperature?

12.  In a perturbed angular distribution experiment there are three fundamental times involved: the mean life $\tau$ of the emitting state, the period $T$ of the interaction (i.e., the reciprocal of the interaction frequency), and the time resolution $t$ of the detector and electronics. Discuss the relationships among these times necessary for the experiment to be possible. Consider such limits as the effect of $t \gg T$ or $\tau \ll T$.

# 17

# MESON PHYSICS

In Chapter 4, we discussed the role of the $\pi$ meson as the carrier of the strong nucleon–nucleon interaction. Exchange of $\pi$ mesons from one nucleon to another is responsible for a major component of nuclear binding. As a part of the model, we regard the nucleon as surrounded by a "cloud" of virtual $\pi$ mesons that are continually being emitted and absorbed. The maximum distance these mesons can travel before they must be absorbed (so as not to violate conservation of energy for a time longer than allowed by the uncertainty relationships) determines the range of the nuclear force and the "size" of a nucleon. Other mesons, including $\rho$ and $\omega$, contribute to the short-range nuclear interaction, particularly the tensor, spin-orbit, and repulsive core terms.

The $\pi$ mesons (pions) are the lightest members of the *meson* family, one of the three major groupings of particles (*leptons*, which include the electrons and neutrinos, and *baryons*, which include the nucleons, are the other two families). The mesons are particles that have integer spins (leptons and baryons have half-integer spins) and interact with nucleons through the strong force (in addition of course to weak and electromagnetic forces). Free mesons can be produced in nucleon–nucleon collisions, and they decay rapidly to lighter mesons, photons, or leptons either through the strong, electromagnetic, or weak interactions. Typically, the decay lifetimes are $10^{-20}$–$10^{-23}$ s for strong decays, $10^{-16}$–$10^{-18}$ s for electromagnetic decays, and $10^{-8}$–$10^{-10}$ s for weak decays.

On the most fundamental level, mesons are composite particles made up of a quark and an antiquark. This aspect of meson structure, and the relationship to the structures of the lepton and baryon families, is discussed in Chapter 18. In the present chapter, we discuss the properties of the mesons as nuclear particles, including their production in nuclear reactions, their use as nuclear probes in scattering reactions, and the properties of quasibound states of a meson and a nucleon, which are classified as excited states of the baryons.

## 17.1 YUKAWA'S HYPOTHESIS

In 1935, Japanese physicist Hideki Yukawa proposed a mathematical potential to represent the nucleon–nucleon interaction. Already at that time, the need for an exchange-force basis for the nuclear interaction was recognized. Yukawa attempted to find a potential that would describe the exchange of particles giving rise to the nuclear force, in much the same way that the electromagnetic potentials and fields describe the exchange of photons that give rise to the electromagnetic force. One major difference between the electromagnetic interaction and the strong nuclear interaction is the infinite range of the electromagnetic force compared with the 1-fm range of the nuclear force. If $m$ represents the rest mass of the exchanged particle, then a virtual particle can be created and exist for a time $t$ without violating conservation of energy, as long as $t$ is no greater than what is permitted by the uncertainty relationship:

$$t = \frac{\hbar}{mc^2}$$  (17.1)

The greatest distance the particle can move is then

$$x = ct = \frac{\hbar c}{mc^2} = \frac{200 \text{ MeV} \cdot \text{fm}}{mc^2}$$  (17.2)

and thus for a range of 1 fm, the mass of the exchanged particle is of order 200 MeV/$c^2$. Photons, on the other hand, have zero rest mass and infinite range.

The basic equations for the electromagnetic field are Maxwell's equations, which yield wave equations that can be solved for the fields that govern the propagation of real as well as virtual photons. For the nuclear field, the electromagnetic field equations will clearly not work, for they describe massless field particles. Nor is the Schrödinger equation appropriate, for it is nonrelativistic and does not include the correct relativistic mass-energy relationship that is sure to be necessary for pions. Instead, we need a new equation, consistent with the basic relativistic expression relating total energy and rest mass

$$E^2 = p^2c^2 + m^2c^4$$  (17.3)

In quantum theory, the energy is associated with the operator $i\hbar\partial/\partial t$ and the momentum with $-i\hbar\nabla$, and making these substitutions gives a relativistic wavelike differential equation called the *Klein-Gordon equation*:

$$\left(\nabla^2 - \frac{m^2c^2}{\hbar^2}\right)\phi = \frac{1}{c^2}\frac{\partial^2\phi}{\partial t^2}$$  (17.4)

where $\phi$ represents the amplitude of the field. (In the limit $m = 0$, this reduces to the familiar wave equation for the electromagnetic field.) We begin by looking for a static potential from the time-independent solution to

$$\nabla^2\phi - k^2\phi = 0$$  (17.5)

where $k = mc/\hbar$. In radial coordinates the spherically symmetric solution is

$$\phi = g\frac{e^{-kr}}{r}$$  (17.6)

where $g$ is a constant that represents the strength of the pion field, analogous to the electronic charge $e$ representing the strength of the electromagnetic field. In this analogy, the electromagnetic field of a single charge would be represented by $e/4\pi\epsilon_0 r$, and the strength of the interaction of another elementary charge with this field would be $e^2/4\pi\epsilon_0 r$. Similarly, the strong interaction between two "elementary" nuclear particles would be $-g^2 e^{-kr}/r$, which is of the form of Equation 4.54.

According to Yukawa's calculation, the nuclear force should have a range of the order of $k^{-1} = \hbar/mc$, identical with Equation 17.2 derived from arguments based on the uncertainty relationship. Yukawa's proposed particles were called mesons (originally mesotrons, "meso" meaning "middle") because the postulated mass was midway between the light known particles (electrons) and the heavy nucleons. At the time of Yukawa's proposal, accelerators were not yet powerful enough to liberate mesons, so the early searches for mesons were done among particles produced by cosmic rays, using photographic emulsions carried to high altitude where they record the passage of secondary particles created when cosmic rays strike the atmosphere. In the late 1930s and early 1940s, cosmic-ray results showed evidence for a particle with a mass about 100 MeV/$c^2$, close to Yukawa's estimate, but further work showed that these particles had too long a range in solid matter. Such a long range was inconsistent with the assumption of a particle that interacted strongly with nuclei, and thus there was some doubt that this particle was in fact Yukawa's meson. Later work in 1947 by C. F. Powell and his collaborators showed evidence for two distinct mesons from the emulsion tracks (Figure 17.1), a heavier meson ($\sim$ 150 MeV) decaying to the lighter one ($\sim$ 100 MeV). It is the heavier meson that is Yukawa's particle, now known as the $\pi$ meson. The lighter particle is a muon; although it was originally called a $\mu$ meson, it is in fact not a meson at all (it does not interact strongly and it has spin $\frac{1}{2}$), but a true elementary particle of the lepton family.

Very soon after the cosmic-ray results gave evidence for the $\pi$ meson, synchrocyclotrons reached energies beyond the production thresholds for $\pi$ mesons, permitting their properties to be studied in the laboratory. Today, at accelerators known as "meson factories" (LAMPF in Los Alamos, New Mexico; TRIUMF in Vancouver, British Columbia; and SIN in Zurich, Switzerland) $\pi$ mesons are produced by intense proton beams in such great quantities that secondary beams of pions can be extracted, permitting detailed studies of reactions induced by pions.

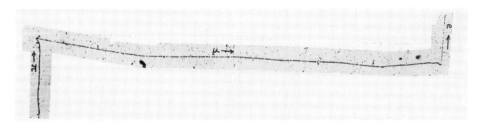

**Figure 17.1**  A $\pi$ meson (left) decays to a muon (center), which in turn decays to an electron (right).

## 17.2 PROPERTIES OF $\pi$ MESONS

### Electric Charge

Pions can carry electric charges of $+e$, 0, or $-e$ and are thus denoted by $\pi^+$, $\pi^0$, or $\pi^-$. The $\pi^0$ is its own antiparticle (like the photon), and the $\pi^+$ and $\pi^-$ are antiparticles of each other. It makes little sense, in the case of $\pi^+$ and $\pi^-$, to identify one with "particle" and the other with "antiparticle." In the case of particles such as electrons or nucleons, we identify the components of ordinary matter as particles, and thus positrons and antinucleons are identified as antiparticles. Ordinary matter, however, is not composed of pions, and so no such identification is possible. The ultimate factor in determining the identification of particles and antiparticles in the case of mesons has to do with the relative statistical weights of the cross sections of various pion reactions, from which it is concluded that the pions must be regarded as a set of three particles $\pi^+$, $\pi^0$, $\pi^-$ whose antiparticles are, respectively, $\pi^-$, $\pi^0$, $\pi^+$.

### Isospin

The understanding of reactions among elementary particles is aided by classifying the isospin of the particles. In Chapter 11 the isospin notation was introduced to represent the nucleons as a $T = \frac{1}{2}$ isospin doublet, with projection $T_3 = +\frac{1}{2}$ representing a proton and $T_3 = -\frac{1}{2}$ representing a neutron. The $\pi$ mesons can be represented with a similar isospin notation. Because there are three pions, the multiplicity $2T + 1$ demands $T = 1$. Again letting the member with the maximum electric charge have the highest projection, we have $T_3 = +1$ for $\pi^+$, $T_3 = 0$ for $\pi^0$, and $T_3 = -1$ for $\pi^-$. The usefulness of this scheme for understanding reactions and decays involving mesons is considered later in this chapter. As in the case of the nucleons, in the absence of electromagnetic interactions, the pion isospin triplet would have identical masses; the small mass difference within the pion triplet can be ascribed to the isospin-nonconserving electromagnetic interaction.

### Mass

The mass of the $\pi^-$ is determined with great precision from the energies of $\pi$-mesic X rays, which are emitted when a $\pi^-$ is captured into an atomic orbit and cascades down toward the nucleus. Photons are emitted in this process, in exact analogy with electronic optical or X-ray transitions in atoms or muonic X rays, which are used to deduce the nuclear charge radius (see Section 3.1). The most precise individual measurement yet reported is that on pionic X rays in phosphorus and titanium, done by Lu et al. (*Phys. Rev. Lett.* **45**, 1066 (1980)). Pions were produced as secondary particles when a high-energy proton beam from a synchrotron struck a thick target. The pions were slowed and finally stopped in the desired material (P or Ti) where the negative pions were captured into atomic-like orbits. As with electrons in atoms, higher excited levels decay to lower levels by photon emission; the major difference is that as the pion moves closer to the nucleus, it has an increasing chance to disappear through nuclear processes such as $\pi^- + p \to n$. It is therefore necessary to study X rays from states of principal quantum number $n = 3–5$, for the pion will not survive to reach the $n = 2$ or $n = 1$ levels. Because the hydrogenic level energy is proportional to $n^{-2}$, this means that experimenters must work with relatively small energy differences.

A rough estimate of the energies of the $n = 4$ to $n = 3$ transition in P ($Z = 15$) and the $n = 5$ to $n = 4$ transition in Ti ($Z = 22$) can be obtained from the

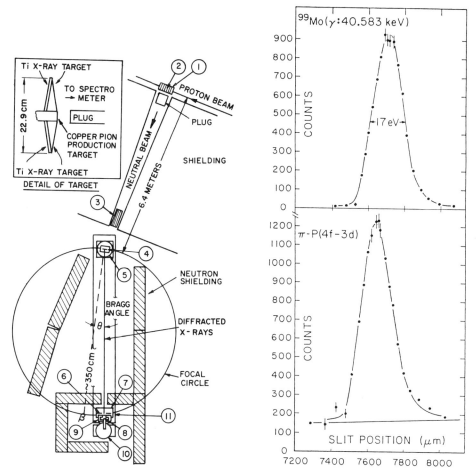

**Figure 17.2** (Left) Apparatus used to observe pionic X rays. The proton beam strikes a pion production target (2) which is surrounded by a P or Ti X ray target to stop the pions; a thick plug prevents radiation from the pion production target from traveling to the detection apparatus. A collimator (3) forms the X rays into a beam which strikes a crystal (4) in a Bragg scattering apparatus; the crystal spectrometer is capable of extremely precise determination of photon energies. Moving a slit (6) in effect varies the scattering angle for observation of the radiation by the detector (8). At right are shown the crystal spectra observed with the $^{99}$Mo radioactive calibration source (top) and the P X-ray target (bottom). Data from D. C. Lu et al., *Phys. Rev. Lett.* **45**, 1066 (1980); apparatus from L. Delker et al., *Phys. Rev. Lett.* **42**, 89 (1979).

hydrogenic formula given in Section 2.4, using the pion mass energy of about 140 MeV: $\Delta E_P = 40.8$ keV and $\Delta E_{Ti} = 40.6$ keV. These energies are very close to a well-known $\gamma$ ray in the radioactive decay of $^{99}$Mo ($40.5835 \pm 0.0002$ keV) and to the ordinary electronic K X rays in Sm ($40.1181 \pm 0.0003$ keV) and Eu ($40.9101 \pm 0.0003$ keV), which can be used for calibration. Figure 17.2 shows a comparison of the measured 4f-3d transition in P with the $^{99}$Mo calibration transition. The resulting measured energies, corrected for small relativistic and non-Coulomb effects (amounting to about 0.1 keV and calculable with great precision) are

$$\Delta E_P = 40.4892 \pm 0.0003 \text{ keV}$$

$$\Delta E_{Ti} = 40.3861 \pm 0.0004 \text{ keV}$$

and the deduced pion mass is

$$m_{\pi^-}c^2 = 139.5675 \pm 0.0009 \text{ MeV}$$

This X-ray method is not applicable to positively charged particles, and so another method must be found for $\pi^+$. In this case we can measure the energies and masses of the final products in the decay

$$\pi^+ \rightarrow \mu^+ + \nu_\mu$$

and deduce the $\pi^+$ mass. In the rest frame of $\pi^+$, the energy balance is

$$m_{\pi^+}c^2 = E_{\mu^+} + E_\nu$$

$$= m_{\mu^+}c^2 + T_{\mu^+} + cp_{\mu^+} \qquad (17.7)$$

where the latter result follows from $E_\nu = cp_\nu = cp_{\mu^+}$ based on massless neutrinos and momentum conservation in the pion rest frame. Measuring the momentum $p_{\mu^+}$, we can calculate $T_{\mu^+}$ and deduce $m_{\pi^+}c^2 - m_{\mu^+}c^2$. From independent results for the $\mu^+$ mass (which is known to greater precision), it is possible to deduce $m_{\pi^+}c^2$.

In an experiment reported by Daum et al., *Phys. Rev. D* **20**, 2692 (1979), the SIN meson factory was used to produce a $\pi^+$ beam that was brought to rest; using a magnetic spectrometer the momentum of the $\mu^+$ emitted in the $\pi^+$ decay could be measured with high precision. Figure 17.3 shows a detail of their spectrometer and some sample results. The deduced muon momentum was

$$p_{\mu^+} = 29.7877 \pm 0.0014 \text{ MeV}/c$$

from which (assuming $m_\nu$ is either zero or negligibly small), Equation 17.7 gives

$$m_{\pi^+}c^2 = 139.5658 \pm 0.0018 \text{ MeV}$$

The mass of the neutral pion can be obtained by studying the reaction

$$\pi^- + p \rightarrow n + \pi^0$$

The initial $\pi^-$ is first slowed to rest and is captured by the target proton from an atomic-like orbit. The $Q$ value is therefore

$$Q = (m_i - m_f)c^2$$

$$= (m_{\pi^-} + m_p - m_n - m_{\pi^0})c^2 \qquad (17.8)$$

$$= T_f - T_i$$

$$\cong T_f \qquad (17.9)$$

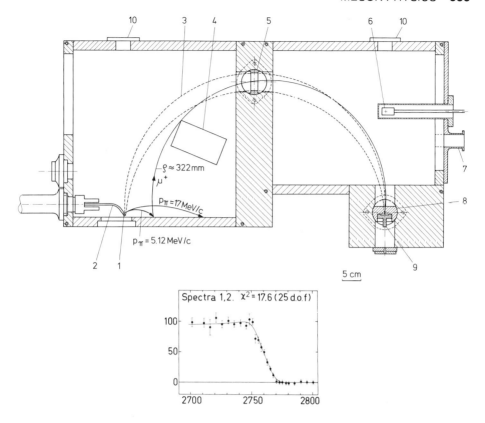

**Figure 17.3** (Top) Spectrometer for measuring momentum of muons from $\pi^+$ decay. Pions enter through the window at left bottom and strike the tip of the scintillator (1), where they stop and decay to muons; the scintillation signal from the pions travels along the light guide (2) to the photomultiplier at left. The muons follow the dashed trajectory (3) through a collimator (5) and eventually to a solid-state silicon detector (9). A magnetic field (the source for which is not shown) bends the muon path into a circular arc. The two-body pion decay should give a monoenergetic muon, but those pions that decay in the interior of the scintillator can result in muons that lose part of their energy within the scintillator; as the magnetic field is increased, eventually the counting rate drops to zero as the trajectories of even the most energetic muons are bent into too small an arc to reach the detector. At bottom is shown a sample data set, in which the counting rate drops to zero above 2750 G. From M. Daum et al., *Phys. Rev. D* **20**, 2692 (1979).

where the last step can be made because the initial kinetic energy is negligibly small. The final kinetic energy is shared by the neutron and the $\pi^0$:

$$(m_{\pi^-} - m_{\pi^0})c^2 = T_n + T_{\pi^0} + (m_n - m_p)c^2 \qquad (17.10)$$

Here $T_n$ and $T_{\pi^0}$ are related by momentum conservation, $p_n = p_{\pi^0}$, since we assumed the initial capture occurred at rest. The final kinetic energy of the $\pi^0$ can be deduced from the energy of the $\gamma$ rays in the decay $\pi^0 \rightarrow 2\gamma$. An equivalent procedure is to measure the angle between the two $\gamma$ rays; in the $\pi^0$ rest frame,

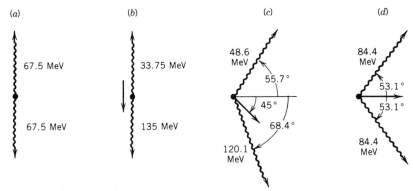

(a)    (b)    (c)    (d)

67.5 MeV    33.75 MeV    48.6 MeV    84.4 MeV

67.5 MeV    135 MeV    55.7°    53.1°

45°    53.1°

120.1 MeV    68.4°    84.4 MeV

**Figure 17.4** (a) Decay of $\pi^0$ at rest gives two 67.5-MeV photons moving in opposite directions. (b) Pions with $v/c = 0.6$ decaying along their direction of motion give two photons again with a relative angle of 180°. (c) If the angle between the laboratory direction of $\pi^0$ and the rest-frame photon direction is 45°, the opening angle is 124.1°. (d) The minimum opening angle, 106.2°, occurs when the pions in the rest frame decay at right angles to their laboratory momentum.

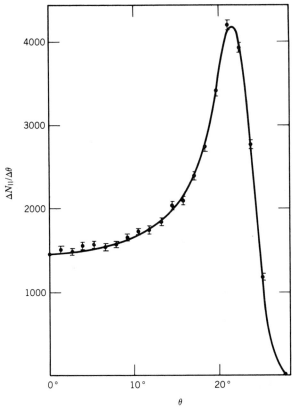

**Figure 17.5** Opening angle (plotted as $180° - \theta$) between two photons in $\pi^0$ decay. The peak corresponds to the maximum angle as in Figure 17.4d. From Vasilevsky et al., *Phys. Lett.* **23**, 281 (1966).

the $\gamma$'s are emitted in opposite directions, but the laboratory momentum $p_{\pi^0}$ results in a range of possible angles, as indicated in Figure 17.4. If the $\gamma$'s are emitted along the (laboratory) direction of $p_{\pi^0}$, the opening angle remains 180°; the minimum opening angle occurs when the $\gamma$'s in the $\pi^0$ rest frame are emitted perpendicular to $p_{\pi^0}$, in which case (see Figure 17.4d)

$$cp_{\pi^0} = m_{\pi^0}c^2 \cot \frac{\theta}{2} \tag{17.11}$$

Figure 17.5 shows the experimental results for the angle between the $\gamma$'s, corresponding to $\theta = 156.6°$ and $T_{\pi^0} = 2.89$ MeV; thus $(m_{\pi^-} - m_{\pi^0})c^2 = 4.603 \pm 0.005$ MeV. Alternatively, the energy of the neutron can be measured directly using time-of-flight techniques. Figure 17.6 shows a sample result, with the neutron observed over a distance of 5.6 m in a time of about 600 ns. The velocity of the neutron is deduced to be $8.9 \times 10^6$ m/s, corresponding to an energy of 0.42 MeV. From this result, the mass difference is deduced to be $(m_{\pi^-} - m_{\pi^0})c^2 = 4.5930 \pm 0.0013$ MeV, a result consistent with (and of uncertainty less than) the value from the $\pi^0 \to 2\gamma$ decay. The net result for $m_{\pi^0}$, using the best value of $m_{\pi^-}$, is

$$m_{\pi^0}c^2 = 134.9745 \pm 0.0016 \text{ MeV}$$

Because the $\pi^-$ and $\pi^+$ are antiparticles of one another, we expect their masses to be equal. This is a consequence of the *CPT theorem*, in which C, P, and T represent the three discrete operations charge conjugation (actually, particle $\leftrightarrow$ antiparticle exchange), parity ($r \to -r$) and time reversal ($t \to -t$). The experimental difference between the $\pi^+$ and $\pi^-$ masses is

$$m_{\pi^+} - m_{\pi^-} = -0.0017 \pm 0.0020 \text{ MeV}$$

consistent with the CPT theorem.

The difference between the masses of the charged and neutral pions must be related to the electromagnetic interaction, a kind of electromagnetic self-energy analogous to that of any finite charge distribution. It is this mass difference that may be responsible for a possible small difference between the nn, np, and pp scattering parameters (Section 4.3).

## Spin and Parity

The decay of $\pi^0$ into two $\gamma$'s and the production of a single pion from nucleon–nucleon collisions such as $p + p \to p + n + \pi^+$ immediately shows that the pion must have integral spin, as do all mesons. The most direct indication of the pion spin comes from the study of the reaction

$$p + p \to d + \pi^+$$

and its inverse

$$\pi^+ + d \to p + p$$

If nature is symmetric with respect to time reversal, the direct and inverse cross sections should be identical, except for statistical factors and kinematics. This is

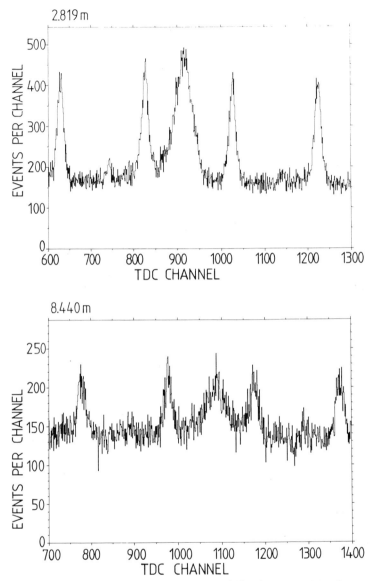

**Figure 17.6** Time-of-flight neutron spectrum following negative pion capture by protons. The top spectrum is taken over a flight path of 2.82 m and the bottom one is over 8.44 m; the shift of the peaks to longer times arises from the additional flight path. The broad peak near the center is the neutrons, and the narrow peaks are photons from $\pi^0$ decay and from $\pi^- + p \rightarrow \gamma + n$. Because the photons travel at $c$, they provide a convenient calibration of the flight path; note the shift of the neutron peak relative to the photon peaks on either side. From J. F. Crawford et al., *Phys. Rev. Lett.* **56**, 1043 (1986).

usually called the principle of *detailed balance*. That is,

$$\sigma \propto \frac{1}{k^2} g \qquad (17.12)$$

where $g$ is the spin-dependent statistical factor, Equation 11.67, and $\hbar k$ is the center-of-mass momentum of the incident particle. Thus

$$\frac{\sigma(\text{pp} \rightarrow \pi\text{d})}{\sigma(\pi\text{d} \rightarrow \text{pp})} = \frac{g(\text{pp} \rightarrow \pi\text{d})}{g(\pi\text{d} \rightarrow \text{pp})} \frac{k_\pi^2}{k_\text{p}^2} \qquad (17.13)$$

$$= \frac{(2s_\pi + 1)(2s_\text{d} + 1)}{\frac{1}{2}(2s_\text{p} + 1)^2} \frac{k_\pi^2}{k_\text{p}^2} \qquad (17.14)$$

$$= \frac{3(2s_\pi + 1)}{2} \frac{k_\pi^2}{k_\text{p}^2} \qquad (17.15)$$

The extra statistical factor of $\frac{1}{2}$ in the denominator of Equation 17.14 comes about because the Pauli principle reduces the number of possible initial pp states by $\frac{1}{2}$. Figure 17.7 shows the cross sections for the two reactions, corrected for the kinematical factor $k_\pi^2/k_\text{p}^2$ and assuming $s_\pi = 0$. The agreement is excellent, indicating the correct choice of $s_\pi$; if we were to have chosen $s_\pi = 1, 2, 3, \ldots$, the two cross sections would have differed by factors of $2s_\pi + 1 = 3, 5, 7, \ldots$, in violation of detailed balance and time-reversal invariance in the strong nuclear interaction.

Evidence for the spin of $\pi^0$ comes from observing its decay into $2\gamma$. In the $\pi^0$ rest frame, the $\gamma$'s are emitted in opposite directions. Photons have spin 1 and are required to have $m_s = \pm 1$ along their direction of motion; $m_s = 0$ is forbidden. (This follows from the *transverse* nature of electromagnetic waves; $\boldsymbol{E}$ and $\boldsymbol{B}$ are transverse to the direction of propagation.) The total $m_s$ of the two photons can thus be 0 or $\pm 2$. The required symmetry of the final state of two photons requires the integer spin of the $\pi^0$ to be even, and zero is the obvious choice in analogy with the charged pion.

The intrinsic parity of the pion can be deduced from studying the reaction

$$\pi^- + \text{d} \rightarrow \text{n} + \text{n}$$

using incident pions of sufficiently low energy that only s-wave capture can occur. The initial parity is

$$\pi_\text{i} = \pi_\pi \pi_\text{d} (-1)^{\ell_\text{i}} = \pi_\pi \qquad (17.16)$$

with $\pi_\text{d} = +1$ and $\ell_\text{i} = 0$. (The deuteron has even intrinsic parity, as discussed in Chapter 4.) Conservation of parity in strong interactions means that the initial and final parities must be equal. The final parity is

$$\pi_\text{f} = \pi_\text{n} \pi_\text{n} (-1)^{\ell_\text{f}} = (-1)^{\ell_\text{f}} \qquad (17.17)$$

Thus $\pi_\pi = +1$ requires $\ell_\text{f} =$ even, while $\pi_\pi = -1$ requires $\ell_\text{f} =$ odd. The total angular momentum of the initial state is

$$\boldsymbol{J}_\text{i} = \boldsymbol{s}_\pi + \boldsymbol{s}_\text{d} + \boldsymbol{\ell}_\text{i} \qquad (17.18)$$

PION LABORATORY ENERGY, MeV, ($\pi^+$+d $\rightarrow$ p+p )

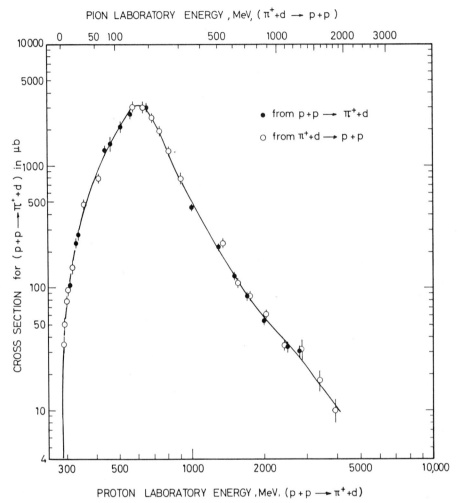

PROTON LABORATORY ENERGY, MeV, (p+p $\rightarrow \pi^+$+d)

**Figure 17.7** Comparison of cross sections of p + p $\rightarrow \pi$ + d and its inverse. The excellent agreement between the two cross sections confirms the spin of zero for the pion. From W. O. Lock and D. F. Measday, *Intermediate Energy Nuclear Physics* (London: Methuen, 1970).

and knowing that $s_\pi = 0$, $\ell_i = 0$, and $s_d = 1$ requires $J_i = 1$. The final angular momentum is

$$J_f = s_{n_1} + s_{n_2} + \ell_f \tag{17.19}$$

Because neutrons are spin-$\frac{1}{2}$ particles, their total wave function must be antisymmetric with respect to the interchange of the neutrons. If the spin part of the combined wave function is symmetric ($s_{n_1} + s_{n_2} = 1$) then the spatial part must be antisymmetric ($\ell_f$ = odd); conversely, if the spin part is antisymmetric ($s_{n_1} + s_{n_2} = 0$), then the spatial part is symmetric ($\ell_f$ = even). The latter possibility is eliminated by the requirement that $J_f = 1$, for if $s_{n_1} + s_{n_2} = 0$, there is no way to couple an even $\ell_f$ to get $J_f = 1$. We therefore conclude $s_{n_1} + s_{n_2} = 1$ and $\ell_f$ = odd, and the only odd value of $\ell_f$ that can satisfy the coupling of Equation

17.19 to give $J_f = 1$ is $\ell_f = 1$. From Equation 17.17 we conclude $\pi_f = -1$, hence $\pi_i = -1$ and the pion has odd parity.

A similar argument can be made for the $\pi^+$ through the reaction $\pi^+ + d \to$ p + p. The $\pi^0$ parity can be deduced either directly from observation of the polarization of the electrons in the decay $\pi^0 \to 2e^+ + 2e^-$ or indirectly from the reaction $\pi^- + d \to 2n + \pi^0$ (see Problem 3).

## Decay Modes

The pion is the lightest meson and therefore the lightest strongly interacting particle. It cannot decay, as heavier mesons can, into lighter strongly interacting particles by the strong interaction (with a characteristic lifetime of the order of $10^{-22}$ s). It must decay by the much slower electromagnetic or weak interaction, with consequently a much longer lifetime.

The $\pi^0$ decays electromagnetically:

$$\pi^0 \to \gamma + \gamma$$

in a time of the order of $10^{-16}$ s. This is too short a time to be detected directly, but it is too long a lifetime to be deduced by a measurement of the width of the energy distribution in the production of $\pi^0$, as is done for other mesons discussed in Section 17.4. The width corresponding to the $\pi^0$ lifetime is only 8 eV, far smaller than typical experimental energy resolutions. Since neither of these usual methods are applicable, it is necessary to employ an unusual technique—the reverse reaction is used, in which $\gamma$ radiations produce $\pi^0$ in the Coulomb field of a heavy nucleus in a process called *photoproduction*:

$$\gamma + {}^A_Z X \to \pi^0 + {}^A_Z X$$

Because this is also an electromagnetic process, the cross section for the $(\gamma, \pi^0)$ reaction is directly related to the reverse process, the $\pi^0 \to 2\gamma$ decay. In the experiment reported by Browman et al., *Phys. Rev. Lett.* **33**, 1400 (1974), beams of high-energy (4–6 GeV) photons were directed against targets of various heavy materials, and the $\pi^0$ was detected by observing the two decay $\gamma$'s. The deduced mean lifetime was

$$\tau_{\pi^0} = (0.82 \pm 0.04) \times 10^{-16} \text{ s}$$

The $\pi^0 \to 2\gamma$ process occurs in the $\pi^0$ decay with a branching intensity of 98.8%. The only other major competing process is the so-called *Dalitz decay mode* $\pi^0 \to \gamma + e^+ + e^-$, with a branching intensity of 1.2%.

The most precise measurement of the lifetimes of the charged pions was done in an experiment reported by Ayres et al., *Phys. Rev. D* **3**, 1051 (1971). Beams of $\pi^+$ or $\pi^-$ were extracted as secondary beams from the Berkeley 184-inch proton synchrocyclotron. A counter was moved along the pion beam and measured the number of pions at various distances. The radioactive decay law $N = N_0 e^{-t/\tau'}$ gives the relative number of pions surviving at time $t$ if the laboratory-frame lifetime is $\tau'$. The beam travels at velocity $v$, and the decay law can be written in terms of distance $x = vt$ as

$$N = N_0 e^{-x/v\tau'} \tag{17.20}$$

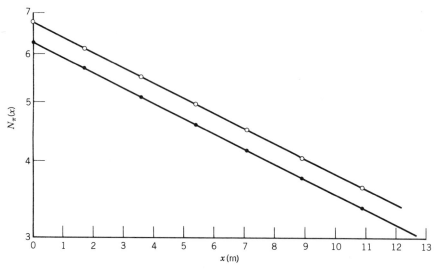

**Figure 17.8** Decay rate vs distance for $\pi^+$ (●) and $\pi^-$ (○) decaying in flight. Data taken from Ayres et al., *Phys. Rev. D* **3**, 1051 (1971).

In the laboratory frame, the lifetime $\tau'$ is not the same as the so-called *proper lifetime* $\tau$ measured in the rest frame of the pion. The relationship is

$$\tau' = \tau/\sqrt{1 - v^2/c^2} \tag{17.21}$$

and thus

$$N = N_0 e^{-Bx} \tag{17.22}$$

where

$$B = \frac{1}{v\tau'} = \frac{\sqrt{1 - v^2/c^2}}{v\tau} \tag{17.23}$$

In the experiment of Ayres et al., the pion detector was moved through a distance of over 10 m and the change in rate due to the decay was observed. Figure 17.8 shows the exponential decay in a semilog plot of the data. The slopes of the $\pi^+$ and $\pi^-$ decays are each 0.0575 m$^{-1}$. Using time-of-flight techniques, the beam momentum was measured to be $p = 311.89$ MeV$/c$, in terms of which

$$\frac{1}{\tau} = B\frac{v}{\sqrt{1 - v^2/c^2}} = \frac{B}{m}\frac{mv}{\sqrt{1 - v^2/c^2}} = \frac{Bp}{m} \tag{17.24}$$

$$\tau = \frac{1}{c}\frac{mc^2}{(B)(pc)} = 26.02 \pm 0.04 \text{ ns}$$

These results of Ayres et al. on the $\pi^\pm$ lifetimes also provide two direct tests of fundamental physics. The CPT theorem, discussed above in connection with the $\pi^\pm$ masses, also requires that the $\pi^+$ and $\pi^-$ lifetimes be identical. The equality of the slopes in Figure 17.8 indicates that the lifetimes are equal, and from separate fits to the two slopes it is possible to deduce their ratio:

$$\frac{\tau(\pi^+)}{\tau(\pi^-)} = 1.00055 \pm 0.00071$$

consistent with CPT. The measured lifetime also indicates directly the validity of the special relativity equation 17.21 for *time dilation*. Ayres et al. observed the decay of pions moving at $v/c = 0.913$; other experiments on the decay of pions at rest give exactly the same lifetime of $26.02 \pm 0.04$ ns. This experiment thus provides one of the most precise direct tests of Einstein's Special Theory of Relativity.

The charged pions decay by the weak interaction (as suggested by the $10^{-8}$-s lifetime), into leptons (electrons or muons and their neutrinos):

$$\pi^- \rightarrow \mu^- + \bar{\nu}_\mu \qquad (\cong 100\%)$$
$$\rightarrow e^- + \bar{\nu}_e \qquad (1.23 \times 10^{-4})$$
$$\rightarrow \mu^- + \bar{\nu}_\mu + \gamma \qquad (1.24 \times 10^{-4})$$
$$\rightarrow e^- + \bar{\nu}_e + \gamma \qquad (5.6 \times 10^{-8})$$
$$\rightarrow e^- + \bar{\nu}_e + \pi^0 \qquad (1.03 \times 10^{-8})$$

A similar set exists for $\pi^+$, with all particles changed to their antiparticle. Each decay conserves lepton number, which is counted separately for muon-type leptons ($\mu^+$, $\mu^-$, $\nu_\mu$, $\bar{\nu}_\mu$) and electron-type leptons ($e^+$, $e^-$, $\nu_e$, $\bar{\nu}_e$). Since there are no leptons in the initial state, the final state must have a lepton number of zero, and must thus include one lepton and one antilepton of the same type. Decays such as $\pi^- \rightarrow e^- + \bar{\nu}_\mu$ are forbidden because they fail to conserve both electron and muon lepton numbers.

The relative branching of the $\pi^\pm$ decays into $\mu^\pm$ or $e^\pm$ can be understood by considering the statistics of the decay process. The decay probability, according to Equation 2.79, is proportional to the square of a matrix element times a density of states function. We assume that electrons and muons have identical weak interaction strengths, and so the differences in the decay branches must be due to the density of final states; for decay into two particles this is (see, for example, Equation 9.15)

$$\rho(E_f) \equiv \frac{1}{V}\frac{dn}{dE_f} = \frac{4\pi p^2}{h^3}\frac{dp}{dE_f} \tag{17.25}$$

Here $E_f$ represents the total final energy in the decay, and $p$ represents the momentum of either the charged lepton ($\mu^\pm$ or $e^\pm$) or the neutrino, since for decays of $\pi^\pm$ at rest momentum conservation requires them to be equal and opposite. Thus, since $E_f$ in the $\pi^\pm$ rest frame is just its mass energy,

$$E_f = m_\pi c^2 = E_\nu + E_\pm \tag{17.26}$$

where $E_\nu$ is the neutrino energy and $E_\pm$ is the energy of the charged lepton

$$E_f = m_\pi c^2 = cp + \sqrt{c^2 p^2 + m^2 c^4} \tag{17.27}$$

$$dE_f = \left[ c + \tfrac{1}{2}(c^2 p^2 + m^2 c^4)^{-1/2} 2c^2 p \right] dp \tag{17.28}$$

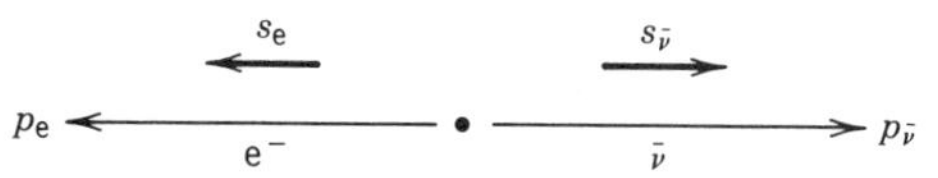

**Figure 17.9** Decay of $\pi^-$ at rest gives $\bar{\nu}$ and $e^-$ with opposite momenta. Because **s** and **p** must be parallel for $\bar{\nu}$, the direction of $\mathbf{s}_{\bar{\nu}}$ must be as shown, and because the pion has no intrinsic spin, $\mathbf{s}_e$ and $\mathbf{s}_{\bar{\nu}}$ must add to zero; thus $\mathbf{s}_e$ must be opposite to $\mathbf{s}_{\bar{\nu}}$.

Here $m$ represents the mass of the charged lepton (e or $\mu$). Using Equation 17.27 to eliminate $p$, it follows that

$$p^2 \frac{dp}{dE_f} = \frac{\left(m_\pi^2 + m^2\right)\left(m_\pi^2 - m^2\right)^2}{8cm_\pi^4} \tag{17.29}$$

One other factor appears in the density of states factor. Figure 17.9 shows a schematic picture of the decay process. The $e^-$ and $\bar{\nu}_e$ are emitted in opposite directions. Because the pion has spin zero, the spins of $e^-$ and $\bar{\nu}_e$ must add to zero. All antineutrinos have **s** and **p** parallel (see the discussion on $\beta$ decay in Section 9.6), and thus the electron spin must be parallel to its momentum. However, $\beta$-decay electrons have helicity $h = \mathbf{s} \cdot \mathbf{p} / |\mathbf{s} \cdot \mathbf{p}|$ equal to $-v/c$, and extremely relativistic electrons, such as those emitted in $\pi^-$ decay, are expected on the basis of weak interaction theory to have helicity $-v/c \approx -1$. The geometry of Figure 17.9, on the other hand, requires positive helicity of about $+1$. The fraction of positive-helicity electrons emitted in the decay is thus $1 - v/c$ and the decay probability is inhibited by that same factor. (Because electrons have $v/c \approx 1$, it is primarily this factor that brings about the extremely small branching ratio into the electron mode relative to the muon mode.) Again using Equation 17.27, it follows that

$$1 - \frac{v}{c} = \frac{2m^2}{m_\pi^2 + m^2} \tag{17.30}$$

Combining these factors, the pion decay probability $\lambda$ depends on the kinematical factors according to

$$\lambda \propto \frac{m^2\left(m_\pi^2 - m^2\right)^2}{m_\pi^4} \tag{17.31}$$

and thus

$$\frac{\lambda(\pi \rightarrow e\nu)}{\lambda(\pi \rightarrow \mu\nu)} = \frac{m_e^2\left(m_\pi^2 - m_e^2\right)^2}{m_\mu^2\left(m_\pi^2 - m_\mu^2\right)^2} \tag{17.32}$$

$$= 1.28 \times 10^{-4}$$

in agreement with the observed ratio of $(1.23 \pm 0.02) \times 10^{-4}$.

## Production

Pions are most conveniently produced (as in the meson factories) from collisions of protons with nuclear targets. The simplest nucleon–nucleon reactions are

$$p + p \rightarrow p + p + \pi^0$$
$$\rightarrow p + n + \pi^+$$
$$p + n \rightarrow p + p + \pi^-$$
$$\rightarrow p + n + \pi^0$$

Note that the initial state and the final state both have two nucleons; this is a

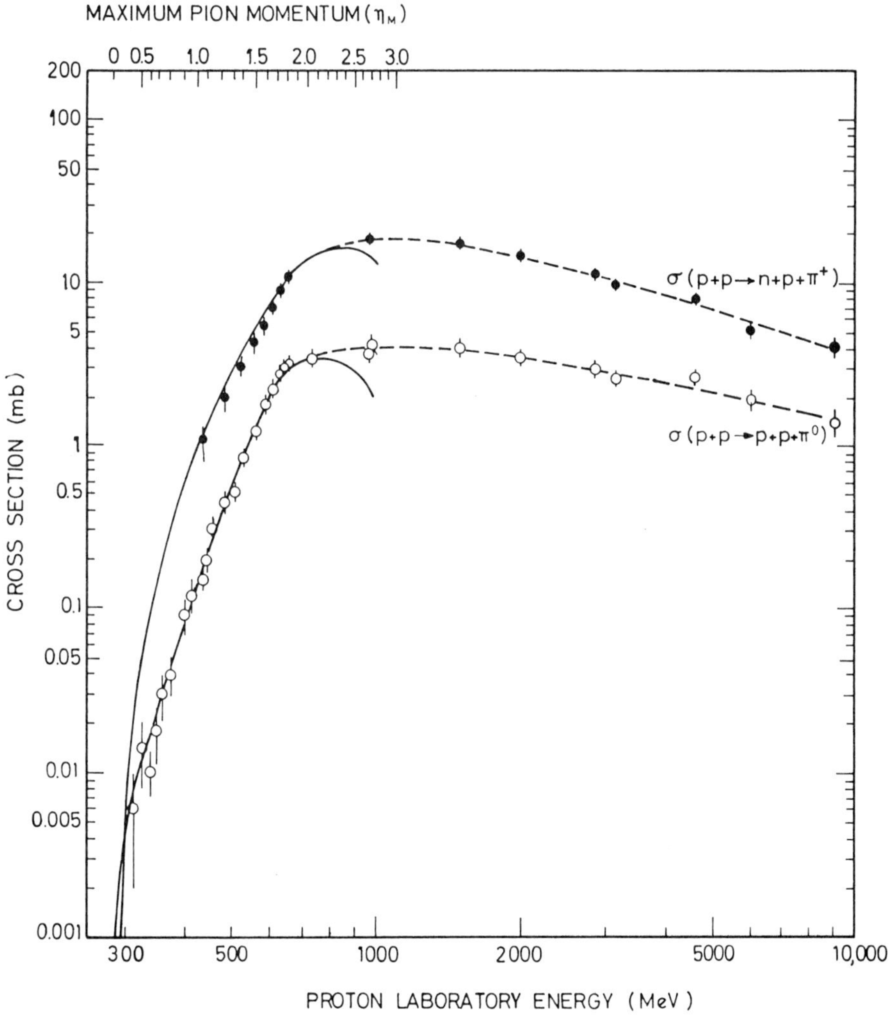

**Figure 17.10** Cross sections for production of pions in proton-proton collisions. From W. O. Lock and D. F. Measday, *Intermediate Energy Nuclear Physics* (London: Methuen, 1970).

consequence of baryon conservation, the nucleons being the lightest members of the baryon family.

The $Q$ value for these reactions is roughly the negative of the pion rest energy, and thus the threshold is, by Equation 15.13, $T_{th} = 290$ MeV. The cross section for single pion production is quite small near threshold, but it rises rapidly to a broad peak near 1 GeV (Figure 17.10), which is why the meson factories operate with proton beams in the range 0.5–0.8 GeV.

Unlike spin-$\frac{1}{2}$ fermions such as leptons and baryons, there is no law requiring conservation of the number of integral-spin particles such as mesons. Therefore, nucleon–nucleon reactions can produce any number of mesons, consistent with energy and charge conservation. At a threshold of about 600 MeV, two-pion production becomes possible:

$$p + p \rightarrow p + p + \pi^+ + \pi^-$$
$$\rightarrow p + p + \pi^0 + \pi^0$$
$$\rightarrow p + n + \pi^0 + \pi^+$$
$$\rightarrow n + n + \pi^+ + \pi^+$$

The cross sections (Figure 17.11) in the 0.5–0.8-GeV region are small compared with those for single pion production.

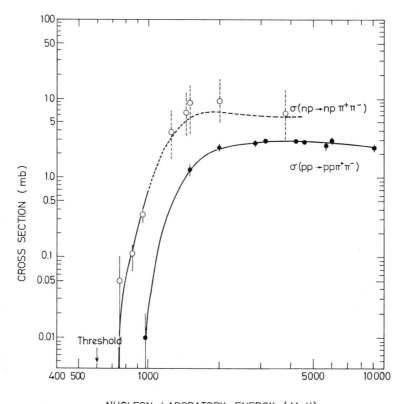

**Figure 17.11** Two-pion cross sections from nucleon-nucleon collisions. From W. O. Lock and D. F. Measday, *Intermediate Energy Nuclear Physics* (London: Methuen, 1970).

Gamma rays incident upon nucleons can also produce pions:

$$\gamma + p \rightarrow n + \pi^+$$
$$\rightarrow p + \pi^0$$

The threshold for this photoproduction process is about 150 MeV. Beams of such high-energy photons are available as secondary beams at electron accelerators, obtained when electrons are made to travel in a circle (synchrotron radiation) or stopped in a solid target (bremsstrahlung).

In the meson factories, the pion production targets are solids of relatively low-$Z$ materials, such as carbon or beryllium. At 0.5–0.8-GeV energies, the incident protons have a Compton wavelength of 0.8–1.1 fm and thus "see" individual nucleons, rather than the nucleus as a whole. Nevertheless, the nucleon–nucleon cross sections are modified considerably in nucleon–nucleus collisions, by the motion of individual nucleons in the nucleus and by the presence of other nucleons. Figure 17.12 shows an example of the pion production cross sections for protons incident on $^9$Be and $^{12}$C. Note that $\pi^+$ production is much more probable than $\pi^-$ production. Only one reaction can produce single $\pi^-$ ($p + n \rightarrow p + p + \pi^-$), while there are two reactions that can produce $\pi^+$ ($p + n \rightarrow n + n + \pi^+$ and $p + p \rightarrow p + n + \pi^+$, the second of which occurs with greater probability). At high resolution, the effect of specific final nuclear excited states can be observed. These also can be quite different for $\pi^+$ and $\pi^-$ final states, as Figure 17.12 illustrates. Despite the mirror-image final states of $^{10}_4\text{Be}_6$ and $^{10}_6\text{C}_4$, the relative cross sections are different.

## 17.3 PION–NUCLEON REACTIONS

Reactions induced by pions incident on nucleons can be classified and analyzed using techniques similar to those developed in Chapter 11 for nucleon-induced reactions. We can identify three types of reactions: elastic scattering, inelastic scattering, and charge exchange. Elastic scattering includes such reactions as

$$\pi^+ + p \rightarrow \pi^+ + p$$
$$\pi^- + p \rightarrow \pi^- + p$$

In the inelastic reactions discussed in Chapter 11, the target nucleus is left in an excited state. In the case of pions, the energy is deposited through the creation of new pions:

$$\pi^+ + p \rightarrow \pi^+ + \pi^0 + p$$
$$\rightarrow \pi^+ + \pi^+ + n$$

Charge exchange reactions are similar to (p, n) reactions:

$$\pi^- + p \rightarrow \pi^0 + n$$

Only in the case of inelastic scattering leading to pion production is there a sizeable threshold (170 MeV); the other reactions will occur even for very low energy pions.

The cross sections for $\pi^+ p$ and $\pi^- p$ reactions, illustrated in Figures 17.13 and 17.14, show a number of broad resonances of the same type as those discussed in

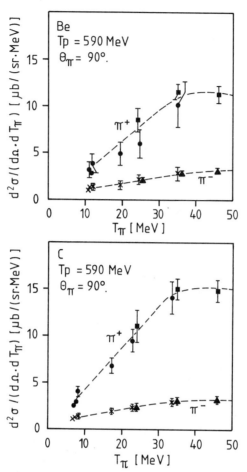

**Figure 17.12** Cross sections for pion production by protons incident on $^9$Be and $^{12}$C. Above are shown comparisons of $\pi^+$ and $\pi^-$ production by 590-MeV protons and at right are shown reactions of 185-MeV protons on $^9$Be leading to individual excited states. Known excited states in $^{10}$Be are at 3.37 MeV, 5.96–6.26 MeV (a multiplet of four states), 7.37–7.54 MeV (two states), and 9.27–9.4 MeV (two states); in $^{10}$C known states are at 3.35 MeV, 5.22–5.38 MeV (two states), and 6.58 MeV. From (above) J. F. Crawford et al., *Helv. Phys. Acta* **53**, 497 (1980), and (right) S. Dahlgren et al., *Nucl. Phys. A* **204**, 53 (1973).

Section 11.12, but somewhat broader. In low-energy nuclear reactions, we regard resonances in much the same way we regard discrete nuclear levels—they have a definite energy, lifetime (or width), spin-parity assignment; we can study their formation and decays into different final states. In short, a resonance is as real as any nuclear bound state.

How then are we to regard these pion-nucleon resonances? From additional experiments, we learn that they too have definite energies, decay lifetimes, spin-parity assignments, and decay modes. Each pion-nucleon resonance forms a structure as definite and real as an "ordinary" proton or neutron, and the fact that these resonances are extremely short-lived should not prejudice us against

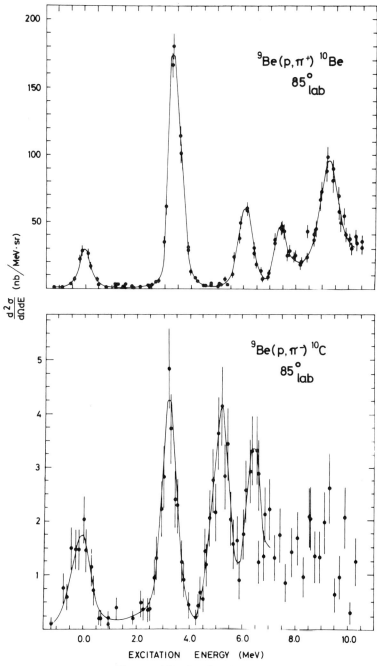

**Figure 17.12**  Continued.

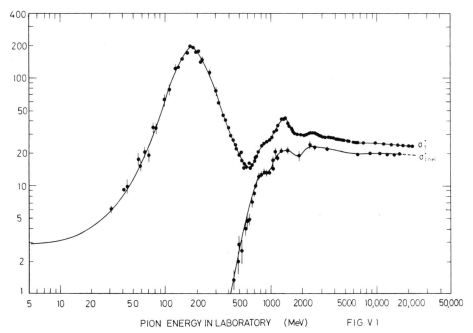

**Figure 17.13** Total and inelastic cross sections for $\pi^+ + p$ reactions. From W. O. Lock and D. F. Measday, *Intermediate Energy Nuclear Physics* (London: Methuen, 1970).

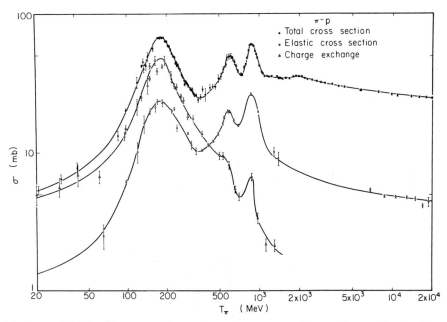

**Figure 17.14** Cross sections for $\pi^- + p$ reactions. From R. J. Cence, *Pion-Nucleon Scattering* (Princeton, NJ: Princeton University Press, 1969).

including them in a list of particles. Such resonance particles play a significant role in the particle classification schemes to be discussed in Chapter 18.

The $\pi^+$p cross section is dominated by a huge resonance at a pion energy of about 200 MeV (corresponding to a center-of-mass energy of 1232 MeV); the same resonance occurs in the $\pi^-$p elastic and charge-exchange cross sections as well. Nor is this resonance restricted to pion scattering—Figure 17.15 shows results of inelastic nucleon–nucleon scattering, inelastic electron scattering from protons, and photoproduction. These cross sections have other weaker resonances in common as well. Some of these resonances, such as 1232 MeV, are called $\Delta$ resonances and always occur in multiplets of four charge states ($+2$, $+1$, 0, and $-1$). Others of the resonances occur only in two charge states ($+1$ and 0) like the nucleons; these are called N (or sometimes N*) resonances. It is quite appropriate to interpret the N resonances, and in a similar way the $\Delta$'s, as excited states of the nucleon. Ignoring for the present the internal structure that may be responsible for the excited states, we can regard the excitation spectrum in the same way that we regard the excitation spectrum of the hydrogen atom. Figure 17.16 shows the spectrum of excited states of N and $\Delta$, some of which are listed in Table 17.1.

Each of the N states is, like the nucleons, a doublet and can thus be assigned isospin $T = \frac{1}{2}$. Each of the $\Delta$ states is a quartet, corresponding to a $T = \frac{3}{2}$ assignment ($2T + 1 = 4$). The masses of the members of the multiplet would be equal to one another in the absence of electromagnetic interactions; the observed mass splittings are the order of a few MeV/$c^2$, small compared with the masses themselves and even with the differences in mass between different multiplets. These isospin identifications are consistent with the interpretation of the N and $\Delta$ resonances as excited states of the pion–nucleon system. Coupling the $T = 1$ pion to the $T = \frac{1}{2}$ nucleon, we expect resultants of $T = \frac{1}{2}$ or $\frac{3}{2}$, according to the usual angular momentum coupling rules of quantum mechanics.

The identification of these resonances is illustrated in the case of the 1232-MeV $\Delta$ in Figure 17.17. The resonance has the Breit-Wigner form of Equations 11.69 and 11.70:

$$\sigma = \frac{\pi}{k^2} \frac{\Gamma\Gamma_{\pi p}}{(E - E_R)^2 + \Gamma^2/4} \frac{2s_\Delta + 1}{(2s_\pi + 1)(2s_p + 1)} \qquad (17.33)$$

where $\Gamma_{\pi p}$ is the partial width of the $\Delta$, $\Gamma$ is the total width, $E_R$ is the resonance energy, and $s$ represents the intrinsic spins of the particles ($s_\pi = 0$, $s_p = \frac{1}{2}$). For the $\Delta$, $\pi^+$p is the only decay mode and so $\Gamma = \Gamma_{\pi p} = 115$ MeV, corresponding to a lifetime of $6 \times 10^{-24}$ s. At the resonance ($E = E_R$),

$$\sigma = \frac{2\pi}{k^2}(2s_\Delta + 1) \qquad (17.34)$$

At a pion laboratory kinetic energy of 200 MeV, $pc = 230$ MeV in the center-of-mass frame, so $1/k^2 = \hbar^2 c^2/p^2 c^2 = 0.7$ fm$^2$ = 7 mb. Thus $\sigma = 45(2s_\Delta + 1)$ mb, and the measured peak cross section of 200 mb is consistent only with $s_\Delta = \frac{3}{2}$. The spin of the $\Delta$ resonance at 1232 MeV is therefore $\frac{3}{2}$, and the only way to couple a spin-0 pion and a spin-$\frac{1}{2}$ nucleon to a spin-$\frac{3}{2}$ resultant is in a state of orbital angular momentum $\ell = 1$. The parity of the $\Delta$ is therefore even; the

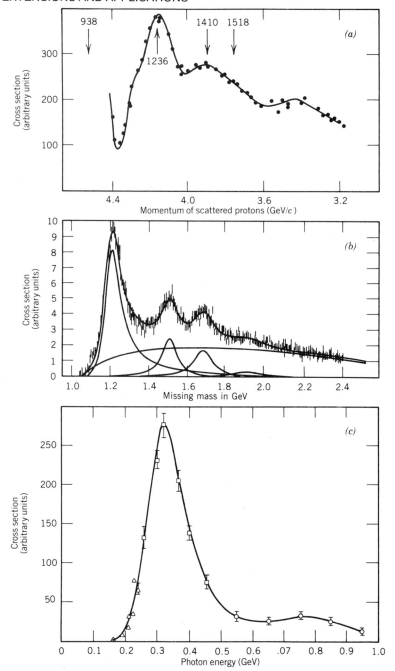

**Figure 17.15** Cross sections for (*a*) proton–proton inelastic scattering at 4.55 GeV / *c* incident momentum, (*b*) electron–proton inelastic scattering at 7 GeV incident energy, and (*c*) photoproduction, $\gamma + p \rightarrow \pi^0 + p$, in which the incident photon energy of 340 MeV corresponds to the 1232-MeV resonance. Data from (respectively) I. M. Blair et al., *Phys. Rev. Lett.* **17**, 789 (1966); K. Gottfried and V. F. Weisskopf, *Concepts of Particle Physics*, Vol. 1 (Oxford: Clarendon, 1984); and E. Segrè, *Nuclei and Particles* (Reading, MA: Benjamin, 1977).

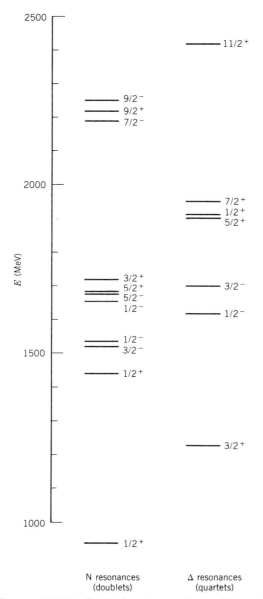

$E$ (MeV)

2500

—— 11/2 $^+$

—— 9/2 $^-$
—— 9/2 $^+$
—— 7/2 $^-$

2000

—— 7/2 $^+$
≡≡≡ 1/2 $^+$
       5/2 $^+$

—— 3/2 $^+$
≡≡≡ 5/2 $^+$
       5/2 $^-$
—— 1/2 $^-$

—— 3/2 $^-$

—— 1/2 $^-$

—— 1/2 $^-$
—— 3/2 $^-$

—— 1/2 $^+$

1500

—— 3/2 $^+$

1000

—— 1/2 $^+$

N resonances
(doublets)

Δ resonances
(quartets)

**Figure 17.16**   Excited-state spectrum of N and Δ.

intrinsic parity of the Δ is the product of the intrinsic parities of the constituent π and p with the relative parity from the orbital motion, thus: $\pi_\Delta = \pi_\pi \pi_{\mathrm{p}} (-1)^\ell = +1$. Through similar arguments, we can work out the orbital angular momentum and spin-parity assignments of the other N and Δ resonances listed in Table 17.1.

The decays of the N's and Δ's occur by the strong interaction in characteristic times of the order of $10^{-23}$ s. The dominant decay modes are into lighter N's and

**Table 17.1** Nucleon Resonances

| Designation | Energy (MeV) | Width (MeV) | Spin Parity | Isospin | $\ell_\pi$ |
|---|---|---|---|---|---|
| N(939) | 939 | — | $\frac{1}{2}^+$ | $\frac{1}{2}$ | 1 |
| N(1440) | 1440 | 200 | $\frac{1}{2}^+$ | $\frac{1}{2}$ | 1 |
| N(1520) | 1520 | 125 | $\frac{3}{2}^-$ | $\frac{1}{2}$ | 2 |
| N(1535) | 1535 | 150 | $\frac{1}{2}^-$ | $\frac{1}{2}$ | 0 |
| N(1650) | 1650 | 150 | $\frac{1}{2}^-$ | $\frac{1}{2}$ | 0 |
| N(1675) | 1675 | 155 | $\frac{5}{2}^-$ | $\frac{1}{2}$ | 2 |
| N(1680) | 1680 | 125 | $\frac{5}{2}^+$ | $\frac{1}{2}$ | 3 |
| N(1720) | 1720 | 200 | $\frac{3}{2}^+$ | $\frac{1}{2}$ | 1 |
| N(2190) | 2190 | 350 | $\frac{7}{2}^-$ | $\frac{1}{2}$ | 4 |
| N(2220) | 2220 | 400 | $\frac{9}{2}^+$ | $\frac{1}{2}$ | 5 |
| N(2250) | 2250 | 300 | $\frac{9}{2}^-$ | $\frac{1}{2}$ | 4 |
| $\Delta$(1232) | 1232 | 115 | $\frac{3}{2}^+$ | $\frac{3}{2}$ | 1 |
| $\Delta$(1620) | 1620 | 140 | $\frac{1}{2}^-$ | $\frac{3}{2}$ | 0 |
| $\Delta$(1700) | 1700 | 250 | $\frac{3}{2}^-$ | $\frac{3}{2}$ | 2 |
| $\Delta$(1905) | 1905 | 300 | $\frac{5}{2}^+$ | $\frac{3}{2}$ | 3 |
| $\Delta$(1910) | 1910 | 220 | $\frac{1}{2}^+$ | $\frac{3}{2}$ | 1 |
| $\Delta$(1950) | 1950 | 240 | $\frac{7}{2}^+$ | $\frac{3}{2}$ | 3 |
| $\Delta$(2420) | 2420 | 300 | $\frac{11}{2}^+$ | $\frac{3}{2}$ | 5 |

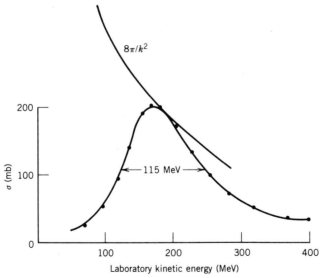

**Figure 17.17** Pion-proton cross section near the $\Delta$ resonance. The peak Breit-Wigner cross section, equal to $8\pi/k^2$ assuming a spin of $\frac{3}{2}$, is also shown. The center-of-mass energy at the peak is 1232 MeV.

$\Delta$'s plus one or more pions:

$$N(1440) \rightarrow N(939) \ + \pi \qquad (\sim 60\%)$$
$$\rightarrow \Delta(1232) + \pi \qquad (\sim 20\%)$$

$$\Delta(1232) \rightarrow N(939) \ + \pi$$

$$N(1650) \rightarrow N(939) \ + \pi \qquad (\sim 60\%)$$
$$\rightarrow \Delta(1232) + \pi \qquad (\sim 10\%)$$
$$\rightarrow N(939) \ + \rho \qquad (\sim 20\%)$$

$$\Delta(1950) \rightarrow N(939) \ + \pi \qquad (\sim 40\%)$$
$$\rightarrow \Delta(1232) + \pi \qquad (\sim 40\%)$$
$$\rightarrow N(939) \ + \rho \qquad (\sim 20\%)$$

Occasionally, a decay may appear to be energetically forbidden, as N(1650) $\rightarrow$ N(939) $+ \rho$, in which the final energy is 1707 MeV. It is the large widths of the initial and final states that permit this decay, for the energy uncertainty of the 1650-MeV resonance easily overlaps 1707 MeV. (The $\rho$ meson is discussed in the next section.)

Differential cross sections have also been measured for pion–nucleon reactions; the results of their analysis are consistent with the $\Delta$ and N descriptions. In the 200-MeV range, the angular distributions are very characteristic of $\ell = 1$; at higher energies, the angular distributions are more complicated, but yield to an analysis in terms of $\ell$ transfers.

It is also possible to characterize the pion–nucleon scattering cross sections in terms of phase shifts, just as was done for the nucleon–nucleon cross sections in Chapter 4. The p-wave phase shift goes through 90° for incident pion energies near 200 MeV. Recall from Equation 11.45 that the cross section depends on the phase shift like $\sin^2 \delta_\ell$, which reaches its maximum value at $\delta_\ell = 90°$. This result is consistent with an $\ell = 1$ resonance at 1232 MeV.

In summary, the results of pion–nucleon scattering give strong evidence for resonances that are interpreted in terms of discrete particle states as excitations of the nucleon. This interpretation is consistent with resonances seen in other studies (pp, $\gamma$p, ep scattering) and with differential cross sections. These $\Delta$ and N resonances are historically called *isobars*, and the analysis of $\pi$p scattering data described above is called the *isobar model*. The internal structure of nuclei may also depend on the presence of isobars, which can be formed in the process of virtual pion exchange between different nucleons in the nucleus.

## 17.4 MESON RESONANCES

The pions are the lightest members of the meson family. As the incident energy is increased, there is the possibility to produce other mesons in proton–proton or pion–proton reactions. All of these mesons have masses greater than twice the pion mass, and because there is no conservation law for the number of mesons, they can decay into two or more pions through the strong interaction on a time scale of the order of $10^{-23}$ s. We have no hope of direct observation of such short lifetimes, but we can infer the existence of the mesons as short-lived resonances by observing their decay products, and from the energy distribution of the decay products we can deduce the width of the resonance and therefore the lifetime of the particle.

As an example of this procedure, we consider the $\rho^+$ meson, which can be formed in pion–nucleon collisions and which decays rapidly into $\pi^+$ and $\pi^0$:

$$\pi^+ + \text{p} \to \rho^+ + \text{p}$$
$$\phantom{\pi^+ + \text{p} \to \rho^+}\hookrightarrow \pi^+ + \pi^0$$

It is also possible to produce a $\pi^0$ directly in the reaction

$$\pi^+ + \text{p} \to \pi^+ + \pi^0 + \text{p}$$

Because we do not observe the $\rho$ meson directly, the particles actually observed in either case are the two pions and the proton; both reactions lead to the same final particles. How is it possible for the experimenter to distinguish the reaction in which a $\pi^0$ is produced directly through a "knockout" process from the case in which the $\rho^+$ resonance is formed and then decays? The solution comes about by analyzing the energies of the emitted pions. If it were possible to measure the energy and momentum of the $\rho^+$ directly, they would be related by the usual relativistic formula:

$$\left[E_\rho^2 - c^2 p_\rho^2\right]^{1/2} = m_\rho c^2 \tag{17.35}$$

If the two pions come from the decay of the $\rho^+$, then their total energy must be equal to the total energy of the $\rho$

$$E_\rho = E_{\pi^+} + E_{\pi^0} \tag{17.36}$$

and the vector sum of their momenta must equal the original momentum of the $\rho$

$$\boldsymbol{P}_\rho = \boldsymbol{p}_{\pi^+} + \boldsymbol{p}_{\pi^0} \tag{17.37}$$

Combining these results,

$$\left[(E_{\pi^+} + E_{\pi^0})^2 - c^2 |\boldsymbol{p}_{\pi^+} + \boldsymbol{p}_{\pi^0}|^2\right]^{1/2} = m_\rho c^2 \tag{17.38}$$

What is so useful about this procedure is that the quantity on the right side of Equation 17.38 is the rest energy of a particle, a true invariant that must have the same value no matter what reference frame we work in. We can therefore measure the two pion energies and momenta in the lab frame, and if a $\rho$ meson is indeed formed, the quantities on the left side of Equation 17.38 will, no matter what their distribution of individual values, always combine to yield the same value of $m_\rho c^2$.

Unfortunately, it is not possible to exclude the direct process $\pi^+ + \text{p} \to \pi^+ + \pi^0 + \text{p}$ from participation in the reaction, but the three-body final state produced directly shares a distinguishable feature in common with the three-body final state of $\beta$ decay—the energy is distributed statistically among the final products. The invariant mass distribution therefore has two components—a continuous background (extending from a minimum at $2m_\pi c^2$ to a maximum determined by the incident energy) and a peak at a unique energy corresponding to the mass of the resonance particle. Figure 17.18 shows an example of the invariant mass plot

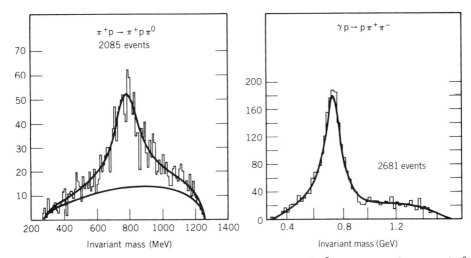

**Figure 17.18**   Invariant mass distribution for (left) $\pi^+\pi^0$ following $\pi^+p \rightarrow p\pi^+\pi^0$ at 2.08 GeV/$c$ incident momentum, and (right) $\pi^+\pi^-$ following $\gamma p \rightarrow p\pi^+\pi^-$ using 2.8 GeV photons. The resonance at left corresponds to the $\rho^+$ and the one on the right to $\rho^0$. The resonance energies are 770 MeV and the widths are 150 MeV. (Left) From F. E. James and H. L. Kraybill, *Phys. Rev.* **142**, 896 (1966). (Right) From J. Ballam et al., *Phys. Rev. D* **5**, 545 (1972).

for this reaction, and the resonance corresponding to the $\rho^+$ can be easily identified at an energy of 770 MeV, with a width of about 150 MeV, typical for a strongly decaying particle. The $\rho^+$ is, like the pion, a member of a $T = 1$ triplet of particles $\rho^+$, $\rho^0$, $\rho^-$. By measuring the relative angular distribution of the two pions emitted in the decay of the $\rho$, it is possible to deduce that it has spin 1. Because the constituent pions have spin zero, they must therefore be in a relative p state ($\ell = 1$) in the $\rho$ meson. The intrinsic parity of the $\rho$ meson can thus be shown to be odd:

$$\pi_\rho = \pi_\pi \pi_\pi (-1)^\ell = -1 \qquad (17.39)$$

It is also possible to produce the $\rho$ meson from $e^+e^-$ annihilations using colliding-beam accelerators. Again, because there is no number conservation law for mesons, and in this case because the initial state of an electron and a positron has a total particle content of zero (one particle plus one antiparticle), the simplest possible process is permitted to occur:

$$e^+ + e^- \rightarrow \rho^0 \rightarrow \pi^+ + \pi^-$$

Figure 17.19 shows the resonance in the $e^+e^-$ cross section corresponding to the $\rho^0$. The cross section at resonance can again be analyzed using the Breit-Wigner form from Equation 17.33:

$$\sigma = \frac{\pi}{k^2}\frac{\Gamma\Gamma_{ee}}{\left(E - E_R\right)^2 + \Gamma^2/4}\frac{\left(2s_\rho + 1\right)}{\left(2s_e + 1\right)^2} \qquad (17.40)$$

and at the resonance

$$\sigma = \frac{\pi}{k^2}\frac{\Gamma_{ee}}{\Gamma}\left(2s_\rho + 1\right) \qquad (17.41)$$

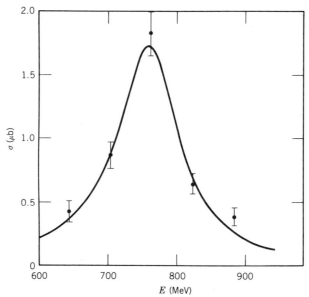

**Figure 17.19** Cross sections for meson production in $e^+e^-$ annihilation. The cross section is observed by measuring the rate of $\pi^+\pi^-$ production as the electron energy is varied. The peak cross section is about 1.5 $\mu$b. Data from J. E. Augustin et al., *Phys. Lett. B* **28**, 508 (1969).

where $\Gamma_{ee}$ is the width for formation of $\rho$ in $e^+e^-$ collisions. We can find $\Gamma_{ee}/\Gamma$ from the relative probability for $\rho$ to *decay* into $e^+e^-$, which is measured to be $(4.6 \pm 0.2) \times 10^{-5}$. In the center-of-mass system (which is identical to the lab system for colliding-beam experiments), the electron energy is one-half of 770 MeV, from which it follows that $\sigma = 0.38(2s_\rho + 1)$ $\mu$b, and the peak cross section is consistent with $s_\rho = 1$.

As a second example of meson resonances, we consider the electrically neutral three-pion final states formed in the reaction

$$\pi^+ + p \rightarrow \pi^+ + p + (\pi^+ + \pi^- + \pi^0)$$

Figure 17.20 shows the three-pion invariant mass distribution; again, there is a continuous background but there appear two resonant peaks, one at 549 MeV ($\eta$ meson) and another at 783 MeV ($\omega$ meson).

Figure 17.21 shows the three-pion combination from the proton–antiproton reaction

$$p + \bar{p} \rightarrow \pi^0 + 2\pi^+ + 2\pi^-$$

The charged final states (such as $\pi^+\pi^+\pi^-$) have no resonant peaks at these energies, so both the $\eta$ and $\omega$ are singlet mesons occurring only in the neutral charge state.

The $\omega$ is found in $e^+e^-$ collisions (Figure 17.21) with nearly the same cross section as the $\rho$. As its electromagnetic decay branch is about the same as that of the $\rho$ ($\Gamma_{ee}/\Gamma = 6.7 \times 10^{-5}$ for $\omega$), the numerical factors in the cross section are nearly the same as those for the $\rho$ and the deduction of the spin gives the same result ($s_\omega = 1$). The width of the $\omega$ resonance is about 10 MeV, indicating a

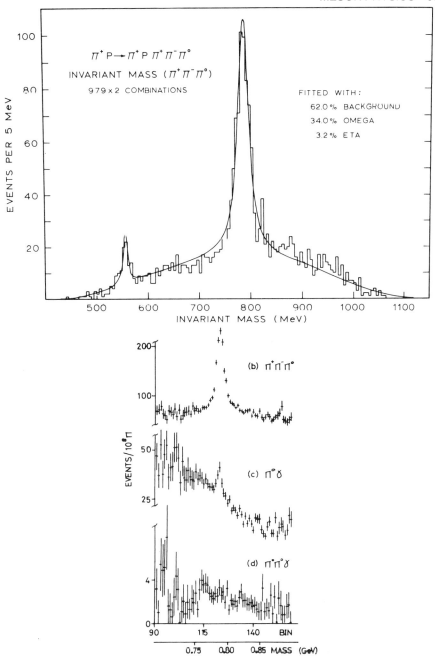

**Figure 17.20** Production of $\omega$ mesons in (top) $\pi^+ p$ reactions and (bottom) $\pi^- p$ reactions. The $3\pi$ resonance at top is at 786 MeV. At bottom are shown various possible final states following the $\pi^- p \rightarrow \omega n$ reaction. There is a $3\pi$ resonance at 782 MeV with a width of 10 MeV. The resonance also appears for $\pi^0 \gamma$ final states, but is considerably weaker; the ratio between the peak areas gives 8.4% for the relative branching ratio of $\omega \rightarrow \pi^0 \gamma$ and $\omega \rightarrow 3\pi$ decays. No evidence is seen for $\pi^0 \pi^0 \gamma$ final states. Data from (top) F. E. James and H. L. Kraybill, *Phys. Rev.* **142**, 896 (1966), and (bottom) J. Keyne et al., *Phys. Rev. D* **14**, 28 (1976).

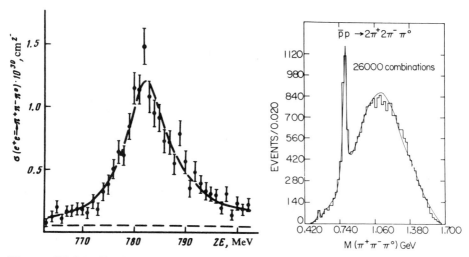

**Figure 17.21** Production of $\omega$ mesons from (left) $e^+ e^- \to 3\pi$ and (right) $p\bar{p} \to 5\pi$. The peak at 783 MeV, with width 10 MeV, is the resonance corresponding to the $\omega$ meson. Data from (left) A. M. Kurdadze et al., *JETP Lett.* **36**, 274 (1982), and (right) A. M. Cooper et al., *Nucl. Phys. B* **146**, 1 (1978).

decay about an order of magnitude slower than the $\rho$. The dominant decay modes are

$$\omega \to \pi^+ + \pi^- + \pi^0 \quad (90\%)$$
$$\to \pi^0 + \gamma \quad (9\%)$$
$$\to \pi^+ + \pi^- \quad (1\%)$$
$$\to e^+ + e^- \quad (6.7 \times 10^{-3}\%)$$

The decay into $3\pi^0$ is forbidden by isospin conservation, which ultimately has to do with the statistics of counting in various charge states. An oversimplified version of this rule requires that decay processes starting with a uniform distribution of all possible charge states of the initial particle must lead to a uniform distribution of all possible charge states of the product particles. Since there is only one charge state of $\omega$, the decay $\omega \to \pi^+ + \pi^- + \pi^0$ satisfies this rule, for it leads to a uniform distribution of the charge states of the pions. The decay $\omega \to 3\pi^0$ does not lead to a uniform distribution and so is forbidden to go by the strong interaction. Notice, however, that the decays $\omega \to \pi^0 + \gamma$ and $\omega \to \pi^+ + \pi^-$ do not lead to uniform distributions and are forbidden to go by the strong interaction. They do occur about 10% of the time, because they proceed by the electromagnetic interaction, which is known not to respect the isospin conservation rule.

Another example of the quantitative application of this rule is the decay of the $\rho$ meson:

$$\rho^+ \to \pi^+ + \pi^0$$
$$\rho^0 \to \pi^0 + \pi^0$$
$$\to \pi^+ + \pi^-$$
$$\rho^- \to \pi^- + \pi^0$$

If we begin with a uniform population of the three $\rho$ mesons (say, 100 of each), then we must finish with a uniform distribution of pions (200 each of $\pi^+$, $\pi^-$, $\pi^0$). The only way to obtain 200 $\pi^+$ is from the 100 $\rho^+$ decays and all 100 of the $\rho^0 \rightarrow \pi^+ + \pi^-$ decays. Thus the $\rho^0 \rightarrow \pi^0 + \pi^0$ decay must be forbidden by this rule and indeed it does not occur. Further details on these counting processes can be found in an article by Charles G. Wohl, *Am. J. Phys.* **50**, 748 (1982).

Isospin can give the same result more directly. The state with two $\pi^0$ must be represented with a symmetric total wave function, because the pions are bosons. The resultant isospin of the combination must be 0 or 2, and thus the $T = 1$ $\rho$ cannot decay into two $\pi^0$ by the strong interaction. The decay could proceed through the electromagnetic interaction, but because the decay into $\pi^+\pi^-$ is permitted by isospin conservation, the decay into two $\pi^0$ is too rare to be observed. Similar arguments forbid the $\omega \rightarrow 3\pi^0$ and $\omega \rightarrow \pi^+\pi^-$ decays through the strong interaction; the decay into $\pi^+\pi^-$ is observed with an intensity of about 1% of the strong decay into $\pi^+\pi^-\pi^0$ which is permitted by isospin. The partial lifetime of the isospin-violating branch is about $10^{-18}$ s, consistent with other electromagnetic decays. (Usually, an electromagnetic process is indicated by the presence of a $\gamma$ in the final state; however, certain decays can occur

**Table 17.2**  Selected Meson Resonances

| Symbol | Mass $(\text{MeV}/c^2)$ | Spin-Parity | Isospin | Width (MeV) | Principal Decay Modes |
|---|---|---|---|---|---|
| $\eta$ | 549 | $0^-$ | 0 | 0.00083 | $\gamma\gamma$ (39%) <br> $3\pi^0$ (32%) <br> $\pi^+\pi^-\pi^0$ (24%) <br> $\pi^+\pi^-\gamma$ (5%) |
| $\rho$ | 769 | $1^-$ | 1 | 154 | $2\pi$ (100%) |
| $\omega$ | 783 | $1^-$ | 0 | 9.9 | $\pi^+\pi^-\pi^0$ (90%) <br> $\pi^0\gamma$ (9%) <br> $\pi^+\pi^-$ (1%) |
| $\eta'$ | 958 | $0^-$ | 0 | 0.29 | $\eta\pi\pi$ (65%) <br> $\rho^0\gamma$ (30%) <br> $\omega\gamma$ (3%) <br> $\gamma\gamma$ (2%) |
| $\phi$ | 1020 | $1^-$ | 0 | 4.2 | $K^+K^-$ (49%) <br> $K_L K_S$ (35%) <br> $\pi^+\pi^-\pi^0$ (15%) <br> $\eta\gamma$ (1%) |
| $f$ | 1270 | $2^+$ | 0 | 178 | $\pi\pi$ (84%) <br> $\pi^+\pi^+\pi^-\pi^-$ (3%) <br> $K\overline{K}$ (3%) |

For more complete data, see the most recent listing of the Particle Data Group, which updates a list of particle properties every year or two. The 1984 listing was published in *Rev. Mod. Phys.* **56**, S1 (1984).

through the electromagnetic interaction without producing a $\gamma$. Decay lifetimes in the range $10^{-16}$ to $10^{-20}$ s usually are characteristic of electromagnetic processes.)

The $\eta$ meson decays as follows:

$$\eta \rightarrow \gamma\gamma \qquad (39\%)$$

$$\rightarrow 3\pi^0 \qquad (32\%)$$

$$\rightarrow \pi^+ + \pi^- + \pi^0 \quad (24\%)$$

$$\rightarrow \pi^+ + \pi^- + \gamma \quad (5\%)$$

The $\eta$ resonance is quite narrow (the width shown in Figure 17.20 actually represents the resolution of the experimental apparatus). The measured value is 0.8 keV, corresponding to a lifetime of about $10^{-18}$ s. This, along with the prominent $\gamma$ branches in the decay, suggests an electromagnetic process. The strong-interaction decay into three pions is forbidden by isospin conservation.

A listing of some of the properties of meson resonances is given in Table 17.2. We return to the elementary-particle properties of mesons in Chapter 18, but for now you should keep in mind these details: mesons are produced in collisions of strongly interacting particles or in $e^+e^-$ collisions; they generally decay rapidly ($10^{-23}$ s) to lighter mesons, unless the violation of some conservation law forces the decay to proceed by the much slower electromagnetic or weak interactions.

## 17.5  STRANGE MESONS AND BARYONS

In the microworld of particle physics, processes occur which have no analog in our ordinary experience. Classifying and then understanding those processes are the challenges faced by nuclear and particle physicists. Fundamental to achieving any sort of theoretical description of reaction and decay processes is a set of rules that aid in the understanding of why certain processes may be suppressed or forbidden, even though they may be allowed on other grounds. (The conservation of parity discussed in Chapter 9 and of isospin mentioned previously in this chapter are examples of such rules. Without the parity rule, we could not understand why a $0^+ \rightarrow 2^-$ $\alpha$ decay is forbidden.) As we approach the most fundamental level of particle interactions more of these rules emerge. The existence of these rules provides valuable clues to the internal structure of particles and is a great aid in forming classification schemes. However, because these empirical rules have no counterparts among ordinary objects, we have no hint of what it is that the rules "really" represent. (In fact, the very use of the word "really" is subject to debate.) As a result, physicists have given arbitrary and often fanciful names to the particle properties, primarily as an aid in remembering the properties. Names such as *strangeness, color, flavor, charm,* and *bottomness* are used to classify particles, but have absolutely no connection with our ordinary use of those descriptions. You should properly regard these names as a mechanism for assigning quantum numbers to particles to explain the observation or nonobservation of various processes.

Historically the first example of this sort of classification scheme was *strangeness*, with a corresponding set of strange particles (that is, particles with a nonzero value of strangeness). The K mesons, with a mass of about 500 MeV, are the lightest strange mesons; strange baryons just heavier than the nucleons exist also. K mesons (or *kaons*) can be formed through reactions similar to pion-producing reactions:

$$\pi^- + p \rightarrow n + K^+ + K^-$$

The cross section for this process is in the range of millibarns, typical for a process involving the strong interaction. Yet the decay of $K^+$ or $K^-$ occurs with a lifetime of about $10^{-8}$ s, characteristic of a weak-interaction process. Moreover, the obvious strong-interaction decay mode ($K^+ \rightarrow \pi^+ + \pi^0$) is suppressed in competition with the process $K^+ \rightarrow \mu^+ + \nu_\mu$, which is a clear signal of a weak interaction. How is it possible for a particle created strongly interacting to decay only weakly interacting?

Another unusual property of strange particles is *associated production*—strange particles are always produced in sets in reactions. The example given above is typical; a positive kaon may be paired with a negative kaon or with one of the strange baryons in associated production.

Let's begin by assigning strangeness quantum numbers $S$ to the strongly interacting particles, in much the same way as we might assign electric charge "quantum numbers" to the electrically interacting particles. All nonstrange particles (p, n, $\pi$, etc.) are assigned $S = 0$. Arbitrarily we choose $S = 1$ for $K^+$. We then postulate the following rule:

> Strangeness is conserved in all strong and electromagnetic processes; strangeness changes in weak interaction processes.

From this rule and the $\pi^- + p \rightarrow n + K^+ + K^-$ reaction we immediately deduce $S = -1$ for $K^-$, to make $S = 0$ on both sides of the reaction. Notice how already the strangeness rule explains the observed phenomenon of associated production—if we start with only nonstrange particles, then the final state must have $S = 0$ and so for every product particle with $S = +1$ there must be a corresponding particle with $S = -1$. Other members of the kaon family are the neutral $K^0$ and its antiparticle $\overline{K}^0$. (We regard the K mesons as a doublet of particles ($T = \frac{1}{2}$) $K^+$ and $K^0$ with antiparticles $K^-$ and $\overline{K}^0$. The K's are therefore different from the $\pi$'s, where $\pi^0$ is its own antiparticle.) The K's decay into two or three pions or into leptons (usually $\mu^\pm$ and its neutrinos). The final state has $S = 0$, and thus the decays are characteristic of the weak interaction which is permitted to change $S$ by one unit. The decay $K \rightarrow 2\pi$ is absolutely forbidden by the strangeness rule from going by the strong or electromagnetic interactions; only the weak interaction remains to cause the decay process.

Beyond the K are many heavier resonances assigned as strange mesons. These can decay into lighter *strange* mesons (usually K's) without violating the $\Delta S = 0$ rule. For example, the 892-MeV resonance called $K^*$ can decay according to $K^* \rightarrow K + \pi$ with $S = +1$ (or possibly $-1$) on both sides of the decay. The decay can therefore go by the strong interaction, and indeed does so with a lifetime of $10^{-23}$ s. Only the K mesons have no lighter strange mesons into which to decay strongly; they alone must decay weakly (and therefore slowly) to the nonstrange $\pi$'s.

Examples of the decay modes of $K^+$ are as follows (for $K^-$, change all particles to their antiparticles):

$$K^+ \rightarrow \mu^+ + \nu_\mu \qquad (63\%)$$
$$\rightarrow \pi^+ + \pi^0 \qquad (21\%)$$
$$\rightarrow \pi^+ + \pi^+ + \pi^- \qquad (6\%)$$
$$\rightarrow \pi^+ + \pi^0 + \pi^0 \qquad (2\%)$$
$$\rightarrow \pi^0 + \mu^+ + \nu_\mu \qquad (3\%)$$
$$\rightarrow \pi^0 + e^+ + \nu_e \qquad (5\%)$$
$$\rightarrow e^+ + \nu_e \qquad (0.0015\%)$$

The $K^0$ decays are similar, but show yet further unusual behavior, which is discussed in the next section.

There is also a set of nucleon-like strange baryons, called *hyperons*. Their strangeness values can be assigned through their production reactions:

$$\pi^+ + n \rightarrow \Lambda^0 + K^+$$

Since this is a strong interaction process, $\Delta S = 0$ and thus $S(\Lambda^0) = -1$. The $\Lambda^0$ is the lightest hyperon, with a rest energy of 1116 MeV. It must decay into a proton or neutron (to conserve baryon number, which, as far as we know, is absolute). Decay channels such as $\Lambda^0 \rightarrow p + K^-$, which would conserve $S$, are forbidden by energy conservation (the final mass energies total more than 1400 MeV, compared with the $\Lambda^0$ at 1116 MeV). The available decay modes that include a nucleon are $p + \pi^-$ or $n + \pi^0$, both of which have $S = 0$. Thus the $\Lambda^0$ decay changes $S$ by one unit and must be a weak interaction process. Its lifetime is observed to be $2.6 \times 10^{-10}$ s, consistent with a weak decay. (The purely electromagnetic decay $\Lambda^0 \rightarrow n + \gamma$ is also forbidden by the strangeness rule.)

The next heaviest baryon is $\Sigma$, a multiplet of three particles $\Sigma^+, \Sigma^0, \Sigma^-$ at a mass energy of about 1190 MeV. An example of the production process is $\pi^- + p \rightarrow \Sigma^- + K^+$, showing that $S = -1$ for $\Sigma^-$ (and for $\Sigma^+$ and $\Sigma^0$). The decays of $\Sigma^\pm$ must go to protons or neutrons plus a pion, and again the $\Delta S = 1$ decay must be through the weak interaction (the observed lifetimes are about $10^{-10}$ s). The decay $\Sigma^0 \rightarrow \Lambda^0 + \gamma$ can proceed without changing strangeness, and does so in a characteristic electromagnetic decay time of $10^{-19}$ s.

The $\Xi^-$ and $\Xi^0$ doublet at 1320 MeV is the next heaviest hyperon. It is formed through $K^- + p \rightarrow \Xi^- + K^+$, and must therefore have $S = -2$. The weak decay can change $S$ by only one unit, and so the decay products must include one of the $S = -1$ baryons, either $\Lambda^0$ or $\Sigma$. There is not enough energy in the $\Xi$ decay to form $\Sigma + \pi$, so the only available decay modes are $\Xi^0 \rightarrow \Lambda^0 + \pi^0$ and $\Xi^- \rightarrow \Lambda^0 + \pi^-$. The lifetimes are about $10^{-10}$ s.

The eight particles n, p, $\Lambda^0$, $\Sigma^{+,0,-}$, and $\Xi^{0,-}$ form a group of spin-$\frac{1}{2}$ baryons with very similar properties, as we see when we examine their structure according to the quark model. Their strangeness values form a progression from $S = 0$ to $S = -2$. Another group of baryons with spin $\frac{3}{2}$ includes the four $S = 0$ $\Delta$'s (1232 MeV), three $S = -1$ $\Sigma^*$ resonances (1385 MeV), two $S = -2$ $\Xi^*$ resonances (1530 MeV), and one $S = -3$ particle, the $\Omega^-$ at 1673 MeV. Notice the

**Table 17.3** Strange Baryons (Hyperons)

| Name | Mass (MeV/$c^2$) | Spin-Parity | Isospin $T, T_3$ | Strangeness | Width or Decay Lifetime | Principal Decay Modes |
|---|---|---|---|---|---|---|
| $\Lambda^0$ | 1116 | $\frac{1}{2}^+$ | 0, 0 | $-1$ | $2.6 \times 10^{-10}$ s | $p\pi^-$ (64%); $n\pi^0$ (36%) |
| $\Sigma^+$ | 1189 | $\frac{1}{2}^+$ | 1, +1 | $-1$ | $0.80 \times 10^{-10}$ s | $p\pi^0$ (52%); $n\pi^+$ (48%) |
| $\Sigma^0$ | 1192 | $\frac{1}{2}^+$ | 1, 0 | $-1$ | $5.8 \times 10^{-20}$ s | $\Lambda\gamma$ |
| $\Sigma^-$ | 1197 | $\frac{1}{2}^+$ | 1, $-1$ | $-1$ | $1.5 \times 10^{-10}$ s | $n\pi^-$ |
| $\Xi^0$ | 1315 | $\frac{1}{2}^+$ | $\frac{1}{2}, +\frac{1}{2}$ | $-2$ | $2.9 \times 10^{-10}$ s | $\Lambda\pi^0$ |
| $\Xi^-$ | 1321 | $\frac{1}{2}^+$ | $\frac{1}{2}, -\frac{1}{2}$ | $-2$ | $1.6 \times 10^{-10}$ s | $\Lambda\pi^-$ |
| $\Sigma^{*+}$ | 1382 | $\frac{3}{2}^+$ | 1, +1 | $-1$ | 35 MeV | |
| $\Sigma^{*0}$ | 1382 | $\frac{3}{2}^+$ | 1, 0 | $-1$ | 35 MeV | $\Lambda\pi$ (80%); $\Sigma\pi$ (12%) |
| $\Sigma^{*-}$ | 1387 | $\frac{3}{2}^+$ | 1, $-1$ | $-1$ | 40 MeV | |
| $\Xi^{*0}$ | 1532 | $\frac{3}{2}^+$ | $\frac{1}{2}, +\frac{1}{2}$ | $-2$ | 9 MeV | $\Xi\pi$ |
| $\Xi^{*-}$ | 1535 | $\frac{3}{2}^+$ | $\frac{1}{2}, -\frac{1}{2}$ | $-2$ | 10 MeV | |
| $\Omega^-$ | 1673 | $\frac{3}{2}^+$ | 0, 0 | $-3$ | $0.82 \times 10^{-10}$ s | $\Lambda K$ (69%); $\Xi\pi$ (31%) |

progression in $S$, in the multiplicity of the particle, and in the mass values (each step is about 150 MeV). Based on this projection, the $\Omega^-$ was predicted at an energy of about 1680 MeV; its observation in 1964 with the predicted mass and strangeness was a great triumph for the theory upon which the prediction is based. The $\Omega^-$ has no strong decay modes available to it; decay into the $S = -3$ combination $\Xi^0 + K^-$ is forbidden by energy conservation. Several $S = -2$ final states are available for the strangeness-changing weak decay: $\Lambda^0 + K^-$, $\Xi^0 + \pi^-$, and $\Xi^- + \pi^0$.

Table 17.3 presents a summary of the strange spin-$\frac{1}{2}$ and spin-$\frac{3}{2}$ baryons. We leave the discussion of the elementary-particle aspects of this topic until the next chapter, but here we consider the nuclear physics of these heavy baryons or hyperons.

At first glance the $\Lambda^0$ appears to be similar to a heavy neutron (although a "strange" neutron), and it makes reasonable sense to study the interaction of $\Lambda^0$ and p in the same way we studied the nucleon–nucleon force through np scattering as described in Chapter 4. Neither beams nor targets of $\Lambda$'s are readily available, so the $\Lambda$p scattering experiment is difficult. In practice, the $\Lambda^0$ are produced by allowing a beam of $K^-$ to enter a liquid hydrogen target:

$$K^- + p \rightarrow \Lambda^0 + \pi^0$$

The $\Lambda^0$ will move through the target, perhaps colliding with a proton as it does. From an analysis of such scattering events, it is concluded that the scattering at relatively low energy (s-wave interactions only) can be characterized by two parameters, the scattering length $a$ and the effective range $r_0$, just as we did for the nucleon–nucleon interaction. The scattering data favor the values $a = -1.8 \pm 0.2$ fm and $r_0 = 3.16 \pm 0.52$ fm; because of the lack of specific results, no allowance is made for the possibility of different parameters for singlet and triplet scattering, even though there is almost certainly a spin dependence in the

$\Lambda$p scattering cross section. The results for the scattering length suggest that the $\Lambda$p interaction is somewhat weaker than the np interaction (recall the discussion accompanying Figure 4.11—the difference in magnitudes between negative scattering lengths does not necessarily suggest a substantial difference between the strength of the interactions.) There is also no bound state if the scattering length is negative, so there exists no analog of deuterium for the $\Lambda$p system.

Of course, we expect the $\Lambda$p interaction to be different from the np interaction. Exchange of a single pion between $\Lambda$ and p is forbidden by statistical (isospin) considerations. Two-pion exchange and K exchange are permitted. Thus the long-range (one-pion) parts of the $\Lambda$p and np interactions should be very different.

An alternative procedure for studying the nuclear interactions of $\Lambda$'s is through *hypernuclei* in which a neutron in a nucleus is replaced with a $\Lambda^0$. This can be accomplished by bombarding nuclei with a beam of $K^-$:

$$K^- + n \rightarrow \Lambda^0 + \pi^-$$

Using relativistic kinematics, it can be shown that if the $\pi^-$ are detected in the forward ($0°$) direction as defined by the incident $K^-$ beam, then the $\Lambda^0$ is produced essentially at rest for $K^-$ of initial momentum 500 MeV/$c$ incident on neutrons at rest. If the $\Lambda^0$ is produced at low momentum, it has a high probability of remaining bound to the nucleus, most likely even in the same orbital state as the original neutron. The $\Lambda^0$ can drop quickly to the 1s shell-model state even though there may already be two neutrons in that state; because $\Lambda^0$ are different particles from neutrons, the Pauli principle does not forbid neutrons and $\Lambda^0$ from occupying the same state. In the 1s state, the $\Lambda^0$ remains until it decays, according to $\Lambda^0 \rightarrow p + \pi^-$ or $n + \pi^0$, or else it undergoes a strangeness-changing weak reaction with one of the nucleons:

$$\Lambda^0 + n \rightarrow n + n$$

$$\Lambda^0 + p \rightarrow n + p$$

The reactions or the decay take place on a $10^{-10}$ s time scale, which is long by

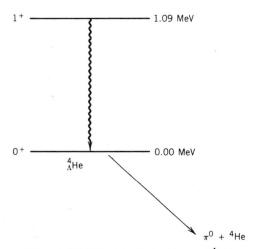

**Figure 17.22** Level structure of $^4_\Lambda$He.

nuclear standards and allows enough time to observe the properties of the hypernucleus.

A typical reaction for forming a hypernucleus is indicated

$$\mathrm{K}^- + {}^A\mathrm{X} \rightarrow {}^A_\Lambda\mathrm{X} + \pi^-$$

For example, a target of $^4\mathrm{He}$ might produce $^4_\Lambda\mathrm{He}$, a nucleus consisting of four baryons (two protons, one neutron, and one $\Lambda^0$). In a fashion similar to ordinary nuclides, the number $A$ to the upper left of the chemical symbol gives the total number of baryons, including $A - 1$ nucleons and one $\Lambda^0$. The structure of $^4_\Lambda\mathrm{He}$ provides an example of the difference between ordinary nuclear physics and hypernuclear physics. In the $^4_\Lambda\mathrm{He}$ ground state, all particles are in the 1s state (as they are in ordinary $^4\mathrm{He}$). However, in ordinary $^4\mathrm{He}$, the Pauli principle requires the two neutrons to have their spins oriented in opposite directions, so that the net spin of $^4\mathrm{He}$ is zero. No such restriction occurs for $^4_\Lambda\mathrm{He}$, so the spins of the neutron and the $\Lambda^0$ can be either antiparallel (for a total of 0) or parallel (for a total of 1). Figure 17.22 shows the deduced level scheme of $^4_\Lambda\mathrm{He}$, showing the $0^+$ ground state and $1^+$ first excited state. (In ordinary $^4\mathrm{He}$, there are no bound excited states.) The 1.09-MeV $\gamma$ ray is detected in coincidence with the $\pi^0$ resulting from the ground-state decay. The level structure of $^4_\Lambda\mathrm{H}$ is very similar.

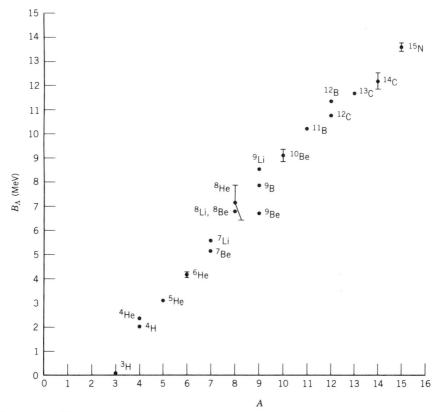

**Figure 17.23** Binding energy of hypernuclei. In most cases the error bars are smaller than the experimental points.

By carefully measuring energies of the particles resulting from the decays of the ground state of a hypernucleus, it is possible to deduce its mass. We can then determine the binding energy of the $\Lambda^0$ in the usual way, by comparing the mass energy of the constituents and of the combination, in analogy with Equation 3.25:

$$B_\Lambda = \left[ m(\Lambda^0) + m(^{A-1}X) - m(^A_\Lambda X) \right] c^2 \tag{17.42}$$

In the case of the $^4_\Lambda$He, for instance, the binding energy of the $\Lambda$ (in the ground state) is 2.39 MeV. Repeating this measurement for several nuclei, we can plot the dependence of $B_\Lambda$ on $A$ as is shown in Figure 17.23. The binding energy increases linearly with $A$ for these light nuclei, in contrast to the neutron binding energy (which is just the neutron separation energy as defined by Equation 3.26). This is a direct result of the action of the Pauli principle in limiting possible neutron interactions, a limitation that does not apply to $\Lambda^0$. For these light nuclei, the $\Lambda^0$ can interact with all of the nucleons, and so $B_\Lambda \propto A$, roughly. For heavier nuclei, which are more difficult to study experimentally, the binding energy saturates at about 23 MeV.


## 17.6   CP VIOLATION IN K DECAY

The decays of K mesons show a number of unusual quantum mechanical effects that have no counterpart in other areas of physics. For example, a $K^0$ produced at time $t = 0$ can later be observed as a $\overline{K}^0$; the spontaneous conversion of a particle into its antiparticle (Figure 17.24) is forbidden for all particles but the $K^0$ and the similar $D^0$ and $B^0$ (see Section 18.6).

The key to understanding the unusual properties of the neutral K mesons is that $K^0$ and $\overline{K}^0$ are not independent particles—because each can decay into two pions, there is a coupling between $K^0$ and $\overline{K}^0$:

$$K^0 \leftrightarrow 2\pi \leftrightarrow \overline{K}^0$$

As a result of this coupling, an initially pure collection of $K^0$ will become, at a later time, a mixture of $K^0$ and $\overline{K}^0$.

The neutral K mesons can be produced by strong reactions, such as

$$\pi^+ + p \rightarrow K^+ + \overline{K}^0 + p$$

$$\pi^- + p \rightarrow K^0 + \Lambda^0$$

which conserve strangeness. The decays of $K^0$ and $\overline{K}^0$ to pions cannot go by the strong or electromagnetic interactions, which must conserve strangeness, but must go by the weak interaction. As we discussed in Section 9.9, nuclear $\beta$ decay violates both the symmetries of parity P and charge conjugation (particle $\rightarrow$ antiparticle) C, but it does so in such a way that the combined CP symmetry remains valid. This argument was made indirectly in Section 9.9 based on a "thought experiment" involving the decay of anti-$^{60}$Co. To test the CP symmetry directly, it is necessary to study the decay of a particle and the decay of its CP image, and to examine the CP properties of the particles they decay into. The weak decays of $\pi^\pm$ into $e^\pm$ or $\mu^\pm$ and neutrinos cannot be used for this test, for the final state has no definite CP properties. (Particles which interact only weakly

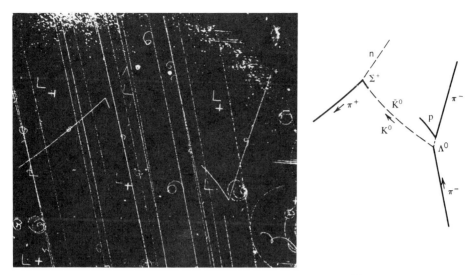

**Figure 17.24**  Bubble-chamber photograph showing $K^0 - \overline{K}^0$ conversion. The sequence of reactions, diagrammed at right, begins with an incident $\pi^-$, causing $\pi^- + p \rightarrow \Lambda^0 + K^0$. The $\Lambda^0$ (leaving no track) travels a short distance and decays according to $\Lambda^0 \rightarrow \pi^- + p$; the proton leaves a short track characteristic of a heavy particle. The $K^0$ moves to the left and converts to $\overline{K}^0$ which can then strike a proton producing yet another reaction: $\overline{K}^0 + p \rightarrow \Sigma^+ + \pi^0$. Finally, the $\Sigma^+$ leaves a short track and decays to a $\pi^+$ and a neutron. From E. Segrè, *Nuclei and Particles*, 2nd ed. (Reading, MA: Benjamin, 1977).

fall into this category.) To make a definitive test, it is necessary that the initial *and* final states consist of strongly interacting particles, and the weak decay of K into pions is the most easily accessible candidate for a study of CP conservation.

Let's consider the possible $2\pi$ final states resulting from $K^0$ or $\overline{K}^0$ decay. Electrically neutral final states are $\pi^0 + \pi^0$ or $\pi^+ + \pi^-$. We will represent combined final wave functions as $\psi(\pi_a^0, \pi_b^0)$ or $\psi(\pi_a^+, \pi_b^-)$, which indicates the first pion is in state a and the second pion is in state b. The basic operations C and P have the following effects:

$$P\psi(\pi) \rightarrow -\psi(\pi) \tag{17.43}$$

$$C\psi(\pi^+) \rightarrow \psi(\pi^-) \tag{17.44a}$$

$$C\psi(\pi^-) \rightarrow \psi(\pi^+) \tag{17.44b}$$

$$C\psi(\pi^0) \rightarrow \psi(\pi^0) \tag{17.44c}$$

where the parity operation includes a minus sign because pions have negative intrinsic parity.

Now consider the effect of CP on the two-pion states:

$$P\psi(\pi_a^0, \pi_b^0) = (-1)^2(-1)^\ell \psi(\pi_a^0, \pi_b^0) \tag{17.45}$$

where the factor of $(-1)^2$ comes from the two intrinsic parities and $(-1)^\ell$ is the usual orbital angular momentum parity factor.

$$C\psi\left(\pi_a^0, \pi_b^0\right) = \psi\left(\pi_a^0, \pi_b^0\right) \qquad (17.46)$$

and thus

$$CP\psi\left(\pi_a^0, \pi_b^0\right) = \psi\left(\pi_a^0, \pi_b^0\right) \qquad (17.47)$$

assuming $\ell = 0$, as it must for decays of spin-0 kaons into spin-0 pions. Similarly

$$P\psi\left(\pi_a^+, \pi_b^-\right) = (-1)^2(-1)^\ell \psi\left(\pi_a^-, \pi_b^+\right) \qquad (17.48)$$

$$C\psi\left(\pi_a^+, \pi_b^-\right) = \psi\left(\pi_a^-, \pi_b^+\right) \qquad (17.49)$$

and

$$CP\psi\left(\pi_a^+, \pi_b^-\right) = \psi\left(\pi_a^+, \pi_b^-\right) \qquad (17.50)$$

In either case, the conclusion is that the CP operation leaves the final state unchanged. If CP is to be a valid symmetry, then the initial decaying state must also be invariant with respect to CP. However, the CP operations on the $K^0$ states do not leave them invariant.

$$CP\psi(K^0) \rightarrow \psi(\overline{K}^0) \qquad (17.51a)$$

$$CP\psi(\overline{K}^0) \rightarrow \psi(K^0) \qquad (17.51b)$$

Therefore, the decays $K^0 \rightarrow 2\pi$ and $\overline{K}^0 \rightarrow 2\pi$ seem to be forbidden by the CP symmetry. We can restore the symmetry by forming instead two states

$$\psi(K_1) = \frac{1}{\sqrt{2}}\left[\psi(K^0) + \psi(\overline{K}^0)\right] \qquad (17.52)$$

$$\psi(K_2) = \frac{1}{\sqrt{2}}\left[\psi(K^0) - \psi(\overline{K}^0)\right] \qquad (17.53)$$

It follows from Equations 17.51a and 17.51b that

$$CP\psi(K_1) \rightarrow \psi(K_1) \qquad (17.54a)$$

$$CP\psi(K_2) \rightarrow -\psi(K_2) \qquad (17.54b)$$

The state represented by $K_1$ is a mixture of $K^0$ and $\overline{K}^0$; the decay of the neutral kaons into two pions is permitted only if the initial state consists of the specified mixture of $K^0$ and $\overline{K}^0$.

The combined effect of CP on possible three-pion final states can be shown to be

$$CP\psi(3\pi) \rightarrow -\psi(3\pi) \qquad (17.55)$$

We therefore have two possible decays permitted by the weak interaction, $K_1 \rightarrow 2\pi$ and $K_2 \rightarrow 3\pi$; each of these decays is permitted by the CP symmetry. We can regard $K_1$ and $K_2$ as particles in the same sense that $K^0$ and $\overline{K}^0$ are. The particles produced by the strong interaction are $K^0$ and $\overline{K}^0$; $K_1$ and $K_2$ cannot be produced in reactions that conserve strangeness, because they are mixtures of $S = +1$ and $S = -1$. The particles that decay by the weak interaction are $K_1$ and $K_2$; $K^0$ and $\overline{K}^0$ cannot decay weakly because of CP conservation. Notice that the $K^0$ and $\overline{K}^0$ parts of $K_1$ and $K_2$ can each participate in allowed $\Delta S = \pm 1$ weak decays to $S = 0$ states of pions. Notice also that $K_1$ and $K_2$ are *not* antiparticles of one another; thus they may have different properties (including different masses and decay lifetimes).

One important factor in determining decay rates is the density of final states, as in Equation 17.25, which includes a factor $p^2$ depending on the momentum of the final particles. In the decay $K_1 \rightarrow 2\pi$, the kinetic energy available is more than 200 MeV, while only about 70 MeV is available for the $K_2 \rightarrow 3\pi$ decay. The density of final states will be much smaller for $K_2 \rightarrow 3\pi$, and it will have a smaller decay constant and a larger lifetime. The observed lifetimes are

$$\tau(K_1) = 0.892 \times 10^{-10} \text{ s}$$

$$\tau(K_2) = 5.18 \times 10^{-8} \text{ s}$$

The difference between the two lifetimes is primarily from the density of final states. The short-lived and long-lived decaying states are usually indicated as $K_S$ and $K_L$, rather than as $K_1$ and $K_2$.

The existence of $K_S$ and $K_L$ helps us to understand the unusual behavior of the neutral K mesons. Suppose we produce $K^0$ in a reaction such as

$$\pi^- + p \rightarrow K^0 + \Lambda^0$$

This reaction cannot produce $\overline{K}^0$, for there is no baryon with $S = +1$ to keep the total strangeness zero. If we form the $K^0$ into a beam, it will initially be pure $K^0$, but within a few $K_S$ lifetimes ($10^{-10}$ s, corresponding to distances of cm to m for beams traveling at nearly $c$), the $K_S$ component has completely disappeared, leaving only $K_L$. At this point the beam in effect consists of equal amplitudes of $K^0$ and $\overline{K}^0$, as can be demonstrated by allowing the beam to undergo strong reactions, such as

$$\overline{K}^0 + p \rightarrow \Lambda^0 + \pi^+$$

Indeed, because of such reactions when the beam is passed through a given thickness of absorbing material, the $\overline{K}^0$ react more strongly and are absorbed. The balance between $K^0$ and $\overline{K}^0$ is upset, and the beam can now be considered a mixture of $K_S$ and $K_L$. The passage through matter has *regenerated* the $K_S$ component of the beam.

These properties may seem unusual, but in fact they are basic manifestations of typical quantum behavior, especially for systems that can exist in two states, for instance linearly polarized light or electron spins passing through a Stern-Gerlach apparatus. In the latter case, a beam of spin-up atoms ($s_z = +\hbar/2$) is passed through a magnet which analyzes its spin in the $y$ direction. There is a 50% chance to measure either $s_y = +\hbar/2$ or $s_y = -\hbar/2$. Furthermore, after selecting one or the other of these outcomes, we could then measure the $z$ component again. We would find a 50% chance to have $s_z = -\hbar/2$, even though no such component existed in the original beam.

To return to our original topic, we expect to observe the decays $K_S \rightarrow 2\pi$ and $K_L \rightarrow 3\pi$ if CP is a valid symmetry. The observed decay modes are consistent

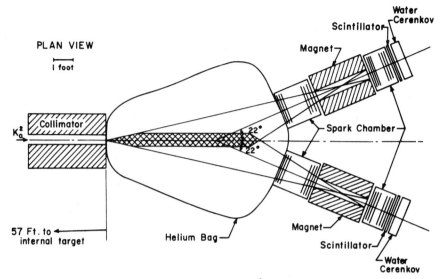

**Figure 17.25** Two-pion spectrometer. The $K^0$ beam enters through the collimator at left; decays that occur in the cross-hatched region are within the view of the two pion spectrometers at right. From J. H. Christenson et al., *Phys. Rev. Lett.* **13**, 138 (1964).

with these expectations:

$$K_S \rightarrow \pi^+ + \pi^- \qquad (68.6\%)$$

$$\rightarrow \pi^0 + \pi^0 \qquad (31.4\%)$$

$$K_L \rightarrow \pi^0 + \pi^0 + \pi^0 \qquad (21.5\%)$$

$$\rightarrow \pi^+ + \pi^- + \pi^0 \qquad (12.4\%)$$

$$\rightarrow \pi^\pm + \mu^\mp + \nu_\mu(\bar{\nu}_\mu) \qquad (27.1\%)$$

$$\rightarrow \pi^\pm + e^\mp + \nu_e(\bar{\nu}_e) \qquad (38.7\%)$$

In 1964, Cronin and Fitch and their co-workers performed an experiment to test the CP symmetry in the decay of the neutral kaon. Figure 17.25 shows their apparatus. A beam of $K^0$ was allowed to travel through a long enough distance that its $K_S$ component was negligibly small. It was then allowed to decay, and the apparatus was set to look for the forbidden $K_L \rightarrow 2\pi$ mode. In a surprising outcome, they found decay into two pions occurred with a branching ratio of about 0.3%. Figure 17.26 shows the care the experimenters took to ascertain that their observed events did indeed originate with a $K_L$ traveling in the beam direction, but the final conclusion is irrefutable—CP is indeed violated at a level of about 0.3%.

It is also possible to observe CP violation by comparing the rates for the decays

$$K_L \rightarrow \pi^- + \mu^+ + \bar{\nu}_\mu$$

$$K_L \rightarrow \pi^+ + \mu^- + \nu_\mu$$

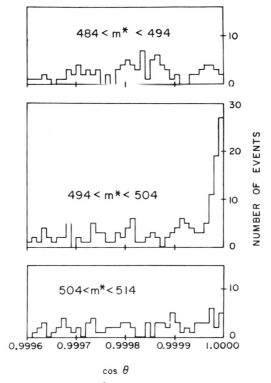

**Figure 17.26** The signal for $K_L^0 \to 2\pi$ decay is an invariant mass in the two spectrometers equal to the $K_L^0$ mass (498 MeV) *and* a total momentum parallel to the original $K_L^0$ direction ($\cos\theta = 1$). In this way it can be determined that the experimenters are not simply counting two of the three particles in the $K \to 3\pi$ decay. The top and bottom figures show no evidence for an invariant mass (called $m^*$) above or below the $K_L^0$ mass, while the middle figure clearly shows a peak in the correct mass range and with the proper direction of the total momentum ($\cos\theta > 0.99999$). From J. H. Christenson et al., *Phys. Rev. Lett.* **13**, 138 (1964).

If CP were an exact symmetry, the rates would be equal, but they are observed to be slightly different, again by about 0.3%.

The issue of CP violation raises a number of fundamental questions that have not yet been resolved despite more than 20 years of active experimental and theoretical effort. Perhaps the most obvious one is: Why is CP violation observed in no other system? The coupling of $K^0$ and $\overline{K}^0$ that is responsible for the $K_L$ (and $K_S$) decays occurs on a time scale determined by the $K_L$ and $K_S$ mass difference. This mass difference is a kind of "self-energy" analogous to the electromagnetic self-energy of a charge distribution. In this case, however, it is a weak self-energy determined from the processes

$$K_S \leftrightarrow 2\pi \leftrightarrow K_S$$
$$K_L \leftrightarrow 3\pi \leftrightarrow K_L$$

Because these modes are slightly different, the masses of $K_S$ and $K_L$ are different, but the difference is extremely small:

$$\Delta mc^2 = \left[ m(\mathrm{K_L}) - m(\mathrm{K_S}) \right] c^2 = 3.52 \times 10^{-6} \text{ eV}$$

that is, about 1 part in $10^{14}$ of the neutral kaon mass. The $\overline{\mathrm{K}}^0$ amplitude in an initially pure $\mathrm{K}^0$ beam oscillates with a frequency determined by this energy difference:

$$\omega = \frac{\Delta E}{\hbar} = \frac{\Delta mc^2}{\hbar} = 5.3 \times 10^9 \text{ s}^{-1}$$

Thus, for $\mathrm{K_S}$, $\omega\tau \approx 0.477$, which makes the oscillations observable on a time scale of the lifetime of $\mathrm{K_S}$. No other known particle decay possesses these unique properties.

A second question: What is the agent responsible for CP violation? Because it has not yet been observed in another system, it is impossible to gather independent evidence on CP violation that might help to answer this question. If we accept the CPT theorem, then CP violation implies T violation, and there are ways to search for evidence of T violation in arbitrary strong, electromagnetic, and weak processes. Many such searches have been done; to date no such evidence for T violation has been observed, but the experimental precision is mostly inferior to that of the CP violations in $\mathrm{K_L}$ decay, and the theoretical interpretations are not as direct. It has been postulated that CP violation is a characteristic of a new type of interaction, a *superweak* interaction, which is completely CP violating and which permits direct $\Delta S = 2$ interactions, allowing $\mathrm{K}^0$ and $\overline{\mathrm{K}}^0$ to mix directly, thereby permitting $\mathrm{K_L} \to \mathrm{K_S} \to 2\pi$ as the origin for the observed $\mathrm{K_L} \to 2\pi$ decay. This incredibly impotent force, at about $10^{-9}$ the strength of the ordinary weak interaction, would be effectively unobservable in all other known situations.

Finally, aside from what must be seen as an oddity in the $\mathrm{K}^0$ system, what do we learn from CP violation? Here we look to the early history of the universe, in the instant after the Big Bang. Any reasonable theory of particle production in the Big Bang must give rise to equal numbers of particles and antiparticles, as $\gamma \to e^+ + e^-$ or $\gamma \to \mathrm{p} + \overline{\mathrm{p}}$. A thoroughly mixed collection of matter and antimatter would not permit large clumps of one kind on the scale of galaxies to form. Yet as best we can discover, our galaxy (and probably all others as well) are made of matter, not antimatter. How did a basically symmetric situation between matter and antimatter turn into such an asymmetric one? Perhaps CP violation may hold a clue. If matter and antimatter were originally created in equal abundances, but at some phase in the cooling and expansion of the universe it passed through an epoch when CP-violating interactions were prevalent, the matter-antimatter equilibrium could have been upset, leading to a slight imbalance of one over the other. If what we call matter became dominant, even by so small a fraction as one part in $10^9$, then after a thorough mixing all of the antimatter would disappear in annihilation reactions with all but one part in $10^9$ of the matter, which in turn would constitute the present universe. This scenario is in fact quite consistent with the observed relative abundance of nucleons and photons in the universe. We will continue with these cosmological speculations in Chapter 19 after we expand our background in fundamental particle physics in the next chapter.

## REFERENCES FOR ADDITIONAL READING

A reading list covering elementary particle physics including mesons is given at the end of Chapter 18. The following list includes only those references that pertain specifically to mesons and meson-nucleus interactions.

An account of the early cosmic-ray work leading to the discovery and study of pions can be found in C. F. Powell, P. H. Fowler, and D. H. Perkins, *The Study of Elementary Particles by the Photographic Method* (New York: Pergamon, 1959); see also C. F. Powell, *Rep. Prog. Phys.* **13**, 350 (1950).

A comprehensive work on nucleon–nucleon interactions involving mesons and on meson–nucleon interactions is W. O. Lock and D. F. Measday, *Intermediate Energy Nuclear Physics* (London: Methuen, 1970). Another similar reference is R. J. Cence, *Pion–Nucleon Scattering* (Princeton, NJ: Princeton University Press, 1969). Both of these books present experimental results and their theoretical interpretation.

A review of meson–nucleus scattering has been presented by M. M. Sternheim and R. R. Silbar, *Ann. Rev. Nucl. Sci.* **24**, 249 (1974), and pion production in proton-induced reactions was reviewed by D. F. Measday and G. A. Miller, *Ann. Rev. Nucl. Part. Sci.* **29**, 121 (1979). A review of experimental techniques relating to pionic atoms and a summary of results can be found in G. Backenstoss, *Ann. Rev. Nucl. Sci.* **20**, 467 (1970).

The experimental and theoretical aspects of hypernuclei are reviewed by B. Povh, *Ann. Rev. Nucl. Part. Sci.* **28**, 1 (1978). A summary describing work with beams of hyperons is J. Lach and L. Pondrom, *Ann. Rev. Nucl. Part. Sci.* **29**, 203 (1979).

An elementary discussion of the general two-state system in quantum mechanics and its application to the $K^0$ can be found in R. P. Feynman, R. B. Leighton, and M. Sands, *The Feynman Lectures on Physics* (Reading, MA: Addison-Wesley, 1965); see especially Vol. 3, Chapters 10 and 11. A general introduction to CP violation is P. K. Kabir, *The CP Puzzle* (London: Academic, 1968). A review of this topic is given by K. Kleinknecht, *Ann. Rev. Nucl. Sci.* **26**, 1 (1976). An account of the relationship between CP-violation and time-reversal invariance is R. G. Sachs, *Science* **176**, 587 (1972).

## PROBLEMS

1. (a) Compute the energies of the pionic M X rays ($n = 4$ to $n = 3$) in Ca, Sn, and Pb. (b) Compare the mean radius of the $n = 3$ pionic state in Ca, Sn, and Pb with the nuclear radius.

2. Compute the threshold kinetic energy for the production in nucleon–nucleon reactions of: (a) single pions; (b) pairs of pions; (c) single kaons.

3. Discuss the permitted angular momentum and parity states in the reaction $\pi^- + d \rightarrow n + n + \pi^0$, given that the pion has intrinsic spin of zero and negative intrinsic parity. Assume very low energy incident $\pi^-$. (*Hint:* The final state has two neutrons in a relative orbital angular momentum state $\ell_{2n}$ and a pion in an angular momentum state $\ell_\pi$ relative to the two neutrons.)

4. Find which of the following reactions are forbidden by one or more conservation laws. Give all violated laws in each case.
   (a) $K^+ + n \to \Sigma^+ + \pi^0$   (d) $\pi^- + p \to \Sigma^+ + K^-$
   (b) $\pi^- + n \to K^+ + \Lambda^0$   (e) $\pi^- + p \to \Xi^- + K^+ + \overline{K}^0$
   (c) $K^- + p \to n + \Lambda^0$   (f) $d + d \to {}^4He + \pi^0$

5. Show how the decays $\omega \to 3\pi^0$ and $\omega \to \pi^+\pi^-$ are forbidden by isospin conservation from proceeding through the strong interaction.

6. (a) Show that the $\pi^+\pi^-$ system must be either in an $\ell =$ even orbital state with total isospin $T = 0$ or 2, or else in an $\ell =$ odd orbital state with $T = 1$. (b) Show that the $\pi^0\pi^0$ system must be in a state of $T = 0$ or $T = 2$, and thus that only $\ell =$ even states are permitted. (c) Discuss the permitted states of the $\pi^+\pi^+$ system.

7. Explain why the decay $\rho \to \eta + \pi$ is forbidden.

8. (a) Show how the decay $\eta \to 2\pi$ is forbidden by conservation of angular momentum and parity. (b) Discuss how the decay $\eta \to 3\pi^0$ violates isospin conservation. (*Hint:* Find the possible values of the total isospin of two pions, and then couple the third to obtain the resultant.) (c) Discuss the decay $\eta \to 4\pi$. Why is this decay not observed?

9. (a) It is desired to produce the $\Omega^-$ particle in proton–proton collisions. Find the reaction satisfying all strong-interaction conservation laws and requiring the smallest possible threshold energy. (b) Find the reaction and the threshold energy for production of $\Omega^-$ by negative kaons incident on protons.

10. Compute the ratio of the decay probabilities for $K^+ \to e^+ + \nu_e$ and $K^+ \to \mu^+ + \nu_\mu$; compare with the experimental ratio given in Section 17.5 (0.0015%/63%).

11. Discuss possible methods for measuring the masses of $K^+$ and $K^-$.

12. (a) Discuss the possible isospin of the kaon-plus-nucleon system. (b) Use the latest review of particle properties (for example, *Rev. Mod. Phys.* **56**, S1 (1984)) to make a list of particles that could be considered as kaon-nucleon resonances. Use the style of Table 17.1 and specify the isospin of each resonance.

13. Show that, in the reaction $K^- + n \to \Lambda^0 + \pi^-$ used to produce hyper-nuclei, a 500-MeV/$c$ kaon incident on a neutron nearly at rest produces a very low energy $\Lambda^0$ if the pion is detected in the forward (0°) direction. In such a case, the $\Lambda^0$ is likely to occupy the same nuclear state as the original neutron. What is the effect of small initial neutron energies on this conclusion?

# 18

# PARTICLE PHYSICS

In particle physics, also called high-energy physics, we deal with the interactions between particles on the most fundamental level. Particle physicists are engaged in the pursuit of the most elementary constituents of matter and in the exploration and elucidation of the rules governing their behavior. In this process, they must study interactions at ever smaller ranges and therefore involving ever heavier particles and ever larger accelerators.

In analyzing these experiments, particle physicists have been able to classify particles and their interactions into a number of easily identifiable categories, and to establish a number of empirical rules that appear to summarize their behavior. These rules have in turn led to a fundamental theory of the properties and structure of the strongly interacting particles based on the *quark model*. They have also led to attempts, not yet completely successful, to combine the treatments of the elementary strongly interacting particles, the quarks, and the elementary weakly interacting particles, the leptons. Parallel attempts are being made to incorporate the mechanics of the strong, weak, and electromagnetic interactions into a single theory. Partial success has been obtained with the combining of the weak and electromagnetic interactions, and several proposals have been advanced for including the strong interactions. Ultimate success will be achieved when gravity can also be included.

A systematic treatment of particle interactions is beyond the level of this text. Although there is some overlap, particle physics is a discipline separate from nuclear physics, and we cannot do it justice with an abbreviated treatment. In the previous 17 chapters however, we have gathered enough background material that we can shorten the preliminaries with reference to analogies and similarities drawn from nuclear physics.

## 18.1 PARTICLE INTERACTIONS AND FAMILIES

Neglecting gravity, which has no measurable effects on the scale of particle interactions, there are three basic types of forces that operate between particles: weak, electromagnetic, and strong. We can assign each of these forces a relative strength parameter and each will operate over a specific distance or time scale.

In the language of modern field theory, each force is governed by the exchange of field particles or quanta. The field quanta can themselves be regarded as

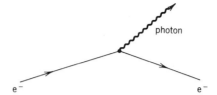

**Figure 18.1** This diagram represents the emission of a photon by an electron. Time increases from left to right, so the initial state consists of only an electron and the final state consists of an electron plus a photon. The vertical dimension represents an arbitrary spatial coordinate.

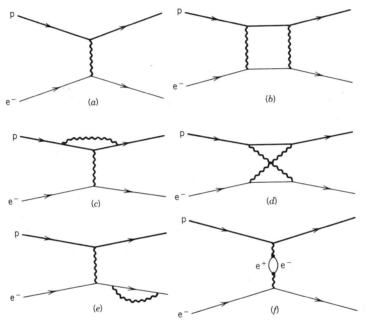

**Figure 18.2** Some of the diagrams that occur in the electron-proton electromagnetic interaction. In calculating the complete interaction, contributions from all such diagrams (and an infinity of similar ones) must be added.

elementary particles with definite sets of properties. A useful graphical way to represent the emission or absorption of field quanta is illustrated in Figure 18.1 for the emission of a photon by an electron. (Such a process cannot occur for free electrons, of course, but we ignore this fact for the purpose of the illustration.) This is an example of a *Feynman diagram*. Not only does the Feynman diagram give a pictorial representation of particle interactions, but it also allows the calculation of the probabilities for various reactions and decay processes by employing a set of rules to go with the diagrams. We will not discuss the rules; we will merely use the diagrams to represent the process. Figure 18.2 illustrates some of the many processes that might occur in the electromagnetic interaction between an electron and a proton. The complete theory involves adding the contributions from all possible processes, of which there is an infinite number.

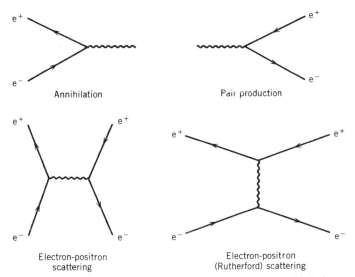

**Figure 18.3** Diagrams for some electromagnetic processes. It is customary to regard positrons (and other antiparticles) as particles traveling backward in time.

Fortunately, there are schemes for either terminating the procedure or removing the infinities that may result.

The electromagnetic interaction is most familiar to us, and to lowest order it is represented by the emission or absorption of a photon. Figure 18.3 shows several examples of diagrams representing electromagnetic processes. Because we are interested primarily in the basic interaction, not in its calculation to high precision, we will ignore the many other contributions such as those shown in Figure 18.2 and deal only with single photon exchange.

Once we have established the basic principles, we can depict other processes using the same scheme. Figure 18.4 shows a diagram for np scattering. The fundamental nature of the np interaction determines how the diagram looks in the area shown as the interaction region. One possibility would be a point interaction (as shown), but we know that not to be the case (nucleons are certainly not point particles). We do know that a major part of the strong nucleon–nucleon interaction is mediated by the exchange of pions, and possible diagrams representing the exchange are shown in Figure 18.4. It is also possible to include the $\Delta$ resonance as an intermediate state in a diagram.

The weak interaction, as indicated in Figure 18.5, poses similar problems for determining what takes place in the interaction region. Because the weak interaction is of very short range, it was originally represented as a point interaction, and indeed the Fermi theory of $\beta$ decay outlined in Chapter 9 is derived on that basis. The decay of a muon can be represented similarly. The present theory of the weak interaction is based on an exchange-force model. The exchanged particles are known as intermediate vector bosons, represented by $W^{\pm}$ and $Z^0$. The neutron $\beta$ decay and the muon decay are shown in Figure 18.5 with the $W^{\pm}$ as carriers of the weak force. In the neutron case, decay into a $W^-$ and a proton occurs at one vertex in the diagram, following which the $W^-$ decays (also weakly)

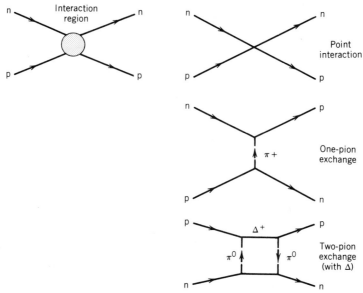

**Figure 18.4** Diagrams in the np interaction. Possible contents of the interaction region are shown at right. The point interaction is not correct, but the others may be possible contributions to the np interaction.

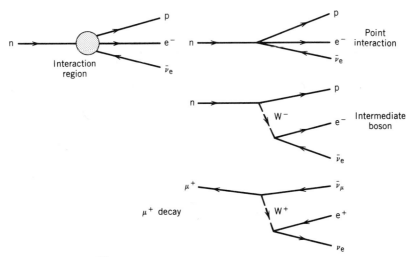

**Figure 18.5** Diagrams of weak decays.

into $e^-$ and $\bar{\nu}_e$. The $\mu^+$ decay behaves similarly. These diagrams remove certain difficulties inherent in the four-particle point vertex.

The existence of the weak bosons was proposed by S. Weinberg and A. Salam who in 1967 separately made the first step toward achieving a unified description of all particle interactions by combining the electromagnetic and weak forces into a single theoretical framework. This *electroweak* theory postulates that at very high energy the weak and electromagnetic forces become completely equivalent. The pure electroweak force would be mediated by four massless spin-1 particles, a triplet (of charges $+1, 0, -1$) and a singlet (uncharged). At lower energies, the symmetry between weak and electromagnetic forces is broken, and three of the four exchanged particles lose their massless character to become the weak bosons $W^\pm$ and $Z^0$. The fourth particle remains massless and is the ordinary photon of electromagnetism.

The Weinberg-Salam theory makes several unique predictions that can be directly tested. The masses of $W^\pm$ and $Z^0$ can be predicted based on the weak interaction coupling constant. Recalling our discussion of $\beta$ decay in Chapter 9, we derived a weak interaction strength constant $g$ with a value of $0.88 \times 10^{-4}$ MeV·fm³. This value was derived from superallowed $0^+ \rightarrow 0^+$ decays, but we can make the hypothesis that this is merely the manifestation of a general property of the weak interaction. According to Equation 9.31, $g$ is related to a general, dimensionless constant $G$ that characterizes the strength of the weak interaction. Equation 9.31 was derived on dimensional grounds, so it does not include constants of order unity. The exact result is

$$G = 4\sqrt{2}\, g \frac{m^2 c^4}{\hbar^3 c^3} \qquad (18.1)$$

According to the Weinberg-Salam electroweak model, the dimensionless weak strength $G$ is directly related to the dimensionless electromagnetic strength, in terms of the fine structure constant $\alpha = e^2/4\pi\epsilon_0 \hbar c$, as

$$G \sin^2 \theta_w = 4\pi\alpha \qquad (18.2)$$

where $\theta_w$ is called the Weinberg angle and is a parameter of the model. The Weinberg angle can be determined from many different experiments, with the result

$$\sin^2 \theta_w = 0.23 \pm 0.01$$

Thus $G = 0.399$, and using the value of $g$ measured in $\beta$ decay (an identical value is obtained in muon decay, lending support to its interpretation as a universal weak coupling constant), we can determine the mass that appears in Equation 18.1. Because the weak interactions used to obtain $g$ involve the exchange of charged $W^\pm$, the mass corresponds to $m_W$ and has the value

$$m_W c^2 = 78 \text{ GeV}$$

A further relationship gives

$$m_Z c^2 = \frac{m_W c^2}{\cos \theta_w} \qquad (18.3)$$

$$= 89 \text{ GeV}$$

Creation of particles of this extremely large mass ($\approx 100 m_p$) requires large accelerators, and no accelerator operating in the fixed-target mode is powerful enough to produce particles in this mass range. Experiments in 1983 at the CERN proton-antiproton collider, in which each beam has an energy of 270 GeV, were able to obtain the first evidence for the $W^\pm$ and $Z^0$, shown in Figure 18.6. The center-of-mass energy (540 GeV) is far greater than that needed to produce $W^\pm$ or $Z^0$, but at threshold the cross section is too small to provide any hope of detecting the particles. The $W^\pm$ and $Z^0$ are not detected directly, but

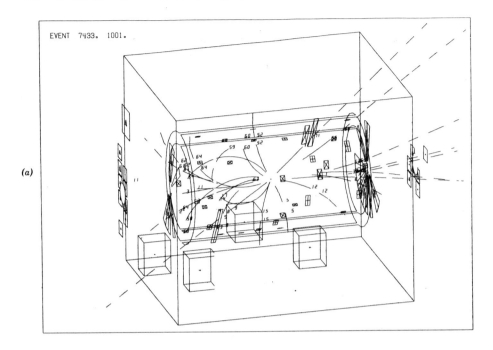

(a)

EVENT 7433. 1001.

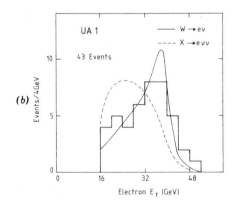

(b)

UA 1

—— W → eν
---- X → eνν

43 Events

Events/4GeV

Electron $E_T$ (GeV)

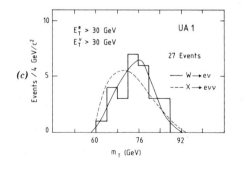

(c)

Events / 4 GeV/$c^2$

$E_T^e$ > 30 GeV
$E_T^\nu$ > 30 GeV

UA 1

27 Events

—— W → eν
---- X → eνν

$m_T$ (GeV)

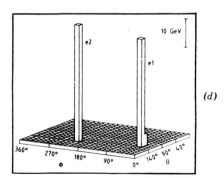

(d)

10 GeV

e2

e1

360°  270°  180°  90°  0°  140° 90° 40°

Φ        θ

**Figure 18.6**

instead through their decays, such as

$$W^{\pm} \rightarrow e^{\pm} + \nu$$
$$Z^0 \rightarrow e^+ + e^-$$

and the experimenters must be able to detect electrons of energy 40–50 GeV. This requires incredibly large detectors, as illustrated in Figure 18.7. The experiments to search for the weak bosons were successful and the deduced masses were

$$m_W c^2 = 80.8 \pm 2.7 \text{ GeV}$$
$$m_Z c^2 = 92.9 \pm 1.6 \text{ GeV}$$

This discovery and its excellent agreement with the predicted values were triumphs for the Weinberg-Salam theory and earned the 1984 Nobel Prize for Carlo Rubbia, the leader of the experimental team, and Simon van der Meer, who developed the principle of stochastic cooling that made operation of the CERN p$\bar{\text{p}}$ collider possible.

The weak interactions discussed so far all involve exchange of the charged W$^{\pm}$. These are called "charged-current" weak interactions. Another testable prediction of the Weinberg-Salam theory is a new type of interaction involving the exchange of Z$^0$ and hence called a *neutral current* weak interaction. An example is neutrino scattering

$$\nu_{\mu} + e^- \rightarrow \nu_{\mu} + e^-$$
$$\nu_{\mu} + p \rightarrow \nu_{\mu} + p$$

as indicated in Figure 18.8. With the availability of intense beams of high-energy ($\sim 100$ GeV) neutrinos in the early 1970s at CERN, it became possible to observe these neutrino scattering events and to confirm the existence of weak neutral currents. This experiment gave the first direct confirmation of the Weinberg-Salam model and as a result, Weinberg, Salam, and co-developer S. Glashow were honored with the 1979 Nobel Prize in physics.

We therefore have the three basic interactions: electromagnetic, involving the exchange of photons; weak, involving the exchange of W$^{\pm}$ and Z$^0$; and strong,

---

**Figure 18.6** (a) Reconstruction of the multitude of particle trajectories leaving the central region of a p$\bar{\text{p}}$ collision at CERN. From G. Arnison et al., *Phys. Lett. B* **126**, 398 (1983). (b, c) In the reaction p + $\bar{\text{p}}$ → W + anything, followed by the decay W → e + $\nu$, the $\nu$ is undetected, but its energy and momentum can be deduced if *all* of the other products of the reaction are detected. Of the million or so collision events, in only 43 is there a clear electron track with a deduced neutrino momentum in the opposite direction (as expected for the two-body decay of the W). The electron energy is shown in (b), and the electron plus neutrino energy in (c). The data in (c) give 81 $\pm$ 2 GeV for the W mass. From G. Arnison et al., *Phys. Lett. B* **129**, 273 (1983). (d) In the reaction p + $\bar{\text{p}}$ → Z$^0$ + anything, the decay of the Z$^0$ gives two electrons, which are both detected. The electrons move in opposite directions ($\phi_1 - \phi_2 = 180°$) as expected, and their total energy (shown on the vertical scale) gives 91 $\pm$ 2 GeV for the Z$^0$ mass. From P. Bagnaia et al., *Phys. Lett. B* **129**, 130 (1983).

**Figure 18.7** The giant UA1 detector at the CERN p$\bar{\text{p}}$ collider, from which the data shown in Figure 18.6*b, c* were obtained. The detector is a 2000-ton electromagnetic and hadronic calorimeter, designed to record the passage of all electrically or strongly interacting products of the reactions, so that energy and momentum balance can be used to deduce the properties of undetected neutrinos. When the two halves of the detector are pushed together, the calorimeter is able to observe particles at transverse angles from 90 to 0.2°. Photo courtesy CERN.

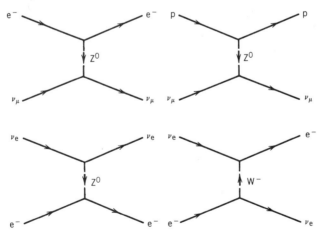

**Figure 18.8** Neutral-current weak interactions are mediated by $Z^0$ exchange. Some processes, such as $\nu_e$e scattering shown at bottom, can occur through both charged-current and neutral-current interactions.

**Table 18.1** Field Particles

| Particle | Mass | Charge | Spin |
|---|---|---|---|
| $\gamma$ | 0 | 0 | 1 |
| $W^{\pm}$ | 81 GeV/$c^2$ | $\pm 1$ | 1 |
| $Z^0$ | 93 GeV/$c^2$ | 0 | 1 |
| Gluons | 0 | 0 | 1 |
| Graviton | 0 | 0 | 2 |

**Table 18.2** Leptons

| Particle | Antiparticle | Mass (MeV/$c^2$) | Charge | Spin | Lifetime(s) | Decay Modes |
|---|---|---|---|---|---|---|
| e | $e^+$ | 0.511 | $\pm 1$ | $\frac{1}{2}$ | Stable | |
| $\nu_e$ | $\bar{\nu}_e$ | < 46 eV/$c^2$ | 0 | $\frac{1}{2}$ | Stable | |
| $\mu^-$ | $\mu^+$ | 105.66 | $\pm 1$ | $\frac{1}{2}$ | $2.20 \times 10^{-6}$ | $e \nu_e \nu_\mu$ |
| $\nu_\mu$ | $\bar{\nu}_\mu$ | < 0.50 | 0 | $\frac{1}{2}$ | Stable | |
| $\tau^-$ | $\tau^+$ | 1784 | $\pm 1$ | $\frac{1}{2}$ | $3.4 \times 10^{-13}$ | $\mu \nu_\mu \nu_\tau (18\%)$; $e \nu_e \nu_\tau (17\%)$; $\pi \nu_\tau (10\%)$; $\rho \nu_\tau (22\%)$ |
| $\nu_\tau$ | $\bar{\nu}_\tau$ | < 164 | 0 | $\frac{1}{2}$ | Stable | |

involving for example the exchange of $\pi$ mesons between nucleons. The first two are truly fundamental processes, while the latter is not, because the nucleons and mesons are composite particles. In Section 18.3 we treat the strong interactions on a more fundamental basis in terms of the quark model, in which the strong interaction between quarks is mediated by the exchange of field particles called *gluons*.

The particles themselves fall into several major categories. First are the *field particles*: $\gamma$, $W^{\pm}$, $Z^0$, gluons (Table 18.1). To this list might be added the graviton, which has been proposed (but not observed) as the particle that carries the gravitational interaction.

Next come the *leptons*, listed in Table 18.2. The third charged lepton, $\tau$, has properties similar to those of e and $\mu$ but a much larger mass, 1784 MeV. The $\tau$ lepton was discovered in 1975 in experiments at SLAC.

As far as we have yet been able to discover, the leptons are true elementary "point" particles. They are totally unaffected by the strong interaction, but in their electromagnetic and weak interactions they appear to be identical to one another. The magnetic moments of e and $\mu$ have been measured and computed using quantum electrodynamics; the difference between the expected $g$ factor of 2 for a pure Dirac particle and the observed value is about $10^{-3}$ and is known to a precision of about $10^{-9}$. The good agreement between theory and experiment in both cases at this incredible level of precision justifies the treatment of the e and $\mu$ as having equivalent properties and interactions. Indeed, the only differences among e, $\mu$, and $\tau$ are in mass and in properties related to mass.

The leptons are readily classified into three pairs of two each: (e, $\nu_e$), ($\mu$, $\nu_\mu$), and ($\tau$, $\nu_\tau$), plus of course the antiparticles. No other types have yet been found, although there certainly could be heavier leptons than we have yet been able to produce with existing accelerators. An upper limit on the number of such groups can be derived from our theories of the evolution of the early universe (Chapter 19)—it is extremely unlikely that there are more than four lepton types, and the upper limit may well be three. There is thus a strong possibility that there are no new types beyond $\tau$.

The next group of particles is the *mesons*, which we have already discussed and partially tabulated in Chapter 17. Mesons are strongly interacting particles with integer spin. The *strange* mesons obey unusual rules in their production and decay that require the introduction of a new property, *strangeness*. Yet other unusual production and decay observations are explained through the introduction of additional properties, named *charm*, *bottomness* (or *beauty*), and *topness* (or *truth*).

The *baryons* are strongly interacting heavy particles with half-integer spins ($\frac{1}{2}, \frac{3}{2}, \ldots$). The lightest members are the proton and neutron. There are many resonances in the pion-nucleon cross section, discussed in Chapter 17, which are assigned to the list of baryons, even though they live for only $10^{-22}$ s or so. There are also strange baryons that display properties similar to those of the strange mesons. Other types of baryons exist as well.

Because the mesons and baryons are composite particles, the number of possible states probably increases without limit, and since we know a good deal about the substructure, it makes little sense to treat them as elementary particles. On the other hand, in nuclear physics at low energy we do not usually observe the substructure, and nucleons or mesons can be treated as fundamental (but spatially extended) units.

## 18.2 SYMMETRIES AND CONSERVATION LAWS

The analysis and interpretation of particle interactions depend on the applicability of a number of symmetries and conservation laws. Some of these are classical in nature and based on our understanding of the implications of elemental properties of space and time. Others are empirical and merely serve to help explain the observation of some processes and the nonobservation of others. Some of these laws are apparently absolute, obeyed in all cases by all processes; some are obeyed in some processes but not in others.

### Energy and Momentum

All decays and reactions obey the conservation of relativistic total energy and momentum. As we have done previously, we can define the $Q$ value of a decay or reaction process, $Q = (m_i - m_f)c^2$, but it is often easier to work directly with the total relativistic energy. In a decay process with a two-particle final state A → B + C, in the rest frame of A, conservation of momentum gives $p_B = p_C$, and with $E_i = m_A c^2$ and $E_f = E_B + E_C$, it is possible to solve uniquely for the

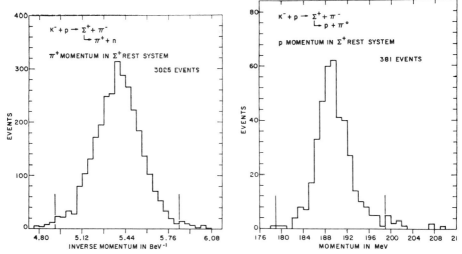

**Figure 18.9** (Left) $\pi^+$ momentum in the decay $\Sigma^+ \to \pi^+ + n$. The width is caused by experimental resolution. The central momentum is $p_\pi = 185.74 \pm 0.12$ MeV$/c$, corresponding to $T_\pi = 92.8 \pm 0.1$ MeV. (Right) Proton momentum in $\Sigma^+ \to$ p + $\pi^0$. The central momentum is $p_p = 189.35 \pm 0.16$ MeV$/c$, which gives $T_p = 18.9 \pm 0.1$ MeV. From P. Schmidt, *Phys. Rev. B* **140**, 1328 (1965).

energies of B and C:

$$T_B = \frac{\left(m_A c^2 - m_B c^2 - m_C c^2\right)\left(m_A c^2 - m_B c^2 + m_C c^2\right)}{2 m_A c^2} \qquad (18.4)$$

$$T_C = \frac{\left(m_A c^2 - m_B c^2 - m_C c^2\right)\left(m_A c^2 - m_C c^2 + m_B c^2\right)}{2 m_A c^2} \qquad (18.5)$$

For example, in the decay $\Sigma^+ \to n + \pi^+$, with $m_\Sigma c^2 = 1189.36$ MeV, $m_n c^2 = 939.57$ MeV, and $m_\pi c^2 = 139.57$ MeV, we find $T_n = 18.0$ MeV and $T_\pi = 92.2$ MeV. Figure 18.9 shows the momentum distributions observed in this decay process, and in the similar decay $\Sigma^+ \to p + \pi^0$, in which we expect $T_p = 18.8$ MeV.

In the case of three-body final states, as in the decay A $\to$ B + C + D, the energies of the final products are not determined uniquely but are shared in a fashion similar to the case of $\beta$ decay. The kinetic energy distributions of B, C, and D extend from zero up to a maximum $T_{\max}$. To conserve momentum in the rest frame of A, the momentum of B must be balanced by one (or both ) of C and D. Particle B will have its maximum energy when as little energy as possible is given to C or D, which means that the recoil momentum should be given to the heavier of C and D, with the other remaining at rest. Assuming $m_C > m_D$, we can then calculate the maximum kinetic energy of B:

$$T_{B, \max} = \frac{\left(m_A c^2 - m_D c^2 - m_B c^2 - m_C c^2\right)\left(m_A c^2 - m_D c^2 - m_B c^2 + m_C c^2\right)}{2\left(m_A c^2 - m_D c^2\right)}$$

$$(18.6)$$

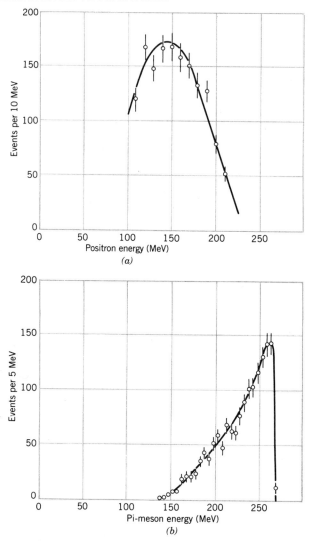

**Figure 18.10** Spectrum of positrons (top) and $\pi$ mesons (bottom) in the decay $K^+ \rightarrow \pi^0 + e^+ + \nu_e$. Data from Botterill et al., *Phys. Lett. B* **31**, 325 (1970).

which can also be obtained directly from Equation 18.4 with the substitution $m_A c^2 \rightarrow m_A c^2 - m_D c^2$. (That is, with D at rest, the available energy to be shared by B and C is reduced by an amount $m_D c^2$.) In the case $K^+ \rightarrow \pi^0 + e^+ + \nu_e$, Equation 18.6 gives $T_{\pi,\,\mathrm{max}} = 130.3$ MeV and $T_{e,\,\mathrm{max}} = 227.9$ MeV. Figure 18.10 shows the distributions of $\pi^0$ and $e^+$ emitted in the $K^+$ decay, and the endpoints are in agreement with the calculated values.

If particle A decays in flight, Equations 18.4–18.6 are no longer valid in the laboratory frame. They remain valid in the rest frame of A, and to transform back to the laboratory frame we can use the Lorentz transformation. Suppose particle A (and therefore the original center of mass) is moving along the $z$

direction at speed $v$. Then, if $E'$ is the total relativistic energy calculated using Equations 18.4–18.6 in the center-of-mass frame ($E' = T' + mc^2$), in the laboratory frame the transformed energy is

$$E = \gamma(E' - \beta p'_z) \qquad (18.7)$$

where $\beta = v/c$ and $\gamma = (1 - \beta^2)^{-1/2}$. The value of $p'_z$ (the $z$ component of the momentum of the particle B, C, or D in the center-of-mass frame) will vary as the direction of emission of B, C, or D vary in the rest frame of A. Thus there will be no unique value of $E$ corresponding to the unique value of $E'$.

In the case of reactions, $A + B \rightarrow C + D + \cdots$, it is again simplest to do the kinematics in the center-of-mass frame and then transform to the laboratory frame. In the lab frame, we will assume particle A to be incident on particle B at rest. The quantity $E^2 - c^2 p^2$ is an invariant for any system of particles, and we may therefore evaluate this quantity in the lab and center-of-mass frames:

$$\left(\sum E\right)^2 - c^2 \left(\sum \mathbf{p}\right)^2 = \left(E_A + m_B c^2\right)^2 - c^2 p_A^2 \quad \text{(lab)} \qquad (18.8a)$$

$$= \left(E'_A + E'_B\right)^2 \quad \text{(c.o.m.)} \qquad (18.8b)$$

Let $E'_0$ be the total energy available in the center-of-mass frame:

$$E'_0 = E'_A + E'_B \qquad (18.9)$$

Then, combining Equation 18.8a with Equation 18.8b,

$$E'^2_0 = m_A^2 c^4 + m_B^2 c^4 + 2 m_B c^2 E_A \qquad (18.10)$$

If there is a two-body final state $C + D$, we can write expressions exactly analogous to Equations 18.4 and 18.5 for the final energies $T'_C$ and $T'_D$ by imagining the decay of a "particle" of mass energy $E'_0$ at rest in the center-of-mass frame:

$$T'_C = \frac{\left(E'_0 - m_C c^2 - m_D c^2\right)\left(E'_0 - m_C c^2 + m_D c^2\right)}{2 E'_0} \qquad (18.11)$$

and similarly for $T'_D$. We can then use Equation 18.7 to transform to the lab frame.

For three-body final states $C + D + E$, we can use the analog to Equation 18.6 in the lab frame to find the maximum energy of C (here D takes the recoil and E is at rest):

$$T'_{C, \text{max}} = \frac{\left(E'_0 - m_E c^2 - m_C c^2 - m_D c^2\right)\left(E'_0 - m_E c^2 - m_C c^2 + m_D c^2\right)}{2\left(E'_0 - m_E c^2\right)} \qquad (18.12)$$

Often our goal in high-energy reactions is the production of new particles, in which case $Q < 0$ and there is a threshold condition on the reaction in the lab frame. (We assume A is incident on B at rest.) At threshold, the product particles $C + D + E + \cdots$ move together as a group. No energy is "wasted" in motion transverse to the original direction of motion of A. In the center-of-mass frame, this corresponds to the group of product particles being formed at rest. Since the product particles are formed as a group, we are not concerned about their identity and can represent their total mass as $M$:

$$M = m_C + m_D + m_E + \cdots \tag{18.13}$$

Then conserving energy and momentum in the laboratory frame, we can derive the threshold laboratory kinetic energy of A:

$$T_{A, th} = \frac{(Mc^2 - m_A c^2 - m_B c^2)(Mc^2 + m_A c^2 + m_B c^2)}{2 m_B c^2} \tag{18.14}$$

or, equivalently

$$T_{A, th} = (-Q) \frac{\text{total mass of all initial and final particles}}{2 \times \text{mass of target}} \tag{18.15}$$

For example, in the production of antiprotons in $p + p \rightarrow p + p + p + \bar{p}$, $Q = -2 m_p c^2$ and

$$T_{A, th} = 2 m_p c^2 \frac{6 m_p c^2}{2 m_p c^2} = 6 m_p c^2 = 5.63 \text{ GeV}$$

### Angular Momentum

All decays and reactions also conserve angular momentum. In the case of two-body decays A → B + C, letting $s_A$, $s_B$, and $s_C$ represent the intrinsic spins of the particles:

$$|s_A| = |s_B + s_C + \ell_{B,C}| \tag{18.16}$$

where $\ell_{B,C}$ is the relative orbital angular momentum of the BC combination. For decay into three particles, we first compute the orbital angular momentum of one pair, say BC. We then evaluate the angular momentum of D relative to the BC pair:

$$|s_A| = |s_B + s_C + \ell_{B,C} + s_D + \ell_{BC,D}| \tag{18.17}$$

In general, we must have knowledge of all but one of the unknowns to use these expressions. Often limits on $\ell$ can be obtained from angular distribution measurements on the decaying particles or from other arguments.

Reactions can be analyzed similarly. For A + B → C + D,

$$|s_A + s_B + \ell_{A,B}| = |s_C + s_D + \ell_{C,D}| \tag{18.18}$$

### Parity

Each elementary particle has an associated intrinsic parity, either odd $(-)$ or even $(+)$. The parity $\pi$ can also describe the behavior of the particle's spatial wave function under the parity operation $r \rightarrow -r$.

$$\psi(r) \rightarrow \psi(-r) = \pi \psi(r) \tag{18.19}$$

Because applying the parity operation twice must bring us back to the original wave function, $\pi^2 = 1$ and $\pi = \pm 1$. The most familiar application of this result is in the case of orbital angular momentum states with parity $(-1)^\ell$. Parities can be deduced in reactions relative to the parity of the proton, which is taken as even by definition.

Parity is absolutely conserved in all strong or electromagnetic interactions, but is violated in weak interactions. The net parity on each side of a decay or reaction process is computed by taking the *product* of the intrinsic parities and the relative spatial parities. Consider, for example, the strong decay of the $\phi$ meson

$$\phi \rightarrow K^+ + K^-$$

The intrinsic parity of $\phi$ is to be determined. Conservation of parity gives

$$\pi_\phi = \pi_{K^+}\pi_{K^-}(-1)^\ell$$

where $\ell$ is the relative orbital angular momentum of $K^+$ and $K^-$. For particles with integral spin, such as $K^+$ and $K^-$, antiparticles have the same parity as particles. (Particles and antiparticles with half-integral spin, such as p and $\bar{p}$, have opposite intrinsic parity.) Thus whatever the intrinsic parity of $K^+$, $K^-$ will be the same and $\pi_{K^+}\pi_{K^-} = +1$. The spins of $K^+$ and $K^-$ are zero, but $\phi$ has spin 1. Conservation of angular momentum, by Equation 18.16, then requires $\ell = 1$ and $\pi_\phi = -1$. Therefore the $\phi$ meson has odd parity.

Consider now the electromagnetic decay process

$$\Sigma^0 \rightarrow \Lambda^0 + \gamma$$

The $\Lambda^0$ is known to have even parity and spin $\frac{1}{2}$. The $\Sigma^0$ also has spin $\frac{1}{2}$. The photon must then carry angular momentum 1 (it is not allowed for photons to have angular momentum of zero) to conserve angular momentum, and given that the photon is observed to be of magnetic dipole character, the parity rules for photon emission, Equation 10.16, require that $M1$ photons do not carry a change in parity. Thus the parity of $\Sigma^0$ must be identical with that of $\Lambda^0$, that is, even.

## Baryon Number

As far as we yet know, all decays and reactions conserve the total baryon number $B$ (but see Section 18.8 for possible violations). If we assign $B = +1$ to baryons, $B = -1$ to the antibaryons, and $B = 0$ to all nonbaryons (mesons, leptons, and field particles), then the sum of the $B$'s must be the same on both sides of the decay or reaction process. For example, antiproton production in proton–proton collisions requires three protons in the final state:

$$p + p \rightarrow p + p + p + \bar{p}$$
$$B = +1 \quad +1 \rightarrow +1 \quad +1 \quad +1 \quad -1$$

and the net value of $B = +2$ is preserved on each side of the reaction. Similarly, baryon number conservation forbids such decays as

$$\bar{\Lambda}^0 \rightarrow p + \pi^-$$
$$B = -1 \rightarrow +1 \quad 0$$

Models proposed for the unification of the strong, weak, and electromagnetic interactions suggest that baryon number conservation may in fact not be absolute, but its violation, if any, occurs only over immensely long time scales

($> 10^{31}$ years), as we discuss in Section 18.8. For all observable particle decays and interactions we will regard $B$ as being absolutely conserved.

## Lepton Number

In analogy with baryon number, for each of the three types of leptons (e, $\mu$, $\tau$) we assign a lepton number of $L = +1$ for leptons, $L = -1$ for antileptons, and $L = 0$ for all nonleptons. Total lepton number is separately conserved for each type of lepton in all particle interactions and decays. For example, consider the decay of the muon

$$\mu^- \;\rightarrow\; e^- \;+\; \bar{\nu}_e \;+\; \nu_\mu$$

$$L_e = \quad 0 \;\rightarrow +1 \quad\;\; -1 \qquad 0$$

$$L_\mu = +1 \;\rightarrow\quad 0 \qquad 0 \quad +1$$

On both sides of the decay process, $L_e = 0$ and $L_\mu = +1$, and so the decay conserves both electron and muon lepton number. The decay $\mu \rightarrow e + \gamma$ is forbidden only by lepton number conservation (it fails to conserve both $L_\mu$ and $L_e$). Searching for this decay provides one of the most stringent tests of this scheme of lepton number assignments: the present upper limit is about $10^{-10}$ relative to the lepton-number conserving decay. Other examples of conservation of lepton number can be derived from neutrino capture reactions:

$$\bar{\nu}_\mu + p \rightarrow \mu^+ + n$$

A positron $e^+$ is never observed in this process.

## Isospin

In Chapter 17, isospin conservation was mentioned as the source of the relative cross sections and decay branches observed for mesons. We will now make those considerations somewhat more explicit. Hadrons (strongly interacting particles) with similar properties and nearly identical masses are grouped into isospin multiplets, thus: (p, n); ($\Sigma^+, \Sigma^0, \Sigma^-$); ($\pi^+, \pi^0, \pi^-$); ($\Lambda^0$); ($\Delta^{++}, \Delta^+, \Delta^0, \Delta^-$); and so on. We use the isospin as a way of labeling the members of the multiplet. We can define the quantity $T_3$ as

$$T_3 = \frac{Q}{e} - \frac{\overline{Q}}{e} \tag{18.20}$$

where $Q$ is the electric charge of a particle and $\overline{Q}$ is the average charge of the multiplet, computed by adding all the charges and dividing by the number of particles. Notice that some multiplets are symmetric about 0 with respect to their charges $(+1, 0, -1)$ and others are asymmetric. This definition of $T_3$ gives to multiplets labels that are always symmetric about $T_3 = 0$. Thus, for the $\Delta$, $\overline{Q}/e = \frac{1}{2}$ and the particles would be labeled as

$$T_3(\Delta^{++}) = +\tfrac{3}{2} \qquad T_3(\Delta^+) = +\tfrac{1}{2} \qquad T_3(\Delta^0) = -\tfrac{1}{2} \qquad T_3(\Delta^-) = -\tfrac{3}{2}$$

and for the nucleons

$$T_3(p) = +\tfrac{1}{2} \qquad T_3(n) = -\tfrac{1}{2}$$

For the multiplets that are already symmetric about zero (the pions or $\Sigma$), $T_3 = Q/e$. Conservation of $T_3$ in decay and reaction processes is thus entirely equivalent to conservation of electric charge.

As was done in Chapter 17, for each multiplet we introduce the isospin quantum number $T$ which characterizes a vector $\mathbf{T}$ in isospin space that has the proper components. Thus $T = 1$ for $\rho$ and the pions (triplets), $T = 0$ for $\Lambda^0$, $\eta$, and $\omega$, $T = \frac{3}{2}$ for the $\Delta$, and so on. In analyzing decays and reactions, we must conserve $T$ in strong-interaction processes, but may violate conservation of $T$ in electromagnetic or weak processes. For example, consider the decay $\eta' \rightarrow \eta + \pi^0$. Both $\eta'$ and $\eta$ have $T = 0$, while $T = 1$ for the pion. Coupling vectors of length 0 and 1 gives a resultant of 1 for the final products, thereby violating isospin conservation in the decay. Similarly, $\eta' \rightarrow \omega + \pi^0$ is isospin forbidden, but the decay $\eta' \rightarrow \eta + 2\pi$ is permitted because the $2\pi$ can be coupled to $T = 0$ to give a net $T$ of zero on the right side.

## Strangeness and Charm

In Chapter 17, we discussed the reasons for assigning the strangeness quantum number to particular mesons and baryons. Certain decay processes that would otherwise be expected to happen in a characteristic strong interaction time of $10^{-22}$ s are slowed to a weak interaction time of $10^{-10}$ s. We account for this slowing by the assignment of strangeness quantum number $S$, and we postulate that strangeness is absolutely conserved in strong and electromagnetic processes, but can change by one unit in weak decays. Then the $\Delta S = 1$ decay $\Lambda^0 \rightarrow p + \pi^-$ is a weak interaction decay, despite the appearance of only strongly interacting particles.

The quantum numbers assigned so far are not all independent; they are related by the Gell-Mann–Nishijima formula:

$$\frac{Q}{e} = T_3 + \frac{B + S}{2} \qquad (18.21)$$

Recent discoveries have suggested yet another quantum number with properties similar to strangeness in inhibiting certain decay processes. This property is called charm, $C$. There are charmed mesons called $D^\pm$, $D^0$ and $\overline{D}^0$, forming a set very similar to the strange K mesons; there are also charmed baryons. As with strangeness, strong and electromagnetic interactions conserve charm; thus there is associated production in the case of charm as there was in the case of strangeness. Weak interactions can change the charm by one unit. The D mesons can decay weakly as

$$D \rightarrow K + \pi$$

changing both $C$ and $S$ by one unit.

There are also particles having both charm and strangeness, for example, the $F^+$ meson with $S = +1$ and $C = +1$.

To account for the additional multiplicity brought about by the charm quantum number, the Gell-Mann–Nishijima relationship must be modified:

$$\frac{Q}{e} = T_3 + \frac{B + S + C}{2} \qquad (18.22)$$

**Table 18.3** Hadrons with Strangeness, Charm, Beauty, and Truth

| Particles | Antiparticles | Spin parity | Isospin | Mass (MeV/$c^2$) | $S$ | $C$ | $B'$ | $T'$ |
|---|---|---|---|---|---|---|---|---|
| Mesons | | | | | | | | |
| $K^+, K^0$ | $K^-, \overline{K}^0$ | $0^-$ | $\frac{1}{2}$ | 495(4) | $+1$ | 0 | 0 | 0 |
| $K^{*+}, K^{*0}$ | $K^{*-}, \overline{K}^{*0}$ | $1^-$ | $\frac{1}{2}$ | 892(7) | $+1$ | 0 | 0 | 0 |
| $D^+, D^0$ | $D^-, \overline{D}^0$ | $0^-$ | $\frac{1}{2}$ | 1866(5) | 0 | $+1$ | 0 | 0 |
| $D^{*+}, D^{*0}$ | $D^{*-}, \overline{D}^{*0}$ | $1^-$ | $\frac{1}{2}$ | 2010(3) | 0 | $+1$ | 0 | 0 |
| $F^+$ | $F^-$ | $0^-$ | 0 | 1971 | $+1$ | $+1$ | 0 | 0 |
| $B^+, B^0$ | $B^-, \overline{B}^0$ | $0^-$ | $\frac{1}{2}$ | 5271(3) | 0 | 0 | $+1$ | 0 |
| Baryons | | | | | | | | |
| $\Lambda^0$ | | $\frac{1}{2}^+$ | 0 | 1116 | $-1$ | 0 | 0 | 0 |
| $\Sigma^+, \Sigma^0, \Sigma^-$ | | $\frac{1}{2}^+$ | 1 | 1190(8) | $-1$ | 0 | 0 | 0 |
| $\Xi^0, \Xi^-$ | | $\frac{1}{2}^+$ | $\frac{1}{2}$ | 1320(6) | $-2$ | 0 | 0 | 0 |
| $\Omega^-$ | | $\frac{3}{2}^+$ | 0 | 1673 | $-3$ | 0 | 0 | 0 |
| $\Lambda_c^+$ | | $\frac{1}{2}^+$ | 0 | 2282 | 0 | $+1$ | 0 | 0 |
| $A^+(?)$ | | | | 2460 | $-1$ | $+1$ | 0 | 0 |
| $\Sigma_c^{++}, \Sigma_c^+, \Sigma_c^0$ | | $\frac{1}{2}^+$ | 1 | 2450 | 0 | $+1$ | 0 | 0 |
| $\Lambda_b^0(?)$ | | | | 5500 | 0 | 0 | $+1$ | 0 |

*Notes.* Evidence for $A^+$ and $\Lambda_b^0$ is currently weak; $\Sigma_c^0$ has not yet been seen. The quantity in parentheses in the mass column is the splitting of the multiplet, in MeV. The last two columns show additional particle attributes beyond charm. They are called either bottomness and topness or beauty and truth. Particles with a nonzero value of $B'$ are thus said to exhibit either bare bottom or naked beauty. (The quantum numbers are labeled $B'$ and $T'$ to prevent confusion with baryon number $B$ and isospin $T$.) Many additional short-lived meson and baryon resonances exist that are not included in this table.

Beyond charm, there is at least one additional attribute with similar properties, and there is strong reason to suspect a second. The evidence for these unusual quantum numbers and their place in the fundamental structure of the hadrons is discussed in the following sections. Table 18.3 gives a summary of some strange and charmed mesons and baryons.

## 18.3 THE QUARK MODEL

The underlying symmetry of the structure of the hadrons can be seen immediately if we make diagrams of the various families on a chart showing isospin component $T_3$ plotted against strangeness. Figure 18.11 shows such diagrams for the spin-0 and spin-1 mesons and the spin-$\frac{1}{2}$ and spin-$\frac{3}{2}$ baryons.

All these diagrams show evidence for a repeated simple structure of three particles: u, d, and s (and their antiparticles $\bar{u}$, $\bar{d}$, and $\bar{s}$), as shown in Figure 18.12. For example, the spin-$\frac{3}{2}$ baryon structure can be reproduced with the diagram shown in Figure 18.13. It is then obvious to make the identification ddd $= \Delta^-$, and so on. These three particles (u, d, and s) are three of the basic

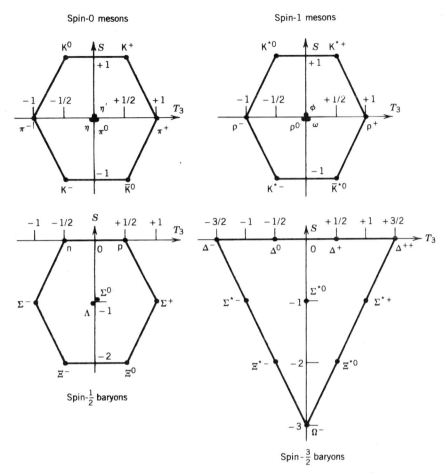

**Figure 18.11** Isospin vs strangeness charts for mesons and baryons.

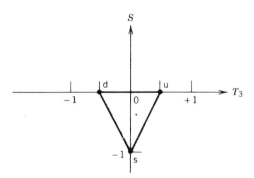

**Figure 18.12** The basic three-quark triplet.

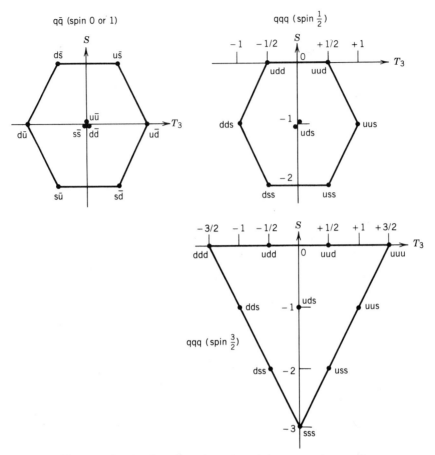

**Figure 18.13** Quark-antiquark and three-quark couplings.

*quarks* of which the hadrons are composed. It has become customary to call the u, d, and s as three *flavors* of quarks. The quark model for the internal structure of hadrons was first suggested in 1964 by Murray Gell-Mann and (independently) George Zweig.

Two features of the quarks are particularly unusual: if we are to have the identifications ddd = $\Delta^-$, uuu = $\Delta^{++}$, and sss = $\Omega^-$, then it is immediately apparent that we must assign fractional electric charges of $+\frac{2}{3}e$ to u and $-\frac{1}{3}e$ to d and s. These charges then reproduce the known charges of the multiplets of Figure 18.11 using the identifications of Figure 18.13 (assigning antiparticles charges opposite to the corresponding particle). Second, if three quarks are to make a baryon, then each quark must have a baryon number of $\frac{1}{3}$. Another property of the proposed couplings is that the quarks must obviously have spin $\frac{1}{2}$; thus a quark and an antiquark in a meson can couple to a total spin of 0 or 1, while three quarks in a baryon can couple to $\frac{1}{2}$ or $\frac{3}{2}$. (We are neglecting the relative orbital angular momentum of the quarks that might contribute to the total intrinsic angular momentum of the mesons or baryons. All of the low-lying states, particularly those shown in Figure 18.11, are $\ell = 0$ states.)

The first question that occurs is whether the quarks actually exist inside the hadrons or whether they are merely a convenient mathematical ingredient leading to the geometrical symmetry (and possibly to other calculable effects). A substantial clue in this direction is obtained in *deep inelastic* scattering from nucleons, in which we use as a probe high-energy ($\sim$ 1–100 GeV) electrons or neutrinos to sample the internal structure of the nucleon. The experiment is somewhat analogous to Rutherford scattering, in that we are searching for compact, massive objects inside the nucleon, much as Rutherford searched for a compact, massive object inside the atom. We use electrons or neutrinos, because we do not want to use a strongly interacting probe that might alter the internal structure, and we use a high-energy probe so that the de Broglie wavelength will be smaller than nucleon dimensions. The probe will thus interact with objects inside the nucleon, not with the nucleon itself. At 10 GeV, for instance, the de Broglie wavelength is 0.1 fm, an order of magnitude smaller than the nucleon dimensions. As was done in the case of Rutherford scattering, the goal is to measure the energy and angular distribution of the scattered particles, from which the properties of the struck object can be inferred.

From these experiments, several conclusions follow: (1) a nucleon contains three point-like objects; (2) the point-like objects have spin $\frac{1}{2}$; (3) the objects have fractional electric charges consistent with $+\frac{2}{3}e$ and $-\frac{1}{3}e$. This direct experimental evidence supports the existence of quarks inside the nucleon.

The baryons and mesons shown in Figure 18.11 can all be described by combinations of the three quarks u (for up), d (for down), and s (for strange). The u and d quarks are members of an isospin doublet ($T_3 = \pm \frac{1}{2}$), while s is a singlet ($T_3 = 0$). The masses of the quarks cannot be determined directly for the binding energy of quarks in hadrons is so strong that the observed total rest energy of a hadron may be small compared with the rest energies of the component quarks. For example, if the u and d quarks had masses of 10 GeV each, a u$\bar{\text{d}}$ combination (a $\pi^+$) could be created with a rest energy of 0.14 GeV if the quark–quark binding energy were 19.86 GeV. Since we know neither the strength of the quark–quark interaction nor the mass of a free quark, these must remain as speculations, but fortunately this lack of knowledge has no effect on the success of the quark theory. In Section 18.7, we consider quark dynamics in terms of an effective mass that the quarks appear to have when bound in hadrons.

## 18.4  COLORED QUARKS AND GLUONS

Consider the three baryons at the corner of the triangle diagram of Figure 18.13; their quark configurations are uuu, ddd, and sss. Here we seem to have systems which are composed of three identical quarks, all with the same quantum numbers—they are each in a state with orbital angular momentum of zero, and to obtain a total spin of $\frac{3}{2}$ for these particles, their individual spins of $\frac{1}{2}$ must all be coupled in the same configuration. This of course violates the Pauli principle, which we would like to believe is an essential part of the description of fermions.

A more direct mathematical way of describing this difficulty is to write the wave function for any composite particle as

$$\psi = \psi_{\text{spin}}\psi_{\text{space}}\psi_{\text{flavor}} \tag{18.23}$$

where each factor in the wave function describes a particular attribute of the particle. In the case of $\Delta^{++}$, $\Delta^-$, or $\Omega^-$, each of the three factors is perfectly symmetric under the exchange of any two quarks. This appears to violate the requirement for fermions to have antisymmetric total wave functions.

We can resolve this difficulty by assigning a new property to the quarks and thereby introducing an additional factor into the combined wave function to describe this new property. We can regard this new property as an additional quantum number that can be used to label the three otherwise identical quarks in the $\Delta^{++}$, $\Delta^-$, and $\Omega^-$. If this additional quantum number can take any one of three possible values, we can restore the Pauli principle by giving each quark a different value of this new quantum number, which is known as *color*. The three colors are labeled red (R), blue (B), and green (G), but these are merely mnemonic devices and have no relation to ordinary colors. The $\Delta^{++}$, for example, would then be $u_R u_B u_G$.

In this scheme, the antiquarks have the "opposite," or in more familiar terminology, complementary color. Thus anti-red ($\overline{\text{R}}$) is cyan, anti-green ($\overline{\text{G}}$) is magenta, and anti-blue ($\overline{\text{B}}$) is yellow.

An essential component of the quark model with color is that *all observed meson and baryon states are "colorless,"* that is, either color-anticolor combinations in the case of mesons, or equal mixtures of R, G, and B in the case of baryons. (Here the analogy with ordinary colors is strong; the overlap of red, green, and blue light falling on a screen gives white, as does the overlap of a color and its complement.)

For baryons, we would write the quark wave function as

$$\psi = \psi_{\text{spin}}\psi_{\text{space}}\psi_{\text{flavor}}\psi_{\text{color}} \tag{18.24}$$

where $\psi_{\text{color}}$ is the total antisymmetric combination of wave functions symbolically represented as R, G, and B:

$$\psi_{\text{color}} = \frac{1}{\sqrt{6}}[\text{RGB} + \text{BRG} + \text{GBR} - \text{RBG} - \text{BGR} - \text{GRB}] \tag{18.25}$$

Notice that the interchange of any two quark labels (the first and second, for instance) produces the effect $\psi_{\text{color}} \rightarrow -\psi_{\text{color}}$, as it should for an antisymmetric wave function. The colorless quark-antiquark combination for mesons is

$$\psi_{\text{color}} = \frac{1}{\sqrt{3}}[\text{R}\overline{\text{R}} + \text{G}\overline{\text{G}} + \text{B}\overline{\text{B}}] \tag{18.26}$$

This is a symmetric wave function, as we expect for integral-spin mesons.

The colorless nature of observed particle states shows immediately why we do not find other quark couplings. For example, we cannot construct colorless two-quark or four-quark systems; with only three available colors, all such states must show a net color and thus are not observed to exist.

The color hypothesis finds experimental confirmation in those effects that depend on the counting of the number of possible quark states. An example is

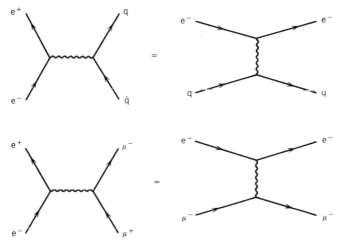

**Figure 18.14** Electron-positron annihilation into quarks (top left) and muons (bottom left) and their Rutherford scattering equivalents.

the production of hadrons in $e^+e^-$ annihilation reactions away from resonances, such as we discussed in Chapter 17. We can visualize the reaction at high energy as

$$e^+ + e^- \rightarrow \gamma \rightarrow q + \bar{q}$$

where q and $\bar{q}$ represent quark and antiquark states. The energetic q and $\bar{q}$ (which are not observed directly) then interact to form many mesons or baryons. Figure 18.14 shows the Feynman diagram for the process, which is identical to the diagram for electron-quark Rutherford scattering. The number of different diagrams of this sort that can be drawn is exactly equal to the number of quarks that can be produced, and the total cross section for all possible $q\bar{q}$ productions is the sum over all such final states:

$$\sigma(e^+e^- \rightarrow \text{hadrons}) = \sum_i \sigma(e^+e^- \rightarrow q_i\bar{q}_i) \qquad (18.27)$$

Each of these diagrams is similar to the diagram for $e^+ + e^-$ annihilation into $\mu^+ + \mu^-$, which is also in turn identical with the diagram for $e^-\mu^-$ Rutherford scattering. We can therefore write, for the ratio of the cross sections at a given energy

$$\frac{\sigma(e^+e^- \rightarrow \text{hadrons})}{\sigma(e^+e^- \rightarrow \mu^+\mu^-)} \equiv R = \frac{\sum_i Q_{q_i}^2}{Q_\mu^2} \qquad (18.28)$$

because the Rutherford cross section is proportional to the square of the electric charge of the scatterer. We use a ratio in this form so that all kinematical factors in the respective cross sections will cancel. If there are three types of quarks (u, d, s) that can be produced, the ratio $R$ is

$$R = \frac{\left(\frac{2}{3}\right)^2 + \left(\frac{1}{3}\right)^2 + \left(\frac{1}{3}\right)^2}{1^2} = \frac{2}{3}$$

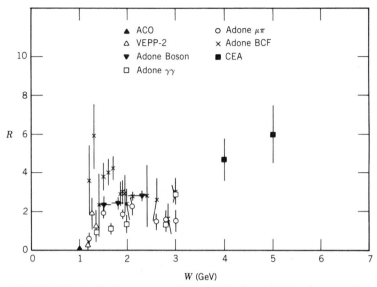

**Figure 18.15** Ratio $R$ of cross section for hadron and muon production from $e^+e^-$ annihilation; $W$ is the total center-of-mass energy of the $e^+e^-$ system. In the range 1–3 GeV, the data cluster about $R = 2$. From R. F. Schwitters and K. Strauch, *Ann. Rev. Nucl. Sci.* **26**, 89 (1976).

But if there are nine types $(u_R, u_B, u_G, d_R, d_B, d_G, s_R, s_B, s_G)$ the ratio is three times as great. Figure 18.15 shows the ratio $R$ in the range below 3 GeV, and the data are certainly in better agreement with $R = 2$ than with $R = \frac{2}{3}$.

Other direct evidence for the additional multiplicity of quarks brought about by the assignment of color comes from the decay $\pi^0 \rightarrow 2\gamma$, for which (based on counting procedures similar to those above), the decay rate calculated without color gives $\frac{1}{9}$ of the observed rate, while including color gives good agreement with the observed rate.

The force between quarks can be modeled as an exchange force, mediated by the exchange of massless spin-1 particles called *gluons*. The field that binds the quarks is a *color field*, and thus color plays a much more fundamental role than merely correcting the counting statistics of quark states. *Color is to the strong interaction between quarks as electric charge is to the electromagnetic interaction between electrons.* It is the fundamental strong "charge" and is carried by the gluons, which must therefore be represented as combinations of a color and a possibly different anticolor. This representation should not be confused with that

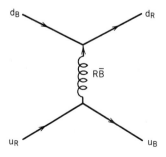

**Figure 18.16** A red up-quark $u_R$ emits a $R\bar{B}$ gluon, which is absorbed by the blue down-quark $d_B$. Note that the gluon does not change the type (flavor) of the quark.

of the mesons, which are quark-antiquark combinations. The gluons are massless and carry their color-anticolor properties just as other particles may carry electric charge. For example, Figure 18.16 shows a gluon $R\overline{B}$ being exchanged by red and blue quarks. In effect the red quark emits its redness into a gluon and acquires blueness by also emitting antiblueness. The blue quark, on the other hand, absorbs the RB gluon, canceling its blueness and acquiring a red color in the process. By simply enumerating the possible color-anticolor combinations, we expect nine possible gluons: $R\overline{R}$, $R\overline{B}$, $R\overline{G}$, $B\overline{R}$, $B\overline{B}$, $B\overline{G}$, $G\overline{R}$, $G\overline{B}$, $G\overline{G}$; the three net "colorless" combinations must be handled carefully for the symmetry properties of the color fields require that they be coupled in three ways:

$$\frac{1}{\sqrt{2}}(R\overline{R} - G\overline{G}) \qquad \frac{1}{\sqrt{6}}(R\overline{R} + G\overline{G} - 2B\overline{B}) \qquad \frac{1}{\sqrt{3}}(R\overline{R} + G\overline{G} + B\overline{B})$$

The first and second of these colorless combinations can transmit color, but the third cannot and it therefore must be excluded as an agent of the color field. This leaves 8 color gluons as the source of the quark-quark interaction.

Evidence for the existence of gluons comes primarily from two experiments. In the deep inelastic scattering of electrons from protons, discussed in the previous section, it is possible to deduce the fraction of the proton's internal momentum carried by the quarks. This fraction is only about 50%, and thus about half of the internal momentum is carried by nonquark systems; these must be the gluons. The second bit of evidence comes from $e^+e^-$ annihilation experiments of the kind discussed above, in which hadrons are produced in the final state. The mechanism $e^+ + e^- \rightarrow \gamma \rightarrow q + \bar{q}$ produces a highly energetic quark-antiquark pair traveling (instantaneously) in opposite directions. In a time of the order of the strong interaction time, the quark and antiquark are transformed into a "shower" of hadrons, but the momentum of the original quark-antiquark pair is preserved. Thus we observe two *jets* of hadrons emerging in opposite directions from the point of the reaction, as shown in Figure 18.17a. A few of the reactions, however, produce 3 (or more) jets, as in Figure 18.17b. Since it is not possible for $e^+ + e^-$ to produce any combination of quarks and antiquarks totalling 3, the accepted explanation is that one of the quarks radiates a gluon, which then forms its own jet. The observed angles between gluon and quark jets are consistent with the spin-1 assignment to the gluon.

## 18.5   REACTIONS AND DECAYS IN THE QUARK MODEL

All reactions and decays of elementary particles can be understood within the quark model, and reaction cross sections and decay lifetimes can be calculated and compared with experiment. In doing so we must follow several rules:

1. Strong interactions cannot change the flavors of quarks; only rearrangements of quarks among particles can occur.
2. Quark–antiquark pairs can be created and annihilated, in exact analogy to positron–electron creation and annihilation with photons.
3. Weak interactions can change the quark flavor through the emission or absorption of a $W^\pm$ weak boson:

$$u \rightarrow d + W^+ \qquad \bar{u} \rightarrow \bar{d} + W^-$$
$$s \rightarrow u + W^- \qquad \bar{s} \rightarrow \bar{u} + W^+$$

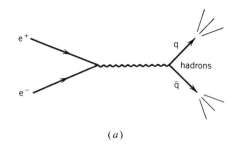

(*a*)

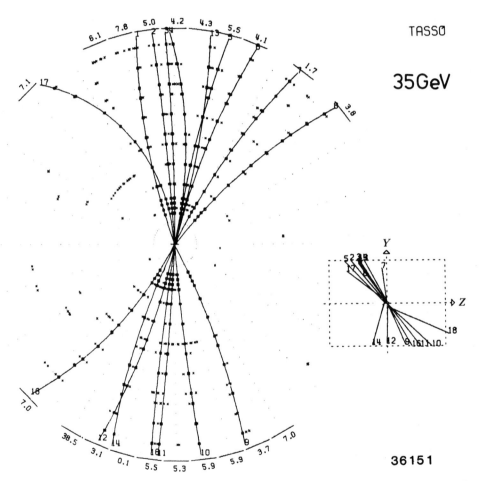

**Figure 18.17** (*a*) A two-jet event, in which q and q̄ each produce a shower of hadrons. (*b*) A relatively less-common three-jet event, in which one of the quarks radiates a gluon, which also creates a shower of hadrons. The visible tracks show the trajectories of the hadrons, bent into arcs by a magentic field. Experimental results are from DESY.

Neutral weak flavor-changing processes (s → d + Z⁰) are forbidden. The only permitted weak neutral processes are those such as Z⁰ → u + ū and variants.

Let's begin by examining some simple reactions for their quark structure. For example, Figure 18.18 shows the reaction

$$\pi^+ + p \rightarrow \pi^0 + \Delta^{++}$$

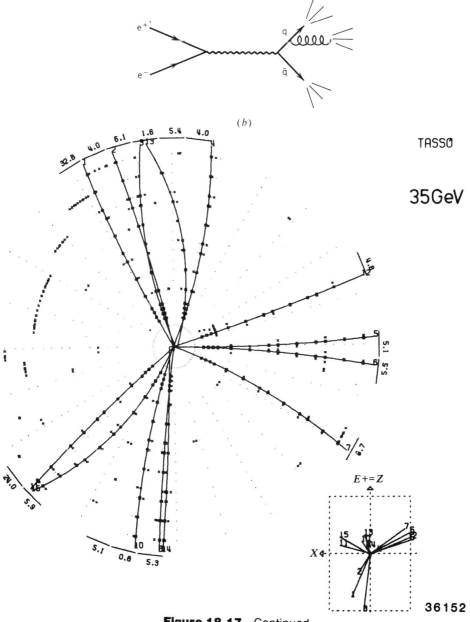

(b)

TASSO

35GeV

36152

**Figure 18.17** Continued.

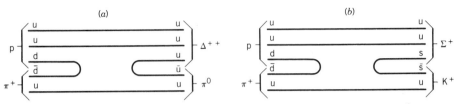

**Figure 18.18** Quark structure of the reactions (a) $p + \pi^+ \rightarrow \Delta^{++} + \pi^0$ and (b) $p + \pi^+ \rightarrow \Sigma^+ + K^+$.

The quark contents are

$$u\bar{d} + uud \rightarrow u\bar{u} + uuu$$

In essence, the process is just $d + \bar{d} \rightarrow u + \bar{u}$, that is, the annihilation of a $d\bar{d}$ pair and the creation of a $u\bar{u}$ pair. Replacing the created $u\bar{u}$ pair by $s\bar{s}$, we would form a different baryon ($uus = \Sigma^+$) and meson ($u\bar{s} = K^+$).

These simple diagrams do not show the gluons that are exchanged among the quarks and antiquarks that participate in the reaction, nor do they show the colors of the various quarks and gluons. A more complete diagram is indicated in Figure 18.19. Gluons are continually being exchanged among the component quarks, and it is gluon exchange that is responsible for the binding of particles and ultimately for the reaction mechanism (which we can imagine in its simplest form as $d + \bar{d} \rightarrow$ gluons $\rightarrow u + \bar{u}$).

The validity of certain empirically derived conservation laws is automatically satisfied by the quark model for reactions. For example, strangeness conservation in the reaction

$$K^- + p \rightarrow \Lambda^0 + \pi^0$$

means simply that we start and end with a single s quark, while associated production, as in Figure 18.18b, always originates with the production of an $s\bar{s}$ pair. The quark model thus provides a direct way to understand the observed phenomenon of associated production. Similarly, the quarks each individually

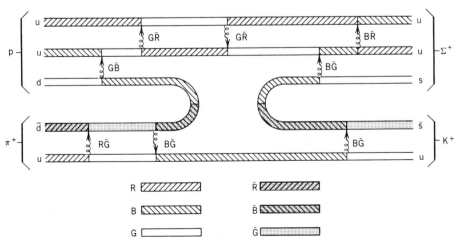

**Figure 18.19** Same as Figure 18.18b, but showing colors of quarks and exchanged gluons.

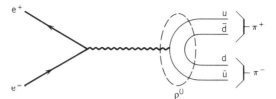

**Figure 18.20**   The reaction $e^+ + e^- \to \rho^0 \to \pi^+ + \pi^-$. The $\rho^0$ exists briefly as the coupling of $u\bar{u}$ and $d\bar{d}$.

satisfy the Gell-Mann–Nishijima relationship, Equation 18.22, and thus any combination of quarks will also satisfy the relationship.

Production of hadrons in $e^+e^-$ collisions involves the formation of one or more $q\bar{q}$ pairs from virtual photon intermediate states, as in Figure 18.20. Here the photon forms $u\bar{u}$ and $d\bar{d}$ pairs, which couple briefly to a $\rho^0$ meson $[\rho^0 = (u\bar{u} - d\bar{d})/\sqrt{2}\,]$ and then decay strongly to $\pi^+$ and $\pi^-$.

Decays of hadrons can occur either by strong, electromagnetic, or weak processes. Strong decays involve either simple rearrangements (as in the $\rho^0$ decay in Figure 18.20) or $q\bar{q}$ production from gluon exchange. Another example of a strong decay is $\Delta^{++} \to p + \pi^+$, which involves the production of a $d\bar{d}$ pair (Figure 18.21).

An observed feature of strong decays is that completely "disconnected" quark-antiquark decays ($q\bar{q} \to$ gluons $\to q'\bar{q}'$) are strongly suppressed relative to quark-antiquark production ($q\bar{q} \to q\bar{q}q'\bar{q}'$). Consider, for example, the decay $\phi \to 3\pi$. The quark content of $\phi$ is $s\bar{s}$. The formation of a $q\bar{q}$ pair must involve no net transfer of color in a diagram of the form of Figure 18.22a, and therefore, since gluons carry a net color, it is necessary that each process involve the exchange of at least two gluons. On the other hand, the continuation of the $s\bar{s}$ pair into the final state, as in Figure 18.22b, allows individual colored gluons to be involved. The suppression of the disconnected decay modes is called *Zweig's rule*.

Based simply on phase-space consideration, we might expect $\phi \to 3\pi$ to be highly favored over $\phi \to K^+K^-$. The density-of-states factor, upon which the decay rate depends, includes momentum-dependent factors that would lead us to expect higher decay rates for decays in which the product particles can take larger momenta. The $Q$ value for the decay $\phi \to 2K$ is only 33 MeV, while for $\phi \to 3\pi$ it is 605 MeV; thus there is 20 times as much kinetic energy available for the $3\pi$ final states as for the $2K$ final states, and the decay $\phi \to 3\pi$ should be highly favored. In fact the observed decays of the $\phi$ meson go about 85% into $2K$ and only 15% into $3\pi$, showing that the favored decay mode is strongly suppressed by the operation of Zweig's rule. This will have important conse-

**Figure 18.21**   Quark structure in the decay $\Delta^{++} \to p + \pi^+$.

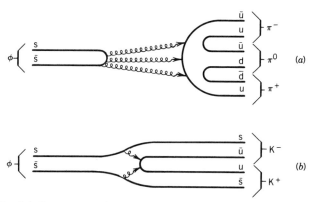

**Figure 18.22** (a) Decay $\phi \to 3\pi$; such "disconnected" quark diagrams correspond to decay processes strongly inhibited by Zweig's rule. (b) Decay $\phi \to 2K$, permitted by Zweig's rule.

quences for studies of the properties of the $\psi$ mesons discussed in the next section.

Electromagnetic decays into hadrons can occur, in the quark model, through annihilation reactions of the form $q\bar{q} \to \gamma$. For example, the $\pi^0$, which in the quark model is represented as $(d\bar{d} - u\bar{u})/\sqrt{2}$, decays to two $\gamma$'s, which we may picture as in Figure 18.23. It is also possible for a quark merely to radiate a $\gamma$, thereby keeping its identity (and its color, since photons carry no strong charge) but possibly changing the internal state of motion of the quarks in the hadron. The decay $\Sigma^0 \to \Lambda^0 + \gamma$, shown in Figure 18.24, is such an example.

The leptonic decays of the mesons can occur electromagnetically, as in $q\bar{q} \to \gamma \to e^+ e^-$ or $\mu^+ \mu^-$. Figure 18.25 illustrates this process, which, like the reverse $e^+ e^- \to q\bar{q}$ reaction shown in Figure 18.14, can be redrawn as an equivalent lepton–quark Rutherford scattering. The decay probability, like the cross section given in Equation 18.28, depends on the total quark charges. In the decay calculation, the quark charges must be weighted by their coefficients in the meson wave functions. Thus, for $\rho^0 = (u\bar{u} - d\bar{d})/\sqrt{2}$

$$\Gamma\left(\rho^0 \to e^+ e^-, \mu^+ \mu^-\right) \propto \left(\sum_i Q_{q_i}\right)^2 = \left[\frac{1}{\sqrt{2}}\left(\frac{2}{3}\right) - \frac{1}{\sqrt{2}}\left(-\frac{1}{3}\right)\right]^2 = \frac{1}{2}$$

while for $\omega = (u\bar{u} + d\bar{d})/\sqrt{2}$

$$\Gamma\left(\omega \to e^+ e^-, \mu^+ \mu^-\right) \propto \left[\frac{1}{\sqrt{2}}\left(\frac{2}{3}\right) + \frac{1}{\sqrt{2}}\left(-\frac{1}{3}\right)\right]^2 = \frac{1}{18}$$

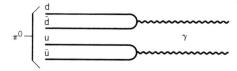

**Figure 18.23** Decay of $\pi^0$ into two photons.

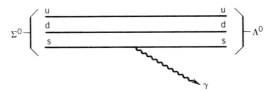

**Figure 18.24** Decay of $\Sigma^0$ to $\Lambda^0 + \gamma$.

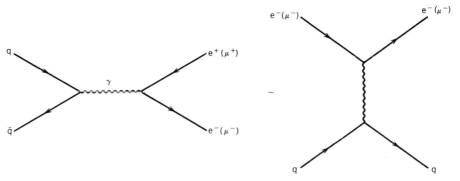

**Figure 18.25** Quark-antiquark annihilation into leptons, and the equivalent Rutherford scattering process.

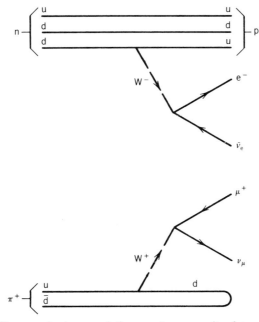

**Figure 18.26** The weak decay of the neutron results from the quark decay $d \rightarrow u + W^-$, followed by $W^- \rightarrow e^- + \bar{\nu}_e$. The $\pi^+$ decay occurs through $u \rightarrow d + W^+$, followed by $W^+ \rightarrow \mu^+ + \nu_\mu$.

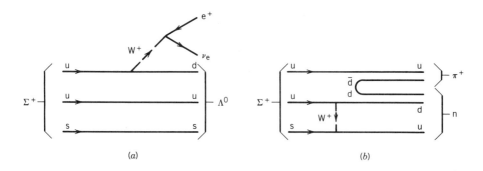

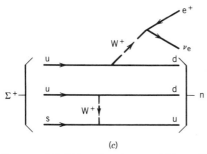

**Figure 18.27** The decays $\Sigma^+ \to \Lambda^0 + e^+ + \nu_e$ (a) and $\Sigma^+ \to n + \pi^+$ (b) each involve only two separate weak interaction vertices, while (c) involves four. The decay $\Sigma^+ \to n + e^+ + \nu_e$ (c) is therefore far less likely to occur than (a) or (b).

and for $\phi = s\bar{s}$,

$$\Gamma(\phi \to e^+e^-, \mu^+\mu^-) \propto \left(\frac{1}{3}\right)^2 = \frac{1}{9}$$

The observed partial widths are related to the total widths by the measured branching fractions and are, approximately, $\Gamma(\rho) = 17$ keV, $\Gamma(\omega) = 1.7$ keV, $\Gamma(\phi) = 2.4$ keV (uncertainties are about $\pm 10\%$). Ignoring the other factors that contribute to the widths, these are in the ratio $10:1:1.5$, in rough agreement with the value $9:1:2$ calculated from the quark charges.

Weak decays of quarks can change quark flavors, and weak decays of hadrons are included in the quark model through processes such as $q \to q' + W^{\pm}$, with the weak boson then decaying into an electron or a muon and its accompanying neutrino or antineutrino. The familiar decay of the neutron is indicated in Figure 18.26. Similar processes will occur in all hadronic decays involving the emission of leptons (called semileptonic decays because only the final state has leptons). Semileptonic decays of mesons can be similarly diagrammed, as shown in Figure 18.26 for $\pi^+$. We can also understand from the quark content why certain decays are suppressed or not observed; the decay $\Sigma^+ \to n + e^+ + \nu_e$ involves the quark decays uus $\to$ udd, and therefore represents a higher-order process involving several weak decays, which must be far less likely to occur than the lower-order weak processes $\Sigma^+ \to \Lambda^0 + e^+ + \nu_e$ or $\Sigma^+ \to n + \pi^+$, as in Figure 18.27.

## 18.6 CHARM, BEAUTY, AND TRUTH

Despite the success of the quark model, all is not quite well with the simple theory described so far. The nonexistence or extreme suppression of neutral (that is, the electric charge of the hadron does not change), strangeness-changing decays is a particular problem, for example, $K^+ \to \pi^+ + \mu^+ + \mu^-$ or $K^0 \to \mu^+ + \mu^-$, as illustrated in Figure 18.28. Upper limits on such processes are $10^{-7}$ or less, relative to other permitted weak processes, even though they appear to be allowed by all known rules for weak decays.

Each of the above processes involves an unobserved intermediate state with a u or $\bar{u}$ quark, as must all neutral, strangeness-changing decays, for s $\leftrightarrow$ d is not permitted through a direct weak process, but only through the indirect process s $\leftrightarrow$ u $\leftrightarrow$ d, involving two W bosons. In 1970, Glashow, Iliopoulis, and Maiani suggested a mechanism whereby these processes could be suppressed. They proposed the existence of a fourth quark, called c or *charmed* quark. It is assigned a new quantum number, charm ($C$), equal to $+1$; all other quarks (u, s, d) are assigned $C = 0$. The c quark is nonstrange ($S = 0$), but in many ways the $C$ quantum number plays a role very similar to strangeness; for example, weak decays can change $C$ by one unit. The c quark has, like the u quark, a charge of $+\frac{2}{3}e$, and can therefore couple s to d in a similar way through two W $^{\pm}$: s $\leftrightarrow$ c $\leftrightarrow$ d. Glashow and colleagues proposed that the two processes s $\leftrightarrow$ u $\leftrightarrow$ d and s $\leftrightarrow$ c $\leftrightarrow$ d exactly canceled one another, leading to the suppression of the decay processes shown in Figure 18.28. Four years after the charm proposal, a very narrow resonance was discovered almost simultaneously at SLAC in $e^+e^-$ collisions (Figure 18.29) and at Brookhaven in proton–proton collisions. The resonance in the cross section of Figure 18.29 shows a width that is characteristic of the instrumental resolution; the actual width of the resonance (called $\psi$ by the SLAC group and J by the Brookhaven group and sometimes known as J/$\psi$, although $\psi$ seems to be gaining favor) is about $1/100$ of the instrumental width shown in the figure.

The width can be deduced with reference to Figure 18.30—even though the natural width is not seen, the area of the resonance is unaffected by the

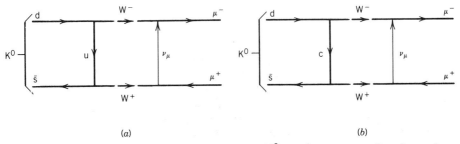

(a)                                                              (b)

**Figure 18.28** Process (*a*) shows the decay $K^0 \to \mu^+\mu^-$ proceeding through an intermediate state containing a virtual u quark. Because the decay is not observed, a competing process, shown in (*b*), must be possible to cancel contributions to the decay rate from process (*a*). Process (*b*) contains a virtual intermediate c quark.

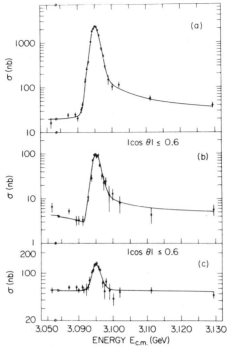

**Figure 18.29** Cross section for $e^+e^-$ showing narrow $\psi$ resonance. Part (*a*) shows the total cross section for production of hadrons in the final state, part (*b*) shows $\mu^+\mu^-$ final states, and part (*c*) shows $e^+e^-$ final states. From A. M. Boyarski et al., *Phys. Rev. Lett.* **34**, 1357 (1975).

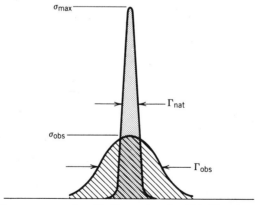

**Figure 18.30** A narrow peak with the natural width $\Gamma_{\text{nat}}$ (shaded) is broadened by instrumental effects into a wider, shorter peak (cross-hatched) of the same area, even though the observed width and cross section are very different from those of the unbroadened peak.

broadening, and the cross sections of Figure 18.29 can be integrated numerically and compared with the integral of the Breit-Wigner resonance for the decay of the resonance back into $e^+e^-$:

$$\int \sigma(E)\,dE = \int \frac{\pi}{k^2} g \frac{\Gamma_{ee}^2}{(E - E_R)^2 + \Gamma^2/4}\,dE$$

$$= \frac{\pi^2}{2k^2}(2s+1)\frac{\Gamma_{ee}^2}{\Gamma} = \frac{\pi^2}{2k^2}(2s+1)\left(\frac{\Gamma_{ee}}{\Gamma}\right)^2 \Gamma \qquad (18.29)$$

where $s$ is the spin of the resonance and where we have assumed $k^2$ varies little over the area of the resonance. Comparison of the $e^+e^- \rightarrow$ hadrons and $e^+e^- \rightarrow$ $e^+e^-$ cross sections (Figures 18.29$a$ and $c$) gives $\Gamma_{ee}/\Gamma \simeq 150/2000 = 0.075$, and making a rough numerical estimate of the area of the resonance in Figure 18.29$c$ (approximating the peak as a triangle) gives about 300 nb · MeV. Subsequent measurements show $s = 1$, and from Equation 18.29 we would deduce $\Gamma \sim 0.03$ MeV. A more careful calculation gives a total width $\Gamma = 0.063$ MeV. This is an extraordinarily narrow width for such a heavy meson; surely a multitude of hadronic decay channels ($\pi^+\pi^-$, $K^+K^-$; $3\pi$, even p$\bar{\text{p}}$) must be available, and we would expect rapid decays by the strong interaction with the customary 100-MeV width. Yet the decay of $\psi$ is slowed by about 3 orders of magnitude.

It was soon realized that the $\psi$ represents the state c$\bar{\text{c}}$ (now called "charmonium" in analogy with $e^+e^-$, called "positronium"). The state is the lowest c$\bar{\text{c}}$ bound state, and other resonances were soon discovered at energies of 3686 MeV ($\Gamma = 0.215$ MeV), 3770 MeV ($\Gamma = 25$ MeV), 4030 MeV ($\Gamma = 52$ MeV), 4160 MeV ($\Gamma = 78$ MeV), and 4415 MeV ($\Gamma = 43$ MeV). The agent responsible for retarding the decay of the lowest two $\psi$ states is Zweig's rule, discussed in Section 18.5. Figure 18.31 shows possible decays of charmonium into $3\pi$ or 2K. Each of these is a "disconnected" diagram and thus suppressed by Zweig's rule. A possible decay mode allowed by Zweig's rule is also shown. The combinations c$\bar{\text{d}}$ and d$\bar{\text{c}}$ are associated with the charmed mesons called D$^+$ and D$^-$. These mesons have rest energies of 1869.4 MeV, and thus the two lowest $\psi$ resonances (3097 and 3686 MeV) are energetically forbidden from the D$^+$D$^-$ decay mode. The lowest resonance has no choice but to decay through a Zweig-suppressed mode, hence its exceedingly narrow width. The first "excited state" of $\psi$ at 3686 MeV is also rather narrow; it can decay electromagnetically to the ground state, as for example $\psi' \rightarrow \psi + 2\pi$ without violating Zweig's rule. The heavier states, at energies above twice the D$^\pm$ energy (3739 MeV), can decay into hadrons through the strong interaction.

The D$^\pm$ mesons and their neutral counterparts D$^0$ and $\overline{\text{D}}^0$ form a set of charmed mesons ($C = \pm 1$) exactly analogous to the set of strange K mesons. The decay of the D$^\pm$ goes by the weak interaction ($\tau = 10^{-12}$ s) into many possible final states, including $K^-\pi^0$, $K^0\pi^-$, etc. The neutral members D$^0$ and $\overline{\text{D}}^0$ decay similarly.

The D mesons represent the couplings of the c quark with u and d; there is also a meson having both charm and strangeness, representing the coupling of c

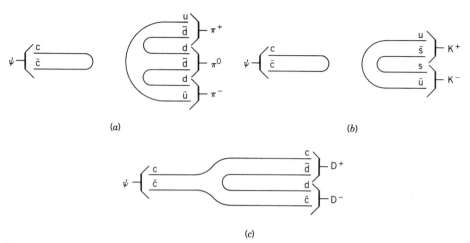

(a)                                                                 (b)

(c)

**Figure 18.31** (a, b) Possible decay modes of $c\bar{c}$ inhibited by Zweig's rule. (c) Decay mode allowed by Zweig's rule but forbidden by energy conservation for the lowest two $\psi$ states.

and s; these are called $F^{\pm}$ (1971 MeV). They also decay weakly ($2 \times 10^{-13}$ s). The strangeness vs isospin diagram of the spin-0 mesons shown in Figure 18.13 must now be modified into three dimensions, with charm along the third axis, as in Figure 18.32. There must also be charmed baryons composed of three quarks including one or more c quarks; discoveries thus far include $\Lambda_c^+$ (udc, analogous to the ordinary $\Lambda^0 =$ uds) at 2282 MeV, decaying weakly primarily to $\Lambda^0$ in a

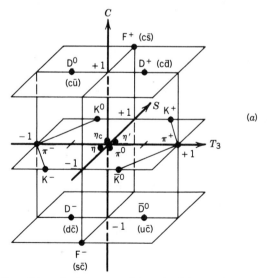

(a)

**Figure 18.32** Possible spin-0 mesons (a), spin-$\frac{1}{2}$ baryons (b), and spin-$\frac{3}{2}$ baryons (c) allowing for all combinations of four quarks u, d, s, c. In each diagram the $C = 0$ plane is identical with Figure 18.11 except for the spin-0 mesons in which the $\eta_c (= c\bar{c})$ must be included. (A similar diagram for the spin-1 mesons would include $\psi (= c\bar{c})$ in the $C = 0$ plane.)

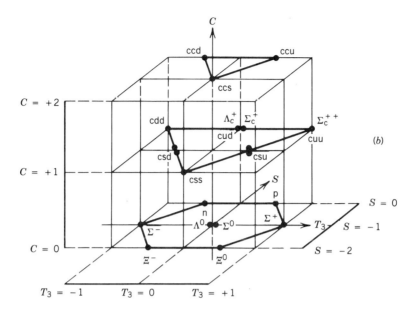

(b)

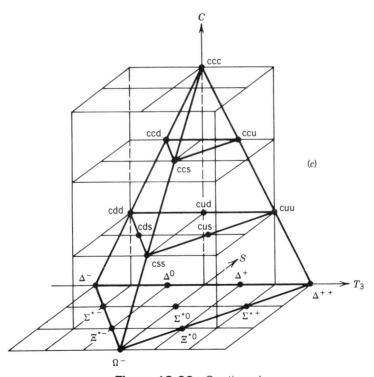

(c)

**Figure 18.32** Continued.

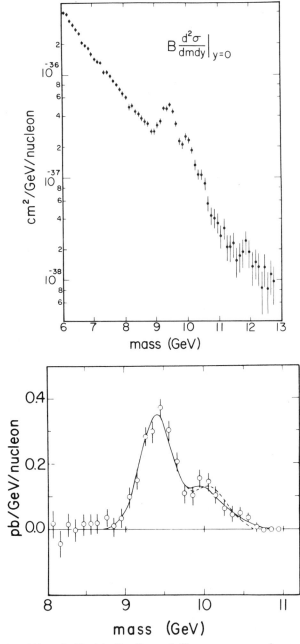

$$B\frac{d^2\sigma}{dmdy}\Big|_{y=0}$$

**Figure 18.33** (Above) First indication of $\Upsilon$ resonances, from proton-nucleus collisions leading to $\mu^+\mu^-$ final states. At bottom is shown the two resonant peaks with background subtracted. The width is large, characteristic of the instrumental resolution. From W. R. Innes et al., *Phys. Rev. Lett.* **39**, 1240 (1977). (Facing page) High-resolution study of the three lowest $\Upsilon$ resonances from the Cornell $e^+e^-$ data. The width of 10 MeV, even though quite narrow, is still far larger than the natural width of 44 keV. From D. Andrews et al., *Phys. Rev. Lett.* **44**, 1108 (1980).

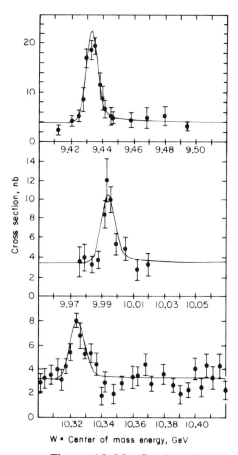

Cross section, nb

W ▪ Center of moss energy, GeV

**Figure 18.33**  Continued.

$\Delta S = -1$, $\Delta C = +1$ mode. There is also evidence for $\Sigma_c$ (uuc) at 2450 MeV, decaying to $\Lambda_c^+ + \pi$.

In 1977, another series of narrow resonances was discovered in the energy range of about 10 GeV in pp experiments at Fermilab. The discovery was quickly confirmed in $e^+e^-$ studies at DESY, and the lowest resonance, now called $\Upsilon$ (upsilon), is shown in Figure 18.33. From here the story proceeds in analogy with that of the $\psi$ resonances. The width of the lowest resonance is exceedingly small (44 keV), and the widths remain small through the excited spin-1 states, in exact analogy with the decays of $\psi$, until the state at 10575 MeV, with $\Gamma = 14$ MeV. The interpretation of these narrow widths is exactly the same as in the case of $\psi$: the $\Upsilon$ is a bound state of yet another new quark called b (for bottom or beauty) and its antiquark $\bar{b}$. The decays are Zweig-suppressed until the energy threshold for production of $B^\pm$ mesons ($b\bar{u}$ and $\bar{b}u$) with a rest energy of 5271 MeV. There are also $B^0$ and $\bar{B}^0$ mesons ($b\bar{d}$ and $\bar{b}d$); thus $B^\pm$, $B^0$, $\bar{B}^0$ form a set similar to D and K. As yet no mesons with quark combinations $b\bar{s}$ or $b\bar{c}$ have been discovered, but there is evidence for a baryon $\Lambda_b^0$ (udb) at 5425 MeV.

The excited spectrum of $b\bar{b}$ states bears a remarkable similarity to the $c\bar{c}$ states, as shown in Figure 18.34 for the spin-1 states (corresponding to parallel spin

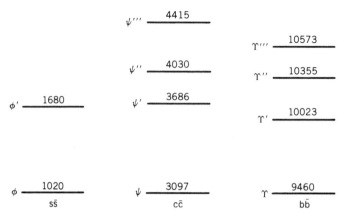

**Figure 18.34** Spin-1 s states of "quarkonium" (q$\bar{q}$ combinations), labeled with their energies in MeV. The excitation energies are not strongly dependent on the quark masses.

alignments of the two quarks, which are in an orbital state $\ell = 0$). There are of course other spin states in both schemes.

Comparison of the measured $\Gamma_{ee}$ width of $\Upsilon$ with that of the other q$\bar{q}$ combinations, as shown in Table 18.4, strongly suggests a charge of $-\frac{1}{3}e$ for the b quark. We expect the width $\Gamma_{ee}$ to be proportional to the squared sum of the quark charges in the meson's wave function, and Table 18.4 shows that $\Gamma_{ee}/(\Sigma Q_q)^2$ is indeed constant as expected.

Another indication of the quark charge comes from the ratio $R$ defined in Equation 18.28. With three flavors (u, d, s) in three colors, the expected ratio is 2, as was shown in Figure 18.15. Above the c$\bar{c}$ threshold, the sum should be carried out over four flavors (u, d, s, c) and three colors, and the expected value of $R$ above 4 GeV should be 10/3. Figure 18.35 shows the ratio $R$ in the 5–7-GeV range, and indeed the data are in excellent agreement with the expected value. Beyond the threshold for b$\bar{b}$ production, it is necessary to include five flavors, and assigning $Q = -\frac{1}{3}e$ to the b quark, we expect the ratio $R$ to rise to 11/3. Figure 18.35 shows that $R$ does increase by about 0.33 above the $\Upsilon'''$ resonance, as expected. Taken together, the hadronic cross sections and e$^+$e$^-$ branching ratios provide convincing evidence for the basic assumptions of the quark model.

Because quarks seem to come in pairs (and because the mathematical theory of the quark dynamics strongly supports such pairings), it is expected that there is a partner to the b quark, called t (for truth, if b = beauty, or top, if b = bottom), with a charge of $+\frac{2}{3}e$. Searches for evidence of the top quark through an increase of the hadronic cross section in the ratio $R$ have not yet been successful

**Table 18.4** e$^+$e$^-$ Widths of q$\bar{q}$ Mesons

| Particle | $\rho$ | $\omega$ | $\phi$ | $\psi$ | $\Upsilon$ |
|---|---|---|---|---|---|
| $\Gamma_{ee}$ (keV) | $6.7 \pm 0.8$ | $0.76 \pm 0.17$ | $1.31 \pm 0.10$ | $4.8 \pm 0.6$ | $1.30 \pm 0.05$ |
| $(\Sigma Q_q)^2$ | $\frac{1}{2}$ | $\frac{1}{18}$ | $\frac{1}{9}$ | $\frac{4}{9}$ | $\frac{1}{9}$ |
| $\Gamma_{ee}/(\Sigma Q_q)^2$ | $13.4 \pm 1.6$ | $13.7 \pm 3.1$ | $11.8 \pm 0.9$ | $10.8 \pm 1.4$ | $11.7 \pm 0.5$ |

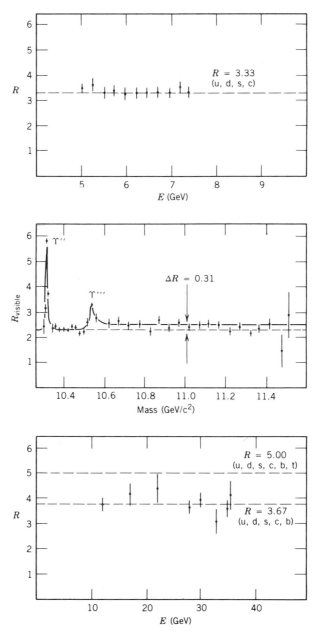

**Figure 18.35**  Ratio $R$ of hadron production to muon production in $e^+e^-$ annihilation. (Top) In the region below the $\Upsilon$ resonances, the measured values are consistent with the presence of four quarks. Data are from SLAC and DESY. (Middle) Above the $b\bar{b}$ threshold ($\Upsilon'''$ resonance), the value of $R$ increases by $0.31 \pm 0.06$, as expected; the data shown have not yet been normalized, but the increase in $R$ is apparent. From measurements at CESR, reported by E. Rice et al., *Phys. Rev. Lett.* **48**, 906 (1982). (Bottom) No evidence is seen for an increase in $R$ that would correspond to crossing the $t\bar{t}$ threshold up to 35.8 GeV, suggesting that the mass of the top quark is at least 18 GeV. Data are from DESY.

(Figure 18.35), and it can be concluded that the mass of the top quark must be at least 18 GeV. In a recent (October 1984) experiment at CERN, two-jet events (as in Figure 18.17) were observed in which the total particle energy carried by the two jets corresponded to the mass of the W boson. The CERN group proposed that they were observing the decay $W^+ \rightarrow t\bar{b}$, and their data would indicate that the mass of the t quark lies in the range 30–50-GeV. Research to confirm this value and its interpretation is presently underway.

## 18.7 QUARK DYNAMICS

No high-energy particle collision has as yet (1987) produced a free quark. With the advent of Fermilab's 1000-GeV collider, this failure to liberate a single free quark leaves us with the philosophically somewhat difficult possibility that the quarks are permanently confined in hadrons and that no amount of energy can liberate a quark from its hadronic environment.

On the other hand, the results of deep inelastic scattering experiments reveal a paradoxically very different property of the quarks—if we examine the quarks at very short distances (through scattering of a nonhadronic probe such as an electron or a neutrino), we find the quarks to move almost freely, as if they are not bound at all.

In the language of quarkologists, these two properties are called *infrared slavery* (confinement of quarks to regions the size of hadrons with large, perhaps infinite energies required to liberate them to larger distances) and *asymptotic freedom* (free movement at short distances). Any successful theory of quark interactions must be able to explain these apparently contradictory properties.

In analogy with quantum electrodynamics (QED), the quantum theory of the electromagnetic field, we have *quantum chromodynamics* (QCD), the quantum theory of the color field. We will not go into the mathematical details of this very abstract theory, but will point out its distinguishing features by comparison with QED.

In QED, electric charges interact through electromagnetic fields propagated through the exchange of real or virtual photons. In QCD, quarks interact by exchanging gluons. The photons are the carriers of the electromagnetic field exactly as the gluons are the carriers of the strong color field. What makes the two theories so different is that the photons themselves carry no electric charge and so are unaffected by electric fields; gluons in contrast carry a net color, and therefore interact directly with the quarks. That is, a quark can emit a gluon and then interact with it and create additional gluons; a photon cannot itself exchange photons with nearby charges. This property of gluons forces QCD into a considerable level of mathematical complexity.

The emission of colored gluons provides a clue for the operation of asymptotic freedom. An electron emitting virtual photons still remains an electron with a charge of $-e$, but a quark emitting a virtual gluon must change its color charge. The color charge of a quark is therefore spread out over a sphere of radius of the order of the size of a hadron (0.5–1 fm). If another quark were to penetrate that sphere, this "smeared-out" color field would cause a considerably reduced quark–quark interaction. If we sample the quark's interactions over a radius

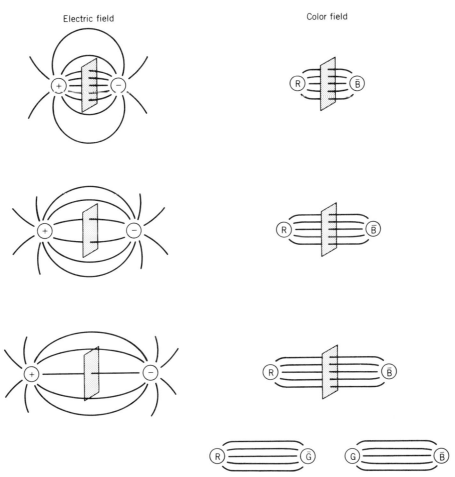

**Figure 18.36** As the distance between two point charges increases, the electric field (corresponding to the density of electric field lines crossing a unit surface area) decreases. The color field remains constant as the distance increases. Eventually, attempts to separate the quarks to large enough distances result in the production of a new $q\bar{q}$ pair.

small compared with 1 fm, we observe only a small fraction of its color charge, and it appears only very weakly bound or nearly free.

The behavior of quark interactions as the separation is increased can also be justified through comparison with QED. Figure 18.36 illustrates one difference between the electric field of two charges and the color field of two quarks. The density of electric field lines through any surface is proportional to the electric field at that location; thus as the charges are separated, the density of the field lines and the electric field between the charges decrease. Regarding the electric field lines as representative of virtual photon exchange between the charges, it is immediately apparent why the QCD field lines behave differently—the exchanged photons do not interact, while the exchanged gluons do. Thus the gluon–gluon interaction forces the color field lines into a narrow tube. The force (again represented as the density of field lines crossing a surface) remains roughly constant as the separation is increased. As we try to separate to large distances, the work will eventually exceed the production threshold for creation of a $q\bar{q}$ pair, resulting in the formation of a meson. Thus putting energy into a nucleus in an attempt to liberate a quark is expected to create new mesons, exactly as is observed.

This simple model does permit some rough calculations of the excitation spectrum of quark-antiquark pairs, such as is found in the $\psi$ or $\Upsilon$ particles. If the force is roughly constant, then the potential varies linearly with the separation distance as $V(r) = kr - V_0$. We can write the radial part of the Schrödinger equation for this potential (assuming s states) as

$$-\frac{\hbar^2}{2m}\frac{d^2u}{dr^2} + [kr - V_0]u(r) = Eu(r) \qquad (18.30)$$

The solution to the differential equation is the Airy function, and the energies are found from mathematical tables of the zeros of the Airy function, the values of which are $a_n$ ($a_1 = -2.3381$, $a_2 = -4.0879$, $a_3 = -5.5206$, $a_4 = -6.7867$, $a_5 = -7.9441, \ldots$):

$$E_n = |a_n|\left(\frac{k^2\hbar^2}{2m}\right)^{1/3} - V_0 \qquad (18.31)$$

and the expected spectrum of s states is shown in Figure 18.37 and compared with the corresponding levels of $\psi$ and $\Upsilon$. The agreement is sufficiently good to lead us to accept the plausibility of the theory.

The masses of the quarks themselves do not appear in any QCD calculation; in fact, if the quarks are permanently confined it makes little sense to discuss the rest mass of a free quark. Instead, we can find the effective mass of a quark when it is found in a hadron; this is usually known as the *constituent* quark mass. The simplest cases are $\Upsilon = b\bar{b}$, $\psi = c\bar{c}$, and $\phi = s\bar{s}$, from which we can estimate

$$m_b c^2 \simeq \tfrac{1}{2}m_\Upsilon c^2 \simeq 4.7 \text{ GeV}$$

$$m_c c^2 \simeq \tfrac{1}{2}m_\psi c^2 \simeq 1.5 \text{ GeV}$$

$$m_s c^2 \simeq \tfrac{1}{2}m_\phi c^2 \simeq 500 \text{ MeV}$$

For the u and d quarks, no pure $u\bar{u}$ or $d\bar{d}$ states exist, although by analogy with the above values we could take $\tfrac{1}{2}m_\rho c^2$ or $\tfrac{1}{2}m_\omega c^2$, since both $\rho^0$ and $\omega$ are combinations of $u\bar{u}$ and $d\bar{d}$. This would give about 380 MeV for the u and d quarks, assuming their masses to be equal. The proton and neutron differ in mass energy by only 1 MeV and differ in quark content by the replacement of d with u; there is thus good evidence that $m_u = m_d$. In fact, an alternative estimate for the u and d quarks would be $\tfrac{1}{3}m_n c^2 \approx 310$ MeV. It therefore seems reasonable to estimate

$$m_u c^2 \approx m_d c^2 \approx 350 \text{ MeV}.$$

As a check on this estimate, we expect the $\Lambda$ ($= uds$) to exceed the nucleon mass by about $m_s c^2 - m_u c^2 \approx 150$ MeV. The observed difference is 170 MeV, in better agreement with this crude estimate than we should expect. The lowest states of the charmed mesons are also quite consistent with these estimates, with D ($= c\bar{u}$)

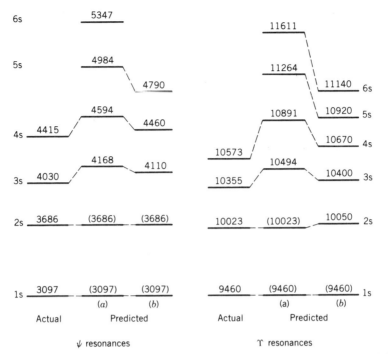

**Figure 18.37** Calculated energies of q$\bar{\text{q}}$ states in simple potential models. The predicted energies are calculated with (a) $V(r) = kr - V_0$ and (b) $V(r) = kr - b/r$. The first potential is somewhat simpler to solve, but the second does a better job of giving the measured $\psi$ and $\Upsilon$ energies. Values in parentheses are inputs for the models and are used to evaluate the parameters ($k$, $V_0$, $b$). Details of calculation (b) can be found in E. Eichten et al., *Phys. Rev. D* **21**, 203 (1980).

at 1870 MeV (estimated value 1850 MeV) and F ($= $c$\bar{\text{s}}$) at 1971 MeV (estimated value 2000 MeV).

Confirming evidence for these effective masses comes from the magnetic moments of the baryons. If we regard the quarks as spin-$\frac{1}{2}$ Dirac particles, then the quark magnetic moments are

$$\mu_q = \frac{Q_q \hbar}{2 m_q} \tag{18.32}$$

where we assume the Dirac $g$ factor $g_s = 2$. The vector coupling of the various quark magnetic moments to the baryon moments gives

$$\mu_p = \tfrac{4}{3}\mu_u - \tfrac{1}{3}\mu_d \tag{18.33a}$$

$$\mu_n = \tfrac{4}{3}\mu_d - \tfrac{1}{3}\mu_u \tag{18.33b}$$

$$\mu_\Lambda = \mu_s \tag{18.33c}$$

Solving these equations simultaneously and using the observed magnetic moments ($\mu_p = 2.79\ \mu_N$, $\mu_n = -1.91\ \mu_N$, $\mu_\Lambda = -0.61\ \mu_N$) gives $\mu_s = -0.61\ \mu_N$, $\mu_d = -0.97\ \mu_N$, $\mu_u = 1.85\ \mu_N$. Because $\mu_N$ is calculated using the proton mass, we get, for example

$$\mu_u = 1.85 \frac{e\hbar}{2m_p} = \frac{\frac{2}{3}e\hbar}{2m_u}$$

$$m_u = 0.36 m_p$$

That is, $m_u c^2 = 340$ MeV, consistent with the previous estimate. Similarly, from the magnetic moments $m_d c^2 = 340$ MeV and $m_s c^2 = 510$ MeV, also consistent with previous estimates. Accepting these values of the three quark magnetic moments, we can then calculate the magnetic moments of the heavier spin-$\frac{1}{2}$ baryons ($\Sigma$ and $\Xi$), and excellent agreement is obtained between the measured and calculated values. If we assume $m_u = m_d$, then Equation 18.32 gives $\mu_u = -2\mu_d$ and Equations 18.33a and b combine to give $\mu_p/\mu_n = -1.50$. The measured ratio is $-1.46$. Thus the quark model has excellent success with the baryon magnetic moments, without resorting to meson clouds or any other artificial structures in the baryons.

Much effort has been put into searches for free quarks, from the debris of extremely high-energy collisions of cosmic rays (which might be energetic enough to free a tightly bound quark) to ocean sediments and other terrestrial matter, in order to search either for cosmic-ray quarks that have accumulated in stable "quark atoms" or else to look for quarks that may have formed in the early universe just following the Big Bang. The most interesting result to date was reported based on experiments done by William Fairbank, in what amounts to a repetition of the classic Millikan oil-drop experiment, but using superconducting niobium spheres magnetically suspended; for a description of the experiment, see G. S. LaRue et al., *Phys. Rev. Lett.* **46**, 967 (1981). Several of the spheres showed charges consistent with $\frac{1}{3}e$, and while it is tempting to associate this result with the presence of a free quark on the sphere, the deduced density of free quarks in terrestrial matter would be far greater than the upper limit deduced in other studies. The reason for this discrepancy is not yet clear, but the existence of free quarks is not at all critical to the success of the model; the discovery of a free quark would require some changes in our ideas about confinement, but it would not be catastrophic to the overall theory.

Of course, the requirement that all observed particles carry no net color forbids the existence of isolated quarks and gluons, but this requirement is not a fundamental postulate of the theory but merely a summary of our present experimental findings. It is interesting to note that while the existence of a free gluon is forbidden, it is possible to construct aggregates of two or more gluons that carry no net color and that therefore might exist. Evidence for such particles, called "glueballs," may have been seen in high-energy collisions.

## 18.8 GRAND UNIFIED THEORIES

The description of elementary particles and their interactions, based on the theories we have outlined so far in this chapter, has been quite successful in analyzing the results of a great variety of experiments. Yet several questions arise which are not addressed by the basic theory: Both leptons and quarks appear to

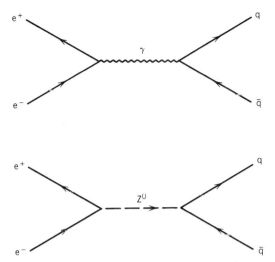

**Figure 18.38** At very high energies, the products of $e^+e^-$ annihilation through photons are similar to those of annihilation through vector bosons.

come in several (probably three) generations of doublets, with all particles having spin $\frac{1}{2}$:

$$\begin{pmatrix} u \\ d \end{pmatrix} \qquad \begin{pmatrix} c \\ s \end{pmatrix} \qquad \begin{pmatrix} t \\ b \end{pmatrix}$$

$$\begin{pmatrix} e \\ \nu_e \end{pmatrix} \qquad \begin{pmatrix} \mu \\ \nu_\mu \end{pmatrix} \qquad \begin{pmatrix} \tau \\ \nu_\tau \end{pmatrix}$$

Is this merely a coincidence, or is it an expression of a more fundamental theory? The weak and electromagnetic interactions have been unified by the electroweak theory. Is it possible to unify all three particle interactions (weak, electromagnetic, and strong) into a single theoretical framework?

As we move to high energy, particle interactions become simpler rather than more complex. For example, we consider the result of electron–positron collisions, indicated schematically in Figure 18.38. At energies well below 100 GeV, $Z^0$ production is not possible and the interaction is purely electromagnetic. At higher energies, the interactions can be mediated by $Z^0$ exchange. If we were to collide electrons at exceedingly high energies, say $10^6$ GeV, exchange of a $10^6$ GeV photon might appear very similar to exchange of a $10^6$ GeV $Z^0$, and there are relatively few experiments that could distinguish the two. The ultimate issue of Grand Unification is whether such an effect also occurs for gluons, so that the strong interaction also loses its individuality at some higher energy, and all three interactions will take on a common character.

If there is a symmetry among the three forces at a high enough energy, why do they appear different in the studies we do at the energies available in our accelerator laboratories? For this answer, we invoke the concept of *spontaneous symmetry breaking*—as we move from the unification energy (whatever it may be) to lower energies, the symmetry of the interactions is spontaneously broken, and they appear to us as the three separate and distinguishable forces. An analogy for this effect is the condensation of hot matter into solids. A collection of gaseous iron atoms will interact randomly (through collisions) with no preferred direction; the interactions have geometrical symmetry. As we cool those atoms and allow them to condense into solid iron, magnetic crystals will form, in which each crystal has a distinct pair of magnetic poles (north and south); the coupling between atoms in the ferromagnetic state introduces a preferred direction in space, and the geometrical symmetry is *spontaneously* broken (that is, without the necessity of introducing any external constraint on a preferred direction). A similar effect might occur in solids that condense into crystals having a single unique symmetry axis; the liquid, from which the crystals form, has no preferred reference axis, yet the individual crystals each acquire an asymmetric form. In either case, the internal dynamics of the system forces a spontaneous breaking of the geometrical symmetry, as the system tries to reach the equilibrium state of lowest energy.

If such an analogy is indeed true, and if the symmetry of the Grand Unified interaction is spontaneously broken, at our "puny" laboratory energies, into strong, electromagnetic, and weak couplings, we may then ask at what energy the unification should occur. Since high-energy probes are also short-distance probes, the question is equivalent to the study of the basic interactions at small distances. Consider the basic interactions of an electric charge, shown in Figure 18.39$a$. Even in the absence of other matter (in a perfect vacuum), the intense electric field at short distances can give rise to (virtual) $e^+ e^-$ pairs, with which the basic charge is surrounded. The pairs are created by virtual photons, the emission of which does not affect the charge of the original particle; as a result the positive member of the virtual pair is attracted to the central negative charge. What we regard as "the" electronic charge is really the effect of many such processes (actually, an infinity of them), but we almost always examine the charge from large distances, where we describe the net effect by the (constant) charge $e$, or equivalently by the dimensionless coupling constant, $\alpha = e^2/4\pi\epsilon_0\hbar c$. If we probe at extremely short distances, however, the electromagnetic coupling constant will increase because we will begin to penetrate the cloud of positive virtual particles with which the "true" electron masks itself. Thus we expect the electromagnetic interaction to increase in strength as we go to higher energy.

The strong interaction behaves differently. Consider a basic red u quark $u_R$, shown in Figure 18.39$b$. It can emit various virtual colored gluons such as $R\overline{B}$ or $R\overline{G}$; for convenience, only $R\overline{B}$ are shown. When $u_R$ emits an $R\overline{B}$ gluon, it becomes a $u_B$ quark: $u_R \rightarrow u_B + R\overline{B}$, to conserve the color charge of the interaction. The strong interactions of the $u_R$ quark are determined by its color field, that is, the spatial distribution of its "redness," which becomes spread out over a volume characteristic of the range of the virtual gluons. As we penetrate to short distances to sample the strong color field, we find it to decrease quite substantially, owing to the distribution of the red color. Thus the strong interaction decreases with decreasing distance. (This distinction in behavior between electromagnetic and strong interactions arises because the photon carries no electric charge, while the gluon does carry a strong charge, that is, color.)

The weak interaction decreases at short distances, like the strong interaction. We can regard an electron as surrounded by a cloud of virtual weakly interacting particles ($e^- \rightarrow W^- + \nu$, $e^- \rightarrow e^- + Z^0$), and in effect the weak interaction is

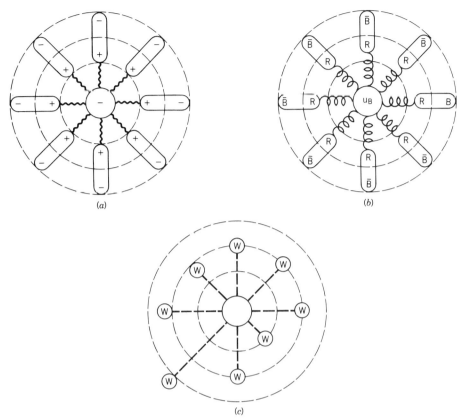

**Figure 18.39** (*a*) An electron surrounded by virtual $e^+e^-$ pairs resulting from virtual photon emission. At short distances, the electric force on a test charge will increase. (*b*) A red quark emitting virtual $R\bar{B}$ colored gluons spreads its red charge over a larger volume of space, thereby reducing the "red" force on a test hadron at small distances. (*c*) An electron emits virtual $W^-$ bosons, spreading its weak interaction and effectively reducing the weak force on a test particle penetrating to small distances.

smeared out. The effect is not quite as dramatic as in the case of the color charge, but the weak interaction does decrease in strength at small distances (high energy).

The actual calculated rate of change of the interactions with distance or energy is shown in Figure 18.40. It is calculated that the increase in the electromagnetic strength and the decrease in strong and weak strengths bring the three interactions together on a distance scale of about $10^{-31}$ m $= 10^{-16}$ fm. If, at this distance, the interactions are carried by an exchanged particle X, to have a range of $10^{-16}$ fm its mass energy would be of the order of $10^{15}$ GeV. We might therefore expect to see evidence of this unification at an energy of $10^{15}$ GeV, which is hopelessly beyond the largest particle accelerators yet built or even contemplated (the largest of which will be in the range of $10^4$ GeV).

There are, nevertheless, calculable and observable consequences of the Grand Unified Theories (GUTs). The first is that they enable the calculation of the

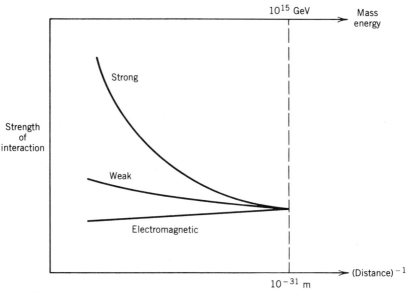

**Figure 18.40** The variation in interaction strength with distance brings the basic interactions together at about $10^{-31}$ m.

mixing amplitude in the electroweak interaction (which was merely regarded as a parameter of the electroweak theory but is now directly calculable, as a function of $m_X$, in GUTs). The value obtained, $\sin^2 \theta_w = 0.22$, is in excellent agreement with the measured value obtained from neutrino interactions (recall, $m_{W\pm}/m_{Z^0} = \cos \theta_w$).

The GUTs can also calculate mass relationships within a generation; for example, the ratio of the masses of the $\tau$ lepton and b quark has been calculated to be about 1 : 3, in excellent agreement with the observed masses.

Perhaps the most distinguishing feature of the GUTs is the direct coupling of leptons and quarks; that is, on an energy scale of $10^{15}$ GeV, leptons and quarks become equivalent. It is then possible for quarks to decay into leptons through the emission of an X particle:

$$q \rightarrow X + l$$

where q stands for a quark and $l$ for a lepton. According to this theory, it should then be possible for the proton to decay:

$$p \rightarrow \pi^0 + e^+$$

according to the diagram shown in Figure 18.41. The estimated lifetime for this process is of the order of $10^{31}$ years (corresponding to $m_X c^2 = 10^{15}$ GeV), but it should nevertheless be accessible to measurement. A cubic container of water 20 m on a side contains about $3 \times 10^{33}$ protons, and thus we should expect to see about one decay per day if the lifetime is $10^{31}$ years. As of this writing, there are several experiments in progress (Figure 18.42) to search for evidence of proton decay. So far, no events have been detected, which has been interpreted to mean that the lifetime may be greater than $10^{32}$ years; this may provide some difficulties for the GUTs to explain. In particular, should the lifetime be of the

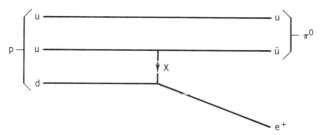

**Figure 18.41** Decay of a proton into $\pi^0 + e^+$ through the emission of an X particle.

**Figure 18.42** An underground chamber, lined with plastic, in the Morton salt mine near Cleveland before it was filled with water; note the worker standing in the corner. The chamber contains about 10,000 tons of water, corresponding to $2.5 \times 10^{33}$ nucleons. The surface of the water is lined with 2048 photomultiplier tubes which look for the light (Cerenkov radiation) emitted by the fast-moving charged particles that would result from proton decay. The experiment is being done by a group from the University of California at Irvine, the University of Michigan, and Brookhaven National Laboratory.

order of $10^{33}$ years or longer (and therefore probably unobservable in any experiment on a realizable scale), the present GUTs would be inconsistent with the calculations of $m_X$ based on the electroweak mixing angle.

These are exciting developments in the history of particle physics; indeed, some prominent theorists have argued that we are approaching the ultimate knowledge of particle physics, and that the next generation of accelerators (the

20-TeV SSC) will discover no new or interesting results. By this argument, the number of generations of quarks and leptons is fixed at three by cosmological constraints (Chapter 19), and so if the t quark is confirmed at about 40 GeV, the spectroscopy of $t\bar{t}$ states can elucidate its nature using the current generation of accelerators. We might thus expect a vast "desert" in which no new properties appear between about 100 and $10^{15}$ GeV. On the other hand, there have been speculations that the quarks and leptons are not fundamental particles, but have substructures that may appear at distances below $10^{-18}$ m (corresponding to energies above 100 GeV). The next generation of TeV accelerators may thus be able to conduct experiments into a completely new spectroscopy of fundamental particles.

## REFERENCES FOR ADDITIONAL READING

References for elementary particle physics fall into two mathematical groups—descriptive (nonmathematical) and advanced; because particle physics cannot be analyzed with nonrelativistic quantum physics, no intermediate-level treatment is possible, and the complete mathematical approach is beyond the level of this text. Elementary descriptive works are useful to provide background material for more advanced study. A reference noteworthy both for its completeness and for its introduction of just enough mathematics to satisfy the advanced reader is J. E. Dodd, *The Ideas of Particle Physics* (Cambridge: Cambridge University Press, 1984). Another elementary work is J. C. Polkinghorne, *The Particle Play* (Oxford: Freeman, 1981), which is distinguished by the elegance of its nonmathematical descriptions of the concepts of modern particle physics. Two descriptive references covering the quark model are: B. McCusker, *The Quest for Quarks* (Cambridge: Cambridge University Press, 1983) and H. Fritzsch, *Quarks: The Stuff of Matter* (New York: Basic Books, 1983).

The history of particle physics is itself a fascinating study. Two references discussing recent history are: *The Birth of Particle Physics*, edited by L. Brown and L. Hoddeson (Cambridge: Cambridge University Press, 1983) and A. Pickering, *Constructing Quarks: A Sociological History of Particle Physics* (Chicago: University of Chicago Press, 1984). A marvelous account featuring the personal aspects in the recent history of the major discoveries in particle physics is R. P. Crease and C. C. Mann, *The Second Creation* (New York: Macmillan, 1986).

Perhaps the best sources of descriptions of recent work in particle physics are the frequent articles in *Scientific American*. Some recent examples are: S. L. Glashow, "Quarks with Color and Flavor" (October 1975); M. Jacob and P. Landshoff, "The Inner Structure of the Proton" (March 1980); H. Georgi, "A Unified Theory of Elementary Particles and Forces" (April 1981); S. Weinberg, "The Decay of the Proton" (June 1981); E. D. Bloom and G. J. Feldman, "Quarkonium" (May 1982); K. Ishikawa, "Glueballs" (November 1982); N. B. Mistry, R. A. Poling, and E. H. Thorndike, "Particles with Naked Beauty" (July 1983); C. Quigg, "Elementary Particles and Forces" (April 1985); J. M. LoSecco, F. Reines, and D. Sinclair, "The Search for Proton Decay" (June 1985). For the most exotic recent fundamental theories of the basic structure of elementary particles, see H. E. Haber and G. L. Kane, "Is Nature Supersymmetric?" (June 1986) and M. B. Green, "Superstrings" (September 1986).

The experimental techniques of particle physics differ substantially from those of nuclear physics. For a comprehensive summary, see R. C. Fernow, *Introduction to Experimental Particle Physics* (Cambridge: Cambridge University Press, 1986).

Textbooks and monographs on particle physics tend to be mathematically abstract and sophisticated. Some works that should partly or mostly be within the grasp of readers of this text are: D. H. Perkins, *Introduction to High Energy Physics*, 2nd ed. (Reading, MA: Addison-Wesley, 1982); E. Segrè, *Nuclei and Particles*, 2nd ed. (Reading, MA: Benjamin, 1977), Chapters 13–20; K. Gottfried and V. F. Weisskopf, *Concepts of Particle Physics*, Vols. 1 and 2 (Oxford: Oxford University Press, 1984 and 1986); H. Frauenfelder and E. M. Henley, *Subatomic Physics* (Englewood Cliffs, NJ: Prentice-Hall, 1974), Chapters 5–13.

Recent advanced textbooks on this topic include: D. C. Cheng and G. K. O'Neill, *Elementary Particle Physics* (Reading, MA: Addison-Wesley, 1979); E. D. Commins and P. H. Bucksbaum, *Weak Interactions of Leptons and Quarks* (Cambridge: Cambridge University Press, 1983); F. Halzen and A. D. Martin, *Quarks and Leptons* (New York: Wiley, 1984); and L. B. Okun, *Leptons and Quarks* (Amsterdam: North-Holland, 1982).

## PROBLEMS

1. Draw Feynman diagrams for the following processes:
    (a) two-pion exchange between $\Lambda^0$ and p (see Figure 18.4, for example, and be sure to identify correctly the particle in the intermediate state);
    (b) $\pi^+ \rightarrow \pi^0 + e^+ + \nu_e$

2. Find the kinetic energy (in the rest frame of the original particle) of each product particle in the following two-body decays:
    (a) $\pi^+ \rightarrow \mu^+ + \nu_\mu$      (c) $\Omega^- \rightarrow \Lambda^0 + K^-$
    (b) $\Lambda^0 \rightarrow p + \pi^-$      (d) $K^+ \rightarrow \pi^+ + \pi^0$

3. Find the maximum kinetic energy (in the rest frame of the original particle) of each final particle in the following decays:
    (a) $\omega \rightarrow \pi^+ + \pi^- + \pi^0$      (c) $K^0_L \rightarrow \pi^- + \mu^+ + \nu_\mu$
    (b) $\Sigma^- \rightarrow n + \mu^- + \bar{\nu}_\mu$      (d) $\psi \rightarrow \omega + K^0 + \overline{K}^0$

4. For the following reactions, find the kinetic energies of the two outgoing particles. In each case, assume the first particle is incident with the given laboratory momentum on the second particle at rest:
    (a) $K^- + p \rightarrow \Sigma^- + \pi^+$      $p_i = 1.2$ GeV/$c$
    (b) $\pi^- + p \rightarrow \rho^0 + n$      $p_i = 2.4$ GeV/$c$

5. Find the threshold kinetic energy for each of the following reactions, assuming the first particle to be incident on the second particle at rest:
    (a) $K^- + p \rightarrow \Xi^- + K^+$
    (b) $\bar{p} + p \rightarrow \Upsilon$
    (c) $\pi^- + p \rightarrow \omega + n$

6. Analyze the following decays or reactions for possible violations of the basic conservation laws. In each case state which conservation laws, if any, are violated and through which interaction the process will most likely proceed (if at all):

(a) $\pi^+ + p \rightarrow p + p + \bar{n}$    (e) $K^+ \rightarrow \pi^+ + \pi^+ + \pi^0 + \pi^-$

(b) $\Sigma^+ \rightarrow n + e^+ + \nu_e$    (f) $K^+ \rightarrow \pi^+ + e^+ + \mu^-$

(c) $K^+ \rightarrow \pi^+ + e^+ + e^-$    (g) $\Lambda^0 + p \rightarrow \Sigma^+ + n$

(d) $\pi^- + p \rightarrow \Lambda^0 + \Sigma^0$    (h) $\Lambda^0 \rightarrow p + K^-$

7. Analyze the following decays according to their quark content:

(a) $\Omega^- \rightarrow \Lambda^0 + K^-$    (c) $\Xi^- \rightarrow \Lambda^0 + \pi^-$

(b) $K^+ \rightarrow \pi^+ + \pi^0$    (d) $\Lambda_c^+ \rightarrow p + \overline{K}^0$

8. Analyze the following reactions according to their quark content:

(a) $K^- + p \rightarrow \Omega^- + K^+ + K^0$    (c) $K^- + p \rightarrow \Xi^- + K^+$

(b) $p + p \rightarrow p + \pi^+ + \Lambda^0 + K^0$    (d) $\pi^- + n \rightarrow \Delta^- + \pi^0$

# 19

# NUCLEAR ASTROPHYSICS

To nuclear and particle physicists, the early universe represents the ultimate particle accelerator, in which energies and densities of particles were beyond what we can ever hope to achieve with artificially constructed accelerators. Reactions occurred with rates and varieties almost incomprehensible, but (perhaps surprisingly) by studying the end products of those reactions, we can infer many details of the reaction processes that cannot be measured on Earth. Essential to the understanding of these evolutionary reactions are known details (cross sections, etc.) of reactions that *can* be measured in our laboratories. What is now called the *standard model* of the hot Big Bang cosmology includes an overall framework based on the General Theory of Relativity, nuclear and particle properties directly measured, inferences from the standard model of elementary particles discussed in Chapter 18, and some extrapolations based on reasonable hypotheses. The input data must yield results in agreement with observations, the primary ones being the relative amounts of various light isotopes produced during the earliest epoch of element formation. The observed abundances can then put severe constraints on the fundamental processes that occurred during the formation epoch.

A much more difficult task is to understand the formation of the heavier elements following fusion reactions and neutron captures in stellar interiors. Not only is the nuclear physics more complicated (and the reactions more difficult to duplicate in the laboratory), but the mechanics and thermodynamics are less well understood. Here the observational evidence consists primarily of astronomical observations, not only with conventional optical telescopes, but more recently with particle and $\gamma$-ray spectrometers carried in orbiting spacecraft. In this case also, nuclear decays and reactions observed in Earth-bound laboratories can determine fundamental limits on the nature and especially the duration of nuclear reactions in stars.

The evolution of the universe can, from our Earthly perspective, be divided into four stages: primordial nucleosynthesis and atomic formation, galactic condensation, stellar nucleosynthesis, and evolution of the Solar System. What we call the age of the universe is merely the sum of the durations of these four sequential periods. The first stage lasts from the Big Bang ($t = 0$) to the formation of stable atomic hydrogen and helium (plus a very small concentration of other products). Although there are many uncertainties about this era (resulting from uncertainties in our knowledge of fundamental nuclear and particle properties), they have very little effect on estimates of the duration of this era, and moreover its duration is so small (about $10^6$ years) that uncertainties are relatively unimportant. Galactic condensation occurs under the influence of gravitational forces alone, and nuclear and particle physics have no influence on this epoch, which has been estimated to last from 1 to 2 Gy (gigayear = $10^9$ y). The era of stellar nucleosynthesis contributes the greatest uncertainty, perhaps $\pm 2$ Gy, but we will see how recent nuclear reaction studies have given results for this era in excellent agreement with independent astronometric deductions. Finally, the duration of the evolution of the Solar System is well known, with little uncertainty.

## 19.1 THE HOT BIG BANG COSMOLOGY

One of the most significant discoveries of twentieth-century physics is the expansion of the universe. This deduction was made by Edwin Hubble following many observations of the absorption line spectra of distant galaxies. Hubble discovered that the spectra are red-shifted; that is, the absorption lines appear closer to the long-wavelength (red) end of the visible spectrum than they do when emitted by terrestrial sources. This red shift is similar to the Doppler shift of electromagnetic radiation, and from the degree of the red shift Hubble was able to deduce the velocities of recession of the galaxies relative to Earth. From independent observations, he knew the distance to the galaxy, and he observed that there was a linear relationship between the distance $d$ and the speed of recession $v$:

$$v = Hd \qquad (19.1)$$

where $H$ is called the Hubble parameter. Figure 19.1 illustrates the linear Hubble relationship. The present best value of the Hubble parameter is about

$$H = 67\frac{\text{km/s}}{\text{Mpc}}$$

where one megaparsec (Mpc) is $3.26 \times 10^6$ light years. The uncertainty in this value is large; the permitted range is about 50–100 km/s/Mpc.

According to the present model, this expansion is a general property of the universe, but one that is likely to vary with time, owing to the effect of gravity. The recession of the galaxies is a result of the general expansion of the universe, but if the universe is infinite we cannot define a radius. Instead, we define a *scale factor* $R(t)$, which gives the time dependence of any typical length. The distance between galaxies, for instance, increases in proportion to the increase of the scale factor. In terms of the scale factor, the Hubble parameter can be written as

$$H = \frac{1}{R}\frac{dR}{dt} \qquad (19.2)$$

If the universe is expanding at a constant rate, then $H$ is a constant, but the mutual gravitational attraction of the galaxies means that the recessional speed will decrease, in which case $H$ is a function of the time.

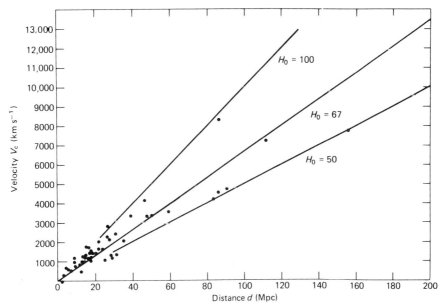

**Figure 19.1**  Velocity-distance relationship for groups and clusters of galaxies. The straight line demonstrates the Hubble relationship. From M. Rowan-Robinson, *The Cosmological Distance Ladder* (New York: Freeman, 1985).

As in the case of any mechanical system, the time evolution of the behavior can be obtained by solving the dynamical equations, in this case the tensor equations of the General Theory of Relativity. The mechanics of the theory are beyond the level of this text, but the result can be expressed in simple form:

$$H^2 = \frac{(dR/dt)^2}{R^2} = \frac{8\pi G}{3}\rho(t) - \frac{kc^2}{R^2} + \frac{\Lambda}{3} \qquad (19.3)$$

where $G$ is the gravitational constant of Newton ($6.67 \times 10^{-11}$ N · m$^2$/kg$^2$) and $\rho(t)$ is the mean mass + energy density of the universe. The geometrical factor $k$ is determined by the fundamental geometry of space-time: $k = 0$ for a "flat" universe, in which the laws of geometry are Euclidian; $k = +1$ for a closed, spherical universe (corresponding to positive curvature); $k = -1$ for a curved universe with a shape analogous to a saddle. The cosmological constant $\Lambda$ is assumed to vanish for the present discussion.

If the expansion had been at a constant rate $dR/dt$ since $t = 0$, then the distance and recession velocity of Equation 19.1 would be connected by $d = vt$, where $t$ is the age of the universe. Thus $t - H^{-1} = 15$ Gy, with an uncertainty of about $\pm 5$ Gy. This is of course an upper limit on the true age of the universe, for we are deducing the Hubble parameter by observing galaxies that have already slowed from their original recession speed. In the past, $H$ would have had a larger value, leading to a shorter deduced age. Figure 19.2 shows how the dependence of various scale factors on time leads to ages of the universe that are less than $H^{-1}$.

In these calculations, the age of the universe is operationally defined as the time since $R = 0$; that is, the expansion of the universe suggests that if we look

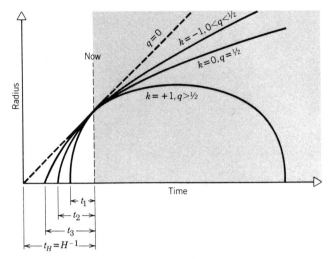

**Figure 19.2** The dependence of the radius or scale factor of the universe on the time for various types of universes: $q = 0$, a universe with a constant rate of expansion; $k = -1$, an open, curved universe; $k = 0$, an open, flat universe; $k = +1$, a closed universe which expands and then contracts. The dimensionless deceleration parameter $q$ is defined as $-R(d^2R/dt^2)/(dR/dt)^2$ and is a measure of the effect of gravitational interactions in slowing (or even reversing) the expansion.

far enough back into time, we find the galaxies so close together that they occupy a state of enormous energy and matter density. Extrapolating further back to $t = 0$, we come to a mathematical singularity, the Big Bang.

Before we examine the state of the universe just after the Big Bang, we should look at the form of solutions to Equation 19.3. We assume $k = 0$, allowing us to neglect what is basically an uninteresting term in the equation. Following the Big Bang, the universe was dominated by radiation (or by matter moving at such high speed that the radiation-like relationship $E = pc = hc/\lambda$ was obeyed), and the radiant energy density $\rho_R$ is

$$\rho_R = \frac{\text{energy}}{\text{volume}} = \text{energy per quantum} \times \text{quanta per volume}$$

$$\propto \frac{1}{R} \times \frac{1}{R^3}$$

where the $R^{-1}$ factor comes from the $\lambda^{-1}$ factor in the energy quantum and the $R^{-3}$ factor comes from the volume. We can thus take $\rho_R = C/R^4$ (the constant $C$ will disappear) and Equation 19.3 becomes

$$\frac{1}{R}\frac{dR}{dt} = \sqrt{\frac{8\pi GC}{3}}\frac{1}{R^2} \tag{19.4}$$

which integrates to give

$$t = \sqrt{\frac{3}{32\pi G\rho_R}} \tag{19.5}$$

If we base $\rho_R$ on the energy density $u(T)$ of blackbody radiation for a radiating system at temperature $T$:

$$u(T) = \sigma T^4 \tag{19.6}$$

where $\sigma$ is the Stefan-Boltzmann constant, then we have an important fundamental relationship between age $t$ (in seconds) and temperature $T$ (in kelvins) during the radiation-dominated era:

$$T = \frac{1.5 \times 10^{10}}{t^{1/2}} \tag{19.7}$$

The temperature is an important parameter to describe the early universe. If the temperature is high enough, matter and radiation are in equilibrium, and particle-antiparticle creation occurs as often as annihilation; in the case of electrons, for example

$$e^+ + e^- \leftrightarrow 2\gamma$$

For $2\gamma \rightarrow e^+ + e^-$ to occur, the radiant photons must have an energy of at least 0.511 MeV; if on the average the photons have an energy of $kT$ (here $k$ is the Boltzmann constant of thermodynamics), the energy corresponds to a temperature of $T = 6 \times 10^9$ K. That is, when $T \lesssim 6 \times 10^9$ K (corresponding to $t > 6$ s), the radiation field is no longer energetic enough to balance $e^+ + e^- \rightarrow 2\gamma$ reactions with $2\gamma \rightarrow e^+ + e^-$, and the former reaction begins to dominate.

This is then the overall scheme of the Big Bang cosmology: The present universe is created from a space-time singularity at essentially infinite temperature and density. It consists of a mixture of the most fundamental particles and their antiparticles plus radiation. As the universe expands and cools, each species of particles in turn goes out of equilibrium with the radiation. We observe today the relics of the particle interactions plus the residual radiation which has had negligible interactions with matter since the final decoupling within the first $10^6$ y.

The experimental discovery which, along with Hubble's expansion, provided dramatic confirmation of the Big Bang theory was the universal microwave background radiation, immediately associated with the residual radiation from the Big Bang. The original discovery was made by Arno Penzias and Robert Wilson in 1964. They were attempting to use a radio receiver, tuned to 7.35 cm wavelength, which had been built for contact with an early communications satellite. In the process, they observed a background "noise" that, despite their best efforts, could not be eliminated from the system. They finally identified the noise as a real signal, and discovered that it came uniformly from all directions and at all times of day and night. By measuring the energy density of the radiation at $\lambda = 7.35$ cm, they were able to deduce the blackbody temperature of $3.1 \pm 1.0$ K. Other measurements subsequently made at different wavelengths demonstrate that the radiation does indeed have the expected blackbody spectrum (Figure 19.3); the present best temperature is $2.7 \pm 0.1$ K.

The essential features of the Big Bang cosmology are thus confirmed, but many details remain to be filled in. The exact mechanism that triggers galaxy condensation from what must have originated as a homogeneous mixture of particles is not yet clear, nor is the mechanism that guarantees that opposite sides of our

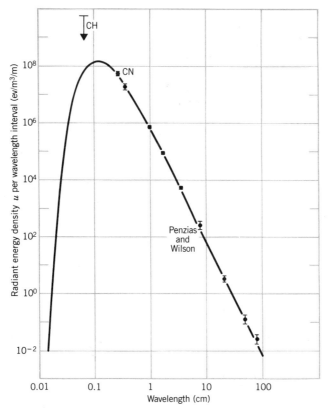

**Figure 19.3** The wavelength spectrum of the cosmic microwave background radiation. The dots are data points and the solid curve is a blackbody spectrum for $T = 2.7$ K.

observable universe, though not causally connected (i.e., a light beam could not have traveled from one end to the other within their present age), appear to have similar properties. Exotic particles may have been available at the large energies and densities of the early universe and may have played an important role; on the other hand, the list of particles generated in Chapter 18 may be nearly complete, as some high-energy physicists suspect, leaving no place for exotics.

## 19.2 PARTICLE AND NUCLEAR INTERACTIONS IN THE EARLY UNIVERSE

Although we do not yet understand how to extrapolate back to the earliest instants after the Big Bang, many of the fine details of the structure of the universe in that era are irrelevant for the particular part of the story in which we are interested.

For example, we consider the times around $t = 10^{-12}$ s. At that point the temperature is, from Equation 19.7, about $10^{16}$ K, corresponding to a mean energy quantum $kT$ of 1000 GeV. With the possible exception of free quarks, whose masses we do not know (and which may not be permitted to exist in a free state), all known particles can be readily created at this energy. All species of particles must exist in equilibrium concentrations; for example, the lepton-production reactions

$$\gamma + \gamma \rightarrow e^+ + e^-$$
$$\gamma + \gamma \rightarrow \mu^+ + \mu^-$$
$$\gamma + \gamma \rightarrow \tau^+ + \tau^-$$

will occur with a probability determined by the density of states available for the final particles, but since the rest energies of all particles are far below 1000 GeV, the available momentum is about the same for each and the birth rates are identical. If there are additional generations of leptons, these also would be produced in equilibrium with the photons.

The number density of photons can be found from the basic expressions for blackbody radiation. The energy density is

$$u(E)\,dE = \frac{8\pi E^3}{(hc)^3} \frac{1}{e^{E/kT} - 1}\,dE \tag{19.8}$$

and the number per energy interval is

$$n(E)\,dE = \frac{u(E)}{E}\,dE = \frac{8\pi E^2}{(hc)^3} \frac{1}{e^{E/kT} - 1}\,dE \tag{19.9}$$

Integrating Equations 19.8 and 19.9 over all energies and evaluating the constants gives the total energy density and number density of photons at temperature $T$:

$$\rho_\gamma = 4.7 \times 10^3\, T^4\, \text{eV}/\text{m}^3 \tag{19.10}$$

$$N_\gamma = 2.0 \times 10^7\, T^3\, \text{photons}/\text{m}^3 \tag{19.11}$$

where $T$ is in kelvins. At the present temperature of 2.7 K, the number density is $4 \times 10^8/\text{m}^3 = 400/\text{cm}^3$. These photons have a present mean energy of less than 0.001 eV, which is the reason we do not detect them under ordinary circumstances; each cubic centimeter has a mass energy of about 0.25 eV in primordial photons.

The present matter density of the universe is difficult to estimate. The best estimates for visible (luminous) matter suggest $\rho_0 \sim 3 \times 10^{-31}$ g/cm$^3$; clusters of galaxies, however, seem to be bound by additional nonluminous matter, and the matter density may be uncertain by as much as one order of magnitude. We will arbitrarily increase the value by a factor of 2 to account for nonluminous matter and take

$$\rho_0 \sim 6 \times 10^{-31}\ \text{g}/\text{cm}^3$$

$$\sim 0.4\ \text{nucleon}/\text{m}^3$$

The present density of nucleons is only about $10^{-9}$ the present density of photons. Because there is no mechanism by which nucleons can be destroyed on a time scale of the age of the universe, our cosmological model must account for the great imbalance of photons over nucleons.

We also believe that the universe is made almost exclusively of matter, rather than antimatter. This is a difficult assumption to test, for galaxies made of antimatter would differ in no observable way from galaxies made of matter. If there were a rough balance of matter and antimatter, we might expect occasional encounters between one and the other in interstellar dust and gas, resulting in annihilation reactions with the emission of high-energy photons. Observational limits on such photons are very low, suggesting little antimatter-matter annihilation. We will therefore assume our universe to be composed of ordinary matter.

If we look back to the time when $T > 10^{13}$ K ($t < 10^{-6}$ s), photons had enough energy to create nucleon-antinucleon pairs, as

$$\text{photons} \leftrightarrow p + \bar{p}, n + \bar{n}$$

We would therefore expect *exactly* equal numbers of nucleons and antinucleons. Even if free quarks could have been created in a much earlier epoch, we would still expect a balance:

$$\text{photons} \leftrightarrow q + \bar{q}$$

The balance between matter and antimatter is upset by CP-violating decays (see Section 17.6), which *do* favor one type over the other. The era of the massive X particles, associated with the Grand Unified Theories (GUTs, Section 18.8) distinguishes between matter and antimatter, and the relative rates of

$$X \rightarrow q + q$$

and

$$\overline{X} \rightarrow \bar{q} + \bar{q}$$

may not be equal. If this is the case, there will be a slight imbalance of matter over antimatter; after $10^{-6}$ s, when nucleons and antinucleons can no longer be created by the radiation field, the antimatter will annihilate with an equal quantity of matter, and the remaining matter is what we observe today. The imbalance must have been very small, about 1 part in $10^9$, for us to observe the present $10^{-9}$ nucleon-to-photon ratio. With the unification mass estimated to be $10^{15}$ GeV, the temperature at which this imbalance becomes established must be about $10^{28}$ K, corresponding to an age of about $10^{-36}$ s.

Let's now jump to the era of $t > 10^{-6}$ s, where $T \lesssim 10^{13}$ K ($E \lesssim 1$ GeV). Nucleon-antinucleon annihilation occurs, but no longer in equilibrium with the reverse creation process. An abundance of leptons and neutrinos is present, so weak interactions can occur to convert protons to neutrons:

$$p + \bar{\nu}_e \leftrightarrow n + e^+$$
$$n + \nu_e \leftrightarrow p + e^-$$

If $e^{\pm}$ and $\nu_e$, $\bar{\nu}_e$ are plentiful, these reactions will go in either direction, and at this early time the number of neutrons and protons should be equal.

At $t = 10^{-2}$ s, $T = 10^{11}$ K ($E = 10$ MeV). Now electrons are the only remaining charged leptons; $\mu$ ($mc^2 = 105$ MeV) and $\tau$ ($mc^2 = 1784$ MeV) are no longer produced (and have long since decayed to electrons or annihilated to photons), but through neutral weak interactions, all species of neutrinos can be made:

$$e^+ + e^- \leftrightarrow Z^0 \leftrightarrow \nu_e + \bar{\nu}_e$$
$$e^+ + e^- \leftrightarrow Z^0 \leftrightarrow \nu_\mu + \bar{\nu}_\mu$$
$$e^+ + e^- \leftrightarrow Z^0 \leftrightarrow \nu_\tau + \bar{\nu}_\tau$$

If the neutrinos are massless (or if their masses are small compared with 10 MeV), then these reactions will be at equilibrium, and the relative numbers of neutrino species will remain constant.

Associated with each kind of particle is a statistical weight $g_i$ ($i = e^+, e^-, \nu_e, \bar{\nu}_e,$ $\nu_\mu, \bar{\nu}_\mu, \nu_\tau, \bar{\nu}_\tau, \ldots$), which essentially gives the permitted number of spin or polarization states. For convenience we group together $e^+$ and $e^-$ (as e) and each $\nu_i, \nu_i$ pair (as $\nu_i$), so that

$$g_e = 4 \quad (2 \text{ spin states for each of } e^+ \text{ and } e^-)$$
$$g_{\nu_i} = 2 \quad (\nu_i \text{ and } \bar{\nu}_i \text{ states for } i = e, \mu, \tau, \ldots) \tag{19.12}$$

The number and energy densities associated with each species are governed by the Fermi-Dirac distribution:

$$u(E)\, dE = g_i \frac{4\pi E^3}{(hc)^3} \frac{1}{e^{E/kT} + 1}\, dE \tag{19.13}$$

with an analogous result for the number density $n(E)$ as in Equation 19.9.

Carrying out the required integrals gives the total volume density of particles and energy for electrons

$$N_e = \tfrac{3}{2} N_\gamma \tag{19.14}$$

$$\rho_e = \tfrac{7}{4} \rho_\gamma \tag{19.15}$$

where $N_\gamma$ and $\rho_\gamma$ are the photon total number and energy densities given by Equations 19.10 and 19.11. The corresponding contribution from neutrinos will be

$$N_\nu = \tfrac{3}{4} n_\nu N_\gamma \tag{19.16}$$

$$\rho_\nu = \tfrac{7}{8} n_\nu \rho_\gamma \tag{19.17}$$

where $n_\nu$ is the number of different neutrino species ($n_\nu = 3$ for e, $\mu$, $\tau$ only).

Neutrons are slightly more massive than protons, so there are slightly fewer of them in an equilibrium mixture at temperature $T$, owing to the Boltzmann factor:

$$\frac{N_n}{N_p} = e^{-(m_n - m_p)c^2/kT} \tag{19.18}$$

The mass difference is about 1.3 MeV, and at $kT = 10$ MeV (at $t \sim 0.01$ s), $N_n/N_p \simeq 0.88$. Maintaining an equilibrium condition requires the presence of sufficient $e^\pm$ and $\nu_e, \bar{\nu}_e$ to allow the n $\leftrightarrow$ p conversion.

At $t = 1$ s ($T = 10^{10}$ K; $E = 1$ MeV), neutrino interactions are no longer important; neither the charged nor the neutral interactions of neutrinos contribute to the reactions. This is the era of *neutrino decoupling*, from which time the neutrinos expand freely with the universe, virtually unaffected by nuclear reactions.

Immediately following the decoupling, $e^+ + e^- \rightarrow 2\gamma$ is possible, but not the reverse. The photon energy density therefore increases somewhat, and in effect the temperature of the photons decreases less rapidly than that of the freely expanding neutrinos. The effect can be shown to be

$$\frac{T_\gamma}{T_\nu} = \left(\frac{11}{4}\right)^{1/3} \cong 1.4 \tag{19.19}$$

for three families of leptons. The present neutrino temperature is thus about 2 K.

Following the annihilation of the positrons and electrons, we are left with a large collection of photons, about $10^{-9}$ as many electrons and protons (equal numbers, to keep the net electric charge zero), and a slightly smaller number of neutrons (from the Boltzmann factor). This ends the era of particle interactions and begins the time of nuclear reactions. After the stable, light nuclei have formed ($^2$H, $^3$He, $^7$Li), the universe continues to expand and cool until the photons decouple; that is, neutral atoms can form when the photons lack the energy necessary to ionize them. This occurs at a temperature of perhaps 3000 K and an age of about 700,000 y. With the continuing expansion of the universe, the photons continue to cool, independent of the interactions of matter (which are restricted to gravity since the other interactions can no longer operate on the large scale) until they reach the presently observed temperature of 2.7 K.

## 19.3 PRIMORDIAL NUCLEOSYNTHESIS

To begin production of heavy nuclei, the first reaction that must occur is

$$n + p \rightarrow d + \gamma$$

At high temperatures, the reverse reaction occurs as quickly as deuterium production, and there is no accumulation of deuterium nuclei. The photon energy necessary for photodissociation is 2.225 MeV (the binding energy of deuterium), but it must be remembered that there are about $10^9$ times as many photons as protons or neutrons. The photons have a blackbody spectrum, given by Equation 19.9, which of course extends to very large $E$ (Figure 19.4). When the number of photons in the high-energy tail above the energy 2.225 MeV is less than the

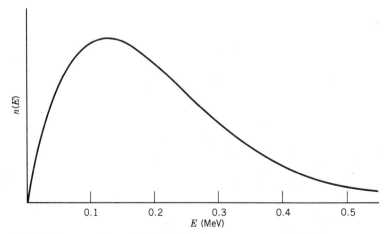

**Figure 19.4**  The number of blackbody photons at energy $E$ for a temperature of $T = 9 \times 10^8$ K.

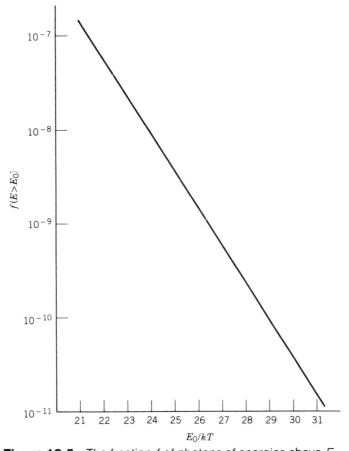

**Figure 19.5** The fraction $f$ of photons of energies above $E_0$.

number of nucleons participating in deuterium formation, there will be too few photons to inhibit deuterium production. We can establish the temperature at which this will occur by approximating the tail as an exponential

$$n(E)\, dE \cong \frac{8\pi E^2}{(hc)^3} e^{-E/kT}\, dE \qquad (19.20)$$

and integrating for energies above $E_0$ gives

$$N_\gamma(E > E_0) = \frac{8\pi}{(hc)^3}(kT)^3 e^{-E_0/kT}\left[\left(\frac{E_0}{kT}\right)^2 + 2\left(\frac{E_0}{kT}\right) + 2\right] \qquad (19.21)$$

Dividing by the total number density gives the fraction $f$ above $E_0$

$$f(E > E_0) = 0.42\, e^{-E_0/kT}\left[\left(\frac{E_0}{kT}\right)^2 + 2\left(\frac{E_0}{kT}\right) + 2\right] \qquad (19.22)$$

Figure 19.5 illustrates this function in the range $10^{-11} < f < 10^{-7}$, corresponding to $21 < E_0/kT < 31$.

The number of nucleons available for deuterium formation is determined by the number of neutrons because there are fewer neutrons than protons. The ratio

$N_n/N_p$ decreases with decreasing temperature, according to Equation 19.18, only as long as $e^{\pm}$ are sufficiently plentiful and react quickly enough for $n \leftrightarrow p$ conversion to take place. At a certain temperature $T_*$, the ratio $N_n/N_p$ becomes "frozen" when the rate of weak interactions becomes too small. Based on the known cross sections for weak interactions, we can estimate this temperature as $T_* = 9 \times 10^9$ K, corresponding to $N_n/N_p \simeq 0.2$; this occurs at a time of about 3 s.

Neutrons therefore originally constitute a fraction of the total number of nucleons equal to 0.2. If the nucleon-to-photon ratio is $10^{-9}$, then the critical fraction of high-energy photons necessary to prevent deuterium formation will be $0.2 \times 10^{-9}$, corresponding to $T = 9 \times 10^8$ K from Equation 19.22 and thus to $t \simeq 250$ s. As can be seen from Figure 19.5, this estimate is very insensitive to the value of $f$ and thus to the $N_n/N_p$ ratio.

Once we have formed deuterium in sufficient quantity, other nuclear reactions become possible. We can form the mass-3 nuclei:

$$d + n \rightarrow {}^3H + \gamma$$

$$d + p \rightarrow {}^3He + \gamma$$

or by

$$d + d \rightarrow {}^3H + p$$

$$d + d \rightarrow {}^3He + n$$

Finally, ${}^4He$ can be formed:

$$^3H + p \rightarrow {}^4He + \gamma$$

$$^3He + n \rightarrow {}^4He + \gamma$$

The binding energies of all these reaction products are greater than that of deuterium; thus if the photons are cool enough to permit deuterium to form, they will certainly permit the remaining reactions.

Because there is no stable mass-5 nucleus, ${}^4He$ is the primary end product of this process. Also ${}^8Be$ is unstable, so two ${}^4He$ cannot combine directly. There will be a small production of mass-7 nuclei:

$$^4He + {}^3H \rightarrow {}^7Li + \gamma$$

$$^4He + {}^3He \rightarrow {}^7Be + \gamma$$

but the Coulomb barriers for these reactions are about 1 MeV, and the nuclei are well below that value in energy (in equilibrium at $T = 9 \times 10^8$ K, the mean kinetic energy is less than 0.1 MeV). Therefore essentially all of the neutrons end as part of ${}^4He$ nuclei, which has a relative abundance $N_{He}/N_p = 0.081$ (calculated from the "frozen" $N_n/N_p$ ratio after correcting for the radioactive $\beta$ decays of the neutrons between $t = 3$ s and $t = 250$ s). The relative primordial abundance of ${}^4He$ by weight, $Y_p$, is thus about 0.24, and except for additional burning of H and He in stars, this relative abundance should have remained constant in the universe from $t = 250$ s until today.

The observed abundance (by weight) of ${}^4He$ is about $Y_p = 0.24 \pm 0.01$, based on observations from a variety of astronomical systems, including gaseous nebulae, planetary nebulae, and stars (including the Sun). This excellent agree-

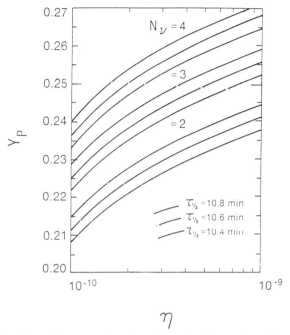

**Figure 19.6** Dependence of primordial helium abundance $Y_p$ on the nucleon-to-photon ratio. The expected dependence is shown for 2, 3, or 4 types of massless neutrino; for each case the three curves show the range corresponding to the uncertainty in the experimental value of the neutron half-life of 10.6 ± 0.2 min. From D. N. Schramm and G. Steigman, *Phys. Lett.* B **141**, 337 (1984).

ment between the calculated abundance and that observed should not be taken as a confirmation of the theory, for the final $^4$He abundance is very sensitive to the assumed frozen $N_n/N_p$ ratio, which in turn is sensitive to the calculated temperature at which the freeze-out occurs. This calculation depends critically on the half-life of neutron decay (which is not extremely well known, $t_{1/2} = 10.6 \pm 0.2$ min) and on the assumed number of species of leptons. Figure 19.6 shows the dependence of $Y_p$ on the number of massless species of neutrinos (at least three, according to the standard model) and on the nucleon-to-photon ratio. You can see that the observed $^4$He abundance allows an additional species of neutrinos (another generation of leptons and presumably also quarks) only for a nucleon-to-photon ratio below $2 \times 10^{-10}$. It permits only two massless neutrinos (allowing, for instance, the $\tau$ neutrino to be massive, which is contrary to the standard model but not disallowed by the presently rather poor experimental limits on its mass) for correspondingly large nucleon-to-photon ratios ($> 6 \times 10^{-10}$).

In addition to $^4$He, there will be a small concentration of primordial $^2$H, $^3$He, and $^7$Li in the present universe. Deuterium in particular is critical for a determination of the nucleon-to-photon ratio—at large nucleon abundances, $^2$H is more rapidly "cooked" into heavy nuclei and its concentration is accordingly reduced. The ratio $N_d/N_p$ can be deduced from the shift in the absorption spectrum of atomic hydrogen caused by the heavier nuclear mass of $^2$H. The observed value is subject to uncertainties arising from the destruction of primor-

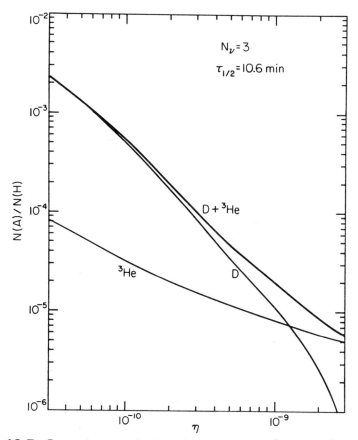

**Figure 19.7** Dependence of the abundance of $^2$H and $^3$He on the nucleon-to-photon ratio. From J. Yang et al., *Astrophys. J.* **281**, 493 (1984).

dial $^2$H in the evolution of galaxies, but the best current value is $N_d/N_p \sim 1$–$3 \times 10^{-5}$.

The isotope $^3$He is, like $^2$H, a result of incomplete primordial processes, and the abundance of $^3$He decreases as the primordial nucleon density increases. Here again, observed present-day abundances may not reflect primordial values, for "new" $^3$He can be produced, in particular from deuterium. The $^3$He abundance observed presently is therefore possibly a measure of the primordial combined $^3$He $+ ^2$H abundance. Based on observed solar abundances, it is suggested that $(N_{^2H} + N_{^3He})/N_p < 6 \times 10^{-5}$.

Figure 19.7 shows the abundances calculated with the standard model (three varieties of massless neutrinos). It is quite clear that the deuterium and $^3$He abundances constrain the nucleon-to-photon ratio to be greater than about $4 \times 10^{-10}$, and returning to Figure 19.6, we see that this value is inconsistent with a fourth neutrino. Although much work, both theoretical and experimental, remains to be done to solidify these conclusions, it appears that these cosmological arguments may indicate that there are no new fundamental particles beyond the present three generations of leptons and quarks.

### 19.4  STELLAR NUCLEOSYNTHESIS ($A \leq 60$)

The dominant process in the formation of the elements with $A \leq 60$ is charged-particle reactions, primarily those induced by protons and $\alpha$ particles. The probability for these reactions to occur will depend on the overlap between the thermal distribution of particle energies and the Coulomb barrier penetration probability, exactly as in the case of fusion reactions discussed in Section 14.2.

Stars begin life as a mixture of hydrogen and 24% (by weight) helium. As the original gas cloud collapses, the individual atoms exchange their gravitational potential energy for kinetic energy, thereby increasing the temperature of the cloud. Eventually, the temperature is high enough that the protons can overcome the repulsive Coulomb energy, and fusion reactions can begin. The outward pressure from the radiation released in fusion effectively halts the gravitational collapse, and the star enters an equilibrium phase (like our Sun) that may last $10^{10}$ y.

The basic reactions of the proton-proton fusion cycle have already been discussed in Section 14.3 and need not be repeated here. When a star's hydrogen fuel is depleted, gravitational collapse can begin again, and eventually a higher temperature is reached (perhaps $1-2 \times 10^8$ K, compared with $10^7$ K for the present Sun) where the Coulomb barrier for $^4$He-$^4$He fusion can be overcome. The increased ignition temperature results in a larger radiation pressure, which expands the outer envelope of the star's surface by a large factor, perhaps by a factor of 100 or 1000. The apparent energy density of the surface decreases, and the star appears at a lower effective surface temperature. This is the *red giant* stage.

As there is no stable mass-8 nucleus, there are no observable end products of the reaction

$$^4\text{He} + {}^4\text{He} \rightarrow {}^8\text{Be}$$

for $^8$Be breaks apart (into two $^4$He again) in a time of the order of $10^{-16}$ s. The $Q$ value is 91.9 keV, and so even at $2 \times 10^8$ K (mean thermal energy of 17 keV) there will be a fair number of energetic $\alpha$'s in the high-energy tail of the thermal distribution to form $^8$Be. There will be a small equilibrium concentration of $^8$Be, of the order of the Boltzmann factor $e^{-91.9 \text{ keV}/17 \text{ keV}} = 4 \times 10^{-3}$, and the reaction rate can be calculated, exactly as we did in Section 14.2 for the D-D or D-T reactions.

We know that $^{12}$C is plentiful in the universe, but the calculated rates of $2\alpha \rightarrow {}^8$Be and $^8$Be $+ \alpha \rightarrow {}^{12}$C are not sufficient to produce the observed large abundances of $^{12}$C, as was first realized by Fred Hoyle in the early 1950s. The $Q$ value for $^8$Be $+ \alpha \rightarrow {}^{12}$C is 7.45 MeV, and Hoyle argued that the large production of $^{12}$C requires this reaction to occur readily, which in turn requires a resonance to account for what must be an increase in the cross section at this energy. He made this suggestion to W. A. Fowler, whose research group at Cal Tech began in the 1950s an extended program concerning the study of nuclear reactions of astrophysical importance. The Cal Tech group discovered the $^{12}$C resonance corresponding to an excited state at 7.65 MeV, just above the energy predicted by Hoyle, but well within the capability of being reached at temperatures of $1-2 \times 10^8$ K. The net $Q$ value for 3 $^4$He $\rightarrow {}^{12}$C* is 285 keV. Subse-

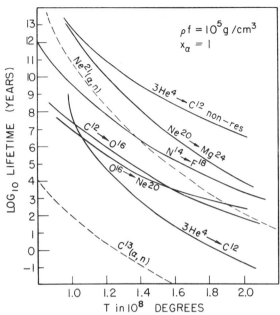

**Figure 19.8** Mean lifetime in years for various $\alpha$-particle reactions as a function of temperature. The reaction rate is the inverse of the lifetime, so reactions near the bottom of the graph are the ones with the highest rates. From Burbidge, Burbidge, Fowler, and Hoyle, *Rev. Mod. Phys.* **29**, 547 (1957).

quently, Fowler and his colleagues were able to populate the 7.65-MeV excited state in $^{12}$C following $\beta$ decay of $^{12}$B. They identified it as a $0^+$ state and observed its decay into three $\alpha$ particles, thereby suggesting that it could indeed be formed by three $\alpha$ particles. The $0^+$ assignment is consistent with the probable s-state couplings in these low-energy reactions.

Once $^{12}$C is formed, other $\alpha$-particle reactions become possible:

$$^{12}\text{C} + {}^4\text{He} \rightarrow {}^{16}\text{O} + \gamma \qquad (Q = 7.16 \text{ MeV}, \ E_B = 3.57 \text{ MeV})$$

$$^{16}\text{O} + {}^4\text{He} \rightarrow {}^{20}\text{Ne} + \gamma \qquad (Q = 4.73 \text{ MeV}, \ E_B = 4.47 \text{ MeV})$$

$$^{20}\text{Ne} + {}^4\text{He} \rightarrow {}^{24}\text{Mg} + \gamma \qquad (Q = 9.31 \text{ MeV}, \ E_B = 5.36 \text{ MeV})$$

As the Coulomb barrier $E_B$ increases for heavier nuclei, it becomes decreasingly likely to continue this sequence of reactions. Figure 19.8 shows the calculated mean lifetime for nuclei participating in these reactions, taken from the classic work on nucleosynthesis by Burbidge, Burbidge, Fowler, and Hoyle. Notice in particular the importance of the $^{12}$C resonance, which effectively increases the reaction rate by about 8 orders of magnitude.

When the helium fuel begins to be exhausted, gravitational collapse sets in again (only if the star is sufficiently massive—otherwise the gravitational force is not strong enough to oppose the "degeneracy pressure" of the electrons that are reluctant to have overlapping wave functions). The star then heats up enough to ignite $^{12}$C and $^{16}$O burning, which permits such reactions as

$$^{12}\text{C} + {}^{12}\text{C} \rightarrow {}^{20}\text{Ne} + {}^4\text{He} \quad \text{or} \quad {}^{23}\text{Na} + \text{p}$$

$$^{16}\text{O} + {}^{16}\text{O} \rightarrow {}^{28}\text{Si} + {}^4\text{He} \quad \text{or} \quad {}^{31}\text{P} + \text{p}$$

at temperatures of the order of $10^9$ K, where the Coulomb barrier can be more easily penetrated.

In addition to these reactions, other $\alpha$-particle and nucleon capture reactions can occur as well. For example, $^{14}$N may be present in second-generation stars, formed originally from $^{12}$C as part of the carbon cycle of proton-proton fusion discussed in Section 14.3. Alpha-capture reactions can produce the chain $^{14}$N $\to$ $^{18}$O $\to$ $^{22}$Ne $\to$ $^{26}$Mg.... Reactions other than $(\alpha, \gamma)$, including $(\alpha, n)$ or $(p, \gamma)$ will occur, with somewhat less probability.

The final stage in the production of nuclei near mass 60 is silicon burning, which is in actuality a complex sequence of reactions that take place rapidly but under nearly equilibrium conditions deep in the hot stellar interiors. The Coulomb barrier is too high to permit direct formation by such reactions as $^{28}$Si $+$ $^{28}$Si $\to$ $^{56}$Ni; instead, what occurs are combinations of photodissociation reactions $(\gamma, \alpha)$, $(\gamma, p)$, or $(\gamma, n)$ followed by capture of the dissociated nucleons:

$$^{28}\text{Si} + \gamma \to {}^{24}\text{Mg} + {}^4\text{He}$$

$$^{28}\text{Si} + {}^4\text{He} \to {}^{32}\text{S} + \gamma$$

and many other similar reactions. In the equilibrium process, the Si that resulted from oxygen burning is partly "melted" into lighter nuclei and partly "cooked" into heavier nuclei. The end products of chains of such reactions are the mass-56 nuclei ($^{56}$Ni, $^{56}$Co, $^{56}$Fe). At this point there is no longer energy released in the capture reactions, and the process is halted.

Confirmation of this scenario can be seen with reference to Figure 19.9; the abundances of the elements formed by $\alpha$ capture ($Z =$ even) are far greater (an order of magnitude or more) than the neighboring odd-$Z$ elements. Notice also

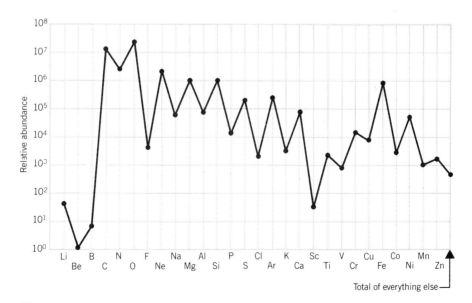

**Figure 19.9**  Relative abundances (by weight) of the elements beyond helium.

the dramatic (6 orders of magnitude) drop for elements below C; these elements are not formed in the stellar processes discussed so far.

To understand the rate of charged-particle reactions in stellar interiors, we must try to duplicate the reaction conditions using accelerators on Earth. The required energies are not large ($\sim$ MeV), but we require beams of the highest possible intensity (because these charged-particle reactions are strongly inhibited at low energy by the Coulomb barrier penetration factors) and best possible energy resolution (to study the behavior near discrete resonances or specific excited states).

The reaction probability in a stellar environment can be calculated in a way similar to the fusion reaction rate, as in Section 14.2. The reacting particles (a + X) are described by a thermal distribution

$$n(E)\,dE \propto e^{-E/kT}\sqrt{E}\,dE \qquad (19.23)$$

while the cross section has the basic form of

$$\sigma(E) \propto \frac{1}{E}\,e^{-2G} \qquad (19.24)$$

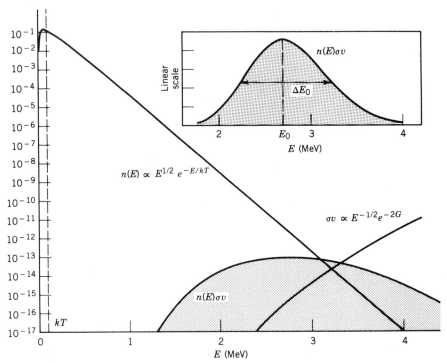

**Figure 19.10** The dependence of $n(E)$ and $\sigma v$ on energy. Their product, which is proportional to the reaction rate, is the shaded region. The inset at the top shows the reaction rate plotted on a linear scale, illustrating the peak energy $E_0$ and the width $\Delta E_0$. Notice that the reaction rate peaks at energies far above $kT$. The curves are drawn for the reaction $^{12}C + {}^{12}C$ at a temperature corresponding to $kT = 0.1$ MeV.

where $G$ is the Gamow factor, as in Equation 8.17:

$$G \cong \frac{e^2}{4\pi\epsilon_0} \frac{\pi Z_a Z_X}{\hbar v} \qquad (19.25)$$

$$2G \cong Z_a Z_X A_{eff}^{1/2} E^{-1/2} \qquad (19.26)$$

All energies and velocities are given in the center of mass. The "effective mass number" $A_{eff}$ is $A_a A_X / (A_a + A_X)$. In Equation 19.26, $E$ is given in MeV.

The barrier penetration factor increases with increasing energy, while the number of particles decreases. Figure 19.10 shows the overlap region between the two functions of energy. The reaction probability is large only in the shaded region where the two distributions overlap, and thus the usual measure of the "effective energy" of a thermal distribution, $kT$, is totally inappropriate to characterize these reactions, which occur exclusively at energies much larger than $kT$.

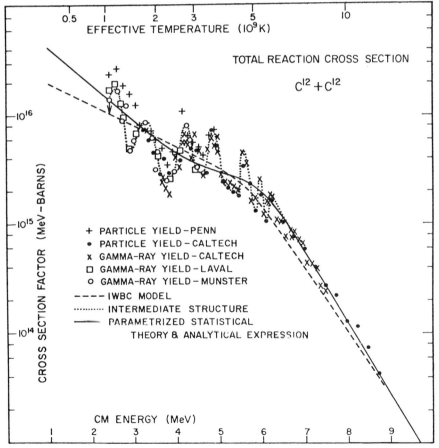

**Figure 19.11**  The cross-section factor $S(E)$ for $^{12}C + ^{12}C$. From W. A. Fowler, *Rev. Mod. Phys.* **56**, 149 (1984).

The reaction rate depends on the product $\sigma v$ and on the number of particles available at a specific energy

$$\text{rate} \propto n(E)\sigma(E)v$$

$$\propto \left(\sqrt{E}\, e^{-E/kT}\right)\left(\frac{1}{E}\, e^{-2G}\right)\sqrt{E} = e^{-E/kT - 2G} \qquad (19.27)$$

which describes the overlap region in Figure 19.10. The characteristic energy of the process can be evaluated in terms of the peak energy $E_0$ and width $\Delta E_0$ of the distribution:

$$E_0 = \left(\tfrac{1}{2}kTZ_a Z_X A_{\text{eff}}^{1/2}\right)^{2/3} \qquad (19.28)$$

$$\Delta E_0 = 2^{7/6}3^{-1/2}\left(Z_a Z_X A_{\text{eff}}^{1/2}\right)^{1/3}(kT)^{5/6} \qquad (19.29)$$

To obtain information on the reaction rate in the interior of a star, we should study reactions with accelerators, not at energies of $kT$, but at energies of $E_0$. In the case of $^{12}C + {}^{12}C$, for example, to duplicate stellar conditions at $kT = 0.1$ MeV (that is, $T = 10^9$ K), we should do the reaction at center-of-mass energy $E_0 = 2.3$ MeV.

Let's rewrite Equation 19.24 as

$$\sigma(E) = \frac{1}{E}\, e^{-2G} S(E) \qquad (19.30)$$

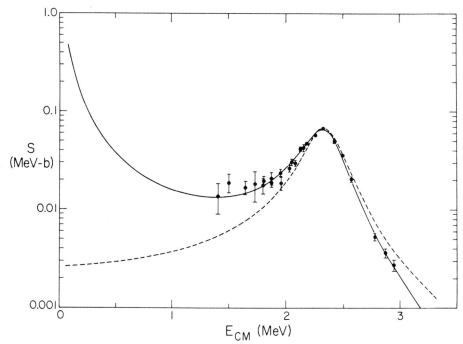

**Figure 19.12** Cross-section factor $S(E)$ for $^{12}C + \alpha \rightarrow {}^{16}O + \gamma$. The dashed curve is a theoretical fit that ignores the contributions of the $^{16}O$ bound states, while the solid curve is a fit that includes the effect of the bound states and gives much better agreement with the experimental data. From S. E. Koonin, T. A. Tombrello, and G. Fox, *Nucl. Phys. A* **220**, 221 (1974).

where $S(E)$ includes all nuclear-structure information other than the barrier penetration factor. For example, if the cross section is to be measured near a resonance, then

$$S(E) = g\Gamma_{aX}\Gamma_{bY}\frac{1}{(E - E_R)^2 + \Gamma^2/4} \qquad (19.31)$$

where $\Gamma_{aX}$ is the noncoulomb contribution to the entrance width (involving only nuclear wave functions). To compute the reaction rate, we must determine the factor $S(E)$, which can be directly obtained from the cross sections:

$$S(E) = E\sigma(E)\,e^{2G} \qquad (19.32)$$

Figure 19.11 shows the $^{12}C + ^{12}C$ cross section factor $S(E)$, which has a smooth nonresonant background along with several peaks illustrating some resonant structure. The $Q$ value for the $^{12}C + ^{12}C \rightarrow ^{24}Mg$ reaction is large (13.9 McV), so the structure corresponds to very highly excited states of the compound system.

Figure 19.12 shows the cross section factor $S(E)$ for $^{12}C + ^4He \rightarrow ^{16}O$. The structure is dominated by a single broad resonance, corresponding to an $\alpha$-unsta-

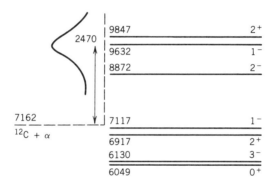

**Figure 19.13** Excited states of $^{16}O$. The broad resonance (which is also shown in Figure 19.12) at 2470 keV above the $^{12}C + \alpha$ threshold includes the $1^-$ and $2^+$ states at 9632 and 9847 keV. The interference of the $1^-$ and $2^+$ states just below the $^{12}C + \alpha$ threshold (7117 and 6917 keV) has a substantial effect on the calculated cross section.

ble state at about 2.5 MeV above the $^{12}$C $+\,^4$He threshold (Figure 19.13). The calculated cross section agrees very well with the measured values, but the two $^{16}$O excited states just below the $^{12}$C $+\,^4$He energy have a substantial influence on the calculated cross section and cannot be neglected.

Careful determinations of charged-particle reaction cross sections for energies in the MeV range, in addition to the most complete knowledge about the structure and properties of excited states, are essential in calculating the reaction rates and interior temperatures for models of stellar interiors. For his work in measuring such reactions, W. A. Fowler was honored with the 1983 Nobel Prize.

## 19.5 STELLAR NUCLEOSYNTHESIS ($A > 60$)

As Figure 3.16 shows, fusion reactions are not energetically favored above about $A = 60$. For these nuclei, neutron capture is the primary production mechanism. Let's consider, for example, $^{56}$Fe, which is the most abundant stable isotope of the most abundant element formed near the end of the chain of fusion reactions. In a flux of neutrons, a sequence of neutron-capture reactions will take place:

$$^{56}\text{Fe} + \text{n} \rightarrow {}^{57}\text{Fe} + \gamma$$

$$^{57}\text{Fe} + \text{n} \rightarrow {}^{58}\text{Fe} + \gamma$$

$$^{58}\text{Fe} + \text{n} \rightarrow {}^{59}\text{Fe} + \gamma$$

The next step in the process depends on the intensity of the neutron flux. The isotope $^{59}$Fe is radioactive, with a half-life of 45 days. If the neutron flux is so low that the probability of neutron capture is far smaller than once per 45 days, then $^{59}$Fe will $\beta$ decay to stable $^{59}$Co, which can then undergo neutron capture leading to radioactive $^{60}$Co. If, on the other hand, the probability of neutron capture is so great that the average time necessary to capture a neutron is small (seconds or less), then the sequence of neutron-capture reactions can continue to $^{60}$Fe ($t_{1/2} = 3 \times 10^5$ y), $^{61}$Fe ($t_{1/2} = 6$ min), $^{62}$Fe ($t_{1/2} = 68$ s), and beyond. When we finally reach an isotope that is so neutron rich that its half-life becomes shorter than the mean life before neutron capture, it will $\beta^-$ decay to an isotope with one higher atomic number. The sequence of neutron captures will begin again, until an extremely unstable isotope of this new sequence is reached, at which time $\beta^-$ decay increases the atomic number by one step again.

These two processes are responsible for the formation of the vast majority of the stable isotopes of the nuclei beyond $A = 60$. The first process, in which neutron capture occurs only on a long time scale, leaving time for all intervening $\beta$ decays to occur, is called the s (for slow) process. The second process, which does not leave time for any but the most short-lived decays, is called the r (for rapid) process. Figure 19.14 shows the s- and r-process paths near $^{56}$Fe.

Before we discuss the neutron-capture processes in detail, let's consider the origin of neutrons. Neutron emission following $\alpha$-particle reactions will be likely only if the original target nucleus is already neutron rich (and thus has a relatively weakly bound neutron). The neutron separation energies of $\alpha$-particle nuclei tend to be quite large ($^{12}$C, 18.7 MeV; $^{16}$O, 15.7 MeV; $^{20}$Ne, 16.9 MeV; $^{24}$Mg, 16.5 MeV; $^{28}$Si, 17.2 MeV . . .) and beyond the range of the incident alphas

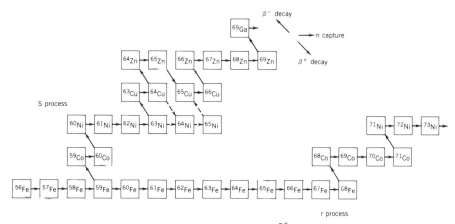

**Figure 19.14** The r- and s-process paths from $^{56}$Fe. The dashed lines in the s process represent possible alternative paths to $^{65}$Cu. Many different r-process paths are possible, as the short-lived nuclei $\beta$ decay; only one such path is shown.

in the stellar environment. More likely candidates are those with less tightly bound neutrons: $^{13}$C$(\alpha, n)^{16}$O ($Q = 2.2$ MeV) and $^{22}$Ne$(\alpha, n)^{25}$Mg ($Q = -0.48$ MeV). These reactions will occur during the helium burning or red-giant phase of stellar evolution, and the reaction rates can be calculated using the methods described in the last section. The results indicate that, in the red-giant temperature range of $1$–$2 \times 10^8$ K, the neutron density $n_n$ is of order $10^{14}/$m$^3$. We can estimate the reaction rate per target atom to be

$$r \simeq n_n \langle \sigma v \rangle \tag{19.33}$$

and at a temperature of $2 \times 10^8$ K, the velocity of a thermalized neutron is about $2 \times 10^6$ m/s. A typical neutron-capture cross section at these energies ($\sim 20$ keV) is about 0.1 b, and so

$$r \sim (10^{14} \text{ m}^{-3})(2 \times 10^6 \text{ m/s})(10^{-29} \text{ m}^2) = 2 \times 10^{-9}/\text{s}$$

or about one per 20 years. This is obviously an s-process situation.

To increase the reaction rate by some 10 orders of magnitude, as is needed for the r process, we obviously require a vast increase in the neutron flux. This is thought to occur in the violent stellar explosions known as supernovae, but there is as yet no general agreement on this hypothesis, nor is there an acceptable theory of the properties of supernovae. Yet the experimental evidence for the production of r-process nuclei is firm, and we will therefore accept that somewhere in the universe is a source (supernova, neutron star, etc.) that is capable of producing the necessary fluxes.

In the production of s-process nuclei, which occurs over a very long time, we expect the establishment of an approximate equilibrium situation. That is, each species has sufficient time to reach its equilibrium abundance, in which the production rates and destruction rates are equal. In a species $A$, with abundance $N_A$, the rate of change of $N_A$ is

$$\frac{dN_A}{dt} \propto \sigma_{A-1}N_{A-1} - \sigma_A N_A \tag{19.34}$$

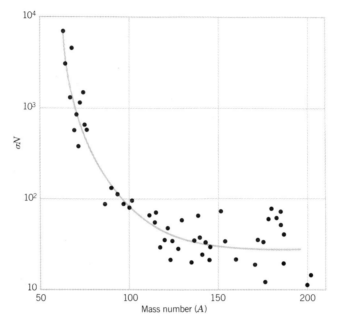

**Figure 19.15** The product $\sigma N$ approaches a constant value well above the Fe peak.

because $A$ is produced through neutron capture by nuclei of mass number $A - 1$, and $N_A$ is depleted by neutron captures of $A$ leading to $A + 1$. In equilibrium, $dN_A/dt = 0$, and we expect

$$\sigma_{A-1}N_{A-1} = \sigma_A N_A = \text{constant} \qquad (19.35)$$

Figure 19.15 shows the product $\sigma N$ plotted for the nuclei beyond $A = 60$. Just above the peak in the abundance at iron, Equation 19.35 is not satisfied, for the Fe abundance is not obtained through the s process. The product $\sigma N$ begins near Fe with very large values, but then decreases to an equilibrium value above $A = 100$. This smooth and gradual behavior indicates that the basic assumptions regarding s-process behavior are valid.

Figure 19.16 shows a small section of the chart of the nuclides, illustrating s- and r-process paths. Certain nuclei are accessible to both r- and s-process paths; trying to account for the abundances of these isotopes means we must be able to separate the two contributions. Where there is a gap in an isotopic sequence of stable nuclei (as between $^{120}$Sn and $^{122}$Sn, or between $^{121}$Sb and $^{123}$Sb), the s process cannot continue along the sequence and must proceed after $\beta$ decay to the next higher atomic number. Thus $^{122}$Sn and $^{124}$Sn can be produced only in the r process. Their abundances are roughly equal, 4.5 and 5.6%, but much smaller than that of $^{120}$Sn, 32.4%. As a first guess, we might estimate that about 5% of the $^{120}$Sn abundance of 32% is from the r process, while the remaining 27% is from the s process. On the other hand, the $\beta$ decays at mass 122, 123, and 124 from r-process nuclei terminate at stable $^{122}$Sn, $^{123}$Sb, and $^{124}$Sn, and are thus unable to reach the Te isotopes $^{122,123,124}$Te, which are *shielded* from the r process. They can be produced only through the s process.

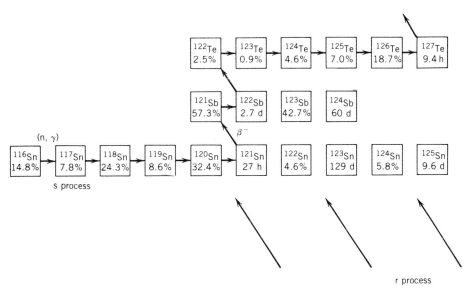

**Figure 19.16**   r- and s-process paths leading to Sn, Sb, and Te isotopes.

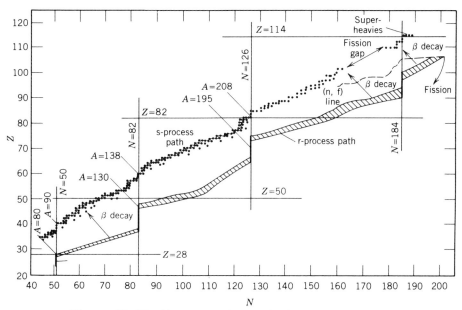

**Figure 19.17**   Neutron capture paths for r and s processes.

Figure 19.17 shows the complete r- and s-process paths leading to the stable isotopes in the chart of the nuclides. The s process proceeds in zigzag fashion through most of the stable isotopes, terminating at $^{209}$Bi, because there are no stable (or metastable) isotopes just above $A = 209$ through which the s process could continue. The r process has no such restriction and can continue until the fission half-lives are as short as the r-process capture times. Near the termination of the r process, it may be possible to produce superheavy nuclei; such a

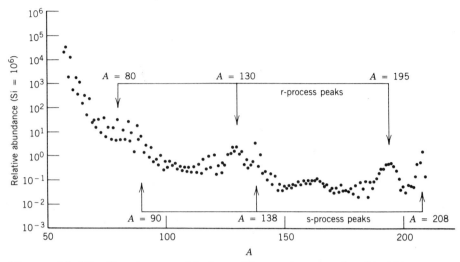

**Figure 19.18** Abundances of isobars. The peaks near $A = 80$, 130, and 195 originate from the $\beta$ decays of r-process progenitors with $N = 50$, 82, or 126. The peaks near $A = 90$, 138, and 208 result from s-process stable nuclei with $N = 50$, 82, or 126. Note the difference in abundance between odd-$A$ and even-$A$ nuclei.

possibility has inspired the search for evidence for superheavy nuclei in natural materials, thus far without success.

Near the magic numbers, the $\beta$ decay times become so short that an added neutron decays to a proton in a time small compared with the r-process capture time. This accounts for the vertical rises in the r-process path at $N = 50$, 82, and 126. As these nuclei subsequently $\beta$ decay toward the stable isobars, a slight overabundance of stable nuclei will result. This will occur near $A = 80$, 130, and 195, as indicated in Figure 19.18.

## 19.6 NUCLEAR COSMOCHRONOLOGY

In Section 6.7, we reviewed the use of radioactive dating methods to determine the age of the Earth (and of other solid objects in the Solar System). By examining the relative numbers of parent and daughter nuclei in decays with half-lives in the proper range ($\sim 1$–10 Gy), it is possible to deduce the time since the condensation of the object. The fundamental assumption of the method is that the "ingredients" of the Earth were thoroughly mixed before condensation, so that the previous decay products of a given species would be unlikely to be found close to a collection of atoms of that species. Since condensation, there was little opportunity for migration or loss of decay products and we can, with suitable care to account for the possibility of such losses, measure the ratio of parent and daughter nuclei and deduce the length of time that the daughters have been accumulating in that particular site. Using a variety of decay processes ($^{40}K \rightarrow {}^{40}Ar$, $^{87}Rb \rightarrow {}^{87}Sr$, $^{232}Th \rightarrow {}^{208}Pb$, $^{238}U \rightarrow {}^{206}Pb$) the oldest materials from the Earth, Moon, and meteorites are dated at 4.55 Gy. The good agreement obtained with decays of differing half-lives, and using materials from differing locations that have differing exposures to conditions that may cause loss of daughter activities, serves to confirm the validity of the method. The age of the Solar System is just a bit older than this, perhaps 4.6–4.8 Gy, representing the additional time necessary for the condensation of the solid material from the "protostar" or nebula.

The sequence of events leading to the present day is then as follows.

1. Big Bang, leading to formation of neutral atoms ($\sim 10^6$ y);
2. Condensation of galaxies and first-generation stars (time interval $\equiv \delta \sim$ 1–2 Gy);
3. Nucleosynthesis in stars and supernovae, leading to the formation of the present chemical elements (time interval $\equiv \Delta$);
4. Condensation of Solar System from debris of earlier stars (time interval $\equiv A_e$ = 4.6 Gy).

The total age of the universe ($A_u$) is then

$$A_u = \delta + \Delta + A_e \tag{19.36}$$

neglecting the first contribution, which is certainly very small compared with the others. The time $\delta$ for the gravitational interaction to bring together original galaxies and stars has been calculated to be in the range of 1–2 Gy. Since $A_e$ is well known, we can determine the age of the universe if it is possible to determine the duration $\Delta$ of nucleosynthesis in first-generation stars.

The basic procedure in this process is to compare observed abundances with those calculated according to a certain model and based on fundamental processes whose rate we can determine independently, primarily through measurements in our laboratories on Earth. Because of the many uncertainties in determining the magnitude of r- and s-process contributions in nuclei that can be produced through both processes, we choose for this procedure nuclei whose production is exclusively through only one process. For the r process, we use heavy nuclei ($A > 209$), which cannot be produced through the s process. We have already alluded to one technique of this sort in Chapter 13. The $^{235}U/^{238}U$ ratio changes with time, owing to their differing half-lives (0.7 Gy for $^{235}U$, 4.5 Gy for $^{238}U$). The present ratio (0.00720) is then considerably different from the value at the time of condensation of the Solar System (0.29). To extrapolate back beyond that time, to formation of $^{235}U$ and $^{238}U$ in the r process, we must know something of the production processes that ultimately result in $^{235}U$ and $^{238}U$. The r-process advance to heavier nuclei is halted when the fission half-lives become sufficiently short, and the heavy nuclei produced (which range up to $A \simeq 300$) then $\alpha$ and $\beta$ decay to produce $^{235}U$ and $^{238}U$. The isotope $^{238}U$ can be formed through $\beta$ decays from the r-process nucleus at mass 238, and from $\alpha$ decays of nuclei at mass 242, 246, and 250 that result (ultimately) from $\beta$ decays of r-process nuclei at those masses. Figure 19.19 illustrates the process. Nuclei above $A = 250$ decay primarily through spontaneous fission and thus do not contribute to the uranium abundances. We can therefore make a calculation of the "original" ratio $^{235}U/^{238}U$, based on these estimates of r-process production. Figure 19.20 is an example of the calculated results, illustrating two extreme cases—a steady-state

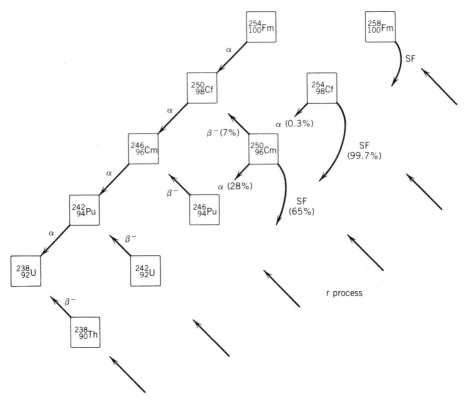

**Figure 19.19** Progenitors of $^{238}$U. When spontaneous fission (SF) becomes the most likely decay process (for $^{254}$Cf, $^{258}$Fm, and beyond), the r process cannot feed the $\alpha$-decay chain leading to $^{238}$U. Production of $^{238}$U from the r process therefore occurs only at $A = 238$, 242, 246, and 250.

production, in which the ratio is constant in time until production ends, following which the ratio decays to its present value over 4.5 Gy, and a sudden production, which produces all of the $^{235}$U and $^{238}$U in their original abundance ratio (1.64), which then decays freely to the present value. In the former case, we can make no prediction of the length of time during which this production occurred; in the latter case, the time $\Delta$ between nucleosynthesis and condensation of the Solar System is 2 Gy.

A more acceptable scenario is that during the interval $\Delta$, r-process events went on gradually and continuously, as also did the relative decays of $^{235}$U and $^{238}$U. Constructing an appropriate model for the production of r-process nuclei, we can then check it against abundances directly measured for stable nuclei produced exclusively through the r process. Finally, we can use the model to compute the unmeasurable abundances of the r-process progenitors of $^{235}$U and $^{238}$U, and we can extrapolate the production-plus-decay rates backward in time until they match the calculated abundance ratio at the beginning of galactic nucleosynthesis. Figure 19.21 shows the result of this calculation, along with a similar result for the $^{232}$Th$/^{238}$U ratio. The deduced most probable value for $\Delta$ is 6 Gy, with a

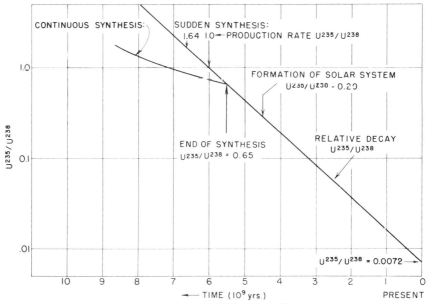

**Figure 19.20** Ratio of abundances of $^{235}$U and $^{238}$U over time, leading to the present value of 0.0072. Various scenarios are shown which may have produced the original nuclides. From Burbidge, Burbidge, Fowler, and Hoyle, *Rev. Mod. Phys.* **29**, 547 (1957).

permitted range (based on the uncertainties inherent in the calculations) from about 4 to 9 Gy.

Such models may also include a "pre-solar spike," a sudden deposit of r-process products just at the time condensation began. (Indeed, there are some theories that suggest a cataclysmic event such as a supernova was responsible for the condensation of the Solar System.) Evidence in support of the spike comes from the abundances of the products of short-lived (on the Gy time scale) decays of r-process nuclei, including $^{129}$I (16 My), $^{244}$Pu (81 My), and $^{26}$Al (0.72 My).

An alternative approach to determining $\Delta$ comes from an s-process argument. Figure 19.22 shows the s- and r-process paths through the Os and Re nuclei. The isotope $^{187}$Re is an r-process nucleus and decays to $^{187}$Os with a half-life of 40 Gy, in just the proper range to be used as a decay chronometer. If we could compare the abundance of $^{187}$Re with that of its $^{187}$Os daughter, we could deduce the interval during which the $^{187}$Re has been decaying. Unfortunately, $^{187}$Os is also produced directly in the s process, so this determination requires a careful computation of the s-process abundance of the Os isotopes. We can use the standard s-process formula, $\sigma_A N_A = \sigma_{A-1} N_{A-1}$, for this purpose if we know the cross sections for neutron capture, given that the abundance of $^{186}$Os is from the s process alone, since it is shielded from the r-process by $^{186}$W. A difficulty with this procedure is that the first excited state of $^{187}$Os is at an energy of only 10 keV; at stellar temperatures, $kT \approx 20$ keV, and the excited state is expected to be strongly populated. Moreover, the excited state has spin $I = \frac{3}{2}$, while the ground state has $I = \frac{1}{2}$; thus the statistical weight $(2I + 1)$ of the excited state is

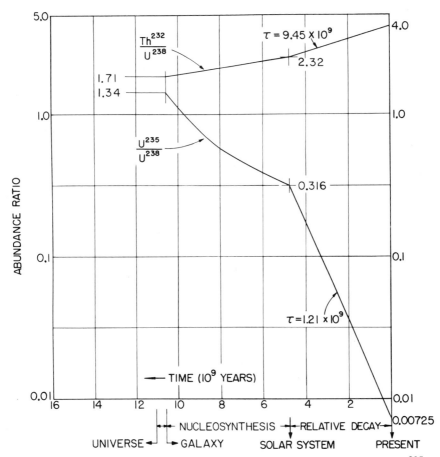

**Figure 19.21** Possible scenario leading to present abundance ratio of $^{235}$U and $^{238}$U. The r process and decay contribute during the nucleosynthesis era ($\Delta$), perhaps resulting from many supernovas ejecting r-process nuclei into the interstellar medium. After the solar system began to condense, no new r-process nuclei were added; the abundance ratio then decayed freely to its present value. From W. A. Fowler, *Quarterly Journal of the Royal Astronomical Society* (1987); courtesy W. A. Fowler.

twice that of the ground state. The analysis therefore requires measurement of the neutron-capture cross section of the excited state as well as the ground state. The measurement of the ground-state capture cross section is relatively easy, but the excited state is not populated in 300 K Earthly laboratories, and it cannot serve as a target for neutron reactions. Two differing approaches have been used to attack this problem. In one, the capture cross section of the $^{189}$Os ground state is measured. This state happens to have the same nuclear structure as the $^{187}$Os excited state, and its measurement can be applied to the determination of the cross section of the $^{187}$Os excited state. In the second method, the inelastic

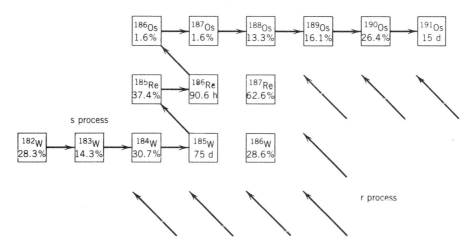

**Figure 19.22**    r- and s-process formation of Re and Os isotopes.

scattering cross section has been measured for $^{187}$Os(n, n')$^{187}$Os*, permitting a determination of the scattering amplitudes and phase shifts, from which the capture cross section can be calculated. Both measurements are done with neutron energies in the astrophysical range (30–60 keV). These methods give values for $\Delta$ in the range of 9–12 Gy, with uncertainties of about $\pm 20\%$. The lower end of this range is consistent with the value of $\Delta$ determined from the $^{235}$U/$^{238}$U ratio, but both methods depend on assumptions that need further theoretical and experimental tests before either calculation is accepted as reliable. If we accept the value $\Delta = 8 \pm 2$ Gy as a fair compromise, then

$$A_{\mathrm{u}} = 14 \pm 2 \text{ Gy}$$

Other model-dependent determinations of the age of the universe may be consistent with this value. The Hubble age $H^{-1}$ is 15 Gy for $H = 67$ km/s/Mpc, but $H^{-1}$ is equal to $A_{\mathrm{u}}$ only in a universe that has been expanding at a constant rate since $t = 0$. Any realistic cosmology includes a gravitational deceleration, and so $A_{\mathrm{u}} < H^{-1}$. If there is just enough matter to "close" the universe (that is, eventually to halt the expansion only as $R \rightarrow \infty$), then $A_{\mathrm{u}} = \frac{2}{3}H^{-1} = 10$ Gy, in rough agreement with the above value deduced from nucleosynthesis. However, the Hubble parameter itself is uncertain by a factor of 2 (the present literature includes values as high as 100 km/s/Mpc), so perhaps the agreement is accidental; there are also claims for values of $\Delta$ from nucleosynthesis that would lead to an age of about 18 Gy. In any case, the determination of the value of $\Delta$ is likely to remain an active pursuit in experimental and theoretical nuclear physics, demonstrating once again the fundamental importance of nuclear physics for understanding astrophysical processes.

## REFERENCES FOR ADDITIONAL READING

There are many excellent and inspiring popular-level books on cosmology that provide interesting background reading for this chapter. Among them are T. Ferris, *The Red Limit* (New York: Morrow, 1977); S. Weinberg, *The First Three Minutes* (New York: Basic Books, 1977); J. Silk, *The Big Bang* (San Francisco: Freeman, 1980).

Other elementary and intermediate (in mathematical level) texts are E. R. Harrison, *Cosmology* (Cambridge: Cambridge University Press, 1981); M. Rowan-Robinson, *Cosmology* (Oxford: Clarendon Press, 1981); P. J. E. Peebles, *Physical Cosmology* (Princeton: Princeton University Press, 1971); M. Berry, *Principles of Cosmology and Gravitation* (Cambridge: Cambridge University Press, 1976); P. T. Landsberg and D. A. Evans, *Mathematical Cosmology* (Oxford: Clarendon Press, 1977); J. V. Narlikar, *Introduction to Cosmology* (Portola Valley, CA: Jones and Bartlett, 1983); H. S. Goldberg and M. D. Scadron, *Physics of Stellar Evolution and Cosmology* (New York: Gordon & Breach, 1981).

The advanced mathematical theory can be found in S. Weinberg, *Gravitation and Cosmology* (New York: Wiley, 1972) and C. W. Misner, K. S. Thorne, and J. A. Wheeler, *Gravitation* (San Francisco: Freeman, 1973).

No detailed study of nucleosynthesis would be complete without a reading of the classic review article known in the literature as $B^2FH$: G. R. Burbidge, E. M. Burbidge, W. A. Fowler, and F. Hoyle, *Rev. Mod. Phys.* **29**, 547 (1957). Additional details on nucleosynthesis can be found in H. Reeves, *Stellar Evolution and Nucleosynthesis* (New York: Gordon & Breach, 1968) and in D. D. Clayton, *Principles of Stellar Evolution and Nucleosynthesis* (New York: McGraw-Hill, 1968). More recent review papers on nucleosynthesis include J. W. Truran, *Ann. Rev. Nucl. Part. Sci.* **34**, 53 (1984), G. J. Mathews and R. A. Ward, *Reports on Progress in Physics* **48**, 1371 (1985), and A. M. Boesgaard and G. Steigman, *Ann. Rev. Astron. Astrophys.* **23**, 319 (1985).

The contributions of George Gamow and William A. Fowler to the fields of cosmology and nucleosynthesis have been celebrated in two volumes of short contributions by their colleagues: *Cosmology, Fusion, and Other Matters*, edited by F. Reines (Boulder: Colorado Associated University Press, 1972) and *Essays in Nuclear Astrophysics*, edited by C. A. Barnes, D. D. Clayton, and D. N. Schramm (Cambridge: Cambridge University Press, 1982). Fowler's contributions to nuclear astrophysics are also summarized in his Nobel Prize address, reprinted in *Rev. Mod. Phys.* **56**, 149 (1984).

The strong relationship between nuclear physics and astrophysics is emphasized in the review by C. Rolfs and H. P. Trautvetter, *Ann. Rev. Nucl. Part. Sci.* **28**, 115 (1978). The similarly strong connection between cosmology and particle physics is discussed by G. Steigman, *Ann. Rev. Nucl. Part. Sci.* **29**, 313 (1979).

For recent work on the determination of the age of the universe from the Os abundances and neutron cross sections, see: J. C. Browne and B. L. Berman, *Phys. Rev. C* **23**, 1434 (1981); R. R. Winters and R. L. Macklin, *Phys. Rev. C* **25**, 208 (1982); R. L. Hershberger et al., *Phys. Rev. C* **28**, 2249 (1983).

## PROBLEMS

1.  (a) In an expanding universe dominated by matter, the mass density should decrease like $R^{-3}$. Use Equation 19.3 with $k = 0$ and $\Lambda = 0$ to find $R(t)$. This case is called the Einstein-deSitter model. (b) Calculate the corresponding relationship between $H^{-1}$ and the age of the universe in this model.

2.  Derive Equation 19.5 from Equation 19.4.

3.  Show that the constants in Equations 19.5 and 19.6 evaluate to give Equation 19.7.

4.  Integrate Equations 19.8 and 19.9, and evaluate the constants to obtain Equations 19.10 and 19.11.

5.  (a) During what era would the universe have been rich in $K^0$ and $\overline{K}^0$ mesons? (b) Could the interactions of these mesons with the existing nucleons and antinucleons have been responsible for the imbalance of matter and antimatter?

6.  On a logarithmic time scale from $10^{-6}$ s to $10^3$ s, make a qualitative sketch showing the time dependence of the relative numbers of free p, n, $^2$H, $^3$H, $^3$He, and $^4$He.

7.  (a) Assuming a nucleon-to-photon ratio of $10^{-10}$, repeat the calculation of Section 19.3 for the age of the universe when deuterium formation is permitted. (b) Assume all of the neutrons combine with an equal number of protons to make $\alpha$ particles. After taking into account the decrease in the number of neutrons during the time calculated in part (a) from their radioactive decay, find the resulting ratio $N_{He}/N_p$.

8.  (a) Calculate the energy $E_0$ at which the stellar reaction rate, Equation 19.27, reaches its peak. (b) Calculate the width $\Delta E_0$ by expanding Equation 19.27 in a Taylor series about $E_0$ and finding the energy $E_{1/2}$ at which the rate drops by half. For this calculation the width $\Delta E_0$ is the full width at half maximum, $2(E_{1/2} - E_0)$.

9.  (a) At what center-of-mass energy should we study the silicon-burning reaction $^4$He + $^{28}$Si to duplicate stellar conditions? (b) If $^4$He are incident in the laboratory on $^{28}$Si at rest, what energy should be chosen?

10. The CNO cycle that may contribute to energy production in stars similar to the Sun begins with the reaction p + $^{12}$C → $^{13}$N + γ. Assuming the temperature near the center of the Sun to be $15 \times 10^6$ K, find the peak energy and width of the reaction rate.

11. Trace the s-process path, in analogy with Figure 19.14, from $^{69}$Zn to $^{90}$Zr. Identify any stable nuclei in this region that cannot be reached in the s process, and note which nuclei are shielded from the r process.

12. Continue the previous problem from $^{90}$Zr to $^{116}$Sn.

# 20

# APPLICATIONS OF NUCLEAR PHYSICS

Throughout this text, we have discussed applications of nuclear techniques, including fission and fusion power, radioactive dating, crystal structure determinations, and so on. Many of the methods and practices of nuclear physics have found their way into still other applications, in particular in the quantitative study of chemical composition and in the diagnosis and treatment of disease.

In this chapter we explore some of these additional applications, with the goal not so much to make a comprehensive study as to indicate the diversity of disciplines to which these techniques may be applied.

## 20.1 TRACE ELEMENT ANALYSIS

Nuclear physics techniques often have a great advantage over traditional chemical techniques for the determination of the elemental composition of materials. Two techniques in particular have become widely used: neutron activation analysis (NAA) and particle-induced X-ray emission (PIXE, pronounced "pixie").

In the NAA technique a small sample of the material to be evaluated is exposed to a flux of thermal neutrons from a reactor. Nuclei of stable elements can become radioactive through the neutron capture $(n, \gamma)$ reaction. Many of these radioactive nuclei decay through $\beta$ and subsequent $\gamma$ emission, and the $\gamma$ rays are characteristic of that particular decay process. By precise determination of the $\gamma$-ray energies, obtained when the irradiated sample is removed from the reactor and counted with a Ge detector, it is usually possible to determine not only *which* isotopes are present, but from the $\gamma$ intensities it can be determined exactly *how much* of the original capturing isotope was present and consequently the amount of that element present in the sample. On-line computers and even microprocessor-based multichannel analyzers can automatically determine energies and identify peaks, and so a large number of samples can be rapidly counted.

NAA is an example of a nondestructive testing method. Aside from a small amount of induced radioactivity (which decreases with time), the sample is essentially unaffected by this process. In the case of a valuable painting or a fragment of ancient pottery, the sample can be restored undamaged in its original condition. In the case of material subject to forensic analysis and used in criminal investigations, it can be saved as evidence or even subjected to reanalysis (or other analysis) at a later time. Contrast this situation with ordinary chemical techniques, in which determination of the elements present in a sample requires that it be vaporized, dissolved, burned, or otherwise permanently altered.

The activity resulting from neutron capture is given in convenient form by Equation 12.23:

$$\mathscr{A} = 0.602 \frac{m}{A} \sigma \phi (1 - e^{-\lambda t}) \tag{20.1}$$

in which we will work in decays per second rather than curies. Here the flux $\phi$ is in neutrons/cm$^2$/s and the cross section for thermal neutron capture is in barns; $m$ is the mass (in grams) of the isotope of mass number $A$ in the sample. We will assume a typical neutron flux of about $\phi = 10^{13}$ n/cm$^2$/s and an irradiation time of $t = 1$ h. A Ge detector may have an efficiency of the order of 10% for $\gamma$ rays of energies normally encountered following $\beta$ emission. Let's consider a sample containing $10^{-12}$ g of Mn, which has only one stable isotope, $^{55}$Mn. The neutron-capture cross section of $^{55}$Mn is 13.3 b, leading to 2.58-h $^{56}$Mn, which emits prominent $\gamma$'s of 0.847 MeV (99% of decays), 1.811 MeV (29%), and 2.110 MeV (15%), among several others in the 1% range. Equation 20.1 then gives 0.3 decays/s, or 0.03 counts/s of the most intense $\gamma$ in a 10%-efficient detector. Counting for 1 h gives more than 100 counts in the 0.847-MeV line, which should be sufficient to identify it and infer the presence of Mn in a sample. The detection limit of Mn is therefore of the order of $10^{-11}$–$10^{-12}$ g with NAA. Table 20.1 shows a sampling of other elements that can be analyzed by NAA. In all, 71 (of the 83 naturally occurring) elements can be analyzed in this method, with widely varying sensitivities (Figure 20.1). Only the eight lightest elements (H-O) and P, S, Tl, and Bi cannot be detected within reasonable limits.

NAA has applications in any area in which we require quantitative knowledge of minute quantities of materials. Atmospheric pollutants, especially particulate

**Table 20.1**  Neutron Activation Analysis

| Element | Isotope (Abundance) | $\sigma$ (b) | Activity (s$^{-1}$) per $\mu$g | $t_{1/2}$ | Prominent $\gamma$ (MeV) | Counts/h | Detection Limit ($\mu$g) |
|---|---|---|---|---|---|---|---|
| Mn | 55 (100%) | 13.3 | $3 \times 10^5$ | 2.58 h | 0.847 (99%) | $1 \times 10^8$ | $10^{-5}$–$10^{-6}$ |
| Ag | 107 (51%) | 35 | $1 \times 10^6$ | 2.4 min | 0.633 (2%) | $4 \times 10^5$ | $10^{-3}$–$10^{-4}$ |
| Au | 197 (100%) | 99 | $3 \times 10^4$ | 2.7 d | 0.412 (95%) | $1 \times 10^7$ | $10^{-4}$–$10^{-5}$ |
| Sc | 45 (100%) | 13 | $6 \times 10^2$ | 84 d | 0.889 (100%) | $2 \times 10^5$ | $10^{-3}$–$10^{-4}$ |
| Cu | 65 (31%) | 2.3 | $7 \times 10^4$ | 5.1 min | 1.039 (9%) | $3 \times 10^5$ | $10^{-3}$–$10^{-4}$ |
| Ir | 193 (61%) | 110 | $8 \times 10^4$ | 17 h | 0.328 (10%) | $3 \times 10^6$ | $10^{-4}$–$10^{-5}$ |
| Sb | 121 (57%) | 6 | $2 \times 10^3$ | 2.8 d | 0.564 (66%) | $4 \times 10^5$ | $10^{-3}$–$10^{-4}$ |
| Ni | 64 (1.2%) | 1.5 | $4 \times 10^2$ | 2.5 h | 1.481 (25%) | $3 \times 10^4$ | $10^{-1}$–$10^{-2}$ |
| Pb | 206 (25%) | 0.03 | $2 \times 10^2$ | 0.8 s | 0.570 (98%) | $2 \times 10^1$ | $10^{-1}$–$10^{-2}$ |
| Zr | 96 (2.8%) | 0.05 | $4 \times 10^0$ | 17 h | 0.747 (92%) | $1 \times 10^3$ | $10^{-1}$–$10^0$ |
| As | 75 (100%) | 4.5 | $9 \times 10^3$ | 26 h | 0.559 (43%) | $1 \times 10^6$ | $10^{-3}$–$10^{-4}$ |
| Hg | 202 (30%) | 4 | $2 \times 10^1$ | 47 d | 0.279 (77%) | $6 \times 10^3$ | $10^{-2}$–$10^{-3}$ |

Activity and counting rate are based on irradiating a 1-$\mu$g sample for one hour in a flux of $10^{13}$ neutrons/cm$^2$/s. The detection limit is based on accumulating 100 counts in one hour using a detector with an efficiency of 10%.

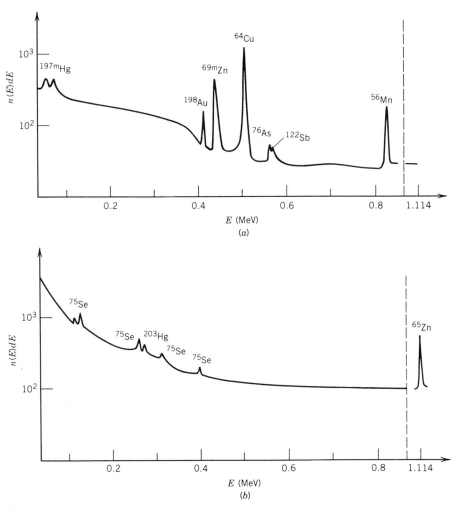

**Figure 20.1** Gamma-ray spectrum following neutron activation of a sample of human hair. Spectrum (*a*) was taken 4 h after the irradiation, and (*b*) was taken 4 weeks later. Trace elements revealed include mercury, gold, zinc, copper, arsenic, antimony, manganese, and selenium. From D. DeSoete et al., *Neutron Activation Analysis* (New York: Wiley-Interscience, 1972).

matter, can be collected and analyzed for the presence of elements that might indicate the origin of the pollution. There are many applications in forensic science where NAA has aided in criminal investigations. Rifle and handgun ammunition contains Ba and Sb compounds, and firing a gun leaves residues of these elements in $\mu$g quantities on the back of the hand. These residues are 1000 times above the detection threshold of NAA for Ba and Sb. Analysis of fragments of bullets recovered in the assassination of President John Kennedy has helped to resolve long-standing doubts about the number of bullets actually fired.

Arsenic and mercury poisoning can be detected in the hair; it is even possible to study preserved samples of the hair of historical figures. Analysis of strands of

Napoleon's hair shows them to be rich in arsenic, which was an ingredient in medicines used in that period. Isaac Newton's hair showed high levels of mercury. Detailed NAA of the mineral content of various ancient pottery fragments shows which pieces have common origins, and systematic studies of fragments found at various archeological excavations can help us to trace the trading routes of ancient cultures. An anomalous concentration of Ir was discovered by NAA at several sites throughout the world at a depth corresponding to the boundary between the Cretaceous and Tertiary periods, about 65 million years ago. It has been hypothesized that a large Ir-rich meteor struck the Earth and that the debris from the impact raised a cloud which blocked sunlight for such an extended period that the extinction of the dinosaurs eventually resulted. Eventually the Ir settled to Earth in the geological strata we observe today.

Particle-induced X-ray emission (PIXE) is also used for analysis of small quantities of elements. In this technique, a thin sample of the material is placed in the target area of an accelerator and bombarded with protons, $\alpha$'s, or even heavy ions. The Coulomb interaction between the incident particle and the target can result in ionization of the target atoms, and the cross section for removal of an inner (K-shell or L-shell) electron can be calculated. Once the vacancy is created, it will be quickly filled because outer electrons will jump to the inner shell with the accompanying emission of an X ray. The K- and L-shell X-ray energies are smoothly varying functions of the atomic number $Z$, and by observing the emitted X rays we can deduce the elements present in the sample. Owing to the short range of charged particles in matter and to the low energies of emitted X rays (10–100 keV), it is necessary to work with thin samples.

The cross sections for K-shell ionization vary smoothly (contrary to the case of capture of thermal neutrons). For light elements, the cross section is of the order 100 b, but it decreases to below 1 b for heavy elements. The sensitivity of PIXE therefore decreases with increasing $Z$ of the target, but it is not impossible to obtain quantitative results all the way up to Pb. Figure 20.2 shows the variation of the K- and L-shell X-ray productions with $Z$. In evaluating the *production* cross sections for X rays, it is necessary first to calculate the *formation* cross sections to produce a vacancy in the appropriate shell, then to calculate the probability that the vacancy will be filled by an X-ray transition of a particular type. The data plotted in Figure 20.2 include these factors. The ionization cross sections also vary with incident energy, but here the variation is well approximated by a universal curve, as shown in Figure 20.3. The cross sections are maximum at about $T_p/\lambda E_K = 1$, where $T_p$ is the incident proton energy and $E_K$ is the ionization energy.

$$T_p = \frac{m_p}{m_e} E_K \tag{20.2}$$

In the standard treatment of the hydrogenic atom, the (negative) total energy of an electron is $E = T + V$, and for the Coulomb interaction it happens that $V = -2T$, so that $|E| = T$. Thus the binding energy of a K-shell electron (or the ionization energy $E_K$) is numerically equal to the kinetic energy, and Equation 20.2 is then equivalent to

$$v_p = v_e \tag{20.3}$$

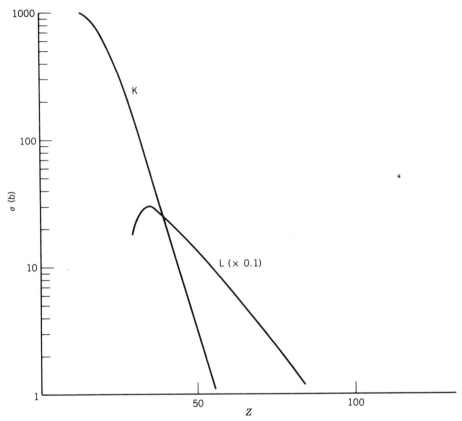

**Figure 20.2** Production cross sections of K and L X rays for incident protons of energy 5.0 MeV.

That is, the cross section reaches a maximum when the incident proton has a velocity equal to the Bohr orbital velocity of the electron. It is of course physically quite reasonable that this should be so.

The theoretical understanding of PIXE cross sections means that the X-ray yield can be used quantitatively to deduce the amounts of various elements present in the sample. We can estimate the counting rates and thereby deduce the sensitivity of the technique to compare with that of NAA. Let's assume a thin target of $10^{-4}$ g/cm$^2$ with an area of 1 cm$^2$, irradiated uniformly with a proton beam of 0.1 $\mu$A $(= 6 \times 10^{11}$ protons/s$)$. The reaction rate, for a cross section of 1000 b, is

$$R = \sigma I n$$
$$= (1000 \text{ b})(6 \times 10^{11} \text{ p/s})(10^{18} \text{ target atoms/cm}^2)$$
$$\simeq 10^9 \text{ reactions/s}$$

The X-ray yield ranges from about 1% for light atoms to 100% for heavy atoms, and while the detection efficiency for X-rays (using a Si(Li) detector) may approach 100%, the reaction geometry requires the detector to be relatively far from the sample, so the detector solid angle will be less than 1% of $4\pi$. The counting rate may thus be only $10^5$/s and using 1 h as the basis (as we did for NAA), we could accumulate $10^8$ counts for our original sample of $10^{-4}$ g. If we

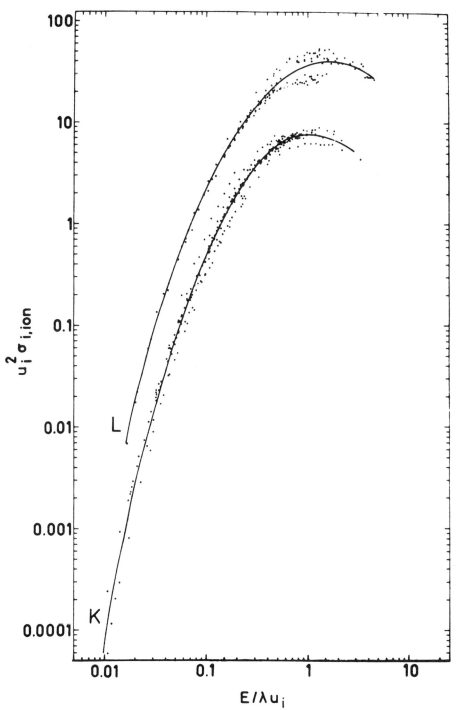

**Figure 20.3**   K- and L-ionization cross sections from proton bombardment. The vertical axis is the cross section $\sigma$ multiplied by the square of the electron binding energy $u_i$. The horizontal axis shows $E/\lambda u_i$, where $E$ is the proton kinetic energy and $\lambda = m_p/m_e$. From S. A. E. Johansson and T. B. Johansson, *Nucl. Instrum. Methods* **137**, 473 (1976).

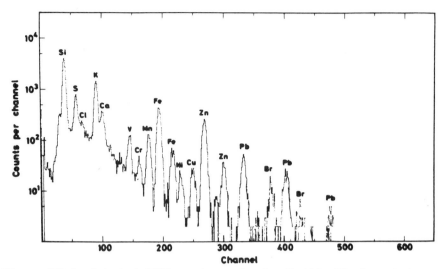

**Figure 20.4** A typical PIXE spectrum, showing the identification of elements present in an aerosol, following bombardment with 2-MeV protons. From S. A. E. Johansson and T. B. Johansson, *Nucl. Instrum. Methods* **137**, 473 (1976).

again require at least 100 counts to make an observable peak in the spectrum, we will be able to detect quantities down to $10^{-10}$ g or $10^{-4}$ $\mu$g, which makes the technique quite comparable to NAA in its sensitivity, and for many elements it may be superior to NAA. If the beam can be focused to a diameter of 1 mm without damaging the target from the heating, the sensitivity increases by a factor of 100.

For precise quantitative analyses, we would need to know exactly the cross sections and the beam current and its profile over the sample. This may limit the ultimate precision of PIXE to of the order of $\pm 10\%$ in quantitative analysis, which is not significantly inferior to NAA. Moreover, NAA is limited in its selectivity by isotopes that may have small cross sections, short radioactive decay times, or no $\gamma$ emissions. PIXE has no such limitations—all elements can be detected with relatively uniform sensitivity. One disadvantage is that the X-ray spectrum contains many components of the K and L lines, which can interfere and make the analysis difficult. For example, the Pb $L_{\alpha 1}$ line has an energy of 10.55 keV, while the As $K_{\alpha 1}$ line is at 10.54 keV; Si(Li) detectors typically have energy resolution of 100 eV and are not able to resolve these two peaks at a 10-eV separation. Such interferences occur in well-known cases, and other components in the spectrum can be used to deduce the presence of the individual elements.

Figure 20.4 shows a PIXE spectrum analyzed to reveal the presence of many elements.

## 20.2 MASS SPECTROMETRY WITH ACCELERATORS

The technique of radioactive dating of ancient materials has, as discussed in Chapter 6, yielded precise information on the ages of minerals and organic materials. The method has a fundamental limitation, however; in processes involving counting long-lived radioactive decays, only a very small fraction of the atoms present in the sample are counted. This means that large samples or long counting times are necessary, and this situation becomes worse with the increasing age of the sample.

Consider for example the $^{14}$C decay used in radiocarbon dating. The $^{14}$C content of present-day organic material is about $10^{-12}$ relative to $^{12}$C, and the decay rate is about 15 per minute per gram of total carbon. The half-life of $^{14}$C is 5730 y; thus in a 1-g sample of age about 10,000 y the decay rate would be about 4/min. A statistically precise observation of 10,000 decays ($\pm 1\%$ statistical uncertainty) would then take about two days, if the counting equipment were 100% efficient. The 1-g sample contains about $5 \times 10^{22}$ atoms of carbon or about $10^{10}$ atoms of $^{14}$C. The counting procedure therefore involves observing only one atom in $10^{6}$ in the sample, a very inefficient procedure.

A reasonable alternative to counting the decays of the $^{14}$C atoms would be to count the atoms directly. A mass spectrometer would be capable of distinguishing $^{14}$C from $^{12}$C, and one might expect it to be a relatively trivial process to introduce a measured amount of carbon into the spectrometer and to determine its $^{14}$C content. A fundamental difficulty arises, however, from contamination from other mass-14 ions that may be present in the sample or in the spectrometer itself; these ions cannot be distinguished from $^{14}$C and therefore contribute a background that may be orders of magnitude more intense than the small concentration of $^{14}$C. Organic matter, for example, would be expected to contain $^{14}$N and $CH_2$, and even a very small amount of residual gas in the spectrometer would be mostly $^{14}$N. The mass differences between the contaminants and $^{14}$C are about one part in $10^{5}$, far too small to be resolved by conventional mass spectrometers.

The nuclear accelerator (cyclotron or tandem Van de Graaff) provides the opportunity for mass and charge selection on the output beams and thus will perform many of the same functions as the ordinary mass spectrometer. The accelerator provides two major advantages: (1) In the ion source, negative ions can be produced. This can result in a substantial reduction in the background, especially in the case of $^{14}$N, which does not form a stable negative ion and can thus be removed from the beam at the source. (2) The analysis system can use $\Delta E \cdot E$ techniques, described in Chapter 7, to determine the charge and mass of the ion, and can also use differential range techniques to reduce or stop the transmission of isotopes of higher atomic number (again, as in the case of $^{14}$N and $^{14}$C). These techniques, combined with conventional momentum or velocity selectors using electric and magnetic fields, give accelerator-based mass spectrometry a considerable advantage over the counting of radioactive decays for dating materials.

Both cyclotrons and tandem accelerators have been used for this sensitive new application of mass spectrometry. The energies available at these accelerators (MeV and above) allow use of particle identification techniques that are not practical at the keV energies of conventional mass spectrometers. In the cyclotron, the resonance condition (Equation 15.4) selects for acceleration only those ions with the proper charge-to-mass ratio, thus immediately eliminating $^{12}$C and $^{13}$C from interfering with the counting of $^{14}$C. The $^{14}$N background can be reduced by use of a negative ion source or by selective analysis of the range of the accelerated particles. The mass-14 molecule $CH_2$ can be eliminated if a multiply ionized beam is extracted from the ion source; if three electrons are removed from $CH_2$, the resulting ion is unstable and the molecule breaks into fragments.

To determine the age of a sample, it is necessary to determine the amount of the rare isotope relative to the stable isotope. To make this comparison with a cyclotron, the resonance frequency must be changed, and corresponding changes in the transmission efficiency of the cyclotron can introduce uncertainties in the determination of the relative isotopic abundances.

The tandem accelerator eliminates the latter difficulty, but lacks the resonant condition that makes the cyclotron such a strongly discriminating mass analyzer. Using a negative ion beam and a multiple stripper in the tandem, both $^{14}N$ and $CH_2$ can be reduced, and electromagnetic analyzers and range filters can accomplish the final separation of the masses. This final analysis is done with what is in effect a separate mass spectrometer, but since the final task is separation of the masses ($^{12}C$ and $^{13}C$ from $^{14}C$) a conventional spectrometer will suffice.

These accelerator-based techniques have been demonstrated to have a sensitivity of $10^{-15}$ in the $^{14}C/^{12}C$ ratio. With only milligram quantities of material, it is possible to determine ages to 100,000 y. This exploratory work has been done in the past few years using conventional cyclotrons and tandems, and the development of accelerators specifically designed for mass spectrometry may bring about significant improvements in sensitivity.

Although $^{14}C$ has been used as an example in this discussion, similar results have been obtained with other isotopes. For example, $^{10}Be$ ($t_{1/2} = 1.6 \times 10^6$ y) is produced by cosmic rays and is found in ocean sediments; the $^{10}Be$ content can be used to study the dynamics of the sea floor. The isotope $^{36}Cl$ ($t_{1/2} = 3.0 \times 10^5$ y) is also produced by cosmic rays in collisions with Ar atoms in the atmosphere; it is carried in rain or snow to the surface water which then seeps gradually through the ground. It may take millions of years to reach deep groundwater, and the $^{36}Cl$ content can be used to determine the length of time it takes the water to travel from the surface. This study has an important application to the problem of the long-term storage of radioactive wastes. Dating techniques using these long-lived isotopes are both practical and precise because of the advances in mass spectroscopic studies using accelerators.

## 20.3 ALPHA-DECAY APPLICATIONS

Alpha decays from long-lived heavy nuclei have two important characteristics that provide for important applications: they are emitted at a unique energy (in contrast to $\beta$ decays) and (if the half-life is long enough) the emission rate is virtually constant.

Let's consider, for example, the $\alpha$ decay of $^{238}Pu$ ($t_{1/2} = 86$ y), which is produced from the decay of $^{238}Np$ following neutron capture by $^{237}Np$. The $Q$ value is about 5.6 MeV and the decay rate of 1 g of $^{238}Pu$ would be

$$\mathscr{A} = \lambda N = \frac{0.693}{t_{1/2}} \cdot \frac{1 \text{ g}}{238 \text{ g/mol}} 6 \times 10^{23} \text{ atoms/mol}$$

$$= (2.6 \times 10^{-10} \text{ decays/s})(2.5 \times 10^{21} \text{ atoms/g})$$

$$= 6 \times 10^{11} \text{ decays/s/g}$$

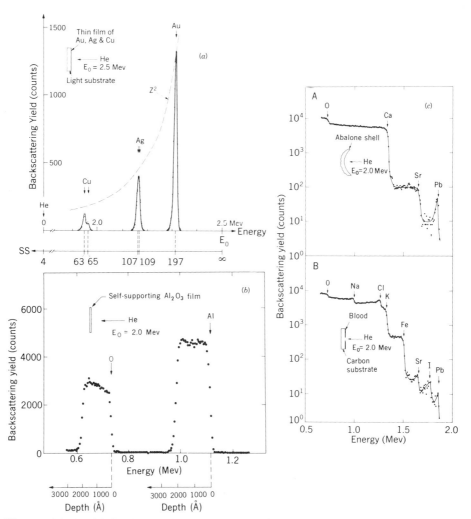

**Figure 20.5**  (*a*) Backscattering spectrum of 2.5-MeV $\alpha$ particles from a thin film of copper, silver, and gold. The dashed line shows the $Z^2$ behavior of the cross section expected from the Rutherford formula. Note the appearance of the two isotopes of copper. (*b*) In a thicker target, some particles may penetrate and lose energy before backscattering. This broadens the peaks into plateaus, the thickness of which depends on the thickness of the scattering target (about 200 nm in this case). (*c*) Scattering from complex substances can reveal the elements of which they are made. Figures from M.-A. Nicolet, J. W. Mayer, and I. V. Mitchell, *Science* **177**, 841 (1972). Copyright © 1972 by the AAAS.

Each decay releases 5.6 MeV, and the power output is therefore

$$P = \left(6 \times 10^{11} \frac{\text{decays/s}}{\text{g}}\right)\left(\frac{5.6 \times 10^6 \text{ eV}}{\text{decay}}\right)(1.6 \times 10^{-19} \text{ J/eV})$$

$$= 0.6 \text{ W/g}$$

Each gram of $^{238}$Pu produces a power output of 0.6 W. While this is not exactly enough to heat or light a house, it is quite sufficient to power simple electrical circuits, and it has the enormous advantage that it continues to operate at a constant rate even under very adverse conditions (in the vacuum and extreme cold of space, for example) and needs replacement only over time scales of the order of its half-life. The need for such stable and reliable power sources has resulted in many applications, from cardiac pacemakers to the Voyager spacecraft that photographed Jupiter, Saturn, and Uranus.

In cardiac pacemakers, the $\alpha$-decay energy is deposited as heat and converted to an electric pulse through a thermoelectric converter. They are capable of producing about 300 $\mu$W of electric power, and do so continuously, limited only by the 86-year half-life of $^{238}$Pu.

Alpha-particle scattering is important for both qualitative and quantitative analysis. Smoke detectors contain a small quantity of the isotope $^{241}$Am ($t_{1/2} = 433$ y) whose $Q$ value is 5.6 MeV. Under normal conditions, the $\alpha$ particles ionize air molecules in the detector; the ions travel to electrodes and establish a small steady-state current in the device. The products of combustion include heavy ionized atoms; when these enter the detector, they collide with the ions responsible for the ambient current, and the resulting decrease in current triggers the alarm. The device is activated not by visible particles of smoke (as in the case of the photoelectric type of smoke detector) but instead by the charged ions resulting from combustion.

Quantitative analysis can be done by observing $\alpha$ particles scattered at large angles, called *Rutherford backscattering*. In our discussion of Rutherford scattering in Chapter 11, we assumed the target nucleus to be infinitely heavy, so the $\alpha$ particle emerges with its original incident energy. In actuality, a small energy is transmitted to the struck nucleus, and for backscattered particles ($\theta \approx 180°$), the loss in energy of the $\alpha$ particle is

$$\Delta T = T\left[\frac{4m/M}{(1 + m/M)^2}\right] \tag{20.4}$$

where $T$ is the incident $\alpha$ energy, $m$ is the $\alpha$-particle mass, and $M$ is the mass of the target. The loss in energy is greatest for light nuclei, but even for heavy nuclei ($m/M \sim 0.02$) the energy loss is of order 0.5 MeV and can easily be detected. Scattering from a target consisting of a variety of isotopes or elements will produce a spectrum of $\alpha$ energies, each corresponding to a unique mass of the struck atom, from which we can infer the composition of the target. The separation between adjacent masses for heavy nuclei is of order 1% or 5 keV, which is somewhat (but not hopelessly) smaller than the resolution of a typical solid-state $\alpha$-particle detector (20 keV), but for light nuclei, the separations between adjacent elements (and even between different isotopes of a single element) can be easily resolved. Figure 20.5 shows an example of a spectrum of

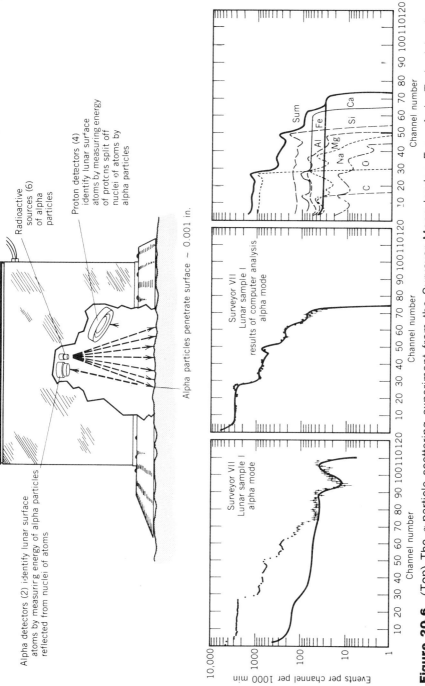

**Figure 20.6** (Top) The $\alpha$-particle scattering experiment from the Surveyor Moon lander. From A. L. Turkevich, *Acc. Chem. Res.* **6**, 81 (1973). (Bottom) Data from the $\alpha$-scattering experiment on the Moon. The figure at left shows the raw data; the solid line is background from a naturally occurring $\alpha$-source. At center is shown the data with background subtracted, and at right is the analysis showing the constituents. The composition is dominated by oxygen and silicon, as are Earthly rocks. From J. H. Patterson et al., *J. Geophys. Res.* **74**, 6120 (1969).

backscattered $\alpha$'s, from which it is possible to deduce the presence of various elements. This technique has been used for analysis of soil samples by the Surveyor spacecraft that landed on the Moon. The Surveyors contained a Rutherford backscattering apparatus including a source of $^{242}$Cm, which decays with a half-life of 163 days. Figure 20.6 shows the Surveyor craft and its backscattering experiment.

## 20.4 DIAGNOSTIC NUCLEAR MEDICINE

Roentgen's discovery of X rays in 1895 was quickly followed with their application to medical diagnosis. Because X rays travel easily through soft body tissue but are strongly attentuated by bone, X-ray photographs can reveal the detailed structure of the human skeletal system and are of course invaluable in medical diagnostic procedures leading to the resetting of broken bones. Two disadvantages of X-ray photographs limit their usefulness: they are not very effective in differentiating between different types of soft tissue (as in locating tumors, for instance), and they produce a flat, two-dimensional image that, even if it reveals an abnormality, would not indicate the depth of that abnormality within the body. Moreover, interesting soft tissue can be shielded or obscured by bone, as for instance the brain within the skull.

The development of techniques in experimental nuclear physics has permitted parallel development in medical imaging: $\gamma$-ray cameras, specialized accelerators for producing medical radioisotopes, and remarkable techniques for obtaining images at specified depths within the body. This branch of research is called *nuclear medicine*, and its practitioners are often nuclear physicists who work in close association with physicians in developing and applying the techniques.

Let's first look at techniques for forming images of specific areas of the body by the introduction of $\gamma$-emitting radioisotopes; the pattern of emitted $\gamma$ rays can then be used to produce an image of that part of the body. (An analogous technique is used in X-ray imaging, in which a high-$Z$ nonradioactive material such as barium or iodine is introduced into a specific organ. Since the body ordinarily contains very little of the high-$Z$ elements, their introduction results in a very detailed and specific image. The use of barium as such a *contrast medium* for studies of the gastrointestinal tract is common.) A very simple (and early) application of this technique is the measurement of iodine uptake by the thyroid gland. Radioactive iodine is orally ingested, and a $\gamma$-ray counter placed near the neck measures the increase in activity with time as the iodine is concentrated in the thyroid. Originally $^{131}$I was used for this purpose; it is a fission product with a half-life of 8 days. Since it is usually desired to observe the thyroid over a period of hours, the half-life is far too long—the activity lingers in the body and results in a large radiation dose to the patient. Moreover, the high-energy $\beta$ emission, which does not contribute to the diagnostic procedure, also increases the radiation dose, which amounts to about 3 rads per $\mu$Ci of $^{131}$I. Normally several $\mu$Ci of $^{131}$I are used for this test, and the absorbed radiation dose can be as high as 30 rads or more. An alternative choice is 2.3-h $^{132}$I, which is a daughter product of 78-h $^{132}$Te. The $^{132}$I can be "milked" from the $^{132}$Te (which is also a fission product) by use of a solvent which dissolves I but not Te. Thus a source of $^{132}$I

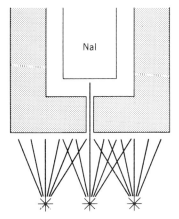 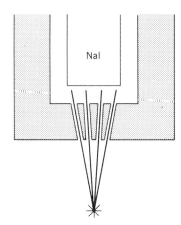

**Figure 20.7** Collimator design for scanning γ-ray camera The design at left allows the scintillation detector to view only radiation coming from the region directly below the hole. The design at right accepts radiation only if it comes from a point at the proper distance below the collimator; such a design is useful to scan at a certain depth within the patient.

has in effect a shelf life determined by the 78-h decay of the parent, but only the 2.3-h activity is introduced into the patient. More recently, the 13-h $^{123}$I has been widely used. This isotope, produced with cyclotron reactions and now readily available from commercial sources, decays by electron capture (thus no radiation dose from $\beta$'s) and emits a single 159-keV γ ray. The radiation dose is only about 2 rads per 100 $\mu$Ci, far less than $^{131}$I or $^{132}$I, and the half-life is also almost ideal, as it is long enough for a 24-h test, but short enough that the activity does not linger in the patient too far beyond the end of the tests.

Kidney function is also studied using a compound labeled with $^{131}$I (sodium iodohippurate) which is administered intravenously. A γ detector views each kidney, and comparisons of the rate of uptake of $^{131}$I by the two kidneys can reveal disorders. The patient receives a relatively small dose of radiation because the compound is almost completely removed from the blood after one passage through the kidneys and is then excreted.

There are now available many pharmaceutical compounds that are labeled with a radioactive isotope. By choosing a compound that tends to accumulate in a specific organ, a local concentration of the radioisotope can be achieved. The spatial pattern of radioactive emissions then gives a complete picture of the organ (as opposed merely to a measure of the quantity of accumulated activity). One possibility is to scan over the area using a scintillation detector; a narrow collimator (Figure 20.7) confines the detector to observe only a small area (2–3 mm in diameter), and scanning back and forth through a rectangular array permits the intensity to be recorded. An alternative collimator design (also shown in Figure 20.7) permits the radiation to reach the detector only if it comes from a certain depth within the patient.

The most frequently used isotope for scans is $^{99m}$Tc (half-life, 6 h), which is the daughter of $^{99}$Mo (66 h) obtained either from fission or from neutron capture by $^{98}$Mo. As in the case of $^{132}$Te, it is possible to design a "generator" in which the shorter-lived $^{99m}$Tc can be periodically extracted from the long-lived parent, by

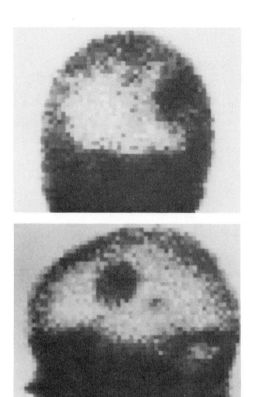

**Figure 20.8** A brain scan obtained by passing a collimated scintillator (such as that shown in Figure 20.7) along the head in a sequence of horizontal lines. The top view is taken from the front of the patient and the bottom view is from the side, with the patient's face to the left. The isotope used was $^{99m}$Tc as pertechnate. The black region near the front of the skull shows a tumor, in which the failure of the normal blood-brain barrier has enabled concentration of the radioisotope. From B. H. Brown and R. H. Smallwood, *Medical Physics and Physiological Measurement* (Oxford: Blackwell Scientific Publications, 1981).

passing a saline solution through an ion-exchange column containing $^{99}$Mo. The single $\gamma$ ray at 140 keV (with no accompanying betas) is also an advantage, for it permits precise design of collimators. If higher energy $\gamma$'s were also emitted, they could blur the image by penetrating the septa that separate the collimation channels. The activity can be introduced into the body in a wide variety of labeled compounds and chemical forms, depending on the organ to be viewed and the diagnosis to be sought.

One of the most common uses of scanning techniques is to produce an image of the brain. The brain scan is possible because of the so-called "blood-brain" barrier—the brain seems to have a very low absorption for impurities in the blood under normal circumstances. If there is disease or a tumor, however, the $^{99m}$Tc can concentrate in the affected region and reveal itself in the scanning image. Figure 20.8 shows a brain scan that clearly reveals a tumor.

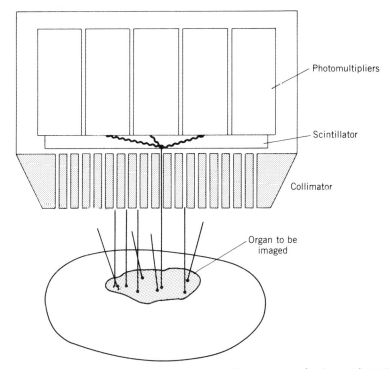

**Figure 20.9**  Gamma-ray scintillation camera. The organ to be imaged contains a radioisotope; emitted γ rays pass through the array of narrow collimators and strike a large scintillator ($\frac{1}{2}$ inch thick and 11 inches in diameter), which is viewed by an array of photomultipler tubes. The relative amounts of light reaching adjacent photomultipliers can be used to deduce the locations of the scintillation event.

A disadvantage of the scanning technique is that only a small portion of the decaying activity is being measured at any single time; the rest is "wasted" and contributes an unnecessary radiation dose to the patient. A device capable of recording a complete image of a large area is the γ-ray camera, illustrated in Figure 20.9. The image is obtained rapidly (~ 1 min), and thus it is possible to do dynamic studies, showing the time evolution of bodily function.

Radiation enters the camera through the many multichannel collimators and strikes a large sodium iodide scintillating crystal. The scintillator is viewed by an array of 19 photomultiplier tubes. A flash of light occurring at a particular place in the scintillator is recorded in part by several tubes, but with relative intensities varying with the distance between the tube and the scintillation event. From the relative intensities recorded in the different tubes it is possible to determine exactly where the original photon struck the scintillator and thus to construct the image after many photons are counted and recorded. Figure 20.10 shows an image of the brain obtained using a γ camera with $^{99m}$Tc.

One of the most remarkable developments in imaging techniques has been the field of tomography, which is capable of imaging a particular "slice" of the internal structure of the body, either through externally incident X rays or internally introduced radioisotopes. Figure 20.11 shows an example of X-ray

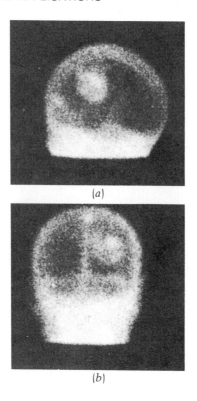

(a)

(b)

**Figure 20.10** Scintillation camera image of the brain, following intravenous injection of 20 mCi of $^{99m}$Tc. Photo (a) shows a side view, with the patient's face to the left; photo (b) is the back view. The bright circular spot shows concentration of blood in a lesion, possibly a tumor. Other bright areas show the scalp and the major veins. Photo courtesy D. Bruce Sodee, M.D., Hillcrest Hospital, Mayfield Heights, Ohio.

linear tomography. The X-ray tube and the film are moved simultaneously in opposite directions so that a fixed point of a section through the body maintains its image on the film. If the motion is linear, all points on the plane retain their locations on the film and all points on other planes are smeared or blurred. The result is a focused image of the selected plane, with the remaining planes blurred but superimposed on the image. This results in a considerable loss of contrast of the image.

Present-day tomographic images are produced by passing a large number of X-ray beams through the region of interest from many different directions. For each beam we can determine the loss in intensity and thus the relative absorption along that particular direction. From these many "one-dimensional" projections of the density profile of the body it is possible to reconstruct a two-dimensional image. A computer is necessary for this reconstruction, and since the X-ray beam normally scans in a circle about the patient, the technique is now known as *computerized axial tomography*, or CAT. Figure 20.12 shows a schematic diagram of a CAT scanner, and a photograph is shown in Figure 20.13. The X-ray source emits a fan of narrow beams that are observed in the detector array. Several

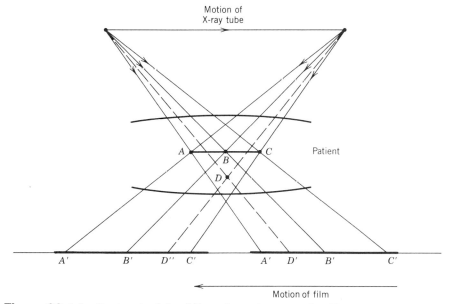

**Figure 20.11** Basic principle of X-ray linear tomography. The motion of the X-ray tube and film are correlated so that the images of points *A*, *B*, and *C* (on a plane within the patient) always appear at the same locations *A'*, *B'*, and *C'* on the film. Points not on the plane, such as *D*, produce smeared images (*D'*, *D''*, etc.) on the film.

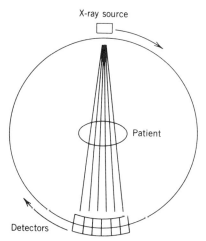

**Figure 20.12** Schematic diagram of early CAT scanner. In modern scanners, the patient is surrounded with a ring of stationary X-ray detectors, and the X-ray source moves around the patient.

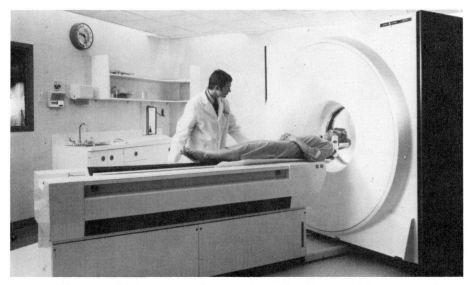

**Figure 20.13** A CAT scanner. The X-ray source and detectors are in the large ring surrounding the patient. Courtesy General Electric Corporation.

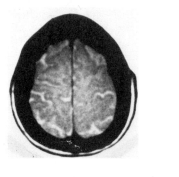
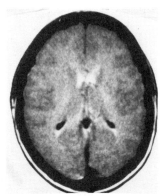

**Figure 20.14** CAT scan of the author's brain, showing two slices about 3 to 4 cm apart. The left photo shows the region near the top surface; folds and fissures in the cortex are evident. The right photo is near the middle of the brain.

hundred X-ray detectors form the array, and a complete image can be obtained in a matter of several seconds. Figure 20.14 shows a typical CAT scan of a section through the brain. The spatial resolution is of the order of mm and a wealth of fine detail is revealed.

Another imaging technique is called *positron-emission tomography* (PET). Positron-emitting isotopes are introduced into the area to be studied, and the two 511-keV photons emitted following electron-positron annihilation are observed in coincidence. Figure 20.15 shows a schematic view of the process. Detection of the two oppositely direct photons serves to identify a line along which the original decay must have occurred. From a large number of such events it is possible to reconstruct the original distribution of radioisotopes and to map an image of the

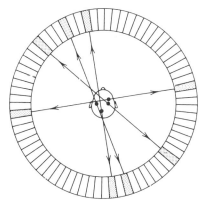

**Figure 20.15** Positron-emitting isotopes within the head result in two 511-keV photons following electron-positron annihilation. The photons move in opposite directions and trigger two of the detectors in the ring.

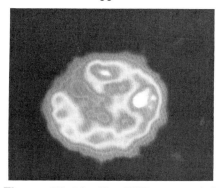 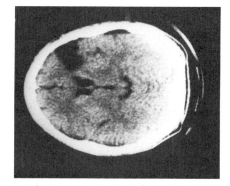

**Figure 20.16** The PET scan at left shows the regions of the brain that emit radiation as a result of the patient's inhaling oxygen labeled with $^{15}O$. A region poor in oxygen as a result of reduced blood flow shows clearly at the upper left. The dark region near the upper left in the CAT scan at right shows the corresponding region of the brain damaged by a stroke. From M. Ter-Pogossian et al., *Scientific American* **243**, 171 (October 1980).

area. Among the isotopes used are $^{15}O$ (2 min), $^{13}N$ (10 min), $^{11}C$ (20 min), and $^{18}F$ (110 min), all of which must be produced with a cyclotron, and because of the short half-lives the cyclotron must be at the site of the diagnostic facility.

In the case of brain scans, PET offers many advantages over CAT. The CAT scan is essentially static, showing only the density of tissue, while the PET scan can reveal dynamic effects, such as blood flow. Figure 20.16 compares CAT and PET scans of the brain of a patient who had suffered a stroke. The PET scan was made after the patient had inhaled oxygen labeled with $^{15}O$ and it clearly shows reduced blood flow to a large part of the brain. Labeling glucose with $^{18}F$ permits observation of brain metabolism because the more active areas will concentrate the activity. Figure 20.17 shows the different areas of the brain that are active for language and music.

Perhaps the newest development in this area is *nuclear magnetic resonance* (NMR) imaging. As we discussed in Chapter 16, in the basic NMR experiment a sample is simultaneously exposed to a large static magnetic field and a time-vary-

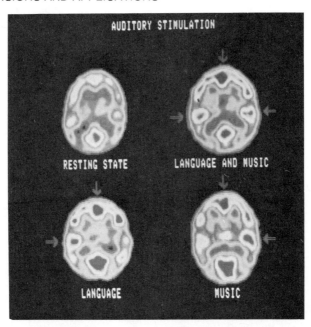

**Figure 20.17** Brain activity in a patient after the introduction of glucose labeled with $^{18}$F. Active areas of the brain metabolize the glucose more rapidly and show increased concentrations of $^{18}$F (lighter shading). The PET image shows different areas of the brain that are active for language and music. The bright area at the bottom of each image is the visual cortex, which is also active because the patient's eyes were open. From Huffman, *Psychology in Action* (New York: Wiley, 1987).

ing radiofrequency (rf) field perpendicular to the static field. The static field in effect gives a Zeeman splitting $\Delta E$ of the adjacent nuclear magnetic substates and the rf field is tuned to the frequency $\nu = \Delta E/h$ so that it induces transitions between the substates, which are observed through the absorption of rf power at the resonance frequency. For medical applications, it is possible to resonate $^1$H, which is of course present in the body in great abundance, but other stable isotopes are possible as well, such as $^{13}$C or $^{17}$O. Each of these nuclei has a unique resonant frequency, determined by the nuclear magnetic dipole moment. A scan through the body is even possible, by taking advantage of the variation in resonant frequency arising from variation in the static field over the volume of the body. The NMR imaging technique has an advantage over X-ray, $\gamma$-ray, and positron-emission imaging techniques, in that the patient is not exposed to ionizing radiations; the direct effects on the body of large static and rapidly varying magnetic fields are believed to be insignificant, but to date there has not been a great deal of research on this subject.

## 20.5 THERAPEUTIC NUCLEAR MEDICINE

The primary use of nuclear radiations in therapy is in the destruction of unwanted or malfunctioning tissue in the body, such as a cancerous tumor or an overactive thyroid gland. This effect originates with the *ionizing* ability of nuclear radiations. In essence, the destruction of tissue proceeds as follows: (1) The incident radiations ionize atoms in molecules of the irradiated material; this *physical* change occurs on a time scale of $10^{-16}$ s or less. (2) The ionized molecules participate in chemical reactions that give rise to free radicals or other excited molecules; this *chemical* change occurs on a time scale of $10^{-15}$ s to perhaps $10^{-3}$ s. (3) These free radicals can then be incorporated into complex biological structures at the molecular level and alter their biological function; it may take hours to years for the effects of this *biological* change to become apparent.

It is possible to have direct action by the radiation on the biologically sensitive molecules, thus bypassing step (2); because the human body is about 80% water, however, it is most likely that the radiation will produce ionizing events with water molecules. That is, a water molecule can be ionized by incident radiation as

$$H_2O \rightarrow H_2O^+ + e^-$$

and the free electron can be captured by another neutral water molecule to produce a molecule with an excess negative charge:

$$H_2O + e^- \rightarrow H_2O^-$$

Both $H_2O^+$ and $H_2O^-$ are unstable ions and can dissociate as follows:

$$H_2O^+ \rightarrow H^+ + OH\cdot$$
$$H_2O^- \rightarrow H\cdot + OH^-$$

In each case, the result is an ion ($H^+$, $OH^-$) and a *free radical* ($H\cdot$, $OH\cdot$). The free radical is an electrically neutral atom or molecule which has a free (that is, unpaired) electron to participate in chemical bonding. Owing to the strong tendency of this unpaired electron to participate in chemical reactions that lead to a more stable paired configuration, these free radicals are extremely reactive. Within about $10^{-6}$ s, they will react along the following possible paths in an environment of pure water:

$$H\cdot + H\cdot \rightarrow H_2$$
$$OH\cdot + OH\cdot \rightarrow H_2O_2$$
$$H\cdot + OH\cdot \rightarrow H_2O$$
$$H_2O + H\cdot \rightarrow H_2 + OH\cdot$$

In the case of organic matter, we can simplify the structure of a complex hydrogen-containing biological molecule as the combination RH of a free radical $R\cdot$ with hydrogen. The free radicals $H\cdot$ or $OH\cdot$ can combine with this molecule:

$$OH\cdot + RH \rightarrow R\cdot + H_2O$$
$$H\cdot + RH \rightarrow R\cdot + H_2$$

In either case, the result is the production of a free radical $R\cdot$ which may be a part of a biologically more complex system (a chromosome, for instance) and may alter the function of that system, possibly causing its death if it is not able to function, or alternatively changing the genetic information that is passed in reproduction so that the structure of the next generation is fundamentally different (a genetic mutation).

As an alternative process, the radiation may interact directly with the molecule RH without the intermediate step of producing free radicals from water. This is accomplished by direct ionization followed by dissociation:

$$RH \rightarrow RH^+ + e^-$$
$$RH^+ \rightarrow R\cdot + H^+$$

again resulting in the free radical $R\cdot$.

If the irradiated material is rich in oxygen, another set of processes is possible:

$$R\cdot + O_2 \rightarrow RO_2\cdot$$

and the organic peroxyradical $RO_2\cdot$ can then interact with another RH molecule

$$RO_2\cdot + RH \rightarrow RO_2H + R\cdot$$

resulting in yet another free radical $R\cdot$, which can initiate a new set of processes. (This is analogous to the chain reaction in neutron-induced fission.) Another process that occurs with oxygen is

$$O_2 + e^- \rightarrow O_2^-$$

because $O_2$ has a large electron affinity. The capture of the electron by $O_2$ not only can initiate an alternative set of chemical reactions but also can prevent the free electron from recombining with the original ions produced by the interaction of the radiation; thus the radiation damage is not able to be "healed" through the recapture of the electrons.

This *oxygen effect* results in highly oxygenated tissue having a greater sensitivity to radiation, and thus irradiated tissue that is rich in oxygen has a smaller survival rate than tissue that is less rich in oxygen. From the standpoint of treatment of tumors with radiation this is somewhat of an unfortunate situation, for tumors generally have an inferior blood supply compared with normal tissue and are thus less well oxygenated. The oxygen effect results in tumors being less sensitive to radiation than the surrounding tissue.

Figure 20.18 illustrates the oxygen effect. In healthy, oxygenated tissue, a radiation dose $D_1$ results in destruction of a certain fraction of the cells. In tissue

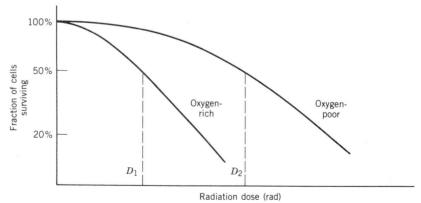

**Figure 20.18** For a given radiation dose $D_1$, the fraction of oxygen-rich cells that survive is smaller than the fraction of oxygen-poor cells. To kill the same fraction of oxygen-poor cells requires a larger dose $D_2$.

that lacks oxygen, a much larger dose $D_2$ is required to produce the same effect. The ratio $D_2/D_1$ for a particular radiation is called its *oxygen enhancement ratio* (OER). Typical values of the OER are 2–3; small values ($\sim 1$) are more desirable, for they indicate that a smaller radiation dose is required to achieve the desired effect.

In comparing the effects of different types of radiation on living tissue, the critical factor is the *linear energy transfer* (LET), which measures the energy deposited per unit distance over the path of the radiation. To the extent that we can neglect bremsstrahlung (which may radiate the energy relatively far from the actual path of the incident radiation), the LET is the same as $dE/dx$ or the linear stopping power defined and discussed in Section 7.1. Examples of high-LET radiations are heavy ions and $\alpha$ particles; these generally have short ranges in matter because they lose their energy quickly in collisions. Dissipating a great deal of energy over a short distance, they have a correspondingly high LET. Typical values for the range might be 0.1–1.0 mm (Figure 7.2) and for the LET, above 100 keV/$\mu$m. At the other extreme, particles with low LET are electrons or photons, with ranges of the order of cm and LET of the order of 1 keV/$\mu$m.

It should be immediately apparent that radiation therapy presents contradictory goals: We should like to concentrate the radiation damage to the area to be treated and simultaneously to minimize radiation exposure to the patient and damage to surrounding tissue. This suggests high-LET radiations. On the other hand, such radiations have a very short range and do not penetrate deep within the body. Primarily because of the need for deep penetration, radiotherapy has been traditionally done with photons (X rays or $\gamma$ rays), which can penetrate to great distances, but which also pose the risk of damage to healthy tissue (especially with the oxygen effect making healthy tissue more sensitive to radiation).

In recent years, studies have been done with beams of neutrons and pions for radiotherapy. Neutrons interact with living tissue through a variety of reactions, the dominant one being elastic scattering from hydrogen. In this process a relatively high-LET proton is created. Another possible interaction is the (n, $\alpha$) reaction on $^{16}$O, which creates a high-LET $\alpha$. Other elastic scatterings from oxygen or carbon can produce heavy recoil nuclei which also have high LET. The neutron, in contrast to charged particles, does not interact through Coulomb effects and is therefore less sensitive to the presence of higher $Z$ atoms such as oxygen; neutrons therefore have a smaller OER, about 1.5–1.9, and may therefore result in destruction of tissue with a smaller radiation dose to the patient.

Pions are slowed as they pass through matter, and are finally stopped and captured by nuclei of the target atoms. When this happens, many nuclear reactions can result, for example

$$\pi^- + {}^{16}\text{O} \rightarrow {}^{16}\text{N}^* \rightarrow {}^{15}\text{N} + \text{n}$$

$$\rightarrow {}^{14}\text{N} + 2\text{n}$$

$$\rightarrow {}^{14}\text{C} + \text{p} + \text{n}$$

$$\rightarrow {}^{12}\text{B} + \alpha$$

The vicinity of the pion capture event is called a "star," after the pattern of tracks radiating from the capture site in a photographic emulsion. Approximately

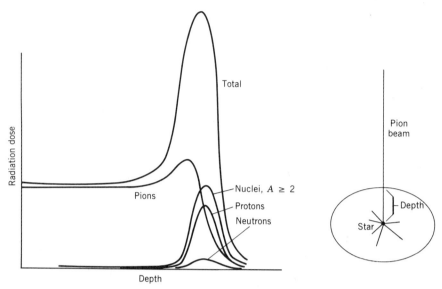

**Figure 20.19** Most of the dose received from a pion beam occurs as a result of nuclear reactions deep within the body. As charged particles, the pions cause ionizing events along their entire path, but the heavy particles and nucleons produced at the "star" (where the pion is captured) give a large dose in a relatively small volume.

35 MeV of charged particles deposit their energy within a few millimeters of the capture site. Figure 20.19 shows the calculated contributions to the radiation dose per incident pion. Notice that the dose is maximized at a specific depth, which is dependent on the energy of the incident pion. In treatment of tumors, the energy would be adjusted so as to deliver the maximum dose at the location of the tumor. As charged particles, pions have an advantage (over neutrons or photons) in that they can be focused to converge at a specific site, further increasing their local effectiveness. The OER of pions is close to 2, but because the nuclear reactions are confined to a relatively small region, the oxygen effect may be less significant for pion radiotherapy than for other radiation.

## REFERENCES FOR ADDITIONAL READING

Neutron activation analysis is reviewed by P. Kruger, *Principles of Activation Analysis* (New York: Wiley, 1971); D. De Soete, R. Gijbels, and J. Hoste, *Neutron Activation Analysis* (London: Wiley, 1972); *Modern Trends in Activation Analysis*, edited by J. R. DeVoe (Washington, DC: U.S. National Bureau of Standards, 1969). Applications of NAA for criminal investigations are reviewed by V. P. Guinn, *Ann. Rev. Nucl. Sci.* **24**, 561 (1974), and other uses of neutrons in science and technology are summarized by D. A. Bromley, *Nucl. Instrum. Methods* **225**, 240 (1984). For a review of the theory and practice of PIXE, see S. A. E. Johansson and T. B. Johansson, *Nucl. Instrum. Methods* **137**, 473 (1976).

For other applications of nuclear techniques to the analysis of materials, see W.-K. Chu, J. W. Mayer, and M.-A. Nicolet, *Backscattering Spectrometry* (New York: Academic, 1976); V. Valković, *Nuclear Microanalysis* (New York: Garland, 1977); M.-A. Nicolet, J. W. Mayer, and I. V. Mitchell, *Science* **177**, 841 (1972); P. Müller and G. Ischenko, *J. Appl. Phys.* **47**, 2811 (1976).

For references to accelerator-based mass spectrometry, see the following introductory and review articles: R. A. Muller, *Science* **196**, 489 (1977); R. A. Muller, *Phys. Today* **32**, 23 (February 1979); H. E. Gove, *The Physics Teacher* **21**, 237 (April 1983); A. E. Litherland, *Ann. Rev. Nucl. Part. Sci.* **30**, 437 (1980); A. E. Litherland, *Nucl. Instrum. Methods Phys. Res. B* **5**, 100 (1984). The latter volume contains in addition several dozen research papers on techniques and applications of accelerator mass spectroscopy.

References on medical applications of nuclear techniques include W. R. Hendee, *Medical Radiation Physics* (Chicago: Year Book Medical, 1979); E. G. A. Aird, *An Introduction to Medical Physics* (London: William Heinemann Medical Books, 1975); J. T. Andrews and M. J. Milne, *Nuclear Medicine* (New York: Wiley, 1977); N. A. Dyson, *An Introduction to Nuclear Physics with Applications in Medicine and Biology* (Chichester, England: Ellis Horwood, 1981); *The Physical Basis of Medical Imaging*, edited by C. M. Coulam, J. J. Erickson, F. D. Rollo, and A. E. James, Jr. (New York: Appleton-Century-Crofts, 1981); T. S. Curry III, J. E. Dowdey, and R. C. Murry, Jr., *Christensen's Introduction to the Physics of Diagnostic Radiology* (Philadelphia: Lea and Febiger, 1984).

For review articles on medical imaging and diagnostic techniques, see R. D. Neumann and A. Gottschalk, *Ann. Rev. Nucl. Part. Sci.* **29**, 283 (1979); W. Swindell and H. H. Barrett, *Phys. Today* **30**, 32 (December 1977); P. R. Moran, R. J. Nickles, and J. A. Zagzebski, *Phys. Today* **36**, 36 (July 1983); M. M. Ter-Pogossian, M. E. Raichle, and B. E. Sobel, *Sci. Am.* **243**, 171 (October 1980).

Neutron radiotherapy is discussed by P. H. McGinley, *The Physics Teacher* **11**, 73 (February 1973), and pion radiotherapy by C. Richman, *Med. Phys.* **8**, 273 (1981).

# Appendix A
# SPECIAL RELATIVITY

In nuclear $\beta$ decay and in most calculations in medium- and high-energy processes, the equations of relativistic kinematics are required. A brief review of these equations is given below.

## A.1 LORENTZ TRANSFORMATION

It is often necessary to convert equations from the laboratory reference frame to another frame moving at constant speed, for example, the center-of-mass frame discussed in Appendix B or the rest frame of a moving particle. The relativistic transformation that accomplishes this is the Lorentz transformation.

Let the motion of a particle be described by coordinates $(x, y, z, t)$ in one frame and $(x', y', z', t')$ in another, which moves relative to the first at speed $u$ along the common $x$ and $x'$ directions. Then

$$
\begin{aligned}
x' &= \gamma(x - ut) \\
y' &= y \\
z' &= z \\
t' &= \gamma(t - ux/c^2)
\end{aligned}
\tag{A.1}
$$

where $\gamma = (1 - u^2/c^2)^{-1/2}$.

The velocity of the particle can be transformed from one frame to another using the above expressions by calculating $dx'/dt'$, $dy'/dt'$, and $dz'/dt'$. The results are

$$
\begin{aligned}
v_x' &= \frac{v_x - u}{1 - uv_x/c^2} \\[2mm]
v_y' &= \frac{v_y}{\gamma(1 - uv_x/c^2)} \\[2mm]
v_z' &= \frac{v_z}{\gamma(1 - uv_x/c^2)}
\end{aligned}
\tag{A.2}
$$

One important consequence of these transformations for nuclear and particle physics is *time dilation*—the duration of a phenomenon measured in one frame of reference differs from the duration of the same phenomenon measured in a relatively moving frame. The duration is longer in all such moving frames and has the smallest value in the frame in which it occurs "at rest"; that is, the events that mark the beginning and the end of the time interval occur at the same coordinates $x$, $y$, $z$. This particular time interval is called the *proper time*; in all other frames of reference in relative motion, the time interval is measured to be longer than the proper time. For example, lifetimes of decaying particles are generally measured in the laboratory frame in which the particles may be in motion. The lifetime we measure in the laboratory is therefore longer than the proper lifetime, which would be measured in the particle's rest frame. The relationship between the proper time interval $\Delta t_0$ and the interval $\Delta t$ measured in the moving frame is

$$\Delta t = \gamma \Delta t_0 \tag{A.3}$$

Note that $\gamma \approx 1$ for small relative velocities, while $\gamma \to \infty$ as the relative speed approaches the speed of light.

## A.2  RELATIVISTIC DYNAMICS

Special relativity also forces us to redefine basic concepts of mass, momentum, and energy to achieve an internally consistent formalism for the analysis of collisions. The relativistic momentum of a particle moving at speed $v$ is

$$p = \frac{mv}{\sqrt{1 - v^2/c^2}} \tag{A.4}$$

Here $m$ is the *rest mass* of the particle—the mass measured in a frame of reference in which the particle is at rest. Often the equations of special relativity are written in terms of a relativistic mass $\gamma m$; such a notation can lead to difficulties because it is in general incorrect merely to replace the mass in classical expressions with this relativistic mass. It is correct, however, to retain the classical notion of momentum conservation if we use Equation A.4 for the momentum. While it is occasionally convenient to use the relativistic mass increase of a moving particle (such as in the discussion of cyclotrons in Chapter 15), in this text we avoid confusion by using only the rest mass $m$. For the magnetic deflection of charged particles, we can write the path radius in terms of the momentum of the deflected particle, thereby eliminating the difficulty.

With the momentum defined as in Equation A.4, the relativistic kinetic energy becomes

$$T = E - mc^2 \tag{A.5}$$

where $E$ is the relativistic total energy

$$E = \sqrt{p^2c^2 + m^2c^4} \tag{A.6}$$

For massless particles (photons, neutrinos), which move at the speed of light,

$$E = pc \tag{A.7}$$

Equation A.7 is also a good approximation for particles whose total energy $E$ is much greater than their rest energy $mc^2$.

Using the binomial expansion we can reduce Equation A.5 to

$$T = \tfrac{1}{2}mv^2\left(1 + \frac{3}{4}\frac{v^2}{c^2} + \cdots\right) \tag{A.8}$$

and thus for motion at speeds small compared with $c$ the classical $\tfrac{1}{2}mv^2$ is of sufficient precision.

In $\beta$ decay and other weak interaction processes, expressions for energy spectra and angular distributions often include the factor $v/c$. By using Equation A.6, it can be shown that

$$\frac{v}{c} = \frac{p}{E} \tag{A.9}$$

In collisions at low energy, we can apply the classical formulas for conservation of linear momentum and energy [assuming that we use the $Q$-value expression $Q = (m_i - m_f)c^2$, which is in effect a relativistic formula]. For collisions at high energy, the relativistic formulas *must* be used. Although it is, of course, correct to use the relativistic formulas even at low energy, they are often cumbersome and complicate the algebraic manipulations. Thus wherever possible, we will use nonrelativistic formulas to analyze reaction and decay processes.

## A.3   TRANSFORMATION OF ENERGY AND MOMENTUM

By using the Lorentz transformation, we can obtain the transformation of the relativistic expressions for the energy and linear momentum:

$$
\begin{aligned}
p_x' &= \gamma\left(p_x - uE/c^2\right) \\
p_y' &= p_y \\
p_z' &= p_z \\
E' &= \gamma\left(E - up_x\right)
\end{aligned}
\tag{A.10}
$$

where $p$ and $E$ are measured in one frame, and $p'$ and $E'$ are measured in another frame moving with speed $u$ in the $x$ direction.

# Appendix B

## CENTER-OF-MASS REFERENCE FRAME

Reactions are most conveniently analyzed in the center-of-mass (CM) reference frame, in which the total linear momentum of the reacting particles is zero. (As we see in the following discussion, the effects of special relativity cause the center-of-mass frame to differ from the center-of-momentum frame. For the low-energy decays and reactions that are discussed in this text, the nonrelativistic equations can be safely applied and the distinction is relatively unimportant. To be strictly correct, the CM frame should be defined as the center-of-momentum frame, as in Equation B.2 below, but with the relativistic expression for momentum, Equation A.4, replacing the classical $mv$.) In the laboratory (L) system, reactions are usually performed by bringing a beam of incident particles against a target of particles at rest. The observed properties of the reaction (energy and momentum of product particles, cross sections) must therefore be transformed from the L to the CM reference frame before the theory and the experiment can be compared.

### B.1 REACTION KINEMATICS

The reaction to be considered will be of the form

$$a + X \rightarrow Y + b$$

in which the beam of particles of type a has a laboratory kinetic energy $T_a$; the target particles X are at rest in the laboratory. To avoid the confusion of multiple subscripts, all L coordinates will be represented by unprimed variables $(T_a, v_a, T_Y, T_b, \theta_Y, \theta_b, \dots)$, while all corresponding CM coordinates will be represented by primed variables $(T'_a, v'_a, T'_Y, \dots)$.

For convenience we choose the incident momentum of a to be along the $x$ direction. Then, for low energy reactions in which nonrelativistic kinematics may be used, the initial linear momenta are

$$p_x = m_a v_a \qquad (B.1)$$

$$p'_x = m_a v'_a + m_X v'_X = 0 \qquad (B.2)$$

in the L and CM frames, respectively. If $v_{CM}$ represents the velocity of the CM as measured in the L frame, then

$$v'_a = v_a - v_{CM} \tag{B.3}$$

$$v'_X = -v_{CM} \tag{B.4}$$

and it can be immediately shown that Equation B.2 is satisfied if

$$v_{CM} = v_a \frac{m_a}{m_a + m_X} \tag{B.5}$$

After the collision, the particles b and Y must move in opposite directions in the CM frame (so that the total linear momentum remains zero). The final momentum in the L frame has $x$ and $y$ components

$$p_x = m_b v_b \cos\theta_b + m_Y v_Y \cos\theta_Y \tag{B.6}$$

$$p_y = m_b v_b \sin\theta_b - m_Y v_Y \sin\theta_Y \tag{B.7}$$

where for convenience the relative signs of the $y$-components are indicated explicitly in Equation B.7 by defining $\theta_b$ and $\theta_Y$ on opposite sides of the $x$ axis.

In the CM frame, the final particles move in the directions $\theta'_b$ and $\theta'_Y = -\theta'_b$. Along the line of motion, the total momentum must vanish, and so

$$m_b v'_b + m_Y v'_Y = 0 \tag{B.8}$$

The manipulation of Equations B.6 to B.8 leads to an interesting result: the velocity of the center of mass after the reaction is not the same as its velocity before the reaction. (This may seem to violate the classical restriction on the uniformity of CM motion under internal forces in an isolated system, but classical physics does not permit the total mass of the system to change. Because the initial mass is not equal to the final mass, their difference being the $Q$ value, the classical equations do not strictly apply.) The relationship between the final and initial CM velocities is

$$v_{CM,f} = \left(\frac{m_a + m_X}{m_b + m_Y}\right) v_{CM,i} \tag{B.9}$$

Note that in elastic collisions, $v_{CM,f} = v_{CM,i}$. The ratio between the masses in Equation B.9 is normally very close to unity; the initial and final masses differ by the $Q$ value, which is typically only $10^{-3}$ to $10^{-5}$ of the total mass energy for reactions at low energy.

The application of conservation of energy in the L and CM frames gives the following result for the energy of particle b:

$$T'_b = \frac{m_b}{m_Y + m_b}\left[Q + \left(1 - \frac{m_a}{m_Y + m_b}\right)T_a\right] \tag{B.10}$$

The corresponding equation for $T_b$ in the L frame was given as Equation 11.5.

A corresponding analysis of the velocity and momentum equations gives the transformation between the angles:

$$\tan\theta_b = \frac{\sin\theta'_b}{\cos\theta'_b + \gamma} \tag{B.11}$$

where

$$\gamma = \frac{v_{CM}}{v'_b} \tag{B.12}$$

$$= \left[ \frac{m_a m_b}{m_X m_Y} \frac{T_a}{T_a + Q(1 + m_a/m_X)} \right]^{1/2} \tag{B.13}$$

## B.2 CROSS SECTIONS

Total cross sections involve only absolute probabilities for reactions to occur, and are therefore unaffected by the transformation between the L and CM frames. Differential cross sections, however, involve angular variables and are therefore dependent on the frame of reference.

The number of particles hitting a small detector is $d\sigma$ in both frames, and writing $d\sigma = (d\sigma/d\Omega) d\Omega \equiv \sigma(\theta) d\Omega$, the relationship between the differential cross sections follows:

$$d\sigma = \sigma(\theta) d\Omega = \sigma(\theta') d\Omega' \tag{B.14}$$

Given a measured laboratory differential cross section $\sigma(\theta)$, the CM differential cross section can be obtained as

$$\sigma(\theta') = \sigma(\theta) \frac{d\Omega}{d\Omega'} \tag{B.15}$$

Integrating over the uninteresting azimuthal variable in both frames, we have $d\Omega = 2\pi \sin\theta \, d\theta$ and $d\Omega' = 2\pi \sin\theta' \, d\theta'$, and from Equation B.11, it follows that

$$\frac{d\Omega}{d\Omega'} = \frac{|1 + \gamma \cos\theta'|}{\left(1 + 2\gamma \cos\theta' + \gamma^2\right)^{3/2}} \tag{B.16}$$

## B.3 THE CM SCHRÖDINGER EQUATION

The Schrödinger equation that governs the mutual interaction of the reacting particles can also be reduced to a CM form. For two particles, the combined equation in Cartesian coordinates can be written

$$-\frac{\hbar^2}{2m_a} \left( \frac{\partial^2 \Psi}{\partial x_a^2} + \frac{\partial^2 \Psi}{\partial y_a^2} + \frac{\partial^2 \Psi}{\partial z_a^2} \right) - \frac{\hbar^2}{2m_X} \left( \frac{\partial^2 \Psi}{\partial x_X^2} + \frac{\partial^2 \Psi}{\partial y_X^2} + \frac{\partial^2 \Psi}{\partial z_X^2} \right)$$

$$+ V(r_a, r_X)\Psi = E\Psi \tag{B.17}$$

where $r_a = (x_a, y_a, z_a)$ and $r_X = (x_X, y_X, z_X)$ represent the coordinates of the particles and $\Psi$ represents the combined wave function. Defining the CM coordinate $R = (X, Y, Z) = (m_a r_a + m_X r_X)/(m_a + m_X)$ and the relative coordinate $r = (x, y, z) = r_a - r_X$, then the partial derivatives can be written in terms of the new coordinates. For example,

$$\frac{\partial \Psi}{\partial x_a} = \frac{\partial \Psi}{\partial X}\frac{\partial X}{\partial x_a} + \frac{\partial \Psi}{\partial x}\frac{\partial x}{\partial x_a}$$

$$= \frac{m_a}{m_a + m_X}\frac{\partial \Psi}{\partial X} + \frac{\partial \Psi}{\partial x}$$

and so on for all first and second partials. The result is

$$-\frac{\hbar^2}{2M}\left(\frac{\partial^2 \Psi}{\partial X^2} + \frac{\partial^2 \Psi}{\partial Y^2} + \frac{\partial^2 \Psi}{\partial Z^2}\right) - \frac{\hbar^2}{2m}\left(\frac{\partial^2 \Psi}{\partial x^2} + \frac{\partial^2 \Psi}{\partial y^2} + \frac{\partial^2 \Psi}{\partial z^2}\right) + V\Psi = E\Psi$$

$$(B.18)$$

where the terms involve respectively the total mass $M = m_a + m_X$ and the *reduced mass* $m = m_a m_X/(m_a + m_X)$. If the potential depends only on the relative coordinate $r$, then the equation is *separable* into one part that depends only on the CM motion and another relative part that involves the interaction. The wave function separates as $\Psi = \psi_{CM}(X, Y, Z)\psi(x, y, z)$. The CM part has the form of a single particle of mass $M$ moving freely (that is, a plane wave). The relative part $\psi$ is the solution of

$$-\frac{\hbar^2}{2m}\left(\frac{\partial^2 \psi}{\partial x^2} + \frac{\partial^2 \psi}{\partial y^2} + \frac{\partial^2 \psi}{\partial z^2}\right) + V(r)\psi = E\psi \qquad (B.19)$$

where $\psi(x, y, z)$ is the wave function representing the interaction of the particles. This simplified form of the two-body equation now only requires that we solve the Schrödinger equation for a single particle of the reduced mass $m$ moving in a fixed potential $V(r)$ that is identical with the two-body laboratory potential. The resulting solution can be used to evaluate cross sections that can then be compared with laboratory cross sections using the transformation Equations B.15 and B.16. Equation B.19 can also be written in spherical polar coordinates; see Equation 2.16.

# Appendix C

## TABLE OF NUCLEAR PROPERTIES

The following table shows some properties of a selection of isotopes. For each element only the stable and relatively long-lived radioactive isotopes are included. Ground-state atomic masses and spin-parity assignments are shown for all isotopes; uncertain spin-parity assignments are in parentheses. Abundances are given for stable isotopes, and for radioactive isotopes the half-life and principal decay mode are shown ($\varepsilon$—electron capture, possibly including positron emission; $\beta^-$—negative beta decay; $\alpha$—alpha decay; f—spontaneous fission). The masses are those of the corresponding neutral atoms and were taken from the 1983 atomic mass evaluation: A. H. Wapstra and G. Audi, *Nucl. Phys.* **A432**, 1 (1985). In the half-life entries, My = $10^6$ y. Uncertainties in the masses are typically $10^{-5}$ u ($10^{-4}$ u for some cases far from stability); uncertainties in the abundances and half-lives are typically at or below the level of the last digit tabulated.

| | Z | A | Atomic mass (u) | $I^\pi$ | Abundance or Half-life | | Z | A | Atomic mass (u) | $I^\pi$ | Abundance or Half-life |
|---|---|---|---|---|---|---|---|---|---|---|---|
| H | 1 | 1 | 1.007825 | $\frac{1}{2}^+$ | *99.985%* | | | 10 | 10.012937 | $3^+$ | *19.8%* |
| | | 2 | 2.014102 | $1^+$ | *0.015%* | | | 11 | 11.009305 | $\frac{3}{2}^-$ | *80.2%* |
| | | 3 | 3.016049 | $\frac{1}{2}^+$ | 12.3 y ($\beta^-$) | | | 12 | 12.014353 | $1^+$ | 20.4 ms ($\beta^-$) |
| | | | | | | | | 13 | 13.017780 | $\frac{3}{2}^-$ | 17.4 ms ($\beta^-$) |
| He | 2 | 3 | 3.016029 | $\frac{1}{2}^+$ | *$1.38 \times 10^{-4}$%* | | | | | | |
| | | 4 | 4.002603 | $0^+$ | *99.99986%* | C | 6 | 9 | 9.031039 | $\frac{3}{2}^-$ | 0.13 s ($\varepsilon$) |
| | | | | | | | | 10 | 10.016856 | $0^+$ | 19.2 s ($\varepsilon$) |
| Li | 3 | 6 | 6.015121 | $1^+$ | *7.5%* | | | 11 | 11.011433 | $\frac{3}{2}^-$ | 20.4 m ($\varepsilon$) |
| | | 7 | 7.016003 | $\frac{3}{2}^-$ | *92.5%* | | | 12 | 12.000000 | $0^+$ | *98.89%* |
| | | 8 | 8.022486 | $2^+$ | 0.84 s ($\beta^-$) | | | 13 | 13.003355 | $\frac{1}{2}^-$ | *1.11%* |
| | | | | | | | | 14 | 14.003242 | $0^+$ | 5730 y ($\beta^-$) |
| Be | 4 | 7 | 7.016928 | $\frac{3}{2}^-$ | 53.3 d ($\varepsilon$) | | | 15 | 15.010599 | $\frac{1}{2}^+$ | 2.45 s ($\beta^-$) |
| | | 8 | 8.005305 | $0^+$ | 0.07 fs ($\alpha$) | | | | | | |
| | | 9 | 9.012182 | $\frac{3}{2}^-$ | *100 %* | N | 7 | 12 | 12.018613 | $1^+$ | 11 ms ($\varepsilon$) |
| | | 10 | 10.013534 | $0^+$ | 1.6 My ($\beta^-$) | | | 13 | 13.005739 | $\frac{1}{2}^-$ | 9.96 m ($\varepsilon$) |
| | | 11 | 11.021658 | $\frac{1}{2}^+$ | 13.8 s ($\beta^-$) | | | 14 | 14.003074 | $1^+$ | *99.63%* |
| | | | | | | | | 15 | 15.000109 | $\frac{1}{2}^-$ | *0.366%* |
| B | 5 | 8 | 8.024606 | $2^+$ | 0.77 s ($\varepsilon$) | | | 16 | 16.006100 | $2^-$ | 7.13 s ($\beta^-$) |
| | | 9 | 9.013329 | $\frac{3}{2}^-$ | 0.85 as ($\alpha$) | | | | | | |

| | Z | A | Atomic mass (u) | $I^\pi$ | Abundance or Half-life |
|---|---|---|---|---|---|
| | | 17 | 17.008450 | $\frac{1}{2}^-$ | 4.17 s $(\beta^-)$ |
| | | 18 | 18.014081 | $1^-$ | 0.63 s $(\beta^-)$ |
| O | 8 | 14 | 14.008595 | $0^+$ | 71 s $(\varepsilon)$ |
| | | 15 | 15.003065 | $\frac{1}{2}^-$ | 122 s $(\varepsilon)$ |
| | | 16 | 15.994915 | $0^+$ | 99.76% |
| | | 17 | 16.999131 | $\frac{5}{2}^+$ | 0.038% |
| | | 18 | 17.999160 | $0^+$ | 0.204% |
| | | 19 | 19.003577 | $\frac{5}{2}^+$ | 26.9 s $(\beta^-)$ |
| | | 20 | 20.004076 | $0^+$ | 13.5 s $(\beta^-)$ |
| F | 9 | 17 | 17.002095 | $\frac{5}{2}^+$ | 64.5 s $(\varepsilon)$ |
| | | 18 | 18.000937 | $1^+$ | 110 m $(\varepsilon)$ |
| | | 19 | 18.998403 | $\frac{1}{2}^+$ | 100% |
| | | 20 | 19.999981 | $2^+$ | 11 s $(\beta^-)$ |
| | | 21 | 20.999948 | $\frac{5}{2}^+$ | 4.3 s $(\beta^-)$ |
| | | 22 | 22.003030 | $(3,4)^+$ | 4.2 s $(\beta^-)$ |
| | | 23 | 23.003600 | $(\frac{3}{2},\frac{5}{2})^+$ | 2.2 s $(\beta^-)$ |
| Ne | 10 | 17 | 17.017690 | $\frac{1}{2}^-$ | 0.11 s $(\varepsilon)$ |
| | | 18 | 18.005710 | $0^+$ | 1.7 s $(\varepsilon)$ |
| | | 19 | 19.001880 | $\frac{1}{2}^+$ | 17.3 s $(\varepsilon)$ |
| | | 20 | 19.992436 | $0^+$ | 90.51% |
| | | 21 | 20.993843 | $\frac{3}{2}^+$ | 0.27% |
| | | 22 | 21.991383 | $0^+$ | 9.22% |
| | | 23 | 22.994465 | $\frac{5}{2}^+$ | 37.6 s $(\beta^-)$ |
| | | 24 | 23.993613 | $0^+$ | 3.4 m $(\beta^-)$ |
| | | 25 | 24.997690 | $(\frac{1}{2},\frac{3}{2})^+$ | 0.60 s $(\beta^-)$ |
| Na | 11 | 20 | 20.007344 | $2^+$ | 0.45 s $(\varepsilon)$ |
| | | 21 | 20.997651 | $\frac{3}{2}^+$ | 22.5 s $(\varepsilon)$ |
| | | 22 | 21.994434 | $3^+$ | 2.60 y $(\varepsilon)$ |
| | | 23 | 22.989768 | $\frac{3}{2}^+$ | 100% |
| | | 24 | 23.990961 | $4^+$ | 15.0 h $(\beta^-)$ |
| | | 25 | 24.989953 | $\frac{5}{2}^+$ | 60 s $(\beta^-)$ |
| | | 26 | 25.992586 | $3^+$ | 1.1 s $(\beta^-)$ |
| | | 27 | 26.993940 | $\frac{5}{2}^+$ | 0.30 s $(\beta^-)$ |
| Mg | 12 | 21 | 21.011716 | $(\frac{3}{2},\frac{5}{2})^+$ | 0.123 s $(\varepsilon)$ |
| | | 22 | 21.999574 | $0^+$ | 3.86 s $(\varepsilon)$ |
| | | 23 | 22.994124 | $\frac{3}{2}^+$ | 11.3 s $(\varepsilon)$ |
| | | 24 | 23.985042 | $0^+$ | 78.99% |
| | | 25 | 24.985837 | $\frac{5}{2}^+$ | 10.00% |
| | | 26 | 25.982594 | $0^+$ | 11.01% |
| | | 27 | 26.984341 | $\frac{1}{2}^+$ | 9.46 m $(\beta^-)$ |
| | | 28 | 27.983877 | $0^+$ | 21.0 h $(\beta^-)$ |
| | | 29 | 28.988480 | $\frac{3}{2}^+$ | 1.4 s $(\beta^-)$ |
| Al | 13 | 24 | 23.999941 | $4^+$ | 2.07 s $(\varepsilon)$ |
| | | 25 | 24.990429 | $\frac{5}{2}^+$ | 7.18 s $(\varepsilon)$ |
| | | 26 | 25.986892 | $5^+$ | 0.72 My $(\varepsilon)$ |
| | | 27 | 26.981539 | $\frac{5}{2}^+$ | 100% |
| | | 28 | 27.981910 | $3^+$ | 2.24 m $(\beta^-)$ |
| | | 29 | 28.980446 | $\frac{5}{2}^+$ | 6.6 m $(\beta^-)$ |
| | | 30 | 29.982940 | $3^+$ | 3.7 s $(\beta^-)$ |
| Si | 14 | 26 | 25.992330 | $0^+$ | 2.21 s $(\varepsilon)$ |
| | | 27 | 26.986704 | $\frac{5}{2}^+$ | 4.13 s $(\varepsilon)$ |
| | | 28 | 27.976927 | $0^+$ | 92.23% |
| | | 29 | 28.976495 | $\frac{1}{2}^+$ | 4.67% |
| | | 30 | 29.973770 | $0^+$ | 3.10% |
| | | 31 | 30.975362 | $\frac{3}{2}^+$ | 2.62 h $(\beta^-)$ |
| | | 32 | 31.974148 | $0^+$ | 105 y $(\beta^-)$ |
| | | 33 | 32.997920 | $(\frac{3}{2}^+)$ | 6.2 s $(\beta^-)$ |
| P | 15 | 29 | 28.981803 | $\frac{1}{2}^+$ | 4.1 s $(\varepsilon)$ |
| | | 30 | 29.978307 | $1^+$ | 2.50 m $(\varepsilon)$ |
| | | 31 | 30.973762 | $\frac{1}{2}^+$ | 100% |
| | | 32 | 31.973907 | $1^+$ | 14.3 d $(\beta^-)$ |
| | | 33 | 32.971725 | $\frac{1}{2}^+$ | 25.3 d $(\beta^-)$ |
| | | 34 | 33.973636 | $1^+$ | 12.4 s $(\beta^-)$ |
| S | 16 | 30 | 29.984903 | $0^+$ | 1.2 s $(\varepsilon)$ |
| | | 31 | 30.979554 | $\frac{1}{2}^+$ | 2.6 s $(\varepsilon)$ |
| | | 32 | 31.972071 | $0^+$ | 95.02% |
| | | 33 | 32.971458 | $\frac{3}{2}^+$ | 0.75% |
| | | 34 | 33.967867 | $0^+$ | 4.21% |
| | | 35 | 34.969032 | $\frac{3}{2}^+$ | 87.4 d $(\beta^-)$ |
| | | 36 | 35.967081 | $0^+$ | 0.017% |
| | | 37 | 36.971126 | $\frac{7}{2}^-$ | 5.0 m $(\beta^-)$ |
| | | 38 | 37.971162 | $0^+$ | 170 m $(\beta^-)$ |
| Cl | 17 | 33 | 32.977452 | $\frac{3}{2}^+$ | 2.51 s $(\varepsilon)$ |
| | | 34 | 33.973763 | $0^+$ | 1.53 s $(\varepsilon)$ |
| | | 35 | 34.968853 | $\frac{3}{2}^+$ | 75.77% |
| | | 36 | 35.968307 | $2^+$ | 0.30 My $(\beta^-)$ |
| | | 37 | 36.965903 | $\frac{3}{2}^+$ | 24.23% |
| | | 38 | 37.968011 | $2^-$ | 37.3 m $(\beta^-)$ |
| | | 39 | 38.968005 | $\frac{3}{2}^+$ | 56 m $(\beta^-)$ |
| | | 40 | 39.970440 | $2^-$ | 1.35 m $(\beta^-)$ |
| | | 41 | 40.970590 | $(\frac{1}{2},\frac{3}{2})^+$ | 31 s $(\beta^-)$ |
| Ar | 18 | 34 | 33.980269 | $0^+$ | 0.844 s $(\varepsilon)$ |
| | | 35 | 34.975256 | $\frac{3}{2}^+$ | 1.78 s $(\varepsilon)$ |
| | | 36 | 35.967546 | $0^+$ | 0.337% |
| | | 37 | 36.966776 | $\frac{3}{2}^+$ | 35.0 d $(\varepsilon)$ |
| | | 38 | 37.962732 | $0^+$ | 0.063% |
| | | 39 | 38.964314 | $\frac{7}{2}^-$ | 269 y $(\beta^-)$ |
| | | 40 | 39.962384 | $0^+$ | 99.60% |
| | | 41 | 40.964501 | $\frac{7}{2}^-$ | 1.83 h $(\beta^-)$ |

| | Z | A | Atomic mass (u) | $I^\pi$ | Abundance or Half-life | | Z | A | Atomic ass (u) | $I^\pi$ | Abundance or Half-life |
|---|---|---|---|---|---|---|---|---|---|---|---|
| | | 42 | 41.963050 | $0^+$ | 33 y ($\beta^-$) | | | 52 | 51.946898 | $0^+$ | 1.7 m ($\beta^-$) |
| | | 43 | 42.965670 | | 5.4 m ($\beta^-$) | | | 53 | 52.949730 | $(\frac{3}{2})^-$ | 33 s ($\beta^-$) |
| | | 44 | 43.965365 | $0^+$ | 11.9 m ($\beta^-$) | | | | | | |
| K | 19 | 37 | 36.973377 | $\frac{3}{2}^+$ | 1.23 s ($\varepsilon$) | V | 23 | 46 | 45.960198 | $0^+$ | 0.42 s ($\varepsilon$) |
| | | 38 | 37.969080 | $3^+$ | 7.61 m ($\varepsilon$) | | | 47 | 46.954906 | $\frac{3}{2}^-$ | 32.6 m ($\varepsilon$) |
| | | 39 | 38.963707 | $\frac{3}{2}^+$ | 93.26% | | | 48 | 47.952257 | $4^+$ | 16.0 d ($\varepsilon$) |
| | | 40 | 39.963999 | $4^-$ | 1.28 Gy ($\beta^-$) | | | 49 | 48.948517 | $\frac{7}{2}^-$ | 330 d ($\varepsilon$) |
| | | 41 | 40.961825 | $\frac{3}{2}^+$ | 6.73% | | | 50 | 49.947161 | $6^+$ | 0.250% |
| | | 42 | 41.962402 | $2^-$ | 12.4 h ($\beta^-$) | | | 51 | 50.943962 | $\frac{7}{2}^-$ | 99.750% |
| | | 43 | 42.960717 | $\frac{3}{2}^-$ | 22.3 h ($\beta^-$) | | | 52 | 51.944778 | $3^+$ | 3.76 m ($\beta^-$) |
| | | 44 | 43.961560 | $2^-$ | 22.1 m ($\beta^-$) | | | 53 | 52.944340 | $\frac{7}{2}^-$ | 1.6 m ($\beta^-$) |
| | | 45 | 44.960696 | $\frac{3}{2}^+$ | 17 m ($\beta^-$) | | | 54 | 53.946442 | $(3,4,5)^+$ | 50 s ($\beta^-$) |
| | | 46 | 45.961976 | $(2^-)$ | 115 s ($\beta^-$) | Cr | 24 | 46 | 45.968360 | $0^+$ | 0.26 s ($\varepsilon$) |
| | | 47 | 46.961677 | $\frac{1}{2}^+$ | 17.5 s ($\beta^-$) | | | 47 | 46.962905 | $\frac{3}{2}^-$ | 0.51 s ($\varepsilon$) |
| | | | | | | | | 48 | 47.954033 | $0^+$ | 21.6 h ($\varepsilon$) |
| Ca | 20 | 38 | 37.976318 | $0^+$ | 0.44 s ($\varepsilon$) | | | 49 | 48.951338 | $\frac{5}{2}^-$ | 41.9 m ($\varepsilon$) |
| | | 39 | 38.970718 | $\frac{3}{2}^+$ | 0.86 s ($\varepsilon$) | | | 50 | 49.946046 | $0^+$ | 4.35% |
| | | 40 | 39.962591 | $0^+$ | 96.94% | | | 51 | 50.944768 | $\frac{7}{2}^-$ | 27.7 d ($\varepsilon$) |
| | | 41 | 40.962278 | $\frac{7}{2}^-$ | 0.10 My ($\varepsilon$) | | | 52 | 51.940510 | $0^+$ | 83.79% |
| | | 42 | 41.958618 | $0^+$ | 0.647% | | | 53 | 52.940651 | $\frac{3}{2}^-$ | 9.50% |
| | | 43 | 42.958766 | $\frac{7}{2}^-$ | 0.135% | | | 54 | 53.938882 | $0^+$ | 2.36% |
| | | 44 | 43.955481 | $0^+$ | 2.09% | | | 55 | 54.940842 | $\frac{3}{2}^-$ | 3.50 m ($\beta^-$) |
| | | 45 | 44.956185 | $\frac{7}{2}^-$ | 165 d ($\beta^-$) | | | 56 | 55.940643 | | 5.9 m ($\beta^-$) |
| | | 46 | 45.953689 | $0^+$ | 0.0035% | | | | | | |
| | | 47 | 46.954543 | $\frac{7}{2}^-$ | 4.54 d ($\beta^-$) | Mn | 25 | 50 | 49.954240 | $0^+$ | 0.28 s ($\varepsilon$) |
| | | 48 | 47.952533 | $0^+$ | 0.187% | | | 51 | 50.948213 | $\frac{5}{2}^-$ | 46.2 m ($\varepsilon$) |
| | | 49 | 48.955672 | $\frac{3}{2}^-$ | 8.72 m ($\beta^-$) | | | 52 | 51.945568 | $6^+$ | 5.59 d ($\varepsilon$) |
| | | 50 | 49.957519 | $0^+$ | 14 s ($\beta^-$) | | | 53 | 52.941291 | $\frac{7}{2}^-$ | 3.7 My ($\varepsilon$) |
| | | | | | | | | 54 | 53.940361 | $3^+$ | 312 d ($\varepsilon$) |
| Sc | 21 | 42 | 41.965514 | $0^+$ | 0.68 s ($\varepsilon$) | | | 55 | 54.938047 | $\frac{5}{2}^-$ | 100% |
| | | 43 | 42.961150 | $\frac{7}{2}^-$ | 3.89 h ($\varepsilon$) | | | 56 | 55.938907 | $3^+$ | 2.58 h ($\beta^-$) |
| | | 44 | 43.959404 | $2^+$ | 3.93 h ($\varepsilon$) | | | 57 | 56.938285 | $\frac{5}{2}^-$ | 1.6 m ($\beta^-$) |
| | | 45 | 44.955910 | $\frac{7}{2}^-$ | 100% | | | 58 | 57.940060 | $3^+$ | 65 s ($\beta^-$) |
| | | 46 | 45.955170 | $4^+$ | 83.8 d ($\beta^-$) | | | | | | |
| | | 47 | 46.952409 | $\frac{7}{2}^-$ | 3.35 d ($\beta^-$) | Fe | 26 | 51 | 50.956825 | $(\frac{5}{2})^-$ | 0.25 s ($\varepsilon$) |
| | | 48 | 47.952235 | $6^+$ | 43.7 h ($\beta^-$) | | | 52 | 51.948114 | $0^+$ | 8.27 h ($\varepsilon$) |
| | | 49 | 48.950022 | $\frac{7}{2}^-$ | 57.0 m ($\beta^-$) | | | 53 | 52.945310 | $\frac{7}{2}^-$ | 8.51 m ($\varepsilon$) |
| | | 50 | 49.952186 | $5^+$ | 1.71 m ($\beta^-$) | | | 54 | 53.939613 | $0^+$ | 5.8% |
| | | | | | | | | 55 | 54.938296 | $\frac{3}{2}^-$ | 2.7 y ($\varepsilon$) |
| Ti | 22 | 43 | 42.968523 | $\frac{7}{2}^-$ | 0.51 s ($\varepsilon$) | | | 56 | 55.934939 | $0^+$ | 91.8% |
| | | 44 | 43.959690 | $0^+$ | 54 y ($\varepsilon$) | | | 57 | 56.935396 | $\frac{1}{2}^-$ | 2.15% |
| | | 45 | 44.958124 | $\frac{7}{2}^-$ | 3.09 h ($\varepsilon$) | | | 58 | 57.933277 | $0^+$ | 0.29% |
| | | 46 | 45.952629 | $0^+$ | 8.2% | | | 59 | 58.934877 | $\frac{3}{2}^-$ | 44.6 d ($\beta^-$) |
| | | 47 | 46.951764 | $\frac{5}{2}^-$ | 7.4% | | | 60 | 59.934078 | $0^+$ | 1.5 My ($\beta^-$) |
| | | 48 | 47.947947 | $0^+$ | 73.7% | | | 61 | 60.936748 | $(\frac{3}{2},\frac{5}{2})^-$ | 6.0 m ($\beta^-$) |
| | | 49 | 48.947871 | $\frac{7}{2}^-$ | 5.4% | | | 62 | 61.936773 | $0^+$ | 68 s ($\beta^-$) |
| | | 50 | 49.944792 | $0^+$ | 5.2% | | | | | | |
| | | 51 | 50.946616 | $\frac{3}{2}^-$ | 5.80 m ($\beta^-$) | Co | 27 | 54 | 53.948460 | $0^+$ | 0.19 s ($\varepsilon$) |

| Z | A | Atomic mass (u) | $I^\pi$ | Abundance or Half-life |
|---|---|---|---|---|
|  | 55 | 54.942001 | $\frac{7}{2}^-$ | 17.5 h ($\varepsilon$) |
|  | 56 | 55.939841 | $4$ | 78.8 d ($\varepsilon$) |
|  | 57 | 56.936294 | $\frac{7}{2}^-$ | 271 d ($\varepsilon$) |
|  | 58 | 57.935755 | $2^+$ | 70.8 d ($\varepsilon$) |
|  | 59 | 58.933198 | $\frac{7}{2}^-$ | 100 % |
|  | 60 | 59.933820 | $5^+$ | 5.27 y ($\beta^-$) |
|  | 61 | 60.932478 | $\frac{7}{2}^-$ | 1.65 h ($\beta^-$) |
|  | 62 | 61.934060 | $2^+$ | 1.5 m ($\beta^-$) |
|  | 63 | 62.933614 | $(\frac{7}{2})^-$ | 27.5 s ($\beta^-$) |
| Ni 28 | 55 | 54.951336 | $\frac{7}{2}^-$ | 0.19 s ($\varepsilon$) |
|  | 56 | 55.942134 | $0^+$ | 6.10 d ($\varepsilon$) |
|  | 57 | 56.939799 | $\frac{3}{2}^-$ | 36.0 h ($\varepsilon$) |
|  | 58 | 57.935346 | $0^+$ | 68.3 % |
|  | 59 | 58.934349 | $\frac{3}{2}^-$ | 0.075 My ($\varepsilon$) |
|  | 60 | 59.930788 | $0^+$ | 26.1 % |
|  | 61 | 60.931058 | $\frac{3}{2}^-$ | 1.13 % |
|  | 62 | 61.928346 | $0^+$ | 3.59 % |
|  | 63 | 62.929670 | $\frac{1}{2}^-$ | 100 y ($\beta^-$) |
|  | 64 | 63.927968 | $0^+$ | 0.91 % |
|  | 65 | 64.930086 | $\frac{5}{2}^-$ | 2.52 h ($\beta^-$) |
|  | 66 | 65.929116 | $0^+$ | 54.8 h ($\beta^-$) |
|  | 67 | 66.931570 | ? | 21 s ($\beta^-$) |
| Cu 29 | 59 | 58.939503 | $\frac{3}{2}^-$ | 82 s ($\varepsilon$) |
|  | 60 | 59.937366 | $2^+$ | 23.4 m ($\varepsilon$) |
|  | 61 | 60.933461 | $\frac{3}{2}^-$ | 3.41 h ($\varepsilon$) |
|  | 62 | 61.932586 | $1^+$ | 9.73 m ($\varepsilon$) |
|  | 63 | 62.929599 | $\frac{3}{2}^-$ | 69.2 % |
|  | 64 | 63.292766 | $1^+$ | 12.7 h ($\varepsilon$) |
|  | 65 | 64.927793 | $\frac{3}{2}^-$ | 30.8 % |
|  | 66 | 65.928872 | $1^+$ | 5.10 m ($\beta^-$) |
|  | 67 | 66.927747 | $\frac{3}{2}^-$ | 61.9 h ($\beta^-$) |
|  | 68 | 67.929620 | $1^+$ | 31 s ($\beta^-$) |
| Zn 30 | 61 | 60.939514 | $\frac{3}{2}^-$ | 89 s ($\varepsilon$) |
|  | 62 | 61.934332 | $0^+$ | 9.2 h ($\varepsilon$) |
|  | 63 | 62.933214 | $\frac{3}{2}^-$ | 38.1 m ($\varepsilon$) |
|  | 64 | 63.929145 | $0^+$ | 48.6 % |
|  | 65 | 64.929243 | $\frac{5}{2}^-$ | 244 d ($\varepsilon$) |
|  | 66 | 65.926035 | $0^+$ | 27.9 % |
|  | 67 | 66.927129 | $\frac{5}{2}^-$ | 4.10 % |
|  | 68 | 67.924846 | $0^+$ | 18.8 % |
|  | 69 | 68.926552 | $\frac{1}{2}^-$ | 56 m ($\beta^-$) |
|  | 70 | 69.925325 | $0^+$ | 0.62 % |
|  | 71 | 70.927727 | $\frac{1}{2}^-$ | 2.4 m ($\beta^-$) |
|  | 72 | 71.926856 | $0^+$ | 46.5 h ($\beta^-$) |
|  | 73 | 72.929780 | $(\frac{3}{2})^-$ | 24 s ($\beta^-$) |
| Ga 31 | 64 | 63.936836 | $0^+$ | 2.6 m ($\varepsilon$) |
|  | 65 | 64.932738 | $\frac{3}{2}^-$ | 15.2 m ($\varepsilon$) |
|  | 66 | 65.931590 | $0^+$ | 9.4 h ($\varepsilon$) |
|  | 67 | 66.928204 | $\frac{3}{2}^-$ | 78.3 h ($\varepsilon$) |
|  | 68 | 67.927982 | $1^+$ | 68.1 m ($\varepsilon$) |
|  | 69 | 68.925580 | $\frac{3}{2}^-$ | 60.1 % |
|  | 70 | 69.926028 | $1^+$ | 21.1 m ($\beta^-$) |
|  | 71 | 70.924701 | $\frac{3}{2}^-$ | 39.9 % |
|  | 72 | 71.926365 | $3^-$ | 14.1 h ($\beta^-$) |
|  | 73 | 72.925169 | $\frac{3}{2}^-$ | 4.87 h ($\beta^-$) |
|  | 74 | 73.926940 | $(4)^-$ | 8.1 m ($\beta^-$) |
|  | 75 | 74.926499 | $\frac{3}{2}^-$ | 2.1 m ($\beta^-$) |
| Ge 32 | 66 | 65.933847 | $0^+$ | 2.3 h ($\varepsilon$) |
|  | 67 | 66.932737 | $(\frac{1}{2})^-$ | 19.0 m ($\varepsilon$) |
|  | 68 | 67.928096 | $0^+$ | 271 d ($\varepsilon$) |
|  | 69 | 68.927969 | $\frac{5}{2}^-$ | 39.0 h ($\varepsilon$) |
|  | 70 | 69.924250 | $0^+$ | 20.5 % |
|  | 71 | 70.924954 | $\frac{1}{2}^-$ | 11.2 d ($\varepsilon$) |
|  | 72 | 71.922079 | $0^+$ | 27.4 % |
|  | 73 | 72.923463 | $\frac{9}{2}^+$ | 7.8 % |
|  | 74 | 73.921177 | $0^+$ | 36.5 % |
|  | 75 | 74.922858 | $\frac{1}{2}^-$ | 82.8 m ($\beta^-$) |
|  | 76 | 75.921402 | $0^+$ | 7.8 % |
|  | 77 | 76.923548 | $\frac{7}{2}^+$ | 11.3 h ($\beta^-$) |
|  | 78 | 77.922853 | $0^+$ | 1.45 h ($\beta^-$) |
|  | 79 | 78.925360 | $(\frac{1}{2})^-$ | 19 s ($\beta^-$) |
| As 33 | 70 | 69.930929 | $4^+$ | 53 m ($\varepsilon$) |
|  | 71 | 70.927114 | $\frac{5}{2}^-$ | 61 h ($\varepsilon$) |
|  | 72 | 71.926755 | $2^-$ | 26.0 h ($\varepsilon$) |
|  | 73 | 72.923827 | $\frac{3}{2}^-$ | 80.3 d ($\varepsilon$) |
|  | 74 | 73.923928 | $2^-$ | 17.8 d ($\varepsilon$) |
|  | 75 | 74.921594 | $\frac{3}{2}^-$ | 100 % |
|  | 76 | 75.922393 | $2^-$ | 26.3 h ($\beta^-$) |
|  | 77 | 76.920646 | $\frac{3}{2}^-$ | 38.8 h ($\beta^-$) |
|  | 78 | 77.921830 | $(2^-)$ | 91 m ($\beta^-$) |
|  | 79 | 78.920946 | $\frac{3}{2}^-$ | 9.0 m ($\beta^-$) |
| Se 34 | 71 | 70.932270 | $\frac{5}{2}^-$ | 4.7 m ($\varepsilon$) |
|  | 72 | 71.927110 | $0^+$ | 8.4 d ($\varepsilon$) |
|  | 73 | 72.926768 | $\frac{9}{2}^+$ | 7.1 h ($\varepsilon$) |
|  | 74 | 73.922475 | $0^+$ | 0.87 % |
|  | 75 | 74.922522 | $\frac{5}{2}^+$ | 119.8 d ($\varepsilon$) |
|  | 76 | 75.919212 | $0^+$ | 9.0 % |
|  | 77 | 76.919913 | $\frac{1}{2}^-$ | 7.6 % |
|  | 78 | 77.917308 | $0^+$ | 23.5 % |
|  | 79 | 78.918498 | $\frac{7}{2}^+$ | < 0.065 My ($\beta^-$) |

| Z | A | Atomic mass (u) | $I^\pi$ | Abundance or Half-life |
|---|---|---|---|---|
| | 80 | 79.916520 | $0^+$ | *49.8%* |
| | 81 | 80.917991 | $(\tfrac{1}{2})^-$ | 18.5 m ($\beta^-$) |
| | 82 | 81.916698 | $0^+$ | *9.2%* |
| | 83 | 82.919117 | $(\tfrac{9}{2})^+$ | 22.5 m ($\beta^-$) |
| | 84 | 83.918463 | $0^+$ | 3.3 m ($\beta^-$) |
| Br 35 | 76 | 75.924528 | $1^-$ | 16.1 h ($\varepsilon$) |
| | 77 | 76.921378 | $\tfrac{3}{2}^-$ | 57.0 h ($\varepsilon$) |
| | 78 | 77.921144 | $1^+$ | 6.46 m ($\varepsilon$) |
| | 79 | 78.918336 | $\tfrac{3}{2}^-$ | *50.69%* |
| | 80 | 79.918528 | $1^+$ | 17.6 m ($\beta^-$) |
| | 81 | 80.916289 | $\tfrac{3}{2}^-$ | *49.31%* |
| | 82 | 81.916802 | $5^-$ | 35.3 h ($\beta^-$) |
| | 83 | 82.915179 | $(\tfrac{3}{2})^-$ | 2.39 h ($\beta^-$) |
| | 84 | 83.916503 | $2^-$ | 31.8 m ($\beta^-$) |
| | 85 | 84.915612 | $(\tfrac{3}{2})^-$ | 2.9 m ($\beta^-$) |
| Kr 36 | 75 | 74.931029 | ? | 4.3 m ($\varepsilon$) |
| | 76 | 75.925959 | $0^+$ | 14.8 h ($\varepsilon$) |
| | 77 | 76.924610 | $\tfrac{5}{2}^+$ | 75 m ($\varepsilon$) |
| | 78 | 77.920396 | $0^+$ | *0.356%* |
| | 79 | 78.920084 | $\tfrac{1}{2}^-$ | 35.0 h ($\varepsilon$) |
| | 80 | 79.916380 | $0^+$ | *2.27%* |
| | 81 | 80.916590 | $\tfrac{7}{2}^+$ | 0.21 My ($\varepsilon$) |
| | 82 | 81.913482 | $0^+$ | *11.6%* |
| | 83 | 82.914135 | $\tfrac{9}{2}^+$ | *11.5%* |
| | 84 | 83.911507 | $0^+$ | *57.0%* |
| | 85 | 84.912531 | $\tfrac{9}{2}^+$ | 10.7 y ($\beta^-$) |
| | 86 | 85.910616 | $0^+$ | *17.3%* |
| | 87 | 86.913360 | $\tfrac{5}{2}^+$ | 76 m ($\beta^-$) |
| | 88 | 87.914453 | $0^+$ | 2.84 h ($\beta^-$) |
| | 89 | 88.917640 | $(\tfrac{5}{2})^+$ | 3.18 m ($\beta^-$) |
| Rb 37 | 82 | 81.918195 | $1^+$ | 1.25 m ($\varepsilon$) |
| | 83 | 82.915144 | $\tfrac{5}{2}^-$ | 86.2 d ($\varepsilon$) |
| | 84 | 83.914390 | $2^-$ | 32.9 d ($\varepsilon$) |
| | 85 | 84.911794 | $\tfrac{5}{2}^-$ | *72.17%* |
| | 86 | 85.911172 | $2^-$ | 18.8 d ($\beta^-$) |
| | 87 | 86.909187 | $\tfrac{3}{2}^-$ | *27.83%* |
| | 88 | 87.911326 | $2^-$ | 17.8 m ($\beta^-$) |
| | 89 | 88.912278 | $(\tfrac{3}{2})^-$ | 15.2 m ($\beta^-$) |
| | 90 | 89.914811 | $(1^-)$ | 153 s ($\beta^-$) |
| Sr 38 | 81 | 80.923270 | $(\tfrac{1}{2}^-)$ | 22 m ($\varepsilon$) |
| | 82 | 81.918414 | $0^+$ | 25.0 d ($\varepsilon$) |
| | 83 | 82.917566 | $\tfrac{7}{2}^+$ | 32.4 d ($\varepsilon$) |
| | 84 | 83.913430 | $0^+$ | *0.56%* |
| | 85 | 84.912937 | $\tfrac{9}{2}^+$ | 64.8 d ($\varepsilon$) |
| | 86 | 85.909267 | $0^+$ | *9.8%* |

| Z | A | Atomic mass (u) | $I^\pi$ | Abundance or Half-life |
|---|---|---|---|---|
| | 87 | 86.908884 | $\tfrac{9}{2}^+$ | *7.0%* |
| | 88 | 87.905619 | $0^+$ | *82.6%* |
| | 89 | 88.907450 | $\tfrac{5}{2}^+$ | 50.5 d ($\beta^-$) |
| | 90 | 89.907738 | $0^+$ | 28.8 y ($\beta^-$) |
| | 91 | 90.910187 | $(\tfrac{5}{2})^+$ | 9.5 h ($\beta^-$) |
| | 92 | 91.910944 | $0^+$ | 2.7 h ($\beta^-$) |
| | 93 | 92.913987 | $(\tfrac{7}{2}^+)$ | 7.4 m ($\beta^-$) |
| Y 39 | 84 | 83.920310 | $(5^-)$ | 39 m ($\varepsilon$) |
| | 85 | 84.916437 | $(\tfrac{1}{2})^-$ | 2.7 h ($\varepsilon$) |
| | 86 | 85.914893 | $4^-$ | 14.7 h ($\varepsilon$) |
| | 87 | 86.910882 | $\tfrac{1}{2}^-$ | 80.3 h ($\varepsilon$) |
| | 88 | 87.909508 | $4^-$ | 106.6 d ($\varepsilon$) |
| | 89 | 88.905849 | $\tfrac{1}{2}^-$ | *100%* |
| | 90 | 89.907152 | $2^-$ | 64.1 h ($\beta^-$) |
| | 91 | 90.907303 | $\tfrac{1}{2}^-$ | 58.5 d ($\beta^-$) |
| | 92 | 91.908917 | $2^-$ | 3.54 h ($\beta^-$) |
| | 93 | 92.909571 | $\tfrac{1}{2}^-$ | 10.2 h ($\beta^-$) |
| | 94 | 93.911597 | $2^-$ | 18.7 m ($\beta^-$) |
| Zr 40 | 87 | 86.914817 | $(\tfrac{9}{2}^+)$ | 1.6 h ($\varepsilon$) |
| | 88 | 87.910225 | $0^+$ | 83.4 d ($\varepsilon$) |
| | 89 | 88.908890 | $\tfrac{9}{2}^+$ | 78.4 h ($\varepsilon$) |
| | 90 | 89.904703 | $0^+$ | *51.5%* |
| | 91 | 90.905644 | $\tfrac{5}{2}^+$ | *11.2%* |
| | 92 | 91.905039 | $0^+$ | *17.1%* |
| | 93 | 92.906474 | $\tfrac{5}{2}^+$ | 1.5 My ($\beta^-$) |
| | 94 | 93.906315 | $0^+$ | *17.4%* |
| | 95 | 94.908042 | $\tfrac{5}{2}^+$ | 64.0 d ($\beta^-$) |
| | 96 | 95.908275 | $0^+$ | *2.80%* |
| | 97 | 96.910950 | $\tfrac{1}{2}^+$ | 16.9 h ($\beta^-$) |
| | 98 | 97.912735 | $0^+$ | 31 s ($\beta^-$) |
| Nb 41 | 89 | 88.913449 | $(\tfrac{1}{2})^-$ | 2.0 h ($\varepsilon$) |
| | 90 | 89.911263 | $8^+$ | 14.6 h ($\varepsilon$) |
| | 91 | 90.906991 | $(\tfrac{9}{2})^+$ | 700 y ($\varepsilon$) |
| | 92 | 91.907192 | $(7)^+$ | 35 My ($\varepsilon$) |
| | 93 | 92.906377 | $\tfrac{9}{2}^+$ | *100%* |
| | 94 | 93.907281 | $6^+$ | 0.020 My ($\beta^-$) |
| | 95 | 94.906835 | $\tfrac{9}{2}^+$ | 35.0 d ($\beta^-$) |
| | 96 | 95.908100 | $6^+$ | 23.4 h ($\beta^-$) |
| | 97 | 96.908097 | $\tfrac{9}{2}^+$ | 72 m ($\beta^-$) |
| Mo 42 | 90 | 89.913933 | $0^+$ | 5.67 h ($\varepsilon$) |
| | 91 | 90.911755 | $\tfrac{9}{2}^+$ | 15.5 m ($\varepsilon$) |
| | 92 | 91.906808 | $0^+$ | *14.8%* |
| | 93 | 92.906813 | $\tfrac{5}{2}^+$ | 3500 y ($\varepsilon$) |
| | 94 | 93.905085 | $0^+$ | *9.3%* |
| | 95 | 94.905841 | $\tfrac{5}{2}^+$ | *15.9%* |

| Z | A | Atomic mass (u) | $I^\pi$ | Abundance or Half-life | | Z | A | Atomic mass (u) | $I^\pi$ | Abundance or Half-life |
|---|---|---|---|---|---|---|---|---|---|---|
| | 96 | 95.904679 | $0^+$ | *16.7%* | | | 108 | 107.903895 | $0^+$ | *26.7%* |
| | 97 | 96.906021 | $\frac{5}{2}^+$ | *9.6%* | | | 109 | 108.905954 | $\frac{5}{2}^+$ | 13.4 h ($\beta^-$) |
| | 98 | 97.905407 | $0^+$ | *24.1%* | | | 110 | 109.905167 | $0^+$ | *11.8%* |
| | 99 | 98.907711 | $\frac{1}{2}^+$ | 66.0 h ($\beta^-$) | | | 111 | 110.907660 | $\frac{5}{2}^+$ | 23 m ($\beta^-$) |
| | 100 | 99.907477 | $0^+$ | *9.6%* | | | 112 | 111.907323 | $0^+$ | 21.0 h ($\beta^-$) |
| | 101 | 100.910345 | $\frac{1}{2}^+$ | 14.6 m ($\beta^-$) | | Ag 47 | 103 | 102.908980 | $\frac{7}{2}^+$ | 65.7 m ($\varepsilon$) |
| Tc 43 | 94 | 93.909654 | $7^+$ | 293 m ($\varepsilon$) | | | 104 | 103.908623 | $5^+$ | 69.2 m ($\varepsilon$) |
| | 95 | 94.907657 | $\frac{9}{2}^|$ | 20.0 h ($\varepsilon$) | | | 105 | 104.906520 | $\frac{1}{2}^-$ | 41.3 d ($\varepsilon$) |
| | 96 | 95.907870 | $7^+$ | 4.3 d ($\varepsilon$) | | | 106 | 105.906662 | $1^+$ | 24.0 m ($\varepsilon$) |
| | 97 | 96.906364 | $\frac{9}{2}^+$ | 2.6 My ($\varepsilon$) | | | 107 | 106.905092 | $\frac{1}{2}^-$ | *51.83%* |
| | 98 | 97.907215 | $(6)^+$ | 4.2 My ($\beta^-$) | | | 108 | 107.905952 | $1^+$ | 2.4 m ($\beta^-$) |
| | 99 | 98.906254 | $\frac{9}{2}^+$ | 0.214 My ($\beta^-$) | | | 109 | 108.904756 | $\frac{1}{2}^-$ | *48.17%* |
| | 100 | 99.907657 | $1^+$ | 15.8 s ($\beta^-$) | | | 110 | 109.906111 | $1^+$ | 24.4 s ($\beta^-$) |
| Ru 44 | 94 | 93.911361 | $0^+$ | 52 m ($\varepsilon$) | | | 111 | 110.905295 | $\frac{1}{2}^-$ | 7.45 d ($\beta^-$) |
| | 95 | 94.910414 | $\frac{5}{2}^+$ | 1.65 h ($\varepsilon$) | | | 112 | 111.907010 | $2^-$ | 3.14 h ($\beta^-$) |
| | 96 | 95.907599 | $0^+$ | *5.5%* | | Cd 48 | 104 | 103.909851 | $0^+$ | 58 m ($\varepsilon$) |
| | 97 | 96.907556 | $\frac{5}{2}^+$ | 2.88 d ($\varepsilon$) | | | 105 | 104.909459 | $\frac{5}{2}^+$ | 56.0 m ($\varepsilon$) |
| | 98 | 97.905287 | $0^+$ | *1.86%* | | | 106 | 105.906461 | $0^+$ | *1.25%* |
| | 99 | 98.905939 | $\frac{5}{2}^+$ | *12.7%* | | | 107 | 106.906613 | $\frac{5}{2}^+$ | 6.50 h ($\varepsilon$) |
| | 100 | 99.904219 | $0^+$ | *12.6%* | | | 108 | 107.904176 | $0^+$ | *0.89%* |
| | 101 | 100.905582 | $\frac{5}{2}^+$ | *17.0%* | | | 109 | 108.904953 | $\frac{5}{2}^+$ | 463 d ($\varepsilon$) |
| | 102 | 101.904348 | $0^+$ | *31.6%* | | | 110 | 109.903005 | $0^+$ | *12.5%* |
| | 103 | 102.906323 | $\frac{3}{2}^+$ | 39.4 d ($\beta^-$) | | | 111 | 110.904182 | $\frac{1}{2}^+$ | *12.8%* |
| | 104 | 103.905424 | $0^+$ | *18.7%* | | | 112 | 111.902757 | $0^+$ | *24.1%* |
| | 105 | 104.907744 | $\frac{3}{2}^+$ | 4.44 h ($\beta^-$) | | | 113 | 112.904400 | $\frac{1}{2}^+$ | *12.2%* |
| | 106 | 105.907321 | $0^+$ | 372 d ($\beta^-$) | | | 114 | 113.903357 | $0^+$ | *,28.7%* |
| | 107 | 106.910130 | $(\frac{5}{2}^+)$ | 3.8 m ($\beta^-$) | | | 115 | 114.905430 | $\frac{1}{2}^+$ | 53.4 h ($\beta^-$) |
| Rh 45 | 98 | 97.910716 | $(2)^+$ | 8.7 m ($\varepsilon$) | | | 116 | 115.904755 | $0^+$ | *7.5%* |
| | 99 | 98.908192 | $(\frac{1}{2}^-)$ | 16.1 d ($\varepsilon$) | | | 117 | 116.907228 | $\frac{1}{2}^+$ | 2.4 h ($\beta^-$) |
| | 100 | 99.908116 | $1^-$ | 20.8 h ($\varepsilon$) | | | 118 | 117.911700 | $0^+$ | 50.3 m ($\beta^-$) |
| | 101 | 100.906159 | $\frac{1}{2}^-$ | 3.3 y ($\varepsilon$) | | In 49 | 110 | 109.907230 | $2^+$ | 69.1 m ($\varepsilon$) |
| | 102 | 101.906814 | $6^+$ | 2.9 y ($\varepsilon$) | | | 111 | 110.905109 | $\frac{9}{2}^+$ | 2.83 d ($\varepsilon$) |
| | 103 | 102.905500 | $\frac{1}{2}^-$ | *100 %* | | | 112 | 111.905536 | $1^+$ | 14.4 m ($\varepsilon$) |
| | 104 | 103.906651 | $1^+$ | 42.3 s ($\beta^-$) | | | 113 | 112.904061 | $\frac{9}{2}^+$ | *4.3%* |
| | 105 | 104.905686 | $\frac{7}{2}^+$ | 35.4 h ($\beta^-$) | | | 114 | 113.904916 | $1^+$ | 71.9 s ($\beta^-$) |
| | 106 | 105.907279 | $1^+$ | 29.8 s ($\beta^-$) | | | 115 | 114.903882 | $\frac{9}{2}^+$ | *95.7%* |
| Pd 46 | 99 | 98.911763 | $(\frac{5}{2}^+)$ | 21.4 m ($\varepsilon$) | | | 116 | 115.905264 | $1^+$ | 14.1 s ($\beta^-$) |
| | 100 | 99.908527 | $0^+$ | 3.6 d ($\varepsilon$) | | | 117 | 116.904517 | $\frac{9}{2}^+$ | 43.8 m ($\beta^-$) |
| | 101 | 100.908287 | $\frac{5}{2}^+$ | 8.5 h ($\varepsilon$) | | Sn 50 | 109 | 108.911294 | $\frac{7}{2}^+$ | 18.0 m ($\varepsilon$) |
| | 102 | 101.905634 | $0^+$ | *1.0%* | | | 110 | 109.907858 | $0^+$ | 4.1 h ($\varepsilon$) |
| | 103 | 102.906114 | $\frac{5}{2}^+$ | 17.0 d ($\varepsilon$) | | | 111 | 110.907741 | $\frac{7}{2}^+$ | 35 m ($\varepsilon$) |
| | 104 | 103.904029 | $0^+$ | *11.0%* | | | 112 | 111.904826 | $0^+$ | *1.01%* |
| | 105 | 104.905079 | $\frac{5}{2}^+$ | *22.2%* | | | 113 | 112.905176 | $\frac{1}{2}^+$ | 115.1 d ($\varepsilon$) |
| | 106 | 105.903478 | $0^+$ | *27.3%* | | | 114 | 113.902784 | $0^+$ | *0.67%* |
| | 107 | 106.905127 | $\frac{5}{2}^+$ | 6.5 My ($\beta^-$) | | | 115 | 114.903348 | $\frac{1}{2}^+$ | *0.38%* |

| | Z | A | Atomic mass (u) | $I^\pi$ | Abundance or Half-life |
|---|---|---|---|---|---|
| | | 116 | 115.901747 | $0^+$ | 14.6% |
| | | 117 | 116.902956 | $\frac{1}{2}^+$ | 7.75% |
| | | 118 | 117.901609 | $0^+$ | 24.3% |
| | | 119 | 118.903311 | $\frac{1}{2}^+$ | 8.6% |
| | | 120 | 119.902199 | $0^+$ | 32.4% |
| | | 121 | 120.904239 | $\frac{3}{2}^+$ | 27.1 h ($\beta^-$) |
| | | 122 | 121.903440 | $0^+$ | 4.56% |
| | | 123 | 122.905722 | $\frac{11}{2}^-$ | 129 d ($\beta^-$) |
| | | 124 | 123.905274 | $0^+$ | 5.64% |
| | | 125 | 124.907785 | $\frac{11}{2}^-$ | 9.62 d ($\beta^-$) |
| | | 126 | 125.907654 | $0^+$ | 0.1 My ($\beta^-$) |
| | | 127 | 126.910355 | $(\frac{11}{2}^-)$ | 2.1 h ($\beta^-$) |
| Sb | 51 | 118 | 117.905534 | $1^+$ | 3.6 m ($\varepsilon$) |
| | | 119 | 118.903948 | $\frac{5}{2}^+$ | 38.0 h ($\varepsilon$) |
| | | 120 | 119.905077 | $1^+$ | 15.8 m ($\varepsilon$) |
| | | 121 | 120.903821 | $\frac{5}{2}^+$ | 57.3% |
| | | 122 | 121.905179 | $2^-$ | 2.70 d ($\beta^-$) |
| | | 123 | 122.904216 | $\frac{7}{2}^+$ | 42.7% |
| | | 124 | 123.905938 | $3^-$ | 60.2 d ($\beta^-$) |
| | | 125 | 124.905252 | $\frac{7}{2}^+$ | 2.7 y ($\beta^-$) |
| | | 126 | 125.907250 | $8^-$ | 12.4 d ($\beta^-$) |
| | | 127 | 126.906919 | $\frac{7}{2}^+$ | 3.85 d ($\beta^-$) |
| Te | 52 | 117 | 116.908630 | $\frac{1}{2}^+$ | 62 m ($\varepsilon$) |
| | | 118 | 117.905908 | $0^+$ | 6.00 d ($\varepsilon$) |
| | | 119 | 118.906411 | $\frac{1}{2}^+$ | 16.0 h ($\varepsilon$) |
| | | 120 | 119.904048 | $0^+$ | 0.091% |
| | | 121 | 120.904947 | $\frac{1}{2}^+$ | 16.8 d ($\varepsilon$) |
| | | 122 | 121.903050 | $0^+$ | 2.5% |
| | | 123 | 122.904271 | $\frac{1}{2}^+$ | 0.89% |
| | | 124 | 123.902818 | $0^+$ | 4.6% |
| | | 125 | 124.904429 | $\frac{1}{2}^+$ | 7.0% |
| | | 126 | 125.903310 | $0^+$ | 18.7% |
| | | 127 | 126.905221 | $\frac{3}{2}^+$ | 9.4 h ($\beta^-$) |
| | | 128 | 127.904463 | $0^+$ | 31.7% |
| | | 129 | 128.906594 | $\frac{3}{2}^+$ | 69 m ($\beta^-$) |
| | | 130 | 129.906229 | $0^+$ | 34.5% |
| | | 131 | 130.908528 | $\frac{3}{2}^+$ | 25.0 m ($\beta^-$) |
| | | 132 | 131.908517 | $0^+$ | 78.2 h ($\beta^-$) |
| | | 133 | 132.910910 | $(\frac{3}{2}^+)$ | 12.5 m ($\beta^-$) |
| I | 53 | 123 | 122.905594 | $\frac{5}{2}^+$ | 13.2 h ($\varepsilon$) |
| | | 124 | 123.906207 | $2^-$ | 4.18 d ($\varepsilon$) |
| | | 125 | 124.904620 | $\frac{5}{2}^+$ | 60.2 d ($\varepsilon$) |
| | | 126 | 125.905624 | $2^-$ | 13.0 d ($\varepsilon$) |
| | | 127 | 126.904473 | $\frac{5}{2}^+$ | 100% |
| | | 128 | 127.905810 | $1^+$ | 25.0 m ($\beta^-$) |
| | | 129 | 128.904986 | $\frac{7}{2}^+$ | 16 My ($\beta^-$) |
| | | 130 | 129.906713 | $5^+$ | 12.4 h ($\beta^-$) |
| | | 131 | 130.906114 | $\frac{7}{2}^+$ | 8.04 d ($\beta^-$) |
| | | 132 | 131.907987 | $4^+$ | 2.30 h ($\beta^-$) |
| Xe | 54 | 121 | 120.911450 | $(\frac{5}{2}^+)$ | 40.1 m ($\varepsilon$) |
| | | 122 | 121.908170 | $0^+$ | 20.1 h ($\varepsilon$) |
| | | 123 | 122.908469 | $(\frac{1}{2}^+)$ | 2.08 h ($\varepsilon$) |
| | | 124 | 123.905894 | $0^+$ | 0.096% |
| | | 125 | 124.906397 | $(\frac{1}{2})^+$ | 17 h ($\varepsilon$) |
| | | 126 | 125.904281 | $0^+$ | 0.090% |
| | | 127 | 126.905182 | $(\frac{1}{2}^+)$ | 36.4 d ($\varepsilon$) |
| | | 128 | 127.903531 | $0^+$ | 1.92% |
| | | 129 | 128.904780 | $\frac{1}{2}^+$ | 26.4% |
| | | 130 | 129.903509 | $0^+$ | 4.1% |
| | | 131 | 130.905072 | $\frac{3}{2}^+$ | 21.2% |
| | | 132 | 131.904144 | $0^+$ | 26.9% |
| | | 133 | 132.905888 | $\frac{3}{2}^+$ | 5.25 d ($\beta^-$) |
| | | 134 | 133.905395 | $0^+$ | 10.4% |
| | | 135 | 134.907130 | $\frac{3}{2}^+$ | 9.1 h ($\beta^-$) |
| | | 136 | 135.907214 | $0^+$ | 8.9% |
| | | 137 | 136.911557 | $\frac{7}{2}^-$ | 3.82 m ($\beta^-$) |
| Cs | 55 | 130 | 129.906753 | $1^+$ | 29.2 m ($\varepsilon$) |
| | | 131 | 130.905444 | $\frac{5}{2}^+$ | 9.69 d ($\varepsilon$) |
| | | 132 | 131.906431 | $2^-$ | 6.47 d ($\varepsilon$) |
| | | 133 | 132.905429 | $\frac{7}{2}^+$ | 100% |
| | | 134 | 133.906696 | $4^+$ | 2.06 y ($\beta^-$) |
| | | 135 | 134.905885 | $\frac{7}{2}^+$ | 3 My ($\beta^-$) |
| | | 136 | 135.907289 | $5^+$ | 13.1 d ($\beta^-$) |
| | | 137 | 136.907073 | $\frac{7}{2}^+$ | 30.2 y ($\beta^-$) |
| | | 138 | 137.911004 | $3^-$ | 32.2 m ($\beta^-$) |
| Ba | 56 | 127 | 126.911130 | $(\frac{1}{2}^+)$ | 12.7 m ($\varepsilon$) |
| | | 128 | 127.908237 | $0^+$ | 2.43 d ($\varepsilon$) |
| | | 129 | 128.908642 | $\frac{1}{2}^+$ | 2.2 h ($\varepsilon$) |
| | | 130 | 129.906282 | $0^+$ | 0.106% |
| | | 131 | 130.906902 | $\frac{1}{2}^+$ | 12.0 d ($\varepsilon$) |
| | | 132 | 131.905042 | $0^+$ | 0.101% |
| | | 133 | 132.905988 | $\frac{1}{2}^+$ | 10.7 y ($\varepsilon$) |
| | | 134 | 133.904486 | $0^+$ | 2.42% |
| | | 135 | 134.905665 | $\frac{3}{2}^+$ | 6.59% |
| | | 136 | 135.904553 | $0^+$ | 7.85% |
| | | 137 | 136.905812 | $\frac{3}{2}^+$ | 11.2% |
| | | 138 | 137.905232 | $0^+$ | 71.7% |
| | | 139 | 138.908826 | $\frac{7}{2}^-$ | 82.9 m ($\beta^-$) |
| | | 140 | 139.910581 | $0^+$ | 12.7 d ($\beta^-$) |
| | | 141 | 140.914363 | $\frac{3}{2}^-$ | 18.3 m ($\beta^-$) |

| | $Z$ | $A$ | Atomic mass (u) | $I^\pi$ | Abundance or Half-life | | $Z$ | $A$ | Atomic mass (u) | $I^\pi$ | Abundance or Half-life |
|---|---|---|---|---|---|---|---|---|---|---|---|
| La | 57 | 135 | 134.906953 | $\frac{5}{2}^+$ | 19.5 h ($\varepsilon$) | | | 145 | 144.912743 | $\frac{5}{2}^+$ | 17.7 y ($\varepsilon$) |
| | | 136 | 135.907630 | $1^+$ | 9.87 m ($\varepsilon$) | | | 146 | 145.914708 | $3^-$ | 5.5 y ($\varepsilon$) |
| | | 137 | 136.906460 | $\frac{7}{2}^+$ | 0.06 My ($\varepsilon$) | | | 147 | 146.915135 | $\frac{7}{2}^+$ | 2.62 y ($\beta^-$) |
| | | 138 | 137.907105 | $5^+$ | 0.089% | | | 148 | 147.917473 | $1^-$ | 5.37 d ($\beta^-$) |
| | | 139 | 138.906347 | $\frac{7}{2}^+$ | 99.911% | | | 149 | 148.918332 | $\frac{7}{2}^+$ | 53.1 h ($\beta^-$) |
| | | 140 | 139.909471 | $3^-$ | 40.3 h ($\beta^-$) | | | 150 | 149.920981 | $(1^-)$ | 2.68 h ($\beta^-$) |
| | | 141 | 140.910896 | $\frac{7}{2}^+$ | 3.90 h ($\beta^-$) | Sm | 62 | 142 | 141.915206 | $0^+$ | 72.5 m ($\varepsilon$) |
| | | 142 | 141.914090 | $2^-$ | 91.1 m ($\beta^-$) | | | 143 | 142.914626 | $\frac{3}{2}^+$ | 8.83 m ($\varepsilon$) |
| Ce | 58 | 133 | 132.911360 | $\frac{1}{2}^+$ | 5.4 h ($\varepsilon$) | | | 144 | 143.911998 | $0^+$ | 3.1% |
| | | 134 | 133.908890 | $0^+$ | 76 h ($\varepsilon$) | | | 145 | 144.913409 | $\frac{7}{2}^-$ | 340 d ($\varepsilon$) |
| | | 135 | 134.909117 | $\frac{1}{2}^+$ | 17.6 h ($\varepsilon$) | | | 146 | 145.913053 | $0^+$ | 103 My ($\alpha$) |
| | | 136 | 135.907140 | $0^+$ | 0.190% | | | 147 | 146.914894 | $\frac{7}{2}^-$ | 15.1% |
| | | 137 | 136.907780 | $\frac{3}{2}^+$ | 9.0 h ($\varepsilon$) | | | 148 | 147.914819 | $0^+$ | 11.3% |
| | | 138 | 137.905985 | $0^+$ | 0.254% | | | 149 | 148.917180 | $\frac{7}{2}^-$ | 13.9% |
| | | 139 | 138.906631 | $\frac{3}{2}^+$ | 137.2 d ($\varepsilon$) | | | 150 | 149.917273 | $0^+$ | 7.4% |
| | | 140 | 139.905433 | $0^+$ | 88.5% | | | 151 | 150.919929 | $\frac{5}{2}^-$ | 90 y ($\beta^-$) |
| | | 141 | 140.908271 | $\frac{7}{2}^-$ | 32.5 d ($\beta^-$) | | | 152 | 151.919728 | $0^+$ | 26.6% |
| | | 142 | 141.909241 | $0^+$ | 11.1% | | | 153 | 152.922094 | $\frac{3}{2}^+$ | 46.8 h ($\beta^-$) |
| | | 143 | 142.912383 | $\frac{3}{2}^-$ | 33.0 h ($\beta^-$) | | | 154 | 153.922205 | $0^+$ | 22.6% |
| | | 144 | 143.913643 | $0^+$ | 284 d ($\beta^-$) | | | 155 | 154.924636 | $\frac{3}{2}^-$ | 22.4 m ($\beta^-$) |
| | | 145 | 144.917230 | $\frac{5}{2}^+$ | 2.98 m ($\beta^-$) | Eu | 63 | 148 | 147.918125 | $5^-$ | 54.5 d ($\varepsilon$) |
| Pr | 59 | 138 | 137.910748 | $1^+$ | 1.45 m ($\varepsilon$) | | | 149 | 148.917926 | $\frac{5}{2}^+$ | 93.1 d ($\varepsilon$) |
| | | 139 | 138.908917 | $\frac{5}{2}^+$ | 4.4 h ($\varepsilon$) | | | 150 | 149.919702 | $0^-$ | 36 y ($\varepsilon$) |
| | | 140 | 139.909071 | $1^+$ | 3.39 m ($\varepsilon$) | | | 151 | 150.919847 | $\frac{5}{2}^+$ | 47.9% |
| | | 141 | 140.907647 | $\frac{5}{2}^+$ | 100% | | | 152 | 151.921742 | $3^-$ | 13 y ($\varepsilon$) |
| | | 142 | 141.910039 | $2$ | 19.2 h ($\beta^-$) | | | 153 | 152.921225 | $\frac{5}{2}^+$ | 52.1% |
| | | 143 | 142.910814 | $\frac{7}{2}^+$ | 13.6 d ($\beta^-$) | | | 154 | 153.922975 | $3^-$ | 8.5 y ($\beta^-$) |
| | | 144 | 143.913301 | $0^-$ | 17.3 m ($\beta^-$) | | | 155 | 154.922889 | $\frac{5}{2}^+$ | 4.9 y ($\beta^-$) |
| Nd | 60 | 139 | 138.911920 | $\frac{3}{2}^+$ | 29.7 m ($\varepsilon$) | | | 156 | 155.924752 | $0^+$ | 15 d ($\beta^-$) |
| | | 140 | 139.909306 | $0^+$ | 3.37 d ($\varepsilon$) | | | 157 | 156.925418 | $\frac{5}{2}^+$ | 15 h ($\beta^-$) |
| | | 141 | 140.909594 | $\frac{3}{2}^+$ | 2.5 h ($\varepsilon$) | Gd | 64 | 149 | 148.919344 | $\frac{7}{2}^-$ | 9.4 d ($\varepsilon$) |
| | | 142 | 141.907719 | $0^+$ | 27.2% | | | 150 | 149.918662 | $0^+$ | 1.8 My ($\alpha$) |
| | | 143 | 142.909810 | $\frac{7}{2}^-$ | 12.2% | | | 151 | 150.920346 | $\frac{7}{2}^-$ | 120 d ($\varepsilon$) |
| | | 144 | 143.910083 | $0^+$ | 23.8% | | | 152 | 151.919786 | $0^+$ | 0.20% |
| | | 145 | 144.912570 | $\frac{7}{2}^-$ | 8.3% | | | 153 | 152.921745 | $\frac{3}{2}^-$ | 242 d ($\varepsilon$) |
| | | 146 | 145.913113 | $0^+$ | 17.2% | | | 154 | 153.920861 | $0^+$ | 2.1% |
| | | 147 | 146.916097 | $\frac{5}{2}^-$ | 11.0 d ($\beta^-$) | | | 155 | 154.922618 | $\frac{3}{2}^-$ | 14.8% |
| | | 148 | 147.916889 | $0^+$ | 5.7% | | | 156 | 155.922118 | $0^+$ | 20.6% |
| | | 149 | 148.920145 | $\frac{5}{2}^-$ | 1.73 h ($\beta^-$) | | | 157 | 156.923956 | $\frac{3}{2}^-$ | 15.7% |
| | | 150 | 149.920887 | $0^+$ | 5.6% | | | 158 | 157.924099 | $0^+$ | 24.8% |
| | | 151 | 150.923825 | $(\frac{3}{2}^+)$ | 12.4 m ($\beta^-$) | | | 159 | 158.926384 | $\frac{3}{2}^-$ | 18.6 h ($\beta^-$) |
| | | 152 | 151.924680 | $0^+$ | 11.4 m ($\beta^-$) | | | 160 | 159.927049 | $0^+$ | 21.8% |
| Pm | 61 | 142 | 141.912970 | $1^+$ | 40.5 s ($\varepsilon$) | | | 161 | 160.929664 | $\frac{5}{2}^-$ | 3.7 m ($\beta^-$) |
| | | 143 | 142.910930 | $\frac{5}{2}^+$ | 265 d ($\varepsilon$) | Tb | 65 | 156 | 155.924742 | $3^-$ | 5.34 d ($\varepsilon$) |
| | | 144 | 143.912588 | $5^-$ | 349 d ($\varepsilon$) | | | 157 | 156.924023 | $\frac{3}{2}^+$ | 150 y ($\varepsilon$) |

| Z | A | Atomic mass (u) | $I^\pi$ | Abundance or Half-life | | Z | A | Atomic mass (u) | $I^\pi$ | Abundance or Half-life |
|---|---|---|---|---|---|---|---|---|---|---|
| | 158 | 157.925411 | $3^-$ | 150 y $(\varepsilon)$ | Yb | 70 | 166 | 165.933875 | $0^+$ | 56.7 h $(\varepsilon)$ |
| | 159 | 158.925342 | $\frac{3}{2}^+$ | 100 % | | | 167 | 166.934946 | $\frac{5}{2}^-$ | 17.5 m $(\varepsilon)$ |
| | 160 | 159.927163 | $3^-$ | 72.1 d $(\beta^-)$ | | | 168 | 167.933894 | $0^+$ | 0.135% |
| | 161 | 160.927566 | $\frac{3}{2}^+$ | 6.90 d $(\beta^-)$ | | | 169 | 168.935186 | $\frac{7}{2}^+$ | 32.0 d $(\varepsilon)$ |
| | 162 | 161.929510 | $1^-$ | 7.76 m $(\beta^-)$ | | | 170 | 169.934759 | $0^+$ | 3.1% |
| Dy | 66 | 153 | 152.925769 | $\frac{7}{2}^-$ | 6.4 h $(\varepsilon)$ | | | 171 | 170.936323 | $\frac{1}{2}^-$ | 14.4% |
| | 154 | 153.924429 | $0^+$ | 3 My $(\alpha)$ | | | 172 | 171.936378 | $0^+$ | 21.9% |
| | 155 | 154.925747 | $\frac{3}{2}^-$ | 10.0 h $(\varepsilon)$ | | | 173 | 172.938208 | $\frac{5}{2}^-$ | 16.2% |
| | 156 | 155.924277 | $0^+$ | 0.057% | | | 174 | 173.938859 | $0^+$ | 31.6% |
| | 157 | 156.925460 | $\frac{3}{2}^-$ | 8.1 h $(\varepsilon)$ | | | 175 | 174.941273 | $\frac{7}{2}^-$ | 4.19 d $(\beta^-)$ |
| | 158 | 157.924403 | $0^+$ | 0.100 % | | | 176 | 175.942564 | $0^+$ | 12.6% |
| | 159 | 158.925735 | $\frac{3}{2}^-$ | 144.4 d $(\varepsilon)$ | | | 177 | 176.945253 | $\frac{9}{2}^+$ | 1.9 h $(\beta^-)$ |
| | 160 | 159.925193 | $0^+$ | 2.3% | | | 178 | 177.946639 | $0^+$ | 74 m $(\beta^-)$ |
| | 161 | 160.926930 | $\frac{5}{2}^+$ | 19.90% | Lu | 71 | 172 | 171.939085 | $(4^-)$ | 6.70 d $(\varepsilon)$ |
| | 162 | 161.926795 | $0^+$ | 25.5% | | | 173 | 172.938929 | $\frac{7}{2}^+$ | 1.37 y $(\varepsilon)$ |
| | 163 | 162.928728 | $\frac{5}{2}^-$ | 24.9% | | | 174 | 173.940336 | $1^-$ | 3.3 y $(\varepsilon)$ |
| | 164 | 163.929171 | $0^+$ | 28.1% | | | 175 | 174.940770 | $\frac{7}{2}^+$ | 97.39% |
| | 165 | 164.931700 | $\frac{7}{2}^+$ | 2.33 h $(\beta^-)$ | | | 176 | 175.942679 | $7^-$ | 2.61% |
| | 166 | 165.932803 | $0^+$ | 81.6 h $(\beta^-)$ | | | 177 | 176.943752 | $\frac{7}{2}^+$ | 6.71 d $(\beta^-)$ |
| Ho | 67 | 162 | 161.929092 | $1^+$ | 15 m $(\varepsilon)$ | | | 178 | 177.945963 | $1^+$ | 28.4 m $(\beta^-)$ |
| | 163 | 162.928731 | $(\frac{7}{2})^-$ | 33 y $(\varepsilon)$ | Hf | 72 | 171 | 170.940490 | $(\frac{7}{2}^+)$ | 12.1 h $(\varepsilon)$ |
| | 164 | 163.930285 | $1^+$ | 29.0 m $(\varepsilon)$ | | | 172 | 171.939460 | $0^+$ | 1.87 y $(\varepsilon)$ |
| | 165 | 164.930319 | $\frac{7}{2}^-$ | 100 % | | | 173 | 172.940650 | $\frac{1}{2}^-$ | 24.0 h $(\varepsilon)$ |
| | 166 | 165.932281 | $0^-$ | 26.8 h $(\beta^-)$ | | | 174 | 173.940044 | $0^+$ | 0.16% |
| | 167 | 166.933127 | $(\frac{7}{2}^-)$ | 3.1 h $(\beta^-)$ | | | 175 | 174.941507 | $\frac{5}{2}^-$ | 70 d $(\varepsilon)$ |
| Er | 68 | 160 | 159.929080 | $0^+$ | 28.6 h $(\varepsilon)$ | | | 176 | 175.941406 | $0^+$ | 5.2% |
| | 161 | 160.929996 | $\frac{3}{2}^-$ | 3.24 h $(\varepsilon)$ | | | 177 | 176.943217 | $\frac{7}{2}^-$ | 18.6% |
| | 162 | 161.928775 | $0^+$ | 0.14% | | | 178 | 177.943696 | $0^+$ | 27.1% |
| | 163 | 162.930030 | $\frac{5}{2}^-$ | 75.1 m $(\varepsilon)$ | | | 179 | 178.945812 | $\frac{9}{2}^+$ | 13.7% |
| | 164 | 163.929198 | $0^+$ | 1.56% | | | 180 | 179.946546 | $0^+$ | 35.2% |
| | 165 | 164.930723 | $\frac{5}{2}^-$ | 10.4 h $(\varepsilon)$ | | | 181 | 180.949096 | $\frac{1}{2}^-$ | 42.4 d $(\beta^-)$ |
| | 166 | 165.930290 | $0^+$ | 33.4% | | | 182 | 181.950550 | $0^+$ | 9 My $(\beta^-)$ |
| | 167 | 166.932046 | $\frac{7}{2}^+$ | 22.9% | | | 183 | 182.953530 | $(\frac{3}{2}^-)$ | 64 m $(\beta^-)$ |
| | 168 | 167.932368 | $0^+$ | 27.1% | Ta | 73 | 178 | 177.945750 | $1^+$ | 9.31 m $(\varepsilon)$ |
| | 169 | 168.934588 | $\frac{1}{2}^-$ | 9.40 d $(\beta^-)$ | | | 179 | 178.945930 | $(\frac{7}{2}^+)$ | 665 d $(\varepsilon)$ |
| | 170 | 169.935461 | $0^+$ | 14.9% | | | 180 | 179.947462 | $1^+$ | 0.0123% |
| | 171 | 170.938027 | $\frac{5}{2}^-$ | 7.52 h $(\beta^-)$ | | | 181 | 180.947992 | $\frac{7}{2}^+$ | 99.9877% |
| | 172 | 171.939353 | $0^+$ | 49.3 h $(\beta^-)$ | | | 182 | 181.950149 | $3^-$ | 115 d $(\beta^-)$ |
| Tm | 69 | 166 | 165.933561 | $2^+$ | 7.70 h $(\varepsilon)$ | | | 183 | 182.951369 | $\frac{7}{2}^+$ | 5.1 d $(\beta^-)$ |
| | 167 | 166.932848 | $\frac{1}{2}^+$ | 9.25 d $(\varepsilon)$ | W | 74 | 178 | 177.945840 | $0^+$ | 21.5 d $(\varepsilon)$ |
| | 168 | 167.934170 | $3^+$ | 93.1 d $(\varepsilon)$ | | | 179 | 178.947067 | $(\frac{7}{2}^-)$ | 38 m $(\varepsilon)$ |
| | 169 | 168.934212 | $\frac{1}{2}^+$ | 100 % | | | 180 | 179.946701 | $0^+$ | 0.13% |
| | 170 | 169.935798 | $1^-$ | 128.6 d $(\beta^-)$ | | | 181 | 180.948192 | $\frac{9}{2}^+$ | 121 d $(\varepsilon)$ |
| | 171 | 170.936427 | $\frac{1}{2}^+$ | 1.92 y $(\beta^-)$ | | | 182 | 181.948202 | $0^+$ | 26.3% |
| | 172 | 171.938397 | $2^-$ | 63.6 h $(\beta^-)$ | | | 183 | 182.950220 | $\frac{1}{2}^-$ | 14.3% |

| Z | A | Atomic mass (u) | $I^\pi$ | Abundance or Half-life | Z | A | Atomic mass (u) | $I^\pi$ | Abundance or Half-life |
|---|---|---|---|---|---|---|---|---|---|
| | 184 | 183.950928 | $0^+$ | *30.7%* | | 198 | 197.967869 | $0^+$ | *7.2%* |
| | 185 | 184.953416 | $\frac{3}{2}^-$ | *15.1 d ($\beta^-$)* | | 199 | 198.970552 | $(\frac{5}{2}^-)$ | 30.8 m ($\beta^-$) |
| | 186 | 185.954357 | $0^+$ | *28.6%* | | 200 | 199.971417 | $0^+$ | 12.5 h ($\beta^-$) |
| | 187 | 186.957153 | $\frac{3}{2}^-$ | 23.9 h ($\beta^-$) | Au 79 | 194 | 193.965348 | $1^-$ | 39.5 h ($\varepsilon$) |
| | 188 | 187.958480 | $0^+$ | 69.4 d ($\beta^-$) | | 195 | 194.965013 | $\frac{3}{2}^+$ | 186 d ($\varepsilon$) |
| Re 75 | 182 | 181.951210 | $2^+$ | 12.7 h ($\varepsilon$) | | 196 | 195.966544 | $2^-$ | 6.18 d ($\varepsilon$) |
| | 183 | 182.950817 | $(\frac{5}{2})^+$ | 71 d ($\varepsilon$) | | 197 | 196.966543 | $\frac{3}{2}^+$ | *100%* |
| | 184 | 183.952530 | $3^-$ | 38 d ($\varepsilon$) | | 198 | 197.968217 | $2^-$ | 2.696 d ($\beta^-$) |
| | 185 | 184.952951 | $\frac{5}{2}^+$ | *37.40%* | | 199 | 198.968740 | $\frac{3}{2}^+$ | 3.14 d ($\beta^-$) |
| | 186 | 185.954984 | $1^-$ | 90.6 h ($\beta^-$) | | 200 | 199.970670 | $1^-$ | 48.4 m ($\beta^-$) |
| | 187 | 186.955744 | $\frac{5}{2}^+$ | *62.60%* | Hg 80 | 193 | 192.966560 | $\frac{3}{2}^-$ | 3.8 h ($\varepsilon$) |
| | 188 | 187.958106 | $1^-$ | 16.9 h ($\beta^-$) | | 194 | 193.965391 | $0^+$ | 520 y ($\varepsilon$) |
| | 189 | 188.959219 | $(\frac{5}{2})^+$ | 24.3 h ($\beta^-$) | | 195 | 194.966640 | $\frac{1}{2}^-$ | 9.5 h ($\varepsilon$) |
| Os 76 | 182 | 181.952120 | $0^+$ | 21.5 h ($\varepsilon$) | | 196 | 195.965807 | $0^+$ | *0.15%* |
| | 183 | 182.953290 | $(\frac{9}{2})^+$ | 13.0 h ($\varepsilon$) | | 197 | 196.967187 | $\frac{1}{2}^-$ | 64.1 h ($\varepsilon$) |
| | 184 | 183.952488 | $0^+$ | *0.018%* | | 198 | 197.966743 | $0^+$ | *10.0%* |
| | 185 | 184.954041 | $\frac{1}{2}^-$ | 93.6 d ($\varepsilon$) | | 199 | 198.968254 | $\frac{1}{2}^-$ | *16.8%* |
| | 186 | 185.953830 | $0^+$ | *1.6%* | | 200 | 199.968300 | $0^+$ | *23.1%* |
| | 187 | 186.955741 | $\frac{1}{2}^-$ | *1.6%* | | 201 | 200.970277 | $\frac{3}{2}^-$ | *13.2%* |
| | 188 | 187.955830 | $0^+$ | *13.3%* | | 202 | 201.970617 | $0^+$ | *29.8%* |
| | 189 | 188.958137 | $\frac{3}{2}^-$ | *16.1%* | | 203 | 202.972848 | $\frac{5}{2}^-$ | 46.6 d ($\beta^-$) |
| | 190 | 189.958436 | $0^+$ | *26.4%* | | 204 | 203.973467 | $0^+$ | *6.9%* |
| | 191 | 190.960920 | $\frac{9}{2}^-$ | 15.4 d ($\beta^-$) | | 205 | 204.976047 | $\frac{1}{2}^-$ | 5.2 m ($\beta^-$) |
| | 192 | 191.961467 | $0^+$ | *41.0%* | Tl 81 | 200 | 199.970934 | $2^-$ | 26.1 h ($\varepsilon$) |
| | 193 | 192.964138 | $\frac{3}{2}^-$ | 30.6 h ($\beta^-$) | | 201 | 200.970794 | $\frac{1}{2}^+$ | 73 h ($\varepsilon$) |
| | 194 | 193.965173 | $0^+$ | 6.0 y ($\beta^-$) | | 202 | 201.972085 | $2^-$ | 12.2 d ($\varepsilon$) |
| Ir 77 | 188 | 187.958830 | $(2^-)$ | 41.5 h ($\varepsilon$) | | 203 | 202.972320 | $\frac{1}{2}^+$ | *29.5%* |
| | 189 | 188.958712 | $\frac{3}{2}^+$ | 13.1 d ($\varepsilon$) | | 204 | 203.973839 | $2^-$ | 3.77 y ($\beta^-$) |
| | 190 | 189.960580 | $(4^+)$ | 11.8 d ($\varepsilon$) | | 205 | 204.974401 | $\frac{1}{2}^+$ | *70.5%* |
| | 191 | 190.960584 | $\frac{3}{2}^+$ | *37.3%* | | 206 | 205.976084 | $0^-$ | 4.20 m ($\beta^-$) |
| | 192 | 191.962580 | $4^-$ | 74.2 d ($\beta^-$) | Pb 82 | 201 | 200.972830 | $\frac{5}{2}^-$ | 9.3 h ($\varepsilon$) |
| | 193 | 192.962917 | $\frac{3}{2}^+$ | *62.7%* | | 202 | 201.972134 | $0^+$ | 0.05 My ($\varepsilon$) |
| | 194 | 193.965069 | $1^-$ | 19.2 h ($\beta^-$) | | 203 | 202.973365 | $\frac{5}{2}^-$ | 51.9 h ($\varepsilon$) |
| | 195 | 194.965966 | $(\frac{3}{2}^+)$ | 2.8 h ($\beta^-$) | | 204 | 203.973020 | $0^+$ | *1.42%* |
| Pt 78 | 187 | 186.960470 | $\frac{3}{2}^-$ | 2.35 h ($\varepsilon$) | | 205 | 204.974458 | $\frac{5}{2}^-$ | 15 My ($\varepsilon$) |
| | 188 | 187.959386 | $0^+$ | 10.2 d ($\varepsilon$) | | 206 | 205.974440 | $0^+$ | *24.1%* |
| | 189 | 188.960817 | $\frac{3}{2}^-$ | 10.9 h ($\varepsilon$) | | 207 | 206.975872 | $\frac{1}{2}^-$ | *22.1%* |
| | 190 | 189.959917 | $0^+$ | *0.013%* | | 208 | 207.976627 | $0^+$ | *52.3%* |
| | 191 | 190.961665 | $\frac{3}{2}^-$ | 2.9 d ($\varepsilon$) | | 209 | 208.981065 | $\frac{9}{2}^+$ | 3.25 h ($\beta^-$) |
| | 192 | 191.961019 | $0^+$ | *0.78%* | | 210 | 209.984163 | $0^+$ | 22.3 y ($\beta^-$) |
| | 193 | 192.962977 | $(\frac{1}{2}^-)$ | 50 y ($\varepsilon$) | | 211 | 210.988735 | $(\frac{9}{2}^+)$ | 36.1 m ($\beta^-$) |
| | 194 | 193.962655 | $0^+$ | *32.9%* | | 212 | 211.991871 | $0^+$ | 10.6 h ($\beta^-$) |
| | 195 | 194.964766 | $\frac{1}{2}^-$ | *33.8%* | | | | | |
| | 196 | 195.964926 | $0^+$ | *25.3%* | Bi 83 | 206 | 205.978478 | $6^+$ | 6.24 d ($\varepsilon$) |
| | 197 | 196.967315 | $\frac{1}{2}^-$ | 18.3 h ($\beta^-$) | | 207 | 206.978446 | $\frac{9}{2}^-$ | 32 y ($\varepsilon$) |

| Z | A | Atomic mass (u) | $I^\pi$ | Abundance or Half-life |
|---|---|---|---|---|
| | 208 | 207.979717 | $(5^+)$ | 0.368 My ($\varepsilon$) |
| | 209 | 208.980374 | $\frac{9}{2}^-$ | *100 %* |
| | 210 | 209.984095 | $1^-$ | 5.01 d ($\beta^-$) |
| | 211 | 210.987255 | $\frac{9}{2}^-$ | 2.15 m ($\alpha$) |
| | 212 | 211.991255 | $1^-$ | 60.6 m ($\beta^-$) |
| Po 84 | 206 | 205.980456 | $0^+$ | 8.8 d ($\varepsilon$) |
| | 207 | 206.981570 | $\frac{5}{2}^-$ | 5.8 h ($\varepsilon$) |
| | 208 | 207.981222 | $0^+$ | 2.90 y ($\alpha$) |
| | 209 | 208.982404 | $\frac{1}{2}^-$ | 102 y ($\alpha$) |
| | 210 | 209.982848 | $0^+$ | 138.4 d ($\alpha$) |
| | 211 | 210.986627 | $\frac{9}{2}^+$ | 0.52 s ($\alpha$) |
| At 85 | 208 | 207.986510 | $6^+$ | 1.63 h ($\varepsilon$) |
| | 209 | 208.986149 | $\frac{9}{2}^-$ | 5.4 h ($\varepsilon$) |
| | 210 | 209.987126 | $5^+$ | 8.3 h ($\varepsilon$) |
| | 211 | 210.987469 | $\frac{9}{2}^-$ | 7.21 h ($\varepsilon$) |
| | 212 | 211.990725 | $(1^-)$ | 0.31 s ($\alpha$) |
| | 213 | 212.992911 | $\frac{9}{2}^-$ | 0.11 $\mu$s ($\alpha$) |
| Rn 86 | 207 | 206.990690 | $\frac{5}{2}^-$ | 9.3 m ($\varepsilon$) |
| | 210 | 209.989669 | $0^+$ | 2.4 h ($\alpha$) |
| | 211 | 210.990576 | $\frac{1}{2}^-$ | 14.6 h ($\varepsilon$) |
| | 212 | 211.990697 | $0^+$ | 24 m ($\alpha$) |
| | 218 | 218.005580 | $0^+$ | 35 ms ($\alpha$) |
| | 222 | 222.017571 | $0^+$ | 3.82 d ($\alpha$) |
| | 224 | | $0^+$ | 107 m ($\beta^-$) |
| Fr 87 | 209 | 208.995870 | $\frac{9}{2}^-$ | 50 s ($\alpha$) |
| | 212 | 211.996130 | $5^+$ | 20 m ($\varepsilon$) |
| | 215 | 215.000310 | $\frac{9}{2}^-$ | 0.12 $\mu$s ($\alpha$) |
| | 220 | 220.012293 | $1$ | 27.4 s ($\alpha$) |
| | 223 | 223.019733 | $(\frac{3}{2})$ | 21.8 m ($\beta^-$) |
| Ra 88 | 222 | 222.015353 | $0^+$ | 38 s ($\alpha$) |
| | 223 | 223.018501 | $\frac{1}{2}^+$ | 11.4 d ($\alpha$) |
| | 224 | 224.020186 | $0^+$ | 3.66 d ($\alpha$) |
| | 225 | 225.023604 | $(\frac{3}{2})^+$ | 14.8 d ($\beta^-$) |
| | 226 | 226.025403 | $0^+$ | 1602 y ($\alpha$) |
| | 227 | 227.029171 | $(\frac{3}{2}^+)$ | 42 m ($\beta^-$) |
| Ac 89 | 224 | 224.021685 | $(0^-)$ | 2.9 h ($\varepsilon$) |
| | 225 | 225.023205 | $(\frac{3}{2}^-)$ | 10.0 d ($\alpha$) |
| | 226 | 226.026084 | $(1^-)$ | 29 h ($\beta^-$) |
| | 227 | 227.027750 | $\frac{3}{2}^-$ | 21.77 y ($\beta^-$) |
| | 228 | 228.031015 | $(3^+)$ | 6.1 h ($\beta^-$) |
| Th 90 | 228 | 228.028715 | $0^+$ | 1.91 y ($\alpha$) |
| | 229 | 229.031755 | $\frac{5}{2}^+$ | 7300 y ($\alpha$) |
| | 230 | 230.033128 | $0^+$ | 75,400 y ($\alpha$) |
| | 231 | 231.036299 | $\frac{5}{2}^+$ | 25.52 h ($\beta^-$) |

| Z | A | Atomic mass (u) | $I^\pi$ | Abundance or Half-life |
|---|---|---|---|---|
| | 232 | 232.038051 | $0^+$ | *100 %* |
| | 233 | 233.041577 | $(\frac{1}{2}^+)$ | 22.3 m ($\beta^-$) |
| Pa 91 | 229 | 229.032073 | $(\frac{5}{2}^+)$ | 1.4 d ($\varepsilon$) |
| | 230 | 230.034527 | $(2^-)$ | 17.7 d ($\varepsilon$) |
| | 231 | 231.035880 | $\frac{3}{2}^-$ | 32,800 y ($\alpha$) |
| | 232 | 232.038565 | $(2^-)$ | 1.31 d ($\beta^-$) |
| | 233 | 233.040243 | $\frac{3}{2}^-$ | 27.0 d ($\beta^-$) |
| U 92 | 233 | 233.039628 | $\frac{5}{2}^+$ | 0.1592 My ($\alpha$) |
| | 234 | 234.040947 | $0^+$ | 0.245 My ($\alpha$) |
| | 235 | 235.043924 | $\frac{7}{2}^-$ | *0.720%* |
| | 236 | 236.045563 | $0^+$ | 23.42 My ($\alpha$) |
| | 237 | 237.048725 | $\frac{1}{2}^+$ | 6.75 d ($\beta^-$) |
| | 238 | 238.050785 | $0^+$ | *99.275%* |
| | 239 | 239.054290 | $\frac{5}{2}^+$ | 23.5 m ($\beta^-$) |
| Np 93 | 236 | 236.046550 | $(6^-)$ | 0.11 My ($\varepsilon$) |
| | 237 | 237.048168 | $\frac{5}{2}^+$ | 2.14 My ($\alpha$) |
| | 238 | 238.050941 | $2^+$ | 2.117 d ($\beta^-$) |
| | 239 | 239.052933 | $\frac{5}{2}^+$ | 2.36 d ($\beta^-$) |
| Pu 94 | 237 | 237.048401 | $\frac{7}{2}^-$ | 45.3 d ($\varepsilon$) |
| | 238 | 238.049555 | $0^+$ | 87.74 y ($\alpha$) |
| | 239 | 239.052158 | $\frac{1}{2}^+$ | 24,100 y ($\alpha$) |
| | 240 | 240.053808 | $0^+$ | 6570 y ($\alpha$) |
| | 241 | 241.056846 | $\frac{5}{2}^+$ | 14.4 y ($\beta^-$) |
| | 242 | 242.058737 | $0^+$ | 0.376 My ($\alpha$) |
| | 243 | 243.061998 | $\frac{7}{2}^+$ | 4.96 h ($\beta^-$) |
| Am 95 | 240 | 240.055278 | $(3^-)$ | 50.9 h ($\varepsilon$) |
| | 241 | 241.056824 | $\frac{5}{2}^-$ | 433 y ($\alpha$) |
| | 242 | 242.059542 | $1^-$ | 16.0 h ($\beta^-$) |
| | 243 | 243.061375 | $\frac{5}{2}^-$ | 7370 y ($\alpha$) |
| | 244 | 244.064279 | $(6^-)$ | 10.1 h ($\beta^-$) |
| Cm 96 | 246 | 246.067218 | $0^+$ | 4700 y ($\alpha$) |
| | 247 | 247.070347 | $\frac{9}{2}^-$ | 16 My ($\alpha$) |
| | 248 | 248.072343 | $0^+$ | 0.34 My ($\alpha$) |
| | 249 | 249.075948 | $\frac{1}{2}^+$ | 64 m ($\beta^-$) |
| Bk 97 | 246 | 246.068720 | $2^-$ | 1.8 d ($\varepsilon$) |
| | 247 | 247.070300 | $(\frac{3}{2}^-)$ | 1380 y ($\alpha$) |
| Cf 98 | 251 | 251.079580 | $\frac{1}{2}^+$ | 898 y ($\alpha$) |
| | 252 | 252.081621 | $0^+$ | 2.64 y ($\alpha$) |
| Es 99 | 252 | 252.082944 | $(4^+,5^-)$ | 472 d ($\alpha$) |
| | 253 | 253.084818 | $\frac{7}{2}^+$ | 20.5 d ($\alpha$) |

| Z | A | Atomic mass (u) | $I^{\pi}$ | Abundance or Half-life | | Z | A | Atomic mass (u) | $I^{\pi}$ | Abundance or Half-life |
|---|---|---|---|---|---|---|---|---|---|---|
| Fm 100 | 256 | 256.091767 | $0^+$ | 2.63 h (f) | Lr | 103 | 260 | 260.105320 | | 180 s ($\alpha$) |
| | 257 | 257.095099 | $(\frac{9}{2}^+)$ | 100 d ($\alpha$) | Rf | 104 | 261 | 261.108690 | | 65 s ($\alpha$) |
| Md 101 | 257 | 257.095580 | $(\frac{7}{2}^-)$ | 5.2 h ($\varepsilon$) | Ha | 105 | 261 | 261.111820 | | 1.8 s ($\alpha$) |
| | 258 | 258.098570 | $(8^-)$ | 55 d ($\alpha$) | | | 262 | 262.113760 | | 34 s (f) |
| No 102 | 258 | 258.098150 | $0^+$ | 1.2 ms (f) | | 106 | 263 | 263.118220 | | 0.8 s (f) |
| | 259 | 259.100931 | $(\frac{9}{2}^+)$ | 60 m ($\alpha$) | | 107 | 262 | 262.122930 | | 115 ms ($\alpha$) |

# CREDITS

The following illustrations were taken from the works listed below and are used with the permissions of the author and publisher: from David Halliday, *Introductory Nuclear Physics*, 2nd edition (New York: Wiley, 1955)—figures 3.14, 4.2, 4.7, 4.10, 6.2, 6.5, 7.1, 12.12, 12.24, 12.25, 15.13, 15.26, 15.28, 16.12, 16.14, 16.15; from Kenneth S. Krane, *Modern Physics* (New York: Wiley 1983)—figures 1.1, 3.1, 3.11, 3.12, 3.16, 5.1, 6.1, 7.19, 10.27, 11.7, 11.8, 11.10, 12.20, 13.5, 14.4, 14.15, 14.18, 15.11, 15.31, 18.10, 19.2, 19.3, 19.9, 19.15. All other figures that are not original to the present work are used with the permissions of the authors and publishers listed in the accompanying legends.

# INDEX